The Application of Laser Light Scattering to the Study of Biological Motion

NATO Advanced Science Institutes Series

A series of edited volumes comprising multifaceted studies of contemporary scientific issues by some of the best scientific minds in the world, assembled in cooperation with NATO Scientific Affairs Division.

This series is published by an international board of publishers in conjunction with NATO Scientific Affairs Division

A Life Sciences **B Physics**	Plenum Publishing Corporation New York and London
C Mathematical and Physical Sciences	D. Reidel Publishing Company Dordrecht, Boston, and London
D Behavioral and Social Sciences **E Applied Sciences**	Martinus Nijhoff Publishers The Hague, Boston, and London
F Computer and Systems Sciences **G Ecological Sciences**	Springer Verlag Heidelberg, Berlin, and New York

Recent Volumes in Series A: Life Sciences

Volume 56—Advances in Vertebrate Neuroethology
edited by Jörg-Peter Ewert, Robert R. Capranica, and David J. Ingle

Volume 57—Biochemical and Biological Markers of Neoplastic Transformation
edited by Prakash P. Chandra

Volume 58—Arterial Pollution: An Integrated View on Atherosclerosis
edited by H. Peeters, G. A. Gresham, and R. Paoletti

Volume 59—The Applications of Laser Light Scattering to the Study of Biological Motion
edited by J. C. Earnshaw and M. W. Steer

Volume 60—The Use of Human Cells for the Evaluation of Risk from Physical and Chemical Agents
edited by Amleto Castellani

Volume 61—Genetic Engineering in Eukaryotes
edited by Paul F. Lurquin and Andris Kleinhofs

Volume 62—Heart Perfusion, Energetics, and Ischemia
edited by Leopold Dintenfass, Desmond G. Julian, and Geoffrey V. F. Seaman

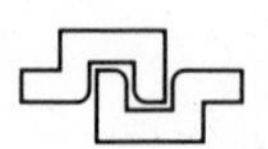

The Application of Laser Light Scattering to the Study of Biological Motion

Edited by

J. C. Earnshaw and
M. W. Steer

The Queen's University of Belfast
Belfast, Northern Ireland

Plenum Press
New York and London
Published in cooperation with NATO Scientific Affairs Division

Proceedings of a NATO Advanced Study Institute on
the Application of Laser Light
Scattering to the Study of
Biological Motion,
held June 20–July 3, 1982,
in Maratea, Italy

Library of Congress Cataloging in Publication Data

NATO Advanced Study Institute on the Application of Laser Light Scattering to the Study of Biological Motion (1982: Maratea, Italy)
The application of laser light scattering to the study of biological motion.

(NATO advanced science institutes series. Series A, Life sciences; v. 59)
"Proceedings of a NATO Advanced Study Institute on the Application of Laser Light Scattering to the Study of Biological Motion, held June 20–July 3, 1982, in Maratea, Italy"—T.p. verso.
Includes bibliographical references and index.
1. Lasers in biology—Congresses. 2. Light—Scattering—Congresses. 3. Biomechanics—Congresses. I. Earnshaw, J. C., 1943– . II. Steer, Martin W. III. Title. IV. Series.
QH324.9.L37N36 1982 574.8′028 83-2327
ISBN 0-306-41268-3

A Division of Plenum Publishing Corporation
233 Spring Street, New York, N.Y. 10013

Printed in the United States of America

PREFACE

Several previous Advanced Study Institutes have concentrated on the techniques of light scattering, while the biological applications were not fully explored. Many of the techniques are now standardised and are being applied to a wide range of biologically significant problems both in vivo and in vitro. While laser light scattering methods are superior to conventional methods, there was a general reluctance among biologists to adopt them because of the complexity of the physical techniques and the accompanying mathematical analysis. Consequently valuable opportunities for advancing the understanding of the biological problems were being missed. Advances in the design and commercial availability of standard light scattering instruments, and the availability of standard computer programs, made the more widespread use of these techniques a practical reality for the biologist. While biologists are unable to cope with the complexities of the physical techniques, physicists are generally unaware of the nature and scale of the biological problems. The meeting at Maratea was an attempt to bring these two groups together and provide an impetus for the application of laser light scattering techniques to biology. This volume differs from the three previous proceedings on laser light scattering in the NATO ASI series (B3, B23, B73), in that it has been published in the Life Sciences series rather than the Physics series, reflecting the shift in emphasis from the development of a new technique to its application in biology.

The contents of this volume have necessarily been organised rather differently from their order of presentation in the meeting. Here we present a linear sequence of contributions from both lecturers and participants, starting with the introductory material. This is followed by techniques, of both experiment and data analysis, and then it progresses from relatively simple macromolecular systems to the complex problems of whole cells.

We are very grateful to both lecturers and participants for their considerable efforts in producing these written contributions, based on their often eloquent and well illustrated verbal presentations.

Our venue contributed substantially to a relaxed, yet enclosed atmosphere during the meeting. It was based in the Hotel Villa del Mare, Maratea. The hotel clings to the cliff face of a narrow bay, trapped between the mountains and the Mediterranean, only accessible by a narrow corniche road. This led to a highly interactive atmosphere, both in the bar and on the beach, that significantly extended the scientific and cultural content of the meeting. The Italian siesta was compulsory, as were the evening sessions, conducted in the knowledge that they were to be followed by a further dose of the enthusiastic hospitality of Sen. Guzzardi and his hotel staff.

The meeting itself, held in a lecture hall with a cliff face for one wall and through the glass of the other a panoramic view of the Mediterranean, was conducted in a friendly yet critically challenging atmosphere. At the outset the biological camp was criticised for being overoptimistic about the possible applications of light scattering to the study of complex biological systems, while the physical sciences camp was accused of being perfectionist, failing to grasp the point that information of almost any quality might assist the biologist. In a sense the following two weeks of lectures and discussions were an attempt to bridge this divide. The success or otherwise of this operation may be judged by the concluding statements.

The running of the meeting and subsequent production of this volume, were due to the efforts of many. Our colleagues on the organising committee provided valuable geographical coverage and some powerful leverage. Much of the labour of turning organisation into practise fell on a small band of research workers, instantly dubbed 'the Belfast mafia' (Grace Crawford, Jill Picton, Jim Crilly and George Munroe). The lecturers responded by providing a continuous stream of lively and inventive entertainment throughout the final banquet, compered by Hyuk Yu. The final production of this volume depended on the skills of our team of secretaries (Anne Clements, Karen Coulter, Dianne Finlay and Norma McAllister), our illustrator (Mrs Simpson) and photographer (George McCartney), to whom we are very grateful.

Belfast

September 1982

J. C. Earnshaw

M. W. Steer

ACKNOWLEDGEMENTS

We are most grateful to NATO, who sponsored this meeting under their Advanced Study Institutes Program.

We wish to express our gratitude to the National Science Foundation and the Company of Biologists (Cambridge, UK) who generously assisted with the travel expenses of certain participants.

We thank the following organizations whose generous donations assisted with organizational costs and enabled us to subsidize a social programme which directly contributed to the success of the meeting:

Langley-Ford Instruments
Malvern Instruments Ltd
Spectra-Physics GmBH

CONTENTS

BIOLOGICAL APPLICATIONS

MUSCLES AND MUSCLE PROTEINS

CTYOPLASMIC STREAMING

MOTILITY

CONCLUDING STATEMENTS

Foreword

FOREWORD : LIGHT SCATTERING NOMENCLATURE

Newcomers to any field are not assisted by variations in nomenclature, terminology and notation. Unfortunately the field of quasi-elastic light scattering abounds in such variations. For example, in 1976 Cummins and Pusey[1] noted that literature searches 'are made particularly difficult by the wide range of titles in use, e.g. light-beating spectroscopy, optical mixing spectroscopy, intensity fluctuation spectroscopy, photon correlation spectroscopy, inelastic light scattering, quasi-elastic light scattering, dynamic light scattering, laser Doppler light scattering, Rayleigh line spectrometry etc, etc'. The situation has not greatly improved since 1976; some of these names may have dropped from common usage, but doubtless others have emerged. In the present volume the term 'elastic light scattering' is used to denote scattering, without change in frequency, from static structures; various titles are used to imply scattering involving changes of the frequency of the light.

The novice will find a great variety of terms used to describe the same thing; sometimes, more confusingly, the same word is used to describe different concepts. Thus 'intensity fluctuation spectroscopy' and 'self-beating spectroscopy' are synonymous. But 'homodyne' may be used for 'self-beating' or for 'heterodyne' (or again, a particular variant thereof), sometimes, but not always, depending on the side of the Atlantic from which the author writes.

Nor are the mathematical equations always reliable land-marks in this sea of shifting sands. While the physical content of a formula is constant, its appearance may vary. We are grateful to Professor Chu for Table 1, showing some of the differences in notation to be found in just three of the lecture courses at this A.S.I. This table is only a partial listing of the differences in these cases; if more authors were considered, the list could doubtless be extended for as long as the investigator's patience held out.

The two basic techniques used to analyse the frequency spectrum of the scattered light (spectrum analysis in the frequency

TABLE 1 : VARIATIONS OF LIGHT SCATTERING NOTATION

Quantity	Degiorgio	Cummins	Chu
Momentum transfer vector	$k = 4\pi (\frac{n_o}{\lambda}) \sin (\theta/2)$	$q = 4\pi (\frac{n}{\lambda}) \sin (\theta/2)$	$K = 4\pi (\frac{1}{\lambda}) \sin (\theta/2)$
Wavelength in medium	λ/n_o	λ/n	$\lambda = \lambda_o/n$
Intensity correlation function	$G_2(\tau) = \langle I_s \rangle^2 (1 + Be^{-2Dk^2\tau})$	$G^{(2)} (\tau)$	$G^{(2)} (\tau)$
Measured intensity correlation function		$C(\tau) = B (1 + a \vert g^{(1)}(\tau)\vert^2)$	
Single-clipped intensity correlation function			$G_k^{(2)} = A (1 + \beta \vert g^{(1)}(\tau)\vert^2)$
Translational diffusion coefficient $(C \to 0)$	D	D_T	D^o (noninteracting)
(at finite but dilute concentration)	D_c (collective)		D
(cooperative diffusion)			(D_c)
Intensity second virial coefficient	k_I		A_2
Hydrodynamic radius	R_H	r_h	r (sphere)
Radius of gyration	R_G		r_g
Normalized linewidth distribution		$F (\Gamma)$	$G (\Gamma)$
Rotational diffusion coefficient		D_R	Θ

domain and photon correlation in the time domain) have usually been considered to yield essentially equivalent information, so that the choice of technique has largely been a matter of experimental convenience.[eg 2] Two contributions in this volume [3,4] show that in some circumstances the two methods can yield complementary information. Such studies may, in the future, cause the various titles cited above to be regarded as less freely interchangeable than they have been to date.

References

1. Cummins, H. Z. and P. N. Pusey in Photon Correlation Spectroscopy and Velocimetry, edited by H. Z. Cummins and E. R. Pike (New York; Plenum, 1977).
2. Ware, B., this volume.
3. Fujime, S., this volume.
4. Volochine, B., this volume.

Introduction

PHYSICAL PRINCIPLES OF LIGHT SCATTERING

Vittorio Degiorgio

Istituto di Fisica Applicata
Università di Pavia
Via Bassi 6, 27100 Pavia, Italy

CONTENTS

INTRODUCTION

A schematic light scattering experiment is sketched in Fig.1. A monochromatic plane wave, linearly polarized, is incident upon a uniform transparent medium. An optical detector placed in the position P reveals the presence of a nonzero light intensity, usually weak, propagating in directions other than that of the reflected and refracted beam. This is what is called scattered light.

The physical origin of light scattering is the spatial inhomogeneity of the dielectric constant ε due to:

i) spontaneous thermal fluctuations (pure fluids)

ii) built-in inhomogeneities (macromolecular solutions).

I will treat only case ii) which is the most relevant for biological applications. I will not consider in these lectures Raman scattering and nonlinear scattering processes. Our light scattering process is linear, i.e. the scattered field E_s is proportional to the incident field E_o, and is due only to single scattering events (no multiple scattering).

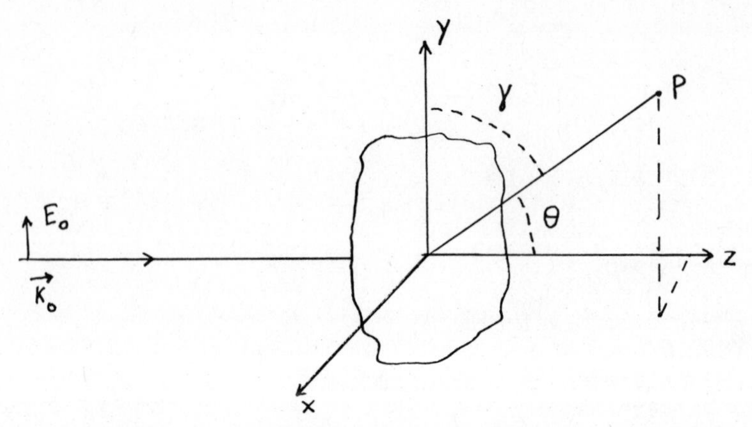

Fig. 1. Schematic light scattering experiment. The incident beam propagates along the z-axis and is linearly polarized along the y-axis. The scattering angle θ is the angle between $\vec{k}_o$ and $\vec{k}_s$. The distance of P is taken to be much larger than the linear size of the scattering volume.

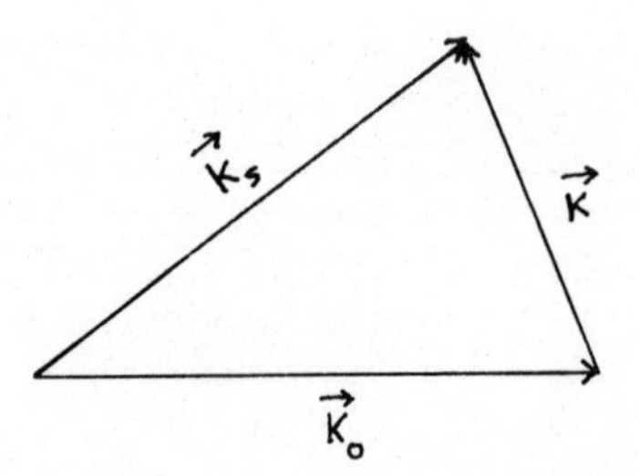

Fig. 2. Triangle of wave vectors which expresses conservation of momentum in the scattering process.

Many excellent books[1-3] and proceedings[4-7] of conferences and schools are available which give an historical introduction, and describe both theory and technique of dynamic light scattering. These volumes also discuss many applications to physics, chemistry, biology and engineering. Exhaustive treatments of static light scattering can be found in the books by Van de Hulst[8] and Kerker[9].

A basic quantity in a light scattering experiment is the vector $\vec{k} = \vec{k}_s - \vec{k}_o$ which represents the difference between the wave vectors of the scattered and the incident beam (see Fig. 2). Since the wavelength of scattered light is very close to that of the incident light, $k_s = 2\pi/\lambda_s \simeq k_o = 2\pi/\lambda_o$, and, therefore, the modulus of k is $\vec{k} \cong 2k_o \sin\theta/2$, where the scattering angle θ is defined in Fig. 1. A scattering particle having linear size a behaves as a point scatterer if $ka \ll 1$.

A light-scattering experiment usually consists of the measurement of the average scattered intensity I_s and (or) of the optical spectrum $S(\omega)$. If ka is not negligible compared with 1, both I_s and $S(\omega)$ will be k-dependent, and therefore have to be measured as function of the scattering angle θ . The scattered intensity I_s contains information on the static properties of the scattering medium (size and shape of the scatterers, thermodynamic quantities). The optical spectrum $S(\omega)$ reflects the dynamics of the scattering medium (particle velocity, transport coefficients, internal motion of scatterers). All the experiments which will be described at this Institute yield optical spectra with a very small broadening (or a very small frequency shift) $\Delta\omega$ with respect to the incident beam frequency ω_o, that is $\Delta\omega/\omega_o \ll 1$. For this reason, such experiments are called quasielastic light scattering (QELS) experiments. Whereas static light scattering can be performed with lamps, QELS requires lasers as optical sources. Since $\Delta\omega$ is often smaller than the frequency resolution of the best optical spectrometers and also smaller than the spectral width of the laser source, the spectrum $S(\omega)$ cannot be measured directly with standard optical methods, but intensity correlation techniques are required.

Light scattering is useful because it is a nonperturbative direct probe of the state and of the dynamics of macromolecules in solution. The perturbation is minimized, of course, if the frequency of the incident light does not fall into an absorption band either of the macromolecule or of the solvent.

It should be mentioned that, instead of using light, one could perform scattering experiments with X-rays or neutrons. It goes beyond the scope of these lectures to make a comparison among different scattering techniques. I will only make a few general comments. Since the wavelength of X-rays and neutrons used is smaller than the light wavelength by a factor 10^3, it is possible to study structures in much greater detail. There are indeed several groups active in X-ray and neutron scattering. In comparison with light-scattering, X-ray scattering has two disadvantages: i) the optical contrast between macromolecule and solvent (usually water) is very low, so that only very concentrated solutions can be studied. This may cause interpretation problems, because of interparticle interactions. ii) no dynamic scattering is feasible. Neutron scattering can be both static and dynamic, using the recently developed neutron spin-echo technique[7]. However, neutron scattering requires a high flux reactor in the next room and the use of deuterated water as the solvent (to enhance the contrast).

2. LIGHT SCATTERING FROM A SINGLE PARTICLE

A. Small Particles

I will start the discussion with scatterers which are very small compared to the wavelength both outside and inside the particle. The first treatment of this problem is due to Rayleigh[10]. The particle is treated as an elementary dipole which is forced to oscillate at the same frequency as the incident field and, in turn, radiates. The radiated intensity follows the well-known dipole radiation pattern.

The simplifications introduced by the small size are that the particle may have arbitrary form and that it may be considered to be placed in a homogeneous electric field E_o.

The electric field of the scattered wave (referring to the geometry of Figure 1) is

$$\vec{E}_s(\vec{R},t) = \frac{\vec{k}_s \times (\vec{k}_s \times \vec{E}_o)}{4\pi\, \varepsilon R}\, \alpha\, e^{i(\omega_o t - \vec{k}_s \cdot \vec{R})}\, e^{i\vec{k}\cdot\vec{r}} \qquad (1)$$

where ε is the relative dielectric constant of the medium in which the particle is suspended, and α is the polarizability of the particle. If the polarizability is isotropic, if the particle is homogeneous and made of a material having index of refraction n, α is given by

$$\alpha = (n^2 - n_o^2)\, V$$

where V is the particle volume and $n_o = \sqrt{\varepsilon}$ is the index of refraction of the medium. The intensity of scattered light at $\vec{R}$ is therefore

$$I_s = I_o \left(\frac{k_o}{n_o}\right)^4 \frac{V^2 \sin^2\gamma}{16\pi^2 R^2} (n^2 - n_o^2)^2 \qquad (2)$$

Eq. (2) shows, as expected, that I_s is proportional to the incident intensity I_o and inversely proportional to R^2. The dependence of I_s on R is such to ensure that the total power of the scattered field through the surface of a sphere with radius R and centered at the scattering volume is independent of R. Furthermore I_s is proportional to k_o^4 that is to the fourth power of the frequency of the incident light. Thus blue light of 450 nm wavelength for example, is scattered more intensely than red light of 670 nm wavelength by a ratio of $(670/450)^4 \simeq 5$. Indeed the blue colour of some mountain lakes is explained by the presence of very small particles, originated by glacier erosion, which are suspended in the lake water. As far as the angular distribution of I_s concerned, Eq. (2) shows that I_s does not depend on the scattering angle θ, but depends on the angle γ. This latter dependence reflects the well known fact that an oscillating dipole does not radiate in the direction of oscillation. Finally I_s depends on the square of the particle volume, that is, for a spherical particle, on the sixth power of the radius. This means, as an example, that a single dust particle having a diameter of 1 μm scatters the same amount of light scattered by 10^{12} proteins having the diameter of 10 nm (if the optical contrast with the solvent is the same of course).

The origin of the phase factor $e^{i\vec{k}\cdot\vec{r}}$ which depends on the position of the particle can be easily understood from Fig. 3. The phase ϕ_s of the scattered field E_s is given by the phase of the incident field at the position $\vec{r}$ plus a fixed phase shift (omitted in Eq. 1 because unimportant for our discussion) introduced by the scattering process plus the phase shift due to the propagation of the scattered field from the position $\vec{r}$ to the position $\vec{R}$. The first phase term is $\omega_o t - \vec{k}_o\cdot\vec{r}$ and the last one is $-\vec{k}_s\cdot(\vec{R}-\vec{r})$. By adding the two terms and rearranging, I obtain

$$\phi_s = \omega_o t - \vec{k}_s\cdot\vec{R} + (\vec{k}_s - \vec{k}_o)\cdot\vec{r} \qquad (3)$$

Note that the dependence on the position $\vec{r}$ of the particle is limited to the phase of the scattered field. The scattered intensity I_s does not depend on $\vec{r}$. Of course, this comes from the fact that I have considered an incident plane wave with infinite wave fronts. In pratical cases the optical source is a laser beam having a Gaussian intensity profile with a beam waist w_o, so that I_s will certainly depend on the position of the particle.

If the particle is moving, $\vec{r}$ is a function of time, and therefore the phase ϕ_s of the scattered field will take a temporal behavior which contains information on particle dynamics. Note that what matters is the product $\vec{k}\cdot\vec{r}$, therefore only movements having a component in the direction of the scattering vector $\vec{k}$ can modify the phase ϕ_s. Furthermore, since $k = 4\pi n_o/\lambda \sin\theta/2$, the particle displacement necessary to change the phase ϕ_s by a fixed quantity, say π, depends on the scattering angle θ. If θ is small, the

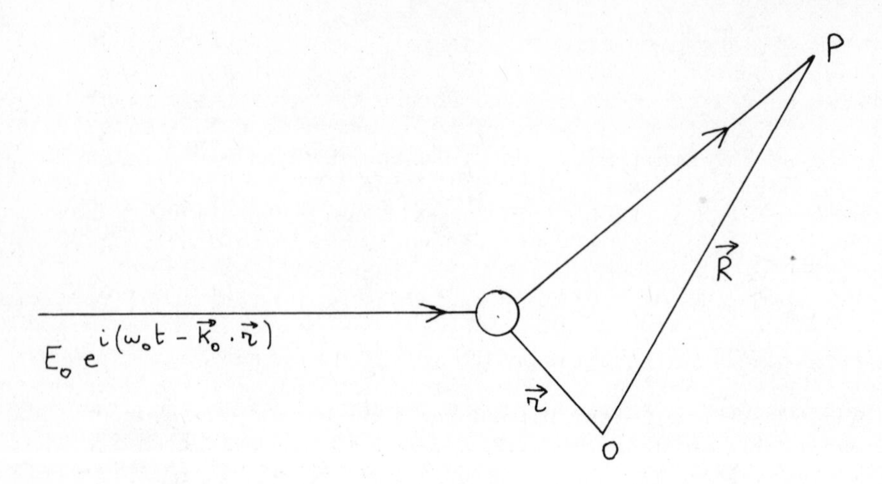

Fig. 3. Scattering from a single particle.

phase ϕ_s is sensitive only to large displacements. If θ is large, a displacement of the order of λ produces already a large effect on the phase ϕ_s. The two important cases of particle dynamics are:

i) the particle is moving with a deterministic uniform velocity $\vec{v}$; that is $\vec{r} = \vec{r}_o + \vec{v}\, t$

ii) the particle is suspended in a medium of much smaller particles (solvent molecules) and is undergoing a Brownian motion; that is $\vec{r}$ is a random variable and the displacement $\Delta\vec{r} = \vec{r} - \vec{r}_o$ is described by a tridimensional Gaussian probability distribution with a variance $<\Delta r^2> = 6Dt$ where D is the translational diffusion coefficient of the particle.

Considering first case i), the phase ϕ_s becomes

$$\phi_s = \omega_o t - \vec{k}_s \vec{R} + \vec{k}\cdot\vec{v}\, t = (\omega_o + \vec{k}\cdot\vec{v})\, t - \vec{k}_s \cdot \vec{R} \qquad (4)$$

The scattered field is monochromatic, because ϕ_s is a linear function of time, but its frequency is shifted from that of the incident field by the quantity $\vec{k}\cdot\vec{v}$. This is nothing else than the Doppler effect. If I consider, for sake of simplicity, the case $\vec{v}$ parallel to $\vec{k}$, the frequency shift is

$$\Delta\nu_D = \frac{2v\sin\theta/2}{\lambda} = \nu_o\, v/c\; 2\sin\theta/2$$

Example: $\nu_o = 6\text{x}10^{14}$Hz (green light)
$v \cong 1$ cm/s
$\theta = 10° \rightarrow \Delta\nu_D = 4\text{x}10^3$Hz; $\theta = 180° \rightarrow \Delta\nu_D = 4\text{x}10^4$Hz

Case ii) is qualitatively different from case i) because now $\vec{r}(t)$ is a random variable, and therefore the scattered field E_s is randomly phase modulated. In order to describe a random variable, one has to measure either probability densities or appropriate averages. The first significant average which contains information on the particle motion is the field correlation function

$$G_1(\tau) = \langle E_s(t)E_s(t+\tau)\rangle \tag{5}$$

which is the average product of the scattered field at time t times the complex conjugate scattered field at the delayed time t + τ. I will discuss only the simple stationary case which allows us to take time averages and yields a correlation function $G_1(\tau)$ which depends only on the delay τ. The same information is contained in the Fourier transform of $G_1(\tau)$ which is the optical spectrum $S(\omega)$,

$$S(\omega) = \frac{1}{2\pi} \int G_1(\tau)\, e^{-i\omega\tau}\, d\tau \tag{6}$$

To gain some insight on the problem it is useful to make some qualitative consideration on the width of the function $S(\omega)$. The spectrum of the incident field is made of an infinitely narrow peak at the frequency ω_o. The scattered field will not be monochromatic because ϕ^o is no longer linearly dependent on t due to the random time-dependence of $\vec{r}(t)$. The spectrum will contain a continuum of frequencies, but will still be centered around ω_o because the particle motion does not contain preferred directions. Clearly, the broadening of the spectrum will depend on the mobility of the particles. If the particles move very fast, the spectrum will contain large frequency shifts with respect to ω_o. Slow particles will produce narrow spectral broadening. The order of magnitude of the broadening $\Delta\nu_B$ can be derived as follows: given a particle in $\vec{r}_o$ at time t = 0, the time τ_c it takes to change the phase ϕ_s by π is given by the condition $k\langle\Delta r_k^2\rangle^{1/2} \simeq \pi$, where $\langle\Delta r_k^2\rangle^{1/2}$ is the root mean square displacement in the direction of $\vec{k}$. Since Δr_k is a variable undergoing a one-dimensional Brownian motion, $\langle\Delta r_k^2\rangle = 2Dt$. I obtain $k(2D\tau_c)^{1/2} \simeq \pi$, and therefore $\tau_c \simeq (Dk^2)^{-1}$. The broadening $\Delta\nu_B$ is given by $(2\pi\tau_c)^{-1} \simeq DK^2/2\pi$. A more formal calculation of τ_c may proceed as follows. The field correlation function $G_1(\tau)$ may be written, according to Eqs.(1) and (5), as

$$G_1(\tau) = I_s e^{i\omega_o\tau} \langle e^{i\vec{k}\cdot[\vec{r}(t)-\vec{r}(t+\tau)]}\rangle \tag{7}$$

The exponential can be formally written as a series expansion:

$$e^{ik[r_k(t)-r_k(t+\tau)]} = e^{ik\Delta r_k(\tau)} = \sum_{\ell=0}^{\infty} \frac{(ik)^\ell\,[\Delta r_k(\tau)]^\ell}{\ell!}$$

The moments of the Gaussian variable Δr_k are:

$$\langle \Delta r_k^{2\ell+1} \rangle = 0$$

$$\langle \Delta r_k^{2\ell} \rangle = (2\ell - 1)(2\ell - 3) \ldots 3\cdot 1 \; \langle \Delta r_k^2 \rangle^{\ell}$$

Recalling that $\langle \Delta r_k^2 \rangle = 2Dt$, I obtain finally

$$G_1(\tau) = I_s \, e^{i\omega_o \tau} \, e^{-Dk^2\tau} \tag{8}$$

The spectrum $S(\omega)$ is easily calculated from Eqs (6) and (8). The result is

$$S(\omega) = I_s \frac{Dk^2/\pi}{(Dk^2)^2 + (\omega-\omega_o)^2} \tag{9}$$

Plots of $G_1(\tau)$ and $S(\omega)$ according to Eqs (8) and (9) are shown in Fig. 4.

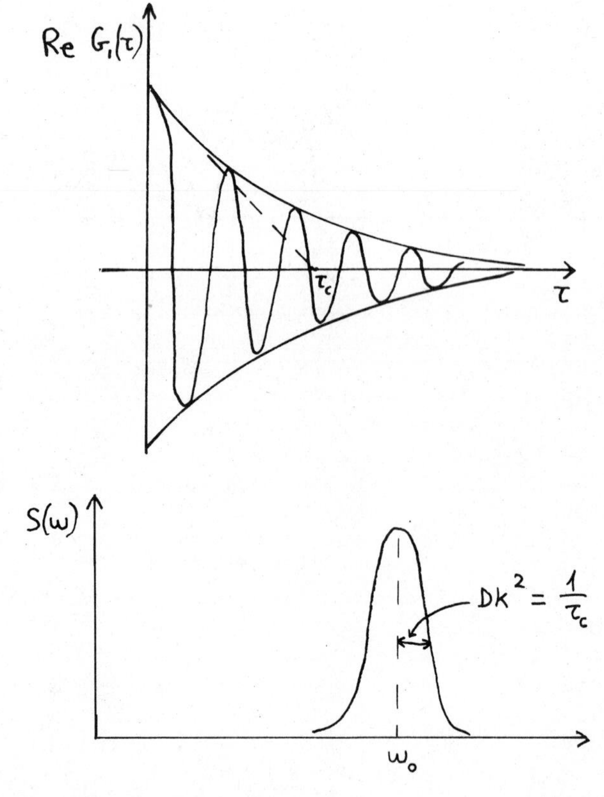

Fig. 4. Field correlation function and optical spectrum of light scattered from a single particle.

The translational diffusion coefficient D is connected to the friction coefficient f through the Einstein relation

$$D = k_B T/f \tag{10}$$

where T is the absolute temperature and k_B is the Boltzmann constant. For a sphere of radius a, $f = 6\pi\eta a$, where η is the dynamic viscosity of the solvent. By substitution into Eq. (10), one obtains the Einstein-Stokes relation.

$$D = k_B T/(6\pi\eta a) \tag{11}$$

More generally, D should be written as

$$D = k_B T/(6\pi\eta R_H) \tag{12}$$

where R_H is the hydrodynamic radius of the particle which is larger than the geometric radius a because the particle also drags solvent molecules in its Brownian motion. For a nonspherical particle, $R_H = f_o R_e$, where R_e is the radius of the sphere having equivalent volume and $f_o \geq 1$ is a form factor which is equal to 1 for a sphere and becomes very large when the shape of the particle deviates considerably from the sphere, as will be discussed in greater detail in Cummins' lectures.

Since it is always instructive to know the limits of validity of the formulas we are using, let me recall that, for very short delays τ, $|G_1(\tau)|$ is Gaussian instead of exponential. Indeed the particle's random walk is such that the particle moves in an arbitrary direction for a step length ℓ_{FP} with a mean velocity v (for a mean duration $\tau_{FP} = \ell_{FP}/v$), then turns around to a randomly selected direction and repeats the same motion. When the spatial coarseness of the light scattering experiment is much larger than the mean free path ℓ_{FP}, that is $k\ell_{FP} \ll 1$, we are in the so-called hydrodynamic regime to which Eq.(8) applies. In the opposite limit (kinetic regime), the particle velocity persists for an interval long enough to give rise, for each mean free path, to a Doppler-shifted scattered light, so that the spectrum $S(\omega)$ will simply reflect the Maxwell velocity probability density. Therefore $S(\omega)$ will be Gaussian and

$$|G_1(\tau)| = I_s\, e^{-k^2 v^2 \tau^2/2} \tag{13}$$

Light scattering experiments of biological interest are in the hydrodynamic regime, so that $G_1(\tau)$ and $S(\omega)$ are given respectively by Eqs (8) and (9). Consider the following numerical example: a spherical particle (protein or micelle) with radius 2.5 nm in aqueous solution at room temperature. Taking $T = 293°k$, $\eta = 10^{-2}$ cp, $\lambda = 0.5$ µm(argon laser), $n_o = 1.33$, $\theta = 90°$, I find:

$$D \simeq 10^{-6} \text{cm}^2/\text{s}$$

$$k^2 \simeq 5\times10^{10}\,\mathrm{cm}^{-2}$$

Therefore the decay time of $|G_1(\tau)|$ is $\simeq 20$ μs or, equivalently, the frequency broadening of scattered light is

$$\Delta\nu_B = \frac{Dk^2}{2\pi} \simeq 8 \text{ kHz}.$$

B. Size Effects

When the linear size of the scatterer is not negligible with respect to k^{-1}, the scattered intensity is generally reduced because of interference effects. These effects may be formally taken into account by including a multiplicative θ-dependent form factor $P(\theta)$ into the scattered intensity formula

$$I_s = I_o\left(\frac{k_o}{n_o}\right)^4 \frac{V^2 \sin^2\gamma}{16\,\pi^2 R^2} (n^2 - n_o^2)^2\, P(\theta) \tag{15}$$

The most precise method of evaluation of $P(\theta)$ is that proposed by Mie, which is rather complex and is usually solved numerically. However, large part of the problems met in light scattering applications to biology can be approached by using the Rayleigh-Debye approximation which is valid when

$$(n - n_o)\, a \ll \lambda/4\,\pi \tag{16}$$

Note that, for small values of $n - n_o$, Eq.(16) can be satisfied also when the size of the particle is comparable to the wavelength. The form factor $P(\theta)$ of a homogeneous particle, within the Rayleigh-Debye approximation, is

$$P(\theta) = \left|\frac{1}{V} \int_V e^{i\vec{k}\cdot\vec{r}}\, dv\right|^2 \tag{17}$$

where V is the volume of the scatterer. The behavior of $P(\theta)$ for some simple shapes, like spheres, discs, rods and random coils can be found in Ref. 11 and works mentioned therein. For any shape of the particle, it can be shown that the small angle limit of $P(\theta)$ is given by

$$P(\theta) \xrightarrow[\theta\to 0]{} 1 - k^2 R_G^2/3 \tag{18}$$

where R_G is the particle's radius of gyration, defined as

$$R_G^2 = \frac{1}{V} \int_V r^2 dv \tag{19}$$

r being the distance of the volume element dv from the center of mass of the particle.

Some examples are: sphere with radius R, $R_G^2 = 3/5\ R^2$; rod with radius R and length L, $R_G^2 = L^2/12$ (for L>>R). The value of $P(\theta)$ for a sphere with radius R = 10 nm, immersed in water, $n_o = 1.33$, taking $\lambda = 0.5$ nm, $\theta = 180°$, is $P(\theta) = 0.96$, that is, the backscattered intensity is 4% lower than the forward scattered intensity.

Eqs (17) and (19) apply only to homogeneous scatterers, but can be easily extended to inhomogeneous ones. I will not enter into a systematic discussion of this point. I will only report about a specific case, discussed few years ago by Kerker [12], which illustrates an extreme situation: a large scatterer which may be invisible. The problem concerns the inclusion of microvoids in paints. Consider the scatterer shown in Fig. 5 which consists of a spherical pigment particle surrounded by a void concentric shell, all encased in a resin. The calculated scattering cross-section is shown in Fig. 6 as a function of f, volume fraction of the core, for the following parameters; $\lambda = 0.546$ µm, n = 2.97 for the pigment core (titania particle), n' = 1.00 for the microvoid (air), $n_o = 1.51$ for the surrounding resin. The parameter ν is the ratio of outer circumference to wavelength. You can see that, when $\nu = 0.3$ and f = 0.42, the scattering cross section decreases by several orders of magnitude.

Next, we consider dynamic light-scattering. If the scatterer cannot be described as a point scatterer and is not spherical, rotational diffusion will give rise to a time-dependent scattering amplitude. Furthermore, if the particle is not rigid (e.g., flexible coils), intramolecular modes of motion will also influence the correlation function of the scattered field. The starting point of the theoretical treatment [13] is to consider the particle as a collection of N identical scatterers each with isotropic, constant polarizability α. The i-th scatterer is at position $\vec{r}_i(t)$ at time t. The scattered electric field at scattering angle θ is then proportional to

$$E_s(t) = \alpha \sum_{i=1}^{N} e^{i\vec{k}\cdot\vec{r}_i(t)} \qquad (20)$$

Let $\vec{r}_o(t)$ be the position of a point within the particle (the center of mass, for instance) and $\vec{b}_i$ be the position of segment $\vec{r}_i$ relative to that point,

$$\vec{r}_i(t) = \vec{r}_o(t) + \vec{b}_i(t)$$

Thus

$$E_s(t) = \alpha\, e^{i\vec{k}\,\vec{r}_o(t)} \sum_{i=1}^{N} e^{i\vec{k}\cdot\vec{b}_i(t)} \qquad (21)$$

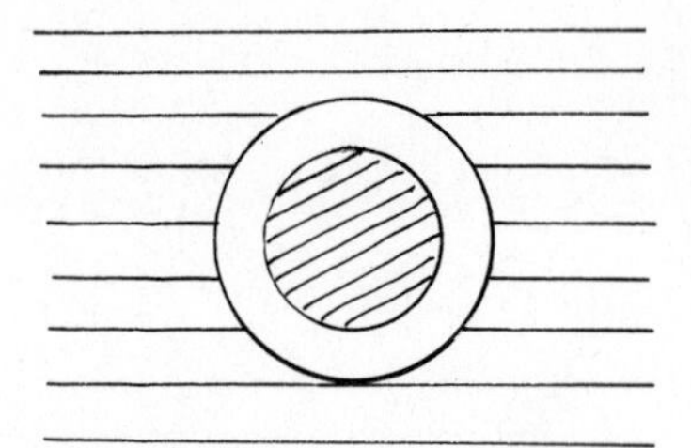

Fig. 5. "Invisible" scatterer: model for pigmented microvoid coating.

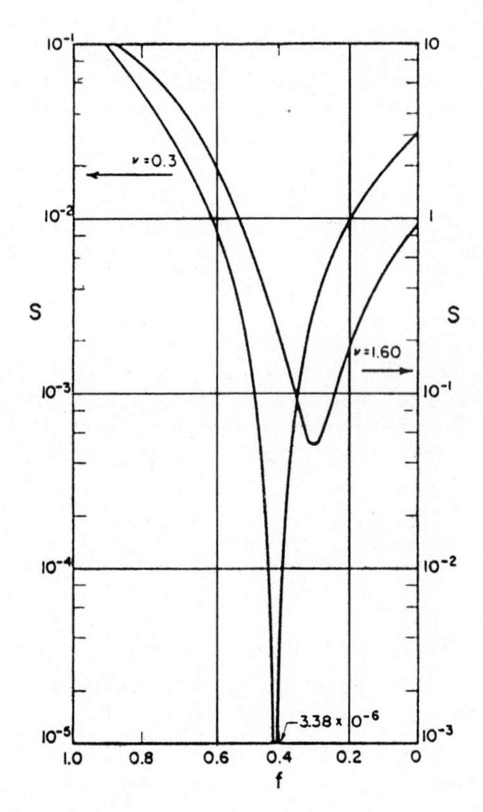

Fig. 6. Scattering cross section of the scatterer of Fig. 5 versus the volume fraction of core.

The scattered field has a time-dependent amplitude, whereas for a point scatterer we found that only the phase of E_s was time-dependent. If the particle is small, that is its size ℓ is such that $k\ell \ll 1$, the summation of the intramolecular exponential terms contributes a constant factor N and we go back to the case of the point scatterer. Starting from eq.(21) and recalling the definition of $G_1(\tau)$ it is possible to compute $G_1(\tau)$ for some simple models. For instance, the case of a rigid rod of length L undergoing independent rotational and translational diffusion gives[13]

$$|G_1(\tau)| = I_s \, e^{-k^2Dt} \, (S_o + S_1 \, e^{-6D_R\tau} + \ldots) \qquad (22)$$

where the omitted terms are negligible for $kL \leqslant 8$, D_R is the rotational diffusion coefficient, and S_o and S_1 are known functions of kL. When $kL \leqslant 3$, $S_o \simeq 1$ and $S_1 \simeq 0$. When $kL \simeq 6$, $S_1 \simeq 0.1$. A second model which was considerably discussed is that of flexible coils. Under several assumptions fully discussed in Ref. 13, $|G_1(\tau)|$ can be put in a form identical to that of Eq.(22), where now the exponential in brackets is $e^{-2\tau/\tau_1}$ instead of $e^{-6D_R\tau}$, τ_1 being the longest intramolecular relaxation time of the chain, and the amplitudes S_o and S_1 depend on kR_G, R_G being radius of giration of the chain.

A final point to be mentioned is the existence of depolarized light scattering. Consider the case of an incident field which is linearly polarized in a direction perpendicular to the scattering plane (the plane containing $\vec{k}_o$ and $\vec{k}_s$). An optically isotropic point scatterer will give a scattered field which is also linearly polarized perpendicularly to the scattering plane. An anisotropic scatterer will originate a second component in the scattered field which is polarized in the scattering plane and is termed the depolarized field. A particle made of optically isotropic material may also give rise to a depolarized scattered field E'_s if it has a nonspherical form. This is called a form-anisotropy effect. The field E'_s will be amplitude modulated by the rotational motion of the particle. For a rigid particle with rotational symmetry, the correlation function of E'_s is

$$|G'_1(\tau)| = I'_s \, e^{-(Dk^2 + 6D_R)\tau} \qquad (23)$$

where I'_s is the depolarized scattered intensity which is usually much smaller than I_s. If $|G'_1(\tau)|$ is measured at small angles, $Dk^2 \to 0$ and the time decay of the correlation function is controlled only by the rotational diffusion coefficient D_R. Just to give an idea of the orders of magnitude involved, $D_R \sim 10^6 \, s^{-1}$ for a small globular protein like lysozyme($M \simeq 15000$), and $D_R \sim 10^2 \, s^{-1}$ for the rodlike tobacco mosaic virus ($M \simeq 40 \times 10^6$).

3. THE LIGHT SCATTERING EXPERIMENT

The light scattering experiment is performed by sending a continuous laser beam into the sample cell. Dynamic light scattering measurements require a small scattering volume (in order to maximize the signal-to-noise ratio [14]), so that the laser beam is focussed down to a spot diameter which is typically in the range 50-100 μm. Although the full path of the beam inside the scattering cell may be 1 cm, the collecting optics selects only a small portion of this path, typically 100 μm, around the focal region. Therefore the actual scattering volume is a cylinder of about 10^{-6} cm^3. Such a volume will usually contain many scatterers. Take, for instance, a water solution of a protein, say bovine serum albumin, having a molecular weight ~60000 daltons, at a concentration of 1 mg/cm^3. It is easy to calculate that 10^{-6} cm^3 will contain 10^{10} proteins! Since the particles are moving, the number N of particles within the scattering volume fluctuates in time. With independent particles one expects Poisson statistics for N, which implies that relative fluctuations are 1/N. When $N = 10^{10}$, fluctuations in the number of scatterers are therefore negligible. I will only consider the case of large N.[15]

The scattered field will be given by the sum of all the contributions from the N particles:

$$E_s(t) = e^{i(\omega_o t - \vec{k}_s \cdot \vec{R})} \sum_{j=1}^{N} A_j \, e^{i\vec{k}\cdot\vec{r}_j(t)} \tag{24}$$

where A_j is the scattering amplitude of the jth particle. The expression of A_j can be derived from Eq. (1). The field E_s is the superposition of a large number of independent contributions, and becomes therefore a Gaussian random variable. An important point to note is the following: the field scattered from a single small particle has a constant amplitude and a randomly modulated phase. The scattered field of Eq.(24) is also amplitude-modulated because of the interference among all the fields scattered by the N particles.

It is useful to briefly discuss the spatial coherence of the scattered field. The problem can be formulated in the following way. Consider a sphere of radius R centered on the scattering volume, with R much larger than the linear size of the scattering volume. Let $E_s(\vec{R},t)$ be the scattered field. One could ask: how far can I move on the sphere surface in order to find a scattered field with an amplitude and a phase different from those found in $\vec{R}$? The area over which E_s is coherent is called the coherence area, A_{coh}. In order to evaluate A_{coh} consider the simple scheme of Fig. 7. The incident beam propagates from left to right hand, the scattering volume is limited to the length d between A and B. Assume that the

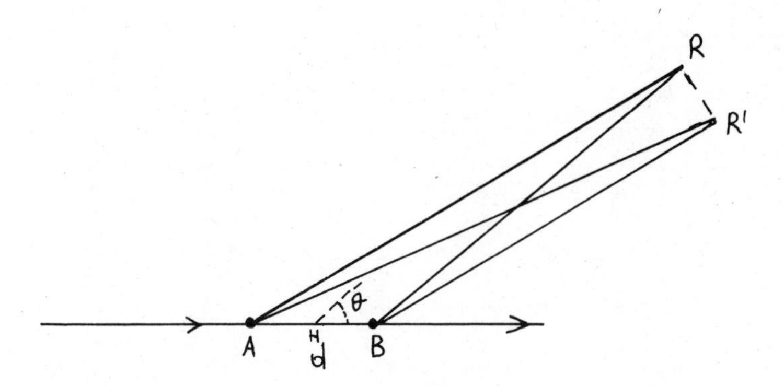

Fig. 7. Elementary calculation of the coherence area.

fields scattered by two scatterers at A and B produce constructive interference in R. Now I move fron R to R'. The phase difference between two scattered fields in R' is

$$\Delta\phi = k_s \left(AR' - AR - BR' + BR\right) = \frac{2\pi}{\lambda} \quad \Delta R \,\frac{d \sin\theta}{R}$$

where ΔR is the distance RR'. The distance ΔR which yields $\Delta\phi = 2\,\pi$ is

$$\Delta R = \frac{\lambda\, R}{d \sin\theta} \tag{25}$$

For a displacement perpendicular to the scattering plane I will find

$$\Delta' R = \frac{\lambda\, R}{w_o} \tag{26}$$

where w_o is the beam diameter. Therefore $A_{coh} = \Delta R \cdot \Delta' R = \lambda^2 R^2 / (w_o d \sin\theta)$. It should be clear that, if we want to study the fluctuations of the scattered field, we should choose a detector area smaller or equal to A_{coh}, otherwise we average out the fluctuations we are interested in. To make A_{coh} large, both w_o and d must be small.

It is important to realize that it is not possible to measure directly by an optical spectrometer the linewidth (or the frequency shift) of the scattered field when the broadening (or the shift) is of the order of 10 kHz as for the examples discussed in the previous section. Indeed what is practically done is to measure not $G_1(\tau)$ (or $S(\omega)$), but the intensity-correlation function

$$G_2(\tau) = \langle I_s(t)\, I_s(t+\tau) \rangle \tag{27}$$

The measurement of G_2 proceeds as follows: the scattered field is collected by a photodetector which yields an output electric current i(t) instantaneously proportional to the incident scattered intensity I_s. The correlation function of i(t) can be obtained by real-time digital correlators, as described in detail in Refs. 16 and 17.

The usefulness of intensity-correlation spectroscopy (also called photon-correlation spectroscopy because the signal i(t) is usually a sequence of single photoelectron pulses) is evident if we recall that G_2 is simply related to G_1 when the scattered field is a Gaussian random process. The relation is

$$G_2(\tau) = G_1^2(0) + |G_1(\tau)|^2 \tag{28}$$

If, for instance, G_1 is given by Eq.(8), we obtain

$$G_2(\tau) = I_s^2 \left(1 + e^{-2Dk^2\tau}\right) \tag{29}$$

Note that we can derive the spectral broadening but information about the central frequency ω_o is lost when we measure $G_2(\tau)$. This is obvious because G_2 is an intensity correlation, and the intensity is not sensitive to the phase of the field. Clearly, the method of direct detectionof the scattered field followed by a correlation measurement is not fit for a laser Doppler velocimetry experiment because in this case the relevant information is contained in the frequency shift and not in the frequency broadening of the scattered field. It is instead necessary to use a method which gives the real part (and not the modulus of $G_1(\tau)$). This is the reference-beam method, schematized in Fig. 8. This method is completely general, that as it does not rely on the assumption that the field is Gaussian. The field E_s we want to study is superposed on a

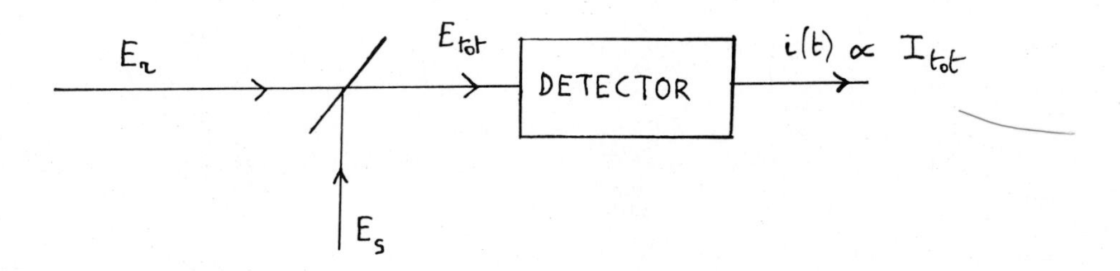

Fig. 8. The reference-beam method.

reference field E_r. The total field impinging on the photodetector surface is

$$E_{tot}(t) = E_s(t) + E_r(t) \tag{30}$$

The intensity is

$$I_{tot} = |E_{tot}|^2 = I_s(t) + I_r(t) + 2\mathrm{Re}\,\{E_s(t)E_r^*(t)\} \tag{31}$$

If the two fields are monochromatic, the third term at right-hand side (interference term) oscillates at the frequency difference between the two fields. Therefore the power spectrum of $I(t)$ will show a peak at the frequency difference. More generally, one can write the intensity correlation function of the total field, by assuming that the two fields are statistically independent, as

$$G_{2tot} = G_{2s} + G_{2r} + 2\langle I_s\rangle\langle I_r\rangle + 2\mathrm{Re}\,\{G_{1s}\,G_{1r}^*\} \tag{32}$$

By putting $E_s(t) = E_{os}(t)\, e^{i[\omega_{os}t + \phi_s(t)]}$

$$E_r(t) = E_{or}\, e^{i(\omega_{or} t + \phi_r)}$$

with E_{or}, ϕ_r deterministic values (the reference field is free from fluctuations), and by writing G_{1s} and G_{2s} as

$$G_{1s}(\tau) = \langle I_s \rangle\, e^{i\omega_{os}\tau}\, g_1(\tau)$$

$$G_{2s}(\tau) = \langle I_s \rangle^2 \left[1 + g_2(\tau)\right],$$

I obtain

$$G_{2tot} = \langle I_s \rangle^2 \left[1+g_2(\tau)\right] + I_r^2 + 2I_r \langle I_s \rangle \{1+\cos(\omega_{os}-\omega_{or})\tau\, g_1(\tau)\} \quad (33)$$

If $I_r \gg \langle I_s \rangle$, the first term can be neglected, and we find that the time-dependent part of G_{2tot} is proportional to the real part of $G_{1s}(\tau)$ shifted around the frequency ω_{or}. The reference-beam technique is therefore useful as a differential technique, and is particularly suited to velocimetry experiments where very small frequency shifts have to be detected. In this case the reference beam is simply a portion of the incident beam, so that the difference $\omega_{os} - \omega_{or}$ coincides with the Doppler shift. To be rigorous, one should note that, if E_r is proportional to the incident beam, E_s is not statistically independent of E_r. However, provided that the incident field has negligible intensity fluctuations, Eq.(33) can still be applied. As discussed in Earnshaw's lectures , what is usually measured is the power spectrum, that is the Fourier transform of G_{2tot}. Spectral analysis is widely used because it gives directly the velocity distribution of the scatterers.

I will close the Section by recalling some basic information about the light-scattering apparatus. The optical source may be a He-Ne laser (λ = 0.6328 μm, output power 1-20 mW) or an ion argon laser (λ = 0.488 and 0.5145 μm, output power 1-200 mW). The laser must be intensity stabilized to better than 1%, operates on a single transverse TEM_{00} mode (Gaussian intensity profile), and can work on many longitudinal modes provided that its coherence volume remains larger than the scattering volume. The scattering cell must be clean and possibly much larger than the scattering volume in order to be able to separate with the collecting optics the useful scattered light from the stray light due to scattering on the cell walls. Dust (or any large unwanted scatterer) is eliminated by microporous filtering of the solution or by centrifugation. The cell may be temperature-controlled. In any case it is important to avoid temperature gradients which could induce convective motions inside the cell. The collecting optics are usually designed in such a way as to select one (or few) coherence areas on the photodetector surface. The photodetector is a low-noise photomultiplier fit for photon counting work (ITT FW 130, EMI 9863 KB, etc.).The electronic processing of the photomultiplier signals requires a fast amplifier, a discriminator, and a digital correlator interfaced with a computer. For further details see Refs 1,16 and 17.

4. INDEPENDENT AND INTERACTING SCATTERERS

If the solution is sufficiently dilute to minimise the interactions among the scatterers, it is easy to show that the scattered intensity is simply the sum of the intensities scattered by the individual particles, i.e.

$$I_s = \sum_{i=1}^{N} I_{si} \tag{34}$$

Polydisperse systems will be discussed in detail by Prof. Chu at this Institute. I will consider, for sake of simplicity, N identical scatterers. In this case Eq.(34) can be put into the form

$$I_s = A \left(\frac{dn}{dc}\right)^2 Mc\, P(\theta) \tag{35}$$

where M is the molecular weight of the particle, c is the concentration of suspended material in g/cm^3, dn/dc is the refractive index increment (measured by a differential refractometer) and A is a calibration constant.

Eq.(35) can be used to derive M from the measured scattered intensity, provided that I_s is extrapolated at zero scattering angle (recall that $P(\theta) \to 1$ as $\theta \to 0$). Since an absolute calibration is difficult to perform, A can be derived by comparison with a reference solution of known scattering properties. I have recently discussed, in a paper written with Corti and Minero [18], the feasibility of using a micellar solution as a light scattering standard. As an example, take the nonionic surfactant $C_{12}E_8$ which forms micelles having M = 65000 in aqueous solution. A solution of 10 mg/cm^3 of $C_{12}E_8$ at 20°C (with dn/dc = 0.13) has a Rayleigh ratio of $30 \times 10^{-6} cm^{-1}$ at λ = 514.5 nm (the intensity of scattered light is 40 times that scattered by pure water).

When the particles are interacting, Eq.(35) can still be used to yield an apparent molecular weight M_{app} which is concentration dependent. To the first order in the concentration c, M_{app}^{-1} can be written in the form

$$M_{app}^{-1} = M^{-1} (1+k_I c) \tag{36}$$

where k_I is proportional to the second virial coefficient, and can therefore be expressed as a function of the interparticle interaction potential. When the potential is repulsive, k_I is positive. As an example, I present in Fig. 9 the behavior of M_{app} as a function of the concentration c for aqueous solutions of sodium dodecyl sulfate (SDS) micelles [19]. The SDS micelles have an electric charge of about 40 electronic charges, and have a globular shape with a hydrodynamic radius around 2.5 nm. The different curves refer to different ionic strengths of the solution. At low ionic strength the excluded volume and electrostatic repulsion effects are predominant ($k_I \simeq 30$ cm^3/g at 0.1 M NaCl). At increasing ionic strength the Coulomb potential is screened more and more effectively and London-van der Waals attraction becomes important. The results may be quantitatively described by the so called Derjaguin-Landau-Verwey-Overbeek theory of colloid stability. Note, incidentally, that for very low ionic strengths the electrostatic interactions may give rise to a long-range ordered system (colloidal crystal). A discussion of this interesting field may be found in several papers by Pusey and others in Ref. 6.

A similar approach applies to dynamic light scattering[19]. For a system of identical, small, non-interacting scatterers, the intensity correlation function is given by

$$G_2(\tau) = \langle I_s \rangle^2 (1 + Be^{-2Dk^2\tau}) \tag{37}$$

where B is an instrumental constant and D is the translational diffusion coefficient of the individual particle. When the particles are interacting, the time-dependent part of $G_2(\tau)$ can usually be fitted with a single exponential, but the obtained diffusion coefficient is the collective diffusion coefficient D_c which is concentration-dependent. For dilute solutions,

$$D_c = D\,(1 + k_D c) \tag{38}$$

where k_D is connected to the interparticle interaction potential by an integral expression which contains also hydrodynamic effects (because the movement of one particle through the fluid generates a velocity field which affects the motion of neighboring particles). The behavior of k_D is qualitatively similar to that of k_I, as shown by the dynamic light scattering measurements on SDS micelles[19] reported in Fig. 10. For hard spheres, $k_I = 8\,\bar{v}$ and $k_D = 1.5\,\bar{v}$, where $\bar{v}$ is the specific volume of the solute particles. A good example of a solution of hard spheres seems to be a dilute solution of nonionic micelles, such as $C_{12}E_8$ in water in the temperature range 10-40°C and in the concentration range 0-30 mg/cm^3 by weight[18].

A last point to be mentioned in connection with interactions is that there are techniques which measure the self-diffusion coefficient D_s (for instance, by radioactive or fluorescent tagging

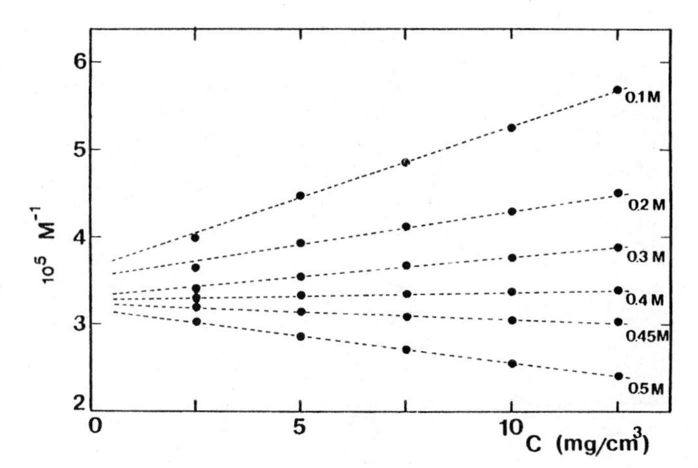

Fig. 9. The reciprocal of the apparent molecular weight M^{-1} versus surfactant concentration for a aqueous solution of ionic micelles at various ionic strengths.

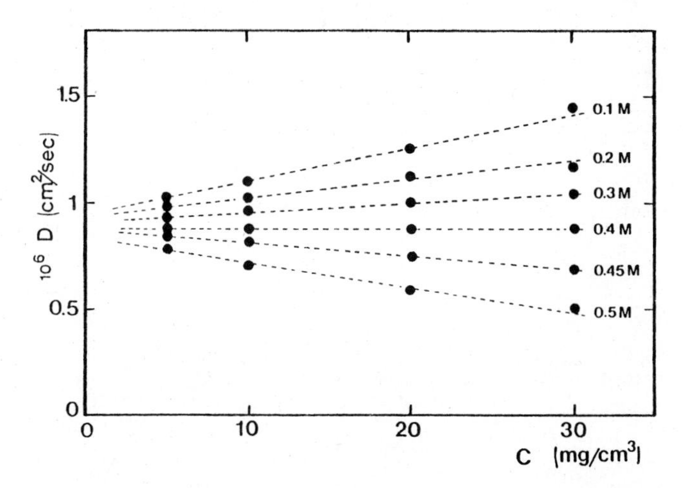

Fig. 10. The diffusion coefficient D versus surfactant concentration for the same system of Figure 9.

of some particles). One should not expect the concentration dependence of D_s to follow that of D_c, so that the two coefficients coincide only in the infinite dilution limit.

REFERENCES

1. B. Chu,"Laser Light Scattering", Academic Press, New York (1974)
2. B. J. Berne and R. Pecora "Dynamic Light Scattering", Wiley, New York, (1976)
3. B. Crosignani, P. Di Porto, and M. Bertolotti,"Statistical Properties of Scattered Light", Academic Press, New York (1975)
4. "Photon Correlation and Light Beating Spectroscopy", H. Z. Cummins, and E. R. Pike, eds.,Plenum, New York (1974)
5. "Photon Correlation Spectroscopy and Velocimetry", H.Z. Cummins, and E. R. Pike, eds.,Plenum, New York (1977)
6. "Light Scattering in Liquids and Macromolecular Solutions", V. Degiorgio, M. Corti, and M. Giglio, eds., Plenum, New York (1980)
7. "Scattering Techniques Applied to Supramolecular and Nonequilibrium Systems", S. H. Chen, B. Chu, and R. Nossal, eds., Plenum, New York (1981)
8. H. C. Van de Hulst,"Light Scattering by Small Particles", Wiley, New York (1962)
9. M. Kerker,"The Scattering of Light and Other Electromagnetic Radiation", Academic Press, New York (1969)
10. A pleasant historical excursion through the physics of light scattering is offered by A.T. Young, Rayleigh Scattering, Phys. Today, p.32 (Jan. 1982)
11. See H. Z. Cummins in Ref. 4
12. M. Kerker, Some Recent Reflections on Light Scattering, J. Coll. Interface Sci., 58:100 (1977)
13. See R. Pecora in Ref. 7
14. See E. Jakeman in Ref. 5
15. For a discussion of the case of small N, see P. N. Pusey in Ref. 5
16. See C. J. Oliver in Ref. 4
17. See V. Degiorgio in Ref. 5
18. V. Degiorgio, M. Corti, and C. Minero, Absolute Calibration of Light Scattering Data from Micellar and Macromolecular Solutions, Nuovo Cimento D (to be published)
19. M. Corti and V. Degiorgio, Quasi-Elastic Light Scattering Study of Intermicellar Interactions in Aqueous Sodium Dodecyl Sulfate Solutions, J. Phys. Chem., 85:711 (1981).

STRUCTURE AND MOVEMENT IN CELLS

Martin W. Steer

Department of Botany
The Queen's University of Belfast
N. Ireland, BT7 1NN

CONTENTS

1. INTRODUCTION

In this introductory chapter I will review basic cellular processes and structures, emphasising the dynamic activity inherent in these biological systems. Our present knowledge of biological structure and functional activity derives from two distinct approaches, structural studies, made on living material in the light microscope and preserved material in the electron microscope; and biochemical studies made on whole living tissues and isolated tissue and cell fractions. While cytochemical studies on cells and microscopical studies on cell fractions represent attempts to bridge the gap between these two approaches there is still a lack of direct information on the dynamic activities of cells below the level of resolution of the light microscope. Our present information is largely derived from extrapolations of biochemical rate processes to the cellular level. These are made over comparatively long time scales and tell us little about the way in which these processes are controlled or modulated in the cell on a biologically significant time scale.

Laser light scattering (LLS) techniques have been successfully used to study _in vitro_ processes and there is now some reason to be optimistic that they can be applied to _in vivo_ situations. This discussion will emphasise the size of the structures involved in living cell activities and the rates at which these structures may move. As far as possible information will be presented in terms useful to those interested in the number and size of potential laser light scatterers within living cytoplasm. However the range of cellular diversity is such that it is essential, with present LLS techniques, to clearly establish the structure of the cells under study using quantitative electron microscopic methods [1,2] From the present survey we will see that cellular systems contain a very broad range of structural sizes and dynamic motions.

2. MEMBRANES

Lipid-protein membranes provide the basic cell architecture, forming the plasma membrane, a barrier separating living intracellular space from the non-living environment, and numerous internal membranes that generate a number of internal cellular compartments, each with a characteristic structure and function.

The component lipid and protein molecules within each half-layer of the membrane[3] undergo thermal diffusion movements within the plane of the membrane layer. These movements have been detected by fluorescence correlation spectroscopy, with diffusion coefficients of 2×10^{-10} cm^2 sec^{-1} for protein and 9×10^{-9} cm^2 sec^{-1} for lipid.[4] Such movements may by restricted by certain lipids, such as cholesterol, and by proteins that span both layers and interact with the underlying cytoplasm;[5,6] they may be enhanced, as in the directed movements of receptor proteins to internalisation sites ('capping') or in the bulk movement of membrane during growth, locomotion etc (Fig 1).

The surface density (area of membrane per unit volume) of these membranes within cells is always high (1-3 μm^2 μm^{-3}) sometimes astonishingly so (12-20 μm^2 μm^{-3}).

Membranes serve two important functions, one is associated with their barrier function, and movement of solutes and water across the barrier, and the second is the enzyme activities of their component protein molecules. Both these functions generate significant flows in the vicinity of the membrane. Plasma membrane fluxes of solutes, low molecular weight sugars, amino acid, nucleotides etc; usually lie in the range 1-20 picomoles cm^{-2} sec^{-1}, although higher values have been reported[7] (Fig 2).

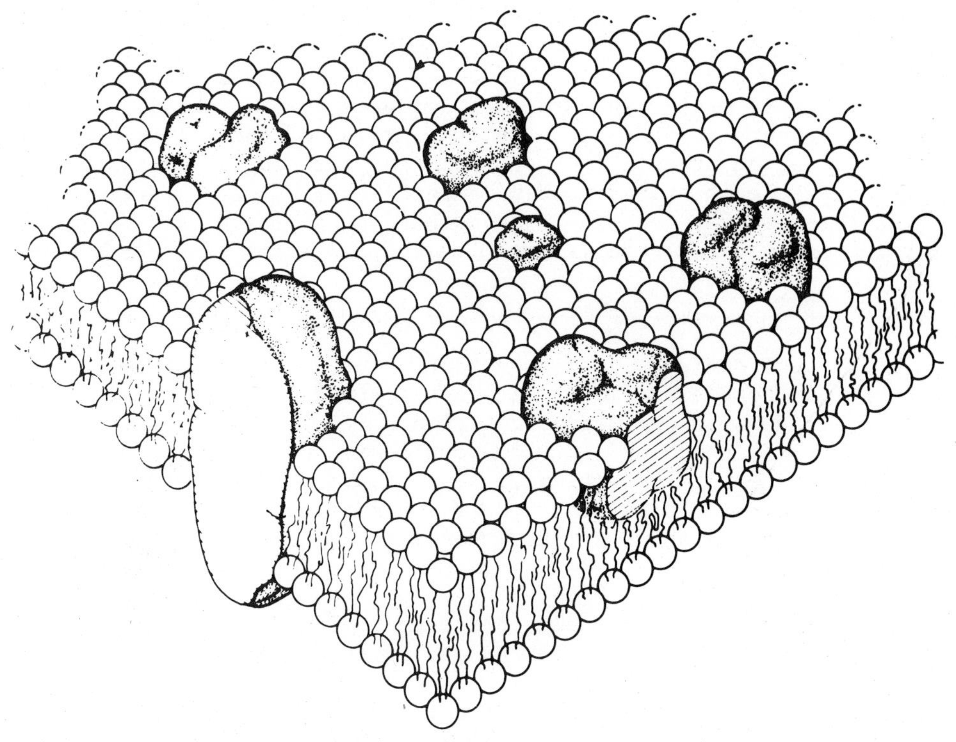

Fig 1. Model of lipid bilayer membrane structure with protein molecules either confined to one lipid leaflet or spanning the entire structure (from ref 3).

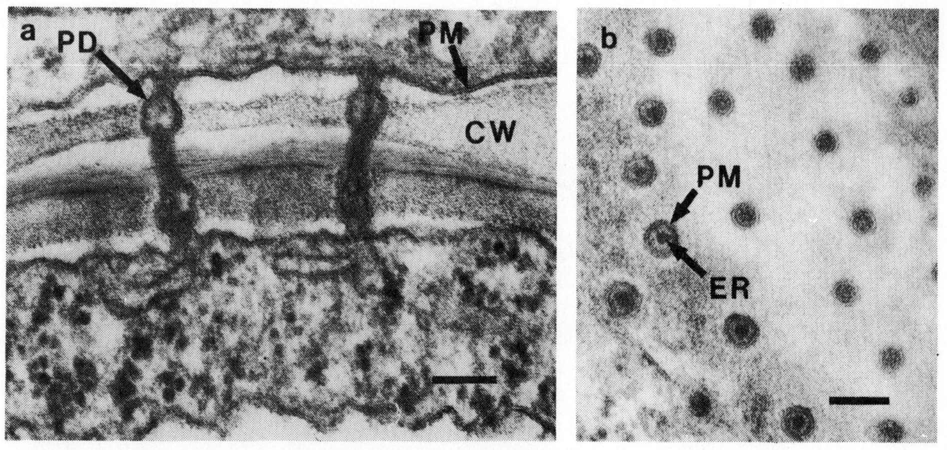

Fig 2. Plasmodesmata (PD) traversing the cell walls of a plant leaf: (a) PD in longitudinal section, note the cell wall (CW) and the plasma membrane (PM) bounding each cell; bar 1 μm; (b) PD in cross section lying in the cell wall, each is bounded by a plasma membrane (PM) and contains a tubule of endoplasmic reticulum (ER), bar 0.1 μm.

Many higher plant cells are interconnected by fine plasmamembrane lined tubes, 40 nm in diameter and about 0.1-1.0 μm long (Fig 2). These tubes usually contain an inner tube of endoplasmic reticulum membrane. While there is much circumstantial evidence that intercellular solute movement occurs in the plasmodesmata, it is not known whether this is by diffusion or bulk flow of solution; both would be capable of sustaining the observed gross movement of molecules. There are plant tissues that contain large numbers of plasmodesmata (> 1,000 per 100 μm^2 of wall) that should be accessible for LLS experiments.[8]

3. NUCLEUS

The cell nucleus contains a high concentration of DNA- and RNA-protein complexes, making it one of the densest cell compartments. It is also the largest, with diameters in the range 5-50 μm. The nucleus is concerned with the synthesis and processing of these large nucleic-protein complexes, some of which are released into the cytoplasm. It also undergoes a cycle of activity culminating in mitosis and movement of chromosomes to different parts of the cell (Fig 3a).

Movements are not usually observed within the nucleus, with the exception of the slow filling and discharge of vacuole like bodies found in some cell types. At the molecular level we know that the nucleus in the centre of intense activity with mRNA synthesis on the chromatin and ribosome synthesis in the nucleoli. Both these products form large macromolecular complexes that must migrate through the nucleoplasm to the nuclear surface and out to the cytoplasm. Some idea of the scale of this movement can be obtained from the rates of synthesis. For example ribosomes are made at the rate of 100-300 sec^{-1} per nucleus.[9]

This flow of large macromolecules crosses the nuclear envelope, which has double membrane construction, via the nuclear pores (Fig 3b). These are open cylinders, 60 nm in diameter and 100-200 nm long, that pass through both lipoprotein membranes.[10] Pore density over the surface of the nucleus varies with the state of activity but usually lies within the range 5-30 pores μm^{-2}. Movement of particles through the pore is known to be restricted to those less than about 13 nm diameter, implying that some obstructions exist within the pore lumen.[11] Ribosomes pass through the pores either as intact particles, 18-20 nm diameter, or as separate large and small subunits at the rate of .01-.03 particles per pore per second (assuming 10^4 pores per nucleus).

This outward flow through the pores must be matched by a corresponding inward flow of the required building blocks for ribonucleoprotein. These will be small metabolites, sugar and nucleotide phosphates, and larger cytoplasmically-synthesised proteins. The rate of inflow of proteins will increase dramatically

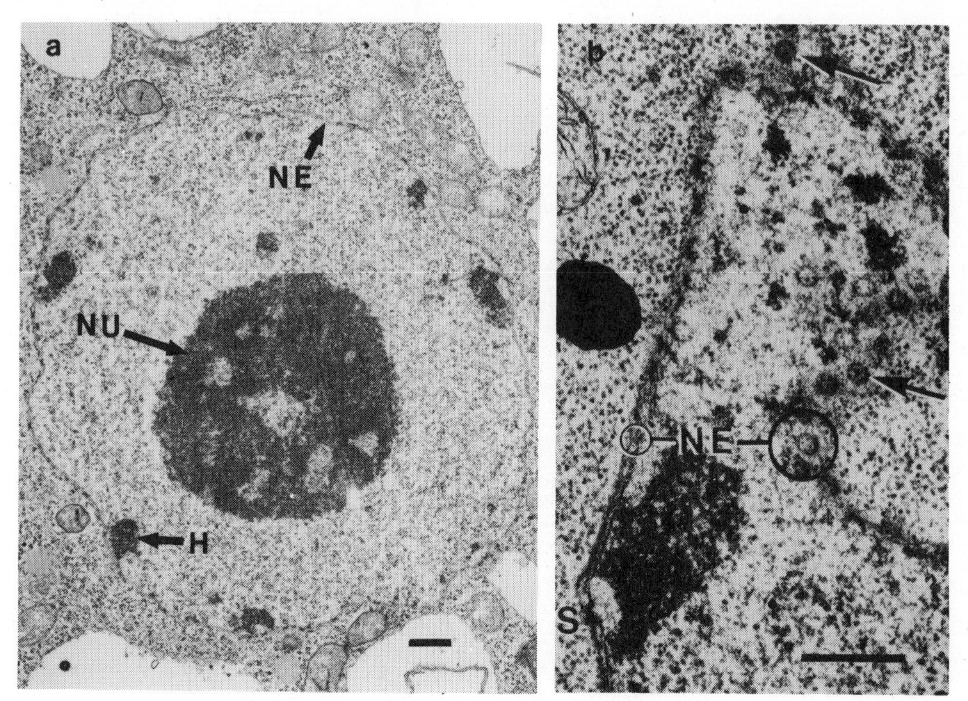

Fig 3. The nucleus: (a) the nuclear envelope (NE) bounds a nucleoplasm containing euchromatin, blocks of heterochromatin (H), and a nucleolus (NU), (b) nuclear pores are seen in side (S) and face (arrows) views according to the angle that the nuclear envelope (NE) makes to the plane of sectioning (from ref 20). Bar 0.5 μm.

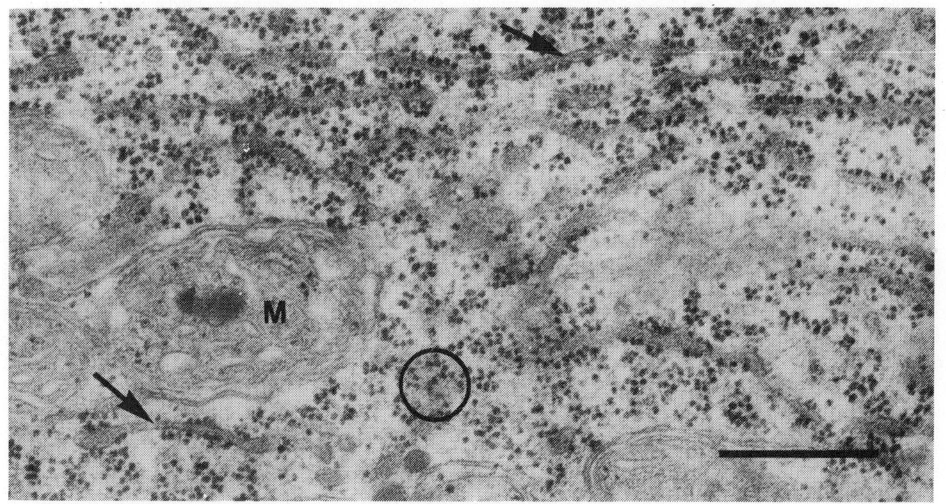

Fig 4. Ribosomes cover the cisternal membranes of the endoplasmic reticulum, seen in face (circled) and sectional (arrows) views. Mitochondria (M) are also present. Bar 0.5 μm.

during the phase of DNA-synthesis when the amount of histone protein in the nucleus is doubled.

4. ENDOPLASMIC RETICULUM

This is a morphologically and functionally diverse system of membranes within the cell that segregate an internal, cisternal space, from the general cytoplasm. This space is usually flattened to a thin layer, about 50 nm thick, between the bounding endoplasmic reticulum (ER) membranes. The surface density of this system within the cytoplasm is usually high, 4-12 $\mu m^2\ \mu m^{-3}$ (Fig 4).

Large molecules, proteins, are synthesised by mRNA-ribosome complexes (polysomes) on the outer membrane surface and the growing polypeptide chains are inserted through the membrane as they are formed. Ribosomes are specifically complexed to the outer surface and probably go through attachment-detachment cycles. The trapped protein molecules migrate through the system to specific sites where they are pinched off in vesicles for transport to other parts of the cell.

Specific information on the movement of molecules other than proteins in the cisternal space is scarce. Synthesis of lipids, steroid hormones and other long chain carbon compounds such as waxes, are known to occur. The cisternal space may also provide a transport route for low molecular weight solutes and ions. Muscle relaxation and contraction cycles are initiated by the movement of Ca^{++} ions into and out of the cisternal spaces, across the ER membrane.[13]

5. GOLGI APPARATUS

The Golgi apparatus is the site of reception and synthesis of macromolecules destined for export to the extracellular environment. This involves synthesis of membrane and formation of vesicles at the margins of the cisternal discs. The vesicles are filled with secretory product from the cisternal spaces which are in turn supplied from the ER or from synthesis within the Golgi cisternae. The vesicles usually enlarge on release from the Golgi and then travel through the cytoplasm, finally fusing with the surface plasma membrane. While the final size of the vesicles may vary in the range 100-500 nm diameter from one cell type to another, the size within each cell is relatively constant (Fig 5).

Vesicle movement through the cytoplasm is effected by microfilaments. There may be a continuous flow of vesicles through the cytoplasm, or the vesicles may accumulate in the cytoplasm until the cell is stimulated to release them en masse. Vesicle movement will be considered in more detail in a later contribution at this meeting.[14]

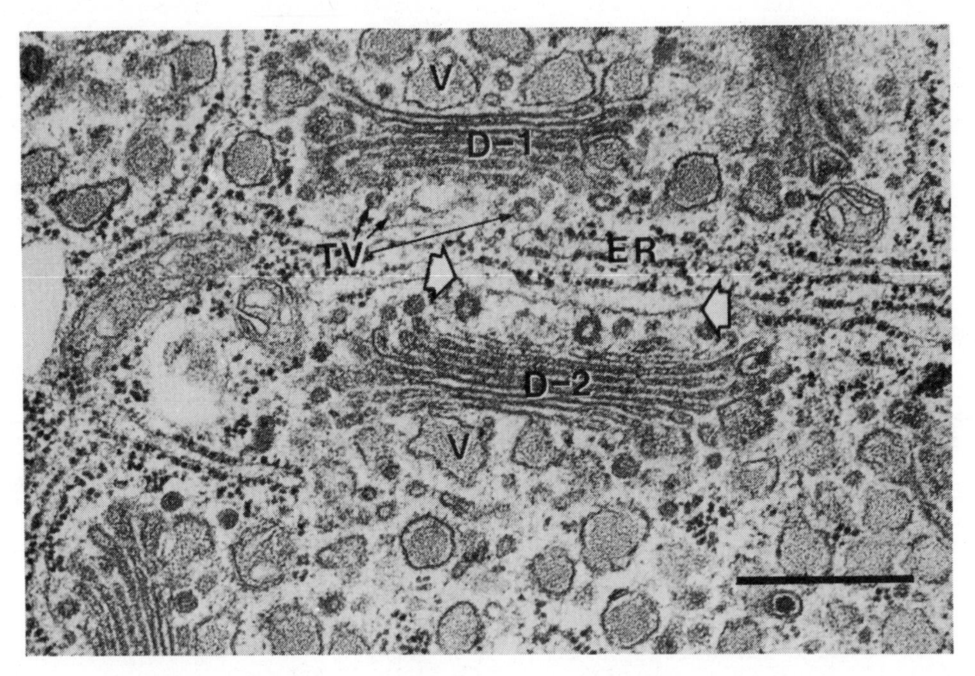

Fig 5. Dictyosomes (D-1, D-2), elements of the Golgi apparatus in plants, giving rise to mucilage containing secretory vesicles (V) in a plant cell. Associated with the forming face of each are cisternae of endoplasmic reticulum which give rise to transitional vesicles (TV, and between white arrows) which contribute to the cisternal stack (from ref 20). Bar 0.5 μm.

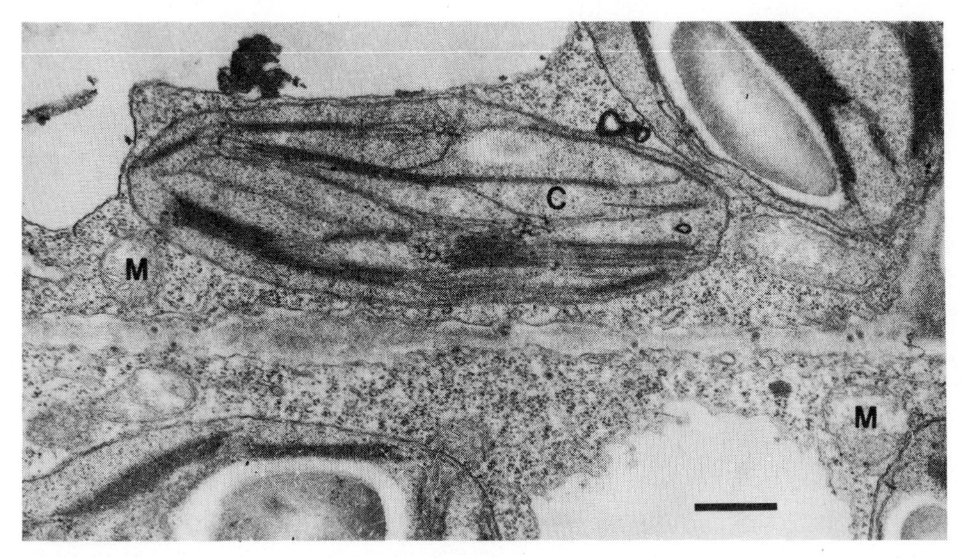

Fig 6. Chloroplasts (C) and mitochondria (M) in a leaf cell. Bar 0.5 μm.

6. MITOCHONDRIA AND PLASTIDS

These organelles are large enough (2-10 μm diameter) to be located and quantified in living cells by light microscopy. They may be either static within the cytoplasm, or caught up in the general cytoplasmic streaming motions. Occasionally they show specific motile properties, as in the shape changes of mitochondria and the light intensity-initiated movements of chloroplasts (Fig. 6).

The metabolic processes within these organelles involve metabolic conversions in the aqueous phase and electron transport in the membranes. There is an intensive exchange of small molecules with the surrounding cytoplasm at the surface of these organelles.

7. STRUCTURE OF THE CYTOPLASMIC MATRIX

A. Ribosomes

Ribosomes not bound to the endoplasmic reticulum ('free' ribosomes) occupy a considerable proportion of the cytoplasmic ground substance, with densities of about 2,000 particles μm^{-3}. They are not necessarily free to move or diffuse independently, since many are associated in helical groups of about 5-30 ribosomes, forming polysomes. Each polysome is held together by a single strand of mRNA and each ribosome will have associated with it a growing polypeptide chain.

B. Microtubules

Microtubules are cylindrical proteinaceous structures, 24 nm in diameter and often many micrometres in length. They frequently lie parallel to each other in arrays of a few to several hundred. Microtubules are found in all eukaryote cells, in a wide variety of situations, such as the cell cytoplasm, the mitotic spindle, just beneath the plasma membrane and the axonemes of cilia and flagella.[15] They are associated with various cell activities ranging from the apparently static maintenance of cell shape to the rather slow movements of chromosomes and the more rapid beating of cilia and flagella. Stationary microtubules may also be associated with the movement of adjacent cytoplasm (Fig 7).

C. Filamentous structures

The cytoplasm of many animal cells contains a great variety of filamentous structures which are only now being clearly defined and understood. Microfilaments (Fig. 7.) are one type of filament that appears to be universally important, occurring in plants and animals, where they were first studied as the thin actin filaments of muscle fibres. In non-muscle cells these microfilaments occur singly or in parallel assays; they are especially evident in cells

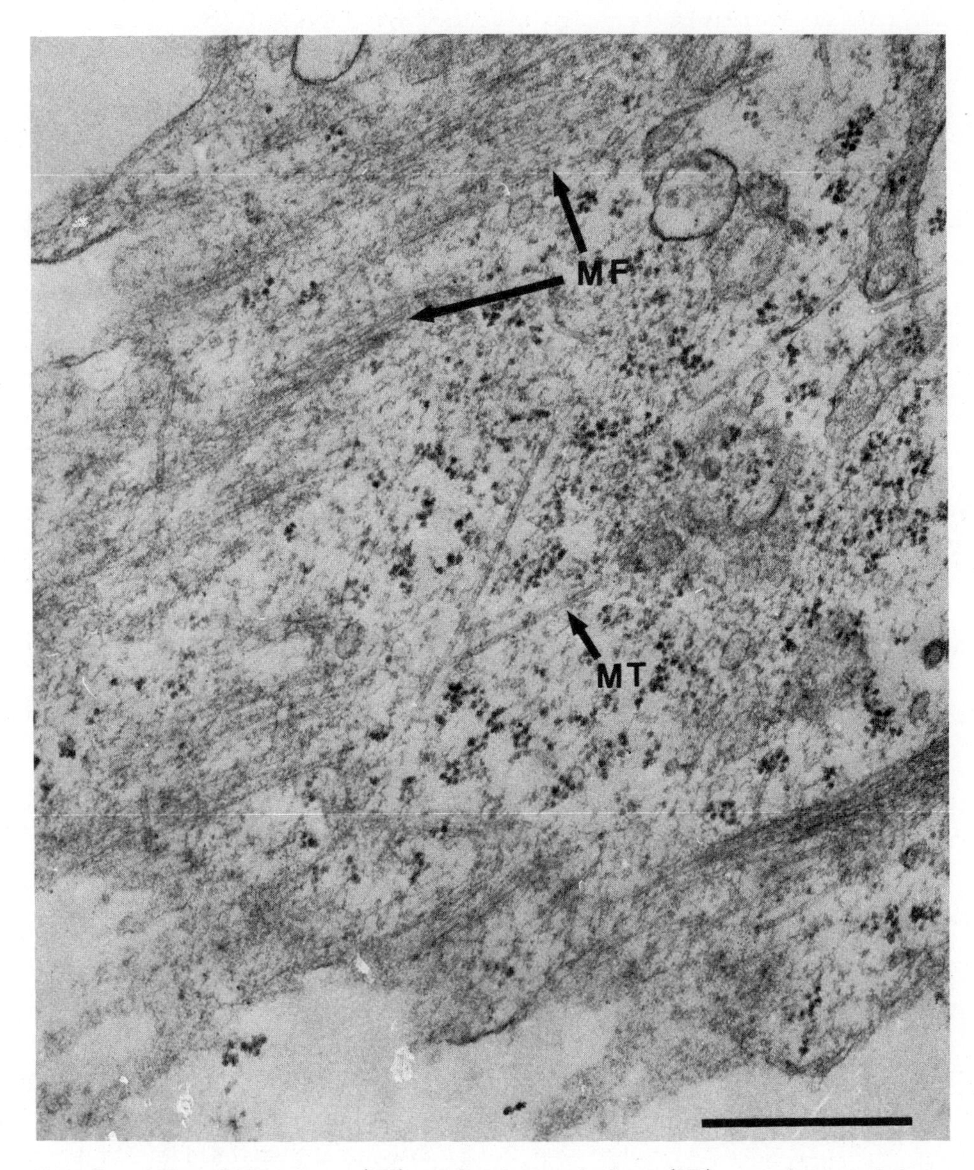

Fig 7. Microfilaments (MF) and microtubules (MT) at the periphery of a tissue culture cell. Bar 0.5 μm.

showing saltatory and cytoplasmic streaming motions[16] and in the advancing edge of moving cells.[17] These cytoplasmic movements are clearly visible at the light microscope level, as a flow of particles that are sufficiently large to be resolved. Typical flow rates range from 1-50 μm sec^{-1}. There has been an intensive search over many years to identify the cellular system that converts cellular energy to cytoplasmic movement.[18] A great deal of circumstantial evidence points to the involvement of the microfilament system in the generation of this flow.

D. Cytoplasmic gel

The ground cytoplasmic substance contains many macromolecules, metabolites and ions in an aqueous environment. Little is known of their specific movements within this gel, although it is assumed that the smaller move in response to concentration gradients, which may be generated by cellular metabolism. Many of the components of this gel are readily soluble in water and ionise or possess surface charges. This forms a gel with complex mechanical and internal surface charge properties that affect the movement of both cell organelles and the simple molecular components of the cytoplasm. In the microscale of the cellular environment we need to be aware that it is very doubtful that random diffusion events can occur; the local viscosity may be in the range 1-10 centipoises while the self-diffusion coefficient of water is reduced by a factor of about 2[19]. Clearly this will effect our interpretation of data from LLS systems.

8. CONCLUSION

This brief survey has shown that cellular systems possess an astonishing range of structural diversity[20], along with a corresponding range of dynamic activity, all contained within a relatively small volume (10-100 μm diameter). The resolution of these activities represents a major challenge to cell biologists, which may be partly met by the application of LLS methods.

REFERENCES

1. E. R. Weibel, Stereological Methods. Volume 1. Practical Methods for Biological Morphometry. Academic Press, London (1979).
2. M. W. Steer, Understanding Cell Structure. Cambridge University Press (1981).
3. S. J. Singer and G. L. Nicolson, Science, 175, 720-731 (1972).
4. J. Schlessinger, D. Axelrod, D. E. Koppel, W. W. Webb, and E. L. Elson, Science, 195, 307-309 (1977).
5. S. E. Lux, Nature, 281, 426-429 (1979).
6. D. E. Koppel, M. P. Sheetz, and M. Schlinder, Proc. Natl. Acad. Sci. USA, 78, 3576-3580 (1981).

7. B. E. S. Gunning, J. S. Pate, F. R. Minchin, and I. Marks, Symp. Soc. exp. Biol., 28, 87-126 (1974).
8. B. E. S. Gunning, in: "Intracellular Communication in Plants : Studies on Plasmodesmata", ed.,B. E. S. Gunning and A. W. Robards. Springer-Verlag, Berlin, Heidelberg, New York, 203-227 (1976).
9. W. W. Franke, Naturwissenschaften, 57, 44-45 (1970).
10. J. R. Harris, Biochem. Biophys, Acta, 515, 55-104 (1978).
11. C. M. Feldherr, in: "Advances in Cell and Molecular Biology, Volume 2", ed., E. J. DuPraw (1972).
12. H. R. Matthews, in: "The Cell Cycle", ed., P. C. L. John, Cambridge University Press, 223-246 (1981).
13. C. F. Louis, P. A. Nash-Adler, G. Fudyma, M. Shigekawa, A. Akowitz, and A. Katz, Eur. J. Biochem., 111, 1-9 (1980).
14. M. W. Steer, in: "The Application of Laser Light Scattering to the Study of Biological Motion", ed., J. C. Earnshaw and M. W. Steer, Plenum Press, London (1982).
15. P. Dustin, "Microtubules", Springer-Verlag, Berlin, Heidelberg and New York (1978).
16. B. A. Palevitz and P. K. Hepler, J. Cell Biol., 65, 29-38 (1975).
17. J. V. Small, J. Cell Biol., 91, 695-705 (1981).
18. Y. Hayashi, J. Theoretical Biol., 85, 469-480 (1980).
19. W. Drost-Hansen and J. Clegg, "Cell Associated Water", Academic Press, New York (1979).
20. B. E. S. Gunning and M. W. Steer, "Ultrastructure and the Biology of Plant Cells", Edward Arnold, London (1975).

APPLICATIONS OF LASER LIGHT SCATTERING TO BIOLOGICAL SYSTEMS

Martin W. Steer

Botany Department
The Queen's University of Belfast
Northern Ireland BT7 1NN

1. INTRODUCTION

I am going to examine the general problems facing biologists who wish to investigate cellular activity using LLS. The physical principles have already been discussed by Degiorgio[1] and it is clear that LLS techniques can provide a measure of diffusional and translational movement in vitro. Extension of these techniques from simple homogenous test systems to complex heterogenous biological systems provides a major challenge to the exponents of this technique. The laser light signals from such systems are very complex and cannot readily be analysed by conventional means. Even when the individual components have been characterised according to their physical parameters we are still left with the problem of assigning each to a specific cell structure and activity. While I am not competent to offer or discuss new methods of signal analysis, I am going to suggest a number of approaches that may be usefully adopted to simplify this complex situation. These will be considered separately, although it is apparent that a combination of all will improve the changes of success. We will conclude that a successful study of cellular systems by LLS will depend on finding methods for isolating and quantifying specific scattered light signals, from within the total signal generated by the system, and on the selection of specific cellular systems that are dominated by only a very limited number of activities, contributing a very large proportion of the total signal.

2. PARTICLE SIZE

The amount of laser light scattered by the specimen is strongly dependent on the size of the particles present, increasing rapidly with size[2]. Hence the signal may be dominated by the movements of the largest particles present. At the cellular level the cytoplasmic particles usually fall into a discrete series of size classes between which numerical density rapidly increases with decreasing size. Thus the single cell nucleus forms a class on its own followed by mitochondria and plastids (3-10 μm diameter, several hundred per cell), dictyosomes (elements of the Golgi apparatus, 0.5-1 μm diameter, $\simeq 10^3$ per cell), secretory vesicles (0.1-0.5 μm diameter, $\simeq 10^4$ per cell), ribosomes (18-20 nm diameter, $\simeq 10^7$ per cell). If the physical technique can be manipulated to give even a very approximate indication of the size and/or numerical density of the scattering particles this will add greatly to the probability of associating an observed motion with a specific cellular particle. In this respect it should be noted that significant changes in the signal strength occur as the detection angle is changed for large scatterers but not for smaller ones.[2]

If studies are carried out on relatively homogeneous populations of cells, then subpopulations can be examined by electron microscopy to determine the numerical density of particles in those parts of the cell examined by a light scattering study. Due to the inherent variability of biological material, the quantitative electron microscopic methods[3] will never yield exact data. These estimates of the relative numbers of each type of particle present can be combined with the distribution $G(\Gamma)$ from the correlation functions to determine the effective viscosity of the cytoplasm. Provided this is found to be consistent over the range of experimental situations to which the cells are to be subjected, it can then be used directly with the distribution $G(\Gamma)$ to determine particle number and size from living cells without the intervention of electron microscopy[4,5]. This means that changes in particle number (for example) could be followed in temporal studies of individual cells undergoing normal growth processes or reacting to a change in environment or metabolite conditions, or to drug treatment.

3. USE OF INHIBITORS AND DRUGS

A very wide range of chemicals are available that have clearly defined and specific primary effects on cellular activities. Less specific secondary effects may occur later, but these should not pose problems over the short time scales that are typical of LLS experiments. These chemicals offer the opportunity of looking for specific differences in the scattered light signals from treated cells, so that a particular cell function may be identified with a specific component of the scattered light.

This approach is not without its problems. A very good quantitative analysis of the signal from normal cells must be available to allow accurate comparison with the treated cells. Interpretation of the observed changes must take into account all aspects of the effects of the chemical on cellular motions. The following examples are not intended to be exhaustive, but to illustrate how this approach could be employed.

In many cells considerable ion fluxes occur across the membranes which should be detectable at the cell surface, even if they are masked at internal membrane surfaces. A considerable enhancement of such transmembrane flow can be achieved by treating cells with ionophores. Molecules of these compounds are incorporated into the cell membranes and create ports for ion transport shuttling backwards and forwards across the membrane with great rapidity.[6]

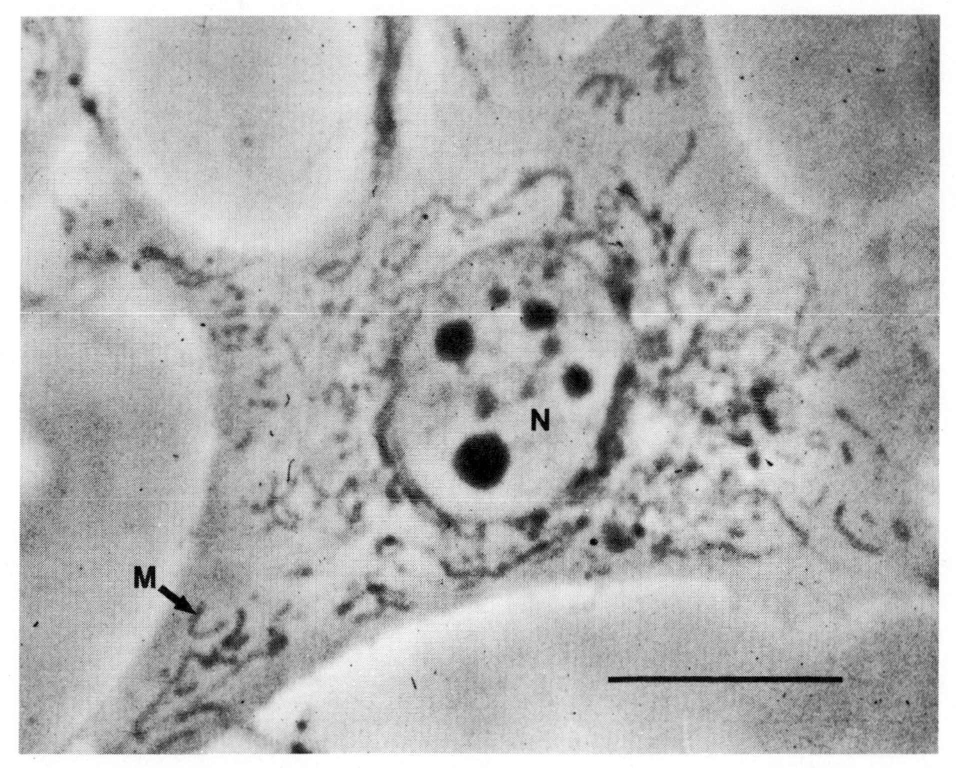

Fig 1. Phase contrast micrograph of living BHK cell growing on glass coverslip. Note the nucleus (N) with dense nucleoli and the mitochondria (M). Bar 10 μm.

Synthesis and subsequent movement of RNA complexes (mRNA and ribosomes) from the nucleus can be inhibited by drugs such as actinomycin D. Since these RNA strands undergo considerable post-synthetic modification before release there is a corresponding time lag on the inhibition of movement by actinomycin D. Nevertheless this inhibitor seems to have considerable potential for use in connection with migration of large components out of the nuclear pores and away from the nuclear surface.

Protein synthesis can be terminated by a number of drugs. Although they all achieve inhibition their mode of action is of importance in our context since the forming polypeptides may remain attached to the polysome complex, or, as with puromycin, be released.[7]

Many microtubular structures (but not cilia and flagella) are disassembled by colchicine to form large pools of monomer tubulin in the cytoplasm. This effect is readily reversible by placing the cells in drug-free medium. The microtubules reassemble, usually by elongation of the protein cylinders. Such cycles of disassembly and reassembly can be induced by other drugs and by application of hydrostatic pressure. These effects have been followed *in vitro* using LLS techniques and there appears to be a good prospect for similar work *in vivo*.[8]

The most conspicuous cell activity is cytoplasmic streaming. This, and many other microfilament related processes, such as secretion, can be effectively inhibited by cytochalasins, which probably inhibit filament elongation.[9] The effects of this group of drugs are probably fully reversible only after short time exposures.

4. SELECTION OF CELL TYPE

Biological research has always concentrated on those organisms, tissues or cells most suitable for the particular problem that is being investigated. The criteria that should be adopted in the selection of cells for examination by LLS will be those which lead to the minimum heterogeneity within the scattering volume. This is most readily achieved in tissues of one or a few cells in thickness, or in single cell systems. Many unicellular systems have a cytoplasmic structure which, while not rigidly stratified, at least allows the selection of areas rich in particular organelles, and areas of activity specifically related to certain functions.

Some of these systems will be considered at this meeting. Amongst those not yet investigated, tissue culture cells appear to offer excellent material for study. The cultures can be grown on coverslips (even $\frac{1}{4}$ λ plates) forming extensively flattened single cells or monolayers in which it is relatively easy to locate the nucleus and nuclear boundary, mitochondrial-rich regions and areas containing only ribosomes and endoplasmic reticulum. The nuclear boundary seems to offer an especially ideal subject for study, since there is only an extremely thin layer of cytoplasm above and below the nucleus, so that nucleocytoplasmic exchange is limited to the nuclear rim. (Figs. 1 and 2)

4. BIOLOGISTS DEMANDS ON LLS TECHNIQUES

Our extremely limited experience of applying LLS techniques to biological systems[10] has highlighted several important problems in the interpretation of the scattered light signal. These are related to defining the type of motion taking place, the detection of minor scattering components and the estimation of the relative and absolute levels of scattered light from each component.

Light scattering theory has usually considered only two types of motion, free diffusion and flow, or translational motion. While many biological systems undergo translational motions, as in the bulk flow of cytoplasmic streaming, there are real doubts about the existence of free diffusive motions in biological systems. The cytoplasm contains densely packed lipoprotein membranes, ribosomes and soluble proteins. All of these interact with the aqueous phase to such an extent that some authorities believe that the proportion of free water molecules, as opposed to those involved in hydration shells, is quite low or even negligible. We will later hear about work on the LLS detection of motions inside artificial gels, these represent an important first step to understanding movements in more complex systems.[11] We need more information on the form of the correlation functions that might be expected from molecules and particles moving in a constrained mode, such as diffusion restricted to certain planes or by gel frameworks, undulating or lateral movements of fibres and particles attached to such fibres, rotational movements of particles undergoing translation etc. Should these motions give rise to complex correlation functions we need to be able to identify and recognise each.

The size of the potential scatterers in a cell ranges from 1 nm to 10 μm. Since the correlation functions will be dominated by light scattered from the larger particles, it is imperative that LLS techniques should be refined to yield the maximum resolution of the smaller components.

Many of the suggested biological applications of LLS techniques demand that some quantitative estimate should be made of the level of scattered light from individual components. This is especially important when comparing normal cells with cells treated to inhibit a specific process, or when observing the normal development of a process with time, perhaps after applying a specific stimulus to the cell. In these cases the signal strength may be as important, or more important, than the actual rate of the process.

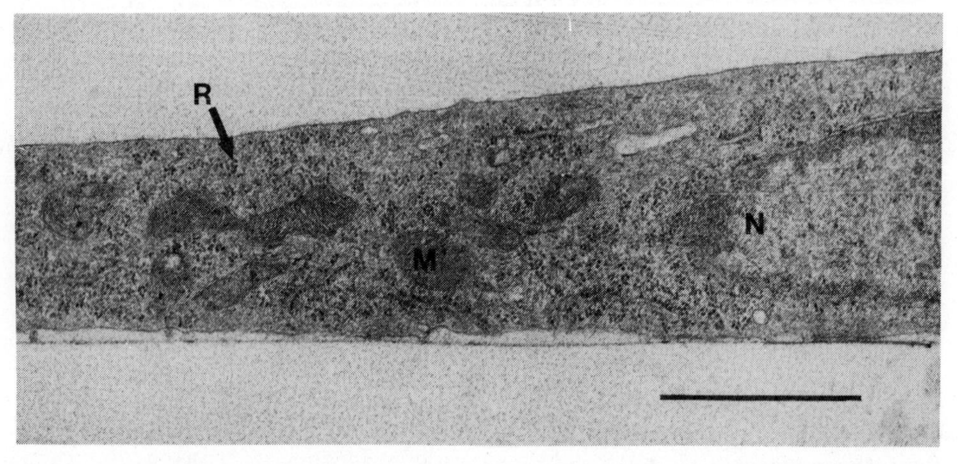

Fig 2. Electron micrograph of section of BHK cell similar to Fig 1 above. Part of the nuclear rim (N) is visible and the cytoplasm, which is about 1 μm thick, contains ribosomes (R) and profiles of mitochondria (M). Bar 1 μm.

REFERENCES

1. V. Degiorgio, in: "The Application of Laser Light Scattering to the Study of Biological Motion," J. C. Earnshaw and M. W. Steer, ed., Plenum, London and New York (1982).
2. H. H. Denman, W. Heller, and W. J. Panyonis, Angular Scattering Functions for Spheres, Wayne State University Press, Detroit (1966).
3. M. W. Steer, "Understanding Cell Structure," Cambridge University Press (1981).
4. B. Chu, this volume, 000-000 (1983).
5. A. N. Lavery, Ph.D. thesis, Queen's University, Belfast (1982).
6. B. C. Pressman, Ann. Rev. Biochem., 45, 501-529 (1976).
7. D. D. Sabatini and B. Kreibich, in: "The Enzymes of Biological Membranes, Volume 2," A. Martonois, ed., pp. 551-580 (1976).
8. P. Dustin, "Microtubules," Springer-Verlag, Berlin, Heidelberg and New York (1978).
9. S. W. Tanenbaum, "Cytochalasins, Biochemical and Cell Biological Aspects," North-Holland, Amsterdam (1978).
10. J. C. Earnshaw and M. W. Steer, Proc. R. Mic. Soc., 14, 108-110 (1979).
11. D. B. Sellen, in: "The Application of Laser Light Scattering to the Study of Biological Motion," J. C. Earnshaw and M. W. Steer, ed., Plenum, London and New York (1982).

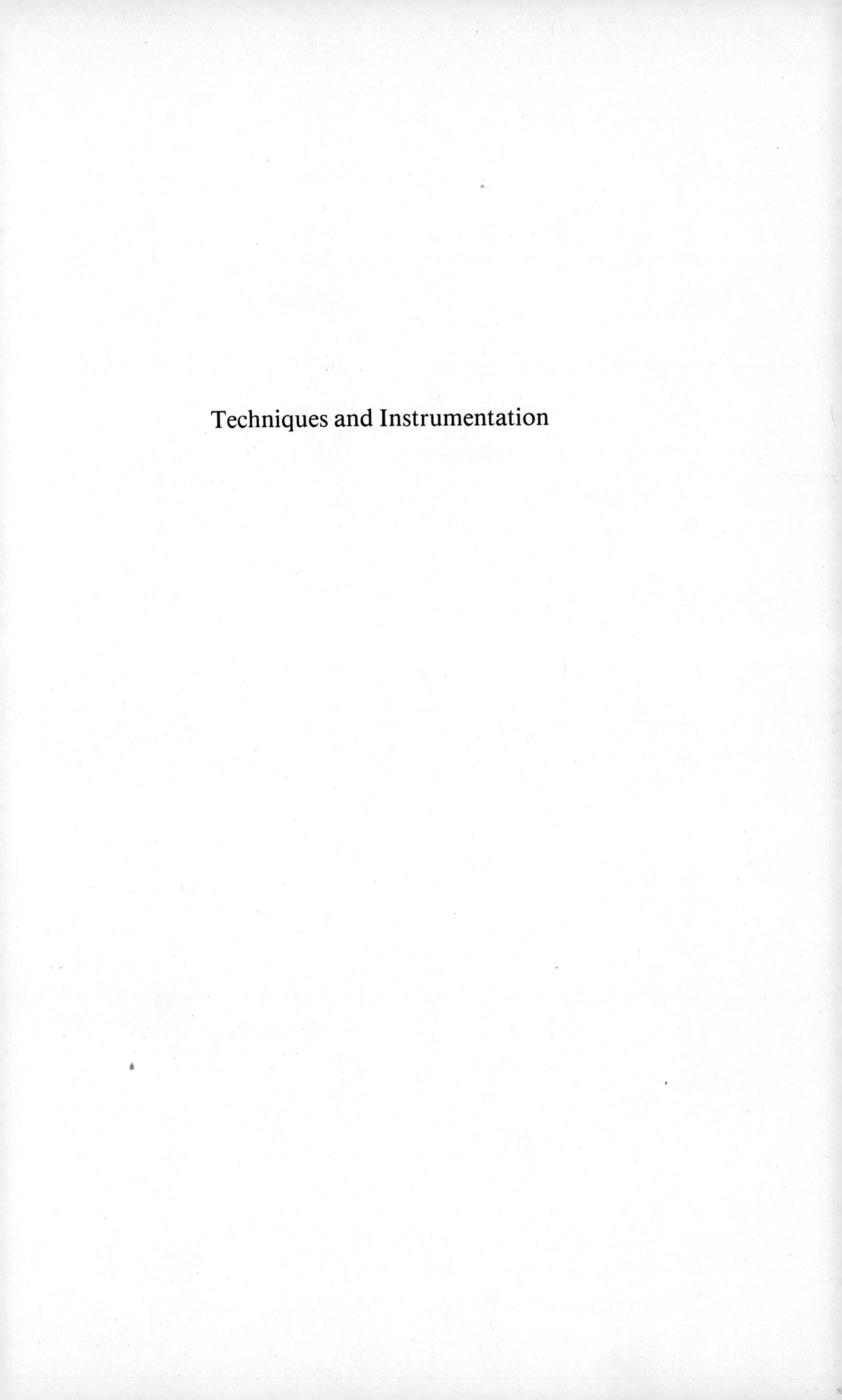

Techniques and Instrumentation

CORRELATION FUNCTION PROFILE ANALYSIS IN LASER LIGHT SCATTERING

I. GENERAL REVIEW ON METHODS OF DATA ANALYSIS

Ben Chu

Chemistry Department
State University of New York at Stony Brook
Stony Brook, New York 11794

ABSTRACT

We present a general review on methods of correlation function profile analysis in photon correlation spectroscopy. The limited objectives are: (1) to emphasize the interactive nature of experimental measurements, data analysis and theory, (2) to provide simple basic equations with a minimum of mathematical details for the different methods of data analysis, and (3) to summarize some useful relations between methods of data analysis and model fitting including molecular polydispersity.

1. INTRODUCTION

Instrumentation in laser light scattering has advanced to the stage whereby we can measure one of the most precise linewidth profiles using either photon correlation or fast Fourier transform (FFT) power spectrum. Having accomplished this experimental achievement, new methods of data analysis are developed to retrieve information on the linewidth distribution function. There now exists a variety of procedures starting from fitting the time correlation function to a single characteristic time over a limited linearly-spaced delay time range to fitting the same function over a finite range of variable delay times using multiexponentials. The delay time range, the number of delay times, the data precision and the measurement technique are related to different methods of data analysis and to the form of the linewidth distribution function. Invariably, it is difficult to prescribe a simple procedure which can maximize the best return in our experimental effort. Furthermore, we are often limited by the type of computing facilities available and the time duration we wish to spend for computations involving multiple variable parameters.

In the three lectures, which are intended to be comprehensible to members of the 'alternate discipline', I shall try to convey to the uninitiated reader the message that some understanding on time correlation function (or power spectrum) profile analysis is essential in the application of laser light scattering to complex macromolecular motions in solution. When we measure a time correlation function containing information on dynamical motions of macromolecules (or particles) ranging from ions to whole cells, we need to know the type of instrument we should use, the setting of the delay time range and spacing, and the measurement precision required, as well as the appropriate method of data analysis, in order to answer a specific question. Therefore, correlation function profile analysis represents a contiguous part of most experiments in laser light scattering. It is advisable, and usually more effective, to incorporate its initial usage in an interactive approach so that we can examine the measured time correlation function profiles during a laser light scattering experiment. Many of us have found the mathematics used in the data analysis to be quite sophisticated. For the mathematically less initiated reader, it may prove to be difficult to understand. In the general review (I), I shall outline the useful fundamental equations in data analysis, molecular polydispersity and model fitting, without making any attempt to justify their usage. Then, (II) I plan to describe a dual delay time correlator and to examine its effect on broad and bimodal linewidth distributions. Many aspects of the contents of my lectures have been reviewed and discussed in previous NATO ASIs on photon correlation and scattering techniques.[1-3] However, in the past two years, significant advances have been made by Pike and his co-workers, including quantitative estimates on the number of multiple exponentials which can be recovered in a given level of noise.

2. TIME CORRELATION FUNCTION

2.1. General Form, Single-Clipped Versus Full

The measured single-clipped photoelectron count autocorrelation function for a detector of finite effective photocathode has the form[4]

$$G_k^{(2)}(\tau) = N_S \langle n_k \rangle \langle n \rangle \left(1 + \beta |g^{(1)}(\tau)|^2\right) \quad (1)$$

where $g^{(1)}(\tau)$ is the first-order normalized correlation function of the scattered electric field, k is the clipping level, τ is the delay time, $\langle n_k \rangle$ and $\langle n \rangle$ are the mean clipped and unclipped counts per sample time, and N_S is the total number of samples with the baseline

$A = N_S \langle n_k \rangle \langle n \rangle$. β is a spatial coherence factor depending upon various experimental conditions, such as coherence and receiver area, and is usually taken as an unknown parameter in the data fitting procedure. Eq. (1) is valid for an extensive class of signals having a Gaussian field probability distribution. If, for example, the number of macromolecules (or particles, such as large cells) in the scattering volume becomes relatively small, the Gaussian field probability distribution of signals may no longer hold. We should then use a full correlator which measures

$$G^{(2)}(\tau) = N_S \langle n \rangle^2 \left(1 + \beta_f |g^{(1)}(\tau)|^2\right) \qquad (2)$$

The value of β_f is comparable to that of β in Eq. (1) if we set k = $\langle n \rangle$ in the single-clipped correlator and the baseline A is equal to $N_S \langle n \rangle^2$ instead of $N_S \langle n_k \rangle \langle n \rangle$. Most commercial correlators are either single-clipped or 4 x n correlators. In a 4 x n correlator, there is an effective limitation of 16 in the number of photocounts during the sampling interval $T(=\Delta\tau)$. Whenever $n(T)$ exceeds 16, scaling[5] is usually used to accommodate high count rates. Otherwise, the full correlation function may be distorted ever so slightly even though $\langle n(T) \rangle$ is less than 16 counts per sample time. In the present discussion, I shall also limit my review to measurements of Brownian motions in the absence of macroscopic movements which imply laser Doppler velocimetry.

For a solution of polydisperse particles (or macromolecules), the first-order electric field correlation function has the form

$$g^{(1)}(K,\tau) = \int_0^\infty G(K,\Gamma)\, e^{-\Gamma(K)\tau}\, d\Gamma \qquad (3)$$

where $G(K,\Gamma)$ is the normalized distribution of linewidth Γ measured at a fixed value of K, defined as the magnitude of the momentum transfer vector. $K = (4\pi/\lambda)\sin(\theta/2)$ with λ and θ being the wavelength of light in the scattering medium and the scattering angle, respectively, and is explicitly shown to emphasize the fact that the dynamic form factor $S(K,\tau)$, which has been expressed simply as $e^{-\Gamma(K)\tau}$, and $G(K,\Gamma)$ are functions of K. The Laplace inversion of Eq. (3) can be performed approximately, depending upon the finite ranges of delay time accessible by experiments, the precision of $g^{(1)}(\tau)$ and the characteristic time ranges of $G(\Gamma)$. In molecular polydispersity analysis, we are concerned mainly with the extraction of estimates of $G(\Gamma)$ due solely to translational motions of different size particles. In reality, we have to take into account particle interactions and complications which may arise when modes of scattering from motions other than translation, such as rotational and internal motions, contribute to the time correlation function. At times, we are concerned mainly with modes of motion in addition to translation. One should always be extra cautious when molecular polydispersity and modes of motions in addition to translation are present in $g^{(1)}(\tau)$.

2.2. Importance of Baseline

In fitting the time correlation function, it is advisable to measure the baseline A and to consider the data analysis in terms of the net unnormalized intensity time correlation $(G_k^{(2)}(\tau)-A)/A = \beta|g^{(1)}(\tau)|^2$ or the net unnormalized electric field time correlation function $\sqrt{\beta}|g^{(1)}(\tau)|$ which is represented by Eq. (3) except for an unknown scaling factor β. If the baseline A cannot be measured accurately, or its value at $\tau\to\infty$ disagrees by more than a few tenths of a percent with the computed baseline, e.g., $A = N_S\langle n_k\rangle\langle n\rangle$ for $G_k^{(2)}(\tau)$, multiple parameter fits of the time correlation function often become unreliable. Therefore, in order to minimize the number of unknown parameters which we need to represent the measured time correlation function, it is worthwhile to take extra pains to establish a reliable measured baseline A.

2.3. Range of Delay Time

The interactive nature of the procedure between data analysis and experimental measurements is again evident when we want to optimize the range of delay times which are necessary to analyze the fastest and the slowest characteristic times of yet to be determined linewidth distribution function $G(\Gamma)$ using the same measured time correlation function. For example, if we have a linewidth distribution which has a spread in the linewidth corresponding to a ratio of the upper to the lower bounds of support $\gamma = b/a = \Gamma^*_{max}/\Gamma^*_{min} \sim 2$ with the subscripts max and min denoting the highest and the lowest frequencies in $G(\Gamma)$, then an average linewidth for $G(\Gamma)$ must fall within $\Gamma^*_{min} < \bar{\Gamma} = 1/\bar{\tau} < \Gamma^*_{max}$. A linearly spaced delay time correlation function, say with 30 channels, should cover a range of delay time intervals corresponding to $\bar{\Gamma}\tau_{max} \approx 3$ because the correlogram which starts at a delay time $\Delta\tau$ corresponding to $\bar{\tau}/10$ and ends at a delay time corresponding to $3\bar{\tau}$ contains the necessary band width for approximating the $G(\Gamma)$. On the other hand, if $\gamma \sim 200$, a correlogram with 30 linearly spaced delay time channels cannot properly cover all the frequency ranges of such a $G(\Gamma)$. Consequently, in order to measure a broader frequency range in the correlogram, we have to use a correlator with a large number of linearly spaced delay time channels. The low frequency components of $G(\Gamma)$ can be covered using more delay channels, while the high frequency components of $G(\Gamma)$ require the use of shorter delay times. Alternatively, we may use a variable delay time so that the start-stop delay channels can span the desired frequency range.

If $\Gamma^*_{max}(=b)$ of $G(\Gamma)$ is much greater than $1/\Delta\tau$, its contribution to $G_k^{(2)}(\tau)$ will be represented by an unusually high value in the first channel $G_k^{(2)}(\Delta\tau)$ and only an averaged value on the distribution of the high frequency components ($\Gamma >> 1/\Delta\tau$) has been measured. Similarly, if $G_k^{(2)}(\tau)$ does not approach A at τ_{max}, we have to consider the possibility of missing some low frequency components of $G(\Gamma)$ where τ_{max} is the maximum delay time in the correlogram.

Several approaches[6] have been developed to correct the deficiency of a linearly spaced delay time correlation function with a limited number of delay channels. For example, for a correlator with 300 channels, instead of 30 channels, the delay time interval is increased by a factor of 10. The increase in the number of channels can be achieved by using straightforward hardware which can become expensive, or by multiplexing a small number of (block) channels over a range of delay time intervals. The latter approach increases the measurement time in multiples of block channels. Aside from the overlapping channels, a correlogram with 300 channels using a correlator capable of multiplexing 30-channel blocks requires the measurement time to be increased by a factor of 10. Unfortunately, each block represents correlation function measurements at different experiment times. For signals which are not stable over the time periods of the multiplexing rate, overlapping of blocks with different baselines in order to form the desired correlogram, introduces additional noises and the normalization procedure may not be simple, depending upon the signal statistics.

An alternative approach which uses the same signal time period to compute the correlogram, is to have a variable delay time option. A dual linearly spaced delay time correlogram represents the simplest step to approximate the desired broader frequency range. Variable delay time and log delay time correlators have become available commercially.

2.4. Location of Delay Channel

After having covered the frequency range of the correlogram, which is appropriate to resolve $G(\Gamma)$, we may also want to ask how many channels we should use, where the channels should be located, and how precise we should make those measurements. As the time correlation function due to thermal Brownian motions of the macromolecules (or particles) decreases monotonically with increasing delay times, there is no need to make measurements of those larger delay time channels whose signals are comparable to the measurement noise. In principle, for a single exponential decay autocorrelation function, we need only a two-channel correlogram to determine the scaling factor β and the characteristic linewidth Γ with $\beta|g^{(1)}(\tau)|^2 = \beta e^{-2\Gamma\tau}$ Clearly, the parameter β can best be computed if we have a value for $G_k^{(2)}(\tau)$ at as small a value of $\tau, G_k^{(2)}(\tau_s)$, as is experimentally feasible, and the slope in a plot of log $\beta|g^{(1)}(\tau)|^2$ versus

τ requires us to have another value of $G_k^{(2)}(\tau)$, $G_k^{(2)}(\tau_e)$, to be sufficiently different from $G_k^{(2)}(\tau_s)$ so that the ratio of $\log \delta G_k^{(2)}(\tau_e)/[\log G_k^{(2)}(\tau_s) - \log G_k^{(2)}(\tau_e)]$ is a minimum. $\delta G_k^{(2)}(\tau_e)$ denotes the uncertainties in the measured time correlation function at some large value of τ, defined as τ_e. The same qualitative reasoning remains valid for finding multiple parameters in a more complex time correlation function. For example, given one point in the time correlation function at a delay time τ, it is usually not very helpful to measure another point in the neighborhood of τ with the difference in $G_k^{(2)}(\tau)$ comparable to their combined statistical uncertainties. Although more channels of the correlator should be used in the range of delay time intervals where the $G_k^{(2)}(\tau)$ curve falls faster, we may reach a point of diminishing returns when too many channels are used in the correlogram, especially when the function is approaching the constant baseline. However, extra delay channels are always good to have because their presence decreases the margin of error. On the other hand, fewer numbers of data points in the correlogram reduces the execution time in the data analysis and permits us to use mainly a microcomputer for such purposes. Some guidelines for obtaining the greatest accuracy in the extraction of spectral parameters have been presented by Oliver.[6]

2.5. Limitations

For complex $G(\Gamma)$, the resolution which may be achieved in photon-correlation spectroscopy in order to differentiate discrete frequency peaks is rather limited, or in terms of the eigen functions and eigen values of the Laplace transform,[7] it is quantized. In practice, it is difficult to distinguish beyond a bimodal distribution. Yet, there are many uses in the methods of data analysis including molecular polydispersity (4.1) which is mostly represented by unimodal distributions and model fitting (4.3) which often involves no more than two characteristic times. Many biomacromolecules of interest have multiple relaxation times. Therefore, it is important that we understand some of the limitations of our data analysis procedure in photon-correlation spectroscopy. We can then design our experiments to extract the information properly.

In data analysis, the first requirement is to perform an inverse Laplace transform on the first-order time correlation function, $g^{(1)}(\tau)$, which has a polydisperse linewidth distribution, or at least to compute the moments of such a $g^{(1)}(\tau)$. However, in addition to being able to determine an approximate form of $G(\Gamma)$, effects of the particle scattering factor $P(K)$ and intermolecular interactions have to be taken into account before we can arrive at the final physical parameters of interest, such as molecular weight or size distributions. In this connection, the method of data analysis is closely related to model fitting. Therefore, we must also be aware of the physical characteristics of the dynamics we are measuring. Again, there is an interactive role between experiment and theory. I have always warned my students not to make too many measurements if they do not have a clear objective. The beginner usually has a tendency to measure too many correlograms of low signal to noise ratio quality or in the less appropriate delay time range. In this review, I shall first summarize the methods of data analysis and then discuss model fitting including the effects of particle scattering factor and intermolecular interactions.

3. METHODS OF DATA ANALYSIS

Three methods of data analysis, including methods of (1) cumulants, (2) multiple exponentials/histograms, and (3) Pearson, are discussed. I shall try mainly to list the essential equations and to provide some simple qualitative guidelines on a reasonable approach to the data analysis.

3.1. The Cumulants Method[8]

In the cumulant expansion,

$$\ln|g^{(1)}(\tau)| = -\bar{\Gamma}\tau + \frac{1}{2!}(\mu_2/\bar{\Gamma}^2)(\bar{\Gamma}\tau)^2 - \frac{1}{3!}(\mu_3/\bar{\Gamma}^3)(\bar{\Gamma}\tau)^3 + \frac{1}{4!}[\mu_4 - 3\mu_2{}^2]\,\tau^4 + \cdots = \sum_{m=1}^{\infty} k_m(\Gamma)(-\tau)^m/m! \qquad (4)$$

where k_m is the mth cumulant with $k_1 = \mu_1 = 0$, $k_2 = \mu_2$, $k_3 = \mu_3$ and $k_4 = \mu_4 - 3\mu_2{}^2$. The average linewidth $\bar{\Gamma}$ is defined by

$$\bar{\Gamma} = \int_0^{\infty} G(\Gamma)d\Gamma \qquad (5)$$

and

$$\mu_i = \int_0^{\infty} (\Gamma-\bar{\Gamma})^i G(\Gamma)d\Gamma \qquad (6)$$

with μ_i being the ith moment of the decay rate. By using the method of nonlinear least squares, we get[9,10]

$$Y(I\Delta\tau) = A\beta|g^{(1)}(\tau)|^2 = A\beta\exp[2(-\bar{\Gamma}I\Delta\tau + \frac{1}{2!}(\mu_2/\bar{\Gamma}^2)(\bar{\Gamma}I\Delta\tau)^2 - \frac{1}{3!}(\mu_3/\bar{\Gamma}^3)(\bar{\Gamma}I\Delta\tau)^3 + \cdots)] \quad (7)$$

where I and $\Delta\tau$ are, respectively, the channel number and the delay time used. The optimum values of the parameters $b_j = \bar{\Gamma}$, μ_2, μ_3, etc. are obtained by minimizing χ^2 with respect to each of the parameters simultaneously,

$$\frac{\partial}{\partial b_j}\chi^2 = \frac{\partial}{\partial b_j}\sum_{I=1}^{N_d}\frac{1}{\sigma_I^2}[Y_m(I\Delta\tau) - Y(I\Delta\tau)]^2 = 0 \quad (8)$$

where N_d is the number of delay channels and σ_I are the uncertainties in the data points $Y_m(I\Delta\tau)$. In our fitting procedure, we terminate the series of Eq. (7) after μ_2, μ_3, etc. terms and refer to the method as second-, third-, etc. order cumulants, respectively. A pertinent point to remember in using the cumulants method is to recall that it is a series expansion in terms of $\bar{\Gamma}\tau$. Table 1 lists the relative contribution of the second term in Eq. (4) for very narrow ($\mu_2/\bar{\Gamma}^2 \sim 0.01$), intermediate ($\mu_2/\bar{\Gamma}^2 \sim 0.1$) and broad ($\mu_2/\bar{\Gamma}^2 \sim 1$) distributions over a range of $\bar{\Gamma}\tau$. At small values of $\bar{\Gamma}\tau$, even over a range of $\bar{\Gamma}\tau$ from 0.1 to 0.5, the contribution of the second term in Eq. (4) for the broad distribution has increased to 25% of $\bar{\Gamma}\tau$. Extrapolation of $\bar{\Gamma}$, which was obtained by force fitting the measured time correlation function to a fixed order of cumulants to zero τ_{max} might not yield a correct $\bar{\Gamma}$(or $\mu_2/\bar{\Gamma}^2$) value. Furthermore, over-fitting the time correlation function with too many terms will render the coefficients ($\bar{\Gamma}$, μ_2, etc.) unreliable. Yet, we cannot limit our analysis only for small $\bar{\Gamma}\tau_{max}$ values often because of frequency range requirements as we have discussed in (2.3) and of limitations in instrumentation, such as after-pulsing. Therefore, as we increase the delay time range for time correlation functions with broad or multimodal linewidth distributions, Eq. (4) must be applied with great care.[11] The same problem has been discussed by Koppel[8] and by

Table 1. Percent Contribution of the Second Term $\frac{1}{2!}(\mu_2/\bar{\Gamma}^2)(\bar{\Gamma}\tau)^2$ with Respect to $\bar{\Gamma}\tau$ for Narrow, Intermediate and Broad Distributions

$\bar{\Gamma}\tau$	0.1	0.5	1.0	3.0
Narrow ($\mu_2/\bar{\Gamma}^2 = 0.01$)	0.05	0.25	0.5	1.5
Intermediate ($\mu_2/\bar{\Gamma}^2 = 0.1$)	0.5	2.5	5.0	15
Broad ($\mu_2/\bar{\Gamma}^2$) = 1)	5	25	50	150

Cummins and Pusey.[2] In practice, it is difficult to extract moments higher than μ_2 and statistical uncertainty limits evaluation of $\mu_2/\bar{\Gamma}^2$ to values greater than about 0.01.

To monitor the quality of fits we compute

(1) the normalized sum of squared errors χ^2

$$\chi^2 = \sum_{I=1}^{N_d} (Y_m(I\Delta\tau) - Y(I\Delta\tau))^2/(N_d-L+1) \tag{9}$$

with L being the number of fitting parameters,

(2) the correlation parameter Q

$$Q = 1 - \{\Sigma(Y_m(I\Delta\tau) - Y(I\Delta\tau))(Y_m((I+1)\Delta\tau) - Y((I+1)\Delta\tau))/ \Sigma(Y_m(I\Delta\tau) - Y(I\Delta\tau))^2\} \tag{10}$$

and

(3) the RSQ

$$RSQ = 1 - \frac{\Sigma(Y_m(I\Delta\tau) - Y(I\Delta\tau))^2}{\Sigma(Y(I\Delta\tau))^2 - ((\Sigma Y(I\Delta\tau))^2/N_d)} \tag{11}$$

The χ-parameter can be compared to the statistical error of measurements as approximated by $\sqrt{A}$. The ratio $R = (A/\chi^2)^{1/2}$ approaches unity for a good fit. A random distribution of errors gives a value of Q near unity. It has been our experience that poor fits will give R and Q < 0.1, while good fits result in values of R and Q greater than 0.8 or 0.9. For correlograms with high-precision data points, the RSQ values are often greater than 0.999. However, values of 0.999 or 0.9999 may not necessarily correspond with good fits. It is advisable to use several quality criteria to examine the goodness of fits. The convergence criteria, R and Q, are difficult to apply for cumulants fits of broad linewidth distributions. In those instances when the cumulants method, even with 3rd or 4th order cumulants fits, provides relatively poor convergence criteria, we still use Eq. (4) in order to estimate appropriate values of $\bar{\Gamma}$ and $\mu_2/\bar{\Gamma}^2$.

3.2. Multiple Exponential Methods

Aside from the cumulants expansion, the basic problem in deriving $G(\Gamma)$ from $g^{(1)}(\tau)$ amounts to inverting the Laplace integral equation

$$g^{(1)}(\tau) = \int_0^\infty \exp(-\Gamma\tau)G(\Gamma)\, d\Gamma. \tag{3a}$$

For discrete spectra, the integral in Eq. (3a) becomes a sum:

$$g^{(1)}(\tau) = \sum_{j=0}^{N_\Gamma} G(\Gamma_j)e^{-\Gamma_j\tau} \tag{12}$$

where N_Γ represents the number of discrete characteristic times, and $G(\Gamma_o)$ accounts for the possibility of an additional unknown baseline with $\Gamma_o = 0$. The discrete linewidth distribution may be represented by a sum of Dirac delta functions:

$$G(\Gamma) = \sum_{j=0}^{N_\Gamma} G(\Gamma_j)\delta(\Gamma - \Gamma_j) \tag{13}$$

and has been solved using damped nonlinear least squares,[12-14] provided that a good enough initial estimate of the set of $(2N_\Gamma + 1)$ parameters $\{G(\Gamma_j),\Gamma_j\}$ is available and N_Γ is known, which is usually not the case. Otherwise, the iterative process can converge to a local minimum which leads to serious errors even though the fitting may be done well enough to be accepted.

In trying to derive the continuous normalized linewidth distribution $G(\Gamma)$, a variety of procedures have been proposed. Gardner et al[15,16] used a Fourier transform solution of Eq. (3a) to develop a procedure that determined N_Γ and the set of $\{G(\Gamma_j),\Gamma_j\}$ parameters provided that a very accurate correlogram over a wide delay time range (beyond most linearly spaced delay time correlators) can be achieved. Provencher[17,18] developed a direct method based on the expansion of the solution of Eq. (3a) in the orthogonal eigen functions of the kernel, while Oștrowsky et al[19] using the eigen functions and eigen values of the Laplace transform discovered by McWhirter and Pike,[7] led to the concepts of exponential sampling. More empirical approaches, such as the histogram method[9,10] and the application of splines,[20] have also been able to give useful practical results. The mathematical details of both the eigen function expansion method[17,18,21,22] and the exponential sampling method[3,7,19,23,24] have been derived and elucidated by the original authors.

3.2.A. Double exponential form. In the double exponential fit, $G(\Gamma)$ is represented by

$$G(\Gamma) = B_1\,\delta(\Gamma-\Gamma_1) + B_2\,\delta(\Gamma-\Gamma_2) = \sum_{i=1}^{2} B_i\,\delta(\Gamma-\Gamma_i) \tag{14}$$

from which we obtain

$$g^{(1)}(\tau) = B_1 e^{-\Gamma_1\tau} + B_2 e^{-\Gamma_2\tau} \qquad (15)$$

with $B_1 + B_2 = 1$. Eq. (14) approximates $G(\Gamma)$ as two Dirac delta functions at Γ_1 and Γ_2 with amplitudes B_1 and B_2. The double exponential form is often used to fit a bimodal linewidth distribution. However, its usage deserves caution because, for a bimodal $G(\Gamma)$ of unequal distribution widths, the amplitude factors B_1 and B_2 as well as the linewidths Γ_1 and Γ_2 will be distorted. In practice, we have found the double exponential form to be a good approximation for a variety of $g^{(1)}(\tau)$ with fairly broad linewidth distributions ($\mu_2/\overline{\Gamma}^2 \sim 0.3$). The aim is to represent $g^{(1)}(\tau)$ by Eq. (15) and then to compute

$$\overline{\Gamma} = (B_1\Gamma_1 + B_2\Gamma_2)/(B_1+B_2) = \sum_{i=1}^{2} B_i\Gamma_i / \sum_{i=1}^{2} B_i \qquad (16)$$

and

$$\mu_2 = [B_1(\Gamma_1-\overline{\Gamma})^2 + B_2(\Gamma_2-\overline{\Gamma})^2]/(B_1+B_2) = \sum_{i=1}^{2} B_i(\Gamma_i-\overline{\Gamma})^2 / \sum_{j=1}^{2} B_i \qquad (17)$$

where we have relaxed the condition $B_1 + B_2 = 1$ indicating that Eqs. (14) and (15) now denote the unnormalized functions. For a broad linewidth distribution, the double exponential form which has four adjustable parameters can often represent the measured time correlation function to within experimental limits and in an iterative process, it can converge to a local minimum for the fitting to be accepted. However, it is very important to emphasize that the linewidth values, Γ_1 and Γ_2, as well as the amplitudes B_1 and B_2 may be without individual physical meaning. Yet, we can still use the double exponential form to compute $\overline{\Gamma}$ and μ_2 according to Eqs. (16) and (17). Those computed values are often better than the corresponding $\overline{\Gamma}$ and μ_2 values determined by using a second-order cumulants fit because the double exponential form tends to simulate the entire correlation function over a broader delay time range than Eq. (4).

In a double exponential fit, we again use a nonlinear least squares method to fit Eq. (15). Proper choice of initial values can often reduce the computation time. In this respect, we usually take $B_1 = B_2$ yielding $\Gamma_1 = \overline{\Gamma} + \mu_2^{1/2}$ and $\Gamma_2 = \overline{\Gamma} - \mu_2^{1/2}$ as the initial values. Both $\overline{\Gamma}$ and μ_2 are estimated by the cumulants method.

3.2.B. Multiple exponential form.

3.2.B.a. Variable characteristic linewidth Γ_j. Eq. (14) may be relaxed to include additional parameters with

$$G(\Gamma) = \sum_{j=1}^{N_\Gamma} B_j \; \delta(\Gamma - \Gamma_j)$$

where Γ_j remains as adjustable parameters. The corresponding net time correlation function is Eq. (12) with $B_j = G(\Gamma_j)$ and the j=0 term neglected:

$$g^{(1)}(\tau) = \sum_{j=1}^{N_\Gamma} B_j \; e^{-\Gamma_j \tau} \tag{12}$$

As we have briefly discussed, if N_Γ is known and if we can provide additional constraints, such as whether $G(\Gamma_j)$ is unimodal, or whether an a priori knowledge of the support of the distribution is known, i.e., whether the upper and the lower bounds of support, b and a, are known, an improvement in the solution can be achieved.

3.2.B.b. Exponentially spaced linewidth $\ln\Gamma_n$. The multiexponential method which I shall emphasize in this section is based on the developments by Pike and his coworkers. Although linearly spaced linewidths can be used to represent narrow linewidth distributions, the exponential sampling follows the natural spacing of the Laplace transform and fixes the Dirac delta functions as a series of exponentially spaced linewidth such that

$$G(\Gamma)\, d\Gamma = G^*(\ln\Gamma) d(\ln\Gamma) \tag{18}$$

and

$$G^*(\ln\Gamma) = \sum_n a_n \; \delta(\ln\Gamma - \ln\Gamma_n) \tag{19}$$

where n denotes the number of delta functions and has an upper limit depending upon the noise level. Attempting to increase the resolution, for points $\ln\Gamma$ closer than

$$\Gamma_n = \exp\left(\frac{\pi}{\omega_{max}}\right) \Gamma_{n-1} \tag{20}$$

or

$$\ln\Gamma_n = \frac{\pi}{\omega_{max}} + \ln\Gamma_{n-1} \tag{21}$$

would amount to extracting information from data below the noise level, an attempt which is bound to yield unreliable results.

Accumulation time sould be long enough to yield data with a noise of $< 10^{-3}$. The optimal cut-off value can be found by successive trials starting with a small ω_{max} value, which allows a low resolution scan of $G^*(\ln\Gamma)$. The resolution limit is attained when the fit parameters become erratic and/or negative. The spacing (π/ω_{max}) is taken such that the values of a_n are all positive and such that the whole range of the linewidth distribution is covered. As the values of a_n tend to go to negative when $\ln\Gamma_n$ is outside the linewidth range of $G(\Gamma)$, we have a practical guideline to approximate the support limits, Γ^*_{min} and Γ^*_{max} in $G(\Gamma)$. We usually select four exponentials (n=4) covering a linewidth range or $4\pi/\omega_{max}$ ($< \ln\Gamma_4 - \ln\Gamma_1$). By substituting Eqs. (18) and (19) into Eq. (3a), we get

$$g^{(1)}(\tau) = \int \sum_n a_n \, \delta(\ln\Gamma - \ln\Gamma_n)\exp(-\Gamma\tau)d(\ln\Gamma) \qquad (22)$$

and for the unnormalized first-order correlation function, we let

$$\beta g^{(1)}(\tau{=}I\Delta\tau) = \sum_n \beta \, a_n \, \exp(-\Gamma_n I\Delta\tau) \qquad (23)$$

where βa_n has absorbed the parameter β and we have retained the $\ln\Gamma$ space with $a_n \neq B_j$. Eq. (23) is used in the least squares method to determine the coefficients a_n where

$$a_n = \frac{\pi}{\omega_{max}} G^*(\ln\Gamma_n) \qquad (24)$$

The reconstruction of $G^*(\ln\Gamma_n)$ can be achieved following two different schemes:

(1) using the interpolation formula

$$\exp(-\tfrac{1}{2}\ln\Gamma)G^*(\ln\Gamma) = \sum_n \exp(-\tfrac{1}{2}\ln\Gamma_n)G^*(\ln\Gamma_n) \frac{\sin[\omega_{max}\ln(\Gamma/\Gamma_n)]}{\omega_{max}\ln(\Gamma/\Gamma_n)} \qquad (25)$$

(2) performing additional fits with different sets of δ-functions having the same exponential spacing π/ω_{max} but with $\ln\Gamma_n$ slightly shifted in increments of $\pi/x\omega_{max}$. The value (x-1) denotes the

additional sets of multiexponentials used in the fitting process. The coefficients a_n obtained from the multiexponential analysis in $\ln\Gamma$-space are <u>not</u> the amplitude factors in Eqs. (12), (13) and (14). From Eqs. (18) and (19), we have

$$G(\Gamma) = \Sigma_n \, (a_n/\Gamma_n)\delta(\ln\Gamma - \ln\Gamma_n) \tag{26}$$

Therefore,

$$G(\Gamma_j) = a_j/\Gamma_j \tag{27}$$

The retrieval of bimodal or trimodal linewidth distributions are possible only if adjacent maxima positions are approximately greater than $2\pi/\omega_{max}$ and if the width of individual peaks are narrow.

3.2.B.c. <u>Exponential sampling of data</u>. Pike et al[24] have extended the eigen value analysis of McWhirter and Pike[7] and showed that the correlation function data points located at a series of exponentially-spaced delay times τ, $e^{\Delta\tau}\tau$, $e^{2\Delta\tau}\tau$, ...τ_n... or τ, $\delta\tau$, $\delta^2\tau$, ... (with the dilation factor $\delta = e^{\Delta\tau}$) together with a given interpolation procedure will allow reconstruction of the correlation function at all values of τ from its values at the points τ_n within the band limit defined by the experimental noise.

The physical significance of this derivation is to show qualitatively that $g^{(1)}(\tau)$ may be fully reconstructed from its values at τ_n using an interpolation formula. Therefore, in the correlogram we can retrieve $G(\Gamma)$ using a set of exponentially-spaced data points which are no closer than $\delta(=e^{\Delta\tau})$ and the integral in Eq. (3a) become truncated. Our earlier qualitative remark on how many correlation data points which we need to measure, has been answered. In practice, we can impose the same criterion on linearly-spaced delay time channels and pay attention to the resolution required to analyze the high frequency components of $G^*(\ln\Gamma)$ provided that there is a sufficient number of delay time channels to extend the range for the low frequency components. The utilization of some excess number of delay time channels will not prevent us from determining the results of interest. As most molecular polydispersity problems involve $(\ln\Gamma^*_{max} - \ln\Gamma^*_{min}) < 5$, a 256 linearly-spaced delay time channel correlator covering a range of ~ 5.5 in the $\ln\Gamma$-frequency space borders on the minimum delay time range needed to resolve Γ^*_{min} and Γ^*_{max} simultaneously. Invariably, a variable delay time approach has the advantage for data analysis of broad (continuous or discrete) linewidth distributions.

3.2.B.d. <u>Resolution of multiexponentials.</u> Pike et al[24] have also made some simple considerations on the resolution of multiexponentials by considering the difference between $e^{-\tau/\tau_o}$ and

$\frac{1}{2} e^{-\tau\delta/\tau_o} + \frac{1}{2} e^{-\tau/\tau_o\delta}$:

$$\Delta g^{(1)}(\tau) = -\frac{1}{2} e^{-\tau\delta/\tau_o} + e^{-\tau/\tau_o} - \frac{1}{2} e^{-\tau/\tau_o\delta} \quad (28)$$

where the three exponentials are located at τ_o/δ, τ_o, and $\tau_o\delta$, respectively. By writing $\delta = 1 + \varepsilon$ and expanding Eq. (28) in powers of ε, they get

$$\Delta g^{(1)}(\tau) = -\frac{\varepsilon^2}{2} e^{-t} (t^2-t) \quad (29)$$

with $t = \tau/\tau_o$. By equating $\partial\Delta g/\partial t = 0$, they find $t = 0.382$ and 2.618 or $\tau = 0.382\ \tau_o$ and $2.618\ \tau_o$ corresponding to $\Delta g = 0.080\ \varepsilon^2$ and $-0.155\ \varepsilon^2$. In Eq. (29), the vanishing of the linear term explains that the two side-band exponentials cannot be distinguished from the central one at their geometric mean position. The value of δ must be sufficiently large that the value of Δg exceeds the noise before we are able to consider whether the exponentials can be resolved.

A single value analysis of the Laplace transform inversion in the presence of noise has been considered by Pike and his coworkers.[25] If we have knowledge of the lower and upper bounds {a,b} [or $\{\Gamma^*_{min}, \Gamma^*_{max}\}$] of the support of $G(\Gamma)$, the exponential components may be recovered at values of Γ_c, $\delta_s\Gamma_c$, $\delta_s^2\Gamma_c$, $\delta_s^3\Gamma_c$, ... within {a,b} where the resolution ratio δ_s has been given as a function of $\gamma(=b/a)$ for various signal-to-noise ratios (S/N). In their calculations, they have shown that recoverable exponential components, even for $S/N \geq 10^3$, are always relatively few. Therefore, photon correlation spectroscopy becomes a poor choice if we want to resolve many peaks in the linewidth (or size) distribution function.

3.2.B.e. Programs for multiexponentials. Aside from the standard least squares techniques, there are other approaches to obtaining the multiexponentials, including the method of non-negatively constrained least squares[26] (NNLS) as proposed by Grabowski and Morrison.[27] They set

$$r_j = |g^{(1)}(\tau)|^2 - \sum_{j=1}^{n} B_j e^{-\Gamma_j\tau} \quad (30)$$

and minimize r_j^2 $(1 \leq j \leq N)$ subject to the constraint $B_j \geq 0$ and $\Gamma_j \geq 0$. Eq. (30) has 2n unknowns. So a large set of decay constants within the band limits of $G_k^{(2)}(\tau)$ is chosen. The NNLS program, as well as the Simplex method , deserves further study as both approaches appear to converge quickly and use less computation time. Finally, extensive programs with documentation have been made available by Provencher.

3.3. The Histogram Method[9,10]

In the histogram method, we assume that $G(\bar{\Gamma}_j)$ is constant over a range corresponding to $\bar{\Gamma}_j - \Delta\Gamma/2$ and $\bar{\Gamma}_j + \Delta\Gamma/2$. Thus, we have

$$g^{(1)}(\tau) = \sum_{j=1}^{n} G(\bar{\Gamma}_j) \int_{\bar{\Gamma}_j-\Delta\Gamma/2}^{\bar{\Gamma}_j+\Delta\Gamma/2} \exp(-\bar{\Gamma}_j\tau)d\Gamma \tag{31}$$

where $\bar{\Gamma}_j$ is the mean linewidth for the jth step and $\Delta\Gamma$ is a constant step width. The computed net signal autocorrelation function $Y(I\Delta\tau)(=\beta|g^{(1)}(\tau)|^2)$ has the form

$$Y(I\Delta\tau) = \beta|g^{(1)}(\tau)|^2 = (\sum_{j=1}^{n} - \frac{P_j^2}{I\Delta\tau} [\exp(-\bar{\Gamma}_{j+1}I\Delta\tau) -\exp(-\bar{\Gamma}_j I\Delta\tau)])^2 \tag{32}$$

where we have set $P_j^2 = B_j$, essentially adding the constraint that $B_j \geqslant 0$. χ^2 is again minimized as in Eq. (8) in order to determine P_j^2. The histogram method computes the difference of exponentials in Eq. (32) instead of the linearly-spaced exponentials in Eqs. (12) and (13). The practical results are comparable to the multiexponential methods, especially when the frequency spacing is changed from Γ to $\ln\Gamma$. We have for linear and exponential spacing:

$$\text{Linear:} \quad \Gamma_j = \Gamma_{j-1} + \Delta\Gamma \tag{33}$$

with $\Delta\Gamma = (\Gamma_{max}-\Gamma_{min})/n$

$$\text{Log:} \quad \ln\Gamma_j = \ln\Gamma_{j-1} + \Delta\ln\Gamma \tag{34}$$

with $\Delta\ln\Gamma = (\ln\Gamma_{max}-\ln\Gamma_{min})/n$. In $\ln\Gamma$-space, $\Gamma_j = \exp[\ln\Gamma_{j-1} + \Delta\ln\Gamma]$. It should be noted that, in $\ln\Gamma$-space,

$\bar{\Gamma} \neq \sum_{j=1}^{n} P_j^2(\Gamma_{j+1} + \Gamma_j)/2/\sum_{j=1}^{n} P_j^2$ but:

$$\bar{\Gamma} = \sum_{j=1}^{n} \frac{1}{2} P_j^2(\Gamma_{j+1}^2 - \Gamma_j^2)/\sum_{j=1}^{n} P_j^2(\Gamma_{j+1} - \Gamma_j) \tag{35}$$

The histogram method corresponds to the zeroth-order splines method.

3.4. Known Distributions - The Pearson Method

If we know the form of the linewidth distribution function $G(\Gamma)$, such as a Gaussian or a log-normal distribution, the number of unknown parameters is known and it becomes quite easy to make a best fit of the data by the method of non-linear least squares. The Pearson type I distribution is one general form which has shown to be quite useful to fit unimodal distributions of unknown form so long as $\gamma = b/a$ is not too large and $\mu_2/\bar{\Gamma}^2 \lesssim 0.4$. It breaks down when long tails exist and involves long computation times because of numerical integration. The Pearson type I equation has the form

$$G(\Gamma) = C(\Gamma/A_1 - 1)^{m_1}(1 - \Gamma/A_2)^{m_2} \quad (36)$$

where C is a normalization constant compensating for the magnitude of β. A_1 and A_2 are the lower and upper limits of Γ, respectively, while m_1 and m_2 govern the shape of the curve.

4. MODEL FITTING

In model fitting, we are referring to the next step, i.e., after we have determined the characteristic decay times and their amplitudes or the form of $G(\Gamma)$, how do we relate Γ to the molecular parameters?

4.1. Molecular Polydispersity

In polydispersity analysis of macromolecules in solution or colloidal particles in suspension, we shall limit our discussions to cases where rotational and internal motions are absent. Experimentally, if we limit our measurements to $KL < 1$ where L is the characteristic length of the particle, rotational and internal motions are negligible. Two effects have to be taken into account before we can extract the molecular parameters from linewidth measurements for polydispersity analysis. These are the particle scattering factor correction and the second virial coefficient correction.

4.1.A. Particle scattering factor and second virial coefficients. For a polydisperse system of macromolecules in solution, the Rayleigh ratio (R_{VV}) for vertically polarized light at finite concentrations has the approximate form

$$\frac{HC}{R_{VV}} = \frac{1}{M_w P_1(K,C)} + 2A_2 \frac{P_2(K,C)}{P_1^{\,2}(K,C)} C + \ldots \quad (37)$$

where $H = 4\pi^2 n_o^2(\partial n/\partial C)^2/N_A\lambda_o^4$ with n_o, C, N_A, λ_o being the respective refractive index, concentration, Avogadro's number, and wavelength of light *in vacuo*; M_w is the weight average molecular weight, $P_1(K,C)$ and $P_2(K,C)$ are the intramolecular and intermolecular interference factors, respectively; and A_2 is the second virial coefficient. At infinite dilution, $P_1(K,C=0) = P(K)$ and $P_2(K,C)/P_1^2(K,C) = 1$. At finite but dilute concentrations, we usually take $P_2(K,C)/P_1^2(K,C) = 1$ and Eq. (39) is reduced to

$$\frac{HC}{R_{vv}} = \frac{1}{M_w P(K,C)} + 2A_2C \tag{38}$$

where we can express the concentration C as

$$C = f_N(M_j)M_j \tag{39}$$

with $f_N(M_j)$ being the number of macromolecules having molecular weight M_j. If we take

$$P^{-1}(K, M_j) = 1 + K^2 r_g^2(M_j)/3 \tag{40}$$

and

$$R_{vv}(M_j) \sim G(\Gamma_j) \tag{41}$$

where r_g is the radius of gyration, and by substituting Eqs. (39)-(41) into Eq. (38), we get

$$f_N(M_j) \sim \frac{G(\Gamma_j)[(1 + K^2 r_g^2(M_j)/3) + 2A_2(M_j)C\,M_j]}{H\cdot M_j^2} \tag{42}$$

which is the basic equation for transforming a multiexponential linewidth distribution function $G(\Gamma_j)$ to the number molecular weight distribution function. In the transformation, we also use the relation

$$\Gamma_j/K^2 = D_j = D_j^o(1 + k_D C) \tag{43}$$

where k_D is the second virial coefficient due to the translational diffusion coefficient and $D_j^o = k_T M_j^{-b}$ with k_T and b being two constants. For a polymer coil in theta solvent, $b = 1/2$, $A_2 = 0$, and

$$k_{D_j} \approx -(\bar{v} + k_f) \approx -(\bar{v}_j + 2.23\, N_A V_h/M_j) \tag{44}$$

where $\bar{v}(=N_A V_1/M_j)$ is the specific volume of the polymer and V_h is the hydrodynamic volume of the polymer molecule. Eq. (44) is based on the Pyun-Fixman theory and has been shown to be valid for polystyrene in cyclohexane at 35°C.[28] The second virial coefficients A_2 and k_D can be determined experimentally and we usually assume A_2 and k_D to be independent of molecular weight. With information on $P(M_j)$, $f_N(M_j)$ can be determined using one linewidth measurement $(G_k^{(2)}(\tau))$ at one scattering angle and one finite concentration.

4.1.B. Molecular weight distribution of rod polymers. For rod polymers, we have $r_g^2 = L^2/12$ yielding

$$1/P(K) = 1 + (K^2/36)(M/M_L)^2 \tag{45}$$

where $L(=M/M_L)$ is the characteristic length of the rod with M_L being the mass per unit length. For rigid rods,

$$D = (k_B T/3\pi\eta_o L)(\ln(2L/d)\text{-less appreciable terms}) \tag{46}$$

If we take $L \sim M$, then $D \sim \ln M/M$. However, when the molecular weight distribution is not too broad, lnM varies slowly when compared with M. So, we have

$$D \sim 1/M \tag{47}$$

With Eqs. (46), (43) and (42), we can again determine $f_N(M_j)$ for rod polymers. In the limit of zero concentration and zero scattering angle we can compute $\bar{\Gamma} \sim 1/M_w$ and

$$\frac{\mu_2}{\bar{\Gamma}^2} \approx -1 + \frac{M_w}{M_n} \tag{48}$$

for rigid rod polymers.

Applications of photon correlation spectroscopy to molecular weight distribution determinations have been reported.[28,29] For $M_z/M_w \lesssim 3$, linearly-spaced linewidth distributions by means of 3.2.B or 3.3 are feasible. For broader distributions, exponential spacing of linewidths becomes essential.

4.2. Polydisperse Analysis of Various Shaped Particles

For a dilute polydisperse suspension of small noninteracting particles, we have

$$g^{(1)}(\tau) = \Sigma N_j M_j^2 e^{-D_j K^2 \tau} / \Sigma N_j M_j^2 \tag{49}$$

When $P(K) \neq 1$, the translational diffusion coefficient measured at K becomes

$$\overline{D} = \Sigma N_j M_j^2 P(M_j) D_j / \Sigma N_j M_j^2 P(M_j) \tag{50}$$

which is no longer the simple z-average translational diffusion coefficient. In most cases, we shall try to avoid using Eq. (50) and assume $P(K) = 1$. If we characterize the rigid particles in terms of their major dimensions, we get for a sphere

$$D = k_B T / 6\pi\eta_o r \tag{51}$$

$$M = \frac{\pi\rho}{6} (2r)^3 \tag{52}$$

where ρ, r, and η_o are, respectively, the mass density of the particle, the particle radius and the solvent viscosity. For a sample with continuous distribution $f(\ell)$ of particle dimensions ℓ, we have $dN = Nf(\ell)d\ell$ where N is the total number concentration of particles. By substituting Eq. (51) into Eq. (50) and taking $P(K) = 1$, we get

$$\begin{aligned}\overline{D} &= (k_B T/6\pi\eta_o)\int_{L_1}^{L_2} \ell^{2x-1} f(\ell)d\ell \Big/ \int_{L_1}^{L_2} \ell^{2x} f(\ell)d\ell \\ &= (k_B T/6\pi\eta_o)\ I(2x-1)/I(2x) \\ &= K^* I(2x-1)/I(2x)\end{aligned} \tag{53}$$

where $I(j) = \int_{L_1}^{L_2} \ell^j f(\ell)d\ell$ with $I(0) = 1$ and $x = 3$ and L_1, L_2 being the lower and upper limits of the particle sizes. Expressions for x and K* for various shaped particles are shown in Table 2. By combining $\overline{D}$ measurements with those of radius of gyration, it is possible to determine the shape parameters for unimodal particle size distributions and for particles whose scattering is described by the Rayleigh-Gans-Debye approximation.

4.3. Particles with Rotational and Internal Motions

For a rigid rod, the first-order time correlation function for vertically polarized incident and scattered light has the form

$$g^{(1)}(\tau) = S_0(KL)e^{-DK^2\tau} + S_1(KL)e^{-(DK^2+6\Theta)\tau} + \ldots \tag{54}$$

Table 2. Expressions for x and K* for Various Shaped Particles[30]

Shape	ℓ	x	axial ratio	K*	Remark
sphere	radius r	3	1	$k_BT/6\pi\eta_o$	
thin rod	rod length L	1	L/d	$(k_BT/3\pi\eta_o)\ln 2(\overline{L/d})$	d=rod diameter
thin disc	disc diameter D	2	D/d	$(k_BT/3\pi\eta_o)$ arctan $(\overline{D/d})$	d=disc thickness
prolate ellipsoid	major axis 2a	3	a/b	$(k_BT/3\pi\eta_o)\frac{\overline{a/b}}{(\overline{a/b}^2-1)^{1/2}}$ $\ln\{(\frac{\bar{a}}{b})+((\frac{\bar{a}}{b})^2-1)^{1/2}\}$	
oblate ellipsoid	major axis 2b	3	b/a	$\frac{k_BT}{3\pi\eta_o}\frac{\overline{a/b}}{(\overline{a/b}^2-1)^{1/2}}$ $\arctan(\overline{a/b}^2-1)^{1/2}$	

where

$$S_0(KL) = (\frac{2}{KL})^2(\int_0^{KL/2}\frac{\sin z}{z}dz)^2 \quad (55)$$

$$S_1(KL) = \frac{5}{(KL)^2}(-3j_1(\frac{KL}{2}) + \int_0^{KL/2}\frac{\sin z}{z}dz)^2 \quad (56)$$

with

$$j_1(z) = \frac{\sin z}{z^2} - \frac{\cos z}{z} \quad (57)$$

The terms $S_0(KL)$ and $S_1(KL)$ are amplitude factors. For $KL \ll 1$, $S_0(KL) \gg S_1(KL)$. When $KL \sim 3$, $S_1(KL)$ becomes noticeable. Therefore, an analysis of $g^{(1)}(\tau)$ requires some knowledge on the behavior of $S_0(KL)$ and $S_1(KL)$ and the use of double exponential form for $g^{(1)}(\tau)$ only at the proper KL ranges.

For a semiflexible worm-like chain,[32]

$$g^{(1)}(\tau) = P_0 e^{-DK^2\tau} + P_2 e^{-(DK^2+2/\tau_1)\tau} + \dots \tag{58}$$

where P_0 and P_2 behave differently from S_0 and S_1; but the form looks exactly alike. Therefore, we need to look at higher order terms over a range of KL in order to distinguish the semiflexible rod from the rigid rod. According to the Harris-Hearst model, we have higher order P_j terms appearing virtually simultaneously at larger values of KL. Such multiple exponentials cannot be resolved using most present methods of data analysis. However, we can always define an average linewidth, such as $\bar{\Gamma}$ from the cumulants method. Indeed, most analysis of complex linewidth distributions is best expressed in terms of some integrated average value, instead of using determined parameters which are not reliable.

5. CONCLUSION

Correlation profile analysis represents a contiguous part of the experiments in laser light scattering. It should be incorporated in an iterative procedure in order to determine the delay time, the delay time range and the accumulation time of an experiment. One procedure which we have found useful in approximating $G(\Gamma)$ is as follows.

1) For a fixed delay time and range, we first compute the $\bar{\Gamma}$ and $\mu_2/\bar{\Gamma}^2$ values using the cumulants method.

2) The $\bar{\Gamma}$ and $\mu_2/\bar{\Gamma}^2$ are checked using the double exponential form.

3) Multiple exponentials with $n \approx 3$-5 are used to define ω_{max} and the bandwidth of $G(\Gamma)$.

4) The form of $G(\Gamma)$ is approximated using $n \approx 5$.

5) The approximate form of $G(\Gamma)$ is further checked using the histogram method.

Sometimes, due to the complex nature of $G(\Gamma)$, computed $g^{(1)}(\tau)$ is compared with measured $g^{(1)}(\tau)$ using a minimum of unknown parameters. In particular, values of $\bar{\Gamma}$ as a function of K are used for comparison purposes. However, it is important to determine $\bar{\Gamma}$ correctly, especially when $\mu_2/\bar{\Gamma}^2 \gtrsim$ say 0.5. A single exponential fit or a second-order cumulants fit is usually a poor practice as those values often depend upon the range of delay times used in the fitting procedure.

ACKNOWLEDGEMENTS

I wish to thank E. R. Pike, N. Ostrowsky, E. F. Grabowski, I. D. Morrison, F. D. Carlson and S. Bott for sending me preprints (24,25,27) and programs, and the National Science Foundation, Polymers Program (DMR 8016521) and the U.S. Army Research Office, Durham, for financial support.

REFERENCES

1. NATO ASI B3, "Photon Correlation and Light Beating Spectroscopy" edited by H. Z. Cummins and E. R. Pike, Plenum Press, New York (1974).
2. NATO ASI B23, "Photon Correlation Spectroscopy and Velocimetry" edited by H. Z. Cummins and E. R. Pike, Plenum Press, New York (1977).
3. NATO ASI B73, "Scattering Techniques Applied to Supramolecular and Nonequilibrium Systems," edited by S. H. Chen, B. Chu and R. Nossal, Plenum Press, New York (1981).
4. B. Chu, "Laser Light Scattering," Academic Press, New York, Chapter 6 (1974).
5. E. Jakeman, C. J. Oliver, E. R. Pike, and P. N. Pusey, J. Phys. A5:L93 (1972).
6. C. J. Oliver, Advances in Physics 27:387 (1978).
7. J. G. McWhirter and E. R. Pike, J. Phys. A11:1729 (1978).
8. D. E. Koppel, J. Chem. Phys. 57:4814 (1972).
9. B. Chu, Es. Gulari, and Er. Gulari, Physica Scripta 19:476 (1979).
10. Es. Gulari, Er. Gulari, Y. Tsunashima, and B. Chu, J. Chem. Phys. 70:3965 (1979).
11. Es. Gulari and B. Chu, Biopolymers 18:2943 (1979).
12. S. L. Laiken and M. P. Printz, Biochem. 9:1547 (1970).
13. M. L. Johnson and T. M. Schuster, Biophys. Chem. 2:32 (1974).
14. W. C. Hamilton, "Statistics in Physical Science," Ronald, New York (1964).
15. D. G. Gardner, J. C. Gardner, G. Laush, and W. W. Meinke, J. Chem. Phys. 31:978 (1959).
16. D. G. Gardner, Ann. N.Y. Acad. Sci. 108:195 (1963).
17. S. W. Provencher, Biophys. J. 16:17 (1976).
18. S. W. Provencher, J. Chem. Phys. 64:2772 (1976).
19. N. Ostrowsky, D. Sornette, P. Parker and E. R. Pike, Optica Acta 28:1059 (1981).
20. G. B. Stock, Biophys. J. 16:535 (1976); ibid, 22:79 (1978).
21. S. W. Provencher, J. Hendrix, L. DeMaeyer and N. Paulssen, J. Chem. Phys. 69:4273 (1978).
22. S. W. Provencher, Makromol. Chem. 180:201 (1979).
23. J. G. McWhirter, Optica Acta 27:83 (1980).
24. E. R. Pike, D. Watson and F. McNeil Watson, "The Analysis of Polydisperse Scattering Data II," presented at First National Aerosol Symposium, Santa Monica, California, 1982.

25. M. Bertero, P. Boccacci and E. R. Pike, Proc. Roy. Soc. A, to be published.
26. C. L. Lawson and R. J. Hanson, "Solving Least Squares Problems," Prentice-Hall, New Jersey (1974).
27. E. F. Grabowski and I. D. Morrison, "Particle Size Distributions from the Analysis of Quasielastic Light Scattering Data," presented at the First National Aerosol Symposium, Santa Monica, California, 1982.
28. Er. Gulari, Es. Gulari, Y. Tsunashima and B. Chu, Polymer 20: 347 (1979).
29. B. Chu and A. DiNapoli, "Extraction of Distributions of Decay Times in Photon Correlation of Polydisperse Macromolecular Solutions," presented at the First National Aerosol Symposium, Santa Monica, California, 1982.
30. V. J. Morris, G. J. Brownsey, B. R. Jennings, JCS Faraday II, 75(1):141 (1979).
31. R. Pecora, J. Chem. Phys. 49:4126 (1968).
32. S. Fujime and M. Maruyama, Macromolecules 6:237 (1973).
33. R. A. Harris and J. E. Hearst, J. Chem. Phys. 44:2595 (1966).

CORRELATION FUNCTION PROFILE ANALYSIS IN LASER LIGHT SCATTERING

II. A HYBRID PHOTON CORRELATION SPECTROMETER*

K. M. Abbey[Δ], J. Shook and B. Chu

Chemistry Department
State University of New York at Stony Brook
Stony Brook, New York 11794

ABSTRACT

We present a 256-channel single-clipped, dual delay time correlator. The PDP 11/35 computer acts as an accumulator for the time correlation function, essentially without limit. Software control allows synchronous start and stop of the correlator front-end, selective delay-time accumulation, and automatic intensity and time correlation measurements as a function of scattering angle. A specific comparison of percent relative error of linewidths and amplitudes of a double exponential time correlation function using single and dual delay times are presented.

1. INTRODUCTION

In (I), I have emphasized the iterative nature in determining time correlation function profiles in laser light scattering. More specifically, results of time correlation function measurements depend upon number and location of delay time channels and accumulation time. As the correlator is a specific purpose computer, another approach, instead of a hardware correlator, is to modify a general purpose computer (or microprocessor) in order to accommodate the fast computation of time correlation functions. Several commercial correlators, including the multiplexing technique, have taken advantage of software and firmware microprocessor control. The two main unique features of our 256-channel correlator are a

---

*Presented at Division of Organic Coatings and Plastics, ACS Symposium on Computer Applications in Coatings and Plastics, New York City, 1981.

[Δ]Present address: NASA Lewis Research Center, Cleveland, Ohio.

dual delay time circuit and direct software control of time correlation function measurements including setting of delay time and length of accumulation time by an iterative procedure where data analysis becomes an integral part of the time correlation function measurement.

2. DUAL CLIPPING CIRCUIT CORRELATOR

Figure 1 shows a block diagram of the dual clipping circuit correlator. A train of TTL pulses enters the digital subtractor which can subtract from the incoming signal a preset value of counts per sample time ranging from 0 to 65535. The digital subtractor is particularly useful in performing homodyne and heterodyne measurements where the scattered light is optically mixed with a local oscillator, such as a small fraction of the incident laser light in a homodyne experiment. The net unclipped signal from the digital subtractor is passed to the clipping circuits and to the correlator cards. A maximum of 256 channels is available and the distribution of channels between the two clipping circuits each with its own separate delay (or sampling) time intervals can be varied. The sampling clocks, whose frequencies correspond to the chosen delay times are generated in the control circuit. Their ratio may vary from 1 to 65535. When both sampling intervals are identical, the instrument functions as a single clipped, 256-channel correlator. Sampling intervals can range from 100 nsec to several seconds in 100 nsec increments, and clipping levels from 0 to 65535 in steps of one are possible. The control circuit also allows complementary clipping for high intensity signals. The total number of counts (denoted by TOTAL COUNTS) from the unclipped signal, the number of clipped counts in each clipping circuit (denoted by CLIPPED COUNTS 1 and 2), and the total number of sampling intervals (denoted by TOTAL SAMPLES) are accumulated in 32-bit counters. Signal counting is initiated by the control pulse which activates the correlator cards. The correlator cards were based on the design of Nossal and Chen.[1] We made only some minor variations. Therefore, the details will not be discussed here. The correlation function, scaled by a factor of 16, is transferred from shift registers on the correlator cards through a 256-channel data buffer to the memory of a PDP-11 minicomputer on a bidirectional data bus through the Tracor Northern 1310 Interface control. Three aspects of the correlator will be discussed in greater detail: (1) the software, (2) the digital subtractor circuit, and (3) the fast 256-channel data buffer.

2.1. Software

A Fortran program consisting of assembly language subroutines provides complete control of the correlator. This program initiates correlation, clears counters, and transfers data to and from the

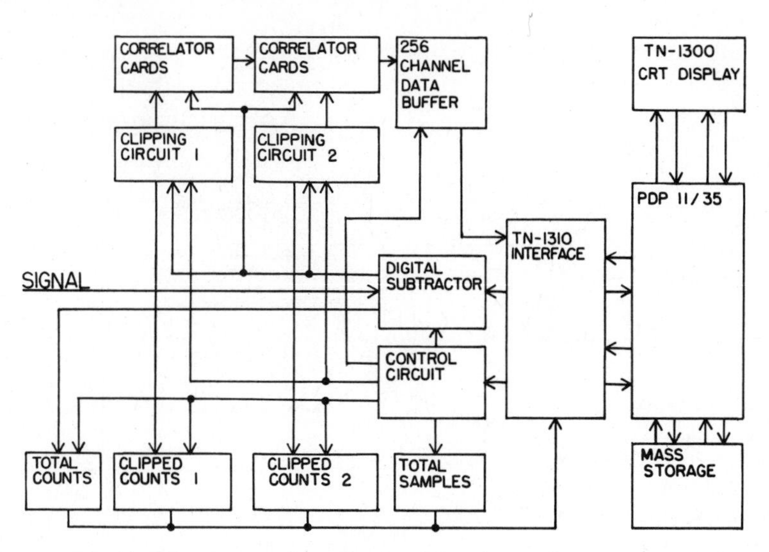

Fig. 1. Block diagram of a dual clipping circuit correlator.

correlator. Both sampling intervals are set independently through the software. The program can set the digital subtractor level, change the number of channels (16-256) in steps of 16, and control the total correlation function accumulation time. The values of the correlation function for each channel are read from the data buffer to the PDP-11 minicomputer and accumulated without limit in multiple precision arrays. The total counts, the clipped counts from both clipping circuits and the number of sampling intervals are also read from the 32-bit counters in the correlator. The software also controls the display on the CRT, storage on floppy disks and any further analysis or manipulation of data.

2.2 Digital Subtractor Circuit[2]

The essential features of our digital subtractor circuit are depicted in Fig. 2. The sampling clock is applied to one input of a flip-flop. The outgoing signal alternates between each output of the flip-flop at a frequency equal to one half of the input frequency. Each output is connected to a different set of counters. Consequently, at alternate sampling intervals different halves of the subtractor circuit are used. Passing the unclipped signal through a series of non-inverting buffers insures that the unclipped signal and the sampling clock are entirely synchronous. This design minimizes dead time effects which become important at short sampling times. The unclipped signal is logically "ANDED" with the sampling clock and the resulting signal passed into one set of counters which performs the subtraction. While one set of counters subtracts the preset value from the incident pulses, the same preset value is loaded into the other set from data buffers accessed by the bidirectional data bus.

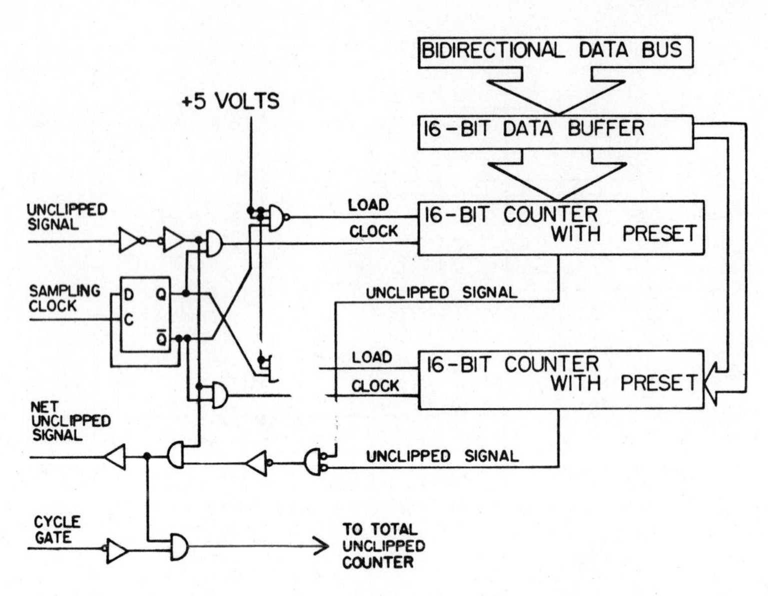

Fig. 2. Block diagram of a dual clipping circuit correlator.

This procedure eliminates the dead time associated with loading of the preset value to the counters. Upon subtraction of the selected preset value and re-synchronization with the sampling clock, the net unclipped signal is transferred to the clipping and correlation circuits. The clipping circuits function in a similar manner with respect to eliminating dead time and synchronizing the clipped signal with the sampling clock and, therefore, will not be shown here.

2.3. 256-Channel Data Buffer

A block diagram of the data buffer is shown in Fig. 3. Clock signals (1.6 MHz) are generated from a 10 MHz quartz oscillator in the control circuit and are used to clear the overflow counters on the correlator cards and to control transfer of the correlated data. Sixteen 256 x 1 read/write memories are connected to form 256 16-bit channels. Data memory address counters allow selected channels to be accessed. Data from individual channels are transferred to counters, added to the correlated data, and returned to their respective locations in the read/write memories. In the control portion of the circuit, a hardware condition is set when addressing the first channel. This condition is reflected in the buffer status register. A data ready flag indicates that the data from the buffer memory can be read onto the data bus from the output latches. The data buffer channels are cleared when read into the output latches. However, if the contents of a buffer channel exceed 65535 before the channel can be read again, an overflow is indicated in the status register. The correlated data is transferred to the buffer from

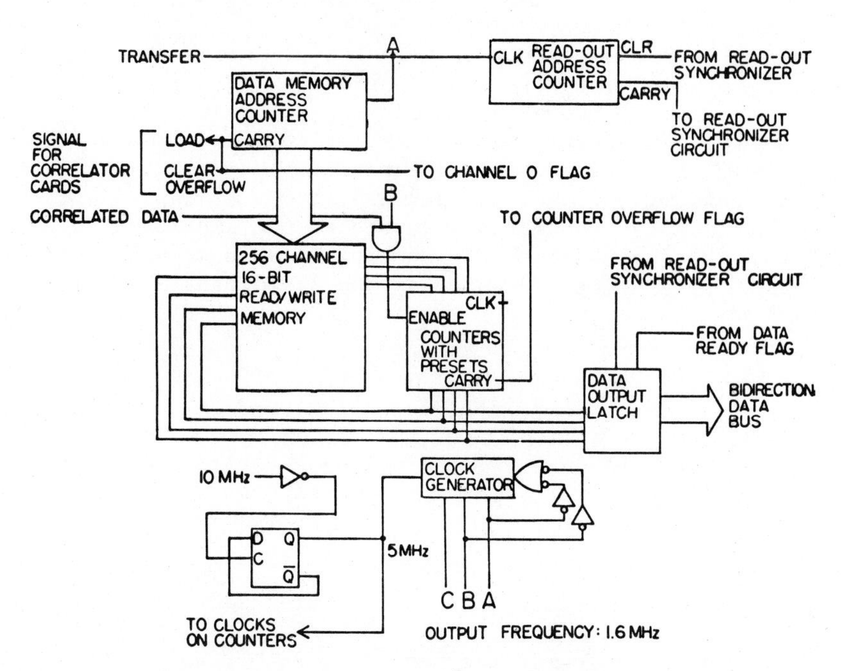

Fig. 3. Block diagram of a 256-channel data buffer.

the correlator cards at a rate of 1.6 MHz. The computer software limits the rate at which the buffer contents can be transferred to the PDP-11 memory through the TN1310. A 256-channel transfer requires less than 5 msec. Consequently, the 256-channel data buffer extends the rate of the data collection by a factor of 10^3.

3. DATA SIMULATION

In order to study the effects of measuring the photon correlation function simultaneously at two different delay times data representing typical double exponential correlation functions were generated. These functions assumed the form

$$g^{(1)}(I\Delta\tau) = A_1 \exp(-\Gamma_1 I\Delta\tau) + A_2 \exp(-\Gamma_2 I\Delta\tau) \quad (1)$$

with $I = 1,2,\ldots 96$; $A_1+A_2 = 1$ and $\Gamma_2 > \Gamma_1$. The baseline A had previously been subtracted. In order to save computation time, only 96 channels were used in the simulated functions. The amplitudes, A_1 and A_2 were chosen such that $g(0) = 1 \times 10^5$, and random noise ($\pm 5 \times 10^3$) was superimposed on each channel of the correlation function.

Data acquired at one delay time were first generated to illustrate the effects of a single discrete sampling time ($\Delta\tau = T$) on the extraction of parameters from a bimodal correlation function. Sampling intervals were selected to optimize the relation $N\bar{\Gamma}T \sim 2\text{-}3$ where N is the total number of channels and $\bar{\Gamma} = A_1\Gamma_1 + A_2\Gamma_2$. A four parameter non-linear least squares algorithm was used to fit the simulated data. The results, expressed in percent relative error in linewidths, Γ_1 and Γ_2, and amplitudes A_1 and A_2 are shown in Table 1. The largest error occurs when the amplitude A_2 corresponding to the larger linewidth species is much greater than A_1. The percent error also increases as the linewidths approach one another.

Depending on the relative amplitudes of the two contributions to the correlation function, $N\bar{\Gamma}T$ may differ greatly from $N\Gamma_1T$ and $N\Gamma_2T$. Delay times for which $N\bar{\Gamma}T \sim 2\text{-}3$ may yield values for $N\Gamma_1T$ and $N\Gamma_2T$ that are far from optimum. Figure 4 gives the percent relative error in amplitudes and linewidths for a specific case with $\Gamma_2/\Gamma_1 = 5$ and $A_1/A_2 = 1/9$.

When $N\bar{\Gamma}T = 2.3$, $N\Gamma_1T = .5$, and $N\Gamma_2T = 2.5$. Therefore, the errors in the determination of A_1 and Γ_1 are high. However, when $N\bar{\Gamma}T = 10.4$, $N\Gamma_1T = 2.3$ and $N\Gamma_2T = 11.3$, the amplitude and the linewidth errors for all four parameters, A_1, A_2, Γ_1 and Γ_2 are minimized In this particular case, errors in the evaluation of A_2 and Γ_2 are not very sensitive to the value of $N\Gamma_2T$ whereas errors in A_1 and Γ_1 (species present in smaller amounts) are very dependent on $N\Gamma_1T$. In Fig. 5 with $\Gamma_2/\Gamma_1 = 3$ and $A_1/A_2 = 1$, the best fit for A_1, A_2, Γ_1 and Γ_2 fall in the ranges $N\bar{\Gamma}T = 2.4 - 5$, with $N\Gamma_1T = 2 - 2.5$ and $N\Gamma_2T = 3.5 - 7.5$. Thus, the choice of an optimum sampling interval is

Table 1. % Relative Error - Linewidths and Amplitudes, Single Delay Time Correlation Function $G(IT) = A_1\exp(-\Gamma_1 IT) + A_2\exp(-\Gamma_2 IT)$, $N\bar{\Gamma}T \sim 2\text{-}3$

Γ_2/Γ_1	9				5				3			
	A_1	A_2	Γ_1	Γ_2	A_1	A_2	Γ_1	Γ_2	A_1	A_2	Γ_1	Γ_2
A_1/A_2												
9	.26	1.4	.29	1.8	.45	1.0	.41	2.3	.55	2.6	.45	.57
1	1.2	4.5	.49	5.4	1.2	1.6	1.7	1.5	2.9	3.3	1.3	2.3
1/9	7.3	1.0	12.6	.72	11.7	1.5	1.8	1.0	12.5	1.6	4.7	1.1

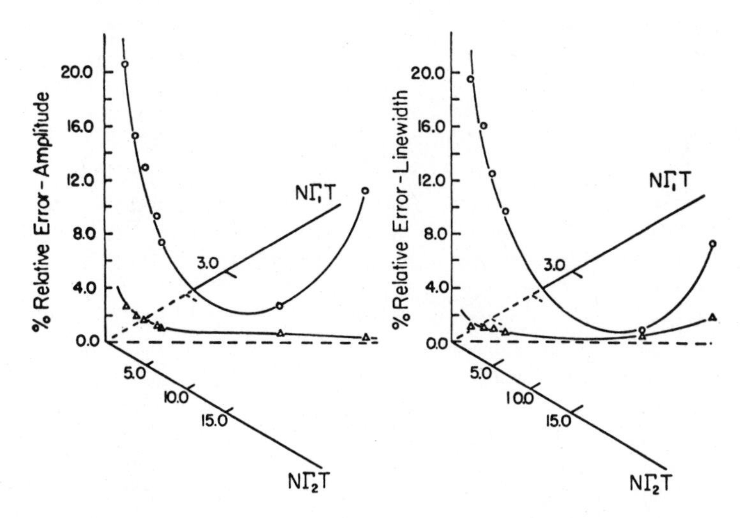

Fig. 4. Percent relative error in amplitudes and linewidths for $\Gamma_2/\Gamma_1 = 5$ and $A_1/A_2 = 1/9$. Hollow circles denote component 1; hollow triangles denote component 2.

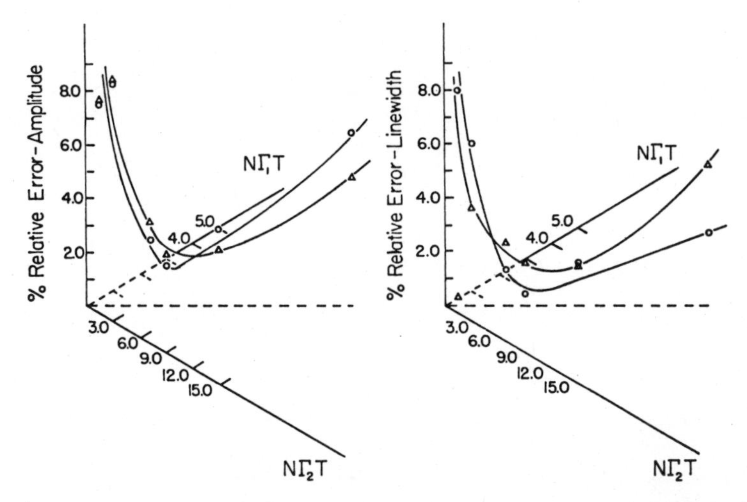

Fig. 5. Percent relative error in amplitudes and linewidths for $\Gamma_2/\Gamma_1 = 3$ and $A_1/A_2 = 1$. Hollow circles denote component 1; hollow triangles denote component 2.

affected by the relative amplitude and coherence time ratio of the species present.

The dual clipping circuit correlator allows simultaneous

optimization of the quantities $N_1\Gamma_1T_1$ and $N_2\Gamma_2T_2$. The 96-channel correlation function is constructed by sampling the signal at two different delay times. A series of correlation functions with different Γ_2/Γ_1 and A_1/A_2 ratios was generated as before; however, by choosing two different sampling intervals and varying the number of samples acquired, it was possible to satisfy both $N_1\Gamma_1T_1 \sim 2\text{-}3$ and $N_2\Gamma_2T_2 \sim 2\text{-}3$. Subsequent linewidth and amplitude errors are listed in Table 2. Substantial reduction in error occurred for those particular values where the deviation was significant using a correlator with a single clipping circuit.

When performing measurements on real systems, Γ_1 and Γ_2 are not known so that we do not have a priori knowledge of $N_1\Gamma_1T_1$ and $N_2\Gamma_2T_2$. We then use the following practical procedure: A correlation function using only one delay time would first be measured. If data analysis by means of the cumulants method and of the multi-exponential/histogram method suggested the presence of two exponentials, estimates of Γ_1 and Γ_2 could be made and the delay times of the dual clipping circuit correlator set accordingly. In contrast, with a single delay time, only one of $N\Gamma_1T$, $N\Gamma_2T$, or $N\bar{\Gamma}T$ could be optimized. The values tabulated indicated that the dual delay time functions yielded better estimates for A_1, A_2, Γ_1 and Γ_2.

4. INTENSITY AUTOMATION

Automation of intensity measurements was achieved with a PDP-11/35 computer and interfacing through Tracor Northern hardware. Figure 6 shows the hardware configuration used. Measurements of

Table 2. % Relative Error of Linewidths and Amplitudes
$G_k^{(2)}(IT) = A_1\exp(-\Gamma_1 IT) + A_2\exp(-\Gamma_2 IT)$
$N_1\Gamma_1T_1 \sim 2\text{-}3$, $N_2\Gamma_2T_2 \sim 2\text{-}3$

Amplitude Ratio $A_1/A_2 = 1/9$	% Relative Error A_1	A_2	Γ_1	Γ_2	Linewidth Ratio Γ_2/Γ_1
Single Delay	11.7	1.5	1.8	1.0	5
Dual Delay	2.0	.44	.23	.41	5
Single Delay	12.5	1.6	4.7	1.1	3
Dual Delay	2.9	.07	.71	.19	3
Single Delay	7.3	1.0	12.6	.72	9
Dual Delay	.29	.26	1.6	.25	9

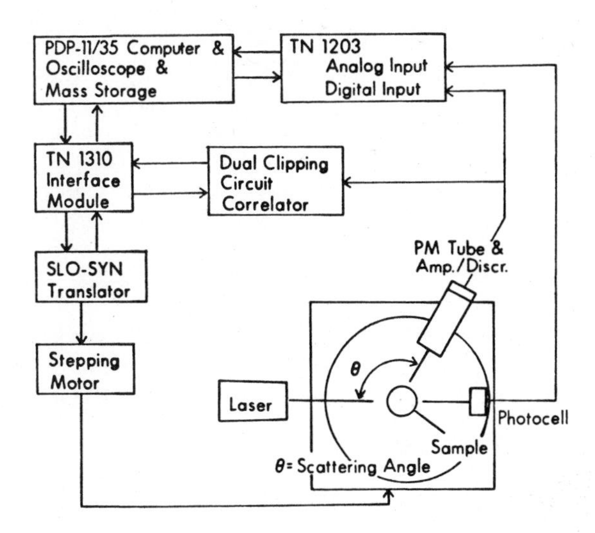

Fig. 6. System diagram of an automated laser light scattering spectrometer.

scattered light intensity are accomplished through a multichannel scaling input on the Tracor Northern 1203 signal processor. Digital pulses from a photomultiplier tube are amplified, discriminated and counted for a programmable length of time using a 16-bit counter. The number of counts accumulated during the measurement time is transferred to a buffer in the PDP-11 interface. Under software control, this number is then deposited in an appropriate memory location in the computer to await further analysis.

The Tracor Northern 1310 hardware module is connected directly to the PDP-11 unibus and facilitates interfacing of several devices to the computer. An axis positioner board (a circuit contained in the 1310), provides control signals to a Slo-Syn translator (Model #ST-1800 BV) which drives a Slo-Syn stepping motor (Model #MO-93 FC 07E). This motor rotates a turntable on which the photomultiplier tube is mounted. Intensity measurements as a function of scattering angle are obtained through rotation of the turntable. For very large or very small angles, proximity-limit-microswitches slow the movement of the turntable. Motion is halted altogether by limit microswitches, positioned at the lowest and highest angles attainable for a particular physical arrangement ($\sim 25°$ to $\sim 150°$).

In a typical experiment, the axis positioner module is programmed to control positioning of the detection optics at a specific angle with respect to the light incident on the sample. The values obtained from measurements of the intensity at one angle are stored in computer memory. The angle is then changed and new measurements are made.

The automated intensity measurements are programmed using Flextran, a higher level language supplied by Tracor Northern, and its associated machine language subroutines. The following steps are included in the program:

(1) All status and control registers for the different devices connected to the TN-1310 are cleared.

(2) The TN-1310 master interrupt is enabled.

(3) Through the Flextran subroutines and the TN-1310 axis positioner module, the turntable is positioned at the first angle. Variables in the program allow one to designate the Slo-Syn motor speed, direction of rotation and number of steps (angle increment) per movement. These conditions are reflected in the axis positioner control and status words.

(4) The intensity is then measured where the control and status registers in the TN-1203 have been programmed to count photoelectron pulses for the length of time desired. The number of samples acquired at a particular angle as well as the addresses in computer memory where the values are to be stored are deposited in specific memory locations before the measurement begins. The multichannel scaling subroutine reads these values before beginning to count pulses.

(5) The turntable is moved to the next angle and the measurement repeated. While the turntable can be rotated in either reverse or forward directions, the motor overshoots the destination in the reverse direction by a slight amount (1° or 2°) and executes the final approach in the forward direction. This technique is used to obtain greater accuracy and reproducibility of the turntable positioning.

It is important to emphasize that measurements of angular distribution of absolute scattered intensity complements measurements of angular distribution of linewidths. Therefore, automation of intensity measurements together with photon correlation should become a powerful tool for studies of static and dynamic properties of macromolecules in solution. Our computer-controlled light scattering system further combines data acquisition and data analysis as an experimental procedure of the spectrometer instrumentation thus making the iterative process in setting the proper delay time ranges, signal-to-noise ratio, methods of data analysis, and model fittings, an integral portion of the experiments.

The 256-channel dual delay time correlator can satisfy most of the needs in our correlation function profile analysis. Our simulated data analysis has addressed only one aspect of the many points we need to consider for real data analysis, namely, the effect of measured delay time range corresponding to a lower frequency limit of the finite bandwidth. In fact, we need to consider the relation between the upper frequency limit of the measured bandwidth and $G(\Gamma^*_{max})$, the measured background, the signal-to-noise ratio of the measured time correlation function to the extent we can approximate the linewidth distribution function $G(\Gamma)$ in terms of a fixed number of histograms or multiexponentials. The more noise we have, the fewer histograms or exponentials we will be permitted to use as otherwise the solution becomes unstable.

ACKNOWLEDGEMENTS

I wish to thank the National Science Foundation, Polymers Program (DMR 801652) and the U.S. Army Research Office, Durham, for financial support.

REFERENCES

1. R. Nossal and S. H. Chen, J. Phys. (Paris) Suppl. 33:C1-172 (1972).
2. K. Schätzel, Appl. Phys. 22:251 (1980).

ELECTROPHORETIC LIGHT SCATTERING:
MODERN METHODS AND RECENT APPLICATIONS
TO BIOLOGICAL MEMBRANES AND POLYELECTROLYTES

B. R. Ware

Department of Chemistry
Syracuse University
Syracuse, New York 13210

CONTENTS

1. INTRODUCTION

Electrophoretic light scattering (ELS) is simply the application of the laser Doppler principle for automated analytical electrophoresis.[1-3] Although the potential power of this approach has been apparent to many workers, both in the fields of electrophoresis and quasi-elastic light scattering (QELS), the technical difficulties of interfacing two complex and unrelated methodologies have discouraged a number of prospective practitioners. The development and publication of a number of technical improvements and successful applications have steadily increased the size of the ELS user community, and in little more than a decade following its invention, the technique is now commercially available and quite widely employed.

Both the applications and the methodology of electrophoretic light scattering (also known as laser Doppler spectroscopy or laser Doppler electrophoresis) have been reviewed in detail recently.[4-8] The objective of this chapter is to summarize the principles of ELS, to outline the important aspects of the experimental methodology, and to present a few recent applications that illustrate the powers of this technique.

A. Electrophoresis

Application of an electric field to a solution or suspension of electrically charged particles causes each particle to move toward the electrode of opposite polarity. The magnitude of the electrophoretic drift velocity (v) is proportional to the field strength (E), and the constant of proportionality (u), called the electrophoretic mobility, is thus given by the expression

$$u = v/E. \tag{1}$$

The relationship between the electrophoretic mobility and more fundamental molecular parameters is inexact because of the effect of the counterions in solution which interact both electrostatically and hydrodynamically with the electrophoresing particle. It is physically reasonable to suspect that the electrophoretic mobility will be proportional to the magnitude of the electric charge on the particle and inversely proportional to the friction constant. With appropriate correction for the effect of counterions,[9] the electrophoretic mobility may be written as

$$u = \frac{Ze}{f}\,\frac{X(\kappa R)}{(1+\kappa R)} \tag{2}$$

where Z is the number of unit charges on the particle, f is the friction constant of the particle, R is the radius of the particle, and κ is the familiar Debye-Hückel constant. The function $X(\kappa R)$, known as Henry's function, varies from 1.0 for small particles to 1.5 for large particles. In work on biological cells and organelles, it can generally be assumed that the particles are much larger than the reciprocal of the Debye-Hückel constant κ so that Eq. 2 can be simplified to the following form:

$$u = \frac{\sigma}{\eta\kappa} \tag{3}$$

where σ is the surface charge density at the hydrodynamic surface of the particle and η is the solution viscosity. Eq. 3 can also be expressed in terms of the electrical potential, generally called the ζ potential, at the hydrodynamic surface of the particle. In this case we can write the familiar Smoluchowski equation:[10]

$$u = \frac{\varepsilon\zeta}{4\pi\eta} \tag{4}$$

where ε is the dielectric constant of the medium.

The classical technique for studying the electrophoretic properties of biological cells and organelles is microelectrophoresis, in which an experimenter views the particle through an optical microscope, applies an electric field, and times the migration of the particles by eye. This is clearly a tedious and time-consuming process which is subject to a number of human errors, including sampling bias by the experimenter. This technique is obviously limited to particles which can be seen in a microscope, and its application to living cells is limited by the fact that cells whose electrophoretic characteristics may be changing in time, or whose electrophoretic distribution may be important to determine, are difficult to characterize because of the long time necessary for a statistically significant number of particles to be measured.

For submicroscopic particles, electrophoretic motion was first detected by the migration of a boundary formed between solution and solvent. The development of the moving-boundary technique for analysis and separation of the plasma proteins led to a Nobel prize for Tiselius, but the moving-boundary method is rarely employed nowadays because it is slow and because the concentration gradients and boundaries can lead to anomalies and difficulties of interpretation. Modern electrophoretic techniques for submicroscopic particles include isoelectric focusing, gel electrophoresis, and paper electrophoresis, none of which provide an accurate value for the solution electrophoretic mobility.

B. Laser Doppler Velocimetry

The Doppler shift is an apparent change of frequency of radiation as a result of relative motion between source and observer. Determination of the velocity of the motion by measurement of the Doppler shift has been common as radar and sonar applications for decades, and optical analogues have been known in astronomy for some time. Use of laser radiation for this purpose is called laser Doppler velocimetry (LDV). The first LDV measurement was accomplished shortly after lasers became commercially available,[11] and LDV technology is now commonplace in a diverse range of research and engineering applications (e.g. [12-15]). I summarize here only the basic principles essential for an understanding of the ELS technique. Other chapters in these proceedings will treat LDV theory and methods in more detail.

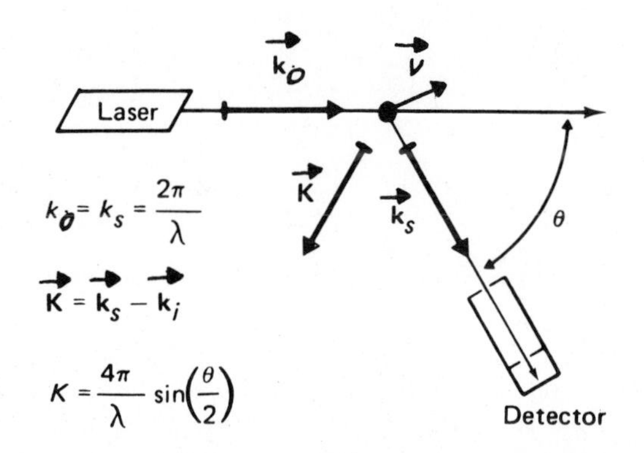

Fig. 1. The geometry of a laser Doppler velocimetry experiment, defining the incident wave vector $\vec{k}_o$, the scattered wave vector $\vec{k}_s$, the scattering angle θ, and the scattering vector $\vec{K}$. The symbol λ is used to indicate the wavelength of the laser light in the medium (λ_o/n), and corrections for refraction at the boundaries of the sample are not indicated.

The geometry of an LDV experiment is defined in Fig. 1. In a laser light scattering measurement of a moving particle there are actually two Doppler shifts. Motion with respect to the laser causes a shift in apparent frequency given by:

$$\Delta\nu_1 = \frac{-\hat{k}_o \cdot \vec{v}}{c} \nu_o \tag{5}$$

where the vector dot product $-\hat{k}_o \cdot \vec{v}$ is just the component of $\vec{v}$ in the direction of the laser. Similarly motion with respect to the detector causes a second Doppler shift given by:

$$\Delta\nu_2 = \frac{\hat{k}_s \cdot \vec{v}}{c} \nu_o \tag{6}$$

The total Doppler shift is then

$$\Delta\nu_{Tot} = \Delta\nu_2 + \Delta\nu_1 = \frac{\hat{k}_s \cdot \vec{v}}{c} \nu_o - \frac{\hat{k}_o \cdot \vec{v}}{c} \nu_o \tag{7}$$

Realizing that $\hat{k}_s - \hat{k}_o = \frac{\lambda}{2\pi} \vec{K}$ and that $c = \nu_o \lambda$, we can simplify Eq. (4) to become:

$$\Delta\nu_{Tot} = \frac{\vec{K} \cdot \vec{v}}{2\pi} \tag{8}$$

Thus the Doppler shift is given by the vector dot product of the particle velocity with the scattering vector. The division by 2π produces a result in Hz rather than angular frequency units.

Two important experimental considerations can be realized by inspection of Eq. (8). First we note that the Doppler shift is proportional to K ($= \frac{4\pi}{\lambda} \sin(\frac{\theta}{2})$) and thus will vanish in the limit of low scattering angle (where the Doppler shifts with respect to laser and detector cancel each other exactly). Secondly we note that the velocity must have some component along the scattering vector in order to produce a Doppler shift and that the shift will be maximized when the velocity to be detected is parallel to the scattering vector.

The Doppler shift magnitude is often too small for direct spectral resolution. Consequently the detection method generally employs a beating principle. Unshifted light from the incident beam (the "local oscillator") is mixed with the Doppler-shifted scattered light at the photodetector, whose output then contains an oscillating component at the difference (beat) frequency. Spectrum analysis of these beat frequencies provides a direct determination of the Doppler shift magnitude.

The shape and width of an LDV spectral peak are determined principally by three sources: velocity heterogeneity, transit-time broadening, and diffusion broadening. Transit-time broadening refers to the uncertainty broadening caused by the transit of the particle across the scattering volume during the measurement; the spectral peak can not be narrower than the reciprocal of the transit time of the scatterer through the scattering volume. The diffusion broadening is given (in Hz) by $DK^2/2\pi$ just as it is in the absence of a directed velocity. If transit-time broadening and diffusion broadening are negligible, then the LDV spectrum can be interpreted as the velocity histogram of the sample within the scattering volume, with each particle being weighted in the histogram in proportion to its light scattering cross-section at the experimental scattering angle.

C. Application of the Laser Doppler Principle to the Detection of Electrophoretic Motion

The theory and the first successful experiments of electrophoretic light scattering were reported in 1971 by Ware and Flygare.[1] It was apparent from this paper that the new technique held the potential to make important advances for two previously independent scientific communities. For the electrophoresis community, ELS provided a spectroscopic approach to analytical electrophoresis. The complete electrophoretic histogram of a sample of particles can be obtained in a few seconds without employing concentration gradients or boundaries and without the need for tedious, time-consuming, and subjective visual clocking by a technician. For the laser light scattering community, the introduction of an electric field represents an experimental means for dealing with the ubiquitous problem of polydispersity as well as for providing an additional experimental parameter (the electrophoretic mobility) for each resolvable species.

Calculation, by the standard means, of the theoretical expression for the ELS spectrum of a monodisperse solution of particles of mobility u and diffusion coefficient D yields:[1,3,5,8]

$$S(\nu) = \frac{DK^2/2\pi}{(\nu - u\vec{E}\cdot\vec{K}/2\pi)^2 + (DK^2/2\pi)^2} \tag{9}$$

Thus the spectrum is predicted to be Lorentzian with a half-width determined by D and a center frequency determined by u. A sample with n distinct species will produce an ELS spectrum that is the superposition of n spectra. Resolution of these peaks will be determined by the relative magnitudes of the shift differences to the respective half-widths, which in turn depends on the scattering angle, as will be discussed later. It is clear from consideration of the theoretical predictions that ELS has a great potential for characterization of complex samples and for rapid electrophoretic measurements. Realization of this potential requires careful consideration of a number of aspects of the experimental method, which I now discuss in some detail.

2. EXPERIMENTAL METHODS

A. Electrophoretic Light Scattering Chambers

For the scientist with a working laser light scattering

apparatus, the highest hurdle in extending its capabilities to electrophoretic measurements is the design and construction of an electrophoretic light scattering chamber. ELS chambers are not readily available as separate items for purchase at this time. The selection of a design and the construction of a working chamber are major efforts which will occupy a good deal of time and will be prime determinants of eventual success or failure.

An ELS chamber must serve simultaneously the objectives of laser light scattering and electrophoresis. There must be a clear path for entry and exit of the light. The chamber must have provision for application of a uniform electric field of known magnitude, as large as possible, and efficient heat transfer for removing the resulting Joule heat generated by the passage of the current. The chamber geometry should be designed to minimize convection in the scattering region while still permitting temperature regulation and stabilization. It is desirable to reduce or eliminate electroosmosis and to avoid artifacts which may result from electrode reactions. Sample volume should be minimized. The materials of which the chamber is made must be chemically compatible with the samples being studied. Finally, it is a desirable objective that the chamber should be convenient to use - i.e., easy to clean, assemble, and align, with a fast and simple method for removing old samples and inserting new ones. Needless to say, the ideal chamber has not yet been developed. A number of compromises must be made, and each experimenter must decide which set of advantages and disadvantages of the major options seems most suitable for his application.

The two major options in chamber design are distinguished by the placement of the electrodes. The more conventional approach separates the electrodes by a channel, in which the LDV measurement is made. A second approach is to employ narrowly spaced parallel-plate electrodes and to make the LDV measurement by introducing the laser through the small gap between the electrodes.

The early ELS channel-type chambers were straightforward adaptations of a Tiselius or microelectrophoresis chamber. The general feature of such designs is that the two electrodes are immersed in electrolyte pools on either side of the channel. If desired, the solution in the electrode pools may be different from the sample, perhaps buffer medium only, and semi-permeable membranes may be used to isolate the electrolyte pools from the sample. This design has the advantage that it permits maximum separation of electrodes from sample, essentially eliminating the undesirable effects of reaction of sample particles at the electrodes; production of bubbles or particles at the electrodes is generally not consequential in this design, since the bubbles or particles produced are kept remote from the scattering volume. The electrodes may be as large as desired, so that current density at the electrodes is not a limiting factor. These advantages bring two corresponding problems. First, this design may require either large sample volumes or the added nuisance of maintaining separate electrolyte pools. The second, more serious problem, is that this design does not allow for efficient removal of heat. The thermal time constant of design may be as high as 1 min or more.

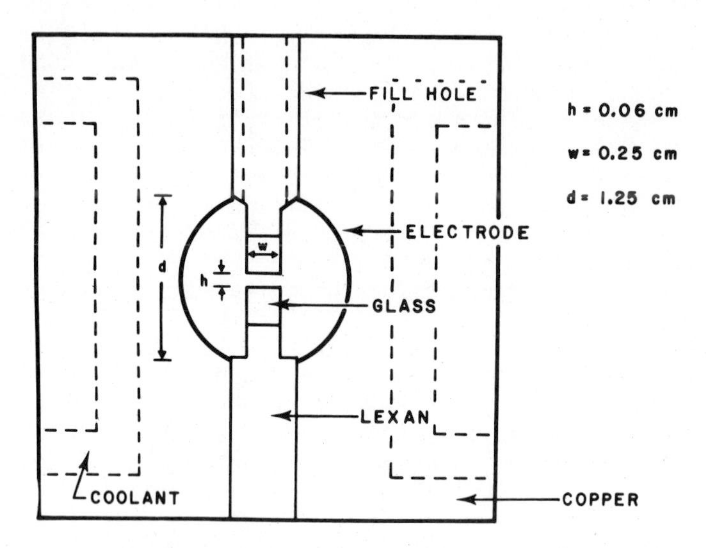

Fig. 2. The modified channel design for an ELS chamber using reduced sample volume and hemicylindrical electrodes. Details of this design are given in references 4 and 16.

A useful modification of the channel design is shown in Fig. 2. Introduced by Haas and Ware,[16] this chamber design reduces the liquid volume and the thermal time constant by placing the two electrodes just outside a narrow channel formed by dielectric spacers. The electrodes are hemicylindrical in order to obtain a uniform current density. The electric field and current density are effectively "focused" into the narrow channel. The flat, narrowly-spaced surfaces of this gap inhibit convection, while the curved walls of the electrode region encourage convection. Thus heat is conducted out of the gap without convection and then carried by convection to the electrodes, which are thermally conducting and in thermal contact with the large metal chamber halves, whose temperature is maintained by circulating coolant. Experimental details of the electric field configuration and the thermal properties of this design have been published.[4,8,16] The thermal time constant is on the order of ten seconds, depending of course on the choice of dimensions and material. Although the two metal halves of the chamber can act directly as electrodes, it is more common to plate them with an inert metal such as silver and to fix the electrode material into place with electrically-conducting epoxy. Adaptations of design B have been described by Smith and Ware[4] and by Schmitz.[17]

One major drawback of any channel-type chamber is the possible distortion due to electroosmotic flow. Electroosmosis is an electrokinetic effect, akin to electrophoresis, which may complicate electrophoretic measurements. Most materials are electrically charged when placed in contact with water; generally the charge on the material is negative. The glass of a cuvette or electrophoretic chamber attracts positive ions to it. If an electric field is applied, those positive ions move toward the negative electrode, drawing water with them. If the system is closed, the small pressure head built up at the negative electrode will cause water to flow back through the chamber, establishing a parabolic flow profile. The electroosmotic flow velocities are of the same order as electrophoretic velocities and are therefore a significant source of absolute error and velocity broadening if not dealt with by one of the following methods:

i) Calibrate the electroosmotic velocity at the point of measurement and correct for it. This approach suffers from the fact that the electroosmotic velocities may change in time if the wall charge changes (e.g. by adsorption of matter).

ii) Perform the measurements at the "stationary layer", i.e., the point in the chamber at which the electroosmotic and back flow velocities cancel. This approach has the drawback that the velocity gradient is steepest at the stationary layer, so that broadening effects are greatest. The scattering volume must therefore be kept unacceptably small. In addition, the stationary layer may move if matter adsorbs to different extents to each wall or if particles sediment to the bottom.

iii) Coat the walls with a neutral substance. This approach is successful, but it requires that the coating be stable and that there be no significant adsorption of charged material to the wall during the measurement.

The second major option for an ELS chamber design has been pioneered by Uzgiris.[5] Two parallel plates spaced about 1 mm apart are immersed in the sample and a potential is applied to them. Because the volume through which current passes is small, the amount of heat generated is small, and the thermal time constant is very short. Moreover, since the field is largely contained between the parallel plates and is essentially zero at the walls of the cuvette, electroosmosis is negligible. Both in concept and in construction, this design is simple, and its obvious advantages are extremely attractive. However, three essential features of this design raise problems which are not encountered in channel-type designs. The first is that the current density in the observation region must be the same as that at the electrodes. The second is that observation of the motion must be conducted very near the electrodes, so that particles, bubbles, or pH changes generated at the electrodes appear in the scattering volume very quickly. The third is that the conducting medium must be rendered significantly non-uniform by the passage of the current. Consequently, it has generally been necessary to reverse the direction of the electric field frequently during the measurement when employing this design, a procedure that can introduce artifactual sidebands or broadening effects into the spectrum. Other important considerations for this design, including electrode non-parallelism, electric field edge effects, and sample convection, have been discussed in detail by Uzgiris.[5]

B. Application of the Electric Field

The electric field that induces the electrophoretic motion must be applied by connecting the electrodes of the chamber to a power supply. Choice of power supply is not critical except that it is far preferable to employ a power supply that delivers constant current. There are two major reasons for this choice. First, if a constant voltage is applied to the electrodes, the conduction of ions to and from the vicinity of the electrodes changes the local conductivity, particularly in the parallel-plate design. The potential drop per unit length, which is how the electric field will be calculated, becomes nonuniform and indeterminate, so that an error in the field strength will result. A constant-current power supply automatically corrects for these local changes; it is rigorously true that if a constant current i is passed through a region of constant cross-sectional area A and conductivity σ, the electric field E in the region will be constant and given by:

$$E = \frac{i}{A\sigma} . \tag{10}$$

One may well object at this point to the notion of a constant conductivity, since the generation of Joule heat will change the conductivity in proportion to the change in viscosity (ca. 2% per C°). This observation leads us to the second motivation for employing a source of constant current. As the solution inevitably warms due to the passage of current, the resulting fall in viscosity leads to an increase in both the electrophoretic mobility and the conductivity in the same proportion. The increase in conductivity causes the electric field, at constant current, to drop (Eq. (10)) in the same proportion as the mobility is rising. Thus, to first order, the measured velocity remains the same despite the rise in temperature when one uses a constant-current power supply. It is only essential that the electrophoretic mobility and the conductivity be quoted at the same temperature. Rarely does nature afford us such a fortuitous cancellation of physical variables, and it would be a pity not to take advantage.

If the constant current is passed through the chamber for a time, a steady state condition will eventually be reached at which heat dissipation and heat generation are equal. The temperature in the sample will have risen and there will be thermal gradients from the center of the sample to the heat sinks. If the temperature at steady state is tolerable and if the steady-state thermal gradients are acceptably low in that they do not produce distorting convection flows, then the current may be left on through the measurement unless this leads to unacceptable electrode or solution effects. If, as is often the case, one is pushing the electric field to the maximum, it is generally possible to achieve a greater field strength under stable conditions by pulsing the field. The duration of the field pulse should be as long as the minimum measurement time specified by the sampling theorem. Thus if a spectral resolution of 0.5 Hz is desired, one switches the current source on, triggers the autocorrelator or spectrum analyzer to collect data for one complete record (2 sec), then switches the current source off to permit dissipation of the accumulated Joule heat. This dissipation time depends upon the thermal time constant of the chamber, generally on the order of 10 sec for the optimal channel designs. Successive pulses are usually alternated in polarity in order to prevent macroscopic accumulation of matter on one side of the chamber due to long-term electrophoresis. Because of the Doppler ambiguity in absolute direction, these successive pulses of alternating polarity produce equivalent signals that may be signal averaged for data accumulation. It has been rightly observed that this pulsing process reduces the efficiency of the method, but under some conditions the loss is minimal. For example, when analyzing suspensions of large particles, spectral quality is determined less by the Doppler signal-to-noise ratio, which is often adequate after a single pulse, but rather more by the time to measure a large number of different particles. The entry of new particles is accomplished by stirring or slow sedimentation into the sample volume, so the off-time is utilized. Some workers who are inclined to computerization may find the off-time useful for direct calculation of the spectrum or autocorrelation function by computer or for data manipulation and analysis. There is no question, however, but that it is best to minimize the off-time or eliminate it entirely, since the greatest single advantage of ELS is its rapidity.

When a parallel-plate chamber is being employed it is also necessary to switch the field, but for a different set of reasons. The high current density at the electrodes induces substantial changes in local conductance and solution conditions that must be reversed periodically. At low current density it may be possible to maintain constant field during the minimum measurement time specified by the sampling theorem. Generally, however, it is necessary to reverse the field several times during a sampling period. The effect of this velocity reversal is to produce in the spectrum a series of peaks at multiples of the switching frequency. The intensity envelope of these peaks is determined by the Doppler spectrum. A number of variants of this technique have been shown to reduce the pronounced sidebands and to produce a spectrum that approaches the true Doppler spectrum.[5,18] Schmitz[19] has proposed to utilize the modulation technique to study intramolecular degrees of freedom in ELS spectra from flexible macromolecules such as DNA. In principle the sidebands may be removed from the ELS spectrum if the functional form of the spectrum may be assumed, but for most of the interesting applications the form of the spectrum is at least as interesting as the peak mobilities. Thus far in the literature the sideband modulations have been dealt with most often by drawing the envelope of the artifactual peaks and thus suffering the broadening and loss of spectral detail which are the most serious drawbacks of the parallel-plate configuration. Recent improvements in electrode preparation may make it possible to reduce the necessary switching frequency to a level which will cause only minimal spectral broadening.

Magnitudes of electric field that are applied for ELS experiments vary widely. In the best cases a field strength of a few volts/cm suffices, and no serious heating problems will result. Field strengths as high as hundreds of volts/cm have been employed under conditions of low ionic strength, short pulse length, and low duty cycle. More typical ranges are 20 V/cm to 60 V/cm, which in terms of current is on the order of milliamperes for most chambers. Much depends on the ionic strength, as described in the next section.

C. Ionic Strength

The ionic strength μ of a solution is calculated from the concentration C and charge Z of the mobile ions in solution from the equation

$$\mu = \frac{1}{2} \sum_i C_i Z_i^2 \tag{11}$$

where the sum is over both positive and negative species. Usually macroions are excluded from the calculation because of their large friction coefficients. For a uni-univalent salt such as NaCl the ionic strength is equal to the concentration. As the ionic strength increases, the electrophoretic mobility decreases (through the counterion screening effects) and the conductivity increases, thus increasing the heating rate for a given electric field strength and the required current density at the electrodes. All these effects make the measurement more difficult. Consequently, the electrophoresis experimenter will always prefer a solvent medium of low ionic strength, but for biological samples the choice of optimal solution conditions is often dictated by the specimen. There is no lower limit of ionic strength for ELS, but loss of buffering capacity is appreciable below 10 mM and severe below 1 mM.

For large particles (>1 μm) it is routinely possible to make ELS measurements at physiological ionic strength (.157 M). However, for smaller particles this measurement becomes increasingly more difficult because these measurements must be done at lower scattering angles, for which the Doppler shifts are so small as to be difficult to measure with acceptable accuracy.

D. Choice of Scattering Angle

In the very first paper in this field, Ware and Flygare[1] defined the analytical resolution of this new technique to be the ratio of the Doppler shift to the diffusion-broadened half-width:

$$r = \frac{\vec{K}\cdot\vec{v}}{DK^2} \tag{12}$$

If $\vec{K}$ and $\vec{v}$ are parallel, the relation can be simplified to:

$$r = \frac{v}{DK} = \frac{uE\lambda_0}{D\ 4\pi n\sin(\theta/2)} \tag{13}$$

In practical terms Eq. (13) tells us that the analytical resolution of a diffusion-broadened ELS spectrum can be improved either by increasing the electric field strength or by decreasing the scattering angle. Note that decreasing the scattering angle actually decreases the Doppler shift, rendering it even more difficult to measure with desired precision, but decreases the diffusion width as the second power, so that the line becomes sharper.

Electrophoretic mobilities do not depend much on particle size, in fact do not vary greatly at all, but diffusion coefficients vary over several orders of magnitude, roughly as the linear characteristic dimension of the particle. For proteins and other monodisperse macromolecular systems in the size range 1-100 nm, diffusion broadening is the primary limitation of electrophoretic resolution by ELS. It is therefore essential to work at low scattering angle. The limitations on how low an angle one should select are placed by the ability to measure very small Doppler shifts with desired precision and by the ability to define the scattering angle with the required precision and accuracy. We have worked at angles as low as 1.5°, measuring Doppler shifts as low as 5 Hz. However, in order to make such measurements with precision it is necessary to maintain the electric field without switching for at least 4 sec. This in turn limits the ionic strength at which one can work. At present it would be impractical to attempt ELS experiments on small macromolecules (1-5 nm) unless the ionic strength can be kept at or below about 20 mM.

For larger particles such as living cells, the diffusion constant is very small, and the principal source of line broadening is electrophoretic heterogeneity. (These sources of broadening are easily distinguished by the facts that heterogeneity broadening is dependent on the first power of the scattering vector and the first power of the electric field strength, whereas diffusion broadening is dependent on the square of the scattering vector and is independent of the electric field strength.) In such cases it is desirable to work at higher scattering angles in order to increase the magnitude of the Doppler shift. How high an angle to choose is limited either by the design of the electrophoretic chamber or by the appearance of a diffusion component to the linewidth. If one can assume that there is an instrumental broadening component which is independent of scattering angle, then an optimum angle for resolution can be calculated[5] with the result that diffusion broadening should be allowed to equal the instrumental component. This optimal angle does not take into account other advantages which may be the result of measuring a shift of greater magnitude, such as the ability to employ shorter pulses of electric field. In practice the angular range 30° to 60° is best for analyzing large particles, and the resolution for such large particles as blood cells (~10 μm) is not, in our experience, very sensitive to scattering angle in this range.

Clearly for particles between the limits we have discussed, such as viruses and vesicles in the 50 nm to 1 μm range, the choice of optimal scattering angle will be intermediate as well. The extent of diffusion broadening tolerable or desirable is the operative criterion. It should be emphasized in this context that the minimization of diffusion broadening is desirable only if one's primary goal is to optimize electrophoretic resolution. The ability to measure simultaneously the diffusion constant and the electrophoretic mobility is a prime advantage of ELS and a major motivation for its invention. It is often desirable to choose a scattering angle for which diffusion broadening is a substantial component of the linewidth, so that changes in particle size by conformational change or aggregation can be detected.

E. Apparatus Components

Having discussed the special electrophoretic requirements, we proceed now to brief mention of the other major components of the apparatus. In many ways ELS is a quite specialized adaptation of laser light scattering, particularly because the data are in such a low frequency range. Thus an apparatus designed exclusively for ELS may be quite different from an optimal QELS apparatus, but it is quite straightforward to adapt a QELS apparatus for electrophoretic applications. A general block diagram of an ELS apparatus is shown in Fig. 3.

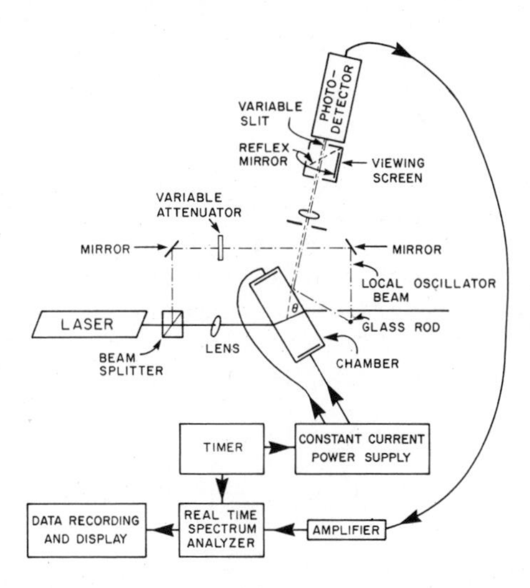

Fig. 3. Block diagram of an ELS apparatus.

The laser is not a critical choice in designing an ELS apparatus. Any optical wavelength will do, unless an intended specimen is colored, in which case minimization of absorption is desirable to avoid convection and photochemical damage. Low power lasers (5-50 mW) are usually sufficient unless one is particularly intending to study very dilute solutions of small macromolecules. Inherent noise on the laser light can be a problem. Fluctuations in the intensity of the light at a frequency within the spectral range being measured will appear as noise in the spectrum. Power-line-frequency ripple in the laser intensity is particularly troublesome in heterodyne applications such as ELS. Helium-neon lasers are generally the best of the commonly available types with respect to noise, reliability, convenience of operation, and price. Argon-ion lasers are also quite suitable, possibly superior, for ELS, but the difference in price and reliability is generally not justified for ELS applications.

Focusing and detection optics for ELS applications follow the same criteria as other LDV and QELS applications, with the inclusion, of course, of provision for a local oscillator. If possible, it is highly desirable to include in the detection optics provision for viewing a real image of the scattering volume and the local oscillator through the same optics that will be viewed by the photomultiplier tube.[4] This configuration permits the experimenter to select visually the illuminated region that will be the scattering volume and to ascertain visually that the local oscillator is spatially superimposed over the entire scattered signal. The maximization of the scattering volume to include as large a region as possible while avoiding the windows and the regions very near the window is an important objective, particularly for suspensions of large particles for which the number of particles in the scattering volume can be the limiting aspect of spectral quality. The positioning of the local oscillator and selection of its relative intensity are critical to the optimization of signal quality, particularly at high scattering angles. It is wise to view a signal from a test sample while making this adjustment. The local oscillator mixing efficiency and ratio will not affect the magnitude of the Doppler shift, but it will greatly affect the signal-to-noise ratio and the relative magnitude of interference from homodyne signals and laser noise.

Photomultiplier tube selection is not an important consideration for the ELS apparatus, particularly because the signals are strong and at low frequency. It is advisable to provide very good rf shielding for the photomultiplier housing; rf pickup will introduce ripple interference at multiples of the line frequency. It is important to know the linear response region of the photomultiplier system and to stay within it.

In contrast to most other QELS applications, ELS measurements are not well suited for photon counting. Because the ELS measurement is a low-frequency heterodyne measurement, often with high signal-to-noise ratio, the typical photon flux per measurement time is far in excess of the counts which can be handled by the counting device, and the absolute photon flux is often far above the level that will saturate the photomultiplier and/or the discriminator-amplifier. If one is constrained by prior decisions or other considerations to employ photon counting, it will generally be necessary to attenuate the laser by a large factor in order to arrive at a count rate that the equipment can handle. The result will be a lower signal-to-noise ratio and a longer time to achieve a satisfactory spectrum. In this regard, autocorrelators that include some sort of baseline subtraction feature are often preferable for handling heterodyne signals, since they can presubtract the uncorrelated local oscillator photons.

The preferred method for detection is to treat the photocurrent as an analog signal, passing it to ground through a dropping resistor (typical magnitude, 10 kΩ). The voltage above the resistor should then be amplified by a relatively low-gain (10X-100X) amplifier with variable high- and low-frequency cutoffs which are set to exclude signals outside the frequency region of interest. The output of the amplifier should then be processed to obtain its frequency spectrum. For low-angle measurements the frequency range of interest is perhaps 0.5 Hz-50 Hz; for higher scattering angles the region of interest may extend up to 200 Hz, but rarely beyond. Commercial autocorrelators and spectrum analyzers are thus greatly overdesigned for the needs of ELS measurements in that much of the expense and effort has been put in to the high speed, high frequency capabilities of these instruments.

Of the commercial equipment currently available, the best choice for ELS experiments is one of the Fourier-transform real-time spectrum analyzers. Unless some other application is also contemplated, it is probably best to purchase the cheapest such unit available provided it is real-time efficient up to at least 200 Hz and provided its baseline flatness and frequency accuracy specifications are better than 1%. It is also desirable for the spectrum analyzer to respond to an external trigger to collect a single spectrum so that it can be synchronized with the pulses of the electric field if desired.

Autocorrelators are less convenient for velocimetry applications because one does not view the data immediately in the form in which it is most readily interpretable. If an autocorrelator is employed, provision should be made for rapid transfer of the result to a computer for Fourier transformation to obtain the frequency spectrum. If a spectrum analyzer is employed, it is desirable to record both a trace and a digital record of the data. Whether the data will require computer analysis depends to a great degree upon the application, as will be discussed later.

F. Sample Preparation

Although it may seem like a Herculean task to assemble a working ELS apparatus with optimized capabilities, once this has been accomplished it is common experience that the greatest challenge in ELS experimentation is the design of the experiment and the preparation of proper samples, much as it is in other QELS applications. Techniques of sample preparation are not different from those for other QELS experiments as already covered. A few special considerations are discussed here.

The effect of dust and particulate contamination is a special consideration for ELS experiments. Although the presence of dust in a heterodyne QELS measurement may not be a serious problem, the dust does not hide harmlessly under the local oscillator signal when an electric field is turned on. Like virtually everything else, dust bears an electrical charge in water and will consequently produce a Doppler-shifted peak. Very often the peak will be at or near the position of the peak due to the intended sample, because the smaller sample particles tend to adsorb to larger particles and thus impart to the larger particles their own charge density. This procedure has even been used as a classical technique for measurement by microelectrophoresis of the electrophoretic mobilities of submicroscopic particles. Thus the principal effect of dust may be on the shape and width of the observed spectrum, a fact which must be considered in data analysis. The presence of dust and aggregates becomes much less serious when one is analyzing large particles such as blood cells. For samples of smaller particles, removal of dust and particulate contamination by the methods already described is an essential aspect of sample preparation.

As the electrophoretic mobility is a function of both the solution ionic strength and pH, control of these variables is rather more important than for most diffusion measurements. Particles that can be washed by centrifugation should receive several washings; smaller particle samples must be submitted to extensive dialysis. The conductivity of the sample must be known precisely to determine the electric field strength (Eq. (10)); it should be measured before the experiment and after the sample is retrieved from the chamber for most careful work. For measurements on living cells, cell viability should also be determined before and after the measurement.

One may generally be assured that the rather mild magnitudes of electric field and the short and slight temperature rises which are common in ELS measurements are not injurious to the biological material being studied, but it is a point to keep in mind. If it is suspected that the sample is coming to unnatural harm in the chamber, the obvious before and after tests should be done, and, if necessary, separate tests can be tried using other generators of temperature and electric field, respectively. I am not aware of cases where this type of damage has been demonstrated, but it is conceivable for such samples as weakly bound supramolecular complexes.

3. APPLICATIONS

A. Overview

The first successful ELS experiments were reported in 1971 by Ware and Flygare,[1] and independent reports from two other groups appeared in the following year.[20,21] In the ensuing decade there have appeared over sixty publications describing methods and applications in this field; the most comprehensive summary and bibliography up to the present is to be found in the review by Ware and Haas.[8] Much of the early work, and even some of the more recent work, has been motivated more by a goal of feasibility demonstration than by a fundamental interest in the investigated system, so the literature of true applications is less extensive than it may appear. Nevertheless there have already been a diverse range of biological, natural, and synthetic systems whose electrokinetic properties have been successfully investigated using ELS, and there is a growing list, of impressive scope, of examples for which ELS has eclipsed the capabilities of alternative techniques.

In surveying the past or potential future applications of ELS, particularly with regard to its efficacy vis-a-vis other electrophoretic techniques, it is useful to distinguish on the basis of the size domain of the particles to be investigated. ELS, like QELS, is very difficult to apply to particles of low molecular weight ($\lesssim$10,000) because of the low light scattering cross-section. In the molecular weight range of most proteins, ELS is feasible but difficult, largely because of the substantial contribution of diffusion to the ELS linewidth. In this range ELS is superior to the moving boundary technique for low-ionic-strength applications but probably inferior for applications at physiological ionic strength and above. In cases for which electrophoretic resolution is the desired result and a determination of the absolute magnitude of the electrophoretic mobility is not essential, the combined electrophoretic-chromatographic separation techniques using gels and other supporting media are generally far superior to ELS or moving boundary methods. An interesting exception is the case of associating or dissociating systems, for which ELS has the analytical advantage over separative techniques in that it does not perturb the equilibrium in order to make the measurement. For example, the spontaneous dissociation of hemoglobin at high pH is of interest because of the relationship of the tertiary and quaternary state of the hemoglobin tetramer to its oxygen binding capacity. Haas and Ware[22] used ELS and photon correlation spectroscopy to study the electrophoretic mobilities and diffusion coefficients of hemoglobin at high pH. The photon correlation spectroscopy results confirmed the dissociation of hemoglobin from tetramers to dimers above pH 10 and provided new estimates of the dissociation equilibrium constants in this range. The ELS data revealed that the electrophoretic mobilities of tetramers and dimers are indistinguishable in this range to within at most 7%, implying an increase in the electrical charge of the dimer of at least 2.8 to 4.4 net negative charges upon dissociation. A more recent ELS application in this size range of small macromolecules is given in part E of this section.

For the electrophoretic characterization of large particles (>1μm) such as living cells, the classical technique is the optical cytopherometer, in which the experimenter selects individual particles in the field of view of a microscope and visually clocks the motion in a known electric field. ELS has proved to be superior to the optical cytopherometer, largely because of the ability to analyze many particles simultaneously. In fact most of the reported applications of ELS have been in the area of electrophoretic investigation of living cells. Two recent reviews discuss this area comprehensively;[6,8] an illustrative series of experiments of this type will be discussed in part B of this section.

There is an intermediate size range (~.1 μm-1 μm) for which the application of either moving boundary or microscope techniques is difficult. Applications of ELS in this area have been quite successful. ELS spectra for particles in this size range are generally broadened both by diffusion and by electrophoretic heterogeneity, so that both charge density distributions and size effects (aggregation, changes in conformation) can be studied simultaneously. Common specimens in this size range include viruses, vesicles, and certain polyelectrolytes. Parts C and D of this section are illustrative of this region.

The reader is referred to the recent comprehensive reviews[6,8] for a survey of published applications. The remainder of this section will consist of four areas of recent applications in my own laboratory which should serve to illustrate the diverse powers of the ELS experiment.

B. Endocytosis and Exocytosis

ELS and other methods of cell electrophoresis can be used to characterize living cells on the basis of their average surface charge density. This characterization is generally more interesting when done in correlation with reactions on the cell surface whose effects are manifest by alterations in the surface charge. A class of reactions that we have studied in this context are the endocytic reactions, or more specifically, soluble pinocytosis. In the latter process, reactions on the external face of the cell plasma membrane, generally resulting in the cross-linking of certain cell-surface receptors, induce vesiculation of regions of the plasma membrane, followed by internalization of the endocytic vesicle. A major question to be posed about the mechanism of this process is whether the region of membrane that becomes the endocytic vesicle is chemically distinct from the remainder of the plasma membrane - i.e., are specific chemical groups partitioned into or out of the portion of the membrane that becomes the endocytic vesicle? A direct form of evidence for this process could be a concomitant change in the electrophoretic mobility of the cell, indicating specific partitioning of charged vs. uncharged groups. A distinct advantage of ELS for such studies is the ability to measure rapidly the complete electrophoretic histogram of a total population of cells, so that shifts in the surface charge properties of any significant fraction of the population can be detected readily.

Two reports from my lab in 1979 provided the first evidence that the endocytic reaction may be accompanied by alterations in the cell surface charge density. In the first of these studies,[23] macrophages and eosinophils from the guinea pig peritoneal cavity were incubated with soluble IgG immune complexes (insulin plus anti-insulin). Although neither insulin nor anti-insulin alone produced a major electrokinetic alteration, the combination caused macrophage electrophoretic mobilities to decrease by about 60%. Eosinophils showed a bimodal response: one population of eosinophils showed a negligible electrophoretic alteration, while the major population decreased in mobility by about a factor of three. The ELS spectra, shown in Fig. 4, provide a graphic illustration of the power of the technique that results from the ability to measure the complete electrophoretic distribution of a population of cells.

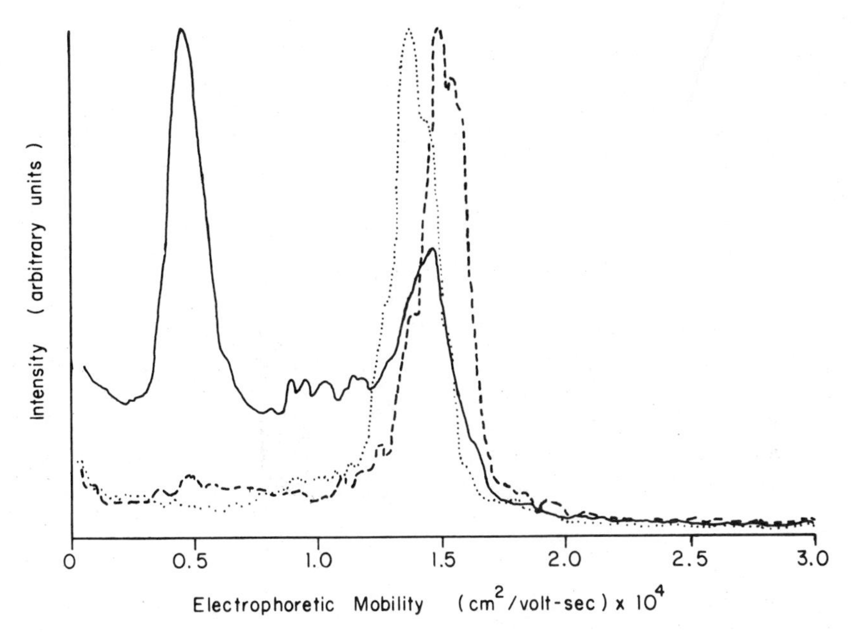

Fig. 4. A comparison of the electrophoretic mobility distributions (at 0.015 M ionic strength) of guinea pig resident peritoneal eosinophils treated with (a) bovine insulin (---), (b) guinea pig antibovine insulin (···), and (c) bovine insulin plus guinea pig antibovine insulin (—). From reference 23.

In the second study of this type,[24] we observed the response of peritoneal macrophages to the tetrameric lectin concanavalin A (Con A). Although binding of Con A to macrophage surfaces had no electrophoretic effect at the concentrations employed, subsequent incubation of the washed cells for 90 min produced a greatly altered electrophoretic distribution, whose mode mobility was about 15% lower but whose width increased three-fold. This effect was attributed to cross-linking of cell surface receptors by the following experiment: reaction of the same cells with dimeric succinyl-Con A (which has similar metabolic effects but induces little surface cross-linking) produced no electrokinetic alteration, but a dramatic electrokinetic alteration of the same form could be regenerated by incubation with anti-Con A, which would cross-link the succinyl Con A dimers on the cell surface. Illustrative ELS spectra are shown in Fig. 5. These electrokinetic effects on resident peritoneal macrophages were then correlated with alterations of surface

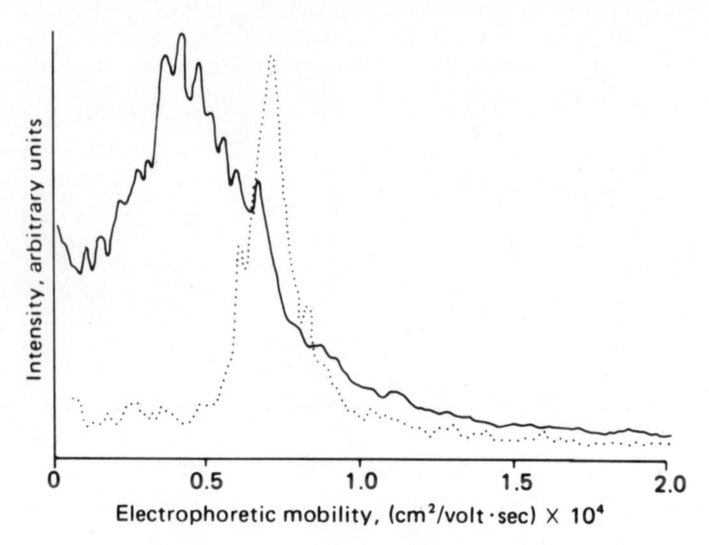

Fig. 5. ELS spectra (at 0.015 M ionic strength) of macrophages treated with succinyl-Con A for 30 min, washed, resuspended in the presence of 100 μg/ml of rabbit anti-Con A and incubated for 90 min at 37°C. The dotted line is the control, for which α-methylmannoside, a hapten sugar of Con A, was included in the final incubation.

morphology.[25] Con A was shown to decrease the number of surface folds and ruffles and to be internalized via endocytic vesicles. Succinyl Con A caused no morphological change nor was it internalized unless it was cross-linked by anti-Con A. Similarly it was shown that anionic sites, as visualized by cationized ferritin labelling, were redistributed on the macrophage surface following treatment with Con A or succinyl Con A plus anti-Con A, but not succinyl Con A alone, corroborating the ELS results.

We have performed a similar series of ELS experiments on human granulocytes incubated with Con A and succinyl-Con A, and we found an electrokinetic effect of the same form and magnitude as observed in the resident macrophage.[26] However, more surprising to us were a similar series of experiments on inflammatory (oil-elicited) guinea-pig peritoneal macrophages.[27] The electrokinetic effects observed previously for resident cells were completely absent for inflammatory cells. Scanning electron micrographs showed that the morphological change in the surface folds of the membrane was essentially identical, and transmission electron micrographs using ferritin-conjugated Con A showed vesicular internalization of the same form as for resident cells. However, the anionic site redistribution observed in the resident cells did not occur upon crosslinking of the Con A receptors in the inflammatory cells, in corroboration of the ELS results. It appears that resident macrophages specifically partition anionic groups into endocytic vesicles, whereas inflammatory macrophages do not. This difference in endocytic mechanisms is one of several newly discovered distinctions between these two cell types; its fundamental significance is not yet clear, but it may be related to a difference in digestive mechanism in the phagolysosome whereby the presence of a higher surface charge density on the interior vesicle surface serves to stabilize a lower pH or steeper ionic distribution in the resident cells.

The inverse of endocytosis, generally called exocytosis, is a fundamental aspect of certain types of vesicular secretion. We have used ELS to characterize the electrophoretic alterations accompanying stimulated secretion by rat serosal mast cells.[28] We observed a dramatic increase in electrophoretic mobility of stimulated cells that went through a bimodal distribution suggestive of an all-or-none process. Sample data are shown in Fig. 6. The interpretation of these experiments (with associated controls) was that the interior surface of the secretory vesicles (granules) bears a high net

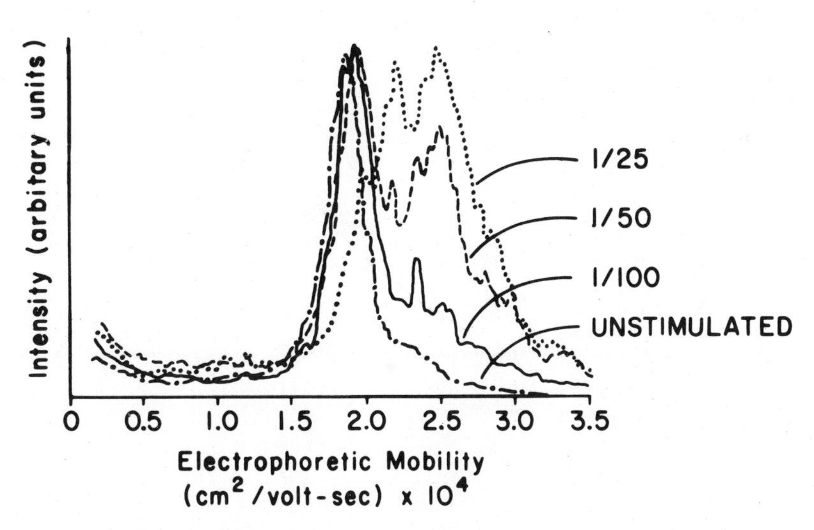

Fig. 6. Effect of immunological stimulation upon rat serosal mast cells. The fractions represent dilution factors of rabbit anti-rat $F(ab')_2$ antiserum. Increasing concentration of this antiserum clearly increases the proportion of cells in the higher mobility population.

negative charge, which adds to the negative surface charge density of the mast cell when the exocytic process transforms the mast cell granule interior to become part of the mast cell exterior. The selective asymmetry of the granule membrane, with a higher interior charge density, is the same as that inferred for the endocytic vesicles of the resident macrophage and circulating granulocyte. Our awareness of the potential significance of membrane surface charge asymmetry through these studies led us to consideration of the next series of experiments described.

C. Inside-Out and Rightside-Out Vesicles from Red Blood Cell Membranes

It is always important to bear in mind that the electrophoretic mobility of a particle is determined by its electrostatic potential (with respect to bulk solvent) at the hydrodynamic surface of shear. This potential in turn depends primarily on the surface density of electrically charged chemical groups on the external surface of the particle, the frictional contributions of surface structure, and the concentration and electrovalence of mobile counterions in solution. Because of the shielding effects, charge distributions within the internal structure of a particle do not affect its electrophoretic mobility. This insensitivity of electrophoretic mobility to internal charges can be put to good use in certain cases, and we have recently completed a study of the charge asymmetry of the red blood cell membrane that clearly illustrates this point.[29,30]

The red blood cell has been perhaps the most common of all specimens for electrophoretic study, so that the electrokinetic properties of its external surface are quite well known.[31] The cytoplasmic side of the red cell plasma membrane is not revealed in these measurements. However, recent development of methods for the selective preparation and characterization of vesicles which have inverted sidedness (IOV) or normal sidedness (ROV) makes the electrophoretic comparison of the two surfaces feasible.

We have used ELS for the characterization of the electrokinetic behavior of IOV and ROV of red blood cell plasma membranes. At neutral pH, ROV have a (~25%) higher electrophoretic mobility than IOV, and the two peaks can be resolved in the ELS spectrum to provide a quantitative estimate of the IOV/ROV ratio. Two examples are shown in Fig. 7. The ROV peak coincides with the electrophoretic mobility of fresh red blood cells and of resealed ghosts. The IOV/ROV ratio determined from the integrated intensity of the two peaks was generally consistent with the ratio determined by assay of the enzyme acetylcholinesterase. The agreement of these two methods indicates that the vesicles are sealed; otherwise the enzyme assay would count leaky IOV as ROV.

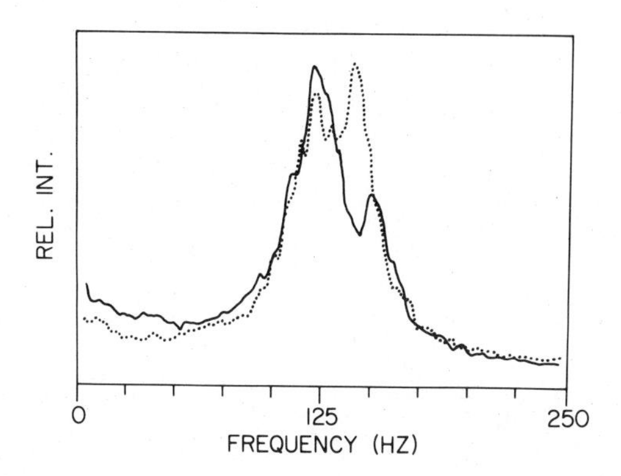

Fig. 7. ELS spectra of two preparations of vesicles made to be predominantly IOV (solid line) or ROV (broken line). For both spectra E = 60 V/cm and $\theta = 36.7^\circ$. From reference 30.

The presence and exposure of chemical groups on the surface of the membrane can be studied by reaction of the vesicles with specific enzymes. For example, neuraminidase, which cleaves charged sialic acid groups, reduced the ROV mobility by a factor of 2.6 but had little or no effect on the IOV peak. These observations are consistent with the prevailing view that most of the charge on the external surface of the red cell membrane is due to sialic acid and that sialic acid is absent from the cytoplasmic side of the membrane. Treatment of a mixed vesicle sample with trypsin resulted in a single narrow peak at about 60% of the mobility of ROV. The narrowness of the peak leads us to conclude that the vesicles inter-aggregated after trypsin treatment. Treatment of IOV with phospholipase C (which hydrolyzes the P-O between the phosphate and glycerol moieties) leaves the electrophoretic mobility unaltered, which is surprising since cleavage of phosphatidyl serine by this enzyme would have been expected to reduce the net negative charge of IOV. Phospholipase D (which hydrolyzes the other P-O bond, leaving an exposed phosphatidic acid) increased the mode mobility of IOV by 22%, presumably by removal of the $-NH_3^+$ of phosphatidylethanolamine.

The electrophoretic mobility titration curve of IOV from pH 2 to pH 10 is shown in Fig. 8. Three distinct inflection points are discernible at pH 4, 6.3, and 9.2, which we attribute respectively to carboxyl, phosphate, and amino groups on the cytoplasmic surface. About 30% of the peripheral proteins remain attached to the IOV in these preparations, so that some of these charged groups may be found on the proteins as well as the phospholipids. We estimate that at physiological pH, about two-thirds of the negative charge on the cytoplasmic surface is attributable to carboxyl groups.

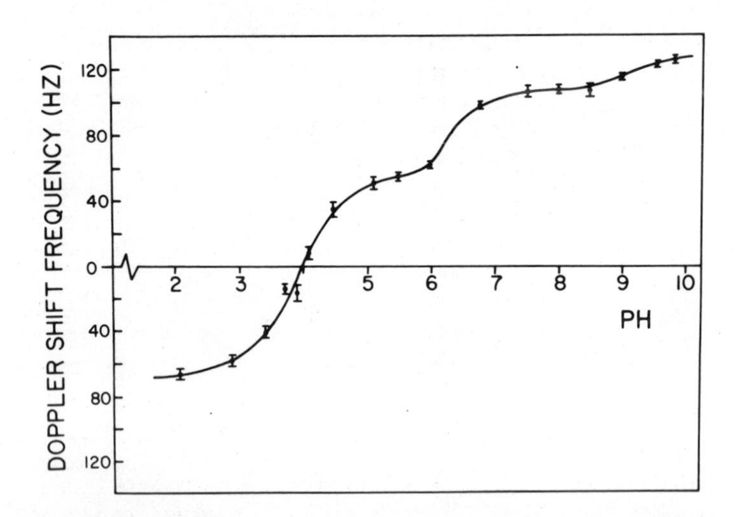

Fig. 8. Titration curve of a preparation of IOV (IOV/ROV = 4.1) represented as Doppler shift vs. pH. The Doppler shift point was taken as the maximum of the IOV peak. Each point represents at least three independent determinations, and error bars indicate SEM. The Doppler shift frequencies may be converted to electrophoretic mobility in the usual units (10^{-4} cm^2 sec^{-1} V^{-1}) by multiplying by 0.013.

Methods for preparation of ROV and IOV of plasma membranes from other cell types are appearing in the biochemical literature. These vesicles preserve many of the major physiological characteristics of their parent membranes while maintaining advantages of relative simplicity and stability as well as selected sidedness. Electrokinetic investigation of such vesicle systems may be a fruitful area of investigation in the years to come.

D. Counterion Condensation onto DNA

The association of counterions with a linear polyelectrolyte in solution can be described quite simply as a condensation process that occurs so long as the gain in electrostatic energy exceeds the thermal energy.[32,33] The fraction of charges on the polyelectrolyte that are effectively neutralized by condensation depends on the charge spacing on the polyelectrolyte and on the electrovalence of the counterion. Clearly the reduction in the electrophoretic mobility of the polyelectrolyte effected by counterion condensation is a direct experimental measure of the condensation phenomenon, and the work of Drifford and coworkers has illustrated the effectiveness of ELS for this type of investigation.[34,35]

We (Yen and Ware, to be published; Yoo and Ware, to be published) have recently conducted a series of experiments on the condensation of counterions onto DNA in solution. The general protocol has been to start with solutions of DNA in uni-univalent salt solutions of about 2 mM concentration (double-stranded DNA is not stable in distilled water) and to add incremental aliquots of multi-valent ions or to dialyze against fixed multi-valent ion concentrations. The resulting reductions in electrophoretic mobility are then compared with the predictions of counterion condensation theory. Data of this form may also be used to detect specific interactions of counterions with the nucleic acid. A striking example is illustrated in Fig. 9. In this experiment solutions of DNA from bacteriophage λ plasmid dvl in a solvent of 2 mM NaCl were dialyzed against solutions containing 2 mM NaCl and controlled concentrations of Mg^{+2}, Co^{+2}, or Hg^{+2}. (In the case of Hg^{+2}, the actual free divalent ion concentration is buffered by the formation of complexes with Cl^-.) The effect of increasing concentrations of Mg^{+2} or Co^{+2} is to reduce the mobility due to counterion condensation and increased screening, though the effects of the two are distinguishable. The Hg^{+2} curve has an inflection at around 10μM, which is roughly the concentration of DNA base pairs. This observation is consistent with the proposal that Hg^{+2} binds to the bases of DNA with the expulsion of two protons and with concomitant changes in the molecular conformation of the double-stranded polyion.[36] Detailed analysis of data such as these will be described in a forthcoming publication.

Although the counterion condensation experiments in our laboratory have not yet been completed nor thoroughly analyzed, it is possible at this point to make a few definitive statements and draw some tentative conclusions. The addition of multi-valent cations, in the absence of specific interactions (as for Hg^{+2}), reduces the mobility of the DNA molecule in a manner that is qualitatively consistent with counterion condensation theory. Ions of higher valence have a much greater effect. The extent of reduction of the mobility is determined primarily by the ratio of the multi-valent counterion concentration to the univalent counterion concentration and is

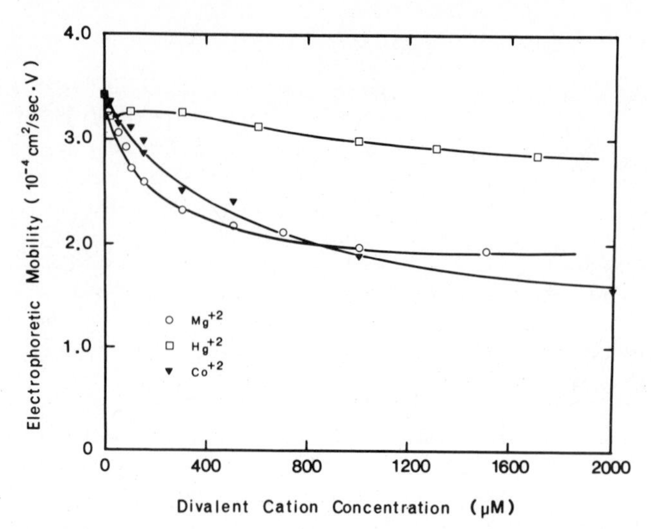

Fig. 9. Plot of the electrophoretic mobility of DNA (from bacteriophage λ plasmid dvl) as a function of the divalent ion concentration against which it had been dialyzed. The effects of Mg^{+2} and Co^{+2} are anticipated from counterion condensation, though the extent and form are distinct for the two ions. The effect of Hg^{+2} is consistent with earlier reports that Hg^{+2} separates the DNA strands and binds to the bases with expulsion of protons.[36] Because of the buffering effect of complex formation with Cl^- ions, the free Hg^{+2} concentration is much lower than the total concentration represented on the axis. (From Yoo and Ware, to be published.)

relatively insensitive to polyion concentration. Back-dilution experiments have verified that the counterion condensation is reversibl with no apparent hysteresis. High degrees of counterion condensation are often followed by collapse of the polyion conformation (as revealed by an increase in the diffusion component in the ELS linewidth), which in turn may be followed by aggregation of the polyion chains (as revealed by subsequent narrowing of the ELS linewidth). The distinction between condensation, collapse, and aggregation is vital to maintain, and ELS is an ideal technique for this purpose. Quantitative comparisons of our data with counterion condensation theory are limited by the indeterminacy of the theoretical parameter V_p, the volume about the polyion within which a counterion is considered to be condensed. If this parameter is treated as a variable, theory and experiment are quantitatively consistent, but different experiments on the same polyion do not yield constant values of V_p.

Further analysis on this point may elucidate trends of physical significance. It is clear to us that analysis of polyelectrolyte behavior using ELS is a rich area for further experimentation.

E. Extraordinary Phase of Poly-L-lysine

Polyelectrolyte solutions become even more complex, particularly at low ionic strength, as the polyion concentration is increased to the point that polyion-polyion repulsions become significant. Long-range position correlations can be expected to increase, and concentration fluctuations are generally suppressed because of the electrostatic energy increases that accompany them. A number of studies of these effects have been described in recent reports. Perhaps the most bizarre are the reports of Schurr and coworkers on the so-called "extraordinary phase" of solutions of poly-L-lysine.[37,38] If one measures the apparent diffusion coefficient, as determined by QELS, as a function of polyion concentration or ionic strength, one finds a precipitous decrease of more than an order of magnitude as the salt concentration is decreased and/or the polyion concentration is increased. Schurr and coworkers have described these observations as indicating a "dramatic rise in the friction factors of the isolated polyions",[38] which is attributed primarily to electrolyte friction.[39]

It has been suggested that the electrolyte frictional effects would lead to a low electrophoretic mobility of poly-L-lysine, and a value of 1.12×10^{-4} cm^2 s^{-1} V^{-1} has been predicted.[39] Accordingly we have measured the electrophoretic mobility by performing ELS experiments on solutions of poly-L-lysine in the extraordinary phase (40). We verified the extraordinary phase by performing QELS experiments using the same apparatus and obtained values that are consistent with those reported by Schurr and coworkers. Representative ELS spectra are shown in Fig. 10. Although the linewidth of the ELS peaks are anomalously narrow to a degree consistent with the QELS measurements, the electrophoretic mobilities calculated from the positions of the ELS peaks are 4.6×10^{-4} cm^2 s^{-1} V^{-1} in 5 mM Tris buffer and 3.5×10^{-4} cm^2 s^{-1} V^{-1} in 5 mM Tris buffer plus 20 mM KCl, which are characteristic values for ordinary polyelectrolytes of high charge density.

It is important to realize when performing light scattering measurements that one determines only the magnitudes and dynamics of the scattering elements - which in general are the Fourier components of the concentration fluctuations of the solution with wavelength $2\pi/K$. It is well known that intermolecular interactions can cause substantial deviations of the magnitudes and dynamics of such fluctuations from the properties that would be predicted from the characteristics of the individual particles alone. Fluctuation mobilities may be substantially different from molecular mobilities,[41] and it is not appropriate to calculate molecular friction factors from fluctuation mobilities in such cases.[40] To investigate this point in the

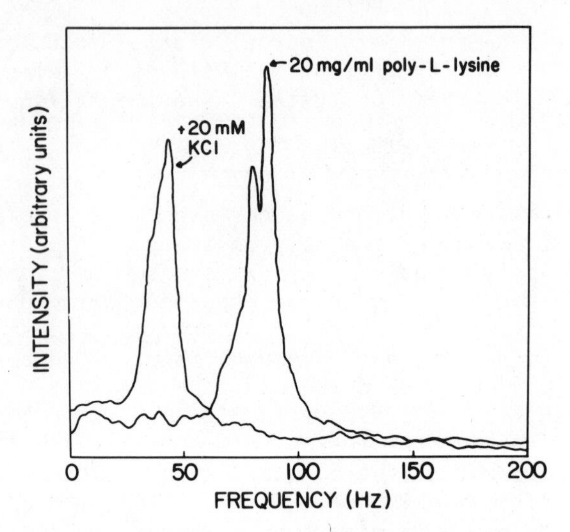

Fig. 10. Representative ELS spectra for the study of the ionic strength dependence of the electrophoretic mobility of poly-L-lysine. For both spectra the poly-L-lysine concentration was 20 mg/ml, the scattering angle was 20°, the temperature was 25°C, and the constant electrophoretic current was 2.0 mamp. Because of differences in conductivity of the two samples, the electric field was 24.1 V/cm in the absence of added salt and 15.2 V/cm with 20 mM KCl. The respective electrophoretic mobilities in the two cases are 4.6×10^{-4} $cm^2/Vsec$ and 3.5×10^{-4} $cm^2/Vsec$. Both samples contained 5 mM Tris buffer at pH 8.

case of the extraordinary phase of poly-L-lysine, we have performed measurements of the molecular (tracer) diffusion coefficients of labelled poly-L-lysine molecules using the technique of fluorescence photobleaching recovery (FPR), which is described in detail elsewhere in these proceedings. Our variation of the experimental methodology has been published.[42] The poly-L-lysine molecules were labelled by reaction with fluorescein isothiocyanate to an average degree of labelling of less than one fluorescein molecule per 50 residues. The FPR measurements yielded a tracer diffusion coefficient that varied with increasing salt from 4×10^{-7} to 6×10^{-7} cm^2/sec. Thus the tracer diffusion coefficient, which is directly related to molecular friction, is not reduced in the extraordinary phase. Our experiments therefore indicate that the effective mutual diffusion coefficient from QELS is indeed more than an order of magnitude lower in the extraordinary phase, but that the electrophoretic mobility and tracer diffusion coefficient remain "ordinary". We conclude that the extraordinary phase of poly-L-lysine is characterized by long-lived concentration fluctuations but not by reduced molecular mobilities. Further experimental and theoretical work will be required for a thorough understanding of this fascinating system.

4. PROGNOSIS

The maturation of ELS into an accepted technique with numerous practitioners and a wide range of successful applications has been accomplished in one decade. It may be appropriate at this time to set down some fundamental considerations for selecting future applications and to speculate on the potential of the technique in its second decade.

It seems likely that ELS will replace the optical cytopherometer for most applications in particle electrophoresis. The advantage of ELS with regard to speed, accuracy, automaticity, and ability to construct the complete electrophoretic histogram are generally more important considerations than the one advantage of the optical cytopherometer - i.e., the ability to select visually the particles to be analyzed. Most of the published applications of ELS in particle electrophoresis are studies of the surfaces of living cells. The biological and medical potential of cell electrophoresis must be analyzed critically. Cell surfaces are exceedingly complex. The measurement of a single average parameter of such a complex system - particularly when the normal variation in that parameter among all cells is hardly more than a factor of two - is unlikely to answer the detailed questions at the forefront of modern membrane research. Experiments in this field must be carefully designed in order to yield unambiguous information of true importance, and corroborating experiments using other physical and biological techniques will generally be required before hard conclusions can be drawn. A number of workers in the cell electrophoresis community are becoming increasingly conscious of the need for more thorough experimentation and continued activity in this field can be anticipated. Related efforts are in progress for the development of clinical tests based on cell electrophoresis. If a straightforward protocol - probably involving multiple treatments with specific enzymes, antibodies, or other reagents - can be worked out for an important problem, ELS is an ideal technique for automated on-line electrophoretic analysis.

Although most of the published applications of ELS in particle electrophoresis have been studies of living cells, there is already an equally vigorous effort in the study of particles and materials for industrial applications. Most of this work remains unpublished, but the growth in activity has been rapid and this remains one of the most promising areas for future growth. Synthetic particles for coatings, photocopying, and structural applications constitute a multi-billion dollar industry, and the measurement of surface charge can be important both for development and for routine processing. Another large industry for which automated electrophoresis can be an important tool is the design and production of polymeric flocculating agents.

The applications of ELS to systems in the small and intermediate size ranges will probably be less numerous, but well designed ELS experiments for these systems have the greatest potential for utilization of the full power of the technique. It should always be kept in mind that ELS experiments are progressively more difficult for smaller particles and macromolecules, particularly if the measurements can not be made at low ionic strength. Many new investigators enter the field to perform difficult experiments on small particles at high salt only to be frustrated by the technical difficulties. This area will probably be confined to research applications by relatively few expert investigators, at least for the immediate future.

One of the promising aspects of the maturation of ELS as a technique is that it is no longer a full-time job to maintain and develop ELS apparatus; ELS investigators are thus now free to combine ELS measurements with other types of measurements for a more complete characterization of the system under investigation. We have found that structural techniques such as scanning and transmission electron microscopy, and tracer techniques, such as fluorescence photobleaching recovery, are excellent complementary techniques for ELS and for light scattering in general. The structural and tracer information make the hydrodynamic and electrokinetic information much more meaningful and lead the experimenter to a much better ability to draw meaningful conclusions.

In conclusion, I believe that ELS, like other light scattering techniques, is a valuable addition to the technical arsenals of the physical chemists and biophysicists, provided that the technique is used within its limitations, that other techniques are used to provide essential independent information, and that the combined data are analyzed critically to pursue an important problem to a meaningful conclusion.

ACKNOWLEDGEMENTS

I am indebted to my students and co-workers whose names appear in the respective citations for the original work. Electrophoretic light scattering projects in my laboratory are supported by Grant #GM 27633 from the National Institutes of Health and by

Grant #PRF-12865-AC7-C from the Petroleum Research Fund, administered by the American Chemical Society.

REFERENCES

1. B. R. Ware and W. H. Flygare, Chem. Phys. Lett. 12: 81 (1971).
2. B. R. Ware, "The Invention and Development of Electrophoretic Light Scattering" (Ph.D. Thesis, The University of Illinois at Urbana-Champaign, 1972).
3. B. R. Ware, Advan. Colloid Interface Sci. 4: 1 (1974).
4. B. A. Smith and B. R. Ware, in "Contemporary Topics in Analytical and Clinical Chemistry", Vol. 2 (D. M. Hercules, G. M. Hieftje, L. R. Snyder, and M. A. Evenson, eds.), pp. 29-54 (Plenum, New York, 1978).
5. E. E. Uzgiris, Prog. Surface Sci. 10: 53 (1981).
6. E. E. Uzgiris, Advan. Colloid Interface Sci. 14: 75 (1981).
7. B. R. Ware, in "Biomedical Applications of Laser-Light Scattering" (D. B. Sattelle, B. R. Ware, and W. L. Lee, eds.) (Elsevier, Amsterdam, 1982).
8. B. R. Ware and D. D. Haas in "Fast Methods in Physical Biochemistry and Cell Biology" (R. Sha'afi and S. Fernandez, eds.) (Elsevier, Amsterdam, 1982).
9. D. C. Henry, Proc. Royal Soc. A203: 514 (1931).
10. M. Smoluchowski, Z. Physik. Chem. 92: 129 (1918).
11. Y. Yeh and H. Z. Cummins, Appl. Phys. Lett. 4: 176 (1964).
12. B. M. Watrasiewicz and M. J. Rudd, "Laser Doppler Measurements" (Butterworths, London, 1976).
13. L. E. Drain, "The Laser Doppler Technique" (Wiley, New York, 1980).
14. B. R. Ware, in "Chemical and Biochemical Applications of Lasers" (C. B. Moore, ed.), pp. 199-239 (Academic Press, New York, 1977).
15. H. Z. Cummins and E. R. Pike (editors), "Photon Correlation Spectroscopy and Velocimetry", NATO Advanced Study Institute, Series B: Physics, Vol. 23 (Plenum, New York, 1977).
16. D. D. Haas and B. R. Ware, Anal. Biochem. 74: 175 (1976).
17. K. S. Schmitz, Chem. Phys. Lett. 63: 259 (1979).
18. J. D. Harvey, D. F. Walls, and M. W. Woolford, Optics Commun. 18: 367 (1976).
19. K. S. Schmitz, Chem. Phys. Lett. 42: 137 (1976).
20. T. Yoshimua, A. Kikkawa, and N. Suzuki, Japan. J. Appl. Phys. 11: 1797 (1972).
21. E. E. Uzgiris, Optics Commun. 6: 55 (1972).
22. D. D. Haas and B. R. Ware, Biochemistry, 17: 4946 (1978).
23. H. R. Petty, R. L. Folger, and B. R. Ware, Cell Biophys. 1: 29 (1979).
24. H. R. Petty and B. R. Ware, Proc. Natl. Acad. Sci. USA 76: 2278 (1979).
25. H. R. Petty, Exptl. Cell Res. 128: 439 (1980).

26. H. R. Petty and B. R. Ware, Cell Biophys. 3: 19 (1981).
27. H. R. Petty and B. R. Ware, J. Ultrastruct. Res. 75: 97 (1981).
28. H. R. Petty, B. R. Ware, and S. I. Wasserman, Biophys. J. 30: 41 (1980).
29. W. S. Yen, R. W. Mercer, B. R. Ware, and P. B. Dunham, in "Scattering Techniques Applied to Supramolecular and Non-equilibrium Systems", NATO Advanced Study Institute, Series B: Physics, Vol. 73 (Plenum, New York, 1981), pp. 861-864.
30. W. S. Yen, R. W. Mercer, B. R. Ware, and P. B. Dunham, Biochim. Biophys. Acta, in press.
31. G. V. F. Seaman, in "The Red Blood Cell, Vol. II) (D. Surgenor, ed.) (Academic Press, New York, 1975), pp. 1135-1229.
32. G. S. Manning, Quart. Rev. Biophys. 11(2): 179 (1978).
33. G. S. Manning, Accounts Chem. Res. 12: 443 (1979).
34. H. Magdelénat, P. Turr, P. Tivant, M. Chemla, R. Menez, and M. Drifford, Biopolymers 18: 187 (1979).
35. J. P. Meullenet, A. Schmitt, and M. Drifford, J. Phys. Chem. 83: 1924 (1979).
36. T. Yamane and N. Davidson, J. Am. Chem. Soc. 83: 2599 (1961).
37. W. I. Lee and J. M. Schurr, J. Polym. Sci. 13: 873 (1975).
38. S.-C. Lin, W. I. Lee, and J. M. Schurr, Biopolymers 17: 1041 (1978).
39. J. M. Schurr, Chem. Phys. 45: 119 (1980).
40. B. R. Ware, Donna Cyr, Sridhar Gorti, and Frederick Lanni, in "Measurement of Suspended Particles by Quasi-Elastic Light Scattering" (B. Dahneke, editor) (John Wiley and Sons, New York, 1982).
41. M. B. Weissman and B. R. Ware, J. Chem. Phys. 68: 5069 (1978).
42. Frederick Lanni and B. R. Ware, Rev. Sci. Instrum., in press.

LASER DOPPLER VELOCIMETRY IN A BIOLOGICAL CONTEXT

J.C.Earnshaw

Department of Pure and Applied Physics
The Queen's University of Belfast
Belfast BT7 1NN, Northern Ireland

CONTENTS

1. INTRODUCTION

Since the pioneering experiments of Yeh and Cummins[1] laser Doppler velocimetry (LDV) has developed into a mature subject which has been well reviewed.[2] No attempt will be made to cover this field fully but attention will be focussed upon the particular problems which are posed by biological flows. LDV has largely been developed in the context of fast, turbulent flows of gases which are tenuously populated by small particles from a more or less homogeneous population. In the biological situation however, we are likely to be confronted with slow laminar flows of dense fluids incorporating particles ranging from macromolecules to organelles. Each difference leads to differences in the implementation of LDV or the interpretation of the observed data.

In this chapter, after a brief summary of the basic principles of the technique, the difficulties and limitations arising in applications to biological systems are systematically considered. In the main-stream development of LDV many ideas and technical devices have appeared which may be usefully adapted to solve these difficulties. Unfortunately much of the literature is not addressed to the aims and problems specific to flows of a biological nature. Thus finding the relevant ideas in the extensive literature of the subject may not be easy.

2. PRINCIPLES

Light scattered by a particle with translational velocity $\underline{v}$ is shifted in frequency due to the Doppler effect. The basic equation for the Doppler shift is

$$\Delta\omega = \underline{k} \cdot \underline{v} \tag{1}$$

where $\underline{k}$, the scattering vector, depends upon the angle of scattering (θ) as $4\pi n/\lambda \sin \theta/2$. Light scattering is only sensitive to the projection of $\underline{v}$ on to $\underline{k}$. If the angle between $\underline{k}$ and $\underline{v}$ is ϕ equation (1) becomes:

$$\Delta\omega = \frac{4\pi n}{\lambda} v \sin \theta/2 \cos \phi \tag{2}$$

showing explicitly that $\Delta\omega$ is maximum for 180° back-scattering. Using He-Ne lasers and assuming flow in aqueous media, the maximum $\Delta\omega$ (for $\theta = 180^{\circ}$, $\phi = 0^{\circ}$) is 4.20 Hz/μm s^{-1}. Such frequency shifts are low, but measureable. To detect a shift in frequency the scattered light falling upon the detector must be mixed with a reference beam. This can be achieved in several ways; various typical experimental arrangements are sketched in Figure 1. The differential Doppler arrangement (1b) involves the mixing of light scattered out of two beams incident upon the scattering volume. This arrangement has certain advantages - notably the frequency of the detector output depends only upon the angle between the incident beams so that light can be gathered over large solid angles - which have led to its widespread use in aerodynamics. Unfortunately, as the density of particles within the flow medium increases the light scattered by different particles interferes, leading to a decrease in the signal strength. In such denser suspensions the reference beam arrangements are most useful. In all of these systems, the sensitive volume of the velocimeter is principally defined by the focussing and detecting optics. For example, in Figure 1a, the scattering volume is basically set by the region of overlapping of the two beams. Note that ideally the laser beam should be focussed before it is split into two beams, but in the

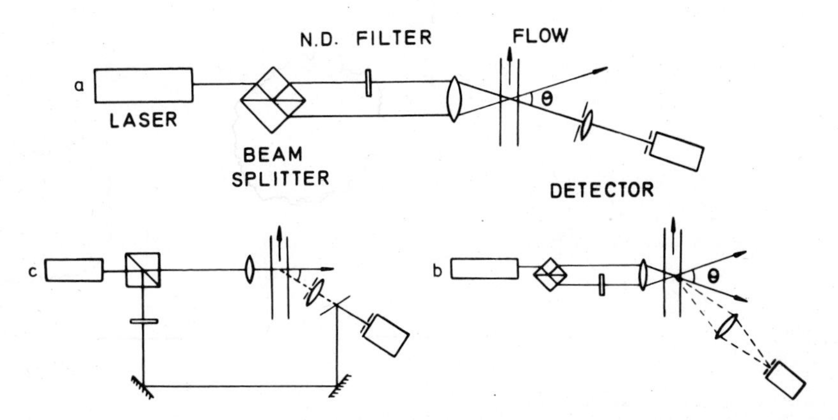

Figure 1 : Representative LDV systems: (a) reference beam arrangement, (b) differential Doppler system, (c) 'external reference beam' arrangement.

biological context the requirement for a small scattering volume may dictate focussing with microscope objectives when arrangements such as those shown are unavoidable. The laser beams will then be focussed beyond their cross-over point (as laser beams are divergent, whereas the beam splitters should produce parallel output beams) causing minor distortions of the Doppler spectrum.[3]

The Doppler spectrum of the scattered light directly reflects the velocities of all the particles which contribute to the detected light. Spectrum analysis in the frequency domain is thus the most transparent technique, providing data which immediately map to velocities. Spectrum analysis is also the technique of choice when the rate of detection of scattered photons is high, as is likely in many biological applications of LDV. However photon correlation has been successfully used in such situations.

In most LDV systems signals due to the transits of several particles through the scattering volume are averaged. In this averaging, each particle makes a contribution which is weighted according to the intensity of light which it scatters during the transit. This depends upon the intensity distribution within the scattering volume. Usually the incident laser beams will have Gaussian intensity profiles and the intensity distribution within the scattering volume will be three-dimensional Gaussian:[2]

$$I(r) = I_o \exp \{- \frac{(x - x_o)^2}{2\sigma_x^2} - \frac{(y - y_o)^2}{2\sigma_y^2} - \frac{(z - z_o)^2}{2\sigma_z^2} \}, \quad (3)$$

the scattering volume being centred at $\underline{r}_o$. The various σ's are determined from the equations governing the propagation of the laser beams.[4]

Now there is a directional ambiguity in LDV, as $\Delta\omega$ is independen of the sense of $\underline{v}$. While in many situations this is not a grave disadvantage, there are cases in which determination of the direction of motion is important or where motion periodically reverses. Various techniques have been devised to circumvent this limitation. The basi principle involves generating a constant frequency difference (ω_o) between the incident beam and the reference beam, so that the Doppler spectrum is shifted from $\Delta\omega$ to ($\omega_o \pm \Delta\omega$), the sign depending upon the sense of $\underline{v}$. Provided $\omega_o > |\Delta\omega|$, motion in both directions is clearly identifiable. Originally the frequency difference ω_o was produced by modulating one beam by Bragg diffraction from travelling ultrasonic waves.[1] While this leads to values of ω_o which are likely to be excessive for biological problems, combinations of two Bragg cells can yield lower effective values of ω_o. Electronic down mixing has also been used to give effective frequency shifts as low as 10 kHz. An alternative approach, generating a linearly ramped phase shift between the two beams, is equivalent to imparting a constant frequency shift.[5] A moving diffraction grating can be used to generate two diffracted beams whose frequencies ($\pm\ \omega_o$) are related to the velocity of the grating. Rotational motion of circular gratings has been useful.[6] Other methods used have included amplitude modulation[7] and moving mirrors;[8] some of these devices are commercially available.

Finally it can be noted that for complex flow fields LDV systems sensitive to 2 or 3 orthogonal velocity components have been devised. The basic idea is to set up 2 (or 3) independent systems - using laser beams of different colours for example - having scattering vectors arranged at right angles to each other. Such systems may find applications in detailed studies of biological motions. A fibre-optic Doppler anemometer[9] has been developed which may be useful in studying flows in inaccessible situations *in vivo* with less disturbance than would be necessary in gaining access for conventional LDV systems.

3. BROADENING OF THE DOPPLER SPECTRUM

Equation (1) implies that for uniform translational motion (ie single $\underline{v}$) the spectrum of the scattered light is a δ function at $\Delta\omega$; the correlation function is just $\cos \Delta\omega t$. Two effects - transit-time broadening and diffusive motion - cause the observed

Doppler spectrum to deviate from this ideal, even for the case of a unique flow velocity. The relative significance of the two effects varies with both $\underline{k}$ and $\underline{v}$.

(a) Transit time broadening

As a particle traverses the scattering volume it will contribute to the observed signal for a time (the transit time) limited by the spatial extent of the scattering volume. Taking the Gaussian ellipsoid intensity distribution of equation 3 (setting $\underline{r}_o = 0$) and if $\underline{v}$ lies parallel to the x axis:

$$I(t) = I_o \exp\left\{-\frac{x(t)^2}{2\sigma_x^2} - \frac{y^2}{2\sigma_y^2}\right\} e^{i\Delta\omega t} \quad (4)$$

$$\propto \exp\left(-\frac{v_x^2 t^2}{2\sigma_x^2}\right) e^{i\Delta\omega t}$$

assuming $\underline{k}$ is parallel to x.

Thus the photon correlation function is given by

$$G(t) \propto \exp\left(-\frac{v_x^2 t^2}{2\sigma_x^2}\right) \cos \Delta\omega t. \quad (5)$$

The damping term effectively broadens the spectrum of the scattered light. The significance of the transit time broadening depends upon the relative magnitudes of $\underline{k}\cdot\underline{v}$ and v_x/σ_x. For low velocities the broadening is unlikely to be important unless either $\underline{k}$ or σ_x is very low. Except in such cases the Gaussian in equation 5 can reasonably be approximated by unity. When $\underline{k}$ *is* very low, as for example in electrophoresis, then the spectral width may have finite contributions due to transit time effects.

(b) Diffusive broadening

The Doppler spectrum reflects all motions within the scattering volume. Even if no overall translation occurs the scattering particles are still liable to be undergoing diffusive motion. The consequent spectral broadening is treated by Cummins in this volume.[10] The general case, incorporating both directed motion and diffusion, has been analysed by Edwards et al[11] in the frequency domain. Their treatment incorporates the effect of particles diffusing into and out of the scattering volume. This finite volume effect can be neglected provided that $K\sigma >> 1$ (ie large sample volumes or large $\underline{k}$), which is likely to be the case in most practical situations. The Doppler spectrum is then a convolution of a Lorentzian of half-width DK^2 with a Gaussian (centred at frequency $\Delta\omega$) due to transit-time broadening. Thus the correlation function takes the form:

$$G(t) \;\alpha\; \exp(-DK^2 t)\exp\left(-\frac{v_x^{\,2}\,t^2}{2\sigma_x^{\,2}}\right)\cos\Delta\omega t \qquad (6)$$

which is characterised by two damping times, $\tau_d = (DK^2)^{-1}$ and $\tau_t = \sqrt{2}\,\sigma_x/v_x$. Such a spectrum is an example of a class of functions, the Voigt functions, characterised by a parameter $a = \tau_d/\tau_t$. For low a, the Voigt function reduces to a Gaussian, whereas at large a it tends towards a Lorentzian.[11] In practice, with reasonable light scattering data we have found that the inequalities $a \lesssim 0.1$ and $a > 1$ satisfy these limits reasonably well.

Thus if the diffusion time τ_d is less than the time the flowing particles reside in the scattering volume (τ_t), the diffusional broadening of the spectrum will dominate. For example, taking $D = 10^{-8}$ cm^2/sec and using $\sigma_x = 10$ μm and $k = 10^5$ cm^{-1} ($\theta \sim 45°$), the two effects are comparable for $v_x = 1.4$ mm/s. Clearly transit time broadening will be negligible for most intra-cellular flows. This is in sharp contrast to LDV applications in aerodynamics where a reduction in τ_d (higher D) is more than offset by the fall in τ_t (higher v_x). The diffusion sets a lower limit to the velocity values perceptible by any technique. If $DK^2 \sim \Delta\omega$, the correlation function of equation (6) will be heavily damped; at some stage overdamping will occur and no oscillations will be visible. Clearly we require $v_x \gtrsim DK$, so that, for the combination of parameters just cited, v_x must exceed some 10 μm/s. In practical biological situations this lower limit is liable to be reduced by the lower value of D appropriate to the viscous cytoplasm. However, it seems apparent that only in particular artificial cases where diffusion is negligible will velocities considerably lower than 1 μm/s be measureable.[13]

It bears noting that both of the effects discussed here broaden the spectrum symmetrically about $\Delta\omega$. Thus correlation functions of the form of equation (6) yield mean Doppler shifts appropriate to the (unique) velocity present in the scattering volume.

Note that the transit time broadening is, to first order, independent of K whereas the diffusive broadening scales as K^2. Thus the variation of the width of the Doppler spectrum with K will serve to check the predominant source of broadening. If - as considered below - the spectral width arises from the presence of a spread of velocities within the scattering volume, then the width should scale as K.[14,15]

4. VELOCITY AVERAGING

Laminar fluid flows normally exhibit a variation of velocity across the flow, rather than a uniform translational motion. The analysis above will still be useful if the scattering volume covers a small portion of the flow profile, within which the velocity is nearly constant.

Many interesting flows (including biologically significant cases) are characterised by small length scales. In such cases it will not be possible to probe the flow with a scattering volume very much smaller than the scale of the flow profile: the effect of the spatially varying velocity field must be allowed for. The Doppler spectrum reflects the velocities of particles which pass through the scattering volume. If there is a variation of velocity within that volume, it is often useful to ascertain a characteristic velocity which can be ascribed to the centre of the scattering volume, to permit the flow profile to be mapped out. The LDV system perceives a velocity averaged over the scattering volume (or that portion within the flow, if it is curtailed by the walls containing the flow). If the flow profile is constant, or substantially linear across the scattering volume, the mean perceived velocity just equals that at the centre of the scattering volume (v_o). For non-linear flow profiles the mean velocity differs from this value, being shifted to lower velocities for physically reasonable, smoothly varying flow profiles.

Differences arise in the averaging of the velocities within the scattering volume for the two cases of tenuous and dense seeding. The two cases are treated explicitly by Kreid.[16] In most biological flows the flowing medium is sufficiently dense that several individual scattering particles are present within the scattering volume at any instant of time. This gives rise to a 'continuous-wave' signal.[16,11] However, if the passage of a scattering particle through the sensitive volume is a rare event (the 'individual realisation' case), there is a biassing towards the *higher* velocities present.[17,18] The extra complication of this 'velocity biassing' will not be necessary here.

The cw case arises from simultaneous scattering from several particles. These particles are assumed to be uniformly distributed in space (with density ρ). Considering a single instantaneous measurement of the average Doppler frequency, we have, following Kreid:[16]

$$\overline{\Delta\omega} = \frac{\sum_i n_i \, \Delta\omega_i \, w_i}{\sum_i n_i \, w_i} , \qquad (7)$$

the sum extending over all the particles present simultaneously in the scattering volume. In equation (7), n_i particles are at a point where the velocity is such that the Doppler shift is $\Delta\omega_i$. The weigh factor w_i reflects the intensity scattered by these particles, which derives from the intensity distribution within the scattering volume (equation (4)). The various possible configurations of particles within the scattering volume occur at random, so that equation (7) may be written (using equation (1)) as

$$\bar{v} = \frac{\int \rho\, v(r)\, w(r)\, dr}{\int \rho\, w(r)\, dr} \tag{8}$$

The results of this averaging are most clearly seen by considering a one-dimensional flow with $\underline{v}$ everywhere normal to the y-z plane and with only one gradient of v (dv/dy) different from zero. Then equation (8) becomes:

$$\bar{v} = \frac{\int v(y)\, w(y)\, dy}{\int w(y)\, dy} \tag{9}$$

where the integrals are truncated at the boundaries of the flow or by the finite extent of the scattering volume. Expanding v(y) as a Taylor series, the first term affecting $\bar{v}$ is the second differential:[9]

$$\bar{v} \simeq v_o + \tfrac{1}{4}\, \frac{d^2v}{dy^2} \tag{10}$$

Thus curvature of the flow profile across the scattering volume is required if a velocity shift is to occur. The spectrum is broadened, the width depending upon the local gradient of the flow profile. Edwards et al[11] explicitly derive the skewing of the Doppler spectrum which causes the shift in the perceived velocity. Such skewed spectra have been observed experimentally,[19] the spectra being more affected for narrower flow vessels, as expected.

Interpretation of the velocity $\bar{v}$ as the velocity appropriate to the centre of the scattering volume yields erroneous flow profiles In an experimental examination of this effect, Cochrane and Earnshaw[2] studied one-dimensional flows in narrow rectangular glass channels. The density of seeding of the flow medium ensured that the cw analysis presented here was appropriate. The perceived average velocities ($\bar{v}$) mapped across several different channels using a scattering volume of constant size (70 μm in length) are shown in

Figure 2. Where the dimensions of the flow are comparable to those of the scattering volume, the observed profiles are notably blunted. Such blunted profiles have been reported for blood flow in very narrow channels[20] although the relevant size of the scattering volume is not clear.

The shape of the Doppler spectrum reflects the velocity profile. If the flow profile is analytically well-behaved and if the shape of the Doppler spectrum is dominated by the spatial variation of velocity then, in principle, a spectrum observed with the scattering volume at a single (known) position upon the profile should permit $v(y)$ to be evaluated. Cochrane[22] has shown that the moments of the spectrum can yield some knowledge of $v(y)$. In particular, assuming a parabolic flow profile, the perceived $\bar{v}$ and $\overline{v^2}$ enable a flow profile to be reconstructed which is reasonably close to theoretical prediction. To be sure, velocity profiles encountered in biological applications are unlikely to be so simple. However, higher moments, such as the skewness, would assist in the parameterization of more general profiles.

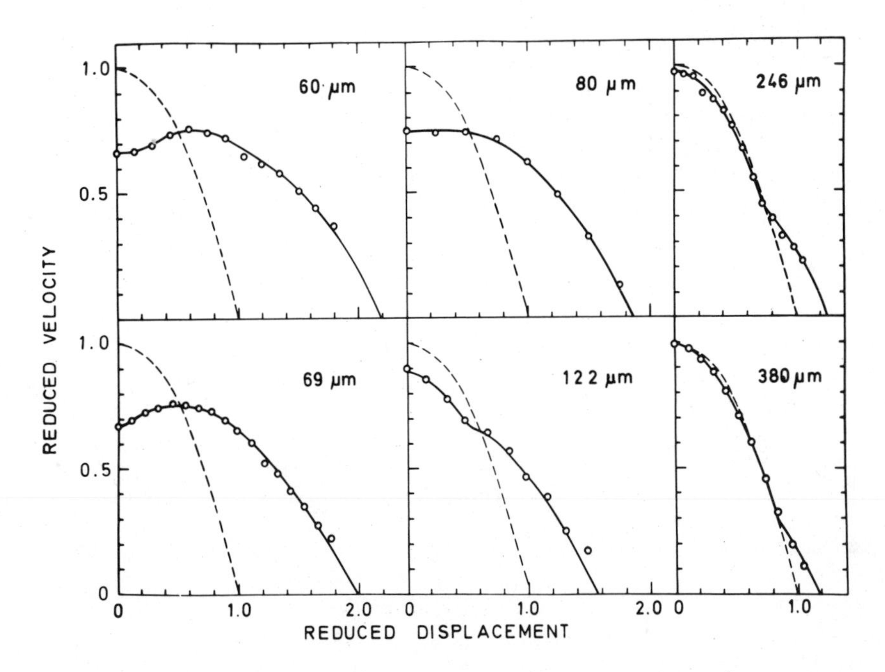

Figure 2 : Experimentally measured flow profiles in glass channels of the widths shown. The true profiles are dashed; the full lines represent the effect of averaging over the scattering volume as suggested by equation (9). (From reference 20.)

5. DENSELY POPULATED FLOWS

Biologically interesting fluids are densely populated with different types of particles which will scatter light. A beam of light incident upon such a fluid will be attenuated at a rate dependent upon the turbidity of the medium. In extreme cases the light will be multiply scattered and interpretation of the Doppler spectrum will become less simple. Multiple scattering will not occur for sufficiently narrow flow channels, but for very turbid media in wide vessels it is unavoidable. For example, for venous flow of whole blood the Doppler spectrum is not displaced from 0 Hz[23]; this is only comprehensible in terms of multiple scattering. For flows of uniform velocity it appears that $\Delta\omega$ is unchanged, but the situation is much less clear where there is a spatial variation of velocity. There are as yet no theoretical analyses directly relevant to this problem. Some experimental guidelines can be given and the present theoretical situation summarised.

Experimentally, with very tenuous flows the differential Doppler LDV (Figure 1b) yields a perceived $\Delta\omega$ which can be interpreted in terms of a flow velocity at the scattering volume. As the density of scatterers increases so that more than one is present within the scattering volume, a reference beam system (Figure 1a) becomes more useful than the differential arrangement. At larger particle densities the signal derived from such an LDV deteriorates, likely due to mixing of the scattered light with low-angle scattering arising from the reference beam. If the reference beam is led to the detector by an optical path which avoids the flowing medium (Figure 1c), substantially denser media can be investigated.[22] This is only a temporary alleviation; eventually double and multiple scattering must complicate the spectrum. Figure 3 is an indication of the effect of increasing the concentration of the flow medium in an 'external reference beam' LDV. For the denser suspension the diffuse appearance of the scattering volume associated with multiple scattering was clearly evident. The increased period in Figure 3b may result from an increased viscosity of the suspension, but the distortion of the spectrum is clear. The suspensions involved were quite dense and some inter-particle interactions may have occurred, further complicating detailed analysis. While no theoretical justification is available, it seems that the perceived Doppler frequency retains some validity as an indication of the 'average' velocity involved even in the presence of some degree of multiple scattering. In such cases the 'scattering volume' and hence the appropriate point to which the velocity may be assigned will be very uncertain.

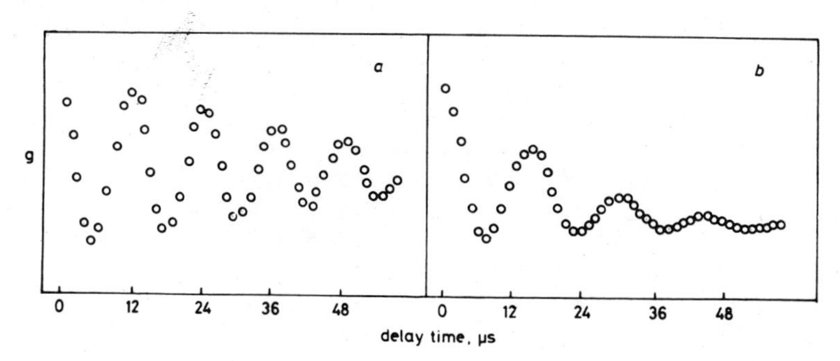

Figure 3 : The effect of increasing density of scatterer. The concentration in b is 6 times that in a (1 part milk to 20 parts water). (From reference 22.)

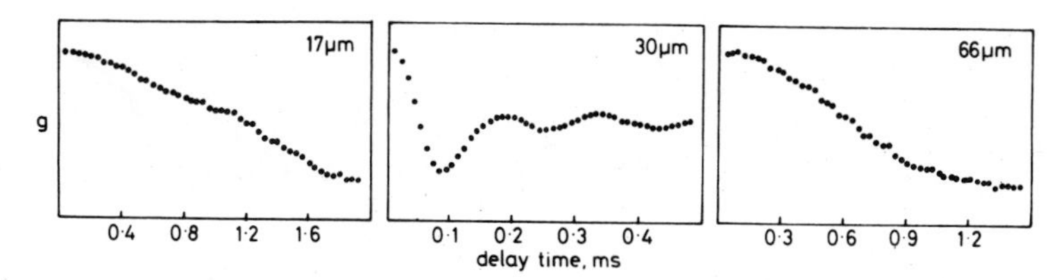

Figure 4 : Correlation functions observed for blood flow in the web of a frog's foot. The vessels of diameter 17 and 66 μm were surrounded by a layer of fat, which diffusely scattered the light. (From reference 22.)

For situations where the light diffuses from the source to the detector (as in wide vessels of very dense fluids or for the case of flow in vessels embedded in diffusing media), it has proved possible to relate the width of the spectral line (which here would be centred on 0 Hz) to the average flow speed in the illuminated volume (the scattering volume is no longer a useful concept). Bonner and Nossal[24] analyse scattering by blood flowing in many narrow vessels embedded in tissue. The intensity of light scattered m times-I_m (t) - is just $|I_1 (t)|^m$ if the positions of the separate scattering events are uncorrelated.[25] This fact is used to relate the correlation function to that for single scattering from particles moving in random directions with an appropriate velocity distribution. They relate the time for G(t) to decay to half of its initial value-G(0)-to the r.m.s. velocity. Experiments on model systems verify their predictions. The general picture is also confirmed by studies on single narrow blood vessels.[22] Where such a vessel was surrounded by a diffusing layer of fat no oscillatory correlation functions were observed (Figure 4). Bonner and Nossal suggest that their model may also be useful for light diffusing in large vessels carrying dense fluids (cf. ref 23) if the scattering vectors in successive scattering events are sufficiently uncorrelated.

6. LARGE PARTICLES

In aero or hydro-dynamics it is usual to have to seed the flow media to make the flow 'visible' to LDV. The particles used are spheres which are small enough to ensure that they will faithfully follow the flow patterns. In the biological case there is usually no need to seed the flow (exceptions may occur, such as the case of phloem transport) and the motions of the particles define the velocities of interest - there is no longer the problem of ensuring that the particle velocities are representative of the flow. While particles of different sizes may move at different rates, this likely reflects significant differences in the biological system. However the motion of large particles may be impeded by the static network of structural elements within the cytoplasm.

The presence of a large range of particle sizes in biological flows, extending to organelles larger than the wavelength of light, may cause complications. Berne and Nossal [26] have shown how the spectrum of light scattering by large or non-spherical particles can be modified by the particle structure. In fact, from this analysis it appears that the Doppler shift is independent of particle size or shape provided that the particles pass through the scattering volume with no change of orientation. The only effect upon the Doppler spectrum of particles which cannot be treated as point scatterers (or spheres) arises from rotational motions. Such motions are known to exist in cytoplasmic streaming,[27] contributions arising from passive spinning of particles due to velocity differences across their diameter and also, in some cases, from autonomous rotations. Rotational diffusion may also occur and will add further diffusive broadening to the spectra. Thus the interpretation of a Doppler spectrum derived from flowing cytoplasm is likely to be non-trivial. Observations of the effects of inhibition of directed motion may permit at least some of the complicating factors to be identified.

A comparatively novel laser velocimetry technique which would be of use in suppressing Doppler frequencies due to rotation is the two-spot system.[28] Here two laser beams are focussed to separate points, close to each other and aligned with the direction of flow.

The time taken for scattering particles to travel the distance between these points is measured. Clearly only translational motion will contribute to the signal in such a system.

7. REFINEMENTS

A. Time Dependent Flows

In many biological flows the velocity varies with time. Similar situations have been extensively studied in engineering contexts and so technology more than competent to cope with biological situations is well established. In cases where the flow is periodic, a mean flow velocity is only occasionally likely to be useful. More often, the variation of velocity through the cycle will be of interest. In such cases the detector output can be gated prior to signal analysis; the signal is only analysed over a pre-determined section of the cycle, data being averaged over several cycles. In a study of flow in a model bronchial system, pulsating air flows in narrow glass tubes have been measured in this way,[29] good agreement with theory being found.

Non-periodic velocity variations may also be of interest. If a strong signal is available, frequency trackers are ideal signal processing systems for such investigations. Indeed cyclic variations in cytoplasmic streaming in *Physarum* have been studied in this way.[30] In noise limited situations, photon correlation will, however, be the optimal technique. In this case it would be desirable to record several correlation functions of brief duration and to extract from each an estimate of the flow velocity averaged over the relevant interval. Brief intervals will be most useful; in a study of the shuttle streaming in *Physarum*,[31] the velocity variations were somewhat smoothed out by the 5s data acquisition cycles. Again, this procedure parallels certain engineering studies. A major problem is data interpretation of the inevitably noisy correlation functions: model fitting requires a reasonable initial guess for $\Delta\omega$. Recently Cole and Swords[32] have developed a rapid algorithm for extracting the mean Doppler frequency from a noisy correlation function, specifically for such brief experiments. Simulations showed that $\Delta\omega$ can be recovered to within 10% even for correlation functions with noise (standard deviation) as high as 30% of the peak amplitudes (assuming Doppler spectra having standard deviations up to 0.4 $\overline{\Delta\omega}$). Great improvements over straightforward Fourier transformation methods were observed.

B. Specific Flow Components

Biological systems are complex and inhomogeneous and the transport of a specific component within such a system may be of interest. A tool which could potentially be used is Fluorescence Correlation Spectroscopy (see Koppel[33]). Madge et al[34] have extended FCS to systems undergoing translation through the laser beam exciting the fluorescence. They contrast pure diffusion with the cases of uniform translation and laminar flow both with and without diffusion. In all cases, characteristic correlation functions are deduced. For example, for uniform translation in the absence of diffusion a Gaussian correlation function

$$G(t) \ \alpha \ \exp \left| - \left(\frac{vt}{\omega}\right)^2 \right|$$

is predicted, v being the translational velocity and ω the standard deviation of the intensity profile of the exciting beam. The practicality of the method was demonstrated by experimental studies where the entire velocity profile of laminar flow (with diffusion) was sampled.

There do not seem to have been any biological applications of this technique. Clearly, though, with fluorescent labelling of specific cell components, it would form a powerful extension of LDV for biologists.

8. DATA INTERPRETATION

The problems of inversion of a correlation function to yield the parent velocity distribution are well-known.[2] For a reference beam LDV, a distribution of velocities P(v) - where v is the component of $\underline{v}$ parallel to $\underline{k}$ - will yield a correlation function

$$G(t) \ \alpha \int P(v) \cos(\underline{k} \cdot \underline{v}t) \, dv \ . \qquad (11)$$

The inversion of this integral equation is subject to the same problems as attend the inversion of the Laplace transform (see Chu[35]); the problems are slightly less intractable due to the well-behaved nature of the cosine kernel. But the sensitivity of P(v) to noise on the observed G(t) remains, and if the measured G(t) does not fully decay within the time range sampled, Fourier transformation of equation (11) will yield spurious peaks in P(v).

This problem has been studied extensively in the context of laser anemometry, where the main goal is improving the velocity resolution to the limit. In the biological context this is unlikely to be required, but data may contain information which is not revealed by conventional Fourier transformation. Model fitting is only reliable when an accurate description of the processes occurring is available and amenable to description in terms of a limited range of parameters. In general one of the several techniques used in anemometry may be useful in determining the spectrum of velocities (or rather frequencies, to cover cases where rotation occurs) *ab initio*. Some of these[36,37] are restricted in their applicability by the assumptions and approximations employed. Two methods may be useful; both largely avoid the ill-conditioning of the inversion of equation (11) by demanding smoothness of the resulting P(v).

McWhirter[38] has developed the eigen-value approach of McWhirter and Pike[39,40] to incorporate a cubic b-spline representation of P(v). Such a model naturally enforces a smooth P(v); sampling and truncation can be taken into account by restricting the number of degrees of freedom of P(v). This approach basically represents the fitting of a very general and flexible model, the implementation of which is governed by the information-theoretic considerations of McWhirter and Pike. Tests of the method[38] on simulated and real data for turbulent air flows were presented.

Davies et al[41] have reported an approach which principally aims to overcome the problems of truncation, notably the false peaks in P(v). Various methods[eg 42] of extending the observed G(t) to reduce these peaks have been proposed. Unfortunately there is an inherent uncertainty involved in extrapolating any observed data and this must reflect in uncertainty in the resultant P(v). This situation parallels the variation (with choice of the regularizer) in G(Γ) obtained by inversion of the Laplace transform[35,43]. Davies et al effectively choose a particular extrapolation procedure which yields unambiguous results; the utility of the procedure is established by tests on simulated data. To enforce non-negativity on P(v) - effectively suppressing spurious oscillations - P(v) is written as $[q(v)]^2$. Then the autoconvolution of x(t), the Fourier transform of q(v), corresponds to the measured autocorrelation function. Least squares fitting of $x(t) * x(t + t_m)$ to G(t) yields $x(t_m)$ at M + 1 points (m = 0 ... M) where $M \leq N$, the number of samples of G(t). In practice Davies et al took M = N. Then the autoconvolution

$$y(t_n) = \sum_{m=-M}^{M} x(t_m)\, x(t_{n-m})$$

is defined over twice the original time domain (for M = N). Now G(t) can be replaced in equation (11) by this extended and smoothed $y(t_n)$. Inversion of this y(t) yields a positive approximation of P(v) with the effects of the original truncation much reduced. The replacement of G(t) by $y(t_n)$ removes the noise upon G(t). The basic assumptions of the method are non-negativity of P(v) and the time-limited nature of G(t). If the time limit falls between t_N and t_{2N}, the transformation should yield an accurate P(v). This method has been developed further (Davies, to be published). Examples of inversions of real and simulated data appropriate to biological systems show the power of the approach.[41] In particular the increased resolution over normal Fourier transformation shows up two flow velocities in blood flow in a narrow vessel, a situation where spontaneous changes in flow velocity are known to occur

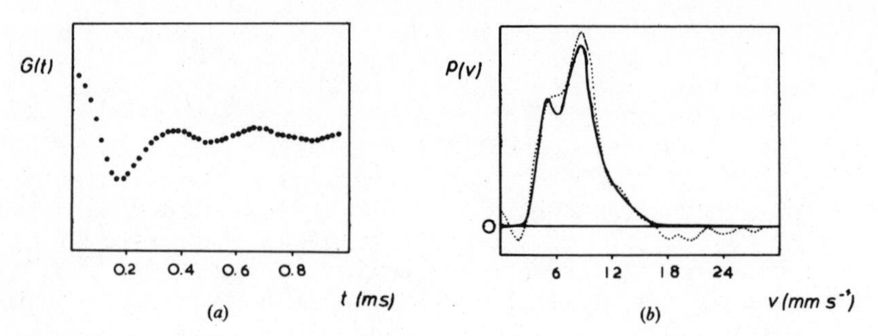

Figure 5 : Blood flow in narrow vessel (as Figure 4). The inverted P(v) is shown as the direct Fourier transform (dotted) and as derived by Davies et al (full line). (Reprinted from reference 41 by permission.)

(Figure 5). This illustrates the need to extract the maximum information in biological systems.

Note that P(v) will incorporate false frequencies if any rotational motions occur in the scattering volume. The interpretation of the data must, therefore, be guided by other observations of the processes.

9. BIOLOGICAL APPLICATIONS

The main areas of application of LDV of present interest are electrophoretic light scattering, blood flow and cytoplasmic streamin While the first of these is the most widespread biologically relevant application,[44] it is extensively discussed by Professor Ware in this volume and so will not be discussed further.

Many reports of studies of blood flow by LDV have appeared; only a brief selection will be cited as an indication of the possibilities. Flow in single blood vessels has been examined over a wide range of vessel sizes. Blood is a highly turbid medium and in large vessels multiple scattering will heavily modify the signals. For vessels of diameter below 100-200 μm, the Doppler spectra of the scattered light are sufficiently clear that flow profiles can be examined.[19,45] At the top end of this range some double scattering will doubtless occur, but the mean perceived $\Delta\omega$ probably remains meaningful. For larger vessels it seems unlikely that the light will be able to propagate through the blood to the far side of the vessel. Back-scattered light from vessels in the retina has been observed;[46] such data probably refer to flow in the proximal portions of the flow profiles. Such measurements are, of course, still of considerable clinical significance. Finally, while flow in major vessels can be examined (eg by fibre-optic catheter[23]) the spectra are very distorted by multiple scattering and other effects and only an indication of some ill-defined average velocity is obtained. Laser Doppler methods have been used to monitor sub-cutaneous blood flow;[47] indeed clinical instruments have been developed for such purposes.[48] The theoretical analysis of Bonner and Nossal[24] has served to validate such applications where the light diffuses to and from the scattering event in the flow.

Cytoplasmic streaming in a few species has been investigated using LDV methods.[49] Species studied have included the algae Nitella[50,14,51,13,52] and Chara,[53] the slime mold Physarum[54,31] and the higher plant Elodea.[55] These studies have been of a preliminary nature, but demonstrate the potential of this approach. To date little biologically novel information has been obtained. The validity of interpreting frequency shifts in light scattered by intra-cellular processes as Doppler shifts has been established by the linear dependence of $\Delta\omega$ upon κ.[49,51] As the relative contribution to the scattered light intensity due to small particles increases as the scattering angle (and hence K) is raised, this linear dependence has further been used as evidence that particles of all sizes move with the same velocity.[49] The streaming velocities inferred are comparable with those found by classical techniques. However, the Doppler spectra contained observable intensity at frequencies corresponding to velocities somewhat greater than those previously reported (up to 3 mm s^{-1} in Physarum[54]). It is possible that such high frequencies may arise from rotation of the translating particles[27] (cf a rolling wheel, whose top travels faster than the centre of mass). For Elodea the streaming velocities are much smaller than in the other organisms cited and rotational motion seriously affected LDV observations.[55] For Nitella, where the Doppler spectrum had a clearly defined width, this linewidth varied linearly with K.[14,15] suggesting that the broadening arose predominantly from the spread of velocities within the scattering volume (effects due to transit time or diffusion have not been firmly established). The two reports cited found opposite deviations from this K scaling at large K; it seems safest to consider instrumental origins for such deviations.

Some new information on streaming in Physarum has emerged[54,15]; large velocities transverse to the veins have been observed[56] as have variations of scattered intensity at the shuttle streaming frequency (though out of phase with it). The latter observation has been interpreted in terms of rhythmic wall contractions. Both of these phenomena have been independently observed by other techniques.[57]

In none of these studies of streaming has the potential spatial resolution (∿ 5 μm) of the LDV method been exploited. It should be possible to measure detailed flow profiles. Such detailed measurements, preferably repeated on cells under varying conditions affecting their streaming rates would form a most significant contribution to the analysis of streaming phenomena and their understanding.

REFERENCES

1. Y. Yeh and H.Z. Cummins, Appl. Phys. Letts. 4: 176 (1964).
2. E.R. Pike in Photon Correlation Spectroscopy and Velocimetry edited by H.Z. Cummins and E.R. Pike (New York, Plenum, 1977). p 246.
3. J.B. Abbis, T.W. Chubb and E.R. Pike, Optics and Laser Tech. 6:249 (1974).
4. H. Kogelnik and T. Li, Appl. Optics, 5:1550 (1966).
5. R. Foord, A.F. Harvey, R. Jones, E.R. Pike and J.M. Vaughan, J. Phys. D, 7:L36 (1974).
6. J. Oldengarm, A.H. van Krieken and H.W. van der Klooster, J. Phys. E, 8:203 (1975).
7. P.H.Y. Lee, Appl. Phys. Letts. 25:737 (1974).
8. T. Yoshimura, Y. Syoji, N. Wakabayashi and N. Suzuki, J. Phys. E 11:777 (1978).
9. R.B. Dyott, IEE J. Microwaves, Optics and Acoustics, 2:13 (1978).
10. H.Z. Cummins, this volume.
11. R.V. Edwards, J.A. Angus, M.J. French and J.W. Dunning, J. Appl. Phys. 42:837 (1971).
12. H.C. van de Hulst and J.J.M. Reesinck, Astrophys. J. 106:121 (1947).
13. D.A. Jackson and D.S. Bedborough, J. Phys. D. 11:L135 (1978).
14. R.V. Mustacich and B.R. Ware, Biophys. J. 16:373 (1976).
15. K.H. Langley, R.W. Piddington, D. Ross and D.B. Sattelle, Biochim. Biophys. Acta, 444:893 (1976).
16. D.K. Kreid, Applied Optics, 13:1872 (1974).
17. D.K. McLaughlin and W.G. Tiederman, Phys. Fluids 16:2082 (1971).
18. L. Shemer and S. Einav, Rev. Sci. Instrum., 50:879 (1979).
19. H. Mishina, T. Ushizaka and T. Asakura, Optics and Laser Tech. 8:121 (1976).
20. T. Cochrane and J.C. Earnshaw, J. Phys. D. 11:1509 (1978).
21. S. Einav, H.H. Berman, R.L. Fuhro, P.R. DiGiovanni, S. Fine and J.D. Fridman, Biorheology, 12:207 (1975).
22. T. Cochrane, J.C. Earnshaw and A.H.G. Love, Med. & Biol. Eng. and Computing, 19:589 (1981).
23. T. Tanaka and G.B. Benedek, Applied Optics, 14:189 (1975).
24. R. Bonner and R. Nossal, Applied Optics 20:2097 (1981).
25. C.M. Sorensen, R.C. Mockler and W.J. O'Sullivan, Phys. Rev. A 17:2030 (1978).
26. B.J. Berne and R. Nossal, Biophys. J. 14:865 (1974).
27. M. Kwiatkowska, Protoplasma, 75:345 (1972).

28. A.E. Smart and W.T. Mayo, Jr, in 'Proceedings from the 4th International Conference on Photon Correlation Techniques in Fluid Mechanics' ed. W.T. Mayo Jr and A.E. Smart (Stanford, Stanford University) pII.1 (1980).
29. T. Mullin and C.A. Greated, J. Phys. E, 11:643 (1978).
30. R.V. Mustacich and B.R. Ware, Rev. Sci. Instrum. 47:108 (1976).
31. S.A. Newton, N.C. Ford Jr, K.H. Langley and D.B. Sattelle, Biochim. Biophys. Acta, 496:212 (1977).
32. J.B. Cole and M.D. Swords, J. Phys. D. 14:1731 (1981).
33. D. Koppel, this volume.
34. D. Magde, W.W. Webb and E.L. Elson, Biopolymers, 17:361 (1978).
35. B. Chu, this volume.
36. J.B. Abbiss, Physica Scripta, 19:388 (1979).
37. P.R. Sharpe, Physica Scripta, 19:411 (1979).
38. J.G. McWhirter, Optica Acta, 28:1453 (1981).
39. J.G. McWhirter and E.R. Pike, J. Phys. A, 11:1729 (1978).
40. J.G. McWhirter and E.R. Pike, Physica Scripta, 19:417 (1979).
41. A.R. Davies, T. Cochrane and O.M. Al-Faour, Optica Acta, 27:107 (1980).
42. P.P. Stone, Physica Scripta, 19:402 (1979).
43. S.W. Provencher, Makromol. Chem. 180:201 (1979).
44. B.R. Ware, this volume.
45. G.V.R. Born, A. Melling and J.H. Whitelaw, Biorheology, 15:163 (1978).
46. G.T. Feke and C.E. Riva, J. Opt. Soc. Am. 68:526 (1978).
47. M.D. Stern, Nature, 254:56 (1975).
48. G.E. Nilsson, T. Tenland and P.Å. Öberg, IEEE Trans. Biomed. Eng, BME27:12, 597 (1980).
49. J.C. Earnshaw and M.W. Steer, Pestic. Sci. 10:358 (1979).
50. R.V. Mustacich and B.R. Ware, Phys. Rev. Lett. 33:617 (1974).
51. R.V. Mustacich and B.R. Ware, Biophys. J. 17:229 (1977).
52. D.B. Sattelle, D.J. Green and K.H. Langley, Physica Scripta, 19:471 (1979).
53. D.B. Sattelle and P.B. Buchan, J. Cell. Sci. 22:633 (1976).
54. R.V. Mustacich and B.R. Ware, Protoplasma, 91:351 (1977).
55. J.C. Earnshaw in 'Photon Correlation Spectroscopy and Velocimetry' edited by H.Z. Cummins and E.R. Pike (New York, Plenum, 1977) p461.
56. J. Picton, unpublished observations.
57. K.E. Wohlfarth-Bottermann, this volume.

IMPLEMENTATION OF TWO DIFFERENT TECHNIQUES FOR MEASURING LATERAL DIFFUSION AND OPTIMISATION OF THE FLUORESCENCE CORRELATION SPECTROSCOPY CONCEPT

Hugo Geerts

Department of Mathematics, Physics and Physiology
Limburgs Universitair Centrum
3610 Diepenbeek Belgium

INTRODUCTION

The experimental implementation of a computer-based spectroscopic instrument is presented. The design is primarily aimed at the measurement of the lateral diffusion coefficient of labelled molecules. It can be used in the Fluorescence Correlation Spectroscopy mode. This approach has been presented at another lecture in this summer school.

The second concept is based on Continuous Fluorescence Microphotolysis (Peters et al, 1980). In this technique, one uses a slightly more intense laser beam to create a balance between photobleaching and diffusion. The decreasing fluorescence intensity finally gets down to a quasi steady-state level and reflects the subtle balance between these two processes. Upon solving the particular two-dimensional reaction-diffusion equation, one gets a value for the diffusion coefficient (assuming first order photobleaching kinetics).

The main difference between these two concepts is the introduction of the photobleaching as an additional process for studying diffusion processes. So we started a series of experiments where we applied the FCS and CFM techniques to the same samples. The FCS-concept is assumed to introduce a minimal amount of bleaching, due to the very small intensities (microwatt range).

Table 1. Diffusion coefficient of 80 nM Rodamine in a 90% glycerol-water mixture $\mu m^2/sec$)

F.C.S.	C.F.M.
1.82	1.78
	1.70
	1.75
	1.85

These initial results showed no systematic difference between the two techniques.

We are actually studying a whole range of systems, including proteins in solution and motion of lipid probes on multilayer systems

OPTIMISATION OF THE F.C.S. APPARATUS

The second part concerned a study of the distortion of the statistics at high photon count rates. This is often necessary for achieving a good S/N ratio in FCS.

We measured a reasonable electronic dead-time by studying the distribution function p(n), the number of photon counts in a sample window. Table 2 shows the values, extracted from an appropriate analytical expression for p(n) (Bedard, 1967).

Table 2. Dead-time measured for Malvern EMI 9558B detector

Freq(Mhz)	Dead-time (ns)
2.35	43.1
2.0	50.1
1.38	70.9

A further investigation concerned the value of the modified factorial moments:

$$m_k = \langle n!/(n-k)! \rangle = \sum_{n=k}^{\infty} n(n-1) \ldots (n-k+1) \; p(n)$$

For the incoherent light, m_k should be equal to $\bar{n}^2$(Saleh, 1978). The deviation of the quantity $Q = \sum_{i=1}^{6} (m_i/\bar{n}_i^{-2} - 1)$ from the expected zero value shows some kind of distortion (Table 3).

Table 3. Value of the quantity Q at high photon rates

Photon freq(Mhz)	Q (10^{-4})
.75	5.73
1.05	16.2
1.35	18.4
1.65	19.9
2.00	24.5

EFFECT OF EXPERIMENTAL CONDITIONS

Finally the optimisation of the experimental conditions was an extension of an earlier work on the S/N ratio of FCS (Koppel, 1974). This included the effect of sample window T, the number of samples taken N, and the maximum channel number for the autocorrelation function (Geerts, 1982).

The theoretical predictions were tested against experimental values for the S/N ratio, for the case of a simple diffusion process. We showed that the variance of the measured diffusion coefficient was inversely proportional to the number N of samples taken.

Furthermore, as expected, we discovered an optimum time window T. It has been shown that the critical parameter for the S/N ratio was the number of photocounts per sample window per molecule. It is of great interest, therefore, to increase the sample time. However at a certain "Nyquist" frequency one gets into trouble.

The dependence of the variance on the maximum channels in the autocorrelation function showed a weak minimum, probably due to increasing noise in the later channels.

Finally, for time constrained conditions (NT=constant), one approaches asymptotically a saturation value, characteristic for continuous sampling ($T \ll 1$ and $N \gg 1$).

REFERENCES

1 G. Bedard (1967) Proc. Phys. Soc. 90: 131.
2 D. Koppel (1974) Phys. Rev. A. 10: 1938.
3 M. Saleh (1978) Photoelectron Statistics (Springer Verlag, Berlin).
4 R. Peters, K. Schulten, A. Brunger (1980) Bioph. Struct. Mech. 6: 105.
5 H. Geerts (1982) Journ. of Stat. Phys. 28: 173.

LASER DOPPLER MICROSCOPY: ESPECIALLY AS A METHOD FOR STUDYING BROWNIAN MOTION AND FLOW IN THE SIEVE TUBES OF PLANTS

Richard P.C. Johnson

The Botany Department
University of Aberdeen
Old Aberdeen, Scotland, AB9 2UD

CONTENTS

ABSTRACT

Laser Doppler velocimetry and photon correlation spectroscopy are well established as methods for measuring the flow and diffusion of particles in fluids. The diameters of the laser beams used are usually too wide to be placed precisely within plant or animal cells, capillaries, or other specimens of interest to biologists. Several groups of workers have described laser Doppler microscopes, either with crossed beams or with single beams. These instruments allow scattered laser light to be detected from volumes with diameters of 10 μm or less and enable the scattering volume to be seen and placed accurately in specimens viewed at high magnification. Some designs and principles of laser Doppler microscopes are discussed, especially of a single beam instrument and its application to the study of Brownian motion and flow in sieve tubes, the food transport channels in higher plants.

INTRODUCTION

Since the demonstration of laser Doppler anemometry by Yeh and Cummins in 1964 many theoretical and practical studies of laser light scattering from moving particles have established the analysis of Doppler shifts of light scattered from specimens as a method for measuring flow and diffusion in them. In most practical studies the beams applied to specimens have been about the same diameter as, or greater than, the diameters of most plant or animal cells, which are typically less than about 100 μm across. However, in biology there are many problems of flow and diffusion which require movement to be examined on a smaller scale. For example, it may be of interest to measure movement at various positions across small vessels in the vascular systems of plants or animals, or in selected parts of living cells. To do this it is necessary to collect laser light scattered from small defined volumes in specimens and to see exactly where in the specimen the scattering volume is. Some adaptation of a microscope is likely to be needed.

The first description of a laser Doppler microscope (LDM) was published by Maeda and Fujime in 1972[1]. Since then various kinds of laser Doppler microscopes have been described by about ten groups of workers world-wide. About forty papers have appeared describing LDMs and results obtained with them, the greatest number having come from Mishina, Koyama, Asakura and coworkers at the Research Institute of Applied Electricity, Hokkaido University.

Generally, two main kinds of optical arrangements have been used in LDMs. In some, two beams have been projected through a lens to produce a pattern of interference fringes in the specimen[2-10]. In others a single beam has been used, with a reference signal derived, either from light scattered by the elements of the objective lens, the microscope slide, coverslip and stationary parts of the specimen itself[11-14], or from a separate reference beam diverted from the laser to shine directly on the surface of the detector[3]. Other arrangements have also been used[3,10,15] including single fibre-optics light guides in the Fibre Optic Doppler Anemometer (FODA)[14,16]. An excellent review of laser Doppler microscopy, its principles and potential for the study of movement in cells was published by Earnshaw and Steer in 1979[17], see also Johnson and Ross[14].

In this paper I discuss some designs of laser Doppler microscopes, especially my single beam version and its potential for use in relation to studies of structure and function in sieve tubes, the food-transport channels in higher plants.

THE STRUCTURE AND FUNCTION OF THE SIEVE TUBE

In higher plants the sugars and other food substances made

in the leaves by photosynthesis are transported, "translocated", to other parts of the plant, the shoots, roots and fruits, mainly in the sieve tubes in the phloem tissue[18,19]. The speed of translocation, as measured by the movement of radioisotopes, is usually around 50 to 100 cm/hr. This is far too fast to be explained by diffusion alone; translocation seems to be by mass flow, usually, but not always, of sucrose solution at a concentration of between 5% and 20% w/v, together with amino acids, adenosine triphosphate and other substances. How translocation is driven remains a classic problem in the study of plants.

The phloem tissue lies within the vascular bundles, the veins, where it runs alongside the xylem, the tissue which conducts water from the roots[20]. The phloem contains conduits, the sieve tubes, which are made up of long tubular cells connected end to end. These cells, the sieve elements, start life separately as typical undifferentiated plant cells. They are surrounded by a cellulose cell-wall and, when young, contain a nucleus and normal cytoplasm. Cytoplasmic streaming is sometimes seen in them, but not when they are mature. The sieve elements elongate as they develop so that, generally, they come to be about 100 to 500 μm long and 20 to 50 μm in diameter. Their nuclei disintegrate, their vacuolar membranes break down and their cytoplasm becomes thinner. It comes to reside as a narrow parietal layer against their cell membranes, which remain semipermeable and pressed against the cell wall by osmotic pressure. The parietal layer usually contains a few small mitochondria, stacked membranes and also, as we are coming to realise, a net-like pattern of other structures. During this process of development the disk-shaped end walls of the sieve elements, the sieve plates, become perforated with pores, the sieve pores, which are usually between 0.5 and 5 μm in diameter. The sieve pores connect the sieve elements end to end to form a sieve tube which conducts food substances from source to sink in the plant.

Calculations show that there is probably enough pressure, around 10 atmospheres, from osmosis at the source end of the sieve tubes to drive the observed rates of flow, provided that the sieve pores remain unobstructed[19]. However, in the electron microscope the pores often appear to contain many fine filaments of protein, known as P-protein (P for phloem). The exact arrangement of this material and, indeed, whether it really is in the pores during translocation, has caused great controversy amongst plant physiologists during the past 25 years[18,19,21]. The quantities and packing of P-protein filaments seen in sieve pores in some electron micrographs suggest that they would cause so much drag that pumps would be necessary at places along the sieve tubes to drive the known rates of flow[19,22]. However, it now seems unlikely that the electron microscope can provide a firm answer to the critical question "what is the arrangement of P-protein filaments in sieve pores during translocation". It cannot do so because sieve tubes seem very sensitive to disturbance and because during even the fastest conceivable time to fix intact sieve tubes for microscopy, the Brownian and other kinds of motion in them might move filaments further than the diameter of a sieve pore[21].

P-protein is liberated into the lumens of developing sieve elements from bodies in their cytoplasm[18]. In the electron microscope it usually appears as thin filaments and sometimes tubules, 20 nm or less in diameter[23]. Sometimes these appear more or less uniformly distributed in the lumen. Sometimes they appear bundled together to form strands, possibly caused by flow[24] or, conceivably, causing it. Often the filaments seem to have been blown into the sieve pores by release of turgor when sieve tubes were cut. They may thus help to prevent excessive bleeding when a plant is injured. The P-protein filaments do not seem to occur in more primitive plants In pine trees, for example, their place appears to be taken by aggregations of membranes and small vesicles[18]. Curiously, P-protein differs, chemically, from one species of plant to another; the subunits in it differ in their electrophoretic mobilities, even slightly from one cultivar of cucumber to another[25]. It was thought at one time that P-protein might be a "contractile protein" like actin and might thus be able to drive a flow in the same way that cytoplasmic streaming appears to be driven[26]. But it has now been shown that P-protein is not like actin[27,28]. Nevertheless, it seems worthwhile to continue to investigate the properties of P-protein. Perhaps, in some way, it is able to reduce drag during flow, though because Reynolds number for sieve pores is about 0.003 the flow is unlikely to be turbulent.

One approach to the enigma of the sieve tube is to continue to investigate the ultrastructure and chemistry of their contents. Another is to try to measure rates of flow and diffusion in sieve tubes directly, near P-protein bundles for example and perhaps also in working models, under various conditions. It is here that laser Doppler microscopy may help.

THE MOTION OF PARTICLES IN SIEVE TUBES

In 1971 Lee et al.[29] reported that the root mean square (r.m.s.) displacement of particles in sieve tubes might be up to three times greater than expected for normal random thermal (Brownian) motion. They obtained their results by analysing the displacement of particles from frame to frame of ciné film of sieve tubes viewed under Nomarski differential interference contrast. It now seems certain that most of these particles, which are usually less than one μm in diameter, are starch grains, liberated from plastids in sieve tubes disturbed during dissection for microscopy[30]. However, if particles in disturbed sieve tubes can show greater than Brownian motion then perhaps they are driven by some mechanism which could, in an undisturbed state, convert chemical to mechanical energy to drive translocation. Even a damaged system might be useful for experiment. For this reason Barclay and Johnson[31] repeated Lee et al.'s observations using the same methods. They reported r.m.s. displacements of about one half to one quarter of those to be expected for normal Brownian motion; either the motion of the starch grains was restricted, perhaps by P-protein or by adjacent cell walls, or the viscosity of the solution in the sieve tubes was higher than plant physiologists had supposed.

Johnson[13] found that particles in sieve tubes examined by laser Doppler microscopy showed less than half the r.m.s. displacement expected for particles of that size, at that temperature, in 20% sucrose. My more recent measurements of the Brownian motion of particles in sieve tubes also suggest that they are restrained, but suggest that conditions governing their motion and the variation in intensity of laser light scattered from them are more complex than expected.

LASER DOPPLER MICROSCOPY

A. Basic requirements

The main requirements for a laser Doppler microscope are:

(i) The means to aim a narrow beam or two crossed beams of light from a laser into the specimen. It is necessary to be able to define the scattering volume and to know the angle at which flow intersects it.
(ii) A detector, usually a photomultiplier, to measure the intensity of light scattered from the specimen. For single beam microscopes it is necessary to know the scattering angle; the angle between the illuminating beam and the direction of the detector. Separate lenses may be used to illuminate the specimen and to detect light scattered from it[4,5,10,12], or a single lens may serve both functions[2,3,9,11,13]. A single beam LDM is shown in Figure 1.
(iii) Provision to illuminate the surface of the detector with a reference beam to beat with light that is Doppler shifted by motion in the specimen.
(iv) An optical arrangement to view an image of the specimen, to judge its condition and to see where the laser beam is placed in it, but with no danger of blinding oneself.
(v) A thermometer to record the temperature of the specimen.
(vi) An apparatus to analyse the signal produced by the detector, either a spectrum analyser, a signal correlator or some form of frequency tracker. The output of this can be recorded, processed and displayed conveniently with the aid of a microcomputer[14]. The mathematics and computing required were discussed by Johnson and Ross[14] and will be discussed more fully by others during this meeting.

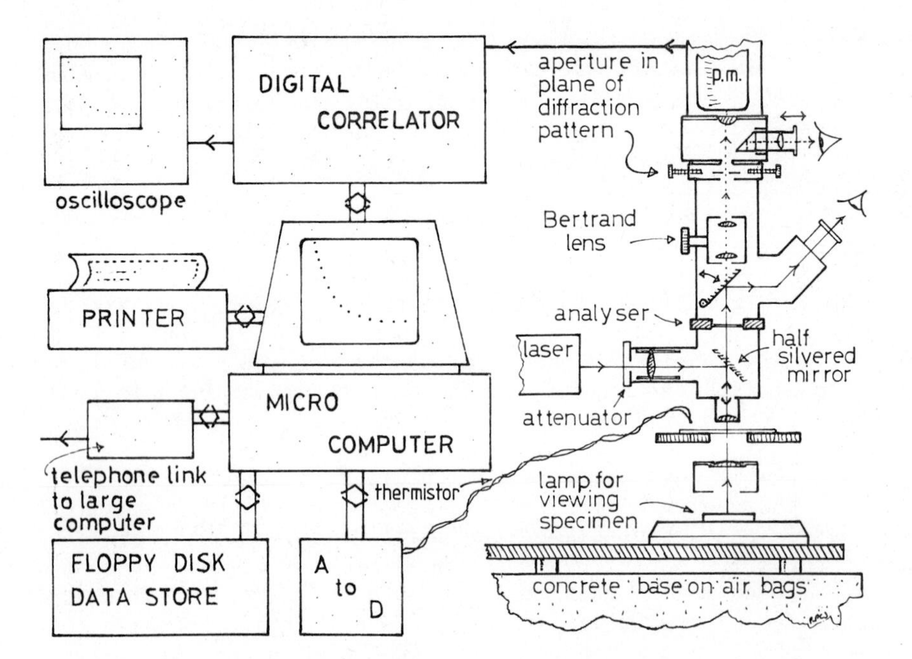

Fig. 1. The optical and signal analyzing arrangements of the laser Doppler microscope.

B. To define the scattering volume

In laser Doppler microscopes with crossed beams the velocity of flow is measured by timing the flashes of light produced when particles move through a pattern of interference fringes formed in the specimen. Strictly speaking the frequency of flashing is the beat frequency between the Doppler shifts given to each of the two beams[32]. In this kind of arrangement it is not necessary to know the angle at which the detector views the specimen. Thus a detector with a large numerical aperture may be used to increase the signal strength. The image plane of the microscope may be focussed on the detector in a crossed-beam microscope, but not for single beams. It remains necessary to know the angle at which the flow intersects the fringes in the specimen.

The pattern of fringes fills the volume where the two beams intersect. The shape of this volume depends on the angle at which the beams cross and their shapes and diameters where they meet. The spacing of the fringes may vary within this volume, especially if the refractive index of the specimen is not uniform[33]. Irregular fringe spacing can cause errors by causing a spread in the detected frequencies. This effect may be minimised by arranging for the beams to cross as nearly as possible at their waists, where they are focussed to their smallest diameter[34]. Also, the mean frequency of the Doppler beat signals may vary with the ratio of the diameter of the scattering particles to the fringe spacing; errors arise as their diameters become similar to the distance between the fringes. These problems were discussed by Abbiss et al.[34], and by Mishina et al.[35] and Kawase et al.[36] who showed that the maximum error of the frequency deviation from the most probable mean frequency reaches about 7% when the diameter of the scattering particle is the same as the fringe spacing; "the frequency deviation of Doppler beat signals produced by sharply edged circular scattering particles varies periodically with their size and gradually decreases with the number N of interference fringes produced in the probing area". However, in a typical crossed-beam microscope the number of fringes seems usually to be about ten[35]. Special analysing apparatus may be needed to identify, track and measure the frequency of the short trains of regular pulses produced by particles as they pass through this small number of fringes[2,8,32,33,37]. In spite of these problems it seems likely that microscopes using crossed beams are the most convenient for measuring flow. Certainly, their use has been developed to a greater extent than that of single-beam microscopes.

The single-beam laser Doppler microscope is best suited to measuring Brownian motion, though flow may be measured with it too. It does not use a pattern of interference fringes and some of the problems of defining the scattering volume differ from those described above for the double-beam microscope.

When the laser beam is projected into the specimen along the axis of an objective lens it will cause light to be scattered from a volume whose shape depends on the diameter of the beam at various depths in the specimen. The beam diameter will increase above and below a waist where it is focussed. If the detector views the specimen also along the axis of the same lens then it will detect light scattered from the full length of the beam where it passes down through the thickness of the specimen. Moving particles anywhere in the beam will cause the detected intensity to vary, even if they are not resolved in the image plane of the microscope[38]. The intensity of light scattered from particles within the beam will depend also on the variation, usually Gaussian, of the intensity of the beam across its diameter and on its divergence.

The scattering volume in a single-beam microscope may be defined if the beam is arranged to leave the objective lens at an angle to its axis. This can be done by aiming the beam so that it enters the back of the lens at some distance from its centre. A length of the beam from which scattered light is to be passed to the detector may then be defined by a movable aperture in the image plane of the microscope.

The best size for the scattering volume will depend to some extent on the size of the particles to be examined. This is because as the diameter of the particles approaches that of the scattering volume, or of the image of the aperture which defines it, the fluctuation in the numbers of particles in the scattering volume will become more significant. The intensity of the scattered light will fluctuate not only because its frequency is being Doppler shifted but because the number of scatterers is varying. This effect was discussed by Herbert and Acton[12] who point out that "If the number of particles in the scattering volume is small, number fluctuations will add an extra term to the autocorrelation function and confuse the interpretation of the phase data". Thus, although it is possible to define very small scattering volumes, of around 1 μm in diameter or even, theoretically, down to a size limited by diffraction, it may not necessarily be useful to do so. As described below, this limitation has arisen in my measurement of the Brownian motion of particles in sieve tubes.

C. To define the scattering angle

In the double beam microscope the rate of flashing does not depend on the angle between the pattern of the fringes and the detector. It is necessary to know only the angle between the flow and the fringes. But, in the single-beam microscope the scattering wave number, Q, varies with the sine of the scattering angle θ_s, which is the angle between the illuminating beam, as it leaves the specimen, and a line drawn between specimen and detector:

$$Q = \frac{4\pi n}{\lambda} \sin(\theta_s/2) \qquad (1)$$

where n is the refractive index of the medium round the particles and λ is the wavelength of the light in metres.

The scattering angle in a single-beam microscope may be defined by projecting the diffraction pattern of the specimen, found in the back focal plane of the objective lens (Fig. 2a), into the plane of a movable aperture in front of the detector (Fig. 2b). The further the aperture is placed from the centre of the pattern the smaller the scattering angle selected. This arrangement may be calibrated by focusing a replica of a diffraction grating under the objective lens[13,14]. The aperture may then be positioned to accept the zero, first, second or third order diffraction spots. By Bragg's law, with a diffraction grating of 15,000 lines per inch, these spots define scattering angles of 180 degrees (backscatter), and about 164, 146, and 123 degrees respectively.

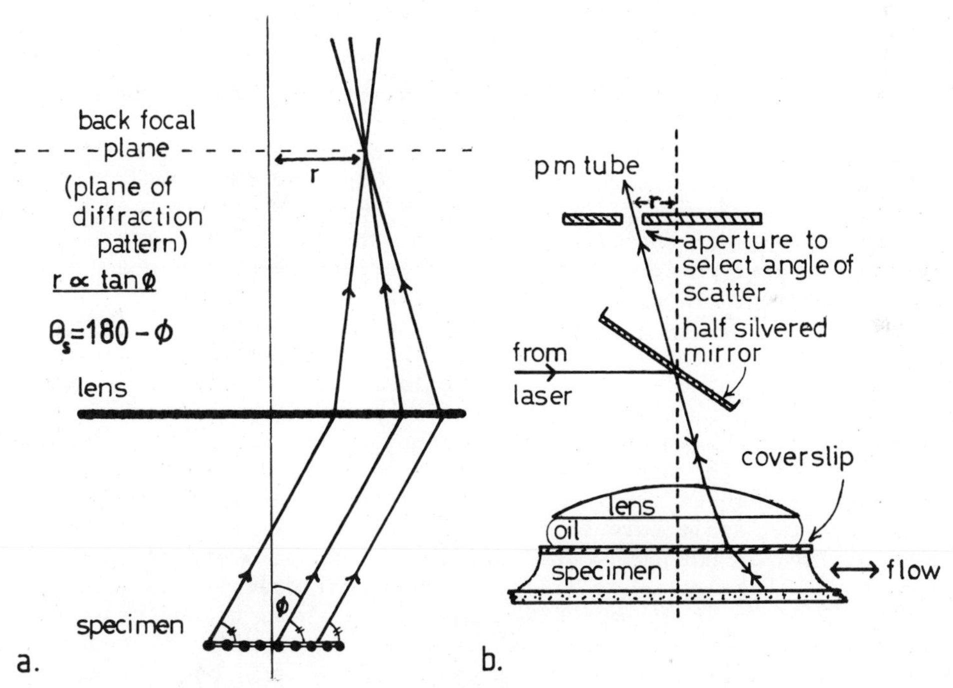

Fig. 2. (a) Light scattered from the specimen at a given angle is brought to a focus at a related distance r from the axis in the back focal plane of the objective lens.

(b) An aperture in the projected back focal plane of the objective lens is positioned to select light scattered at the required angle, here 180°. The Bertrand (projector) lens (Fig. 1) is not shown.

However, if the aperture is made smaller, to define the angle of scatter more closely, the light reaching the detector is reduced and the signal intensity falls. Errors due to the random arrival of photons may then become excessive. This difficulty may be overcome by extending the time during which the autocorrelation function is accumulated or by increasing the intensity of the illuminating beam. For experiments in which the condition of the specimen is changing rapidly it may be necessary to keep the accumulation time short. Conversely, for specimens which are affected by high intensities of laser light it may be necessary to keep the illumination to a minimum. Clearly, there is a set of interacting factors here for optimum performance of the microscope; there is a system of trade-offs to be quantified and balanced.

The arrangement described above allows a range of angles of backscatter to be defined. If detection of forward scattered light is required then the specimen must be illuminated through a lens on the other side of the specimen from the detector. The condenser lens of a microscope can be used for this[2,3,12].

Many flows of interest are likely to be parallel to the specimen stage of the microscope. If, in a single beam microscope, the laser light emerges and returns parallel to the axis of the objective lens it will be at right angles to this flow which will not then have a component of motion along the beam. The frequency of the scattered light will not then be Doppler shifted. There are two ways round this problem. The first, described by Earnshaw and Steer[11], is to incline the specimen on the microscope stage. This is possible if the lens has an adequate working distance. It is not suitable for high resolution lenses with a short working distance. An alternative method[13] is to aim the laser beam so that it emerges at an angle to the axis of the objective lens (Fig. 2b). Unfortunately, if the beam enters a specimen obliquely it and the returning scattered light may be refracted through various angles where the refractive index changes at the surface of the specimen or inside it; the angle between flow and beam may be changed uncontrollably and remain unknown. This difficulty can be overcome if the microscope is set up to collect light backscattered at 180 degrees to the tilted incident beam (Fig. 2b). The angles of incidence and detection are then the same as their bisector. Under this condition, according to Snell's law, any changes in angle caused by changes in refractive index along the beam are cancelled out exactly in equation 1 by the changes in wavelength caused by the changes in refractive index; refractive index throughout the specimen may be taken to be that of the lens or of the immersion fluid in contact with it. Conveniently, the effects of different refractive indexes within the specimen may be ignored. A procedure for achieving this condition was set out by Johnson[13].

A disadvantage of single beam LDMs for measuring flow is that in the autocorrelation function of the intensity of the scattered light (see equation 2 below) the flow term appears multiplied by the diffusion terms. Thus the flow component will fall to zero with the diffusion component. This means that, if flow is to be measured, the diffusion constant of the scattering particles must be arranged to match the component of flow along the scattering wave vector. In other words, the exponential part of the curve, due to diffusion, must include enough cycles of the cosine wave for its period and thus the flow rate to be measured. This can be done by adjusting the scattering angle.

C. Heterodyne versus self beating detection

For a single beam microscope (Fig. 1.) the autocorrelation function G, for time shift t, of the detected intensity of the scattered light from one or more particles together, all of one size and in Brownian motion in a flowing fluid may be represented by:

$$G(t) = C_0 + C_1 \exp(-Q^2 Dt)\cos(2\pi\Delta ft) + C_2 \exp(-2Q^2 Dt) \qquad (2)$$

where C_0 is a constant which represents a baseline due to the intensity of a background of reference light scattered from stationary objects. C_0 is removable by normalising the data. The second term, with coefficient C_1, represents the scattered field autocorrelation of light from moving particles beating with a reference beam scattered or reflected from stationary parts of the microscope and specimen; described as a "heterodyne" signal. $1/Q^2D$ is the time constant of the exponential part of this term where Q is the scattering wave number and D is the coefficient of translational diffusion of the particles[14]. The cosine part of this term contains the mean Doppler shift Δf due to flow. The third term, with coefficient C_2, represents the scattered field autocorrelation of light scattered from moving particles beating with light scattered from other moving particles; a self-beating or "homodyne" signal. The self beating term is squared, i.e. the time constant is halved, because at any instant, on average, the particles are moving relative to each other twice as fast as they are moving relative to stationary scatterers.

The process of fitting this function to the data is simpler, faster and less prone to error if either the heterodyne or the self beating term can be ignored. This can safely be done if the intensity of either of the components of the signal, represented by C_1 and C_2, can be made about 100 times less than the other. Depending on the nature of the specimen it may therefore be an advantage to detect light backscattered from lens elements, coverslips, microscope slides and stationary parts of the specimen or to shield the detector from it. In the microscope shown in Figure 1 some control over this, heterodyne, component of the signal is given by a Zeiss

"Antiflex" lens, which contains a rotatable quarter wave plate, and a rotatable polaroid analyser above it. For some specimens it might be an advantage to arrange an adjustable stationary reference scatterer in the beam to increase the heterodyne component of the detected light. This could perhaps be done by arranging a form of Michelson interferometer just above the objective lens.

E. A practical laser Doppler microscope

The microscope I am developing (Fig. 1.) is based on a Vickers Instruments' M74 metalurgical microscope. It uses a single beam from a 5 mW, plane-polarised helium-neon laser. This is directed into the specimen by the movable half-silvered mirror of a conventional epi-illuminator attachment. The beam passes out through a Zeiss, Neofluar (X63,N.A.1.25) oil immersion, Antiflex objective lens, used to control the intensity of the heterodyne component of the signal (see above). Back-scattered light returns from the specimen into the lens and passes through the half-silvered mirror towards an EMI 9863B100 low dark-count photomultiplier tube. The angle of scattered light seen by this detector is selected by means of a movable aperture in a diffraction pattern of the specimen, projected from the back focal-plane of the objective lens (Fig. 2a) by a Bertrand lens (Fig. 1). The aperture and diffraction pattern may be viewed through a retractable telescope.

The specimen and the position of the laser beam in it may be viewed or photographed by diverting the light path from the specimen towards the eyepieces with a hinged mirror. A movable aperture in the image plane just below this mirror would enable small volumes of the specimen to be selected so that only light from them reached the detector. A microscope lamp and condenser lens below the specimen stage permit normal viewing. The microscope is also equipped for viewing the specimen under Nomarski differential interference contrast which can be adjusted not to conflict with the Antiflex lens

Specimens are mounted on microscope slides or other suitable supports, on a rotatable mechanical stage. A water-immersion lens may be used, but with the oil-immersion Antiflex lens the specimens must be under a coverslip or within a tubular "Microslide". Drops of suspensions of specimens, for example of Dow latex beads in water, may be mounted to hang beneath a coverslip over a cavity in a microscope slide. Drops with a diameter less than 100 μm may be examined.

Pulses from the photomultiplier tube pass via a preamplifier and discriminator to a Malvern Instruments K7025 real-time multibit correlator with 128 channels. 16 channels are replaced by 128 channels of time-delay, followed by the last eight data-channels to define the baseline of autocorrelation functions. Pulses may also be diverted to a loudspeaker which gives a useful guide to count-rate and signal quality[13]. The autocorrelation data from the correlator are passed to a Commodore "PET" 3032 microcomputer and floppy disc unit (Fig. 1).

It is essential to be able to plot graphs and fit model equations to the autocorrelation data quickly if experiments are to be controlled as they proceed. For this reason the microcomputer is fitted with an MTU K-1000-6 high resolution graphics board with a resolution of 200 by 320 picsels. Graphs can be displayed on a separate television screen. The microcomputer can be programmed as a terminal to send data to a large time-sharing computer via a telephone link. But we find that the immediate availability of the microcomputer offsets its comparative slowness in curve-fitting. The input and output processes of the large computer tend to limit progress. The microcomputer takes about 25 minutes to calculate the diameter and standard deviation of particles by means of a least-squares curvefitting program written in BASIC[14].

A. Results

The instrument determines the diameter of Dow polystyrene latex spheres with diameters in the range 0.088 μm to 1.0 μm to within ± 5% of measurements by electron microscopy[13,14]. I am not yet able to comment on the components of this error.

When set up so that the beam emerges at an angle to the axis of the objective lens as described above (Fig. 2b), the microscope measures flow in glass capillaries to within 15% of measurements made by timing the passage of ink drops[13]. I have not calibrated flow rates in the scattering volume by a more precise independent method.

The microscope has provided autocorrelation functions representing Brownian motion and flow from small regions of living cells, for example from streaming cytoplasm, in cells of *Elodea canadensis* leaves, as investigated by Earnshaw[39] and from streaming cytoplasm in the stamenal hair cells of flowers of *Tradescantia virginiana*.

Measurements with the microscope of the movement of particles in sieve elements in dissected vascular bundles from petioles (leaf stalks) of *Nymphoides peltata* (the Fringed Water Lily (Figs. 3a & b) and of *Heracleum mantegazzianum* (Giant Hogweed) suggest that the r.m.s. displacements of the particles are less than to be expected for particles with a diameter of 1 μm in a fluid with the viscosity of 20% sucrose solution. However, conditions in the sieve elements did not seem to be uniform, either from place to place or with time. The time constants of the auto correlation functions and the intensities of the signals obtained sometimes varied during the time taken to accumulate them (100 s). It is not yet clear whether any of the oscillations recorded (Figs. 3a & b) are due to changes

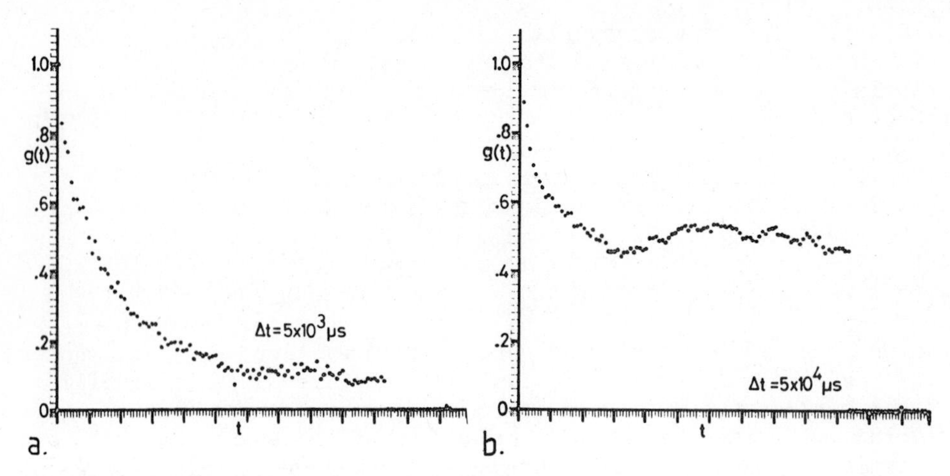

Fig. 3 (a & b) Autocorrelation functions of intensity of light scattered from particles near the same sieve plate in a sieve tube of *Nymphoides peltata*. Oscillations appear, possibly due to flow, possibly to variation in the number of particles in the beam.

in the numbers of particles in the scattering volume, to changes in the amplitude of Brownian motion of the particles or, more interestingly, to transient flows in the sieve tube.

5. CONCLUSIONS

The single beam laser Doppler microscope described here (Fig. 1) measures the diameters of monodisperse suspensions (from the same production lot) of Dow latex beads in the range 0.088 to 1 μm to within ± 5% of sizes determined by electron microscopy. It will do so for very small drops of suspension, one nanolitre or less.

Curve fitting is improved if the microscope is set up to produce a predominantly heterodyne or predominantly self-beating signal.

The microscope will measure flow velocity, though the limits of its accuracy have not yet been determined.

The relative importance of a set of interacting factors which govern the accuracy and efficient use of the LDM described here need to be better quantified.

Applied to the study of the motion of particles in sieve tubes the microscope has provided results which suggest that the particles in the regions examined are undergoing r.m.s. displacements less than expected for the generally accepted viscosity of the fluid round them (about that of 20% sucrose). Two uncertainties are evident. First, the scattering volume is not sufficiently well defined along the length of the illuminating beam; particles out of focus above and below the plane of the image may, nevertheless, be contributing Doppler frequencies to the detected signal. This difficulty may be overcome when an aperture is fitted in the image plane. Second, although the diameter of the scattering volume may be made as small as 2 to 5 µm, depending on refraction and diffraction in the specimen, the signal may then vary in intensity when large particles move in and out of the beam. Oscillations in some correlograms obtained from sieve tubes might be due to this effect or, more interestingly from a biological point of view, these oscillations might be due to transient flows.

ACKNOWLEDGEMENT

I am grateful to the Science and Engineering Research Council for grant GR/B/1712.9 and to Dr. Douglas A. Ross and Dr. Richard J. Hobbs for discussion and help.

REFERENCES

1. I. Maeda and S. Fujime, Quasielastic Light Scattering under Optical Microscope, Rev. scient. Instrum., 43:566-567 (1972).
2. G.V.R. Born, A. Melling and J.H. Whitelaw, Laser Doppler microscope for blood velocity measurements, Biorheology, 15:163-172 (1978).
3. T. Cochrane and J.C. Earnshaw, Practical laser Doppler microscopes, J. Phys. E: Sci. Instrum., 11:196-198 (1978).
4. P.R. DiGiovanni, B. Manoushagian, S. Einav and H.J. Berman, An improved laser Doppler microscope for measurement of *in vivo* velocity distributions in the microcirculation, Proc. 6th New England Bioengineering Conf., Pergamon Press, New York, 113-116 (1978).
5. M. Horimoto and T. Koyama, Measurements of Blood Flow Velocity in Pulmonary Microvessels with laser Doppler Microscope and Investigation of several Factors Affecting Blood Flow Velocity, Biorheology, 18:77-78 (1981).
6. M. Horimoto, T. Koyama, Y. Kikuchi, Y. Kakiuchi and M. Murao, Effect of Transpulmonary Pressure on Blood-flow Velocity in Pulmonary Microvessels, Respiration Physiology, 43:31-41 (1981).
7. T. Koyama, M. Horimoto, H. Mishina, T. Asakura, M. Horimoto and M. Murao, Laser Doppler Microscope in an oblique-backward mode and pulsatile blood flow velocity in pulmonary arteriole, Experentia, 35:65-67 (1979).

7. T. Koyama, M. Horimoto, H. Mishina, T. Asakura, M. Horimoto and M. Murao, Laser Doppler microscope in an oblique-backward mode and pulsatile blood flow velocity in pulmonary arteriole, Experentia, 35:65-67 (1979).
8. H. Mishina, T. Ushizaka and T. Asakura, A Laser Doppler Microscope: Its optical and signal analysing systems and some experimental results of flow velocity, Optics and laser Technology, 121-127 (June 1976).
9. S. Rahat, D.C. Howard, S. Einav and H.J. Berman, Reflectance, fringe mode laser Doppler microscope developed and used to determine velocity profiles of red cells in microvessels, Fed. Proc., 37:214 (1978).
10. B.S. Rinkevichyus, A.V. Tolkachev, V.N. Sutoshin and V.L. Chudov, Laser Doppler Microscope, Radio Eng. and Electron Phys. (USA), 24:114-116 (1979).
11. J.C. Earnshaw and M.W. Steer, Laser Doppler Microscopy, Proc. R. micr. Soc., 14:108-110 (1979).
12. T.J. Herbert and J.D. Acton, "Photon correlation spectroscopy of light scattered from microscopic regions", Applied Optics, 18:588-590 (1979).
13. R.P.C. Johnson, A laser Doppler microscope for biological studies, in: Proc. Conference on Biomedical applications of laser light scattering, Cambridge, England, Sept. 8-10, 1981, B.R. Ware, W.L. Lee and D.B. Sattelle eds., Elsevier North Holland (1982).
14. R.P.C. Johnson and D.A. Ross, Laser Doppler Microscopy and Fibre Optic Doppler Anemometry, in: The Analysis of Organic Surfaces, P. Echlin, ed., Wiley, New York, In Press.
15. A. Koniuta, M.T. Dudermel and P.M. Adler, A laser Doppler anemometer with microscopic intersection volume, J. Phys. E: Sci. Instrum., 12:918-920 (1979).
16. H.S. Dhadwal and D.A. Ross, Size and Concentration of Particles in Syton using the Fibre Optic Doppler Anemometer, FODA, J. Colloid and Interface Science, 76:478-489 (1980).
17. J.C. Earnshaw and M.W. Steer, Studies of cellular Dynamics by laser Doppler microscopy, Pestic. Sci., 10:358-368 (1979).
18. S. Aronoff, J. Dainty, P.R. Gorham, L.M. Srivastava and C.A. Swanson, eds., Phloem Transport, NATO Advanced Study Institute Series A, Volume 4, Plenum Press, New York and London (1975).
19. P.E. Weatherley and R.P.C. Johnson, The form and function of the sieve tube: A problem in reconciliation, Int. Rev. Cytol., 24:149-192 (1968).
20. R.P.C. Johnson, Can cell walls bending round xylem vessels control water flow? Planta, 136:187-194 (1977).
21. R.P.C. Johnson, The Microscopy of P-protein Filaments in Freeze-etched Sieve Pores; Brownian Motion Limits Resolution of their Positions, Planta, 143:191-205 (1978).

22. P.E. Weatherley, Translocation in sieve tubes. Some thoughts on structure and mechanism, Physiol. Veg., 10:731-742 (1972).
23. D.M. Lawton and R.P.C. Johnson, A superhelical model for the ultrastructure of 'P-protein tubules' in sieve elements of *Nymphoides peltata*, Cytobiology, 14:1-17 (1976).
24. R.P.C. Johnson, A. Freundlich and G.F. Barclay, Transcellular strands in sieve tubes; what are they? J. exp. Bot., 27: 1117-1136 (1976).
25. D.D. Sabnis and J.W. Hart, Heterogeneity in Phloem Protein Complements from Different Species, Planta, 145:459-466 (1979).
26. E.A.C. MacRobbie, Phloem Translocation. Facts and Mechanisms: A Comparative Survey, Biol. Rev., 46:429-481 (1971).
27. D.D. Sabnis and J.W. Hart, Studies on the possible occurrence of Actomyosin-like proteins in phloem, Planta, 118:271-281 (1974).
28. B.A. Palevitz and P.K. Hepler, Is P-protein actin-like? - Not yet, Planta, 125:261-271 (1975).
29. D.R. Lee, D.C. Arnold and D.S. Fensom, Some microscopical observations of Heracleum using Nomarski optics, J. exp. Bot., 22:25-38 (1971).
30. G.F. Barclay, K.J. Oparka and R.P.C. Johnson, Induced disruption of sieve element plastids in *Heracleum mantegazzianum* L. J. exp. Bot., 28:709-717 (1977).
31. G.F. Barclay and R.P.C. Johnson, Analysis of Particle motion in sieve tubes of *Heracleum*, Plant, Cell and Environment, 5:173-178 (1982).
32. P. Le-Cong and R.H. Lovberg, Signal-to-noise improvement in laser Doppler velocimetry, Applied Optics, 19:4222-4225 (1980).
33. L.E. Drain, The Laser Doppler Technique, Wiley, Chichester (1980).
34. J.B. Abbiss, T.W. Chubb and E.R. Pike, Laser Doppler Anemometry, Optics and Laser Technology, 249-261 (December 1974).
35. H. Mishina, Y. Kawase, T. Asakura, Frequency Error of Doppler Beat Signals due to Extended Scattering Particles, Japanese Journal of Applied Physics, 15:633-640 (1976).
36. Y. Kawase, H. Mishina and T. Asakura, Frequency Error of Doppler Beat Signals due to Extended Scattering Particles: II. Some Additional Considerations and Experimental Verification, Japanese Journal of Applied Physics, 15:2173-2179 (1976).
37. H. Mishina and T. Asakura, Measurement of Velocity Fluctuations in laser Doppler Microscope by the New System Employing the Time-to-Pulse Height Converter, Appl. Phys., 5:351-359 (1975).
38. M. Baker and H. Wayland, On line volume flow rate and velocity measurement for blood in microvessels, Microvascular Research 7:131-143 (1974).
39. J.C. Earnshaw, Cytoplasmic motion in *Elodea*, in: H.Z. Cummins and E.R. Pike, eds., Photon Correlation Spectroscopy and Velocimetry, NATO Advanced Study Institute Series B; Volume 23, Plenum Press, New York and London (1977).

STUDIES OF NEUROTRANSMITTER RECEPTOR INTERACTIONS USING QUANTITATIVE VIDEO INTENSIFICATION MICROSCOPY (V.I.M.)

D.G. MacInnes, D.K. Green*, A.Harmar, E.G. Nairn and G. Fink

MRC Brain Metabolism Unit, University Department of Pharmacology, Edinburgh
* MRC Clinical and Population Cytogenetics Unit Western General Hospital, Edinburgh

INTRODUCTION

Studies of interactions between neurotransmitters and their receptors would be greatly facilitated by a method to obtain cell fractions enriched with cells that contain a high density of receptors specific for the neurotransmitter. Here we report the use of a fluorescence activated cell sorter to prepare fractions of pituitary cells that contain a high density of high affinity, specific receptors for the decapeptide, luteinizing hormone releasing hormone (LHRH). Since no enzymatic procedures are employed, the dispersed cells permit both binding and biological responses to be analysed immediately without requiring the usual time of 16-48hr for recovery.

METHODS

An agonist of LHRH, (D-Lys^6)-LHRH was coupled to rhodamine isothiocyanate (Rhod-D-Lys^6) LHRH (Hazum et al 1980) and purified by reversed phase, high performance liquid chromatography. The probe was 75% as potent as unlabelled LHRH in releasing LH from dispersed anterior pituitary cells suspended in a Biogel column according to Speight & Fink(1981). A purely mechanical method was used for the dissociation of cells from rat anterior pituitary glands. Rhod-D-Lys^6 LHRH was now added to this suspension at a final concentration of 10^{-7}M at 4°C. After 1h the cells in suspension at 4°C were measured and sorted with a fluorescence activated cell sorter. A conventional arrangement of an 'in air' sample stream, argon-ion laser beam excitation, and charged droplet deflection by electrostatic field form the basic components of the machine (Green D.K. et al., 1980). The laser (Spectra-Physics 164-05) was set at 514nm wavelength and 1.3watts Power. A bandpass 570-650nm filter was placed in front of the fluorescence detecting photomultiplier. A logarithmic amplifier was used to detect light scatter. The fluorescence intensity of each cell was measured in parallel with the forward light scatter. Signal coincidences that occurred within the interval of two drops forming were not recorded.

The distribution of cell size (scatter) versus fluorescence intensity is shown in figure 1. There was a discrete population of large cells which fluoresced brightly.

These fluorescing cells were placed in a microslide (Camlab Ltd) with a fixed pathlength of 0.1nm. They were examined using phase contrast with an oil immersion planapo 63/1.25 N.A. lens on a Zeiss IM35 inverted microscope. The fluorescence was excited using epi-illumination with a 100W mercury lamp with a 560/10nm band pass filter and a 580nm long pass barrier filter. The distribution of fluorescent label on the cell was visualised using a silicon intensified target (SIT) T.V. camera. The image was recorded on a u-matic recorder and viewed directly or by way of a T.V. frame store (Arlunya TF 4000) on a black and white T.V. monitor. This real time digital store incorporates very efficient noise reduction. The picture is digitized in 512 x 512 pixels with 256 grey levels, and each pixel is imaged with quantitative accuracy. The input gain can be changed for different samples. This is a very important feature when investigating the distribution of very low concentrations of fluorescent label.

Photon Counting

The fluorescing cells were viewed in the fluorescence microscope under similar conditions as described above.

To define the area of interest on the fluorescing cells a 0.25mm aperture was placed in the image plane in front of a photon counting photomultiplier tube (EMI 9893 B/350 Bialkali). The shutter in front of the photomultiplier was opened and the slightly defocussed fluorescence image as defined by the aperture was projected onto the photomultiplier end window. The output from the photomultiplier was fed into the C601 amplifier discriminator unit, (EMI Ltd.) and the frequency of the pulsed

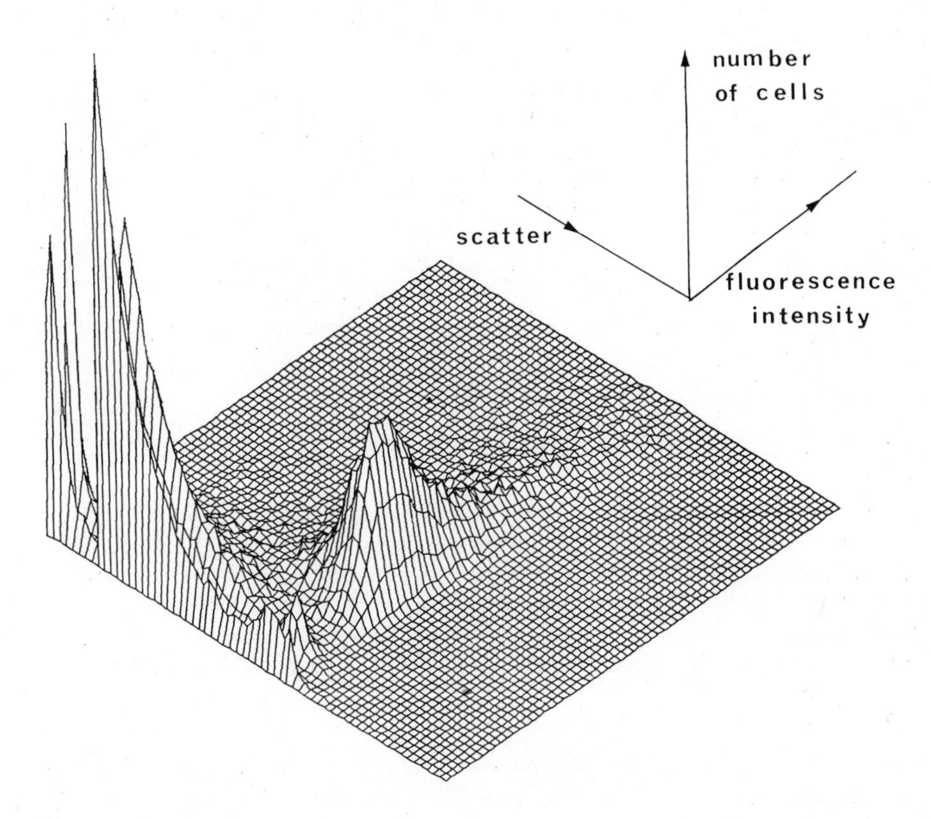

Figure 1
An isometric plot showing the distribution of fluorescence intensity and cell size (scatter) of mechanically dispersed rat pituitary cells as they pass through a fluorescence activated cell sorter. The cells were incubated with rhodamine labelled LHRH at a final concentration of 10^{-7}M.

output from this unit was measured with a PFM 200 (Thandar Ltd.) frequency meter using a gate time of 1s.

The fluorescing cells were examined using the SIT TV camera on the fluorescence microscope. We have observed that the initial even distribution of fluorescent label on pituitary cells changes with time as clusters appear followed by apparent internalization of the receptor-ligand complex.

Preliminary studies were carried out to determine the kinetics of LHRH binding to the fluorescing cell population. For this purpose the output from the photomultiplier was calibrated using stock solutions of rhodamine D Lys^6 LHRH and was found to be linear with concentrations from 10^{-8} to 4×10^{-6} M. The concentrations of free ligand and the total binding was measured on single cells in the presence of varying concentrations of ligand with no calcium present. The non-specific binding was then measured by displacing the ligand with a 100-fold excess of LHRH. The specific binding to LHRH receptors on these pituitary cells was about 75%.

In summary we have described a technique which enables us to collect fractions of viable cells that contain a high density of specific LHRH receptors. This technique is now being used to carry out extensive studies on the kinetics of LHRH binding to single cells and the factors such as gonadal steroids and previous exposure to LHRH (Fink 1979) which may affect this binding. This technique can be extended to investigate a wide range of neurotransmitter receptor interactions in different cell systems.

References

Fink, G. 1979 Neuroendocrine control of gonadotrophin secretion. British Medical Bulletin, 35: 2, 155-160

Green, D.K., Malloy, P., Steel, M. 1980. The recovery of living cells by flow sorting machine. Flow Cytometry IV; 103-106.

Hazum, E., Cuatrecasas, P., Marian, J., Conn, P.M. 1980. Receptor-mediated internalization of fluorescent gonadotropin-releasing hormone by pituitary gonadotropes. Proc. Natl. Acad. Sci., U.S.A. 77: 6692-6695

Speight, A., Fink, G. 1981. Changes in responsiveness of dispersed pituitary cells to luteinizing hormone releasing hormone at different times of the oestrous cycle of the rat. J. Endocr. 89: 129-134.

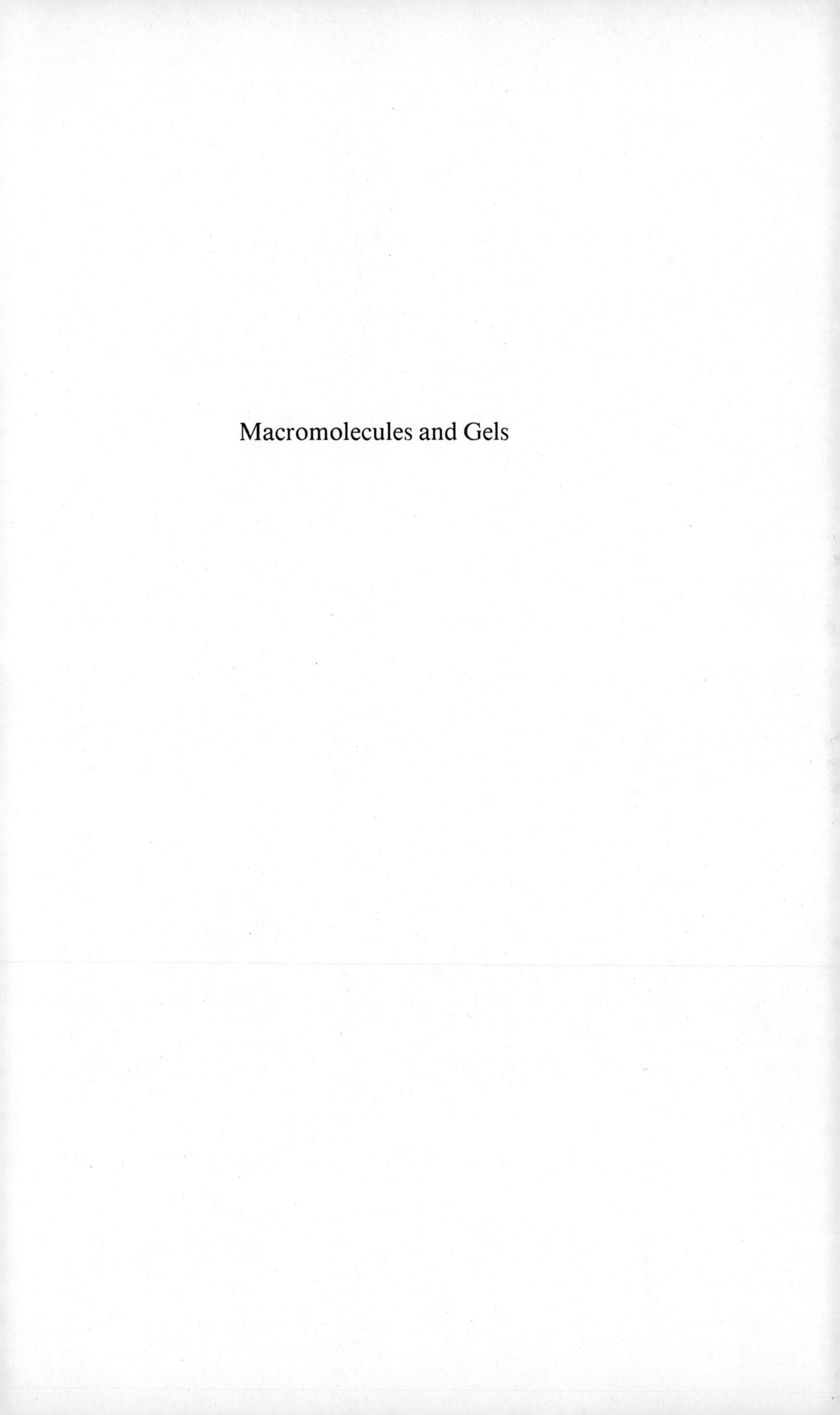

Macromolecules and Gels

ANALYSIS OF DIFFUSION OF BIOLOGICAL MATERIALS BY QUASIELASTIC LIGHT SCATTERING

Herman Z. Cummins

Department of Physics
City College of THE CITY UNIVERSITY OF NEW YORK
New York, New York 10031, U. S. A.

CONTENTS

ABSTRACT

The investigation of translational and rotational diffusion coefficients represents a principal application of laser light scattering in biology. Diffusion coefficients of numerous biologically important molecules as well as viruses, membrane vesicles and biopolymers have been obtained by this technique.

A brief review of the experimental and theoretical aspects of diffusion measurements will be presented including the analysis of polydispersity and time dependent studies of aggregation. Recent experimental investigations of fibrinogen polymerization and of the properties of membrane and phospholipid vesicles, micelles and microemulsions will be described. The emphasis will be on problems related to biological self-assembly.

1. INTRODUCTION

Measurements of translational and rotational diffusion coefficients of biological molecules, polymers and viruses have long provided an important approach to the determination of the size and shape of these objects in solution. Twenty years ago, diffusion coefficient measurements were extremely complicated and time consuming. Translational diffusion coefficients were determined primarily by the free diffusion method in which a sharp boundary is established between a solution of uniform concentration and the pure solvent. After removal of the partition, the time dependence of the concentration profile is followed, and analyzed by fitting against a standard integral equation which contains the translational diffusion coefficient D_T as a parameter. Rotational diffusion coefficients were determined primarily from birefringence measurements or, in a few cases, by transient electric birefringence. The theory and methodology of these techniques as well as a survey of the diffusion coefficients of a number of proteins and viruses was given by Tanford in 1961.[1]

The development of lasers in the early 1960s opened an entirely new approach to the analysis of diffusion coefficients. When a laser beam of constant intensity is focussed in a solution of small particles, light is scattered by the particles, and the spectrum of the scattered light is modified by their translational and rotational motion; analysis of the scattered light can recover the values of the diffusion coefficients.

The first theoretical analysis of the relation between the optical spectrum of the scattered light and the diffusion coefficients was carried out by R. Pecora as his thesis research (completed in 1962) and was published in 1964.[2] The first experimental observation of broadening of the spectrum of laser light scattered from diffusing particles (polystyrene latex spheres) was reported in 1964;[3] the first application to biologically important materials (bovine serum albumen, ovalbumin, lysozyme, tobacco mosaic virus and DNA) was reported in 1967.[4] The first determination of both the translational and rotational diffusion coefficients - for tobacco mosaic virus - was reported in 1969.[5]

The speed and accuracy of diffusion coefficient determinations improved dramatically with the replacement of the early light beating methods of spectral analysis by digital correlation techniques pioneered by E. R. Pike and coworkers at the Royal Radar Establishment at Malvern. At the time of the first Capri ASI on photon correlation and light beating spectroscopy in 1973, the diffusion constants of some 45 biomolecules and viruses had been determined by quasielastic light scattering.[6] By 1976 when the second Capri ASI took place, P. Pusey found that 194 articles on this topic had appeared during the period 1973-76, most of which involved the determination of the diffusion coefficients of particular proteins, viruses or polymers.[7] In just over a decade light scattering had become the principal method for the determination of diffusion coefficients.

Since this topic has been treated extensively in a large number of books, review articles and journal articles, I will only summarize its salient features here and discuss some applications of current interest. In section 2, I present a brief historical survey of the experimental techniques which will be covered more thoroughly in the lectures of V. Degiorgio and B. Chu. In section 3, I will review the fundamental equations relating the measured properties of the scattered light to the diffusional dynamics of the scatterers. Section 4 is devoted to the application of photon correlation spectroscopy to aggregation reactions and section 5 to the study of micelles, microemulsions and vesicles.

2. HISTORICAL SURVEY OF EXPERIMENTAL TECHNIQUES

The properties of laser light which are relevant for quasielastic light scattering experiments can be represented in an idealized fashion as shown in the upper part of Fig. 1: the intensity $I_o(t)$ is constant in time, and the spectrum $I_o(\omega)$ is extremely narrow, represented here as a single frequency "spike" at ω_o. The light which is scattered by the illuminated particles, when viewed at a point sufficiently far from the scattering volume, can be represented as shown in the lower part of Fig. 1: the intensity $I_s(t)$ fluctuates wildly in time, and the spectrum $I_s(\omega)$ is broadened around the incident frequency ω_o. (The structure of $I_s(t)$ and $I_s(\omega)$ will be discussed further in the following section.)

The width of the scattered spectrum $I_s(\omega)$ is typically ~1kHz (hence "quasielastic") which is far too narrow to measure with traditional spectroscopic techniques employing grating spectrometers or Fabry-Perot interferometers. Thus, new techniques were required in order to measure the extremely narrow linewidths produced by quasi-elastic light scattering.

In March 1961, the Second International Conference on Quantum Electronics at Berkeley opened with a paper by C. H. Townes entitled "Some Applications of Optical and Infrared Masers".[8] Looking ahead to future applications for these new light sources, Townes said: "There is a still more intriguing and higher resolution Raman technique available with masers. This involves detection of a mixture

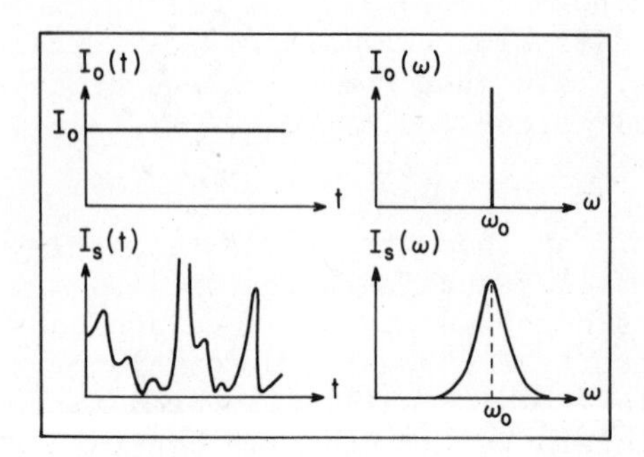

Figure 1. Top: The intensity $I_o(t)$ and spectrum $I_o(\omega)$ of incident laser light. Bottom: The intensity $I_s(t)$ and spectrum $I_s(\omega)$ of laser light scattered by particles in solution.

of the Raman-scattered light and that of the original frequency with a fast photodetector. The two different frequencies would beat together and yield a difference frequency which can be detected and measured with ordinary electronic techniques if their frequency difference is not too great". This technique, initially called light beating spectroscopy or heterodyne spectroscopy, provided the first successful means of analyzing the spectrum of quasielastically scattered light.

There has been some controversy about the historical development of the light beating technique. Usually, the original concept is attributed to well known radio techniques which were then adapted to optics. However, there is an important historical precedent for light beating from the 1870s which precedes the invention of the radio! In 1883, A. Righi published a paper in the Journal de Physique entitled "On the Changes in Wavelength Obtained by a Rotating Polarizer, and on the Phenomenon of Beats Produced with Light Vibrations".[9] In this paper (which also referred to Righi's earlier work published in 1878) he said "if one produces interference between two rays whose frequencies are slightly different, one obtains ...... fringes in uniform movement ...... such that at a point on a diaphragm there will pass each second a number of bright fringes equal to the difference in frequencies. One would therefore have a phenomenon identical to that of beats obtained with sound waves in the air". Righi suggested that the two rays could be obtained from a single source, with the frequency difference introduced by passage through a rotating nicol prism. I have been told that Righi was the teacher of Marconi, suggesting that heterodyne detection may have found its way into radio from optics rather than the other way around![10]

In 1955 Forrester, Gudmundsen and Johnson observed light beats between two Zeeman components of a mercury lamp placed in a magnetic field.[11] The Zeeman splitting was ~10GHz and the beating was detected with a specially constructed microwave phototube. This experiment was extremely difficult beause of the large bandwidth and low intensity of the mercury spectral lines. By contrast, the extremely narrow spectral width and high intensity of lasers makes the observation of light beats with lasers very easy. In typical early experiments, mixing took place in an ordinary photomultiplier tube whose output current was analyzed with an electronic spectrum analyzer.[12]

Two closely related methods of light beating spectroscopy evolved during the 1960s. In one - called heterodyne spectroscopy - some unscattered laser light (which can be shifted in frequency) is mixed with the scattered light. The shape of the spectrum of the photocurrent then exactly matches that of the scattered light, but the center frequency is shifted to a convenient rf frequency so that the spectrum can be analyzed with an electronic spectrum analyzer. In the second method - called homodyne spectroscopy or self-beating spectroscopy - the scattered light effectively mixes with itself at the photomultiplier. The optical and photocurrent spectra for these two techniques are illustrated schematically in Fig. 2. These two techniques were initially described by Forrester[13] and have been widely discussed in the literature.[14]

In the late 1960s, measurements of the photocurrent spectrum $I_s(\omega)$ were largely superceded by measurement of the photocurrent autocorrelation function $C_i(\tau)$. Since the photocurrent consists of digital information in the form of a series of photoelectric pulses which can be shaped and amplified with standard fast pulse electronics, autocorrelators employing digital electronic circuits can extract the maximum amount of information from the photomultiplier output with maximum efficiency and without the analog errors associated with electronic spectrum analysis. Although digital spectrum analyzers are now available as well, the great majority of diffusion measurements conducted during the past decade have used autocorrelators. The photocurrent autocorrelation technique has often been designated as photon correlation spectroscopy or PCS. However, the analysis of quasielastically scattered light by electronic measure-

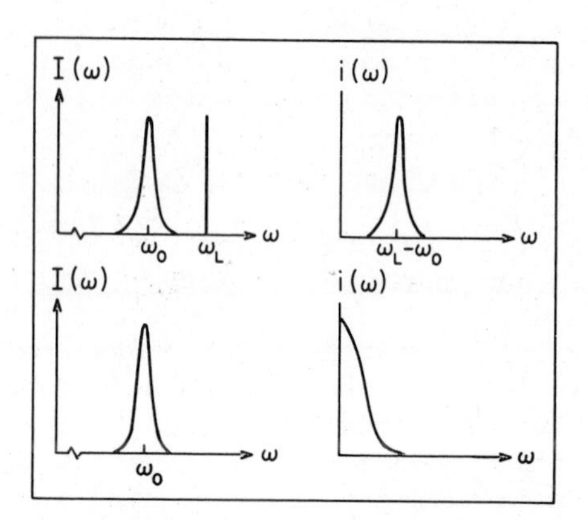

Figure 2. Optical spectrum (left) and photocurrent spectrum (right) for light beating experiments. The upper figures represent heterodyne detection with the local oscillator frequency ω_L shifted from the laser frequency ω_o. The lower figures represent homodyne detection where no local oscillator is used.

ment of the photoelectric current, whether by spectral analysis or by autocorrelation analysis, is also referred to by various authors as dynamic light scattering, quasielastic light scattering or intensity fluctuation spectroscopy.

The principal of photon correlation spectroscopy is illustrated in Figs. 3 and 4. Fig. 3 shows a schematic version of a light scattering spectrometer consisting of a laser source, a scattering region and a photodetector. The scattered light intensity is shown next exhibiting the large fluctuations which would be seen if the photocurrent were viewed directly on an oscilloscope. Finally, the stream of photoelectric pulses which appear when the photocurrent is examined with better time resolution is shown. In Fig. 4 the essential elements of a correlator are displayed. The shift register produces a series of delayed replicas of the incoming sequence of pulses. The delayed replicas are constantly multiplied by the incoming pulse sequence and the products $n(t)n(t+\tau_i)$ for each delay time τ_i are stored in a set of digital scalers. At the end of the run, these scalers contain N points of the autocorrelation function $C(\tau)$. Typically N is between 16 and 100. This is a single clipped correlator. The one-bit shift register can only store a zero or a one. The theory and operation of correlators used in quasielastic light scattering spectroscopy have been discussed exhaustively in several previous ASIs[15,16,17] and will not be further pursued here.

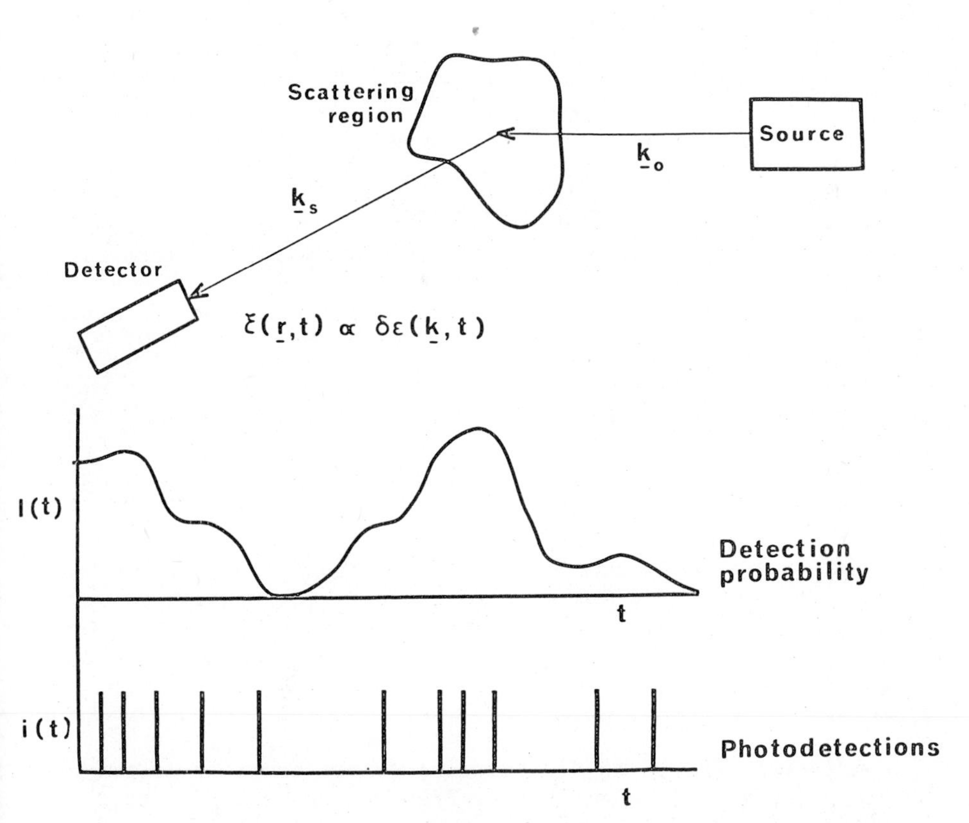

Figure 3. Top: Schematic light scattering experiment. Center: Intensity I(t) of the scattered light. Bottom: Stream of photoelectric pulses that consitute the photocurrent i(t) (from C. J. Oliver, Ref. 15, p. 152).

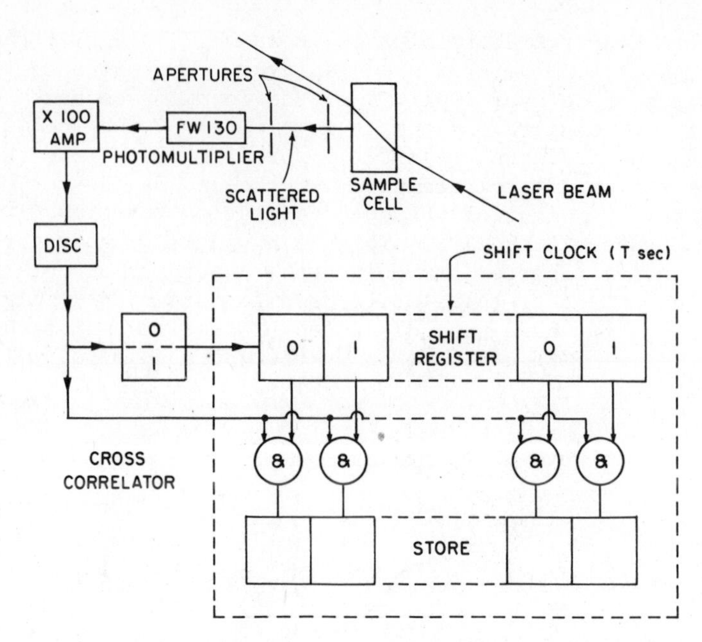

Figure 4. Schematic diagram of a single-clipped photon correlation spectrometer showing the scattering optics and digital correlator based on a one-bit shift register (from E. R. Pike, Ref. 18).

3. LIGHT SCATTERING THEORY

The theory of quasielastic light scattering by particles undergoing translational and rotational diffusion in solution has been reviewed in several previous NATO ASIs,[15,16,17] in several books (e.g. Chu,[19] and Berne and Pecora[20]) and in numerous review articles and research papers. The following discussion is therefore restricted to a brief survey of the most important equations required to extract diffusion coefficients from experimental data. (More extensive discussion are presented in the lectures of Ben Chu and Vittorio Degiorgio.)

Consider a collection of identical scatterers illuminated by a monochromatic laser beam of angular frequency $\omega_o = \nu_o/2\pi$ and wavevector $\vec{k}_o$. Scattered light with wavevector $\vec{k}_s$ is detected in a small area at a large distance from the scattering volume. The electric field of the light scattered from a single particle (j) is

$$E_j(t) = A_j(t)e^{-i\omega_\circ t}\ e^{i\phi_j(t)} \tag{1}$$

$A_j(t)$ is the amplitude of the scattered light, and $\phi_j(t)$ is the phase delay introduced by the difference in optical paths for light scattered from the particle at r_j and one at the origin. This phase shift $\phi_j(t)$ is related to the instantaneous position $r_j(t)$ by

$$\phi_j(t) = \vec{q} \cdot \vec{r}_j(t) \tag{2}$$

where the magnitude of the scattering vector $\vec{q} = \vec{k}_o - \vec{k}_s$ is

$$q \cong 2k_o \sin\frac{\theta}{2} = \frac{4\pi n}{\lambda} \sin\frac{\theta}{2} \tag{3}$$

(n is the refractive index of the solution, λ the wavelength in vacuum of the incident light). The total scattered field is obtained by summing Eq. (1) over all the scatterers:

$$E_s(t) = \sum_j A_j(t) e^{i\vec{q}\cdot\vec{r}_j(t)} e^{i\omega_o t} \tag{4}$$

The principal mathematical functions which relate experimental measurements to $E_s(t)$ are:

1: The field correlation function.

$$G^{(1)}(\tau) = \langle E_s^*(t)\, E_s(t+\tau)\rangle \tag{5a}$$

where <> denotes a time (or ensemble) average, and the normalized field correlation function

$$g^{(1)}(\tau) = G^{(1)}(\tau)/G^{(1)}(0) \tag{5b}$$

2: The optical spectrum.

$$I_s(\omega) = \frac{1}{2\pi} \int_0^\infty G^{(1)}(\tau)\, e^{i\omega\tau} d\tau \tag{6}$$

3. The intensity correlation function.

$$G^{(2)}(\tau) = \langle I_s(t) I_s(t+\tau)\rangle \tag{7}$$

where $I_s = |E_s|^2$

If many independent scatterers are present in the scattering volume, the scattered field will be a Gaussian random process as a consequence of the central limit theorem, and $G^{(2)}(\tau)$ is then related to $g^{(1)}(\tau)$ by

$$G^{(2)}(\tau) = \langle I\rangle\, [1 + |g^{(1)}(\tau)|^2] \tag{8}$$

which is the Siegert relation. The experimentally measured intensity correlation function $C(\tau)$is usually represented as

$$C(\tau) = B[1 + a|g^{(1)}(\tau)|^2] \tag{9}$$

where a, which is typically ~0.7 is a geometrical factor arising primarily from the finite detector size.

A. Spherical Scatterers

For the simplest possible example, a collection of identical spherical scatterers undergoing ordinary Brownian motion in solution characterized by the translational diffusion constant D_T,

$$g^{(1)}(\tau) = e^{-D_T q^2 \tau} e^{-i\omega_o \tau} \tag{10a}$$

$$|g^{(1)}(\tau)| = e^{-D_T q^2 \tau} \tag{10b}$$

The optical spectrum (from Eqs. 6 and 10) is

$$I_s(\omega) = \langle I \rangle_s \frac{D_T q^2 / \pi}{(\omega - \omega_o)^2 + (D_T q^2)^2} \tag{11}$$

which is a Lorentzian of half width $D_T q^2$ centered at ω_o.

The intensity correlation function $C(\tau)$ is

$$C(\tau) = B[1 + ae^{-2D_T q^2 \tau}] \tag{12}$$

Eq. (12) is the conventional starting point for analysis of translational diffusion constants by photon correlation spectroscopy. There are considerable subtleties associated with statistical analysis of such data which have been discussed in detail at previous NATO ASIs,[15,16,17] primarily by E. Jakeman and C. J. Oliver. In practice, correlation functions are usually fit directly to Eq. (12) by subtracting the background B (which is computed by the correlator) and performing a linear regression analysis on

$$\ell n\ (C_s(\tau) - B)/B = \ell n a - 2D_T q^2 \tau \tag{13}$$

Since q is known from λ, θ and n, Eq. (13) yields D_T directly, and from it the hydrodynamic radius r_h via the Stokes-Einstein relation

$$D_T = \frac{kT}{6\pi\eta r_h} \tag{14}$$

where η is the viscosity of the solution.

Fig. 5, from Newman et al,[21] illustrates a typical single exponential fit to a 64 channel correlation function of light scattered

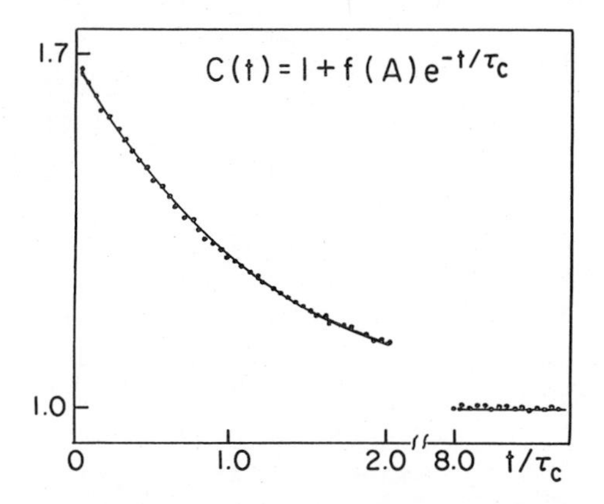

Figure 5. Autocorrelation function of light scattered by a solution of DNA of the fd bacteriophage (from Newman et al., Ref. 21).

from a dilute solution of viral DNA. Note that there is a 128 channel delay between channels 48 and 49. Fig. 6, from Oliver[17] illustrates the linearity of the semilog plot implied by Eq. (13).

B. Nonspherical Scatterers

If the scatterers are identical but nonspherical, the scattering amplitudes $A_j(t)$ in Eq. (4) will change in time due to rotational motion. In the usual approximation that orientation and position are uncorrelated, the field correlation function is

$$G^{(1)}(\tau) = N \langle A(0)A(\tau)\rangle \, e^{-D_T q^2 \tau} e^{-i\omega_0 \tau} \tag{15}$$

The amplitude correlation function $\langle A(0)A(t)\rangle$ for nonspherical particles undergoing rotational diffusion is[2]

$$\langle A(0)A(\tau)\rangle = I_s \sum_{\substack{\ell=0 \\ \text{even}}}^{\infty} B_\ell e^{-\ell(\ell+1)D_R \tau} \tag{16}$$

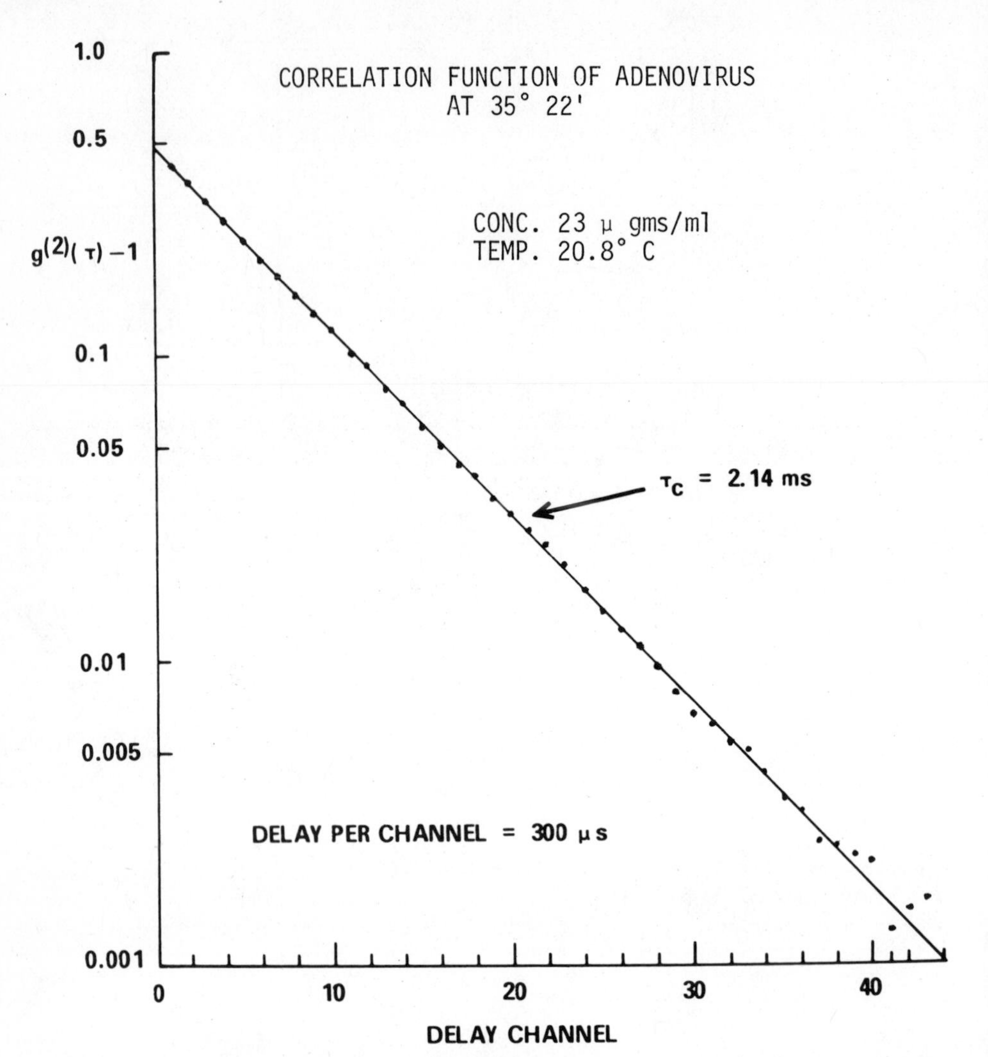

Figure 6. Semilog plot of correlation data of adenovirus illustrating the procedure suggested by Eq. (13) [from C. J. Oliver, K. F. Shortridge and G. Belyavin, Biochem. Biophys. Acta 437, 589 (1976)].

where D_R is the rotational diffusion coefficient and the B_ℓ are coefficients which depend on the size and shape of the scatterers and on the scattering vector q. Combining Eqs. (15) and (16),

$$G^{(1)}(\tau) = I_s e^{-i\omega_\circ\tau}[B_\circ e^{-D_T q^2\tau} + B_2 e^{-(D_T q^2+6D_R)\tau} + B_4 e^{-(D_T q^2+20D_R)\tau} + \ldots] \quad (17)$$

For small particles, or for larger particles at sufficiently small scattering angles, $B_o >> B_2, B_4$ ... so that Eq. (17) reduces to the simpler Eq. (10).

In many cases it is possible to choose a range of scattering angles for which $qL \lesssim 10$ where L is the maximum dimension of the scatterer. Then all terms above B_2 can be neglected. D_T is first determined from small angle scattering data, and B_2/B_o and D_R are found from fits to the approximate equation

$$C(\tau) = B[1+a(e^{-D_T q^2 \tau} + \frac{B_2}{B_o} e^{-(D_T q^2 + 6D_R)\tau})^2] \quad (18)$$

$$\frac{C(\tau) - B}{B} = ae^{-2D_T q^2 \tau}[1 + \frac{B_2}{B_o} e^{-6D_R \tau}]^2 \quad (19)$$

Rotational diffusion coefficients can also be deduced from depolarized light scattering measurements. If the molecular polarizability is significantly anisotropic, then the small-angle depolarized scattering spectrum or correlation function can provide D_R separately.

C. Polydispersity

If the scatterers are spherical but not identical, Eq. (10b) for $|g^{(1)}(\tau)|$ must be replaced by

$$|g^{(1)}(\tau)| = \int_0^\infty F(\Gamma) e^{-\Gamma\tau} d\Gamma \quad (20)$$

where $\Gamma = D_T q^2$ and $F(\Gamma)d\Gamma$ is the fraction of the scattered intensity produced by particles with $D_T q^2$ in the range Γ to $\Gamma + d\Gamma$.

Professor Chu will discuss various methods of deducing $F(\Gamma)$ from $C(\tau)$ in his lectures. I will simply summarize some of the current methods here before proceeding to a discussion of applications.

1. Laplace Inversion. Eq. (20) has the form of a Laplace transform, and can in principle be inverted to find $F(\Gamma)$ exactly if $|g^{(1)}(\tau)|$ is known exactly for all τ. In practice, the limited range of τ for which $|g^{(1)}(\tau)|$ is known and the inevitable presence of statistical noise in the data makes the inversion problem nearly intractable. Nevertheless McWhirter and Pike,[22] Ostrowsky et al.[23] and their coworkers have made substantial progress in obtaining Laplace transforms in the presence of noise based on eigenfunction expansion techniques.

2. Parametrized Distributions. $F(\Gamma)$ can be modelled by an analytic distribution function containing one or more parameters, and a nonlinear least squares fitting procedure can be used to

optimize the parameters. In a recent study of phospholipid vesicles, for example, Ostrowsky and Sornette made use of three model distribution functions.[24]

$$\text{Simple exponential:} \quad F(\Gamma) = \delta(\Gamma-\Gamma_o) \tag{21a}$$

$$\text{Bimodal:} \quad F(\Gamma) = A\delta(\Gamma-\Gamma_1) + (1-A)\delta(\Gamma-\Gamma_2) \tag{21b}$$

$$\text{Gaussian:} \quad F(\Gamma) = A\exp\left[-(\Gamma-\bar{\Gamma})^2/2v\bar{\Gamma}^2\right] \tag{21c}$$

This is the easiest procedure, but the choice of the functional form for $F(\Gamma)$ inevitably biases the results (v is a normalized variance).

3. Moments or Cumulants.[25] From Eq. (9), $[C(\tau) - B]/B = a|g^{(1)}(\tau)|^2$. This quantity is determined directly from the experimental data. One can easily show that[16]

$$\ln[\sqrt{a}|g^{(1)}(\tau)|] = \frac{1}{2}\ln a - \langle\Gamma\rangle\tau + \frac{1}{2!}\mu_2\tau^2 - \frac{1}{3!}\mu_3\tau^3 + \frac{1}{4!}[\mu_4 - 3\mu_2^2]\tau^4 + \ldots \tag{22}$$

where $\mu_n = \int_0^\infty (\Gamma - \langle\Gamma\rangle)^n F(\Gamma)d\Gamma$ is the nth moment of $F(\Gamma)$ about the mean $\langle\Gamma\rangle$.† Thus if $\ln[\sqrt{a}\, g^{(1)}(\tau)]$ vs τ is fit to a low order polynomial, the initial slope gives $\langle\Gamma\rangle$, and the curvature gives the dispersion μ_2. Although in principle $F(\Gamma)$ could be reconstructed completely from all its moments, in practice only the first two (or sometimes three) moments can be found with any confidence.

4. <u>Splines</u>. $F(\Gamma)$ can be represented by a (small) set of N points $F_i(\Gamma_i)$ connected by straight lines (linear splines) or by simple curves. This function is inserted in Eq. (20) and the integration is performed analytically. The N parameters $F_i(\Gamma_i)$ are varied to produce a best fit of Eq. (20) to the data. This method was developed by G. B. Stock[26] for analyzing correlation data obtained with motile microorganisms, and has been applied in Professor Carlson's laboratory at Johns Hopkins to the polydispersity problem also. Although all of these methods have been applied with some success, analysis of polydispersity remains a major unsolved problem in this field.

In this brief survey I have considered only the case of independent non-interacting particles. If the concentration of diffusing scatterers is sufficiently high for particle interactions to become important, the meaning of the diffusion coefficients D_T and D_R must be modified. This problem, which has been discussed extensively in

† The coefficients of $\frac{1}{m!}(-\tau)^m$ are the cumulants or semi-invariants K_m of $F(\Gamma)$.

the literature and is discussed in the lectures of Professor Degiorgio, will not be considered here. We will make use of the theory reviewed here in the following sections which will explore the application of PCS to some problems of current interest in biology.

4. POLMERIZATION REACTIONS: THE FIBRINOGEN-FIBRIN TRANSFORMATION AND FIBRIN AGGREGATION

In section 3 we reviewed the equations which govern the autocorrelation function $C(\tau)$ of light scattered by particles in solution for the usual stationary situation in which the properties of the individual scatterers are assumed not to vary in time. Time dependence may be incorporated into the theory to accommodate conformational changes, chemical reactions, polymerization, etc. The effect of such time dependence on the correlation function is primarily determined by the characteristic time τ_C over which the changes occur relative to the correlator bin time (or delay time per channel) τ_B, and the experiment time τ_E. If $\tau_C < \tau_B$, the time dependence is averaged out and has no effect. If $\tau_B < \tau_C < \tau_E$, there will be additional terms in $C(\tau)$ resembling the extra terms due to rotational diffusion contained in Eq. (17) along with shifting of the apparent background level. If $\tau_c > \tau_E$, the system is quasistationary. The equations of section 3 apply, but the parameters characterizing the scatterers (D_T, D_R, $F(\Gamma)$) will change from one experiment to another. Analysis of a sequence of correlation functions can thus provide the time dependence of these parameters. In this section we consider the application of photon correlation spectroscopy to a polymerization reaction of considerable biological importance: the fibrinogen-fibrin transformation and fibrin polymerization.

Fibrinogen is a protein molecule found in the blood plasma consisting of three pairs of polypeptide chains; its molecular weight is ~330,000. There is controversy over the molecular shape of fibrinogen but a reasonably successful model is a rod about 800Å long, between 40 and 100Å in diameter.[27,28,29] When the body sustains a wound, the enzyme thrombin is produced through a series of proteolytic reactions. Thrombin converts stable fibrinogen to an unstable form called fibrin by cleaving four small terminal peptide groups from fibrinogen, and the activated fibrin then polymerizes to form a clot, a polymer network of fibrin aggregates which prevents the penetration of blood cells, electrolytes and other small molecules as well as water, sealing the wound. Failure of this process to occur properly leads to clotting disorders (e.g., hemophilia or dysfibrinogenemia)) while its occurrence at inappropriate locations causes the pathological disorder of thrombosis, the production of blood clots within the circulatory system. The venom of certain snakes can also cleave peptides from fibrinogen, initiating the potentially lethal formation of clots.

Because of the medical importance of this process, it has been studied intensively for many years. Although the biochemical steps in the processes are well understood, the kinetics of fibrin polymerization remain incompletely understood and controversial, and are the subject of considerable current research. Classical light scattering - measurement of the time averaged scattering intensity $I_s(q)$ - has been used in numerous studies of this system.[27,29,30] The technique of photon correlation spectroscopy has also been used in a number of recent studies of fibrinogen, the most extensive of which was conducted by P. Wiltzius and his coworkers in Zurich.[28,31,32,33] These experiments will be reviewed below.

One possible model for the clotting process suggests the following sequence of steps.[30] The proteolytic enzyme thrombin removes two fibrinopeptides A (FPA) from the Aα chains of fibrinogen and subsequently two fibrinopeptides B (FPB) from the Bβ chains. The fibrin monomers polymerize end to end with an overlap of ~450Å (one half the monomer length) to form fibrin protofibrils which then associate laterally to form fibers consisting of 100 or more protofibrils. The fibers join together to form a polymer network which eventually becomes a gel. The structure is covalently stabilized by the action of another enzyme, factor XIIIa (fibrin stabilizing factor). The correlation of fibrinopeptide release and fibrin polymerization was demonstrated in 1958 by Blomback and Laurent.[29] Substitution of the enzyme reptilase (derived from snake venom) which releases FPA but not FPB has shown that FPA release is sufficient to initiate end to end polymerization of fibrin.

The first fundamental question which Wiltzius and Hofmann investigated by PCS is the relation between the size and shape of fibrinogen and fibrin.[28,33] Purified human fibrinogen was incubated with thrombin. The resulting clot was collected on a glass rod and redissolved in 5M urea. Fibrin devoid only of FPA was prepared similarly using reptilase instead of thrombin. Translational diffusion constants D_T were determined by PCS of polarized scattered light, and rotational diffusion constants D_R from depolarized scattered light. Fibrinogen and both forms of fibrin gave essentially identical results: $D_T \cong 1.94 \times 10^{-7} cm^2/sec$, $D_R \simeq 50{,}000/sec$. These results suggest that fibrinogen and both forms of fibrin are elongated molecules about 800Å in length and 100Å in diameter.[31]

Next Wiltzius et al. investigated the early stages of fibrin aggregation initiated by low concentrations of reptilase. Light scattering measurements were begun 5 minutes after the addition of the enzyme. Both scattered intensity and PCS experiments were performed for ~60 minutes, with individual measurements taking 30 seconds each. The result, illustrated in Fig. 7, shows that the scatterered intensity I_s increases and the mean linewidth $\bar{\Gamma}$ decreases throughout the experiment, and that the time dependence of both $\bar{\Gamma}$ and I_s is linear during the early stages of aggregation. The mean

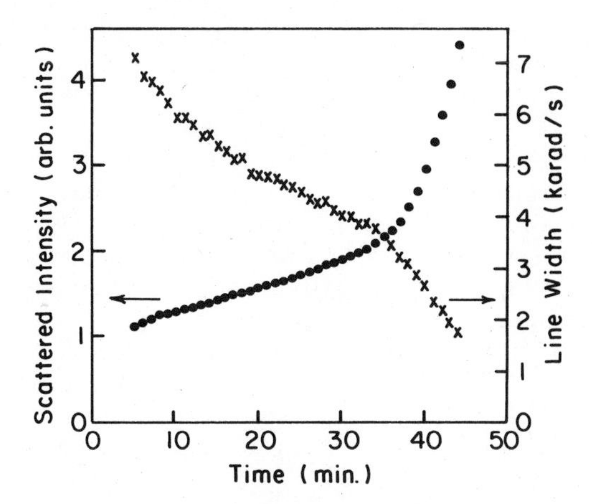

Figure 7. Time dependence of the scattered intensity and mean linewidth $\bar{\Gamma}$ of reptilase-treated human fibrinogen (from Wiltzius et al., Ref. 31).

linewidth $\bar{\Gamma} = \int \Gamma F(\Gamma) d\Gamma$, (which is the initial slope of $\ell n|g^{(1)}(\tau)|$ or equivalently the initial decay rate of $|g^{(1)}(\tau)|$) was found from a cumulant analysis of the data - see Eq. (22).

Wiltzius et al. noted that these results disagree with an earlier dynamic light scattering study of fibrin aggregation by Brass et al.[34] which had shown an initial induction period exhibiting no apparent fibrin aggregation followed by a period of rapid polymerization (we shall return to this point below).

The data of Fig. 7 were analyzed under the assumption that only monomers and dimers are present during the short-time linear regime. The analysis showed that end-to-end association was the most likely dimerization process rather than the staggered overlap association mentioned earlier, although the end-to-end dimers may be somewhat flexible.

Subsequently, measurements were extended to the later stages of fibrin aggregation prior to the sol-gel transition.[32,33] In this study, fibrinogen solutions of 2 mg/ml concentration were incubated with very small quantities of reptilase or thrombin (0.01 NIH units per ml). The evolution of the intensity $I_s(t)$ and initial decay rate (or mean linewidth) $\bar{\Gamma}(t)$ was so slow that the polymerization process could be followed for several hours. In analyzing the data, the earlier results, that the fibrin monomer is a rod about 750Å long and that the initial polymerization reaction is end-to-end dimerization, were explicitly included.

Fig. 8 shows the measured time dependence of $\bar{\Gamma}(t)$ for the reptilase-activated fibrinogen for nine different scattering angles. The initial data points were taken ~3 minues after mixing, and the last, after 6 hours, were taken before the occurrence of gelation. These results (which are very similar to those obtained with thrombin) indicate that $\bar{\Gamma}$ decreases to about ½ its initial value at the gelation point.

Wiltzius et al. analyzed the time evolution of $I_s(t)$ and $\bar{\Gamma}(t)$ using the Flory-Stockmayer distribution for the concentration n_i of the various i-imers: $n_i/\sum_i in_i = e^{-\lambda i}(\lambda i)^{i-1}/i.i!$ This is a one-parameter distribution with the parameter $\lambda = 0$ corresponding to the monomeric solution, and $\lambda = 1$ corresponding to the gel transition

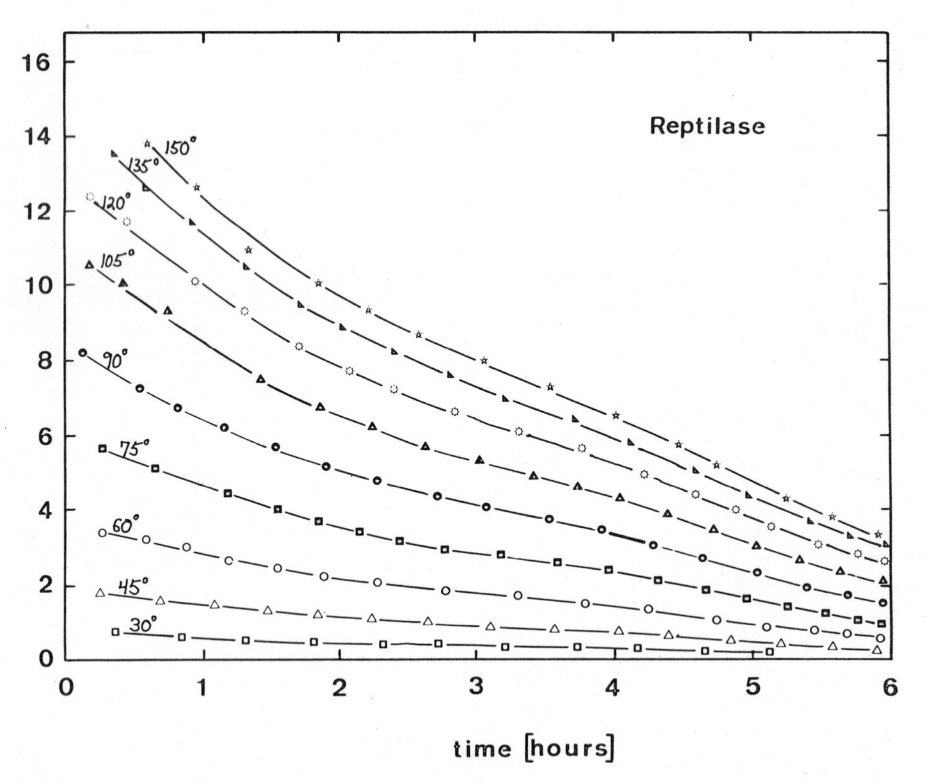

Figure 8. Time dependent $\bar{\Gamma}$ of reptilase-activated fibrinogen (from Wiltzius et al., Ref. 32).

(the experiments correspond to $0 \leq \lambda \leq 0.9$). Wiltzius et al. showed that the Flory-Stockmayer theory predicts that $\bar{\Gamma}$ should decrease to ½ its initial value at the sol-gel transition. Their intensity and linewidth data were in good agreement with the theory, and indicated end-to-end aggregation during the early stage of polymerization ($0 \leq \lambda \leq 0.3$) with staggered overlap association later ($0.3 \leq \lambda \leq 1$). Fig. 9 shows their plot of $I_s(t)/I_s(0)$ vs $\bar{\Gamma}(0)/\bar{\Gamma}(t)$ for both the reptilase and thrombin experiments with three theoretical curves. Up to a value of ~1.8 for both variables, the data shows excellent agreement with a calculation based on the Flory-Stockmayer distribution with end-to-end dimers and staggered overlapped i-mers ($i \geq 3$).

Additional studies of fibrin polymerization, including an investigation of the correlation of fibrinopeptide A release with changes in the static and dynamic light scattering data, have recently been completed by Wiltzius et al. These results - described in Ref. 32(b), were received too late for inclusion in this review.

Drs. H. L. Nossel, A. Hurlet-Jenson, C. Y. Liu and their associates in the Department of Medicine, Columbia University College of Physicians and Surgeons, have investigated the inhibition of thrombin induced fibrinogen polymerization by the synthetic tetrapeptide Gly-Pro-Arg-Pro. In additional to biochemical and optical absorption measurements performed at Columbia, this reaction was investigated by PCS in our laboratory at City College.[35,36]

The synthetic peptide binds to fibrinogen, blocking the polymerization process as monitored by the optical density, with no observable effect on FPA release and no initial effect on FPB release, although the subsequent acceleration of FPB release which is observed without the tetrapeptide is strongly inhibited by its presence (see Fig. 10). These results lend support to the hypothesis that the cleavage of FPB is enhanced by fibrin polymerization.

Light scattering experiments were performed at 37°C with a 60 channel digital correlator interfaced to a PDP-8E minicomputer. All experiments were performed at a scattering angle $\theta = 90°$. Preliminary runs on 1 mg/ml fibrinogen in 0.15M NaCl buffer, pH 7.4, filtered through a 0.45μ millipore filter gave $D_T = 2.39 \times 10^{-7}\ cm^2/sec$ from a single exponential fit, with polydispersity index $\mu_2/\bar{\Gamma}^2 \approx 0.22$ from a cumulant analysis. This D_T is 1.23 times larger than the D_T found by Wiltzius et al., which is somewhat less than the difference expected from the decrease in the viscosity of water between 20° and 37°C (1.44). The difference is probably a result of the different method of data analysis.

Thrombin was added to the filtered fibrinogen to a final concentration of 0.0033 units/ml, gently agitated, and inserted in the thermostat of the light scattering apparatus. Runs began three minutes after mixing and were repeated every 3.5 minutes for 2.5

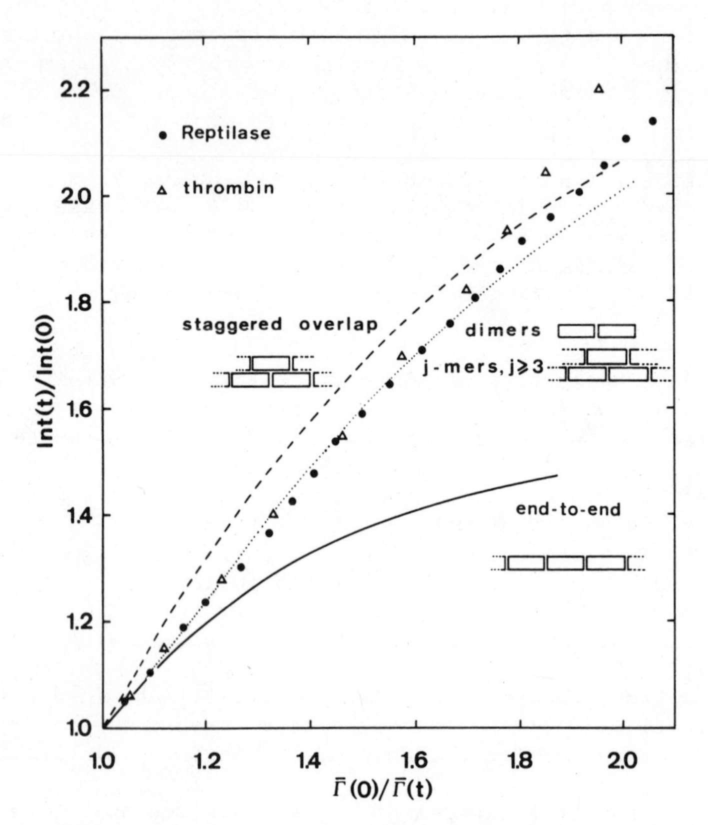

Figure 9. Normalized scattered intensity I(t)/I(0) vs normalized decay rate $\bar{\Gamma}(0)/\bar{\Gamma}(t)$ of fibrinogen activated with reptilase or thrombin. The solid line was calculated for end-to-end aggregation, the broken line for staggered overlap, and the dotted line for a Flory-Stockmayer distribution with end-to-end dimers and staggered overlapped i-mers ($i \geq 3$) (from Wiltzius et al. - Ref. 32).

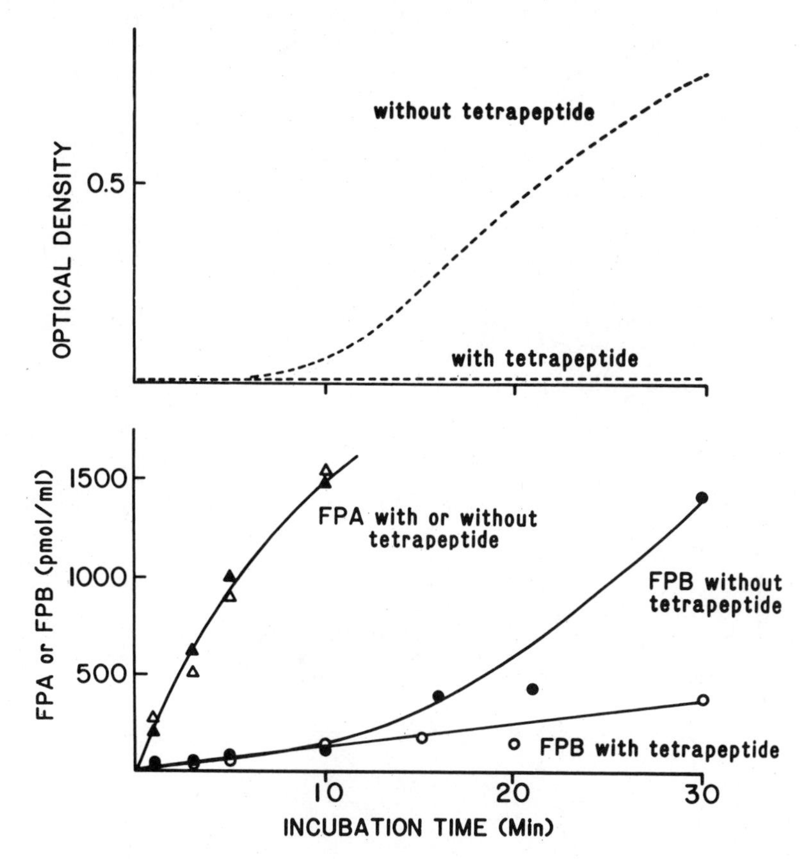

Figure 10. The effect of tetrapeptide Gly-Pro-Arg-Pro on fibrin polymerization (upper curve) and on FPA and FPB release from fibrinogen by thrombin (lower curve). Polymerization was monitored by measuring absorption at 3500Å (from Hurlet-Jensen et al., Ref. 36).

hours. Each run took 200 seconds. By the end of the experiment, the sample had formed a rather rigid gel. In Fig. 11 we show the scattered intensity and the equivalent hydrodynamic radius r_h extracted from the single exponential fit to $C(\tau)$ by use of Eq. (14), taking η as the viscosity of water at 20°C. This procedure is sufficient to show the relative change in r_h. It could easily be refined to give a more quantitatively significant value for the early stages of polymerization, although as the gel point is approached the equivalent hydrodynamic radius no longer describes a one particle property of the system. Note that both $I_s(t)$ and $r_h(t)$ exhibit a latency period of about 20 minutes preceding the period of rapid increase, and that in both the increase is about 10x. The large variations in both I_s

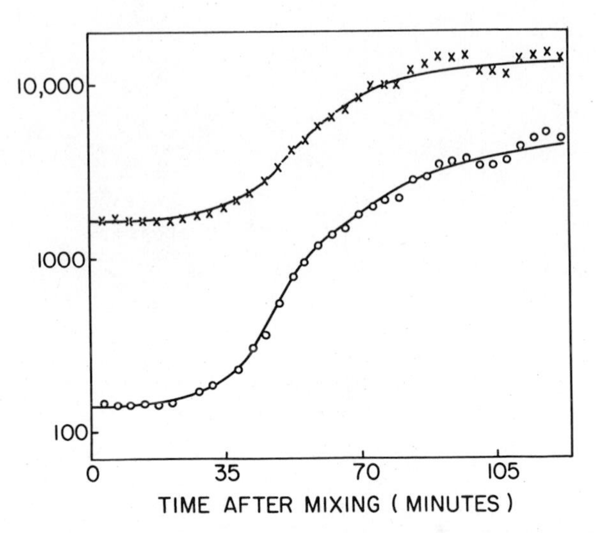

Figure 11. Time evolution of the scattered intensity (x) and equivalent hydrodynamic radius (o) of 1 mg/ml fibrinogen after addition of 0.0033 units/ml of thrombin (T = 37°C, $\theta = 90°$).

and r_h for times $\gtrsim$ 75 minutes are associated with increasing turbidity which precedes the onset of gelation.

The experiment was repeated with 1 mg/ml fibrinogen solution to which again 0.0033 units/ml of thrombin was added, but 1 mg/ml of the synthetic peptide Gly-Pro-Arg-Pro was added simultaneously. Similar 200 second runs were made, this time at 25 minute intervals, over a total span of 32 hours. These results for $I_s(t)$ and $r_h(t)$ are shown in Fig. 12. Comparison of Figs. 11 and 12 shows that the increase of $\bar{\Gamma}$ and I_s is delayed by the inhibitory peptide for several hours, but the time dependence is qualitatively similar on a time scale about 15 times slower. Eventually the second sample also forms a gel.

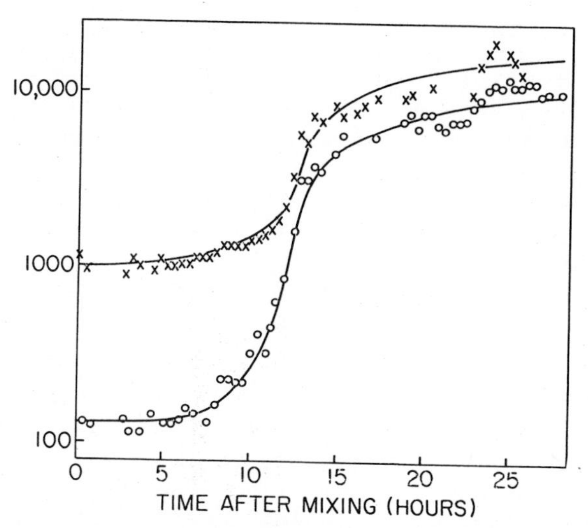

Figure 12. Time evolution of the scattered intensity (x) and equivalent hydrodynamic radius (o) of 1 mg/ml fibrinogen after addition of 0.0033 units/ml of thrombin and 1 mg/ml of the inhibitory peptide Gly-Pro-Arg-Pro.

The studies of fibrinogen discussed here illustrate the utility of PCS in the investigation of polymerization, gelation, and self assembly of proteins. Among the many other examples of biological self assembly studied by PCS are the muscle proteins actin and myosin which form the thin and thick filaments that interact to produce muscle contraction. This topic will be discussed in the lectures of Professor Carlson at this ASI as well as in a number of contributed talks.

5. MEMBRANES, VESICLES, MICELLES AND MICROEMULSIONS

In this section I will review a group of related problems which have been investigated by quasielastic light scattering spectroscopy in recent years. These problems can be loosely grouped under the concept of self-organization driven by the hydrophobic effect, and are all more or less related to the structure of biological membranes.

Biological membranes serve as structural barriers to protect the interior cytoplasm of a cell, and contain proteins which act as conduits or pumps that control the transport of specific ions into and out of the cell. Membranes also surround smaller entities within the cell which play special roles in the cell's metabolism. An important example of such subcellular entities is the mitochondrion which synthesizes adenosine triphosphate (ATP). ATP is the universal cellular energy source, providing bioenergy by virtue of its energy-rich phosphate anhydride bonds. The mitochondria assemble ATP through oxidative phosphorylation, a process which is associated with transport proteins embedded in the mitochondrial membrane.[37]

The current picture of the composite structure of biological membranes which has emerged over the last 50 years is illustrated schematically in Fig. 13.[38] As discussed in Dr. Steer's lectures, the membrane structure consists of a bilayer roughly 50Å thick of amphiphilic molecules called lipids (or phospholipids), a matrix of proteins on the interior side that provides rigidity, and various membrane-bound proteins embedded in the bilayer which mediate the selective ion transport processes.

Amphiphilic molecules, such as detergents or the phospholipids which form the membrane bilayer, consist of two essentially different parts: a water-soluble (hydrophilic) head which is usually (but not always) polar, and a water insoluble (hydrophobic) tail consisting of one (for detergents) or two (for phospholipids) hydrocarbon chains. In the lipid bilayer structure of biological membranes, the hydrocarbon tails form a disordered smectic state which is fluid like, while the polar headgroups form an organized two dimensional sheet. Embedded proteins can move through the bilayer as if it were a two-dimensional liquid. At low temperatures, there is a phase transition of the hydrocarbon chains to a paracrystalline state, in which transport and other functions of the membrane are sharply inhibited.

Lipid bilayers have been widely investigated as a model of biological membranes, a topic which will be discussed at this ASI in the lectures of Dr. Earnshaw. The aspect of amphiphilic molecules which I will consider in this section, is that of aggregation or self assembly. This topic was reviewed in The Hydrophobic Effect: Formation of Micelles and Biological Membranes in 1973 by Charles Tanford.[39]

A. Micelles

When amphiphilic molecules are dissolved in water, they can

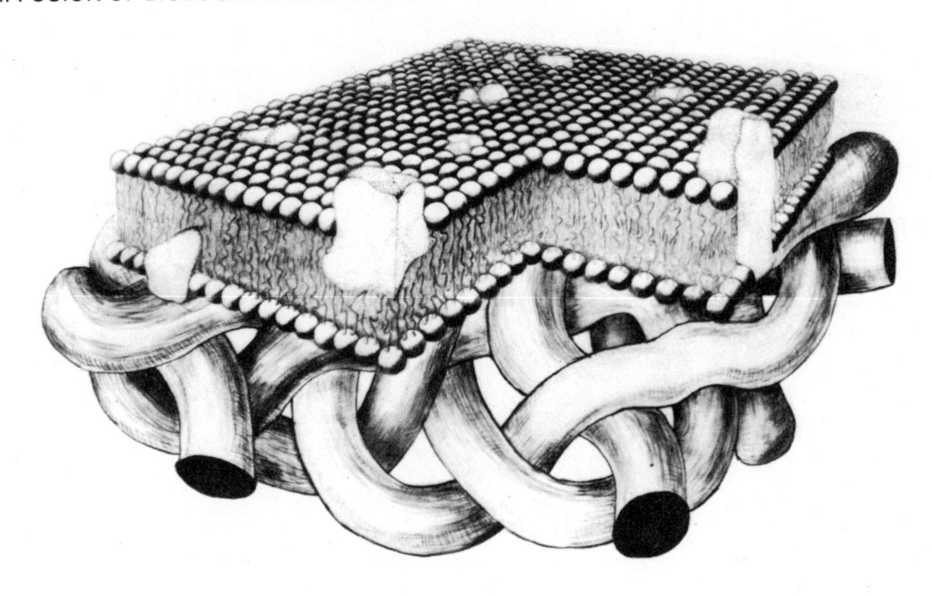

Figure 13. Schematic illustration of the composite structure of a biological membrane showing the lipid bilayer, the embedded proteins, and the random matrix on the cytoplasmic side from which the membrane obtains its solid properties (from Evans and Skalak, Ref. 38).

achieve segregation of their hydrophobic portions from the solvent by self-aggregation to form micelles (see Fig. 14). The molecules do not associate to form micelles until a critical micelle concentration (CMC) of C_0 is reached (the transition actually occurs in a narrow range of concentrations). For concentrations $C > C_0$, virtually all the additional amphiphile is in micelles. The CMC depends primarily on the chain length of the hydrocarbon tail, ranging from ~1 molar for derivatives with hexyl chains to less than 10^{-9} molar for biological phospholipids.[39] Micelle formation is thus a highly cooperative self-association process.

Since the original hypothesis by McBain that amphiphilic molecules in solution at concentration $C > C_0$ form aggregates with the hydrophilic headgroups on the outside and fluid-like interiors composed of the hydrocarbon tails, the structure of micelles has been the subject of intense study and of equally intense controversy.[40] It is generally accepted that just above C_0, about 30 to 150 amphiphilic molecules aggregate to form essentially spherical micelles. With increasing concentration (or decreasing temperature), the micelles enlarge rapidly and ultimately create liquid crystal phases.[40]

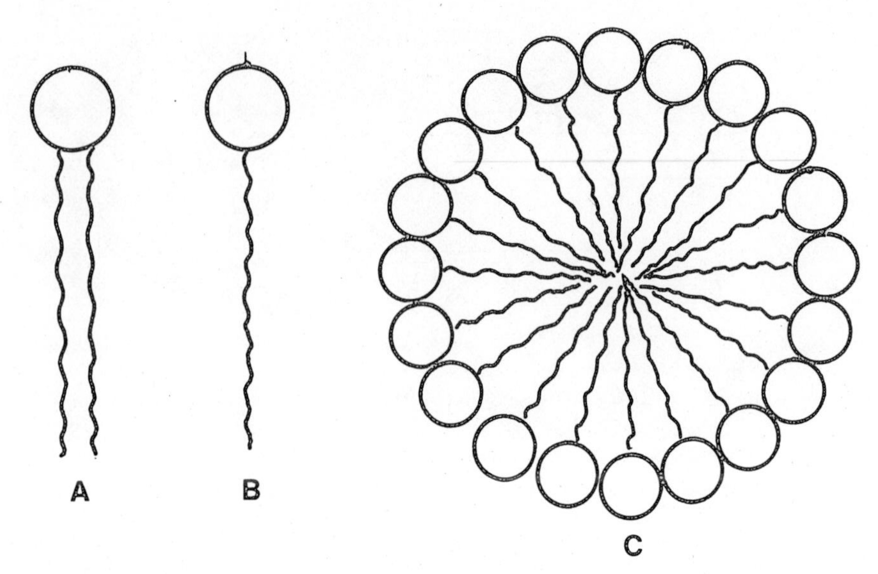

Figure 14. Schematic version of amphiphilic molecules: (a) phospho lipid, (b) detergent, (c) spherical micelle of detergent molecules.

The size and shape of micelles are determined by a combination of geometric and thermodynamic factors. One factor is that one dimension of the micelle core is restricted; it cannot exceed the length of two hydrocarbon chains approaching from opposite sides. According to Tanford, the limit calculated on this basis is ~24Å for dodecyl chains (as in the surfactant SDS) and ~30Å for the chain in the range C_{16} to C_{18} normally found in phospholipids. A spherica micelle of this size of amphiphilic molecules with dodecyl chains could accommodate no more than 20 monomers, which is much less than the aggregation number frequently observed. Since larger spheres are prevented by the fixed dimension constraint, growth must imply a change of shape from spherical to ellipsoidal, either rodlike or disclike. Tanford concluded that the disclike structure is thermodynamically more stable, and that the available measurements of hydrodynamic properties of micelles support the disclike geometry.[39]

In an early study of quasielastic light scattering by micelles, McQueen and Hermans[41] investigated solutions of sodium lauryl sulfat and decyltrimethyl ammonium bromide in water. Their data (obtained with an analog spectrum analyzer) showed a systematic decrease of D_T from 9.5×10^{-7} to $6.8 \times 10^{-7} cm^2/sec$ as C was increased from 5 to 30 mg/ml, demonstrating an apparent increase in micelle size with increasing concentration. More recently, there have been a number of detailed investigations of micelles by PCS which have sought to resolve the problem of micellar shape, and to elucidate the self assembly process. One of the most serious complications which has been explored recently is the effect of micellar interactions on the apparent diffusion constants.[42] This is discussed in the lectures of Dr. Degiorgio.

In an extensive series of publications, Mazer, Benedek and Carey and their coworkers at MIT have reported investigations of quasi-elastic light scattering from aqueous solutions of sodium dodecyl sulfate (SDS).[43-47] In their initial study, correlation data were fit to second order cumulant expansions to yield the mean diffusion constant $\bar{D}_T$ and mean hydrodynamic radius $\bar{r}_h$ of the micelles.[43] SDS concentration, salt concentration, and temperature were all varied. At high temperatures (~60°C), the data for all salt concentrations indicated spherical micelles with a hydrated radius of ~25Å. At lower temperatures, $\bar{r}_h$ increased rapidly, the increase being greatest for the highest salt concentration (0.6M NaCl), reaching 167Å at 18°C. The measured intensity was plotted against $\bar{r}_h$ and compared with theoretical predictions for spheres, prolate ellipsoids (rods) and oblate ellipsoids (discs). The data showed far closer agreement with the rodlike model predictions than with the others (Fig. 15a) . The authors concluded that the SDS micelle is a prolate ellipsoid with a semiminor axis of 25Å and a semimajor axis that approaches 675Å in 0.6M NaCl.

Subsequently, this model was further refined to describe an elongated cylinder with hemispherical caps as illustrated in Fig. 15b, which had been proposed earlier by Debye and Anacker.[44] A thermodynamic theory for micelle formation was developed which was shown to give excellent fits to the data. The theory was subsequently elaborated,[45] and its conclusions were further supported by an angular dissymetry study.[46] Finally, the effect of the hydrocarbon (alkyl) chain length on the sphere-to-rod transition was investigated and again the results were found to agree well with the theory which contains a single parameter K, a measure of the difference in chemical potential between the spherical and cylindrical portions of the micelle.[47]

A similar analysis favoring rod-shaped micelles for dodecyldimethyl ammonium chloride was reported by Flamberg and Pecora,[48] while Nicoli et al. have extended the analysis of Mazer et al. and investigated the effects of hydrostatic pressure on micelle aggregation.[49]

In all of these studies, the analysis followed the procedure outlined in section 3 which is appropriate to non-interacting scatterers. Mazer et al. noted that there could be electrostatic interactions between micelles because of the charge carried by the polar

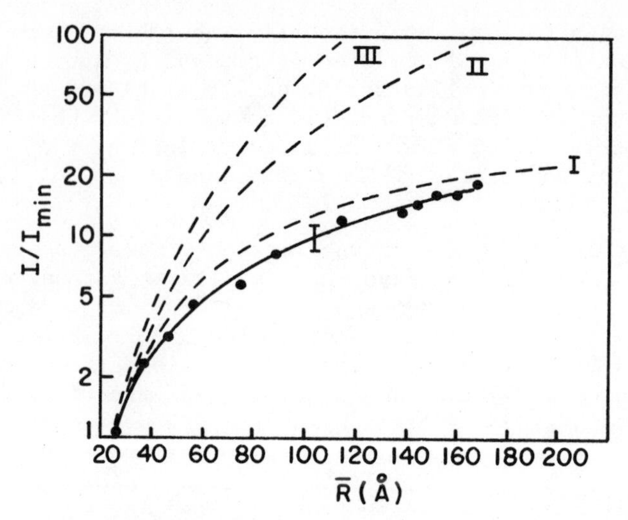

Figure 15a. Scattered intensity I vs mean hydrodynamic radius r_h of 6.9 x 10^{-2}M SDS in 0.6M NaCl. The points represent different temperatures in the range ~18 to 60°C. The dashed curves represent model calculations for: I-prolate ellipsoids; II-oblate ellipsoids; III-spheres (from Mazer, Benedek and Carey, Ref. 43).

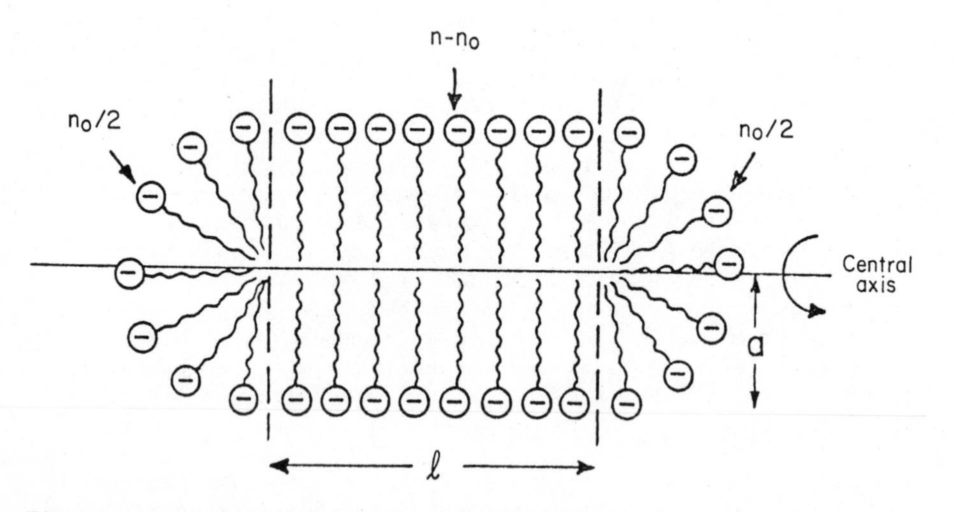

Figure 15b. The Mazer, Carey and Benedek model of SDS micelles. The total aggregation number is n (from Missel et al., Ref. 47).

headgroups, but concluded that the salt concentrations used in their experiments were high enough to effectively screen any such interactions.

M. Corti and V. Degiorgio have also pursued a program of investigation of micelles by quasielastic light scattering.[50-54] As one part of their investigation they have explored the consequences of inter-micelle interactions on the intensity and correlation function of the scattered light[42] as discussed in Dr. Degiorgio's lectures. They have also investigated the critical behavior of nonionic amphiphile micelle solutions which exhibit phase separation at a lower consulate point.[53]

Corti and Degiorgio also investigated the SDS system studied by Mazer et al., but at lower salt and SDS concentrations (C_o for SDS is ~0.45 mg/ml). Their results[52] favor the oblate micelle geometry in agreement with the conclusion of Tanford.[39] Their results, together with those of Mazer et al., suggest that at concentrations close to C_o the initial distortion of the spherical micelle may be oblate, while with increasing amphiphile concentration the shape becomes prolate instead. Such an interpretation appears to imply another phase transition at which the disc-like micelle would reorganize itself into a rod-shaped structure.

B. Microemulsions

Microemulsions are much like micelles except that the amphiphile molecules surround droplets of oil dispersed in water (oil in water microemulsion) or, with the amphiphile heads pointing inward, droplets of water dispersed in oil (water in oil microemulsion). The droplets, sometimes called "swollen micelles", are typically 100Å to 1000Å in diameter. Although the droplets are not permanent and microemulsions are not thermodynamically equilibrated, they are in a state of metastable equilibrium which may be stationary essentially indefinitely. Microemulsions are transparent with low viscosity, and have many practical applications in the food and pharmaceutical industries as well as in tertiary oil recovery. They have been studied extensively since their discovery some forty years ago by Shulman and coworkers. Transparent microemulsions can be prepared with the volume fraction of the droplets as high as 70%.

During the last few years, several groups have applied PCS to various microemulsions. In a particularly interesting series of experiments, Cazabat and Langevin and their coworkers investigated quaternary mixtures of water, SDS, oil (cyclohexane or toluene) and alcohol (butanol, pentanol or hexanol) as cosurfactant.[55-59] The water also contained salt whose concentration provides an additional degree of freedom. Fig. 16 shows the diffusion constant D_T for two oil in water microemulsions as a function of the volume fraction of

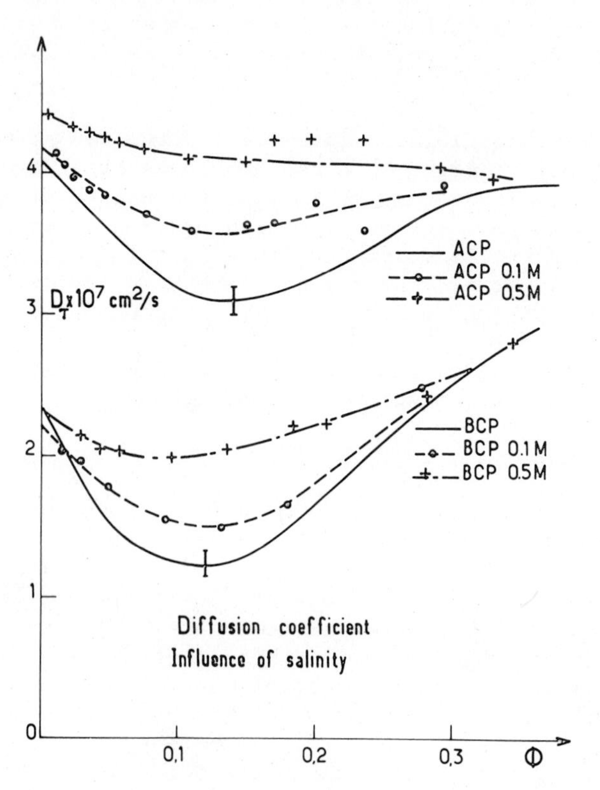

Figure 16. Diffusion coefficients D vs droplet volume fraction Φ of cyclohexane in water microemulsions with SDS surfactant and pentanol cosurfactant (from Cazabat and Langevin, Ref. 55).

the droplets Φ.[55] They are both cyclohexane in water microemulsions with pentanol as the cosurfactant, and a water-to-soap ratio of 0.5 cm^3 (ACP) and 1.0 cm^3 (BCP). The salt concentrations are 0, 0.1M and 0.5M. These experiments have also provided considerable insight into the interactions between the droplets in a microemulsion.

The saltwater-toluene-SDS-butanol system also exhibits a dramatic series of phase transitions as the salt concentration S is varied. For a 47.95% saltwater, 46.35% toluene, 1.95% SDS, 3.75% butanol mixture, the following phases are observed for salinity S between 4 and 10%.

$4 \le S < 5.8\%$: the lower phase is an oil in water microemulsion, the upper phase is a continuous oil rich phase.
$5.8 < S < 7.8$ there is a microemulsion in the center with an oil rich continuous phase on the top and a continuous aqueous phase on the bottom.
$7.8 < S \le 10$ there is a water in oil microemulsion on the top and a continuous aqueous phase on the bottom..

Langevin and coworkers have also examined the interfaces separating these phases by quasielastic light scattering from the interfacial capillary waves or "ripplons".[58,59] They find an extremely small interfacial tension γ at the interfaces near the limit of stability signaling the onset of each phase transition, as shown in Fig. 17.

C. Bilayers and Vesicles

Aqueous solutions of amphiphilic molecules which contain one hydrocarbon chain per headgroup form micelles when their concentration exceeds the critical micelle concentration C_0. Phospholipid molecules which contain two hydrocarbon chains per headgroup tend to aggregate in bilayers which typically form spherical closed vesicles with an internal water filled cavity.[39] Fig. 18 illustrates the chemical structure of a particular phospholipid amphiphile (DPL) and a schematic representation of a vesicle.[60] Typical vesicles formed by sonication of aqueous solutions of phospholipids (e.g. egg yolk lecithin) have outer diameters on the order of 250Å and bilayer thickness on the order of 50Å. The dimensions depend on the particular phospholipid as well as on the method of preparation.

As Tanford points out, the formation of vesicles in response to hydrophobic forces represents the first essential step in biological organization and is, in effect, a beginning in the definition of a living cell.[39] Beyond this fundamental biological importance of phospholipid vesicles, there is also practical interest in their use as targetted drug carriers able to deliver medication to specific sites. Phospholipd vesicles therefore present an attractive subject for investigation by quasielastic light scattering.

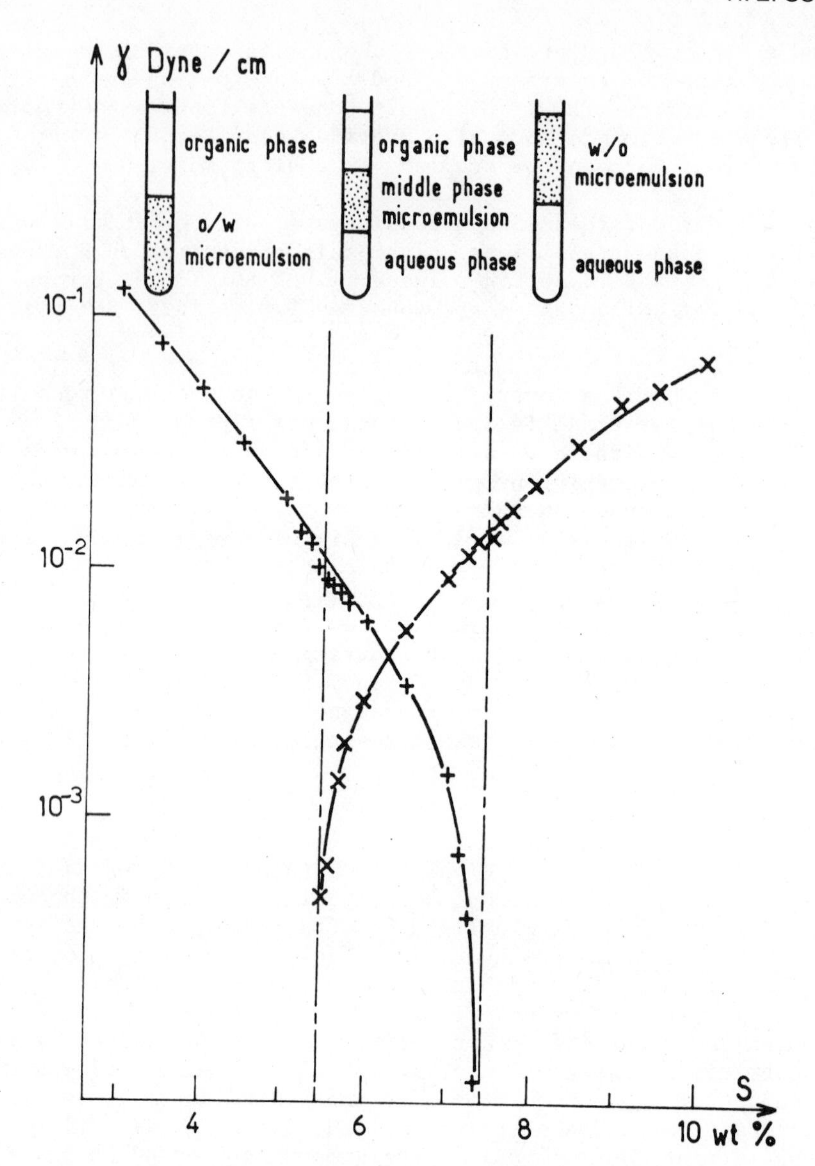

Figure 17. Interfacial tension γ of the saltwater-toluene-SDS-butanol system as a function of salinity S in weight percent (from D. Chatenay et al., Ref. 59).

Ostrowsky and Sornette have applied PCS to study the formation, size and stability of phospholipid vesicles, and I will briefly review some of their results here.[24,60] Sornette and Ostrowsky prepared vesicle solutions by sonication of dimyristoyl phophatidylcholine (DPL) in aqueous buffer solution at 40°C, followed by centrifugation. The phase transition temperature T_{Φ} for DPL from the

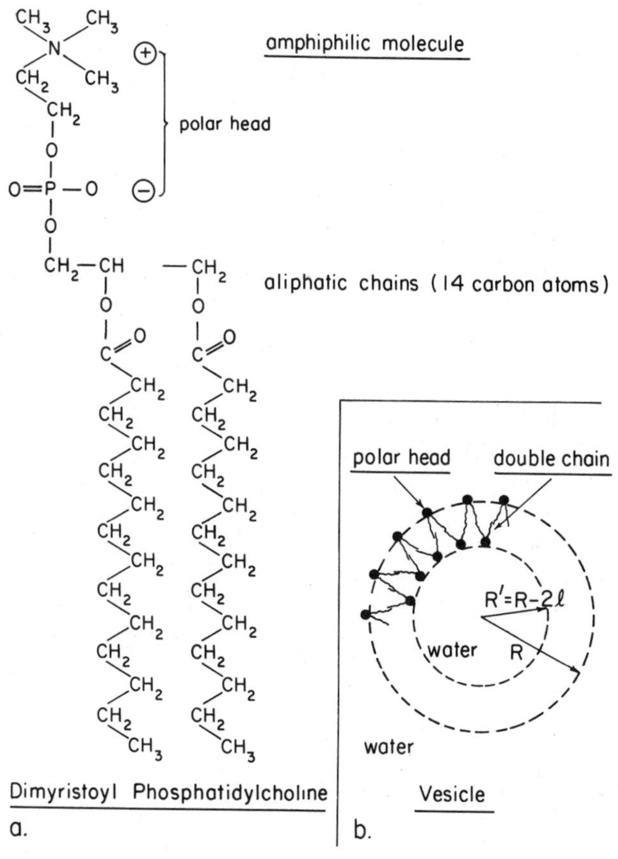

Figure 18 (a) Chemical structure of the phospholipid molecule DPL. (b) Schematic representation of a phospholipid vesicle (from Sornette and Ostrowsky, Ref. 60).

fluid-like to paracrystalline state occurs at ~23°C for planar bilayers and moves down to ~18°C for highly curved bilayers. Measurements at 24°C showed that most of the vesicles have radii r_h ~110Å with a Gaussian size distribution (see Eq. 21C). The vesicles were found not to be stable, however, and the size distribution broadened and moved towards larger mean r_h with time, changing more rapidly for temperatures close to T_Φ.[24]

Subsequently, Sornette and Ostrowsky analyzed the time evolution of the size distribution for two samples, one for T = 19°C ($T < T_\Phi$) and one for T = 32°C ($T > T_\Phi$).[60] In this analysis, $F(\Gamma)$ in Eq. 20 was not modelled by a Gaussian, but was reconstructed from a set of delta functions by a numerical inversion procedure discussed at this Institute in the lectures of Professor Chu. Their results are shown in Fig. 19.

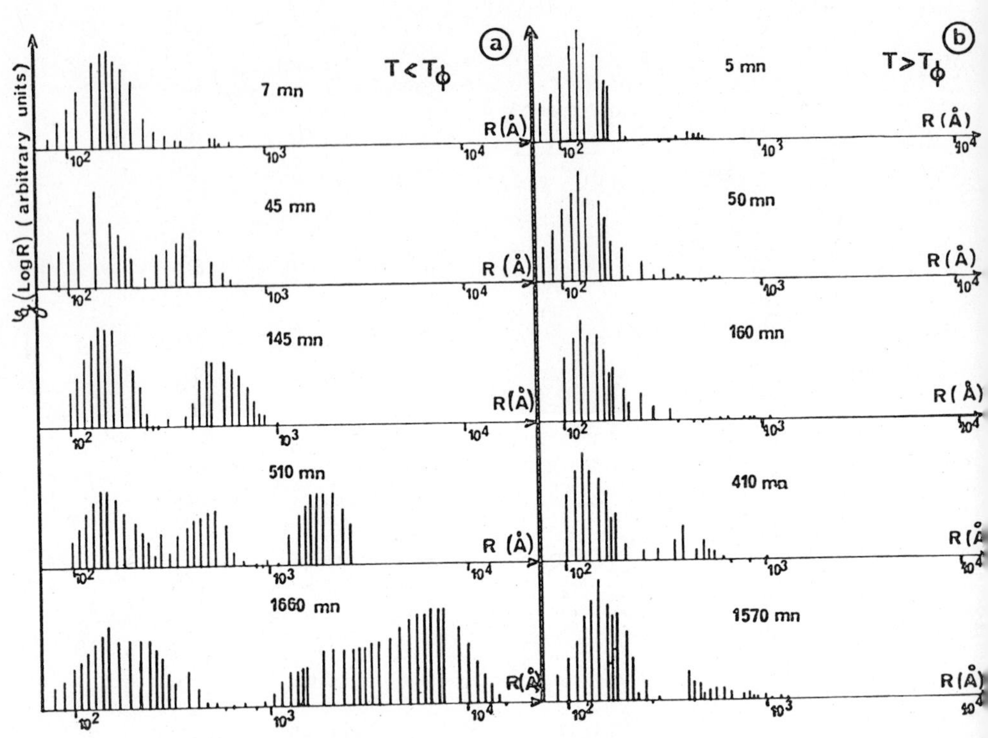

Figure 19. Time evolution of the size distribution of DPL vesicles (a) at T = 19°C ($T < T_\Phi$), and (b) at T = 32°C ($T > T_\Phi$) (from Sornette and Ostrowsky, Ref. 60).

The size distribution at 32°C is quite stable for about one day with some appearance of larger vesicles. At 19°C, however, there is a dramatic change in the size distribution during one day. The authors compare this time dependence to predictions based on two mechanisms, aggregation and fusion. They conclude that an aggregation model cannot explain the data and that the appearance of larger vesicles with time is due to a fusion process triggered by the phase transition of the hydrocarbon tails.

CONCLUSION

The technique of PCS has been applied with great success to the determination of diffusion coefficients of independent scatterers in solution. PCS experiments have also been extended to more concentrated solutions in which particle interactions become important. The problems discussed in this review represent extremé forms of particle interaction leading to polymerization or self assembly. The great biological importance of these processes suggest that such experiments will become a major area of application of laser light scattering in biology.

REFERENCES

1. C. Tanford, Physical Chemistry of Macromolecules, Wiley, New York (1961)
2. R. Pecora, Ph. D. thesis, Columbia University, 1962; J. Chem. Phys. 40, 1604 (1964).
3. H. Z. Cummins, N. Knable and Y. Yeh, Phys. Rev. Lett. 12, 150 (1964).
4. S. B. Dubin, J. H. Lunacek and G. B. Benedek, in Proceedings, Nat. Acad. Sci. U. S., 57, 1164 (1967).
5. H. Z. Cummins, F. D. Carlson, T. J. Herbert and G. Woods, Biophys. J. 9, 518 (1969).
6. H. Z. Cummins, in Photon Correlation and Light Beating Spectroscopy, edited by H. Z. Cummins and E. R. Pike, Plenum, New York (1973) p. 312.
7. P. Pusey, in Photon Correlation Spectroscopy and Velocimetry, edited by H. Z. Cummins and E. R. Pike, Plenum, New York (1976) p. 187.
8. C. H. Townes, in Advances in Quantum Electronics, edited by J. R. Singer, Columbia Univ. Press, New York (1961) p. 8.
9. M. A. Righi, J. Phys (Paris), 2, 437 (1883).
10. O. Righi (private communication).
11. A. T. Forrester, R. A. Gudmundsen and P. O. Johnson, Phys. Rev. 99, 1691 (1955).
12. c.f., H. Z. Cummins and N. Knable, in Proceedings IEEE, 51, 1246 (1963).
13. A. T. Forrester, J. Opt. Soc. Am. 51, 253 (1961).
14. c.f., H. Z. Cummins and H. L. Swinney, in Progress in Optics, Vol. VIII, edited by E. Wolf, North Holland, Amsterdam (1970) p. 133.
15. Photon Correlation and Light Beating Spectroscopy, edited by H. Z. Cummins and E. R. Pike, Plenum, New York (1973).
16. Photon Correlation Spectroscopy and Velocimetry, edited by H. Z. Cummins and E. R. Pike, Plenum, New York (1976).
17. Scattering Techniques Applied to Supramolecular and Nonequilibrium Systems, edited by S. H. Chen, B. Chu and R. Nossal, Plenum, New York (1981).
18. E. R. Pike, Riv. Nuovo Cimento 1, 227 (1969).
19. B. Chu, Laser Light Scattering, Academic Press, New York (1974).
20. B. J. Berne and R. Pecora, Dynamic Light Scattering, Wiley, New York (1976).
21. J. Newman, H. L. Swinney, H. Z. Cummins, S. A. Berkowitz and L. Day, Bull. Am. Phys. Soc. 18, 671 (1973).
22. J. G. McWhirter and E. R. Pike, J. Phys. A11, 1729 (1978); J. G. McWhirter, Opt. Acta. 27, 83 (1980).

23. N. Ostrowsky, D. Sornette, P. Parker and E. R. Pike, Opt. Acta, 28, 1059 (1981).
24. N. Ostrowsky and D. Sornette, in Light Scattering in Liquids and Macromolecular Solutions, edited by V. Degiorgio, M. Corti and M. Giglio, Plenum, New York (1980) p. 125.
25. D. E. Koppel. J. Chem. Phys. 57, 4814 (1972).
26. G. B. Stock, Biophys. J. 16, 535 (1976).
27. M. Muehller and W. Burchard, Biochim. Biophys. Acta 537, 203 (1978).
28. P. Wiltzius and V. Hofmann, Thrombosis Research 19, 793 (1980).
29. B. Blomback and T. C. Laurent, Arkiv fur Kemi 12, 137 (1958).
30. R. R. Hantgan and J. Hermans, J. Biol. Chem. 254, 11272 (1979).
31. P. Wiltzius, W. Känzig, V. Hofmann and P. W. Straub, Biopolymers 20, 2035 (1981).
32. (a) P. Wiltzius, G. Dietler, W. Känzig, V. Hofmann, A. Häberli and P. W. Straub, Biophys. J. 38, 123 (1982);
(b) P. Wiltzius, G. Dietler, W. Känzig, A. Häberli and P. W. Straub, Biopolymers (in press).
33. P. Wiltzius, An Investigation of the Fibrinogen to Fibrin Transition by Means of Light Scattering, dissertation, Swiss Federal Inst. of Technology ETH, Zurich (1981).
34. E. P. Brass, W. B. Forman, R. V. Edwards and O. Lindau, Thrombosis and Haemostasis, 36, 37 (1976).
35. A. Hurlet-Jensen, H. Z. Cummins, H. L. Nossel and C. Y. Liu, Thrombosis and Haemostasis, 46, 182 (1981).
36. A. Hurlet-Jensen, H. Z. Cummins, H. L. Nossel and C. Y. Liu, Thrombosis Research (in press).
37. F. M. Harold, Science 202, 1174 (1978).
38. E. A. Evans and R. M. Hochmuth, J. Membr. Biol. 30, 351 (1977). E. A. Evans and R. Skalak, Mechanics and Thermodynamics of Biomembranes, C.R.C. Press, Boca Raton, Fla. (1980).
39. C. Tanford, The Hydrophobic Effect: Formation of Micelles and Biological Membranes, Wiley, New York (1973); Science 200, 1012 (1978).
40. F. M. Menger, Accounts of Chemical Research, 12, 111 (1979).
41. D. H. McQueen and J. J. Hermans, J. Colloid, Interface Sci. 39, 389 (1972).
42. M. Corti and V. Degiorgio, (Ref. 24, p. 111).
43. N. A. Mazer, G. B. Benedek and M. C. Carey, J. Phys. Chem. 80, 1075 (1976).
44. N. Z. Mazer, M. C. Carey and G. B. Benedek, in Micellization, Solubilization and Microemulsions, Vol. 1, edited by K. L. Mittal, Plenum, New York (1977) p. 359.
45. P. J. Missel, N. A. Mazer, G. B. Benedek, C. Y. Young and M. C. Carey, J. Phys. Chem. 84, 1044 (1980).
46. C. Y. Young, P. J. Missel, N. A. Mazer, G. B. Benedek and M. C. Carey, J. Phys. Chem. 82, 1375 (1978).
47. P. J. Missel, N. A. Mazer, G. B. Benedek and M. C. Carey, in Solution Behavior of Surfactants, edited by K. L. Mittal and E. J. Fendler, Plen, New York (1981).

48. A. Flamberg and R. Pecora (Ref. 17 - p. 803).
49. D. F. Nicoli, R. Ciccolello, J. Briggs, D. R. Dawson, H. W. Offen, L. Romsted and C. A. Bunton (Ref. 17 - p. 363).
50. M. Corti and V. Degiorgio, Opt. Cummun. 14, 358 (1975).
51. M. Corti and V. Degiorgio, Ann. Phys. 3, 303 (1978).
52. M. Corti and V. Degiorgio, Chem. Phys. Lett. 53, 237 (1978).
53. M. Corti and V. Degiorgio, Phys. Rev. Lett. 45, 1045 (1980).
54. M. Corti and V. Degiorgio, Ref. 17, p. 337).
55. A. M. Cazabat and D. Langevin, J. Chem. Phys. 74, 3148 (1981).
56. A. M. Cazabat and D. Langevin, (Ref. 24, p. 139).
57. A. M. Cazabat, D. Chatenay, D. Langevin and A. Pouchelon, (Ref. 17, p. 787).
58. A. Pouchelon, J. Meunier, D. Langevin and A. M. Cazabat, J. Phys. Lett. 41, 239 (1980).
59. D. Chatenay, D. Langevin, J. Meunier and A. Pouchelon, (Ref. 17, p. 795).
60. D. Sornette and N. Ostrowsky (Ref. 17, p. 351; p. 755).

THE DIFFUSION OF COMPACT MACROMOLECULES THROUGH BIOLOGICAL GELS

D. B. Sellen

Astbury Department of Biophysics
The University, Leeds LS2 9JT, U.K.

CONTENTS

ABSTRACT

Experimental procedures for determining diffusion coefficients of compact macromolecules within biological gels and partition coefficients between gels and surrounding solutions are described. Results are presented for dextran fractions diffusing within calcium alginate and agarose gels. A gel concentration - hydrodynamic diameter superposition principle is shown to apply, and a method for calculating the molecular weight per unit length of the fibrous gel structure from the results is discussed, together with the significance of the measured partition coefficients.

1. INTRODUCTON

In recent years measurement of the Rayleigh linewidth of scattered light has been extended to include the light scattered from gels[1-13]. In the case of gels of cross-linked flexible polymers, it has been shown that the width of the broadened component varies as $\sin^2\theta/2$. Thus a diffusion coefficient may be calculated in the same way as for a dilute macromolecular solution. According to the theory of Tanaka et al[1], this corresponds to freely diffusing fluctuations in polymer segment density, and is equal to the ratio of the longitudinal elastic modulus to the force per unit volume required to maintain unit relative velocity between the polymer network and the solvent. In most cases the degree of spectral broadening has been found to be small, most of the scattered light arising from long range stationary spatial fluctuations in polymer segment density, so that effectively heterodyne spectra have been investigated. (This does not appear to be the case for swollen polymeric networks at theta conditions[9]). Diffusion coefficients have been found to be of the same order as those of diffusing macromolecules, with relaxation times in the autocorrelation function of intensity fluctuations of the order of hundreds of microseconds for light scattered at 90°.

Gels formed from stiff polysaccharide chains are, however, thought to have an entirely different structural morphology from those of cross linked flexible polymers. Electron microscopy of beaded agarose for instance[14], indicates a randomly branched filamentous structure in which each filament is thought to consist of an association of up to hundreds of agarose chains. A similar structure has been observed with calcium alginate gels but with much thinner filaments[15]. The degree of spectral broadening of light scattered from calcium alginate[10,11] and agarose[13] gels, like that from most gels of cross linked flexible polymers, is small, but the autocorrelation functions are non-exponential and the relaxation times are much longer - of the order of tens of milliseconds. Also the width of the broadened component does not vary as $\sin^2\theta/2$, much less variation with angle being observed. It has been tentatively suggested[10,11] that the broadened component is associated with a slight instability in the filamentous structure, transient changes in scattered amplitude occurring as chains become dissociated from, or re-associated with, the filamentous structure which is otherwise thought to be stationary, at least in the short term.

The small degree of spectral broadening and the long relaxation times associated with light scattered from polysaccharide gels makes possible the measurement of diffusion coefficients of compact macromolecules within the gels by investigating the heterodyne beat spectra. In this way the sizes of interstitial spaces within the gels may be investigated. In experiments of this kind the autocorrelation function consists essentially of two components resulting from heterodyne spectra, with the unbroadened component of the light scattered from the gel as local oscillator. One corresponds to the diffusing macromolecule and the other to the broadened component of the light scattered from the gel. The former consists, unless the diffusing macromolecule is highly polydisperse, of the single exponential $\exp(-K^2D\tau)$, where $K = 4\pi\sin(\theta/2)/\lambda$, and may be separated from the latter, which usually has a longer relaxation time, by suitably adjusting the concentration of the diffusing macromolecule. To date, experiments have been carried out with calcium alginate[10,11] and agarose[13] gels using dextran fractions as diffusing macromolecules.

2. EXPERIMENTAL

For the following reasons it is advantageous to make measurements with each gel cast as a cylinder, supported with its axis vertical and surrounded by a solution of the diffusing macromolecule.

(i) The gel is observed under conditions of equilibrium swelling.
(ii) The gel may be reduced in diameter sufficient to avoid secondary scatter without introducing troublesome reflections of the incident laser light from its surface.
(iii) Measurements may be made on gels into which the dextran has been allowed to diffuse after the gel has been formed, as well as on gels where the dextran has been introduced before gelation.
(iv) Partition coefficients of the diffusing dextran between the gel and the surrounding solution may be determined from the zero time value of the autocorrelation function.
Cylinders of gel may be formed either in glass tubing in cases where gelation takes place on cooling a hot solution, e.g. agarose, or in dialysis tubing where gelation takes place by ion exchange, e.g. calcium alginate. In general gels so formed are very slightly smaller than the original volume of solution, thus facilitating the easy removal of the cylinders of gel. In view of the time taken to carry out experiments on gels, it is desirable to add a small quantity of bacteriocide (0.02% sodium azide). Solutions should be cleaned by filtering through millipore filters before gelation.

At low concentrations where a true gel does not form (e.g. below 0.3% in the case of agarose of molecular weight 10^6), the gelatinous material has to be allowed to fill the light scattering cell. In passing it should be noted that no abrupt change in either the turbidity or the nature of the spectral broadening occurs at the concentration at which a gel first appears to form, indicating that there is no structural change as such, but that at low concentrations the structure is not coherent throughout the specimen.

The minimum concentration of diffusing dextran necessary is determined by the degree of spectral broadening of the light scattered from the gel itself, and relative decay times of the two components of the autocorrelation function. In the case of a gel such as calcium alginate, where dextrans of molecular weight in the range 10,000 to 500,000 were used, the component due to the diffusing dextran effectively dies away before the decay of the component due to the gel itself starts, so that the two components are easily separated. This is not the case for gels with larger interstitial spaces such as agarose, where dextrans of molecular weight in the range 500,000 to 8,000,000 were used. Here the concentration has to be such that the component of the autocorrelation function due to the diffusing macromolecule is at least an order of magnitude greater than that due to the gel. At the same time, in order to observe a heterodyne spectrum, the scattered intensity due to the diffusing dextran must be much less than the total scattered intensity. Fortunately the degree of spectral broadening of light scattered from agarose gel is very small, falling at 90° scatter, from 1% to 0.01% as the agarose concentration is increased from 0.1% to 4%.

It is essential that the diameter of each gel specimen be small enough to avoid secondary scatter. In the work with agarose, gel diameters as small as 2mm had to be used at the higher agarose concentrations. One method for checking for the absence of secondary scatter is to measure the scattered intensity of incoherent light as a function of angle of scatter at two wavelengths (this is most conveniently done in a conventional light scattering apparatus using mercury green and blue light), and to plot $\lambda^4 R_\theta$ against $\sin^2\theta/2)/\lambda^2$. The plots should be the same for both wavelengths. Another method is to scan the stationary random "speckle" pattern associated with the largely unbroadened laser light scattered from the gel over a few degrees of angle of scatter. The ratio of the variance to the mean value of these fluctuations in intensity should be $\gamma/\sqrt{2}$, where γ is the coherence factor which may be found from laser light scattering measurements on a solution using as nearly as possible the same geometrical optics. The effect of secondary scatter is to reduce these stationary fluctuations as light is then received from outside the coherence area.

Standard equipment may be used to obtain autocorrelation functions. However, the mounting of the apparatus is important. If the gel is set in oscillation due to external vibrations, the otherwise stationary random speckle pattern changes, and oscillations, usually at the frequency of oscillation of the gel, appear in the autocorrelation function. This phenomenon has been used to investigate the mechanical properties of gels[16], but in the type of work being discussed here it is important to avoid these oscillations. This may be achieved most effectively by mounting the apparatus on a heavy base which is then placed on an air bed.

3. DIFFUSION COEFFICIENTS WITHIN GELS

Fig. 1 shows diffusion coefficients of dextran within agarose gel, D, relative to corresponding diffusion coefficients in water, D_o, as a function of hydrodynamic diameter, d, and agarose concentration, C. d is calculated from D_o using the Stokes-Einstein equation. D/D_o has been plotted against log d for each agarose concentration and the plots displaced horizontally relative to the data at 0.7% concentration, (C_o), to yield a single "master" plot.

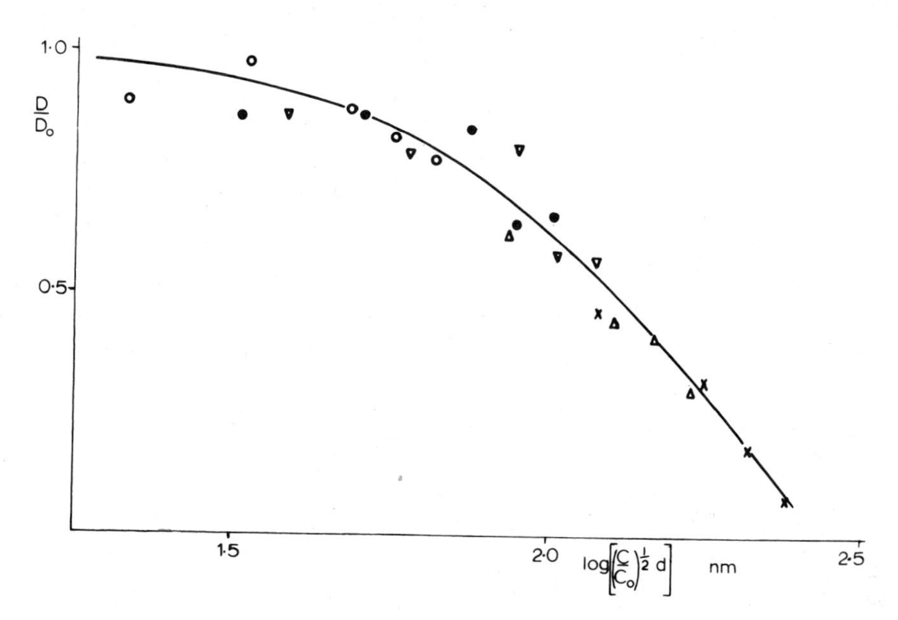

Fig. 1. Diffusion coefficient, D, relative to that in dilute solution, D_o, of dextran fractions within agarose gels of various concentration, C, as a function of dextran hydrodynamic diameter, d. The plots at the various concentrations have been superimposed by displacing horizontally, (C_o = 0.7%, see text). The dextrans were allowed to diffuse in after gelation.
O 0.3% ● 0.7% ▽ 1% Δ 2% X 4%

Fig. 1 shows data for the case where dextran was allowed to diffuse in after gelation. When the dextran was mixed in before gelation, the same unified plot was obtained at a reference concentration of 0.7%, but the displacements required were different. The fact that the plots at different gel concentrations can be superimposed in this manner shows that:

$$D/D_o = f\,[g(C)d] \tag{1}$$

g(C) is a function which can be found from the displacements and f is the function generating the master curve. Double logarithmic plots of the displacement against gel concentration show that:

$$g(C) = a\,C^b \tag{2}$$

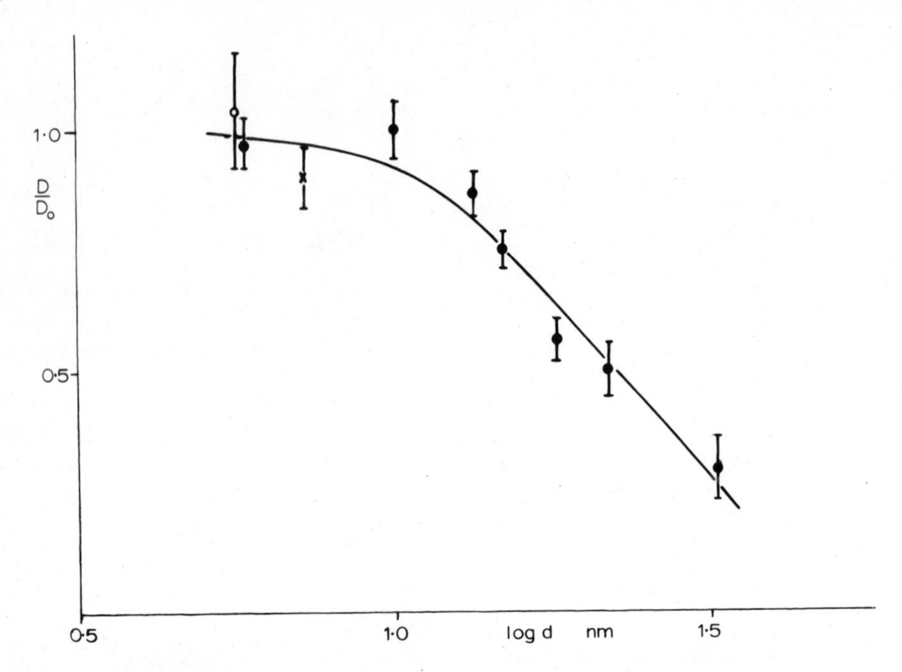

Fig. 2 Diffusion coefficient D, relative to that in dilute solution D_o, of compact macromolecules within 0.5% calcium alginate gels. D is hydrodynamic diameter. ● dextran fractions, O chymotrypsinogen, X bovine plasma albumin.

with b equal to 1/3 and 1/2 for the cases where the dextran was mixed in before gelation and allowed to diffuse in afterwards respectively. Since by dimensional analysis D/D_o must be a function of $\rho_1{}^{1/2}d$ where ρ_1 is the length of fibre per unit volume, and $C = \rho_1 m$ where m is the mean mass per unit length of fibre, b = 1/2 is just the value expected if m were independent of concentration. It would appear that the effect of the presence of dextran during gelation is to reduce m at concentrations below 0.7% and to increase it at higher concentrations.

Fig. 2 shows a similar plot for calcium alginate except that here only one gel concentration has been investigated (0.5%). Within experimental error the curves in Figures 1 and 2 are the same, except that they are displaced relative to one another on the log d scale. If the curve in Fig. 1 is displaced slightly to obtain the curve at 0.5% concentration (using the empirically determined displacement function - equation 2) then it can be shown that equivalent values of d are some seven times larger for agarose gel than for calcium alginate gel, and therefore the mass per unit length of gel fibre is some fifty times higher.

The only available theory for the function f is due to Ogston et al [17], who assume that the diffusion process is a succession of directionally random unit steps, and that the unit step does not take place if a collision with the gel network occurs. The diffusing macromolecule is taken to be a hard sphere, and the model adopted for the gel network is a random array of long straight fibres of negligible width. The unit step is taken to be twice the root mean square value of the radius of "spherical spaces" within the network. The expression derived for the diffusion coefficient is:

$$D/D_o = \mathrm{Exp}\left[-\pi^{1/2}\rho_1{}^{1/2}d\ /\ 2\right] \qquad (3)$$

If plotted as a function of log d, an exponential function is similar in shape to the curves shown in Figures 1 and 2, but the transition from the freely diffusing to the completely immobilized state takes place much more slowly. If the approximately linear regions are extrapolated to $D/D_o = 0$ and $D/D_o = 1$, then an exponential function predicts a transition of 1.2 decades, whilst a transition of 0.6 decades is observed experimentally. An exponential function can be fitted to the data corresponding to the linear region but only with a factor of 1.4 in front of the exponential. Laurent et al. [18,19] obtained a similar result for the sedimentation of compact macromolecules within hyaluronic acid solutions, although in this case Ogston et al[17] attribute the factor in front of the exponential to sedimentation of the hyaluronic acid. A similar explanation cannot apply to diffusion measurements made by laser light scattering.

It might be thought that the model adopted for the gel by Ogston et al is not a realistic representation of a randomly branched fibrous network. An analysis has therefore been carried out[13] with a randomized cubic network, assuming the physics of the diffusion process is that postulated by Ogston et al. The expression obtained for D/D_o however is numerically very similar to equation 3, so that it seems likely that the disparity with experimental data arises from factors affecting the physics of the diffusion process.

Dextran fractions are not monodisperse and the possible effects of polydispersity on the experimental results must be considered. The values of D and D_o obtained by laser light scattering are Z averages[20,21], and the hydrodynamic diameter d, is calculated from the Z average value of D_o. It is possible to obtain an analytical expression corresponding to equation 3 for the case of a Schulz-Zimm function with $M_w/M_n = 2$ where the diffusion coefficient is inversely proportional to the square root of molecular weight[13]. The expression obtained differs little numerically from equation 3 provided it is assumed that the dextran fractions are not significantly refractionated by the gel.

4. PARTITION COEFFICIENTS

The zero time value of that part of the autocorrelation function which relates to the diffusing dextran is given by $2SR_{D\theta}/R_\theta$, where R_θ is the total Rayleigh ratio of the scattered light, and $R_{D\theta}$ is the Rayleigh ratio of the light scattered from the dextran. S is the zero time value of the autocorrelation function which would be expected if the spectral broadening were complete, and may be found from the total photoelectric signal if the coherence factor has previously been determined by experiments on solutions. $R_{D\theta}$ may be taken as being approximately proportional to the concentration of dextran diffusing within the gel. It may be compared with the Rayleigh ratio of light scattered from a dextran solution of the same concentration as that surrounding the gel to obtain the partition coefficient. A correction must however be made for the fact that the light reaching the dextran within the gel and scattered by it, is attenuated by the overall turbidity of the system.

It would be expected that partition coefficients should follow the same scheme of gel concentration - hydrodynamic diameter superposition as D/D_o. Ogston[22] using the same model as for the analysis of diffusion coefficients derives the expression:

$$r = \mathrm{Exp}\left[-\pi\rho_1 d^2 / 4\right] \tag{4}$$

where r is the partition coefficient given by the relative free volume available within the gel to the diffusing macromolecules. In fact for agarose gels the scheme of superposition does not work[13]. It is found that for a given hydrodynamic diameter, r shows a minimum near 1% gel concentration. The picture becomes clearer if r is plotted against D/D_o (Fig.3). The solid line represents the equation obtained by combining equations 3 and 4.

$$r = \mathrm{Exp}\left[-(\ln D/D_o)^2\right] \tag{5}$$

Fig. 3 contains data both from experiments in which the dextran was mixed in before gelation and allowed to diffuse in afterwards. It can be seen that there is fair agreement with equation 5 at the higher concentrations (2% and 4%), but that at lower concentrations r is consistently low. It must be emphasized that r is the relative mobile concentration of dextran. It is possible that at the lower agarose concentrations dextran molecules become trapped in the network. This may be due to the higher mobility of agarose chains which the higher degree of spectral broadening of light scattered from the gel itself indicates. With gelatinous agarose of 0.1% concentration where the material filled the light scattering cell so that the dextran could not diffuse out (data not shown in Figure 3), the relative mobile concentration was still only 70%.

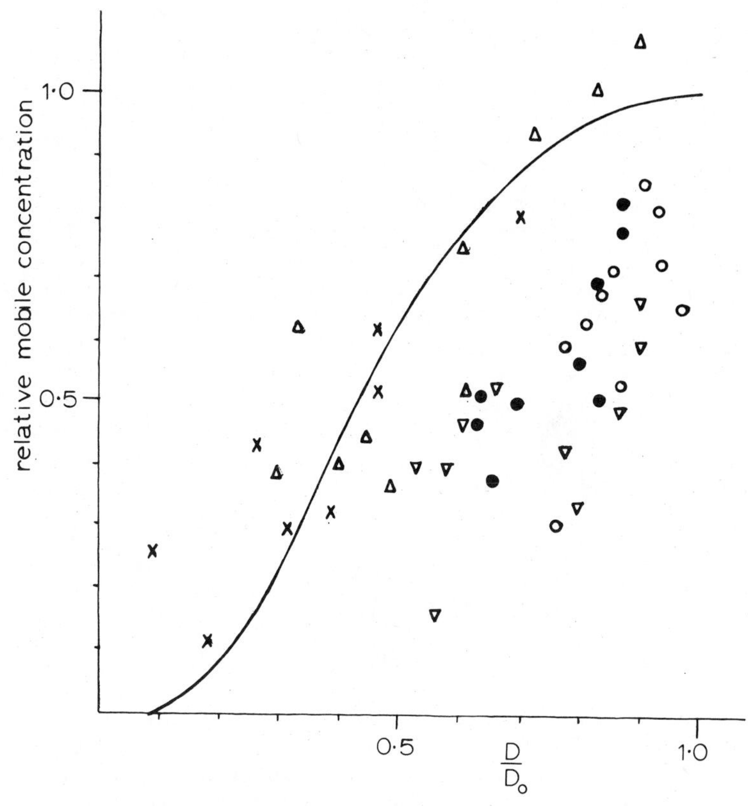

Fig. 3 Mobile concentration of dextran fraction diffusing within an agarose gel relative to concentration in surrounding solution, as determined by laser light scattering. D/D_o is the corresponding diffusion coefficient within the gel relative to that in solution. The solid line is the variation predicted by Ogston et al[17,22] (equation 5, see text).
Agarose concentrations O 0.3% ● 0.7% ∇ 1% Δ 2% X 4%

5. GEL STRUCTURE

Measurements of D/D_o as a function of gel concentration and hydrodynamic diameter of diffusing dextran may be used to determine relative values of the mean mass per unit length of the fibrous network, and its variation, if any, with gel concentration. These relative values are proportional to the square of the hydrodynamic diameter of dextran which produces a given value of D/D_o. It should however be checked that the function f remains the same. To date this has only been done for agarose and calcium alginate.

Absolute determination of the mass per unit length of the fibrous structure is difficult in that theories for the diffusion process do not appear to work. However regardless of the physics of the diffusion process, it would be expected that the theory of Ogston et al would apply near complete immobilization. The point most easily identified is obtained by extrapolating the linear region of the logarithmic plot to $D/D_o = 0$. A compromise between the gel model of Ogston et al and the randomised cubic model yields $\rho_1 d = 3.5$ for this point, whence m may be calculated using C_8. In this way molecular weights per unit length of 25,000 and 598 g mol^{-1} nm^{-1} have been calculated for agarose and calcium alginate respectively. The latter figure is in excellent agreement with the model proposed for calcium alginate gels by Morris et al[23] which envisages polyguluronate chains dimerized into junction zones in association with calcium ions, these junction zones being interconnected by single polymannuronate chains. On the basis of the double helix fibrous structure proposed for agarose by Arnott et al[24], there would be on average some 44 single agarose chains in the fibrous gel association. However, on the basis of the extended single helix model more recently proposed by Atkins and Foord (work to be published) this figure would be 69.

Qualitatively the degree of spectral broadening of the light scattered from the gel itself is an indication of the degree of mobility of the polymer chains of the gelling material. In the case of agarose this decreases with increasing concentration. In addition to the point at which a gel first appears to form when the structure becomes coherent throughout the system, there appears to be some sort of transition at around 1% concentration. The effect of the inclusion of dextran during gelation, in addition to increasing the mass per unit length of the gel fibre at high concentrations and decreasing it at low concentrations, also increases the overall turbidity of the gel at high concentrations and surprisingly, decreases it at low concentrations[13]. Further, it is thought that at low concentrations dextran becomes trapped in the structure due to mobile agarose chains. The fluorescence polarization measurements of Hayashi et al[25] also showed a transition at 1% concentration as well as at the gelling point. It would appear (Atkins and Foord - work to be published) that crystalline X-ray diffraction patterns cannot be obtained with fibrous specimens derived from agarose gels formed at concentrations below 1%. It is possible that at low concentrations the fibres of the gel structure are largely amorphous and slightly unstable, and stabilize into an ordered molecular arrangement only at concentrations above 1%.

Measurement of the diffusion coefficients of compact macromolecules within biological gels promises to become a useful technique for evaluating their structural morphology.

REFERENCES

1. T. Tanaka, O. Hocker, and G. B. Benedeck, J. Chem. Phys. 59:5151 (1973).
2. T. Tanaka, S. Ishiwata, and C. Ishimoto, Phys. Rev. Letters 38:771 (1977).
3. J. P. Munch, S. Candau, R. Duplessix, C. Picot, J. Herz, and H. Benoit, J. Polym. Sci. Polym. Phys. Ed. 14:1097 (1976).
4. J. P. Munch, S. Candau, and G. Hild, J. Polym. Sci. Polym. Phys. Ed. 15:11 (1977).
5. J. P. Munch, S. Candau, J. Herz, and G. Hild, J. Phys. (Paris) 38:971 (1977).
6. J. P. Munch, P. Lemaréchal, S. Candau, and J. Herz, J. Phys. (Paris), 38:1499 (1977).
7. E. Geissler and A. M. Hecht, J. Phys. (Paris), 39:955 (1978).
8. S. J. Candau, C. Y. Young, T. Tanaka, P. Lemaréchal, and J. Bastide, J. Chem. Phys. 70:4694 (1979).
9. S. Candau, J. P. Munch, and G. Hild, J. Phys. (Paris), 41:1031 (1980).
10. W. Mackie, D. B. Sellen, and J. Sutcliffe, J. Polym. Sci. Polym. Symp. 61:191 (1977).
11. W. Mackie, D. B. Sellen, and J. Sutcliffe, Polymer 19:9 (1978).
12. D. B. Sellen, Polymer 19:1110 (1978).
13. P. Y. Key, and D. B. Sellen, J. Polym. Sci. Polym. Phys. Ed. 20:659 (1982).
14. A. Amsterdam, Z. Er El, and S. Shaltiel, Arch. Biochem. & Biophys. 171:673 (1975).
15. O. Smidsrød and O. Skipnes, Norwegian Inst. Seaweed Research Report 34:44 (1973).
16. S. L. Brenner, R. A. Gelman, and R. Nossal, Macromolecules 11:202 (1978).
17. A. G. Ogston, B. N. Preston, and J. D. Wells, Proc. R. Soc. London 333:297 (1973).
18. T. C. Laurent and A. Pietruszkiewicz, Biochim. Biophys. Acta 49:258 (1961).
19. T. C. Laurent, I. Bjork and A. Pietruszkiewicz, Biochim. Biophys. Acta 78:351 (1963).
20. D. B. Sellen, Polymer 14:359 (1973).
21. D. E. Koppell, J. Chem. Phys. 57:4814 (1972).
22. A. G. Ogston, Trans. Far. Soc. 54:1754 (1958).
23. E. R. Morris, D. A. Rees, D. Thom, and J. Boyd, Carbohydrate Research 66:145 (1978).
24. S. Arnott, A. Fulmer, W. E. Scott, I. C. M. Dea, R. Moorhouse and D. A. Rees, J. Mol. Biol. 90:269 (1974).
25. A. Hayashi, K. Kinoshita and M. Kuwano, Polymer Journal 9:219 (1977).

CORRELATION SPECTROSCOPY AND STRUCTURAL PROPERTIES OF MACROMOLECULAR SOLUTIONS

R. Giordano and N. Micali

Istituto di Fisica - Universita' di Messina
Via dei Verdi. 98100 Messina, Italy

In this communication we shall speak about some results concerning the investigation of structural properties in macromolecular solutions, as well as in fused salts in which polymerization occurs. The present results are a part of many experimental investigations carried out in liquids, liquid solutions and so on. The interest in the field of macromolecular solutions is twofold: a macromolecular solution can be treated as a "model", the large size of molecules allowing the use of light scattering instead of X rays or neutrons. On the other hand structural properties of solutions of macromolecules of biological interest can be interesting in themselves, because of a possible connection between structure and biological function. We report the results obtained by using light scattering in two systems, each being characterized by some sort of local order. The first one consists of solutions of macromolecules (both B.S.A. and lysozyme), in which a long-range order seems to occur. The second one is a melted salt (antimony trichloride) in which polymerization takes place. In the first case a main role is played by water itself, which is a strongly - associated liquid because of the existence of hydrogen bonds. In the second-case, chlorine bonds allow the polymerization of the liquid. We use both elastic and quasi-elastic light scattering. We also investigate the coherence properties of the scattered light. A summary of the published papers in this field is reported in the bibliography[1,2,3].

As far as the macromolecular solutions are concerned, we briefly summarize some results obtained with viscosity measurements both in lysozyme[4] and B.S.A. solutions, at very small concentrations (from 10^{-4} to 10^{-2} by weight). The viscosity of such solutions measured with a usual capillary flow viscometer turns out to be practically coincident with that of pure water, and behave regularly both as a function of concentration and of temperature. However if the viscosity is measured at a very low shear rate, using a rotating cylinder viscometer, a peculiar behaviour becomes evident: first of all the relationship between stress and shear rate is no longer linear. The derivative decreases as the shear rate increases and tends to the usual viscosity only for high enough values of the shear rate. In addition the system shows a residual non-zero shear stress at zero shear rate. It is to be noted that all these properties develop gradually in time, reaching a stationary condition only after several hours. Moreover, if the sample is shaken, the initial conditions are restored. All these properties can be explained by assuming the building up of a structure that is stable against small disturbances and produces a solid-like behaviour (thixotropic) in the solution. It is therefore quite obvious to try to investigate the structural properties of the system by using light scattering techniques. We report on the results obtained in a B.S.A. solution. The results are quite similar in the case of lysozyme.

In fig.1 we show the elastic scattering results (full line) i.e. the scattered intensity as a function of the angle: it is quite evident that some kind of structure is present in the system. Although a unique interpretation cannot be given because of the non-uniqueness of the inverse problem in optics, the shape of the curve definitely shows that a simple monodisperse system is

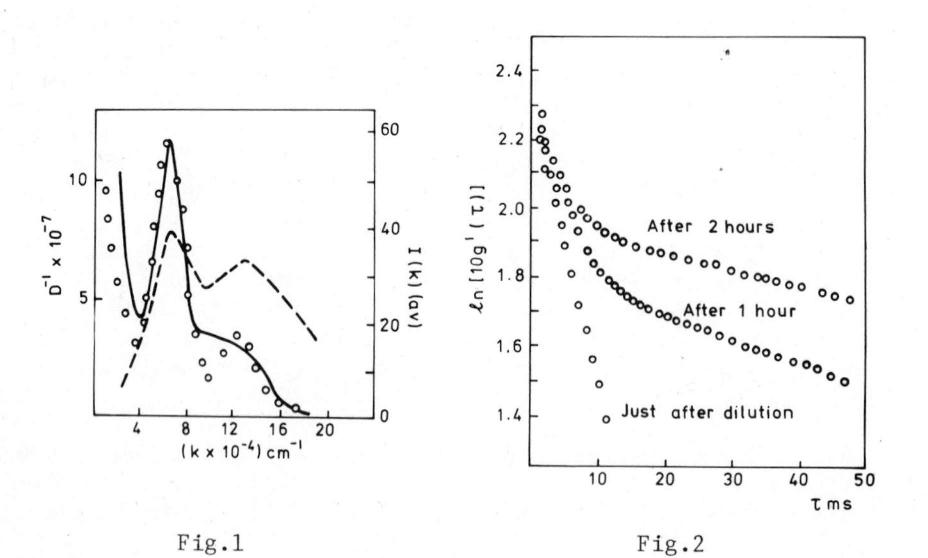

Fig.1 Fig.2

not present. As we shall see later, a suitable model structure can be hypothesized to explain such a behaviour.

In fig.2 we show the results obtained in a lysozyme solution, with correlation spectroscopy (i.e. quasi-elastic light scattering). It is well-known that for a purely Brownian system, the autocorrelation function of the scattered intensity would behave in an exponential way. The decay-time, γ^{-1}, of the autocorrelation function can be, in turn, related to the diffusion coefficient. In the case of our solutions such a behaviour is shown only just after shaking the sample. In such a case the correct value for the diffusion coefficient is found. As the rest time increases, however, the autocorrelation function shows a long tail that can be interpreted either as a failure of the simple Brownian model, or as the appearance of large objects whose diffusion coefficient becomes smaller and smaller. We will see that, in fact, both these circumstances occur. In the case of a structured system, a description can be obtained by allowing a dependence of an "effective" diffusion coefficient upon the exchanged wave-vector:

$$D = D_o/S(k)$$

Where S(k) is the static structure factor. Therefore information concerning the structure can be obtained by plotting the inverse diffusion coefficient as a function of the exchanged wave vector. We performed this kind of measurement in a 1% B.S.A. solution. The value of the "effective" diffusion coefficient is calculated from the long-tail slope of the curve: such a value turns out to be k-dependent, as expected. In fig.1 we show (dashed line) the behaviour of D^{-1}. The results can be compared with the full line, and again indicate the existence of a structure in the liquid.

Let us reconsider the experiments of elastic light scattering. It is trivially known that such an experiment consists of the detection of the intensity scattered at a given angle; i.e. essentially in a "far field" measurement. Now it has been demonstrated that the far field distribution of intensity from a plane source can be related to the coherence properties of the field on the source. More precisely the far field intensity, apart from a cosine factor, is the Fourier transform of the mutual coherence onto the source:

$$I(\theta) \propto \cos^2\theta \int_{source} \Gamma_{12}(r)\, e^{ikr}\, dr$$

Now, assuming that the incoming beam is fully coherent, the far field intensity distribution can be related to a loss of coherence of the beam produced by its interaction with the sample. In turn the coherence properties of the emerging beam could be related to coherence properties of the system, like spatial structure and/or dynamical correlations. From the experimental point of view the arrangement is quite similar to the usual elastic scattering measurements. The only differences consist of the screening of the sample in order to delineate the emerging beam as a plane source. The results are shown in fig.1 (dotted line). Again the same kind of structure is made evident. It can be noted that in such a case the structure is more pronounced. In other words, the interpretation of the elastic light scattering in terms of coherence, enhances the evidence of the structural effects of the system. It can be shown that the behaviour of the curves is the same: in particular the kind of structure revealed by the elastic scattering experiments is reproduced in the dynamical measurements involving the correlation technique.

The features exhibited by the structure factor S(k), as well as the behaviour of the autocorrelation function, can be explained by assuming the following model:
The macromolecules dissolved in water tend to cluster, giving rise to highly correlated regions, whose size can be estimated as some thousands of Å. In turn the clusters interact with each other in such a way that at least two nearest neighbouring clusters are dynamically correlated. We believe that precisely such an interaction between clusters gives rise to the observed thixotropic properties. In the frame of such a model, the observed shape for the scattered intensity comes from the superposition of the shape factor of a simple cluster, modulated by the structure factor that arises because of the interaction between clusters. More research is in progress in order to clarify this point.

As a second example of the autocorrelation technique, we report some results concerning melted antimonium trichloride. The measurements are performed both in the true liquid and in the undercooled region. From X-ray measurements it has been argued that the melted salt tends to give rise to polymeric chains, through the formation of chlorine bridges. Such a polymerization can also be deduced from Raman measurements. In fig.3 we report two autocorrelation curves referring to the melted and undercooled liquid. As a common feature we would like to emphasize the very long tail, that is of the order of a second. The shape of the curve is clearly not exponential, nor can it be decomposed as the sum of two exponentials. Obviously such curves could be decomposed as the sum of enough exponentials, or treated in terms of the Laplace transformation, as the continuous superposition of exponentials. The decomposition of the autocorrelation function as a sum of exponentials implies the existence of a distribution of relaxation times. In our case such a distribution will be very spread out, because of the existence of very short (10^{-4} s) and very long (0.5 s) times. In turn the

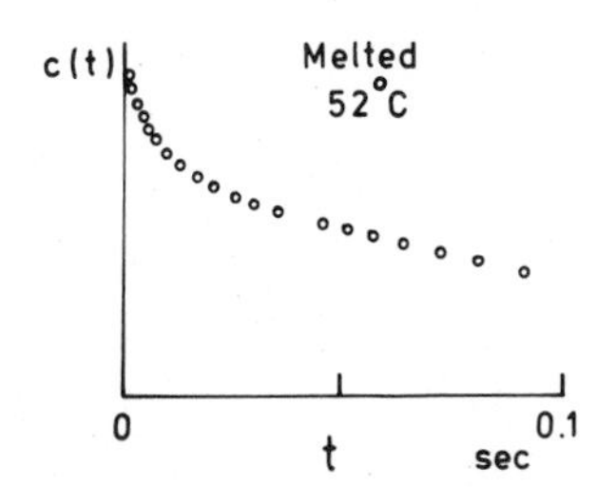

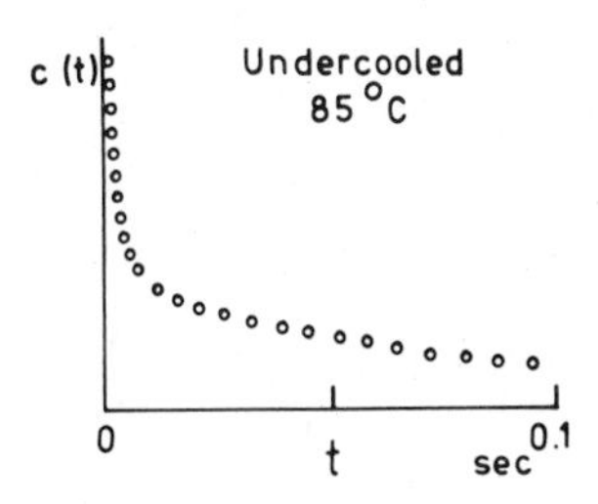

Fig.3

actual meaning of each time constant and its relationship with a suitable dissipation process would be recognized.

However it is possible to treat the data in an alternative way, in which only one non-linear dissipative process is supposed. The scattered intensity at a given instant can be written as proportional to the range R(k,t) of the spatial autocorrelation function of the fluctuations of electrical susceptibility. If such a correlation length fluctuates in time, the scattered intensity will also fluctuate. The autocorrelation function is therefore to be related to the dynamics of the correlation length. Assuming a purely stochastic process with a simple relaxation time, an exponential shape is obtained[5]. However one can assume a non-linear process, in which the time constant depends on the actual size of the fluctuation:

$$\delta R(\underline{k}, t) = \delta R(\underline{k}, o)\, e^{-\gamma\{R(t)\}.t}$$

A suitable choice of such a dependence can reproduce the observed results. In our case we find that the behaviour indicated in fig.4 reproduces the experimental results.

From the physical point of view one can think of the following picture:

1. In the equilibrium state there is a correlation length, that could be associated with the mean length of a polymeric chain.
2. There is however another correlation length that, although being a "non equilibrium" quantity, tends to be stabilized, being characterized by an enormously long relaxation time (metastable state).

Clearly, such a picture is only a tentative one. Its purpose is mainly to show the interpretation of experimental results from the autocorrelation technique.

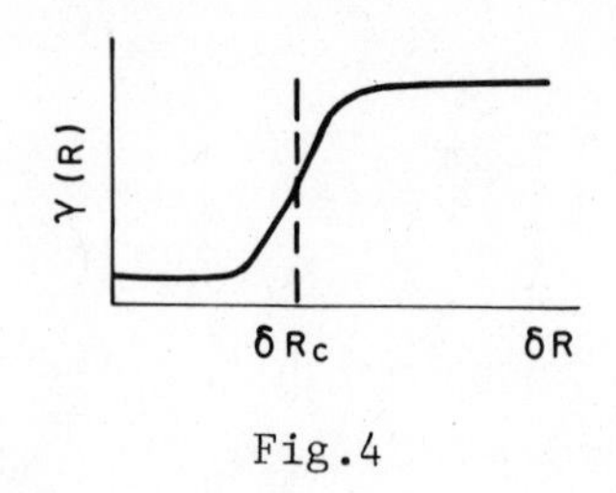

Fig.4

REFERENCES

1.- R. Giordano, M.P. Fontana and F. Wanderlingh, J. Chem. Phys. 74:2011 (1981).
2.- R.Giordano,F.Mallamace,A.Salleo,S.Salleo and F.Wanderlingh, Opt.Acta 27:1965 (1980).
3.- R.Giordano,G.Maisano,F.Mallamace,N.Micali and F.Wanderlingh, J.Chem.Phys.75:4770 (1981).
4.- R.Giordano,A.Salleo,S.Salleo and F.Wanderlingh, Phys Letters 70A:64 (1979).
5.- L.Tisza,"Generalized thermodynamics", The M.I.T.Press (1977), p.307.

DEPOLARIZED RAYLEIGH SPECTRA OF DNA IN SOLUTION

A. Patkowski[+], G. Fytas and Th. Dorfmüller

University of Bielefeld
Faculty of Chemistry
Postfach 8640
D-4800 Bielefeld 1
West Germany

INTRODUCTION

The spectral analysis of scattered light has been extensively used to study dynamical processes in a variety of systems. All the processes which result in a modulation of either the isotropic or the anisotropic part of the optical polarizability of molecules cause shifts of the frequency of the incident light and may be monitored by the polarized or depolarized light scattering spectrum. Different dynamic processes in the scattering medium can be characterized by their characteristic relaxation times (and corresponding frequency shifts) which cover several orders of magnitude in the time (frequency) domain, thus requiring different experimental approaches in order to obtain the dynamical information [1]. Photon correlation spectroscopy, which is based on post detection processing of the electronic signal, giving the intensity correlation function by means of a correlator or a spectrum analyser, may be used to monitor processes with characteristic times in the range of 1 to 10^{-6} s. This time range may be extended down to about 10^{-8} s by analysing the statistics of times of arrivals of scattered photons [2]. Processes faster than 10^{-8} s cannot be monitored by post detection processing of photocounts in the time domain. The analysis of processes with shorter relaxation times (higher frequencies) may be performed in the frequency domain using optical methods. The fre-

[+]On a leave of absence from Quantum Electronics Laboratory, Institute of Physics, A. Mickiewicz University, 60-780 Poznan, Poland

quency shifts of scattered light in the range of 10 MHz to 1000 GHz (corresponding to characteristic times of 0.2 ps to 20 ns) can be measured by means of Fabry-Perot interferometers, while frequency shifts higher than 10 GHz are easily accessible to measurements by means of grating monochromators. A Fabry-Perot interferometer analyses the optical field itself, it acts as a pre-detection frequency filter and produces a spectrum, i.e. a plot of intensity versus frequency.

Different relaxation processes occurring in the same time range can sometimes be separated by the proper choice of polarization of the incident and scattered light relative to the scattering plane defined by the direction of the corresponding wave vectors. With the polarisation vector vertical to the scattering plane the scattered light observed in the vertical plane gives rise to the so-called VV or polarized spectrum, whereas horizontal polarization of the scattered light (i.e. within the scattering plane) provides the VH component or depolarized spectrum.

In this paper we report measurements of the depolarized light scattering spectrum of DNA in solution using Fabry-Perot interferometry. The general formalism of the dynamic depolarized scattering is given in the next section and after describing the experimental part, the data are discussed in terms of localized motion of the anisotropic bases in DNA. The dynamics of the DNA molecule have been extensively studied under various conditions and by means of several experimental techniques (see references in 3). There have been however no experimental studies on the localized dynamics of the optically anisotropic bases in the short time range by measuring a spectrum of the depolarized component of scattered light by means of a Fabry-Perot interferometer. This technique has been used before in this field only to determine the rotational relaxation time of lysozyme [4].

THEORETICAL BACKGROUND

Depolarized light scattering arises mainly from local fluctuations in the intrinsic anisotropy of the polarizability tensor [5]. The spectrum of depolarized scattered light $I_{VH}(\omega)$ is given by the Fourier transform of the corresponding polarizability correlation function C(t) according to

$$I_{VH}(\omega) = \frac{A}{2\pi} \int_{-\infty}^{\infty} C(t) \cdot \exp(-i\omega t)dt \qquad (1)$$

where A is a constant.

In the case of macromolecules by identifying the scatterer with the appropriately defined repeat segment the correlation function C(t) has the form:

$$C(t) = \langle \sum_{m,n}^{N} \sum_{i,j}^{N_s} a_{yz}^{(n)}(j,t) \cdot a_{yz}^{(m)}(i,0) \exp [i\underset{\sim}{k} \cdot (\underset{\sim}{r}_j^{(n)}(t) - \underset{\sim}{r}_i^{(m)}(0))] \rangle \quad (2)$$

where N is the number density of the polymer, N_s the number of segments in the chain, $a_{yz}^{(n)}(j,t)$ represents the yz component of the polarizability tensor of segment j of polymer n at time t in the laboratory fixed coordinate system, $\underset{\sim}{k}$ is the scattering vector and $\underset{\sim}{r}_j^{(n)}(t)$ is the position of the jth unit of polymer n at time t. The brackets in Eq. (2) denote an ensemble average.

The first factor $a_{yz}^{(n)}(j,t) \cdot a_{yz}^{(m)}(i,0)$ in the double sum of Eq. (2) is more affected by orientational motions, whereas the second factor is sensitive to the center of mass motion. For small molecules in the liquid state these two modes are considered to be statistically independent but this assumption cannot always be taken for granted. The polarizability correlation function C(t) of polymeric systems can be affected by several relaxation processes: a) orientational motion of the entire rigid polymer molecule [4], b) long wavelength intramolecular Rouse-Zimm modes in the case of flexible polymer chains [6], c) segmental motions of the polymer backbone [7], and d) local motions of optically anisotropic side groups, which can be either rotations about the main chain axis following the conformational changes of the backbone [6,8] or rotations about the bond connecting the side group with the polymer backbone.

For dilute polymer solutions pair correlations between different polymer molecules in Eq. (2) are negligible. In the case of polymers such as polystyrene or DNA, where the optical anisotropy is due mainly to optically anisotropic side groups, C(t) may be written in the form:

$$C(t) = N \cdot N_s \langle a_{yz}^{(m)}(j,t) \cdot a_{yz}^{(m)}(j,0) \rangle + N N_s (N_s - 1) \langle a_{yz}^{(m)}(j,t) a_{yz}^{(m)}(i,0) \rangle \quad (3)$$

The terms in Eq. (3) represent the self and the cross orientational correlation functions, respectively. The integral intensity of the depolarized component of scattered light (Eq. 3), for t=0 amounts to

$$I_{VH} = A \cdot f(n) \cdot N \cdot N_s \, \beta_s^2 \, (1+F_{intra}) \tag{4}$$

where

$$F_{intra} = (N_s - 1) \frac{\langle a_{yz}^{(m)}(j,0) \cdot a_{yz}^{(m)}(i,0) \rangle}{\langle |a_{yz}^{(m)}(i,0)|^2 \rangle} \tag{5}$$

is a measure of intramolecular static pair correlations of the side groups in the polymer chain and $f(n)$ [$= \frac{1}{n^2} (\frac{n^2+2}{3})^4$] is the product of the geometrical correction $(\frac{1}{n^2})$ and the local field, n being the refractive index of the medium.

F_{intra} for symmetric top scatterers is proportional to the 2-nd Legendre polynomial

$$\langle P_2[\hat{u}_i(0)\hat{u}_j(0)] \rangle = \frac{1}{2} \langle 3 \cos^2 \Theta_{ij}(0) - 1 \rangle \tag{6}$$

with $\Theta_{i,j}(0)$ being the angle between the directions $\hat{u}_i(0)$, $\hat{u}_j(0)$ of the symmetry axis of the scatterers i,j. Thus, according to Eq. (1), (3) and (4), the spectrum $I_{VH}(\omega)$ and the integrated intensity I_{VH} of the depolarized Rayleigh spectrum are related to the dynamics of the anisotropic side groups and the average local conformations of the chain, respectively.

EXPERIMENTAL

The depolarized Rayleigh spectra $I_{VH}(\omega)$ were taken using the apparatus shown in Fig. 1. The light source was an Argon-Ion laser operating in a single mode at 488.0 nm with a power of 400 mW. A reference beam propagated along the optical axis of the scattered beam. This reference beam was used for the alignment of the optical elements of the spectrometer. Light scattering from the sample contained in a rectangular cell (Hellma 10 mm) was spectrally analyzed using a piezoelectrically scanned Fabry-Perot interferometer (FPI). The plate separation of the interferometer amounted to 2.50 cm which corresponds to a free spectral range of 6.00 GHz. The scattered light after passing through the interferometer was detected by a cooled PM tube (RCA C 31024) with photon counting electronics. The depolarized Rayleigh spectra were transfered to a HP-9830 A calculator for further computation. The spectrometer was operated with a finesse of about 60. Depolarized

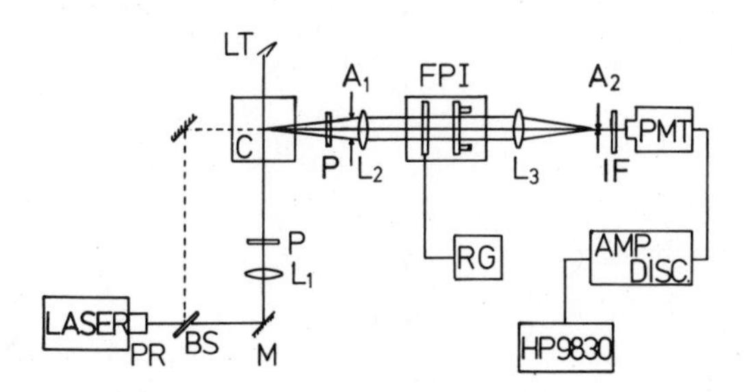

Fig. 1. Schematic diagram of the experimental setup: PR - polarization rotator, L_1, L_2, L_3 - lenses of focal length amounting to 300, 200 and 500 mm, respectively; M-mirror, P - polarizer, BS - beam splitter, C - scattering cell, LT - light trap, FPI - Fabry-Perot interferometer (Tropel 350), IF - interference filter, A_1, A_2 - apertures of the size of 8 mm and 0.3 mm, respectively, PMT - photomultiplier tube, RG - ramp generator, AMP-DISC - amplifier-discriminator, HP 9830 A - calculator.

scattering spectra were measured at a scattering angle of 90°. The incident laser light was polarized horizontally (H) and the scattered light was collected in the H and in the V (vertical) plane relative to the scattering plane. The instrumental function was measured experimentally by recording the spectrum of the polarized component $I_{VV}(\omega)$ of light scattered by the solution under study. A Lorentzian curve was fitted to the experimental spectra and the full width at half height Γ as well as the integrated depolarized intensity I_{VH} were obtained. VH-spectra were taken for four identical DNA solutions in the temperature range of 25 to 97 °C. Typical experimental spectra of the depolarized Rayleigh component $I_{VH}(\omega)$ of light scattered by DNA solution at 25 to 97 °C are shown in Fig. 2.

Solutions of calf thymus DNA (high molecular weight; Boehringer cat. No. 104167) of DNA concentration of 0.2 mg/ml in 0.1 M NaCl; and 0.1 M sodium-citrate buffer pH 7.5 were measured. The samples were passed through a 4µ Millipore filter into the dust free scattering cells before the measurements.

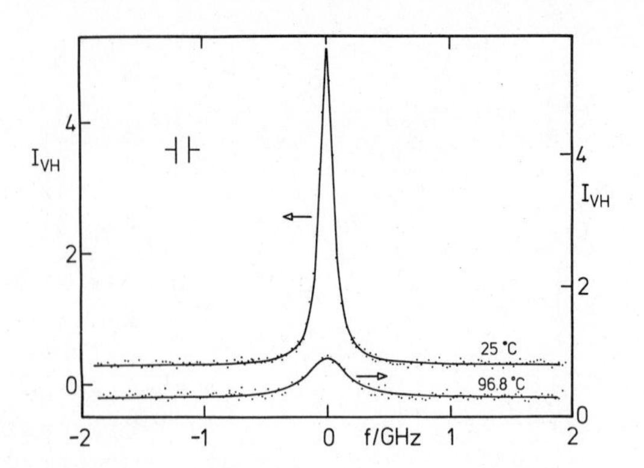

Fig. 2. Depolarized spectra of light scattering by DNA solution at temperatures of 25 and 96.8 °C. The width of the instrumental function is also shown.

RESULTS AND DISCUSSION

The depolarized component of light scattered by DNA solutions originates mainly from the optically anisotropic bases. Therefore, in principle, it should be possible to study the dynamics of DNA bases by measuring the spectrum of the depolarized component of scattered light. The following two parameters of the spectrum may be discussed in terms of the DNA structure: The integral intensity, which according to Eq. (6) is connected to the ordering of the orientation of the bases. This is however a time averaged value and therefore its interpretation in connection with DNA structure is not straightforward. For example a decrease of the depolarized intensity may be interpreted as a result of a decreasing order in DNA, i.e. a more randomized orientation of bases in a relatively static structure, in which the position of each base is fixed with respect to the backbone. On the other hand, the same effect may result from increased mobility of the bases around their equilibrium positions, without significantly destroying the order within the DNA molecule. The width of the depolarized spectrum helps us to distinguish among the two above mentioned possibilities, supplying us with information concerning the dynamics of the bases. For example, the decrease of the depolarized scattered intensity and a broadening of the spectrum (decrease of the relaxation time) means that the bases are moving faster, i.e. more freely. It is, however,

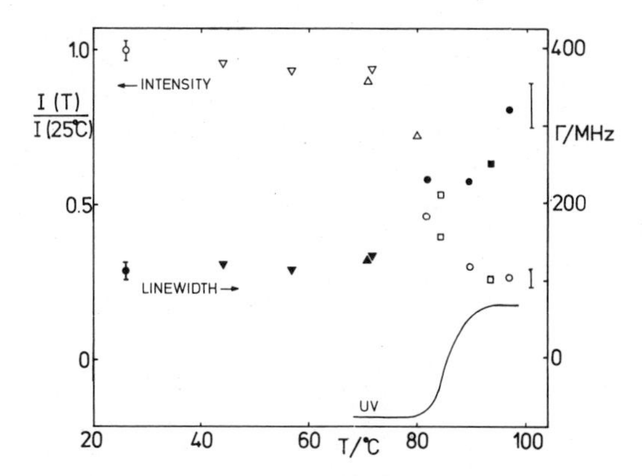

Fig. 3 Plot of the relative intensity (open symbols) and of the line-width (filled symbols) of the depolarized component of scattered light versus temperature for identical DNA solutions (different symbols). The UV-absorption melting curve is also shown for comparison.

impossible to say to what extent this increase of free movement results from breaking of hydrogen bonds or stacking interactions in the double-stranded DNA, without performing further studies on model single- and double-stranded DNA samples. This problem is currently under study. The aim of this work was to check whether the intensity and the spectrum of the depolarized component of light scattered by DNA in solution, as measured by means of a Fabry-Perot interferometer, can supply us with dynamic information on the relaxation of DNA bases, which can be related to conformational changes of the DNA molecule, especially in connection with the helix-coil transition.

Our measurements performed on a high molecular weight calf thymus DNA led to the following conclusions. The relaxation of the bases in a double-stranded DNA is too slow to be measured by means of a plane Fabry-Perot interferometer, which covers the frequency range of 100 MHz to 200 GHz. The same relaxation in melted DNA structure was accessible to our measurements. Melting of the double-stranded DNA caused a significant increase of the width of the depolarized spectrum and decrease of its intensity (Fig. 2). The temperature dependence of both intensity and line-width of the depolarized spectrum, as well as a UV-absorption melting curve are shown in Fig. 3. As can be easily seen, the intensity and line-width changes follow exactly the UV-absorption melting curve in the whole temperature range studied. The value of the fastest relaxation time at 96.8 °C amounts to 1.5×10^{-9} s and is similar to the values obtained in a geometrically somewhat similar systems, for the rotation of the phenyl group in polyphenylmethylsiloxane [8] and polystyrene [6]. It is however not as yet possible to assign the observed relaxation process to a better specified kind of motion in the DNA molecule.

The above experimental findings suggest that this method can be successfully used for conformational studies on double-stranded DNA for all the processes in which unwinding of a double helix occurs. Further studies on synthetic and natural DNA are needed in order to characterize precisely and to assign the relaxation process observed in our experiment to a particular kind of motion in DNA as well as to assess the applicability of this new method to further studies on other systems of biological relevance.

ACKNOWLEDGMENT

We gratefully acknowledge the financial support of the "Minister für Wissenschaft und Forschung des Landes NRW" and the Fonds der Chemischen Industrie.

REFERENCES

1. J.M. Vaugham, Correlation Compared with Interferometry for Laser Light Scattering Spectroscopy, in: "Photon Correlation and Light Beating Spectroscopy", ed. H.Z. Cummins and E.R. Pike, Plenum, New York (1974)
2. A. Patkowski, S. Jen and B. Chu, Intensity-Fluctuation Spectroscopy and tRNA Conformation. II. Changes of Size and Shape of tRNA in the Melting Process, Biopolymers 17: 2643 (1978)
3. A. Patkowski, G. Fytas and Th. Dorfmüller, Thermal Denaturation of DNA: Interferometric Depolarized Light-Scattering Study, Biopolymers 21: 1473 (1982)
4. S.B. Dubin, N.A. Clark and G.B. Benedek, Measurement of the Rotational Diffusion Coefficient of Lysozyme by Depolarized Light Scattering: Configuration of Lysozyme in Solution, J. Chem. Phys. 54: 5158 (1971)
5. B.J. Berne, R. Pecora "Dynamic Light Scattering", Wiley, New York (1976)
6. D.R. Bauer, J.I. Brauman and R. Pecora, Depolarized Rayleigh Spectroscopy Studies of Relaxation Processes of Polystyrene in Solution, Macromolecules 8: 443, (1975)

7. D.R. Jones, C.H. Wang, Depolarized Rayleigh Scattering and Backbone Motion of Polypropylene glycol, J. Chem. Phys. 66: 1659 (1977)
8. Y.-H. Lin, G. Fytas, B. Chu, Depolarized Rayleigh Spectra of Siloxane Polymers, J. Chem. Phys. 75: 2091 (1981)

DOUBLE SCATTERING IN A STRUCTURED SYSTEM OF PARTICLES

T.W. Taylor and B.J. Ackerson

Department of Physics
Oklahoma State University
Stillwater, Oklahoma 74074

INTRODUCTION

Many problems arise in the analysis of dynamic light scattering data. Among those most often encountered in biological systems are polydispersity and multiple scattering. Multiple scattering is a problem due to the relatively high concentrations used and the sizes of the particles under study. A number of papers concerning multiple scattering from suspensions of particles undergoing Brownian motion have appeared over the past few years[1-6]. These papers have for the most part been concerned with non-interacting point particles. Sorensen et al.[1,2] have developed a procedure for determining the double scattered field for such a system. We will extend this procedure to systems with structure due either to interactions or particle size.

THEORY

We will consider the sample to be composed of many correlated regions which are statistically independent from one another. The double scattered field results from two successive single scattering events in different regions. We neglect any double scattering within a single region. The procedure is then similar to the one used by Sorensen et al.[1], where we have exchanged correlated regions for their independent particles. The different wave vectors and scattering angles are shown in Figure 1.

Under these assumptions the double scattered intensity is

$$I_{xy} = \int d\bar{K}'' \, S(\bar{K}-\bar{K}'') \, S(\bar{K}''-\bar{K}')\varepsilon_{xy}^{2}$$

where S(K) is the static structure factor or scattering intensity of a correlated region, ε_{xy} is the final polarization amplitude (y) for a given initial polarization (x)[1,2].

Similarly, the double scattered linewidth or first cumulant is defined by

$$\Gamma_{xy} = \frac{d/dt \int d\bar{K}''\ S(\bar{K}-\bar{K}'';t)\ S(K''-K';t)\,\varepsilon_{xy}^2 \big|_{t=0}}{\int d\bar{K}''\ S(\bar{K}-\bar{K}'')\ S(\bar{K}''-\bar{K}')\,\varepsilon_{xy}^2} \tag{2}$$

If the dynamic structure factor is assumed to have the form[10,11] (at least initially).

$$S(\bar{K}-\bar{K}'';t) = S(K-K'')\ \exp[-D(K-K'')^2 t/S(K-K'')] \tag{3}$$

we may easily perform the time derivative and simplify eq. (2) to find

$$\Gamma_{xy} = \frac{\int d\bar{K}''\ S(\bar{K}-\bar{K}'')(1-\cos\gamma') + S(\bar{K}''-\bar{K}'')(1-\cos\gamma)\,\varepsilon_{xy}^2}{I_{xy}} \tag{4}$$

These equations can be numerically integrated for any given static structure factor to determine the properties of the double scattered field.

RESULTS

The double scattered intensities and linewidths were numerically integrated for both interacting and non-interacting systems of particles. The structure factor for a non-interacting system is shown in Figure 2(a). The double scattered intensities determined by eq. (1) are shown in Figure 2(b). It should be noted that $I_{VV}(90^o) \neq I_{HH}(90^o)$ in contrast to that assumed by Grüner and Lehmann[3]. Both I_{VV} and I_{VH} are constant with depolarization ratio

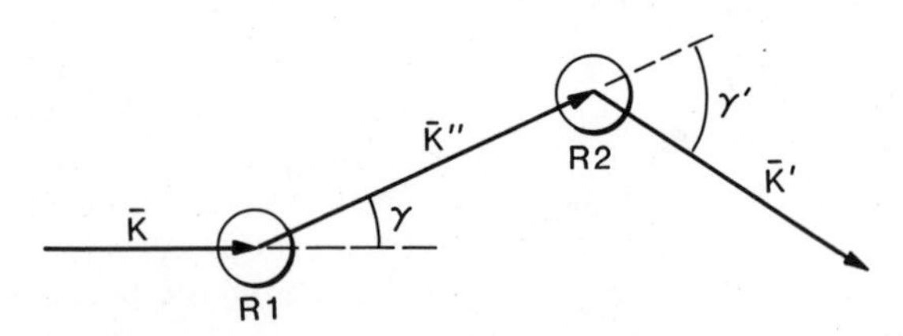

Figure 1. Scattering diagram for a double scattered photon. $\bar{K}$ is the incident wave vector. $\bar{K}''$ is the intermediate scattered wave vector with scattering angle γ. $\bar{K}'$ is the final scattered wave vector with scattering angle γ'. R1 and R2 are two independent correlated regions.

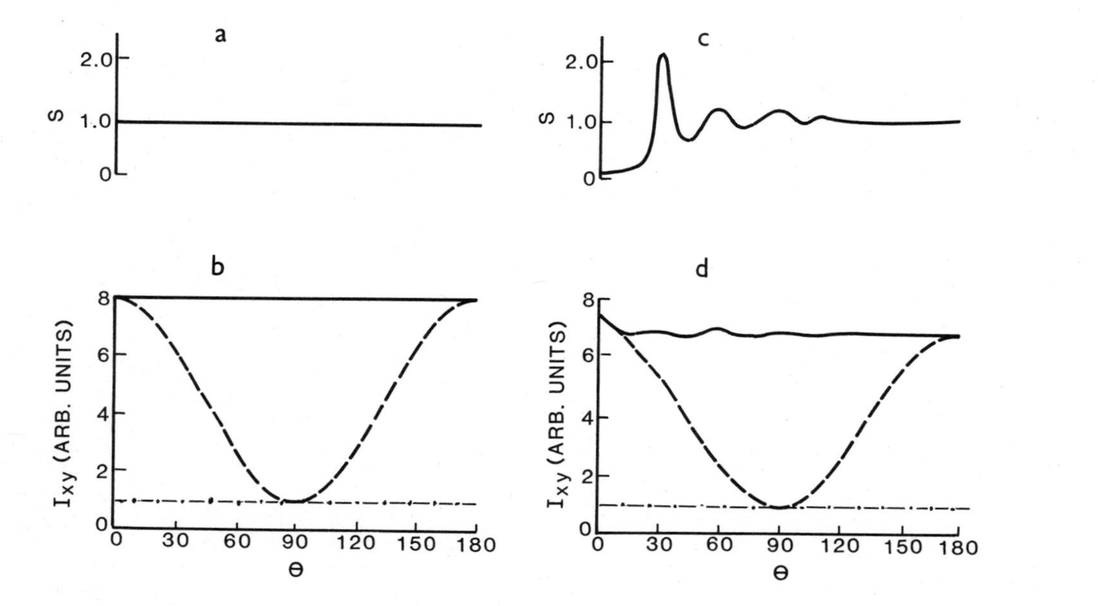

Figure 2. Structure factor and double scattered intensities as a function of scattering angle for two systems of particles (a) is the structure factor for a non-interacting system. (b) is the resulting double scattered intensities. (c) and (d) are for an interacting system. The solid line is I_{VV}, dashed line is I_{HH} and the dotted line is I_{VH}.

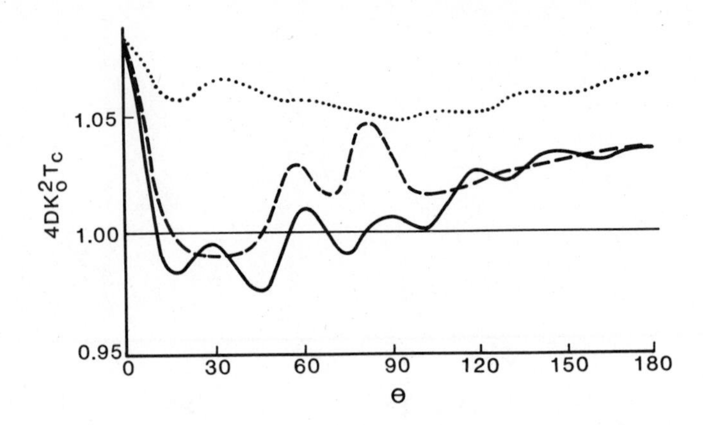

Figure 3. Double scattered correlation time as a function of scattering angle. The straight solid line at $4DK_o^2 T_c=1$ is for all polarizations of the non-interacting system. The curves are from the interacting system with solid curve in the VV polarization, dashed in the HH polarization and the dotted in the VH polarization.

$R_2 = I_{VH}/I_{VV} = 1/8$. I_{HH} has the form

$$I_{HH}(\theta) = (I_{VV}-I_{VH})\cos^2\theta + I_{VH} \tag{5}$$

An experimentally determined structure factor for a system of interacting polystyrene spheres is shown in Figure 2(c) with the resulting double scattered intensities in Figure 2(d). The VV polarized double scattered intensity shows the oscillating features of the structure factor. The highest peak is now at 60°, double the single scattered structure factor's peak position. The VH intensity is effectively constant with a 5% increase at low angles. The HH polarized component follows eq. (5) with the angular dependancies included in I_{VV} and I_{VH}.

The normalized double scattered correlation time determined by eq. (4) is shown in Figure 3. For non-interacting systems the correlation time is independent of angle and polarization. This result is shown as the solid line at $4DK_o^2T_c = 1$. When interactions are turned on there appears both an angular and polarization dependence. This makes correcting the polarized scattered light for multiple scattering difficult. Deducting the VV polarised double scattered component from the VH polarized double scattered component introduces an error of at least 10%. This error increases as higher order scattering enters in.

CONCLUSIONS

Multiple scattering will always remain a problem for light scattering measurements in biological systems. If the system under study shows no structure due to interactions or particle size the multiple scattered components can be resolved[9]. If the system has structure, the problem becomes complex and only a qualitative interpretation can be made. Under these circumstances it would be best to eliminate the multiple scattering by using a smaller scattering volume[5,8] or a two detector method[7].

REFERENCES

1. C.M. Sorensen, R.C. Mockler, and W.J. O'Sullivan, Phys. Rev. A 14, 1520 (1976).
2. C.M. Sorensen, R.C. Mockler, and W.J. O'Sullivan, Phys. Rev. A 17, 2030 (1978).
3. F. Grüner and W. Lehmann, J. Phys. A: Math. Gen. 13, 2155 (1980).
4. A. Bøe and O. Lohne, Phys. Rev. A 27, 2023 (1978).
5. D.B. Siano, B.J. Berne and G.W. Flynn, J. Coll. and Int. Sc. 63, 282 (1978).
6. P.C. Colby, L.M. Narducci, V. Bluemel, and J. Beer, Phys. Rev. A 12, 1530 (1975).
7. G.D.J. Phillies, Phys. Rev. A 24, 1939 (1981).
8. A.J. Hurd, Thesis, University of Colorado.
9. T.W. Taylor and B.J. Ackerson, to be published.
10. B.J. Ackerson J. Chem. Phys. 69, 684 (1978).
11. B.J. Ackerson, J. Chem. Phys. 64, 242 (1976).

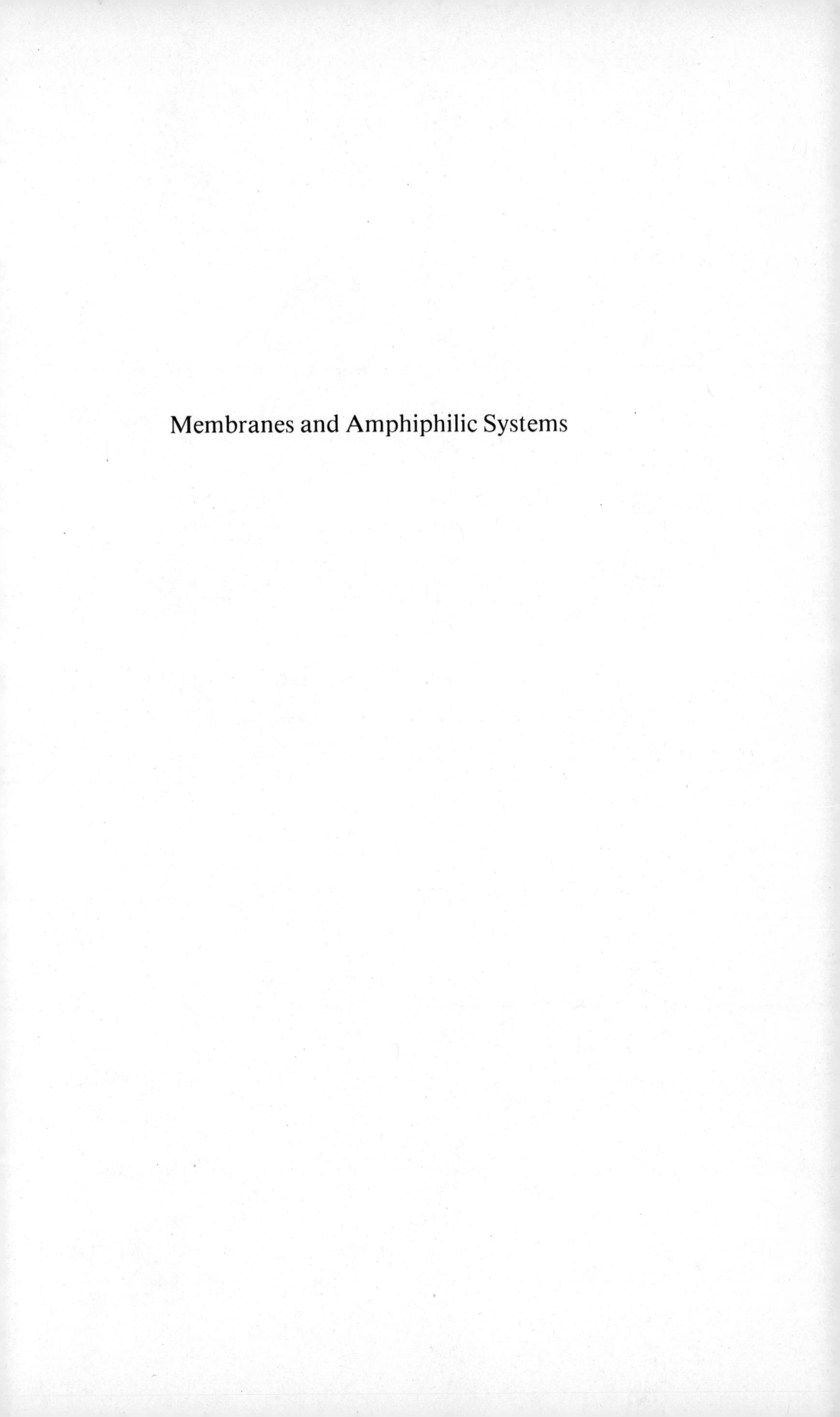

Membranes and Amphiphilic Systems

FLUORESCENCE TECHNIQUES FOR THE STUDY OF BIOLOGICAL MOTION

Dennis E. Koppel

Department of Biochemistry
University of Connecticut Health Center
Farmington, Connecticut 06032, U.S.A.

CONTENTS

ABSTRACT

A variety of fluorescence techniques for the study of biological motion are described and discussed. We consider three basic strategies: time-resolved emission, fluorescence photobleaching, and fluorescence correlation spectroscopy. In the first of these, the sample is excited with pulses of light short compared to the excited singlet-state lifetime. Molecular motions over the time-scale of the excited state lifetime are characterized through their effects on the observed time-resolved fluorescence emission. In the fluorescence photobleaching technique, intense photobleaching pulses, short compared to the time-scale of the motion under study (but very long compared to the excited singlet-state lifetime), are used to selectively deplete the ground-state population. Molecular motions over times long compared to the excited singlet-state lifetime are characterize by measuring the "recovery" of fluorescence after photobleaching monitored with an attenuated CW light source. In fluorescence corre lation spectroscopy, in contrast, the sample is illuminated only wit the CW monitoring beam. Molecular motions over times long compared to the excited-state lifetime are characterized through calculations of the correlation function of the spontaneous, stochastic fluorescence intensity fluctuations about the ensemble average.

1. INTRODUCTION

In recent years, laser light scattering and fluorescence techniques have been applied with great success to the study of motion i biological systems. Recent developments and applications of dynamic light scattering are presented in great detail in other sections of this volume. It is the object of this article to introduce and discuss some of the fluorescence techniques that have been developed, and, wherever possible, to compare and contrast them with laser ligh scattering. Several related reviews have recently appeared.[1-7]

The major advantage of fluorescence emission over light scatter ing can be stated in one word: specificity. All components in a complex system scatter light, with little chemical specificity. Mos light scattering experiments, therefore, have been performed with highly purified samples. Even there, however, minor contaminations of large aggregates or dust can interfere with the measurement, out of all proportion to their weight concentrations.

Fluorescence emission, on the other hand, is a spectroscopic property of exquisite specificity. With the right combination of excitation and detection optics one can selectively study the behavior of a particular chemical component. One can sometimes make use of the intrinsic fluorescence from one of the few naturally fluorescent protein or nucleic acid constituents.[8] More often, it i necessary to introduce a fluorescent analog or fluorescent derivativ of a naturally occurring substance, or to label a cellular component with an extrinsic fluorescent probe. Specific examples are presente below.

The key property of elastically scattered laser light is its coherence. Information on the position of each scattering center is contained in the phases of the scattered waves.[9] This is not true for fluorescence, of course. Spontaneous fluorescence emission (as distinguished from stimulated emission in a dye laser) is a random stochastic process. Thus, whereas molecular motions lead to phase fluctuations in the constituent scattered fields, which can be characterized by autocorrelation or spectral analysis of the resultant scattered intensity,[9,10] fluorescence intensities are simply additive. Different strategies have to be devised to extract information from fluorescence measurements, to take advantage of fluorescence specificity. In the sections below, we shall consider three basic experimental approaches - time-resolved emission, fluorescence photobleaching, and fluorescence correlation spectroscopy - which can be used to study rotational and translational motions in biological systems with characteristic times ranging from nanoseconds to minutes. The principal features of these techniques are summarized in Table I.

2. TIME-RESOLVED EMISSION

A. Basic Principles

Light absorption is accompanied by a transition from the ground state (almost always a singlet state) to a higher singlet electronic energy level.[8] Fluorescence emission occurs from the zero vibrational energy level of the first excited singlet state, which is reached through a rapid (< 10^{-10} sec) internal conversion process from higher energy excited states without emission of light.[8] Decay back to the ground state, with the accompanying emission of a fluorescence photon, occurs as a random stochastic process with a mean excited-state lifetime ranging from less than 10^{-9} to almost 10^{-7} sec. This is long enough so that considerable molecular motion can occur between the absorption and emission transitions.

In "time-resolved emission" techniques, one seeks to characterize these molecular motions through measurements of the response of the sample to a brief (psec-nsec) pulse of light. Linear systems analysis[11,12] assures us that the characteristics of the system are completely defined by such an "impulse response function".

A variety of processes compete with fluorescence emission, leading to a nonradiative deexcitation. In special cases, such processes can be used as reporters of molecular motions. Bimolecular contact quenching[13-15], and nonradiative energy transfer[16-18] are both sensitive to the extent of relative motion between the fluorophore and quencher during the excited state lifetime.

Information on rotational motion can be obtained through consideration of fluorescence polarization. Illumination with linearly polarized light excites an anisotropic ensemble of fluorophores - selectively exciting those molecules with absorption transition dipoles near parallel to the incident polarization. More precisely, if the incident field is parallel to the z-axis, the probability of excitation is proportional to

$$|\hat{\mu}_A \cdot \hat{z}|^2 = \cos^2\theta, \qquad (1)$$

where $\hat{\mu}_A$ is a unit vector along the fluorophore absorption dipole, and θ is the angle between $\hat{\mu}_A$ and the z-axis.[19] The polarization of

Table 1. Summary of Principal Features of Fluorescence Techniques

	Time-Resolved Emission	Fluorescence Photobleaching	Fluorescence Correlation Spectroscopy
Perturbing radiation	nsec laser pulse	msec laser pulse	None
Result	Photoselection	Ground-state depletion	Spontaneous fluctuations
Monitoring radiation	None	Attenuated CW laser	Attenuated CW laser
Time scale	$\simeq \tau_F$	$>> \tau_F$	$>> \tau_F$
General data form	Signal averaged emission transient	One-shot transient recovery	Time-averaged correlation function
Measure of rotational motion	Emission anisotropy	Polarized fluorescence depletion	Polarized emission fluctuations
Measure of translational motion	Bimolecular quenching kinetics	Fluorescence redistribution	Number fluctuations

the subsequent fluorescence emission is governed by the distribution of orientations of the emission transition dipoles ($\hat{\mu}_E$) of the "photoselected" ensemble. In the far (radiation) zone, the time-averaged power radiated per unit solid angle by a classical oscillating dipole moment is proportional to

$$|\hat{n} \times (\hat{n} \times \hat{\mu}_E)|^2 = \sin^2\theta' \tag{2}$$

where $\hat{n}$ is a unit vector along the observed direction of propagation, and θ' is the angle between $\hat{\mu}_E$ and $\hat{n}$[20]. The state of polarization, given by the vector inside the absolute value signs above, is perpendicular to $\hat{n}$ and in the plane defined by vectors $\hat{\mu}_E$ and $\hat{n}$.

In general, for an anisotropic sample, a complete characterization requires the determination of nine intensity functions.[21] These are the functions $I_{ij}(t)$, defined as the fluorescence intensity, with polarization along the jth axis of a laboratory-fixed Cartesian coordinate system, detected a time t after excitation by an incident beam with polarization along the ith axis.

For simplicity, we shall consider only isotropic solutions or suspensions viewed at right angles to the incident polarization. In this case, the problem reduces to just two intensity functions, $I_{||}(t)$ and $I_{\perp}(t)$, defined as the fluorescence intensities with polarizations parallel (||) and perpendicular ($\perp$) to the incident polarization. Although the sample is isotropic initially, photoselection by polarized absorption introduces an anisotropy, so that $I_{||}(t)$ and $I_{\perp}(t)$ are different. This anisotropy is conveniently quantitated as[22]

$$r(t) = [I_{||}(t) - I_{\perp}(t)]/I_T(t), \tag{3}$$

where

$$I_T(t) = I_{||}(t) + 2I_{\perp}(t),$$

the total fluorescence intensity independent of polarization, is introduced to cancel out the trivial effect of fluorescence decay after pulsed excitation. From Eq. 2, it is straightforward to show[23] that if $\hat{z}$ is the axis of incident polarization

$$r(t) = \langle P_2(\hat{\mu}_E(t)\cdot\hat{z})\rangle \tag{4}$$

where

$$P_2(x) = \frac{1}{2}(3x^2-1) \tag{5}$$

is the 2nd order Legendre polynomial, and < > signifies an ensemble average. In a multicomponent system, r(t) is a weighted average, with weight factors proportional to the fluorescence intensity from each component at time t.[19]

Eq. 4 can be analyzed further, in a general way, by repeated application of the addition theorem for spherical harmonics (Ref. 20 p. 67). Consider a set of three unit vectors - $\hat{x}_1$, $\hat{x}_2$ and $\hat{x}_3$. If the azimuthal angle of $\hat{x}_2$ with respect to $\hat{x}_1$ is uniformly distribute and independent of that with respect to $\hat{x}_3$, then from Eq. 3.68 in Re 20

$$\langle P_\ell(\hat{x}_1 \cdot \hat{x}_3)\rangle = \langle P_\ell(\hat{x}_1 \cdot \hat{x}_2)\rangle \langle P_\ell(\hat{x}_2 \cdot \hat{x}_3)\rangle. \qquad (6)$$

Thus, equating $\hat{x}_1$, $\hat{x}_2$ and $\hat{x}_3$ with $\hat{\mu}_E(t)$, $\hat{\mu}_A(0)$ and $\hat{z}$, respectively, it follows from Eqs. 4 and 6 that (neglecting instrumental effects)

$$r(t) = \langle P_2(\hat{\mu}_E(t) \cdot \hat{\mu}_A(0))\rangle \langle P_2(\hat{\mu}_A(0) \cdot \hat{z})\rangle, \qquad (7)$$

where, from Eq. 1, we calculate

$$\langle P_2(\hat{\mu}_A(0) \cdot \hat{z})\rangle = 2/5. \qquad (8)$$

A second application of the general rule of Eq. 6, this time equatin $\hat{x}_1$, $\hat{x}_2$ and $\hat{x}_3$ with $\hat{\mu}_E(t)$, $\hat{\mu}_E(0)$ and $\hat{\mu}_A(0)$, respectively, yields

$$r(t) = (2/5)\ \langle P_2(\hat{\mu}_E(t) \cdot \hat{\mu}_E(0))\rangle \langle P_2(\hat{\mu}_E(0) \cdot \hat{\mu}_A(0))\rangle. \qquad (9)$$

In the simplest case of isotropic rotational diffusion [24] with diffusion coefficient D_{rot}, $r(t)/r(0)$ is a single exponential with characteristic time $(6D_{rot})^{-1}$. This can be seen in analogy to the problem of lateral diffusion (with diffusion constant D) on the surface of a sphere of radius a (Picture the tip of $\hat{\mu}_E$ constrained t diffuse on the surface of a unit sphere). In that case, the general solution for azimuthal symmetry[25] is a sum of Legendre polynomials (of order ℓ) with amplitudes which decay exponentially with time constants $[\ell(\ell + 1)D/a^2]^{-1}$.

For an anisotropic rotator, r(t) assumes the form of a complex sum of five exponentials.[19] For restricted rotation, such as that observed for probes embedded in membranes,[26-28] $r(t)/r(0)$ decays as $t \to \infty$ to a non-zero background, since a totally isotropic distribution cannot be reached. A final application of Eq. 6, equating $\hat{x}_1$, $\hat{x}_2$ and $\hat{x}_3$ with $\hat{\mu}_E(\infty)$, $\langle\hat{\mu}_E\rangle$ (the average of vector $\hat{\mu}_E(t)$), and $\hat{\mu}_E(0)$, respectively, yields

$$r(\infty)/r(0) = S^2 \qquad (10)$$

where S is the time-independent ensemble average,

$$S = \langle P_2(\hat{\mu}_E \cdot \langle\hat{\mu}_E\rangle)\rangle, \qquad (11)$$

commonly known as the order parameter of $\hat{\mu}_E$.[27,28] Note that in a totally "ordered" system, in the absence of motion on the time-scale of the experiment, $\hat{\mu}_E(t) = \langle\hat{\mu}_E\rangle$, so that S = 1. For a "disordered" system, in which $\hat{\mu}_E$ is free to move isotropically about (the now arbitrarily selected) $<\hat{\mu}_E>$, $<\hat{\mu}_E \cdot <\hat{\mu}_E>> = 1/3$, and $S = 0$. If $\hat{\mu}_E$ is assumed to wobble freely and uniformly over a limited range in a cone of semiangle θ_c about $<\hat{\mu}_E>$ (the wobbling-in-cone model[26]), then

$$S = \tfrac{1}{2} \cos \theta_c \ (1 + \cos \theta_c). \quad (12)$$

B. Experimental Methods

Figure 1 presents a schematic diagram of a standard time-correlated single-photon counting spectrofluorimeter[30-32], interfaced to the optics of a fluorescence microscope. The light source is a thyratron-gated low-pressure flash lamp. A time-to-amplitude converter (TAC) coupled to a multichannel analyzer (MCA) generates a distribution function of the time elapsed between the excitation pulse and the detection of the first fluorescence photon. The TAC is gated by the output of a single channel analyzer (TSCA) so that only data arising from pulses generating a single detected fluorescence photon are accumulated. This prevents the biasing of the distribution toward shorter delay times which would result if multiple photon events were included. The optimal data rate is achieved when the probability of detecting a single photon ($<n>e^{-<n>}$) is maximal; i.e. when $<n> = 1$.

Data recorded from such an instrument are the impulse response functions of the fluorescent sample convoluted with an instrumental function; which, in turn, is a convolution of the incident pulse intensity as a function of time with the impulse response function of the measuring system (limited in this case by the transit time spread of electrons in the photomultiplier tubes). Several procedures have been developed[32-36] to perform the deconvolution necessary to extract the true impulse response function of the sample.

The last few years have seen significant technical advances in both illumination and detection systems. Synchrotron radiation[37] provides high intensity, subnanosecond pulses across the entire UV-visible spectrum. Synchronously pumped, cavity dumped dye lasers[38,39] are available which produce high repetition rate psec pulses.

The introduction of streak cameras[40] coupled to silicon intensified target (SIT) optical multichannel analyzers (OMA) provides a detector system with the time-resolution (< 10 psec) and data throughout capacity to take advantage of the new light sources. Unlike the single-photon counting photomultiplier tube detector, streak cameras record complete multiple photon transients, fully utilizing the high intensity pulses.

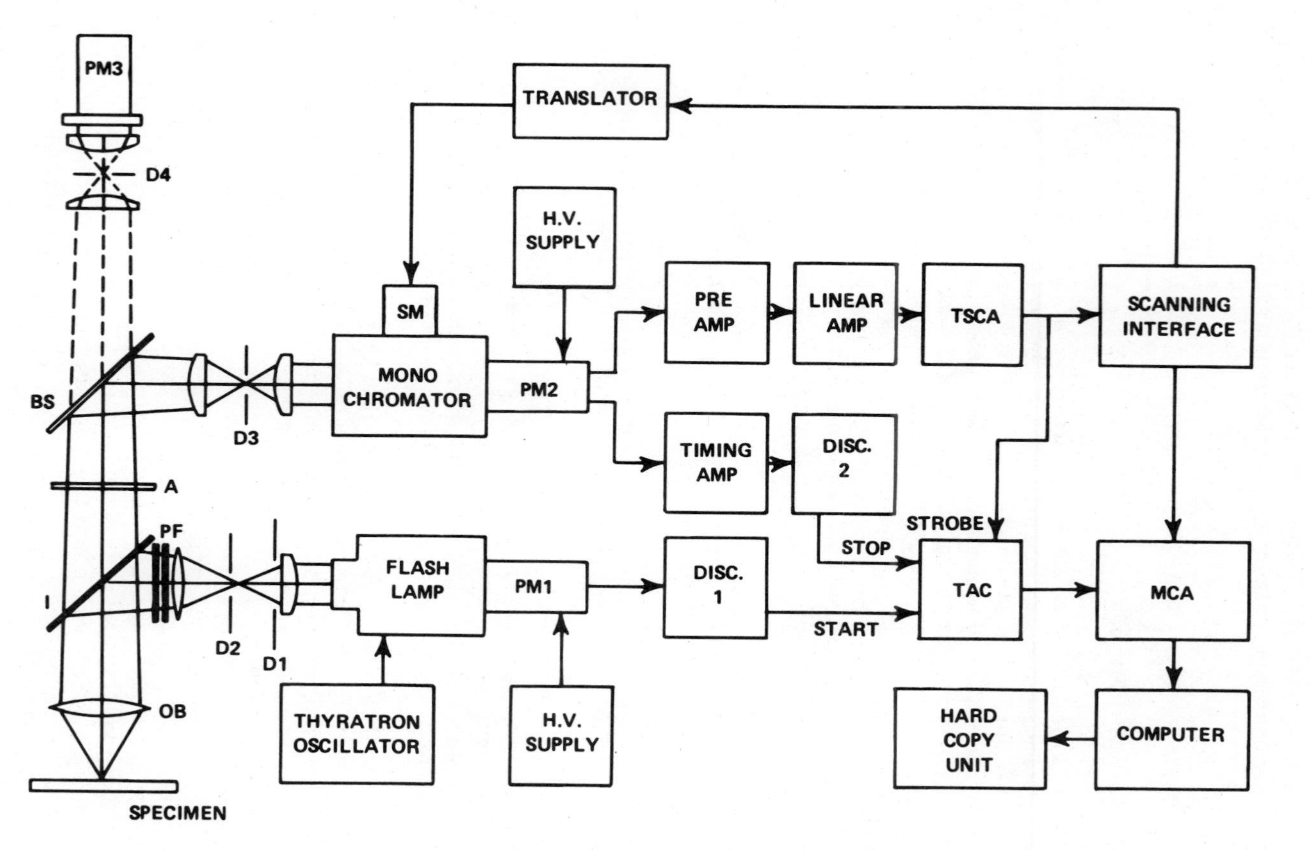

Figure 1. Schematic diagram for time-correlated photon counting spectrofluorimeter. Reprinted with permission from Ref. 29.

C. Applications and Sample Data

In special circumstances, time-resolved emission measurements can be used to study translational motion. In one notable example of this, Haas et al.[17] determined the diffusion coefficients of the ends of oligopeptide chains in solution by analyzing the effect of diffusion on non-radiative excitation energy transfer. The large majority of applications, however, have studied the rotational relaxation of time-resolved emission anisotropy.

Figure 2 reproduces anisotropy data recorded with the pulsed microspectrofluorimeter system diagramed in Fig. 1. The nondeconvoluted emission anisotropy is shown for bovine serum albumin (BSA) non-covalently labeled with the amphipathic fluorescent probe 1-anilino-naphthalene-8-sulfonate (ANS). The inset presents the raw $I_{||}(t)$ (top) and $I_{\perp}(t)$ (bottom) decay curves. The solid line superimposed on the anisotropy data is the convolution of the measured instrumental function with a least-squares best-fit single exponential decay. This analysis yields a rotational correlation time of 41.2 nsec, corresponding to the rotational diffusion of the whole BSA molecule.

Figure 3 presents an example of restricted rotation. The sample under study is a fluorescent coumarin derivative of α-bungarotoxin[41], a low molecular weight protein that binds specifically and irreversibly to acetylcholine receptors. Fluorescently labeled toxin has proven to be an extremely useful probe of the distribution and dynamics of acetylcholine receptors in membranes[42,43]. The anisotropy data in the figure were recorded on solutions in regular buffer (+), and in 75% glycerol (●). In the first case, the data can again be accurately fit to a single exponential decay (solid line), corresponding to rotation of the whole protein molecule. In 75% glycerol, however, the protein motion is essentially frozen, demonstrated by the fact that r(t) levels off to a residual value close to the initial anisotropy obtained for the toxin in buffer. In addition, a new component of the motion appears, corresponding to a restricted localized probe rotation. In buffer, this motion is too fast to be observed with this instrument. By Eq. 11, the measured value of $r(\infty)/r(0)$ in glycerol is consistant with an order parameter of 0.84. In the wobble-in-cone model (Eq. 12) this corresponds to a cone semiangle of 27°.

Recently, the analysis of restricted rotational motion of fluorescent probes in membranes has proved to be of particular interest[26-28,36]. It has been shown[28], for example, that the high steady-state emission anisotropy (measured with CW illumination) observed for diphenylhexatriene in lipid vesicles below the phase transition is not simply a result of slow rotation or high "microviscosity" (as would be deduced assuming isotropic rotation). Rather, it reflects the high degree of order in the system.

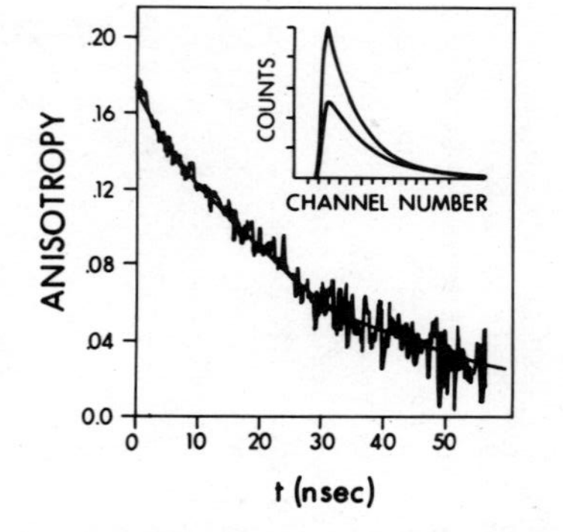

Figure 2. Fluorescence emission anisotropy data for BSA non-covalently labeled with the fluorescent probe ANS. Inset shows the separate decay curves for $I_{||}(t)$ and $I_{\perp}(t)$. Reprinted with permission from Ref. 29.

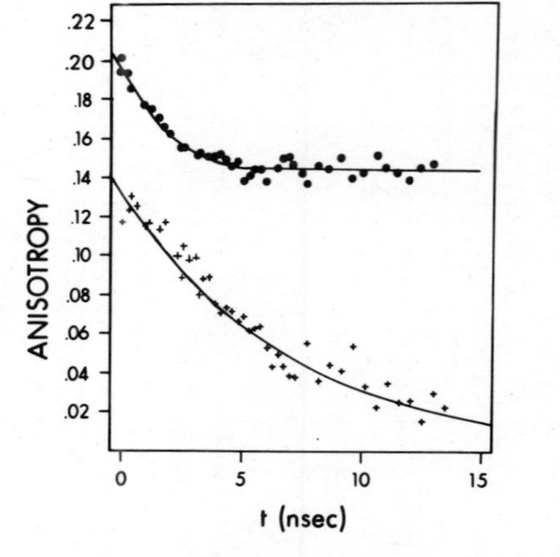

Figure 3. Anisotropy data for coumarin labeled α-bungarotoxin in regular buffer (+), and in 75% glycerol (●). Reprinted with permission from Ref. 41.

2. FLUORESCENCE PHOTOBLEACHING

A. Basic Principles

High intensity laser pulses can promote the transition of a significant fraction of the total population of fluorophores to long-lived spectroscopically distinct states - either excited triplet-states (reached through intersystem crossing from the first excited singlet state), or irreversibly photo-oxidized products. Photobleaching with a localized or nonuniform bleaching beam produces a nonuniform concentration of unbleached fluorophore. Subsequent illumination with an attenuated CW laser beam can be used to monitor the fluorescence redistribution after photobleaching[1-7]. Photobleaching with a polarized bleaching beam produces an anisotropic orientation distribution of unbleached fluorophore. Subsequent illumination with an attenuated, polarized CW laser beam can be used to monitor the time-resolved anisotropy of polarized fluorescence depletion.[7,44,45] In this way, measurements of translational and rotational motion can be extended to time-scales beyond that associated with the ordinary singlet-state lifetime.

Fluorescence photobleaching has been applied most often to the study of lateral diffusion in a plane - the geometry corresponding to extended regions of cell plasma membranes, reconstituted planar membranes, or thin layers of solution or cytoplasm. Two basic approaches have been taken. The first of these uses spatially localized photobleaching pulses, placing primary emphasis on the analysis of distributions in real coordinate space.[46,47] The other employs an extended periodic bleaching pattern[48-51], shifting the analytical emphasis to Fourier transform space. A simple, general analysis of the theory in both coordinate and Fourier space, utilizing the general mathematical formalism of linear shift invariant systems analysis, is presented in Ref. 7.

Lateral diffusion in a plane can be characterized through measurements of $F(\underset{\sim}{r},t)$, the fluorescence intensity excited with a monitoring beam centered at position $\underset{\sim}{r}$ at time t after photobleaching. In general[7,47],

$$F(r,t) = G(\underset{\sim}{r},t) ** F(\underset{\sim}{r},0), \tag{13}$$

the two-dimensional convolution of $F(\underset{\sim}{r},0)$, the fluorescence scan profile at time zero, with $G(\underset{\sim}{r},t)$, the impulse response function of the concentration of fluorescent label. For isotropic diffusion with diffusion coefficient D,

$$G(\underset{\sim}{r},t) = (4\pi Dt)^{-1} \exp(-|\underset{\sim}{r}|^2/4Dt). \tag{14}$$

Explicit forms of $F(\underset{\sim}{r},t)$ can be calculated for particular forms of $F(\underset{\sim}{r},0)$. Bleaching with a focussed laser beam centered at $\underset{\sim}{r} = 0$

produces a localized fluorescence depletion that, to a good approximation, can be represented as a Gaussian profile; ie

$$F(\underset{\sim}{r},0) = F(-)[1-\alpha_o \exp(-|\underset{\sim}{r}|^2/w_o^2)], \quad (15)$$

where α_o is a measure of the extent of bleaching, and w_o, the 1/e radius of the initial scan profile, corresponds to an effective $1/e^2$ laser beam radius. In this case, by Eqs. 13 and 14,

$$F(\underset{\sim}{r},t) = F(-)\{1-\alpha(t)\exp[-|\underset{\sim}{r}|^2/w^2(t)]\}, \quad (16)$$

with

$$\alpha(t) = \alpha_o/(1+t/\tau_D)$$

$$w^2(t) = w_o^2(1+t/\tau_D)$$

where

$$\tau_D = w_o^2/4D$$

is the characteristic time for diffusion.

The problem is simplified significantly in Fourier transform space. Transforming both sides of Eq. 13 gives

$$\tilde{F}(\underset{\sim}{k},t) = \tilde{F}(\underset{\sim}{k},0)\tilde{G}(\underset{\sim}{k},t), \quad (17)$$

where for the $G(\underset{\sim}{r},t)$ of Eq. 14,

$$\tilde{G}(\underset{\sim}{k},t) = \exp(-D|\underset{\sim}{k}|^2 t). \quad (18)$$

Thus, each measured Fourier component decays as a simple exponential. This is analogous to the case of laser light scattering, where the electric field amplitude of light scattered with scattering vector $\underset{\sim}{K}$ is proportional to the spatial Fourier transform of the concentration distribution evaluated at wave vector $\underset{\sim}{K}$. Photobleaching with a periodic pattern of stripes serves to maximize the initial amplitude of the Fourier component corresponding to the fundamental period of the pattern[48-51].

For the analysis of diffusion on the surface of a sphere, the geometry appropriate for the membranes of cells in suspension or for large membrane vesicles, one has to take a different approach. For an azimuthally symmetric concentration distribution as a function of x, the cosine of equatorial angle θ on a spherical surface of radius r, the diffusion equation has the general solution[25]

$$c(x,t) = \sum_{n=0}^{\infty} A_n P_n(x)\exp[-n(n+1)Dt/r^2] \quad (19)$$

In the "normal mode" analysis[52], one computes experimental estimates of

$$\mu_n(t) = \int_{-1}^{1} P_n(x)c(x,t)dx / \int_{-1}^{1} P_o(x)c(x,t)dx, \quad (20)$$

calculated from a series of fluorescence scans across the sphere at times t after photobleaching. Combining Eqs. 19 and 20, applying the orthogonality relation for the Legendre polynomials (Ref. 20, Eq. 3.21), it follows directly that

$$\mu_n(t) = [A_n/(2n+1)A_0]\exp[-n(n+1)Dt/r^2]. \quad (21)$$

Bleaching an "edge" of the sphere serves to maximize the initial value of $\mu_1(t)$.

For the measurement of slow rotational motion[7,44,45], one determines the values of polarized fluorescence depletion, defined as

$$\Delta F_i(t) = F_i(-) - F_i(t), \quad (22)$$

where $F_i(-)$ and $F_i(t)$ are the total fluorescence intensities (for all emitted polarizations) observed with an excitation polarization along the ith axis before bleaching (-) and at time t after bleaching. This approach is directly analogous to earlier studies of transient linear dichroism with naturally occurring membrane chromophores[53], and extrinsic triplet probes[2].

B. Experimental Methods

Figure 4 shows a schematic diagram of an optical system designed for the measurement of lateral motions with the photobleaching technique.[47] As shown, the apparatus is centered about a modified research microscope equipped with a fluorescence vertical illuminator. The light source for both monitoring and bleaching is a CW water-cooled argon ion laser which provides up to ∿ 1 W of power in any of several discrete lines in the blue and green.

The input optics of the microscope perform two principal functions: scanning the laser beam across the sample, and switching between the monitoring and bleaching beams. A servoactivated galvanometric optical scanning mirror (SM) controls the precise input angle of the incident laser beam, and hence the position on the sample of the beam along a scan axis. The combination of diaphragms D1 and D2, uncoated beamsplitters BS1 and BS2, and shutter Sr2 in Fig. 4 work together to switch between monitoring and bleaching. With Sr2 open, a nearly unattenuated beam, transmitted through BS1 and BS2 without reflection, passes through D2, providing the bleaching beam. With Sr2 closed, only that beam reflected four times (once at each glass-to-air interface) is in a position to pass through D2.

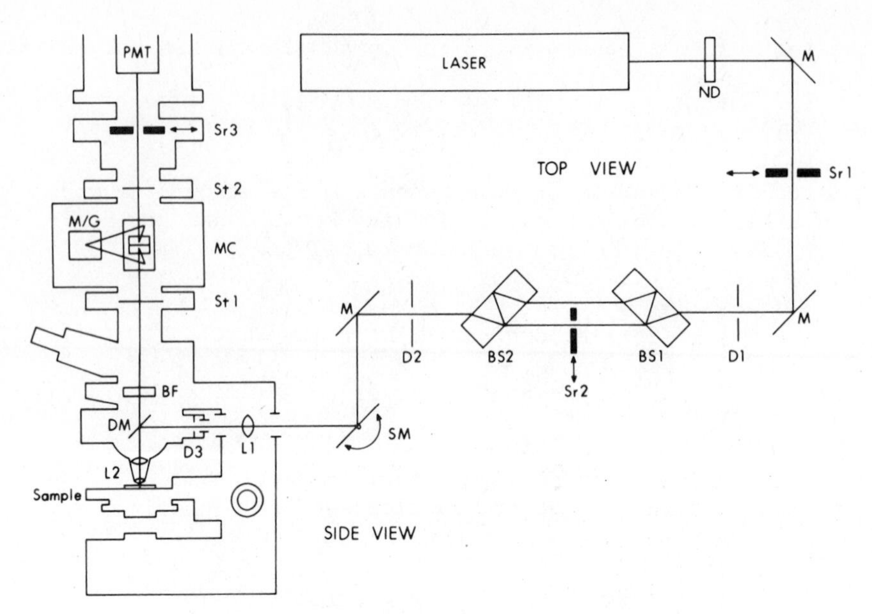

Figure 4. Schematic diagram of optical apparatus for fluorescence photobleaching technique. The symbols used are: ND, neutral density filter; Sr, shutter; M, mirror; D, diaphragm; BS, beam splitter; SM, scanner mirror; L, lens; DM, dichroic mirror; BF, barrier filter; St, slit; MC, monochrometer housing; M/G, mirror or grating; PMT, photomultiplier tube. Reprinted with permission from Ref. 47.

This attenuates the beam by a factor of about 10^4, producing the monitoring beam. Note that in this configuration, with the bleaching and monitoring beams physically separated between BS1 and BS2, either beam can be selectively modified in special cases (profile shaped, polarization rotated) without affecting the other. Further details are presented elsewhere.[7,47,54]

Johnson and Garland[45] have recently described a modified photobleaching apparatus designed to measure rotational motion (see also Ref. 44). An acousto-optic modulator is used to switch between the monitoring and bleaching intensities, with a response time of < 1 μsec. A Pockels cell is used without polarizers to automatically rotate the plane of polarization of the laser beam through 90°, making it possible to determine the anisotropy of fluorescence depletion.

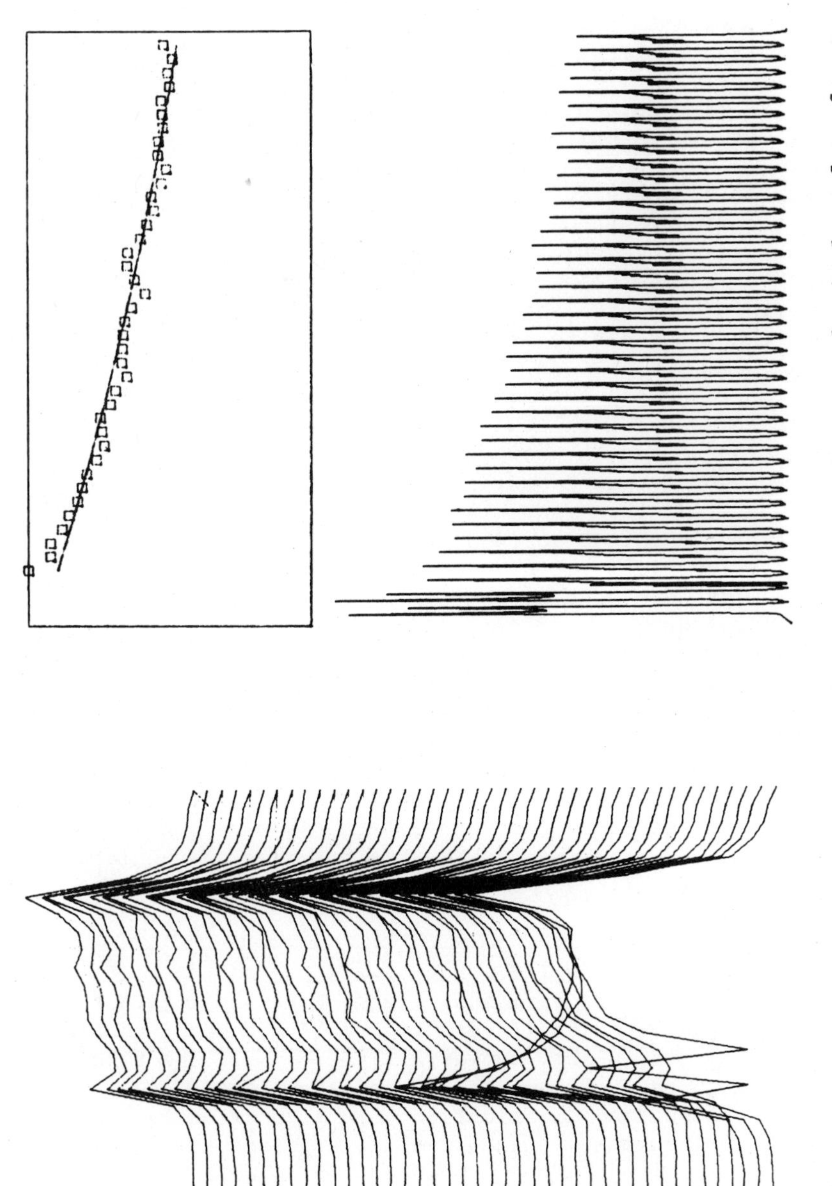

Figure 5. Normal-mode analysis of lateral diffusion on a spherical surface for an osmotically swollen erythrocyte covalently labeled with dichlorotriazinylaminofluorescein. See text.

C. Applications and Sample Data

Fluorescence photobleaching has been widely applied to the study of translational diffusion in membranes. For extensive reviews of the literature, see Refs. 2, 3 and 6. Recently, its use has been extended to the study of translational dynamics in solution[4,55,56] and in cell cytoplasm[57], and to the study of slow rotational motion.[44,45] The data presented below were selected from work in my own laboratory as examples of the capabilities of the technique.

Figure 5 presents an example of the analysis of diffusion on the surface of a sphere, an osmotically swollen human erythrocyte covalently labeled with dichlorotriazinylaminofluorescein (DTAF), a derivative of fluorescein which reacts with amino groups. When intact erythrocytes are labeled with DTAF at pH 9.5, roughly two-thirds of the bound dye is attached to band 3 protein[58].

The distribution of fluorescence on the labeled cell was followed with a series of scans with a focused laser spot. The lower righthand side of Fig. 5 shows the entire linear sequence of 1008 data points (484 sec full scale), displayed as the series of lines connecting the points. The left side of the Figure displays the data again, rearranged this time into a stack of individual 24 point scans. Before photobleaching, the fluorescence distribution is symmetric, with sharp peaks due to the geometrical edge effect. Photobleaching at the leading edge of the cell, during the third scan produces a marked asymmetry, which decays as the labeled protein molecules redistribute over the surface of the sphere.

The values of $\mu_1(t)$ (see Eq. 20), calculated for each individual scan, are displayed in the upper righthand side of the Figure, along with the best least-squares fit to a single exponential decay (see Eq. 21). It is seen that the actual decay is more complex. Here, as with most membrane proteins, a sizeable fraction of the labeled molecules diffuse very much slower (if at all) than the rest of the population[2,3,6].

The next two Figures illustrate two different approaches to the analysis of diffusion in a plane. Figure 6 presents data for a reconstituted planar multibilayer membrane[59] bleached and monitored with a localized circularly symmetric laser spot. The fluorescence intensity, measured at 12 equally spaced locations along a linear scan axis, maps out the distribution of a fluorescently labeled phospholipid, N-4-nitrobenzo-2-oxa-1,3-diazole phosphotidylethanolamine (NBD-PE), incorporated into the membrane. The solid curves are a fit to diffusion theory (see Eq. 16) for the data as a function of position and time.

Figure 7 demonstrates a version of the periodic pattern bleaching method similar to that described by Lanni and Ware[51]. A

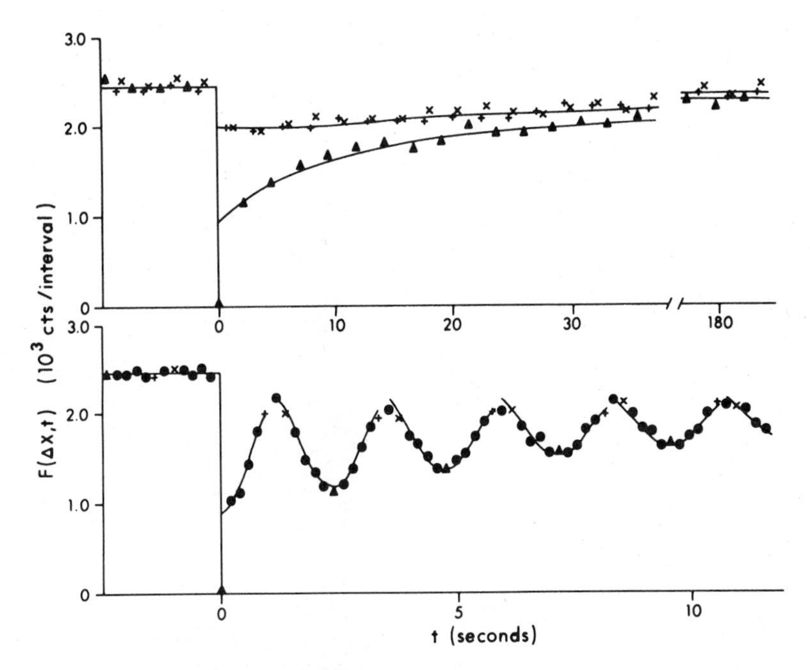

Figure 6. Fluorescence redistribution after photobleaching of NBD-PE in a reconstituted multibilayer membrane (1:1.3:1 by weight phospholipid to matrix protein to lipopolysaccharide), monitored at 12 equally spaced spots (1.0 μm separation) about a single bleach point. Top: Recoveries measured at 3 locations; coincident with the bleach (▲) and 5.0 μm on either side of center (+,x). Bottom: Sequential scans including all 12 locations on expanded time scale. Solid curves are a fit to theory with D=4.71 x 10^{-9} cm^2/sec. Reprinted with permission from Ref. 47.

laser illuminated Ronchi ruling is imaged down onto a 3T3 fibroblast in culture labeled with fluorescein conjugated succinyl-concanavalin A, forming a periodic pattern of stripes (1.8 μm period). Bleaching in this configuration produces a periodic pattern of fluorescent molecules. The modulation amplitude of this concentration pattern is monitored as a function of time after bleaching with a series of fluorescence scans of the monitoring illumination pattern perpendicular to the direction of the stripes.

As in Fig. 5, the total linear sequence of data (484 sec full scale) and the stack of individual 24 point scans are displayed in the lower righthand side and the left side of Fig. 7, respectively. Fourier analysis gives the amplitude and phase of the intensity modulation of each individual scan, with the modulation amplitude (normalized by the average intensity) plotted as a function of time

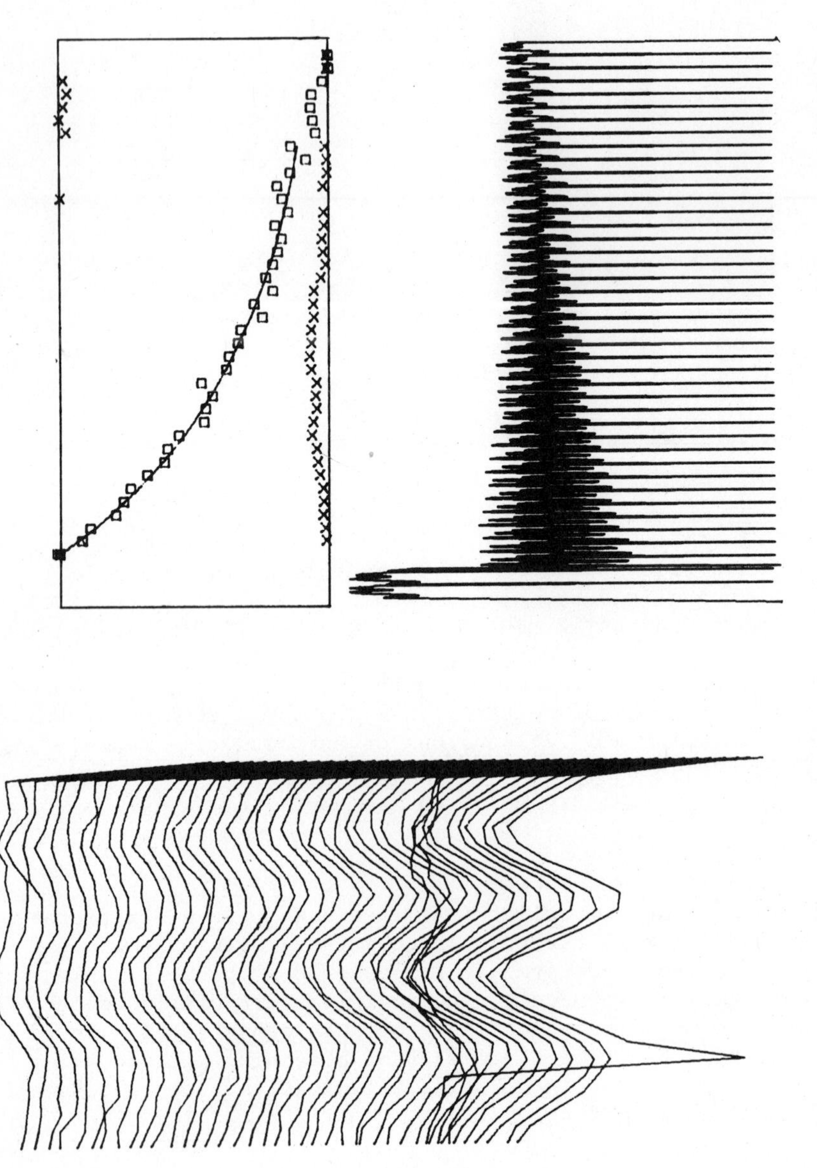

Figure 7. Fourier transform analysis of periodic pattern photobleaching for a 3T3 fibroblast labeled with fluorescein-conjugated succinyl concanavalin A. See text.

in the upper righthand side of the Figure. The solid curve is the best least-squares fit to a single exponential decay (see Eqs. 17 and 18).

Combining spot bleaching (Fig. 6) with periodic pattern bleaching (Fig. 7), we have now developed a promising new approach which allows for measurements over two very different distance scales simultaneously. Briefly, the sample is bleached over a circular area (of radius R) with a pattern of stripes (of period distance $d \ll R$). Subsequently, the redistribution of fluorescence is monitored with a series of scans of the attenuated striped pattern across the circular area. Thus, we simultaneously measure motions over the characteristic distances R (which lead to an increase of the average fluorescence intensity detected within the illuminated region) and d (which lead to a decrease of the amplitude of periodic intensity modulation). For a diffusion process, the characteristic recovery times for the two cases differ by a factor of $(\pi R/d)^2$ (see Eqs. 16-18). This methodology should prove to be extremely useful for the analysis of systems in which one has a wide range of diffusion coefficients, and for the separation of effects due to diffusion and association kinetics.[60]

Figure 8 presents examples of this approach, preliminary data designed to examine the interactions of fluorescently labeled spectrin, the principal component of the erythrocyte membrane skeleton (or matrix), with polymerized actin. The bottom trace, recorded on an extended aggregate of fluorescently labeled erythrocyte membrane skeletons, demonstrates one limiting case. The fluorescence intensity does not recover. Moreover, there is a substantial induced modulation which does not decay. Virtually all of the label is essentially immobile, with a diffusion coefficient $\ll 10^{-11}$ cm^2/sec.

The top data trace was recorded on a solution of fluorescently labeled spectrin alone, with $(\pi R/d)^2 \simeq 250$. In this case, almost all of the fluorescence recovers with a characteristic time of about 35 sec, corresponding to a diffusion coefficient of about 2×10^{-7} cm^2/sec. The corresponding decay time for the modulation amplitude ($\sim$ 0.15 sec) is too short to resolve, and indeed, no modulation above the prebleach background level is observed.

This sets the stage for an evaluation of the middle trace. The observed rate and extent of fluorescence recovery, and the periodic modulation amplitude and decay rate, are all intermediate between the two limiting cases. This demonstrates the effects of the dynamic association between the labeled spectrin and polymerized actin.[60]

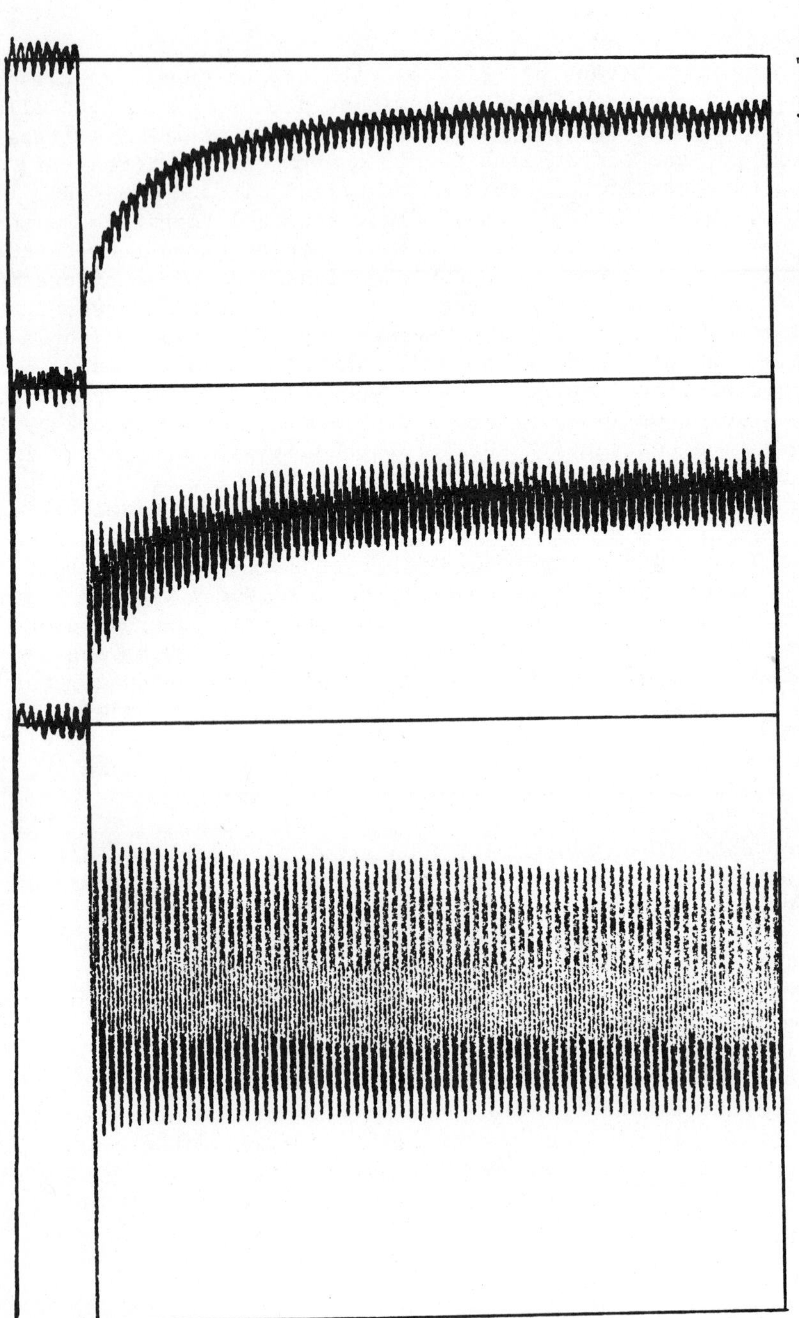

Figure 8. Fluorescence scan data for fluorescein labeled spectrin bleached over a circular area (85 μm diam.) with periodic striped pattern (8.5 μm period). Top: Spectrin alone in bulk solution. Middle: Spectrin interacting with F-actin gel. Bottom: Labeled erythrocyte cytoskeleton aggregates. Full-scale on horizontal time axis equals 420 sec (Sheetz and Koppel, unpublished data).

4. FLUORESCENCE CORRELATION SPECTROSCOPY

A. Basic Principles

Conventional fluorescence measurements on macroscopic samples at equilibrium ordinarily yield equilibrium properties (average concentrations, steady-state absorption or emission anisotropies). If, however, the observation volume is reduced to the point where it contains only a few thousand fluorescent molecules, then, under special circumstances, spontaneous stochastic fluctuations about equilibrium can be observed. These can correspond to concentration or "occupation-number" fluctuations[61,62], as molecules move in and out of the observation volume or undergo chemical transformations[63]. Alternatively, fluctuations can arise from changes is the orientations of the absorption or emission dipoles (see Eqs. 1 and 2). In the fluorescence correlation spectroscopy (FCS) technique[1,4,61-65], measurements of the time correlation function of fluorescence fluctuations are used to characterize the dynamics of these processes.

Figure 9 illustrates the application of FCS to the study of lateral diffusion in a planar membrane. The incident laser intensity is kept constant; but the fluorescence intensity, I(t), changes nevertheless as fluorescent probe molecules move relative to the laser beam, changing the number and distribution of molecules in the illuminated region. Photocount autocorrelation gives experimental estimates of the normalized fluorescence intensity correlation function above background; ie

$$g(\tau) = \langle \delta I(t)\delta I(t+\tau)\rangle / \langle I(t)\rangle^2, \qquad (23)$$

where

$$\delta I(t) = I(t) - \langle I(t)\rangle .$$

For spontaneous fluctuations in the concentration of fluorophore[63],

$$g(\tau) = \int d^3\underset{\sim}{r}\, d^3\underset{\sim}{r}'\, I_o(\underset{\sim}{r}) I_o(\underset{\sim}{r}') \langle \delta c(\underset{\sim}{r},t)\delta c(\underset{\sim}{r}',t+\tau)\rangle \div \left[\int d^3\underset{\sim}{r}\, I_o(\underset{\sim}{r}) \langle c(\underset{\sim}{r},t)\rangle\right]^2, \qquad (24)$$

where $I_o(\underset{\sim}{r})$ is the intensity profile of the incident laser beam.

The extrapolated intercept, $\beta = g(0)$, is of special interest; for it provides an absolute determination of number density. For a Gaussian beam profile with $1/e^2$ radius w,[63]

$$1/\beta = \langle N\rangle, \qquad (25)$$

the average number of independent fluorophores within radius w. For diffusive transport, with lateral diffusion coefficient D[63],

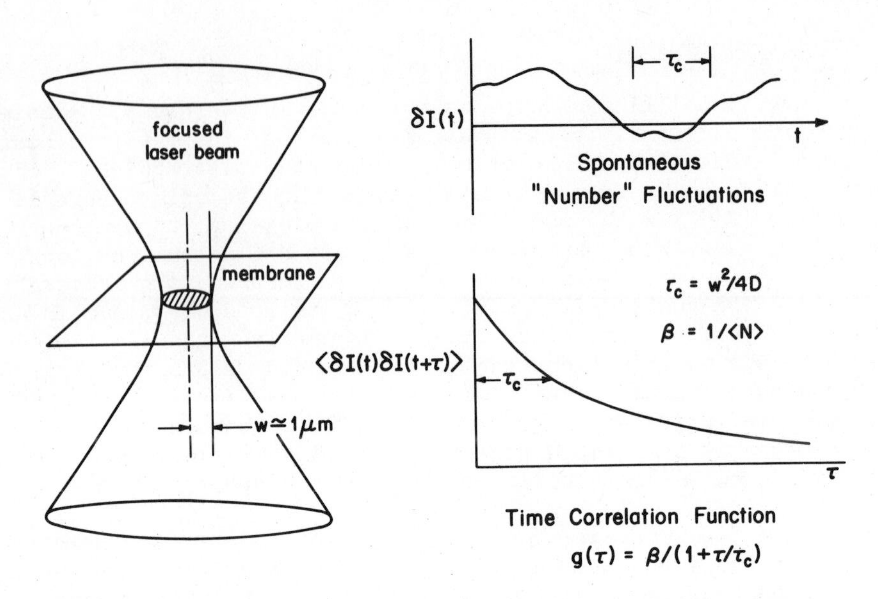

Figure 9. Fluorescence correlation spectroscopy (FCS) illustrated with example of lateral diffusion in a membrane.

$$g(\tau) = \beta/(1+4D\tau/w^2). \tag{26}$$

FCS can be used, in principle, to study rotational motion as well[66-69], providing an additional method of extending measurements of this type beyond the limits of the fluorophore excited-state lifetime. If the sample is illuminated with linearly polarized incident light, the total instantaneous fluorescence intensity is a function of the instantaneous distribution of $\hat{\mu}_A$. Polarized fluorescence components are functions of the distribution of $\hat{\mu}_E$ as well. Autocorrelation functions of the fluorescence intensity, in either case, contain terms, added to that of Eq. 26, which reflect the relaxation of orientation fluctuations.[66-68]

Other physical monitors of spontaneous number fluctuations about equilibrium have been employed.[70-73] The principal advantages of fluorescence detection for this purpose are again its specificity and sensitivity. For a point of reference, in this regard, it is of interest to compare the detected intensities of fluorescence and scattered laser light. For Rayleigh light scattering, the scattering cross section for a single macromolecule of molecular weight M in solution equals[74]

$$\sigma_{LS} = \frac{32}{3} \pi^3 n_o^2 (\frac{dn}{dc})^2 \frac{M^2}{N_A^2 \lambda 4}, \tag{27}$$

where n_o is the index of refraction of the scattering medium, dn/dc is the refractive index increment (in units of cm^3/gm), N_A is Avogadro's number, and λ is the wavelength of light (in units of cm).

The analogous cross section for fluorescence can be defined as the product of the absorption cross section[74] and ϕ_F, the emission quantum efficiency. This gives

$$\sigma_F = 1000 \ \ell n 10 \ \varepsilon \phi_F / N_A, \tag{28}$$

where ε is the molecular extinction coefficient expressed in units of liters/mole/cm. For typical values of n_o (1.33), and dn/dc (0.2 cm^3/g), this gives the ratio

$$\sigma_F/\sigma_{LS} \simeq (6.1 \times 10^{25} \ cm^{-6}) \varepsilon \phi_F \lambda^4 / M^2. \tag{29}$$

Consider, for example, a solution of bovine serum albumin (M = 66,000) labeled with one molecule of fluorescein isothiocyanate ($\phi_F \simeq 0.3$) per molecule, illuminated with an incident laser beam (with λ = 488 nm) near the peak of fluorescein absorption (where ε = 4.2×10^4). By Eq. 29, we calculate that the fluorescence will be brighter than the scattered light by a factor of 1.0×10^3.

Eq. 29, in fact, severely underestimates the relative strength of fluorescence signals. Scattered laser light must be collected in an area small compared to a coherence area[75-77] in order to preserve the field statistics required for coherent interference. One collects over a solid angle of only about $(\lambda/L)^2$ sr, where L is the linear dimension of the scattering volume, with $(\lambda/L)^2$ typically $< 2 \times 10^{-6}$ sr. For fluorescence experiments, high numerical aperture optics can collect the fluorescence over close to 2π sr. Extraordinary detection sensitivity is essentially for a successful FCS experiment. It has been shown[78], that whereas the signal-to-noise figure of merit for the laser light scattering technique is the number of photocounts per coherence area per correlation time, the corresponding parameter for FCS is the number of photocounts per fluorescent molecule per correlation time.

B. Experimental Methods

FCS experiments can be performed with the same basic optical configuration used for fluorescence photobleaching experiments[1]. The raw data consists of a series of numbers corresponding to the numbers of fluorescence photons detected in each of a series of contiguous counting intervals. Correlation analysis is conveniently effected with a dedicated minicomputer. A software correlator allows great flexibility in data processing; permitting, for example, the compensation for slow photochemical drifts.[69]

We are currently implementing a correlation system incorporatin the beam scanning mirror shown in Fig. 4. This will make it possibl to compute photocount correlations as functions of both time and space. The focused beam will jump back and forth between consecutiv positions in two widely separated linear scans. Taking the differences between consecutive data from corresponding positions in the two scans will effect an immediate cancellation of systematic drifts and instrumental fluctuations.

C. Applications and Sample Data

Figure 10 illustrates the FCS technique with data recorded on bulk solutions of fluorescent molecules in thin flat capillary tubes

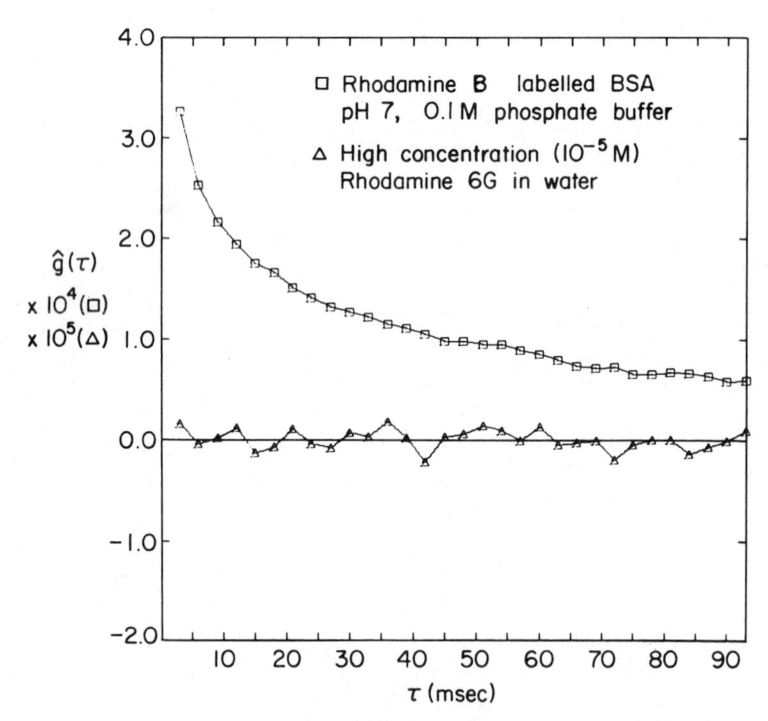

Figure 10. Normalized time autocorrelation functions of fluorescenc intensity fluctuations for solutions of 3×10^{-8} M rhodamine B labeled BSA (□), and 1×10^{-5} M rhodamine 6G (△).

The collection volume was defined optically - limited in two dimensions by the profile of a focused Gaussian laser beam (w = 1.4 μm), and in the third dimension by the fluorescence collection efficiency of the microscope optics[1]. The top trace (□) is data for a dilute solution of rhodamine B labeled BSA. A well defined correlation above background is observed. The calculated intercept gives an <N> of 2500 (see Eq. 25), consistent with an average concentration of 3×10^{-8} M in an effective collection volume of 1.4×10^{-10} ml. The initial slope of the function corresponds to a value of $D = 6 \times 10^{-7}$ cm^2/sec.

As a control, the bottom trace (Δ, on a 10× expanded vertical scale) was recorded on a rhodamine 6G dye solution with a concentration more than 300 times as high. The relative number fluctuations, in this case, are too small to be measured.

Figure 11 presents data recorded on a planar lipid bilayer membrane[79] containing trace amounts of diI carbocyanine dye.[80] As indicated in (a), the three correlation functions correspond to three

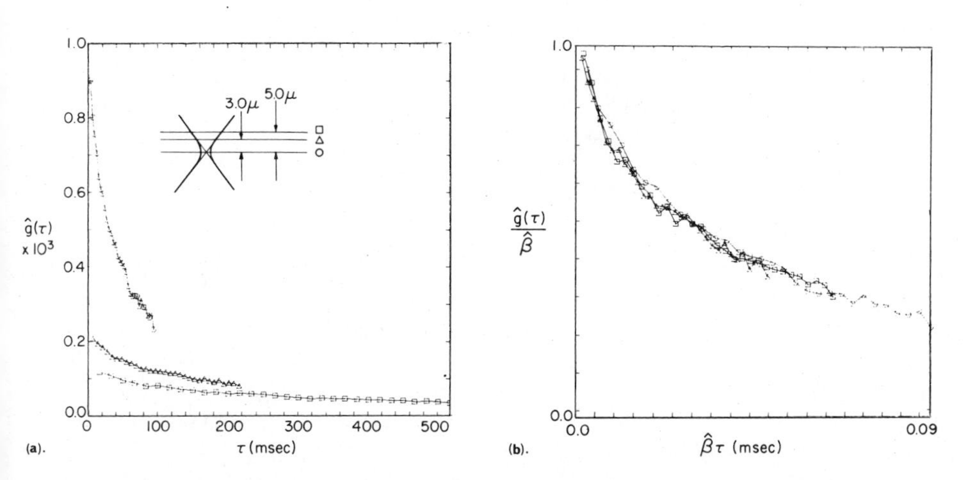

Figure 11. Fluorescence photocount autocorrelation functions of diI carbocyanine dye in a planar lipid bilayer membrane at 12°C formed from a 1:2 molar ratio mixture of egg lecithin:7,dehydrocholesterol in octane. (a) Measured, as indicated, with different laser beam sizes. (b) The same function, each normalized and scaled by $\hat{\beta}$, its computer-fitted intercept. Reproduced with permission from Ref. 1.

different beam sizes, produced by moving the membrane relative to the focused laser beam. In (b), the data are redrawn with the amplitudes and time axes appropriately normalized and scaled to give the universal curve $g(\beta\tau)/\beta$. Combining Eqs. 25 and 26 above, we expect

$$g(\beta\tau)/\beta = 1/[1+4\pi\rho D(\beta\tau)],$$

(where ρ is the average surface density of label), independent of the value of w. Figure 11 bears this relation out.

In several reported cases,[81-83] primary emphasis has been placed on the measurement of the zero time intercept of $g(\tau)$. Weissman et al[81] measured <N> for a sample of fluorescently labeled DNA of known weight concentration, and hence determined the absolute molecular weight. Thompson and Axelrod[82] labeled developing muscle cells with fluorescent α-bungarotoxin, and used the measurement of <N> to demonstrate that acetylcholine receptors diffuse as submicroscopic clusters. In cases where diffusion is extremely slow or absent altogether, systematic sample scanning[65,81,83] can be used to effect an efficient ensemble averaging over the sample.

5. CONCLUSIONS

Three fluorescence techniques have been described. Time-resolved emission can be used in special circumstances to measure short range translational motion; but is more generally useful for the characterization of nanosecond rotations. Fluorescence photobleaching has been applied in a few cases to the study of slow rotational motion; but is most useful for measurements of lateral transport. Similarly, fluorescence correlation spectroscopy has been used to study rotational and translational motion; but its most interesting applications may come from the determinations of absolute number densities. In combination, these techniques provide powerful capabilities for the study of biological motion.

ACKNOWLEDGMENTS

The author would like to thank Drs. S.M. Fernandez and B.A. Herman for the use of Figs. 1-3. This work was supported by NIH Grants GM 23585 and GM 28250.

REFERENCES

1. D.E. Koppel, D. Axelrod, J. Schlessinger, E.L. Elson, and W.W. Webb, Biophys. J. 16:1315 (1976).
2. R.J. Cherry, Biochim. Biophys. Acta 559:289 (1979).
3. K. Jacobson, in "Lasers in Biology and Medicine", F. Hillenkamp, kamp, R. Pratesi, and C.A. Sacchi, eds., (Plenum Press, New York, 1980), p. 271.

4. R. Rigler, and P. Grasselli, in "Lasers in Biology and Medicine", F. Hillenkamp, R. Pratesi, and C.A. Sacchi, eds., (Plenum Press, New York, 1980), p. 151.
5. J. Yguerabide, and M.C. Foster, in "Membrane Spectroscopy", E. Grell, ed., (Springer-Verlag, Berlin, 1981), p. 199.
6. R. Peters, Cell Biol. Internat. Reports 5:733 (1981).
7. D.E. Koppel, in "Fast Methods in Physical Biochemistry and Cell Biology", R.I. Sha'afi and S.M. Fernandez, eds., (Elsevier North-Holland, Amsterdam, in Press).
8. A.J. Pesce, C.G. Rosen, and T.L. Pasby, "Fluorescence Spectroscopy: An Introduction for Biology and Medicine", (Marcel Dekker, New York, 1971).
9. H.Z. Cummins, and H.L. Swinney, in "Progress in Optics", Vol. VIII, E. Wolf, ed. (North-Holland, Amsterdam, 1970).
10. E. Jakeman, in "Photon Correlation and Light Beating Spectroscopy", H.Z. Cummins and E.R. Pike, eds. (Plenum Press, New York, 1974).
11. G.R. Cooper, and C.D. McGillem, "Methods of Signal and System Analysis", (Holt, Rinehart and Winston, New York, 1967).
12. R.N. Bracewell, "The Fourier Transform and its Application", (McGraw-Hill, New York, 1978).
13. H.-J. Galla, and E. Sackmann, Biochim. Biophys. Acta 339:103 (1974).
14. C.S. Owen, J. Chem. Phys. 62:3204 (1975).
15. J.M. Vanderkooi, S. Fischkoff, M. Andrich, F. Podo, and C.S. Owen, J. Chem. Phys. 63:3661 (1975).
16. I.Z. Steinberg, and E. Katchalski, J. Chem. Phys. 48:2404 (1968).
17. E. Haas, E. Katchalski-Katzir, and I.Z. Steinberg, Biopolymers 17:11 (1978).
18. D.D. Thomas, W.F. Carlsen, and L. Stryer, Proc. Natl. Acad. Sci. (USA) 75:5746 (1978).
19. R. Rigler, and M. Ehrenberg, Q. Rev. Biophys. 6:139 (1973).
20. J.D. Jackson, "Classical Electrodynamics", (John Wiley and Sons, New York, 1962).
21. R.A. Badley, W.G. Martin, and H. Schneider, Biochemistry 12:268 (1973).
22. A. Jablonski, Z. Physik. 95:53 (1935).
23. D.D. Thomas, Biophys. J. 24:439 (1978).
24. F. Perrin, Annls. Phys., Paris 12:169 (1929).
25. H.W. Huang, J. Theor. Biol. 40:11 (1973).
26. K. Kinosita, Jr., S. Kawato, and A. Ikegami, Biophys. J. 20:289 (1977).
27. M.P. Heyn, Fed. Eur. Biochem. Soc. Lett. 108:359 (1979).
28. F. Jähnig, Proc. Natl. Acad. Sci. (USA) 76:6361 (1979).
29. B.A. Herman, Doctoral thesis, The University of Connecticut (1980).
30. J. Yguerabide, Methods Enzymol. 26:498 (1972).
31. I. Lewis, and W.R. Ware, Rev. Sci. Instr. 44:107 (1973).

32. I. Isenberg, in "Biochemical Fluorescence", R.F. Chen and H. Edelhoch, eds. (Marcel Dekker, New York, 1975).
33. A. Gafni, R.L. Modlin, and L. Brand, Biophys. J. 15:263 (1975)
34. I. Isenberg, and R. Dyson, Biophys. J. 9:1337 (1969).
35. A. Grinvald, and I.Z. Steinberg, Analyt. Biochem. 59:583 (1974
36. P.K. Wolber, and B.S. Hudson, Biochemistry 20:2800 (1981).
37. K.O. Hodgson, H. Winick, and G. Chu, "Stanford Synchrotron Radiation Laboratory Report 76/100" (Stanford Univ., Stanfo CA, 1976).
38. S.L. Shapiro, ed., "Ultrashort Light Pulses: Picosecond Techniques and Applications" (Springer, N.Y., 1977).
39. K.G. Spears, L.E. Cramer, and L.D. Hoffland, Rev. Sci. Instr. 255 (1978).
40. G.R. Fleming, J.M. Morris, and G.W. Robinson, Aust. J. Chem. 3 2337 (1977).
41. B.A. Herman and S.M. Fernandez, Biochem. Biophys. Res. Commun. 103:1112 (1981).
42. M.J. Anderson, and M.W. Cohen, J. Physiol. 237:385 (1974).
43. D. Axelrod, P. Ravdin, D.E. Koppel, J. Schlessinger, W.W. Webb, and T.R. Podleski, Proc. Natl. Acad. Sci. (USA) 73:4594 (1976).
44. L.M. Smith, R.M. Weis, and H.M. McConnell, Biophys. J. 36:73 (1981).
45. P. Johnson, and P.B. Garland, FEBS Lett. 132:252 (1981).
46. D. Axelrod, D.E. Koppel, J. Schlessinger, E.L. Elson, and W.W. Webb, Biophys. J. 16:1055 (1976).
47. D.E. Koppel, Biophys. J. 28:281 (1979).
48. B.A. Smith, and H.M. McConnell, Proc. Natl. Acad. Sci. (USA) 75:2759 (1978).
49. L.M. Smith, J.W. Parce, B.A. Smith, and H.M. McConnell, Proc. Natl. Acad. Sci. (USA) 76:4177 (1979).
50. B.A. Smith, W.R. Clark, and H.M. McConnell, Proc. Natl. Acad. Sci. (USA) 76:5641 (1979).
51. F. Lanni, and B.R. Ware, Rev. Sci. Instr. 53: in press (1982)
52. D.E. Koppel, M.P. Sheetz, and M. Schindler, Biophys. J. 30:187 (1980).
53. R.A. Cone, Nature (New Biol.) 236:39 (1972).
54. D.E. Koppel, in "Techniques in Lipid and Membrane Biochemistry" J. Metcalfe, and T.R. Hasketh, eds. (Elsevier/North-Holland Amsterdam, 1982).
55. B.G. Barisas, M.D. Leuther, Biophys. Chem. 10:221 (1979).
56. F. Lanni, D.L. Taylor, and B.R. Ware, Biophys. J. 35:351 (1981
57. J.W. Wojcieszyn, R.A. Schlegel, E-S. Wu, and K. Jacobson, Proc. Natl. Acad. Sci. (USA) 78:4407 (1981).
58. M. Schindler, D.E. Koppel, and M.P. Sheetz, Proc. Natl. Acad. Sci. (USA), 77:1457 (1980).
59. M. Schindler, M.J. Osborn, and D.E. Koppel, Nature 283:346 (1980).
60. D.E. Koppel, J. Supramol. Struct. 17:61 (1981).

61. E.L. Elson, and W.W. Webb, Ann. Rev. Biophys. Bioeng. 4:311 (1975).
62. D. Magde, in "Chemical Relaxation in Molecular Biology", I. Pecht, and R. Rigler, eds. (Springer-Verlag, Berlin, 1977) p. 43.
63. E.L. Elson, and D. Magde, Biopolymers 13:1 (1974).
64. D. Magde, E.L. Elson, and W.W. Webb, Biopolymers 13:29 (1974).
65. D. Magde, W.W. Webb, and E.L. Elson, Biopolymers 17:361 (1978).
66. M. Ehrenberg, and R. Rigler, Chem. Phys. 4:390 (1974).
67. S.R. Aragon, and R. Pecora, J. Chem. Phys. 64:1791 (1976).
68. J.T. Yardley, and L.T Specht, Chem. Phys. Lett. 37:543 (1976).
69. J. Borejdo, S. Putnam, and M.F. Morales, Proc. Natl. Acad. Sci. (USA) 76:6346 (1979).
70. G. Feher, and M. Weissman, Proc. Natl. Acad. Sci. (USA) 70:870 (1973).
71. D.W. Schaefer, and B.J. Berne, Phys. Rev. Lett. 29:475 (1972).
72. D.W. Schaefer, and P.N. Pusey, Phys. Rev. Lett. 29:843 (1972).
73. D.W. Schaefer, Science 180:1293 (1973).
74. C.R. Cantor, and P.R. Schimmel, "Biophysical Chemistry, Part II: Techniques for the Study of Biological Structure and Function", (W.H. Freeman, San Francisco, 1980).
75. A.T. Forrester, Am. J. Phys. 24:192 (1956).
76. E. Jakeman, C.J. Oliver, and E.R. Pike, J. Phys. A. 3:L45 (1970).
77. D.E. Koppel, J. Appl. Phys. 42:3216 (1971).
78. D.E. Koppel, Phys. Rev. A. 10:1938 (1974).
79. P.F. Fahey, D.E. Koppel, L.S. Barak, D.E. Wolf, E.L. Elson, and W.W. Webb, Science 195:305 (1977).
80. P.J. Sims, A.S. Waggoner, C.H. Wang, and J.F. Hoffman, Biochemistry 13:3315 (1974).
81. M. Weissman, H. Schindler, and G. Feher, Proc. Natl. Acad. Sci. (USA) 73:2776 (1976).
82. M.L. Thompson, and D. Axelrod, Biophys. J. 37:17a (1982).
83. N.O. Petersen, E.L. Elson, and D.C. Johnson, unpublished.

LIGHT SCATTERING BY MODEL MEMBRANES

J.C.Earnshaw

Department of Pure and Applied Physics
The Queen's University of Belfast
Belfast BT7 1NN, Northern Ireland

CONTENTS

1. INTRODUCTION

It has been known since early in this century[1] that the surface of a liquid is continuously disturbed by thermal molecular agitation. The disturbances can be considered as a dynamically evolving Fourier superposition of capillary waves of all wavelengths, excited according to the classical Boltzmann probability factor.[2] The amplitude of the waves is typically less than a nanometer, but they act as a weak diffraction grating, scattering light. The capillary waves can be considered to constitute 'ripplons' and the scattering process envisaged as

incident photon $\pm$ ripplon $\rightarrow$ scattered photon,

explicitly bringing out the analogy with Brillouin scattering. Early experimental studies[2,3] demonstrated the essential correctness of the theoretical arguments. In this early work the intensity and the polarization of the scattered light were investigated. Further progress awaited the advent of the laser, which was necessar for investigations of the spectrum of the scattered light. The revi of Langevin[4] cites most of the literature prior to 1976; the present paper thus concentrates on work since that date, with particular reference to fluid interfaces or surfaces supporting model biologica membranes.

The thermally excited capillary waves on a fluid interface evolve in space and time; it is a matter of experimental convenience whether one studies the spatial evolution of waves of constant frequency or the temporal evolution of waves of constant wave vector ($q = 2\pi/\Lambda$). The spectrum of light scattered by the waves reflects their temporal evolution. A wave of given (real) wave vector q evolves in time as

$$\zeta = \zeta_q \, e^{\alpha t} \tag{1}$$

where α is the complex frequency:

$$\alpha = i\omega_o - \Gamma .$$

The restoring forces acting upon the surface are those of capillarity: the effects of gravity are negligible[5] within the range of q of interest ($q \gtrsim 100\ cm^{-1}$). The waves are damped by viscous effects. Thus, at least to first order, ω_o is governed by the surface tension (γ) and Γ by the shear viscosity of the liquid (η). A membrane (or other film of surface active molecules) at the surface will introduce further viscoelastic properties which modify the wave propagation[6]. The scattering of light by capillary waves thus provides a technique to study these properties. While other methods to measure γ and η exist, the study of thermally excited capillary waves has the considerable advantage that no perturbation of thermodynamic equilibrium occurs. Similarly while some physical properties of model membranes can be investigated by other techniques, these usually involve disturbance of the membrane or the use of molecular probes which may alter the microscopic environment within the membrane.

Following early studies of light scattering by free liquid surfaces, De Gennes and Papoular[7] suggested that such methods might be applicable to the plasma membranes of living cells. To date, no such work has been reported, but several groups have studied model membranes, including monolayers upon aqueous subphases and bimolecular lipid films immersed in water. Other systems investigated have included liquid crystals[4] and soap films[8]. There are interesting similarities with model membranes, but space prohibits detailed consideration of these systems.

Mechanically generated waves in the capillary regime have long been used to study the properties of liquids, and more recently of films upon fluid surfaces[6]. Such experiments study the spatial evolution of the waves (ie real ω and complex q). Recently very elegant experiments in this class[9] have used light scattering to study waves on fluids and on liquid crystals. Very small wave amplitudes are produced in these experiments by electrocapillarity, reducing possible effects of non-linearity which may arise in mechanical wave generation.

The observable properties of the fluid interface are the macroscopic manifestations of intermolecular interactions. Unfortunately no satisfactory microscopic theory of interfacial rheology exists. Goodrich[10] has demonstrated that the anisotropy of momentum transport within and parallel to a fluid interface entails the existence of four separate interfacial viscosities. This theory is, however, not in a form suited to explicit calculation. Recently Baus[11] has shown that the finite thickness of the interfacial region between two fluids leads to five elastic moduli for the system. Within the fluid phases these reduce to the appropriate bulk and shear moduli. The extra three parameters in the interfacial region comprise the surface tension and two moduli arising from the three-dimensional nature of the transition layer. Baus notes that the dynamic behaviour of such systems should, in principle, permit detailed testing of theoretical models of liquid interfaces. But clearly much development is needed before calculations of the physical properties of interfaces comprising, for example, lipid molecules can confidently be undertaken.

2. THEORETICAL BACKGROUND

The intensity of light scattered by thermally excited capillary waves reflects the mean square amplitude of the fluctuations of the liquid interface, which is determined by the change in energy associated with the deviation from equilibrium[12]. To first order the change in energy per unit area involves the work done in increasing the interface surface area and the gravitational potential

$$\Delta E = \frac{1}{4}\,\zeta_q^2\left[\gamma\, q^2 + (\rho - \rho')\, g\right] \tag{2}$$

where ρ and ρ' are the densities of the two phases separated by the interface. The mean square amplitude is found[12] to be

$$\langle|\zeta_q|^2\rangle = \frac{k_B\, T}{\gamma\, q^2 + (\rho - \rho')\, g}\,, \tag{3}$$

leading to the cross-section for scattering of light:

$$\frac{1}{I_o} \frac{dI}{d\Omega} = \frac{k_o^4}{4\pi^2} \langle |\zeta_q|^2 \rangle \; r_{s,p}^2 \cos^3\theta \qquad (4);$$

where I_o is the intensity and k_o the wavector of the incident light, θ the angle of incidence and $r_{s,p}$ are the Fresnel coefficients for the interface. Equations 3 and 4 clearly show the variation of the intensity of the scattered light with q^{-2}. It is this very sensitive dependence on the angle of scattering which makes the surface scattering perceptible in the presence of the scattering by bulk phonons.

In principle, measurement of I_s should yield information on the surface tension γ. To date no such use has been made of I_s, probably due to the practical difficulties of measuring it absolutely. However in a study of thermal fluctuations of soap films, the amplitudes of oscillatory and exponential contributions to the observed correlation functions have been used to separate the two different mode types present.[13]

The spectrum of the scattered light is just the power spectrum of the fluctuations of the liquid surface. In general, this is related to the dispersion relation connecting the complex frequency α to the wavevector q of the capillary waves. This involves the physical properties of the system (which classically have been restricted to η and γ); for a fluid surface it takes the well-known form[15,16]

$$(\alpha + 2\nu q^2)^2 + g\rho + \frac{\gamma q^3}{\rho} - 4\nu^2 q^3 \left(q^2 + \frac{\alpha}{\nu}\right)^{\frac{1}{2}} = 0 \qquad (5)$$

where $\nu = \eta/\rho$. The solution of equation 5 yields ω_o and Γ as functions of q. To first order Lamb[14] shows that for low damping

$$\alpha \simeq i(g\,\rho + \Gamma q^3/\rho)^{\frac{1}{2}} - 2\nu q^2 \qquad (6)$$

Various higher order approximations to the solutions of equation 5 have been derived[16,17] but are essentially the same as the exact solutions derived numerically[18]. The proportionalities $\omega_o \propto q^{3/2}$ and $\Gamma \propto q^2$ evident from equation 6 for waves in the capillary regime are not significantly modified by such higher order approximations. Now the frequency ω_o and the damping constant Γ are those found experimentally; analytic relationships such as equation 6, or the higher order forms, permit these parameters to be interpreted in terms of the physical properties

(γ and ν) of the system.

Bouchiat and Meunier[19] have derived the theoretical form of the spectrum of the light scattered by the free surface of a liquid. In general the spectrum is complicated but for experimentally useful limits it is approximated quite well by a Lorentzian form. For very viscous liquids the surface disturbances do not propagate and the spectrum is close to a Lorentzian of half-width given by $\gamma q/2\eta$ centred on 0Hz. At the opposite limit of low damping the propagating waves produce a doublet of Lorentzian peaks (of half-width Γ) centred at $\pm\,\omega_o$. Between the two limits there is a region where the spectrum is not well represented by a Lorentzian form[19]. As the condition of critical damping is approached the deviations from the Lorentzian shape become more marked; in particular the peak frequencies and the half-widths of the spectrum fall somewhat below the predicted values ω_o and Γ. It has been shown[20] that critical damping occurs when

$$y = \frac{\gamma\rho}{4\eta^2 q} \simeq 0.145$$

Thus except for situations where interfaces of low tension (such as black liquid membranes or oil-water interfaces) are being studied the Lorentzian approximation is likely to be valid and the observed values for ω_o and Γ will be close to those found from the dispersion equation. For example, one study involved $204 < y < 1140$ (for water) and $8.3 \times 10^{-5} < y < 3.3 \times 10^{-4}$ (glycerine); both cases are very far from the critical region.

More complex physical systems have been investigated, appropriate boundary conditions leading to modified dispersion equations and spectra. Situations which have been analysed include two fluids in contact[21], a fluid surface supporting a thin film[22], thin films in air[8], liquid crystals[23] and a membrane immersed in a fluid[24]. Of these, the second and the fifth are of present concern.

The addition of a thin (eg monomolecular) film to a fluid surface engenders surface viscoelastic effects. The tension (γ) of a surface supporting a monolayer is reduced by a surface pressure (π) exerted by the molecules of the film:

$$\pi = \gamma_\ell - \gamma \tag{7}$$

γ_ℓ being the surface tension of the clean subphase. Now the film is compressible in two dimensions and thus the surface displays elasticity. The surface elasticity - or the modulus of surface dilation - is defined[6] as

$$\varepsilon = -\frac{d\,\pi}{d\,\ell nA} \tag{8}$$

where A is the surface area occupied by each molecule in the film. The variation of the surface pressure and of the elasticity on compression of a myristic acid monolayer upon a aqueous subphase are shown in figure 1. As already noted, fluid interfaces may display up to four separate surface viscosities[10]. These comprise two pairs, each including a shear and a dilational viscosity. One pair (η and κ respectively) describe processes in the plane of the interfac and the other (γ' and κ_N) concern processes normal to the interface plane. Of these, κ_N plays no part in the propagation of capillary waves[10] and η governs surface shear modes which do not couple to the optical field[26]. The remaining two surface viscosities can be incorporated into hydrodynamic theory by expanding the surface tension and elasticity as response functions:

$$\gamma = \gamma_o + i\,\omega\,\gamma' \tag{9}$$

and

$$\varepsilon = \varepsilon_o + i\,\omega\,\kappa \tag{10}$$

where γ_o and ε_o refer to the quantities which would be measured in a truly static process.

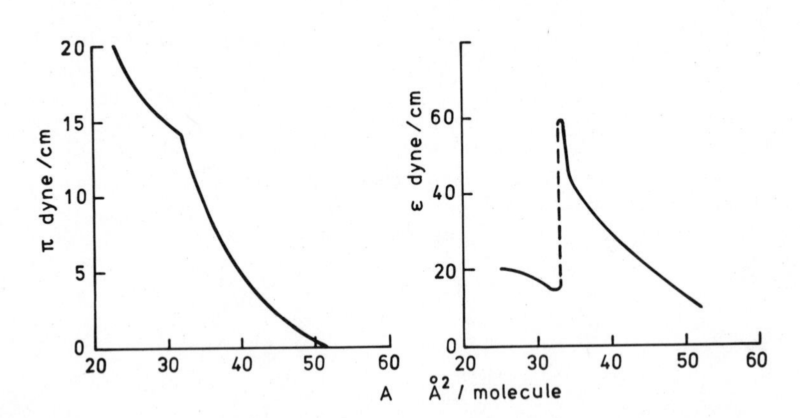

Figure 1: The surface pressure and elasticity of myristic acid on 0.01M NaCl at 20°C adapted from reference 25).

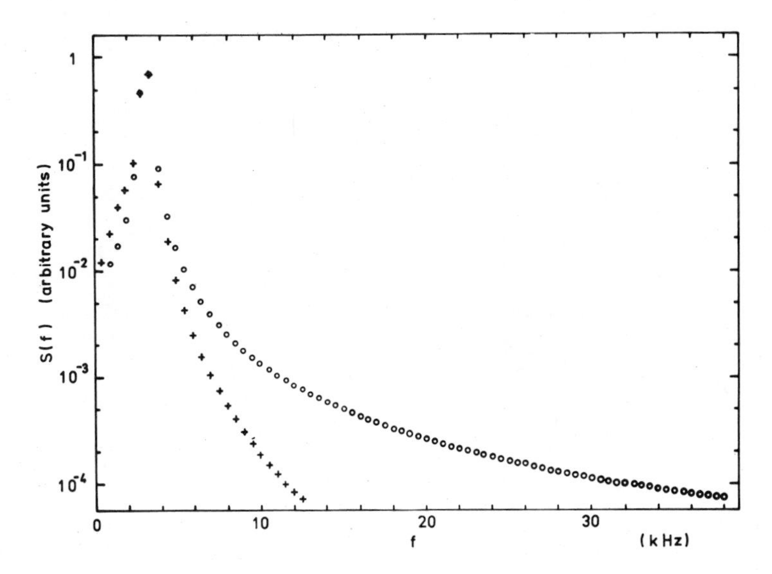

Figure 2: The spectrum of the scattered light at $q = 180\ \mathrm{cm}^{-1}$ for a liquid of $\gamma_o = 72.75$ dyne cm^{-1}, $\varepsilon_o = 3.6$ dyne cm^{-1} and $\kappa = 10^{-5}$ sP. The best fit Lorentzian form is shown (o). (from reference 38).

The theoretical form of the spectrum of light scattered by capillary waves upon a surface bearing a thin film is complicated[22]. However far from y_{crit} the deviations from the Lorentzian form are not large (figure 2). To date experimental data has normally been analysed in terms of such an approximation. No analytical approximations connecting observed values of ω_o and Γ to the physical properties of interest have been found. The interpretation of observed data will be discussed below.

Another interesting case is that of a film bathed by fluid media such as a lipid bilayer membrane immersed in an aqueous fluid. This is of considerable interest as a widely used model of biomembranes. The theoretical analysis again exists[24] and simplifies dramatically for the situation where the fluids separated by the membrane are the same. For such 'symmetric membranes' the effects of ε (both the real and imaginary parts) disappear. In this case only two physical properties of the membrane - γ_o and γ' - affect the scattered light and can be directly related to the observables ω_o and Γ. In fact analytic forms relating ω_o and Γ to γ_o and γ' have been found[24]; their utility is however limited to small values of γ'. Lipid bilayers typically have low γ_o ($\lesssim$ 10% of that for water) and thus the capillary waves are not as far from the point of critical damping as for liquid surfaces. Thus the Lorentzian approximation is somewhat less appropriate in this case.

Recently Loudon[27] has developed a unified theoretical approach to capillary waves upon an interface between two fluids which suppor a thin film. The system is characterised by the properties of the t fluids and by the (complex) tension and elasticity of the interface. The final results of this analysis reduce, with appropriate choices of these properties, to any of the particular cases already mentioned. Further, longitudical acoustic fluctuations, which have to be treated separately in other formalisms, are naturally incorporated. The frequencies of such acoustic waves are much large than those of capillary waves. Here these terms will be eliminated using the appropriate limit ($\omega << \omega_L$). Loudon derives the generaliz dispersion equation

$$D = \left\{\frac{i\rho\omega^2 L}{q^2+TL} + \frac{i\rho'\omega^2 L'}{q^2+T'L'} - \varepsilon q^2\right\}\left\{\gamma q^2 + (\rho-\rho')q - \frac{i\rho\omega^2 T}{q^2+TL} - \frac{i\rho'\omega^2 T'}{q^2+T'L'}\right\} - \omega^2 q^2 \left\{\frac{\eta(T^2 - q^2 - 2TL)}{q^2+TL} - \frac{\eta'(T'^2 - q^2 - 2T'L')}{q^2+T'L'}\right\}^2 \quad (11)$$

where the quantities T and T' are related to the properties of each fluid as

$$T = \left[(i\rho\omega/\eta) - q^2\right]^{\frac{1}{2}}$$

and in the limit of interest,

$$L = L' \simeq i q \quad .$$

From the fluctuation - dissipation theorem the power spectrum of the interfacial fluctuations (and hence of the scattered light) takes the form

$$S(q,\omega) = \frac{k_\beta T}{\pi\omega} \, \mathrm{Im} \, \frac{1}{D} \left\{ \frac{i\rho\,\omega^2 L}{q^2 + TL} + \frac{i\,\rho'\omega^2 L'}{q^2 + T'L'} - \varepsilon^2 q^2 \right\}. \quad (12)$$

Note that if the physical properties of the two fluids are the same the final term of equation 11 vanishes and the spectrum no longer explicitly involves the surface dilational modulus ε, as already noted for symmetric bilayers. The deviations of the spectral form of equation 12 from a Lorentzian form are not large in most practical situations. Neither $S(q,\omega)$ nor the dispersion equation $D = 0$ are amenable to analytic solution.

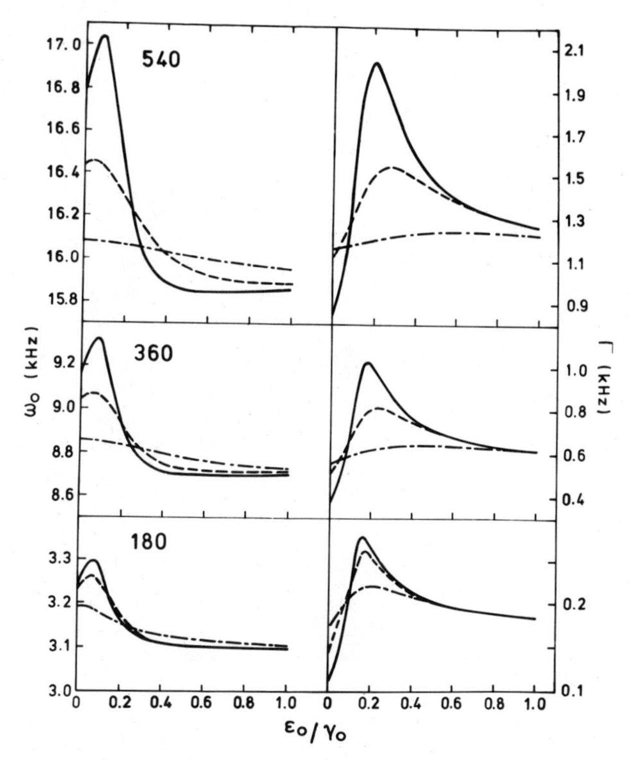

Figure 3: The variation of ω_o and Γ with ε_o and κ for three different q values. The lines represent different κ values: 0sP (——), 10^{-4} sP (----) and 5 x 10^{-4} sP (—— - ——).

The interpretation of the observed values of ω_o and Γ in terms of γ_o, ε_o, γ' and κ depends upon finding combinations of these physical properties which, when substituted into equations 11 or 12, reproduce the observed parameters. The sensitivity of ω_o and Γ to the values of the four properties of the interface determines the utility of the technique.

These effects are explored here for monolayers upon a liquid surface[23]. The effects of ε_o and κ are, to some extent synergetic. Figure 3 shows their effects upon capillary waves of three different wavevectors on a fluid with surface tension of 72.5 dyne cm^{-1}. Note that generally there might be an ambiguity in determining ε_o absolutely. However this is unlikely to be a great practical problem since interest often lies in the gradual change in the physical properties as membranes are manipulated or subjected to varying conditions (eg temperature) and requirements of continuity should resolve the ambiguity. The sensitivity of both ω_o and Γ to the properties varies with q. For example Figure 3 shows how the accessible ranges of ε_o and κ varies with q. The effects of κ tend to saturate as κ increases.

Variation of γ_o causes the curves of Figure 3 to change. The curve for ω_o retains the same form but moves up or down bodily, as suggested by equation 6 (or the second-order approximation), while that for Γ is almost unchanged below the peak but is raised or lowered somewhat at higher ε_o values. Basically γ_o affects ω_o.

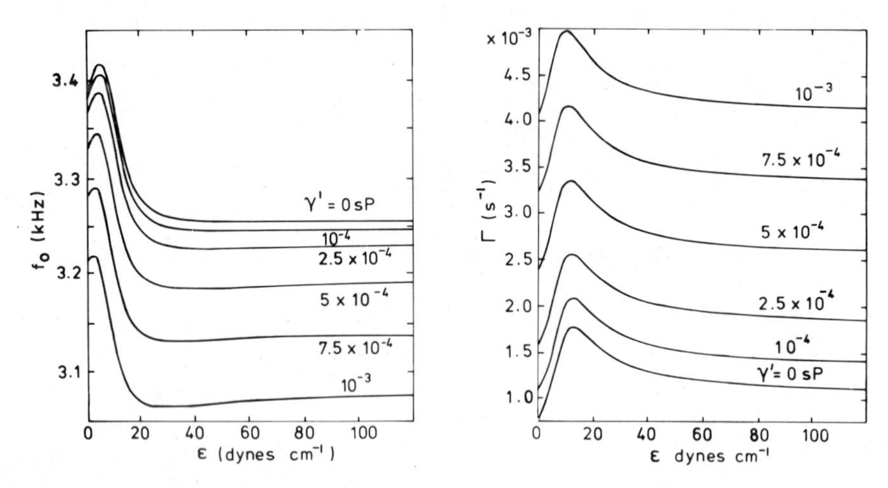

Figure 4: The effect of γ' and ε_o upon ω_o and Γ for waves of $q = 185\ \text{cm}^{-1}$. $\gamma_o = 72.75$ dyne cm^{-1} and $\kappa = 10^{-4}$ sP.

Introduction of γ' slightly depresses ω_o but substantially increases Γ (Figure 4). The effect of increasing γ' upon Γ continues until ultimately the capillary waves become overdamped. At a given q value, the magnitude of γ' necessary to cause over-damping depends upon γ_o; the driving force of surface tension being opposed by the dissipative effects of the transverse shear viscosity. The lower γ_o the more easily overdamping is induced. Formally this may be viewed as a modification of y_{crit} from the value found for a clean surface. For a clean surface below y_{crit} light scattering reveals only one mode. Another surface mode exists[20] but is damped too rapidly to be observed. For an interface with a membrane of γ' sufficient to cause overdamping this mode is slowed down and two modes are observable.[24]

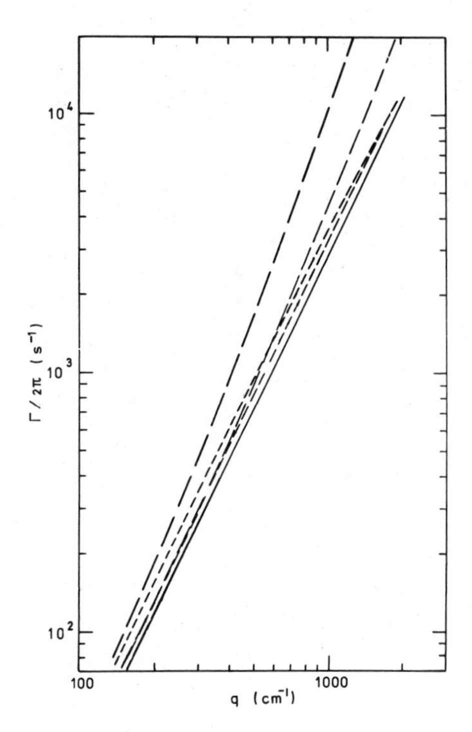

Figure 5: The dependance of Γ upon q for various cases (γ_o = 72.75 dyne cm^{-1}, ε=0): $\kappa=\gamma'=0$ (——), $\kappa = 2 \times 10^{-5}$ and 10^{-4} sP (---) and $\gamma' = 2 \times 10^{-5}$ and 10^{-4} sP (—— ——) (from reference 36).

Only two observable parameters are available - ω_o and Γ - which relate to four physical properties. The extraction of information concerning these properties from the observed data is thus not straight-forward. The changing sensitivity of ω_o and Γ to the various parameters as q is changed has been noted. The variation of ω_o with $q^{3/2}$ and of Γ with q^2 is modified by the inclusion of the various parameters. This is indicated in Figure 5 for the two surface viscosities, κ and γ'. The maxima in ω_o and Γ shown in Figure 3 occur at ε_o values which vary with q, modifying the dispersion behaviour of both parameters substantially. Thus a careful study of the variation of ω_o and Γ over as wide a range of q as possible will be very helpful in determining which of the fourproperties are significant in any situation. Other experiments may yield values for some of the properties (notably γ_o and ε_o) which could usefully be taken into account in interpretation of the light scattering data.

An approach which may permit evaluation of all four physical properties from a light scattering experiment at a single wavevector is that used by Wu and Webb[28] in a study of the critical interface of SF_6. The spectrum is *not* a Lorentzian. While the differences are expected to be small except in the region of critical damping, Wu and Webb found that the full spectrum gave recognizably better fits to their data even in conditions of low damping. They fitted their observed correlation functions to the Fourier transform of the appropriate spectrum to extract the tension and the viscosity of the fluid near the critical point. Returning to the present context of membranes, this approach could be useful if the deviations of equation 12 from a Lorentzian depended in some systematic fashion upon the physical properties of the interface. At present the utility of this idea is not established.

The problems just discussed are much less acute for the case of bilayer membranes where only two membrane properties are accessible. The precision of determination of these properties would, however, be enhanced by observation of light scattered by waves of several q values.

The correlation of the properties accessible via light scattering with those found by other, classical techniques is most important.

Light scattering is uniquely sensitive to the true static and dynamic properties of a film at a liquid interface. Mechanical experiments can measure both γ_o and ε_o (also κ) but at best such experiments involve quasi-static processes and thus values of (say) ε_o may contain some contribution due to the relevant surface viscosity (here κ). This is clearly shown by Hård and Neuman[29] where rapid compression of a monolayer is shown to yield $\varepsilon > \varepsilon_o$. Indeed the response of a monolayer to a step-wise compression can be used to study both the real and imaginary parts of ε[30]. Light scattering does not perturb thermodynamic equilibrium to any significant extent and should thus permit the clear separation of the static and dynamic portions of γ and ε.

The physical properties measured by the light scattering method have been identified with those found by mechanical experiments. This identification is easily seen to be correct. We shall follow the treatment of Landau and Lifshitz[31], modified to a two dimensional system as necessary. The interfacial film is isotropic in the x-y plane and has (constant) thickness d in the z direction. Upon compression the change in film area A is given by

$$A + dA = A(1 + u_{xx})(1 + U_{yy})$$

whence, to first order,

$$u_{ii} = \frac{dA}{A} \quad . \tag{13}$$

The u_{ii} are the diagonal terms of the symmetric strain tensor

$$u_{ij} = \tfrac{1}{2}\left(\frac{\partial u_i}{\partial x_j} + \frac{\partial u_j}{\partial x_i}\right) \quad .$$

The Einstein summation convention is used throughout. We wish now to relate u_{ii} to the stress tensor for compression. The stress tensor is defined by

$$\sigma_{ij} = \left(\frac{\partial F}{\partial u_{ij}}\right) \tag{14}$$

where F is the free energy density of an elastic body. For an isotropic two dimensional body the elastic free energy density involves only two elastic moduli expressed via the Lame coefficients λ and μ:

$$F = \tfrac{1}{2} \lambda u_{ii}^{2} + \mu u_{ij}^{2}$$

The free energy per unit area of the film is thus

$$F = \tfrac{1}{2} \lambda d u_{ii}^{2} + \mu d u_{ij}^{2} \tag{15}$$

As is usual, it is convenient to separate u_{ij} into components expressing pure shear and pure dilation:

$$u_{ij} = (u_{ij} - \tfrac{1}{2} \delta_{ij} u_{kk}) + \tfrac{1}{2} \delta_{ij} u_{kk}$$

which permits equation 15 to be rewritten as

$$F = S(u_{ij} - \tfrac{1}{2} \delta_{ij} u_{kk}) + \tfrac{1}{2} K u_{kk}^{2} \quad .$$

Here the moduli of dilation (or compression) and of shear are defined respectively by $K = (\lambda + \mu) d$ and $S = \mu d$. These definitions are those used in the theoretical analyses of light scattering[22,26]. Using this F in equation 14 we find

$$\sigma_{ij} = K u_{kk} \delta_{ij} + 2 S (u_{ij} - \tfrac{1}{2} \delta_{ij} u_{kk}) \quad ,$$

showing that for a pure hydrostatic compression, σ_{ii} and u_{ij} are related only via the dilational modulus. We require only u_{ij} in terms of σ_{ij}. This follows from the sum of the diagonal terms σ_{ii}:

$$\sigma_{ii} = 2 K u_{ii}$$

where the factor 2 arises from the summation over 2 dimensions. By analogy with hydrostatic compression in 3 dimensions, the stress in a film due to surface pressure $d\pi$ is

$$\sigma_{ij} = - d\pi \, \delta_{ij}$$

so that

$$u_{ii} = - \frac{d\pi}{K}$$

which with equation 13 yields finally

$$K = - A \frac{d\pi}{dA}$$

Thus the modulus of pure dilation accessible via light scattering is exactly identified with the surface elasticity defined for film compression[6] (equation 8).

3. EXPERIMENTAL STUDIES

As previously noted the intensity of the light scattered by capillary waves falls off very rapidly with scattering angle. In this context a scattering vector of 2000 cm^{-1} is large; this corresponds to an angle of scatter rather less than 1° (for light of λ = 488 nm). Spectrometers to study such scattering must permit such low angles of scattering to be precisely defined. A heterodyne spectrometer is also essential to make the frequency shifts of the scattered light manifest. Efficient optical heterodyning requires that the intensities of the scattered light (which varies rapidly with scattering angle) and the reference light should be appropriately matched.

The apparatus sketched in Figure 6 achieves both of these aims by means of a technique introduced by Hård et al[32]. A laser beam is incident upon the fluid interface (with minor modifications the apparatus is equally suited to liquid surfaces or bilayer membranes[33]) where it is specularly reflected and scattere Placed closely to the surface, a coarse diffraction grating intercepts the reflected beam and diffracts a very small proportion

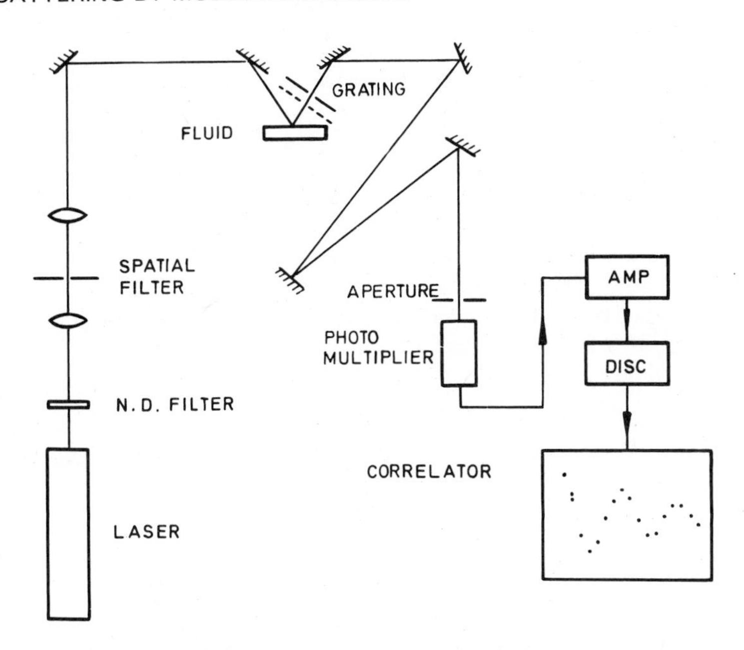

Figure 6: Experimental arrangement for light scattering from fluid interfaces.

of the light incident upon it. Light is diffracted into a series of maxima in the plane of the detector, far from the liquid surface. One diffraction spot is there selected by an aperture and mixes with scattered light which is spatially coherent with this reference beam. The main advantages of this local oscillator are the precise definition of the angle of scattering and the automatic spatial coherence of the reference and scattered fields. It is also relatively easy to adjust the intensity of the reference beam by choosing gratings which diffract varying amounts of light into the higher order (non-zero) intensity maxima.

In such spectrometers the observed spectrum is the convolution of the spectrum of light scattered by waves of the selected q value with an instrumental function. The laser beam, of finite extent, illuminates a limited number of crests of the capillary wave of wavelength Λ ($=2\pi/q$). Thus the beam diffracted by a Fourier component of the surface disturbance forms a spot of finite extent in the detector plane. A given point will be illuminated by light scattered through a range of scattering angles. For a laser beam incident upon the surface with a Gaussian intensity profile, the instrumental function is also a Gaussian[32,34]. The spectrum can be analysed as a Lorentzian convoluted with a Gaussian, but the analysis is more straightforward in the time domain. If photon correlation is used for signal analysis, the appropriate functional form to describe the time varying part of the correlation function is

$$G(t) = \cos(\omega_o t + \phi) \exp(-\Gamma t) \exp(-\beta^2 t^2/4)$$

where β takes account of the instrumental function. The phase term ϕ is included to partially allow for the deviations of the true spectrum from the exact Lorentzian form. We have found that the value of ϕ is very subject to noise on the correlation functions and so it yields no extra useful information.

A Liquid Surfaces

Many model membranes are bounded by aqueous surfaces. Examples are monolayers of amphiphilic molecules on an aqueous subphase or lipid bilayers bathed by aqueous media. It is therefore necessary to ensure that the experimental techniques to be used for studying such membranes yield results for the clean water surface which agree with expectation.

Unfortunately it has been found[18] that the rate at which waves on water are damped (Γ) exceeds that predicted by equation 5 using the accepted values of surface tension and viscosity of water. Briefly, the first order approximations (equation 6) are much closer to the observed data. The loading effect of air upon the waves is insufficient to remove the discrepancies. Systematic errors in the light scattering technique seem unlikely to be the cause of these effects, as similar discrepancies are found for mechanically generated waves of very much lower frequencies[35]. For a while these problems seemed to threaten the use of light scattering to study the properties of membranes.

However, following Goodrich's demonstration that interfacial viscosities may be an intrinsic feature of fluid interfaces[10], it has been shown that the available data on the propagation of capillary waves on water may be understood if one of the interfacial viscosities (γ') assumes a non-zero value[36]. The dispersion equation for capillary waves found by Goodrich[10] is exactly that used in the treatment of waves on a surface supporting a monolayer. Thus a single consistent theoretical model applies to both clean water surfaces and the monolayer case.

B Insoluble monolayers

Several groups have reported studies of the scattering of light by capillary waves at monolayer coated surfaces. These studies indicate the considerable potential of the technique for probing the viscoelastic properties of monolayers. In particular, changes in monolayer properties as the physical or chemical state of the monolayer is modified are perceptible. Applications to phase transitions in model membranes are particularly promising.

The first study of monolayers with this technique was seriously flawed as the analysis took inadequate account of the instrumental effects[37]. Subsequent studies have included insoluble monolayers of fatty acids[38,39,40,29], of propyl stearate[41,40] and of lipids [42,29]. Polymer films have also been studied[40]. The method common to all these studies has involved spreading a monolayer of very low surface coverage and monitoring the changes in the light scattering spectra as the film is compressed. Both correlation techniques [38,40,29], and spectrum analysis[39,40,41] have been used to estimate ω_o and Γ,. From these two observables up to four surface properties have been inferred. To do this, assumptions have been made and other information has been used. In all cases the bulk properties of the subphase have been assumed to have their accepted values. Usually values of γ_o (and hence ε_o) taken from other measurements have been used to guide the interpretation procedure.

Various workers[29,38,41] have assumed that one of the surface viscosities is zero (usually γ'). While this may, of course, be correct in any given situation, there is no *a priori* reason for the assumptions. Similarly, to evaluate all four parameters Langevin and Griesmar[39,40] restrain γ' to be zero until the data is clearly incompatible with this assumption when this constraint is relaxed. This procedure is clearly arbitrary and,if the surface of water does possess a surface viscosity,[36] is unjustifiable.

The values of the dilational modulus and (sometimes) of the surface pressure of the monolayer, evaluated by light scattering, have been found to differ from the values found by classical methods. The physical quantities are identical in both cases: ε_o *is* the pure dilational modulus. Relaxation processes within the monolayer should only affect the dynamic portions of γ and ε, that is to say γ' and κ . Indeed surface viscositieshave been explained in terms of such processes[43]. It appears that such differences may arise from the assumptions made in interpretation of the light scattering data or in the use of erroneous surface pressure versus area data.

For example, Byrne and Earnshaw[38] used published π-A data which, taken together with the assumption that $\gamma' = 0$, was not fully compatible with the light scattering data (ω_o and Γ as functions of A) for myristic acid. As the surface pressure was not independently measured in this work the reliability of the π-A data cannot be checked. However the classical π-A data of Adam and Jessop[25] (Figure 1) are more nearly compatible with the light scattering data provided both κ and γ' are free. For the same material, Langevin and Griesmar[39] find $\varepsilon_o > \varepsilon$ (classical); the variation of ε_o with π found by light scattering seems very similar to that reported by Adam and Jessop. This emphasizes the extreme care which is necessary if surface pressure measurements are to be made reliably[44].

Hård and Neuman find $\varepsilon_o > \varepsilon$(classical). They assume that π has its classically measured value and that $\gamma' = 0$. With these constraints the value of ε_o evaluated from the light scattering data is often greater than that found by differentiation of the π-A curve. Unfortunately the ability to solve the dispersion relation with ε_o and κ as unknowns cannot guarantee the uniqueness of the resulting fit.

These criticisms of the data interpretation used by various authors must not be taken as denigration of some very careful and precise work in an experimentally challenging area. Unfortunately there can be, in general, no unique interpretation of the observed parameters in terms of four variables except in very special cases. If precise values of ε_o and γ_o are available from other experiments (preferably carried out together with the light scattering) these should be used to constrain the data interpretation procedure. Unfortunately such subsidiary experiments are not trivial[44].

These difficulties in data interpretation can be illustrated by a study of fully-compressed monolayers of glycerol mono-oleate[45]. Compression of such monolayers leads to no change in π but to partial collapse of the monolayer, leading to regions of multilamellar structure on the surface. Thus ε_o can be taken as zero. Light scattering data from such a system was interpreted in terms of the remaining three parameters; π, γ' and κ. As π varied over a certain range the values of γ' and κ required to fit the observed ω_o and Γ varied systematically (Figure 7). No unique fits were possible. Further progress was only possible as the surface pressure of such a fully-compressed monolayer was known and was compatible with the *assumption* $\kappa = 0$.

If capillary waves of several different wavevectors can be studied this will certainly reduce the ambiguity of data interpretation. To date this has not yet been fully exploited for studies of monolayers[42]. In a study of a very different fluid surface[46] the damping constants of waves of 13 different q values

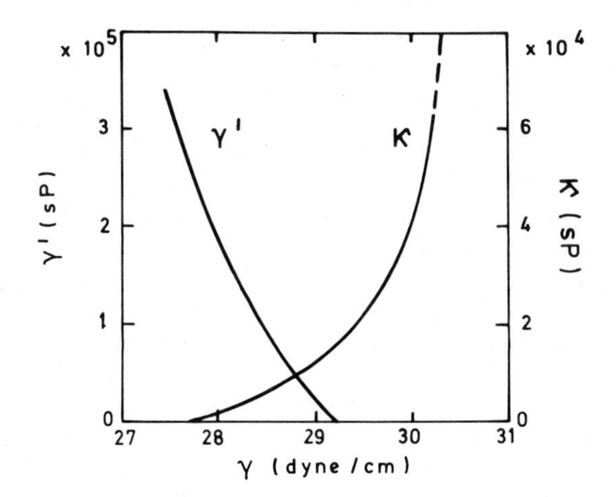

Figure 7: The combinations of γ', κ and γ_o which fit light scattering data (from reference 45).

(210-725 cm^{-1}) sufficed to determine three parameters (γ_o was known). Uncertainties in the values of the parameters (particularly ε_o and κ) are inevitable where large elasticities are involved (see Figure 3): studies over more extended ranges of q may reduce this intrinsic imprecision.

Looking to the future, measurements of such physical properties of monolayers will likely be most significant in investigations of changes in intra-membrane interactions as the chemical or physical state of the membrane is altered. Indeed studies of the compression of monolayers, such as those discussed above, often involve changes of phase of the monolayer. If surface viscosities are regarded as activation processes their temperature variation will yield information on the molecular processes involved in the viscous flow[43]. The study of fully condensed GMO monolayers already mentioned[45] demonstrated the variations of γ_o and γ' with temperature and the latter was interpreted in terms of the free energy, the entropy and the enthalpy of activation. Changes in these thermodynamic variables at a change of phase within the monolayer would be most useful in current attempts to comprehend the nature of these phase transitions. The work on GMO has been extended recently to cover the thermotropic lipid chain melting transition[47].

C Bilayer Lipid Membranes

The application of light scattering to bilayers or black lipid membranes is both easier and harder than is the case for monolayers. Easier because, as already noted, only γ_o and γ' affect the propagation of capillary waves on bilayers, alleviating the problems of data interpretation, and harder because the bilayers do not present large planar surfaces to the incident laser beam, making experiments harder. It is in fact necessary to locate a reasonably flat portion of the bilayer to permit precise definition of the angle of incidence of the light which is necessary to define the q value under investigation. Further, the orientation of a selected portion of the bilayer is liable to change spontaneously, changing q.

Bilayers can be regarded hydrodynamically as two coupled interfaces separated by a distance of a few nanometers. The light scattering technique probes the properties of the interfaces; thus the tension of the interface, not of the membrane, is probed[24]. In such a system, as for soap films, two modes are possible: a bending mode in which the disturbances ζ_q of each interface are everywhere in phase and a squeezing mode in which the ζ_q of the two interfaces are exactly out of phase[8]. For bilayers only the bending mode has been observed.

In an early study of bilayers Grabowski and Cowen[48] analysed their data in terms of a single interface separating two fluids. The damping constants of the capillary waves exceeded those expected from the known properties of the aqueous fluid bathing the bilayers. This was taken as due to a large instrumental function. After taking account of this, a value of membrane tension rather below that accepted for bilayers of oxidised cholesterol was found. More recent work[24] has suggested that the excessive damping may have been due to non-zero transverse shear viscosity, γ'. The reduction of the instrumental correction would tend to raise γ_o, bringing it closer to the accepted value.

Crilly[24] has extensively studied bilayers of glycerol mono-oleate. The interfacial tensions found are in good agreement with independent studies (Figure 8). The Γ values observed for bilayers formed from solutions of GMO in n-decane are entirely consistent with the bathing solution viscosity (Figure 9); no interfacial viscosity is apparent. However such membranes are known to include large proportions of solvent in their structure. In such a system the lipid molecules would likely be very free to move, within the membrane and internally. The exclusion of solvent will restrict these motions. Indeed such 'solvent-free' membranes do exhibit non-zero γ' [24,49]. Cholesterol is a major component of many

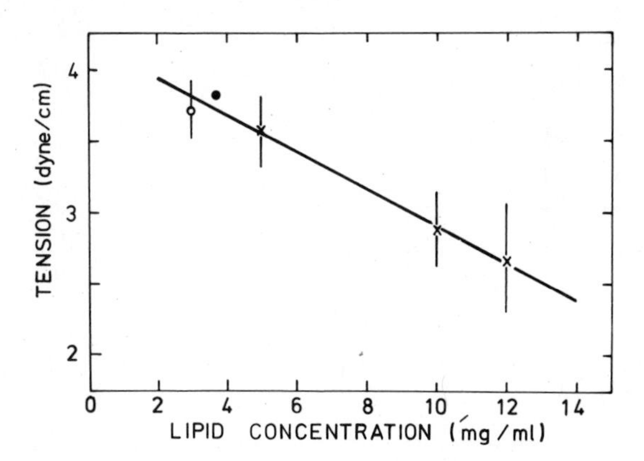

Figure 8: Variation of interfacial tension with GMO concentration in the parent film-forming solution (from reference 24).

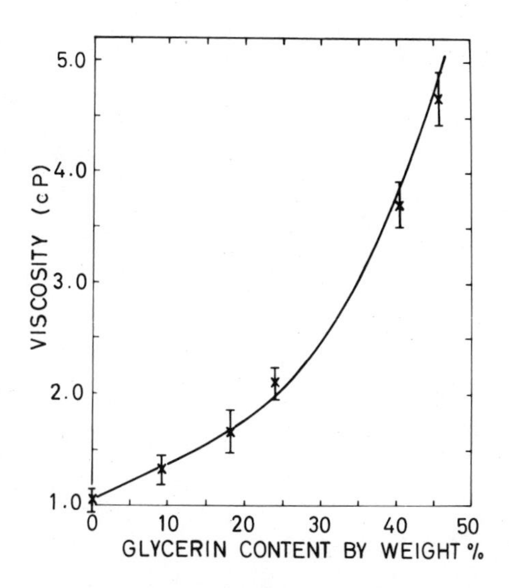

Figure 9: Variation of viscosity with glycerine content in the medium bathing the bilayer. The line indicates the accepted behaviour. (from reference 24).

biomembranes and performs important physiological roles. The effect of cholesterol on the physical properties of simple bilayers is significant in the further understanding of lipid-sterol interactions. Addition of cholesterol to the bilayer-

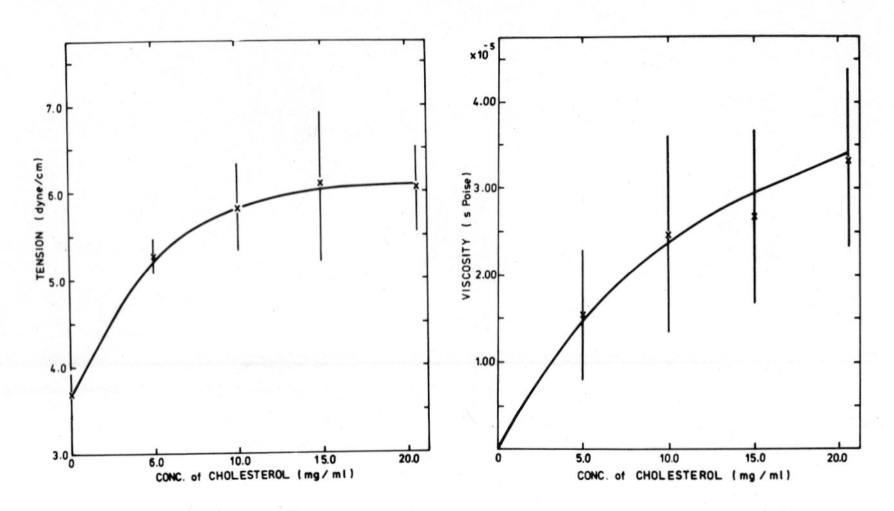

Figure 10: Dependence of γ_o and γ' upon cholesterol content of the film-forming solution (from reference 24).

forming solution induces changes in the membrane properties[24]. As the proportion of cholesterol is increased, the interfacial tension and viscosity both increase and eventually saturate (Figure 10). This saturation behaviour is characteristic of the variation of membrane properties upon addition of cholesterol. The values of γ' found for fully condensed monolayers of GMO, for 'solvent-free' bilayers and for bilayers of GMO saturated with cholesterol are all, within the errors, equal. This remarkable observation is not yet fully explained.

As for monolayers, the most useful applications of laser light scattering to bilayers appear to lie in investigation of temperature dependent changes. Such studies are commencing[33] and thermotropic phase transitions in lipid bilayers have been observed[50].

4 THE PROSPECT FOR CELL MEMBRANES

While De Gennes and Papoular[7] have speculated on the prospect for laser light scattering as a probe of the physical properties of cell membranes, no applications have yet been reported. There are several problems which stand in the way of such studies. Specifically most cell membranes are not large compared to the wave-lengths of capillary waves studied to date. Reduction of Λ by significant factors would lead to experimental difficulties due to the decreasing intensity of the scattered light. Few cells have plasma membranes of macroscopic extent; fewer still adopt the planar geometry considered in the theory summarized above. Amongst the more interesting large cell membranes are the nerve axons. In particular the giant axon in squids can reach 1 mm in diameter. For such cells cylindrical coordinates are appropriate. The theoretical treatment has been suitably modified by Lundström[51]. The variation of the physical properties of such excitable membranes through the time course of the nerve impulse would be most interesting.

It is known that red blood cells exhibit a constant flickering when studied optically. Such cells are not well adapted to laser light scattering because of the small area of the cells. A photometric study[52] has measured the mean square amplitude of the fluctuation of the cell thickness, the frequency spectrum and the spatial correlation length of the fluctuations. Theoretical explanations for this flicker phenomenon have been found[52,53] The key ingredients in these theories are 1) zero surface tension, 2) fluid cytoplasm and 3) refractive indices different inside and outside the cells. The first condition is satisfied because of the closed nature of the cell membrane and the osmotic similarity across the membrane in physiological conditions. Experiments on this flicker phenomenon should thus yield information concerning the membrane elasticity. Unfortunately many cells have very viscous cytoplasms or the membrane is under tension due to osmotic differences, and so the pathological fluctuations of red blood cells are unlikely to be generally encountered.

It is interesting to note that bilayer membranes possess a finite tension just because they are not closed systems, but are in contact with a reservoir of lipid molecules[53].

ACKNOWLEDGEMENTS

This work has been supported by grants from the Science and Engineering Research Council of the UK. I wish to acknowledge the stimulation and the contribution of Drs D Byrne and J F Crilly and Miss G E Crawford. Professor R Loudon kindly sent me a preprint of his theoretical paper.

REFERENCES

1. M. van Smoluchowski, Ann. Physik 25: 225 (1908).
2. L. Mandelstam, Ann. Physik 41: 609 (1913).
3. C.V. Raman and L.A. Ramdas, Proc Roy. Soc. Lond. A. 109 150, 272 (1925).

4. D. Langevin and J. Meunier, in Photon Correlation Spectroscopy and Velocimetry edited by H.Z. Cummins and E.R. Pike (New York, Plenum, 1977) p501.
5. J. Lighthill, Waves in Fluids (Cambridge, CUP, 1978) p221.
6. E.H. Lucassen-Reynders and J. Lucassen, Adv. Coll. Interf. Sci. 2: 347 (1969).
7. P.G. de Gennes and M. Papoular, in Polarization, Matière et Rayonnement (Paris, Presses Universitaires, 1969) p243.
8. A. Vrij, J.G.H. Joosten and H.M. Fijnaut, Adv. Chem. Phys. 48: 329 (1981).
9. C.H. Sohl and K. Miyano, Phys. Rev. A. 20: 616 (1979).
C.H. Sohl, K. Miyano, J.B. Ketterson and G. Wong, Phys. Rev. A. 22: 1256 (1980).
10. F.C. Goodrich, Proc. Roy. Soc. Lond. A. 374: 341 (1981).
11. M. Baus, J. Chem. Phys. 76: 2003 (1982).
12. M.A. Bouchiat and D. Langevin, J. Coll. Interf. Sci. 63: 193 (1978).
13. C.Y. Young and N.A. Clark, J. Chem. Phys. 74: 4171 (1981).
14. H. Lamb, Hydrodynamics (New York, Dover, 1945) p 627.
15. V.G. Levich, Physicochemical Hydrodynamics (Englewood Cliffs, Prentice-Hall, 1962) p603.
16. R.S. Hansen and J.A. Mann Jr., J. Appl. Phys. 35: 152 (1964).
17. E. Mayer and J.D. Eliassen, J. Coll. Interf. Sci. 37: 228 (1971)
18. D. Byrne and J.C. Earnshaw, J. Coll. Interf. Sci. 74: 467 (1980).
19. M.A. Bouchiat and J. Meunier, J. de Phys. 32: 561 (1971).
20. D. Byrne and J.C. Earnshaw, J. Phys. D. 12: 1133 (1979).
21. J.C. Herpin and J. Meunier, J. de Phys. 35: 847 (1974).
22. D. Langevin and M.A. Bouchiat, C.R.A.S. 272B: 1422 (1971).
23. D. Langevin and M.A. Bouchiat, J. de Phys. 33: 101 (1972).
24. J.F. Crilly, PhD. Thesis (Belfast, Queen's University, 1981).
J.F. Crilly and J.C. Earnshaw, to be published.
25. N.K. Adam and G. Jessop, Proc. Roy. Soc. Lond. A. 112: 362 (1926).
26. L. Kramer, J. Chem Phys. 55: 2097 (1971).
27. R. Loudon, in Surface Excitations edited by V.M. Agranovich and R. Loudon (Amsterdam, North-Holland, to be published).
28. E.S. Wu and W.W. Webb, Phys. Rev. A. 8: 2077 (1973).
29. S. Hård and R.D. Neuman, J. Coll. Interf. Sci. 83: 315 (1981).
30. G. Loglio, E. Rillaerts and P. Joos, Coll. and Polymer Sci. 259: 1221 (1981).
31. L.D. Landau and E.M. Lifschitz, Theory of Elasticity (Oxford, Permanon Press, 1970).
32. S. Hård, Y. Hamnerius and O. Nilsson, J. Appl. Phys. 47: 2433 (1976).
33. G.E. Crawford, J.F. Crilly and J.C. Earnshaw, Farad. Symp. of the Chem. Soc. No 16, to be published (1982).

34. D. Byrne and J.C. Earnshaw, J. Phys. D. 10: L207 (1977).
35. J.A. Stone and W.J. Rice, J. Coll. Interf. Sci. 61: 160 (1977).
36. J.C. Earnshaw, Nature 292: 138 (1981).
37. D. McQueen and I. Lundström, J. Chem. Soc. Faraday Trans. I. 69: 694 (1973).
38. D. Byrne and J.C. Earnshaw, J. Phys. D. 12: 1145 (1979).
39. D. Langevin and C. Griesmar, J. Phys. D. 13: 1189 (1980).
40. D. Langevin, J. Coll. Interf. Sci. 80: 412 (1981).
41. S. Hård and H. Lofgren, J. Coll. Interf. Sci. 60: 529 (1977).
42. H. Birecki and N.M. Amer, J. de Phys. 40: C3-433 (1979).
43. W.J. Moore and H. Eyring, J. Chem. Phys. 6: 391 (1938).
44. J. Mingins and J.A.G. Taylor, A manual for the measurement of interfacial tension, pressure and potential at air or non-polar oil/water interfaces (Port Sunlight, Unilever Research, 1970).
45. J.F. Crilly and J.C. Earnshaw, in Biomedical Applications of Laser Light Scattering edited by D.B. Sattelle, B. Ware and W. Lee (Amsterdam, Elsevier/North-Holland, in the press).
46. J.C. Earnshaw, to be published.
47. J.F. Crilly and J.C. Earnshaw, this volume.
48. E.F. Grabowski and J.A. Cowen, Biophys. J. 18:23 (1977).
49. J.F. Crilly and J.C. Earnshaw, Proceedings of the 4th International Conference on Photon Correlation Techniques in Fluid Mechanics (Stanford, Stanford University, 1980) p21.1
50. G.E. Crawford and J.C. Earnshaw, this volume.
51. I. Lundström, J. Theor. Biol. 45: 487 (1974).
52. F. Brochard and J.F Lennon, J. de Phys. 36: 1035 (1975).
53. F. Brochard, P.G. de Gennes and P. Pfeuty, J. de Phys. 37: 1099 (1976).

THE MOVEMENT OF MOLECULES ACROSS MEMBRANES: THE THERMODYNAMIC ANALYSIS OF THE DEPENDENCE ON STRUCTURE, PRESSURE, AND TEMPERATURE

Roger A. Klein

Medical Research Council
Molteno Institute
University of Cambridge
Downing Street
Cambridge CB2 3EE

1. INTRODUCTION

The thermodynamic analysis of biological processes can be a useful phenomenological approach to understanding the molecular interactions which take place at membrane level. The parameters used to describe any particular process and the way in which it changes with alterations in structure, pressure and temperature comprise the free energy, enthalpy, entropy and molar volume change of the system.

There are two possible ways of looking at the thermodynamics of membrane processes. One may be dealing with either a property of the system which depends upon the equilibrium concentration of a molecule within the membrane such as an anaesthetic or other pharmacological agent. Or one may be dealing with the rate at which molecules surmount an energy barrier, passing through an activated state before proceeding to a final state, for example during diffusion or transport across the membrane, and of course in other systems as well; it is important to have some knowledge of the multiplicity of the individual reaction steps which go to make up the overall process together with information on the rate limiting step. This goal may be difficult to achieve in many biological systems because of their inherent complexity.

In this contribution I intend to restrict myself to a consideration of reversible (chemical) thermodynamics without touching on irreversible thermodynamics and the coupling of reactions.

Equilibrium processes may be dealt with using an important series of phenomenological equations. These are the van't Hoff and Clapeyron equations and the Clausius-Clapeyron approximation. The van't Hoff equation (1a)

$$\frac{d(\ln Kp)}{dT} = \frac{\Delta H^{\ominus}}{RT^2} \tag{1a}$$

is often written in the transformed form

$$\frac{d(\ln Kp)}{d(^1/T)} = \frac{-\Delta H^{\ominus}}{R} \tag{1b}$$

since $\frac{d}{dT}(\frac{1}{T}) = \frac{-1}{T^2}$. The Clapeyron equation

$$\frac{dP}{dT} = \frac{\Delta H}{T\Delta V} \tag{2}$$

and the Clausius-Clapeyron approximation, properly valid for ideal gases, since

$$(^{\delta P}/_{\delta T})_V = (^{\delta S}/_{\delta V})_T$$

$$\frac{d(\ln P)}{dT} = \frac{\Delta H_v}{RT^2} \tag{3}$$

These last two equations are of importance when dealing with the pressure dependence of thermotropic lipid phase transitions.

In these equations the concept is embodied that the temperature dependence of an intensive variable such as pressure may be used to obtain a thermodynamic property of the system, in this case the enthalpy ΔH. This equilibrium can in principle at any rate be determined directly and independently by calorimetry.

The equilibrium constant may be expressed in terms of the total free energy difference between the two states in equilibrium, $\Delta G^{\ominus}$,

$$K = \exp(-\Delta G^{\ominus}/RT) \tag{4}$$

where

$$\Delta G^{\ominus} = \Delta H^{\ominus} - T\Delta S^{\ominus} + P\Delta V^{\ominus} \tag{5}$$

since

$$(\delta G^{\ominus}/\delta T)_p = - S^{\ominus} \tag{6}$$

The free energy for an equilibrium may be derived experimentally from measurements of the equilibrium constant

$$\Delta G^{\ominus} = - RT \ln K \tag{7}$$

and the enthalpic and entropic terms estimated from the temperature dependence. The pressure dependence gives a direct measure of the molar volume change $\Delta V^{\ominus}$.

Rate processes can be described phenomenologically using the Eyring activated transition-state formulation[1] for the forward rate constant

$$k_f = K.\frac{kT}{h} . \exp(-\Delta G^*/RT) \tag{8}$$

where

$$\Delta G^* = \Delta H^* - T\Delta S^* + P\Delta V^* \tag{9}$$

with the superscript indicating that the quantities refer to the free energy, enthalpy, entropy and molar volume change of activation. It is important to realize that the activation parameters are related to the equilibrium parameters because at equilibrium the forward and reverse reaction rates are equal, therefore

$$\Delta G^{\ominus} = \Delta G^*_f - \Delta G^*_r \tag{10}$$

In a system with temperature and pressure as independent variables it is convenient to think in terms of the reaction rate, or for that matter an equilibrium, as the dependent variable occurring in pressure-temperature space as a rate or equilibrium surface. A simple example is shown in Figure 1.

The theoretical basis for and interpretation of thermodynamic parameters, and the derivation of the Eyring theory have been summarized recently[2] and a number of examples of its use cited. It is important to stress that the descriptive thermodynamics touched on above is truly phenomenological. It does not imply any particular model for the process being studied, and the thermodynamic values derived from the experimental results are ensemble averages. The difficulties arise in transfering such "averaged" values and any concepts associated with them to individual molecules, or small aggregates of molecular dimensions.

In the remainder of this paper I deal with three recent

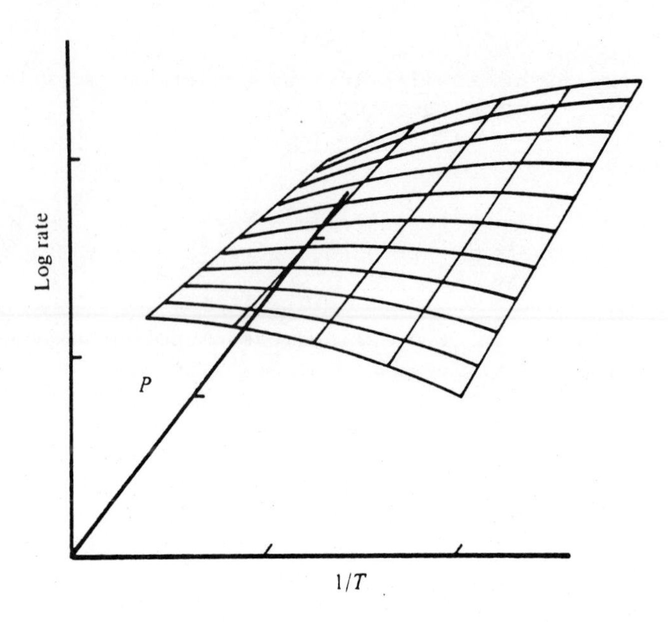

Figure 1. Rate surface in pressure-temperature coordinates, with a temperature-dependent ΔV^* and finite C_p^*. From Klein[2].

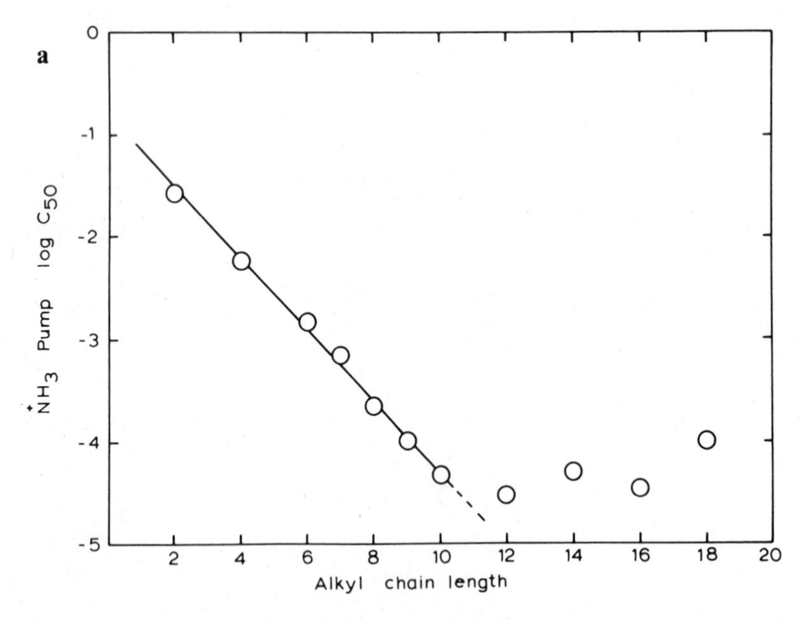

Figure 2. Logarithm of the 50% inhibitory concentration for the K^+ - influx for the sodium pump in human erythrocytes vs. carbon atoms in the alkyl chain for (a) n-alkyl ammonium bromides, (b) n-alkyl trimethylammonium bromides and (c) n-alkyl triethylammonium bromides. From Klein and Ellory[11].

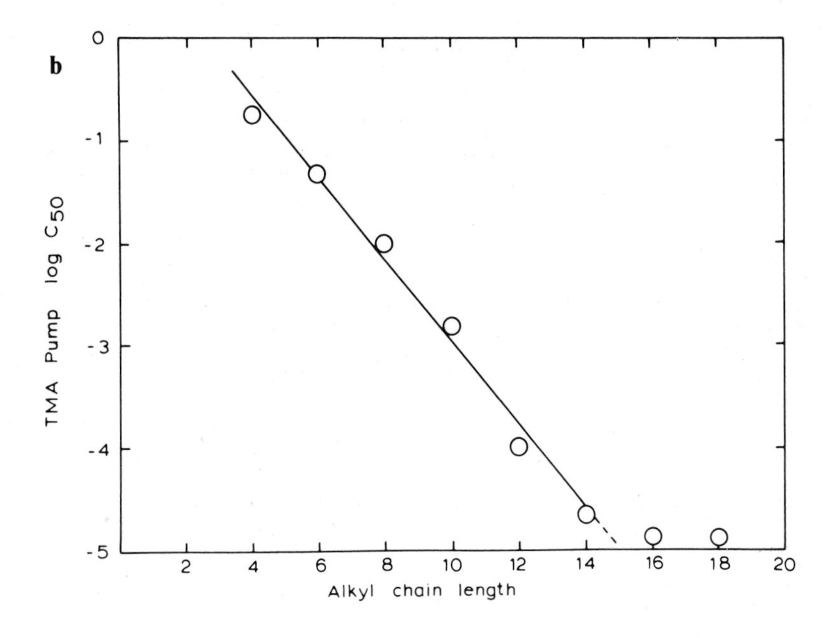
b
TMA Pump log C50
0
-1
-2
-3
-4
-5
2
4
6
8
10
12
14
16
18
20
Alkyl chain length

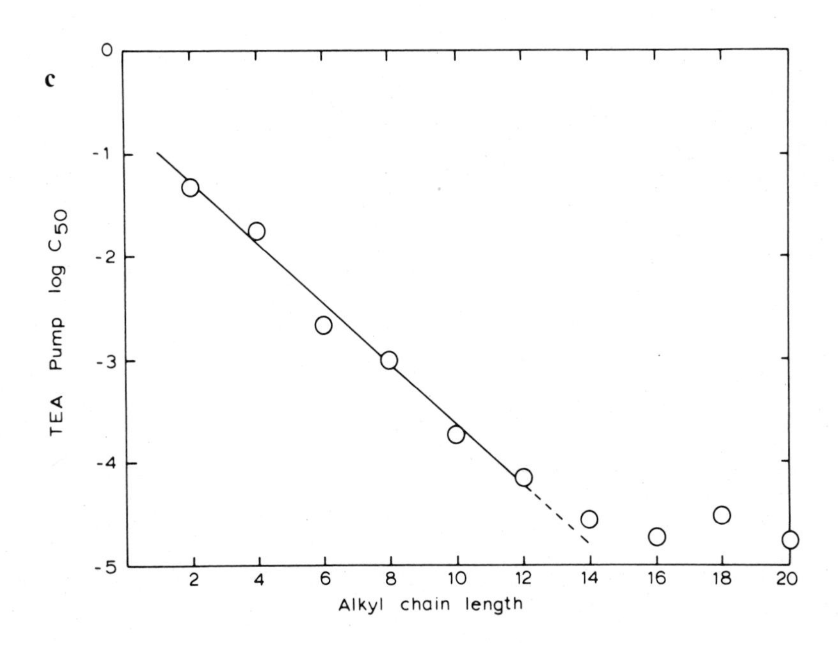
c
TEA Pump log C50
0
-1
-2
-3
-4
-5
2
4
6
8
10
12
14
16
18
20
Alkyl chain length

examples in which we have used thermodynamics to obtain some understanding of the movement and dynamics of molecules as they cross the cell membrane.

2. LINEAR FREE ENERGY RELATIONSHIPS - What is meant by membrane hydrophobicity?

It has been known for a very long time that the rate at which uncharged molecules permeate biological membranes is related to their partitioning behaviour between an aqueous and a non-polar phase, for example olive oil. This is the basis of the so-called Meyer-Overton rule[3].

If partitioning behaviour is examined for an homologous series of related long-chain alkyl compounds, for example fatty acids or their esters, then it is strikingly apparent that there is an excellent correlation between the logarithm of the measured partition coefficient, or some directly related property such as permeability, and the number of carbon atoms in the alkyl chain

$$\ln P = \alpha n_c + \beta = -\Delta G^{\ominus}/_{RT} \qquad (11)$$

One of the clearest examples of this type of behaviour is found in the relationship between the logarithm of the GLC retention times and the chain lengths for a series of homologous fatty acid esters[4].

For homologous alkyl compounds this generally excellent correlation has led to the concept of an <u>incremental free energy per methylene group</u> which is formally equivalent to the slope α in equn. (11). The relationship between the logarithm of the partition coefficient and the sum of suitably weighted additive structural terms has an important place in the Hansch approach to predicting pharmacological activity from structural features[5].

Such linear free energy relationships have been rationalized in terms of solvophobic theory[6]. Solution in aqueous phases is controlled mainly by solvation and entropy effects because of the anomalous structure of water itself[7]. Solution of large apolar molecules in fluid hydrocarbon phases, including the membrane hydrophobic core above its phase transition, is determined rather more by a surface area or cavity free energy term[8]. In macroscopic terms this is equivalent to the product of the surface area of the molecular cavity and the interfacial surface tension. There are considerable difficulties, however, in using a macroscopic concept such as surface tension at a microscopic or molecular level[9].

I refer later to the pitfalls of equating bulk isotropic hydrocarbon behaviour with that of the membrane core which is anisotropic, without taking into account conformational terms.

Dennis Koppel touches on a comparable problem in this volume in the use of the term 'microviscosity' as commonly applied to membranes.

One would expect the cavity free energy term to increase linearly with alkyl chain length. Experimental measurements, statistical mechanical calculations which allow for geometry, and estimates from molecular models indicate this to be the case[10,11]. Values of 33 cal.mol.$^{-1}$Å^2 and around 30Å^2 per methylene group have been reported for short chain alkanes[10]. Somewhat lower values of between 20 and 25 cal.mol.$^{-1}$Å^2 have been reported by Tanford's group[12] with an area increment of 31.8Å^2 per CH_2 group. These authors, however, considered the alkyl chains to be fully extended, an approximation criticized by Hermann[10].

The incremental free energy per CH_2 group for partitioning into a hydrocarbon phase is found to be around 884 cal.mol.$^{-1}$ experimentally in a wide variety of systems for many long chain molecules[13]. It is unusual to obtain figures as high as this for ΔG_{CH_2} in membrane systems[14]. This discrepancy has often been attributed to membrane "hydrophobicity" or "polarity" and used as a measure of this quantity[15].

I show below that this discrepancy is in fact better explained as resulting from the high degree of order found within the membrane hydrocarbon region as compared to bulk solvent. It is not justifiable to think of the membrane at the macroscopic level of normally isotropic bulk solvents. Considerable restraint and caution should be applied in the use of the terms hydrophobic force, hydrophobic effect and hydrophobicity. Some of the semantic difficulties and nuances of thermodynamic interpretation have been argued over recently[16].

We have studied the inhibition of the ouabain-sensitive K^+-influx in human red cells by a series of homologous long chain alkyl ammonium cations, as well as the ability of these compounds to produce haemolysis[11]. Inhibition of K^+-influx was more specific than haemolysis, and occurred at concentrations some 3 to 50-fold lower than for lysis. A good semilogarithmic relationship was obtained between the 50% inhibitory concentration and chain length (Figure 2) for the alkyl ammonium, alkyl trimethyl ammonium and alkyl triethyl ammonium series. Incremental free energies per CH_2 group for the three series (A, TMA & TEA) were clearly distinguishable for inhibition of the K^+-influx: -2.09 ± 0.07 kJ.mol^{-1}; 2.39 ± 0.13 kJ.mol^{-1}; -1.67 ± 0.08 kJ.mol^{-1}. Values for haemolysis were somewhat higher and not significantly different from one another: -2.98 ± 0.20 kJ.mol^{-1}; -2.72 ± 0.14 kJ.mol^{-1}; -2.81 ± 0.28 kJ.mol^{-1}.

We wished to explain why the values for K^+-influx inhibition were so much lower than the typical figures available in Tanford's book[13] as determined using bulk solvent : n-alkanes, -3.70 $kJ.mol^{-1}$; fatty acids, -3.45 $kJ.mol^{-1}$; alcohols, -3.44 $kJ.mol^{-1}$. We found it unreasonable to attribute this difference solely to a doubtful concept of membrane hydrophobicity or polarity.

A significant difference in the treatment necessary for a bulk solvent and the membrane hydrocarbon core lies in the need to consider a configurational energy term which takes into account the probability of trans to gauche interconversions, and the effect on this of membrane order.

We calculated therefore the configurational partition function from the statistical weight matrix for a chain of n bonds following the method of Flory[17], and obtained a (tg)-configurational free-energy term of ≃1.38 kJ.mol CH_2^{-1}. This correction factor explains very satisfactorily, to a first approximation, the discrepancy between 'bulk' incremental free energy values (≃3.5 $kJ.mol^{-1}$) and our experimental values for inhibition of K^+-influx (-1.7 to -2.4 $kJ.mol^{-1}$). The specificity between the series must lie in rather more complex solvation effects related to head group structure which are not independent of the alkyl chain contribution.

It is significant in terms of our analysis that the values for haemolysis are higher (≃2.8 $kJ.mol^{-1}$) and less series specific. This would suggest that the membrane is more disordered, almost certainly brought about by the detergent action of the compounds themselves as part of the process of haemolysis. This is a probable explanation for some of the high values in the literature, as for example in the case of fatty acid permeability in intestinal brush borders[18].

One noticeable feature of our results for the inhibition of K^+-influx is the flattening off of the response above a chain length of C12-C14. This too is likely to be a reflection of the order and average membrane bilayer thickness.

3. PRESSURE EFFECTS

Increased applied pressure may affect the position of an equilibrium process by increasing the $P\Delta V^{\ominus}$ contribution to the overall free energy term. A positive value for the molar volume change will result in a negative pressure coefficient and vice versa

$$\left(\frac{\delta \ln K}{\delta P}\right)_T = \frac{-\Delta V^{\ominus}}{RT} \qquad (12)$$

The pressure-volume product in $J.mol^{-1}$ is given very closely by (P x $\Delta V^{\ominus}$ x 0.1), with P in atmospheres absolute and $\Delta V^{\ominus}$ in $cm^3.mol^{-1}$. Unless $\Delta V^{\ominus}$ is of the order of hundreds of $cm^3.mol^{-1}$ (some protein conformational changes[19]) then the P·V product is not significant compared to kT until pressures in the range of a few hundred ATA are reached.

The thermotropic gel to liquid crystalline phase transition for the alkyl chains of membrane lipids has been shown to have a pressure coefficient of approximately 0.02 K^o ATA^{-1} and is defined by

$$\frac{\delta T_m}{\delta P} = \frac{T_m \Delta V_m}{\Delta H_{VH}} \qquad (13)$$

where ΔH_{VH} is the van't Hoff enthalpy for the transition[20].

Rate processes are also significantly affected by high hydrostatic pressures, and the magnitude of this effect depends again on the value and sign of ΔV^*, the molar volume change of activation for the forward reaction. A broad review of the effects of pressure on many aspects of biological systems may be found in the excellent book by Johnson, Eyring and Stover[21].

We have recently examined the effects of hydrostatic pressure on the three components of K^+ uptake in human red cells[22]. The inhibitors ouabain and bumetanide were used to distinguish the flux mediated by the sodium pump, the chloride-dependent Na^+K^+ cotransport system, and the residual passive leak. We define K^+-passive diffusion across the erythrocyte membrane as that component of the flux that is ouabain and bumetanide-insensitive, and exhibits a linear concentration dependence.

The pump and cotransport components show positive values for ΔV^* with complex non-linear behaviour exhibited for cotransport. The passive leak, however, showed positive pressure dependence with a high negative ΔV^* of around -80 to -90 $cm^3.mol^{-1}$ potassium ion in the presence of chloride (Figure 3). In the presence of other anions different values were obtained for ΔV^*: chloride, -87 $cm^3.mol^{-1}$; bromide, -52 $cm^3.mol^{-1}$; nitrate, -19 $cm^3.mol^{-1}$; iodide, -6 $cm^3.mol^{-1}$. It is highly significant that the order of the ions is the same as their Hofmeister ranking and is directly related to their partial molal volumes of solution (Figure 4). It is generally considered that the water structure-breaking activity of these ions increases in the order $Cl^- < Br^- < NO_3^- < I^-$ [23].

We interpret these results in the following way. If the rate determining step for the formation of the activated transition-state complex for K^+ involves the stripping-off of water from cosphere II, which is energetically very much more likely than

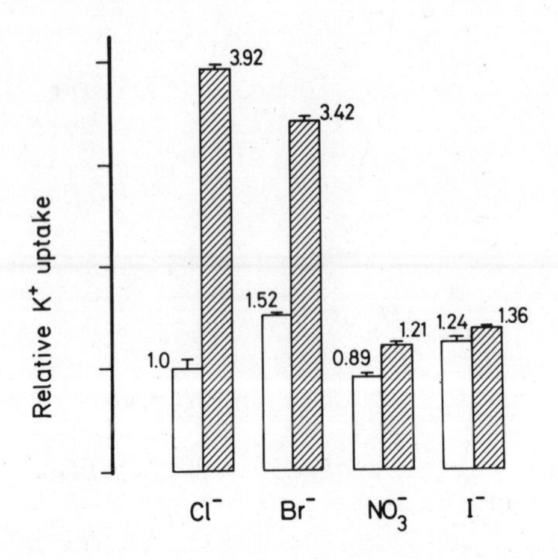

Figure 3. The effect of different anions on the pressure sensitivity of ouabain-sensitive (passive) K^+ uptake in human red cells. Uptakes at 1 and 400 ATA are shown. From Hall, Ellory and Klein[22].

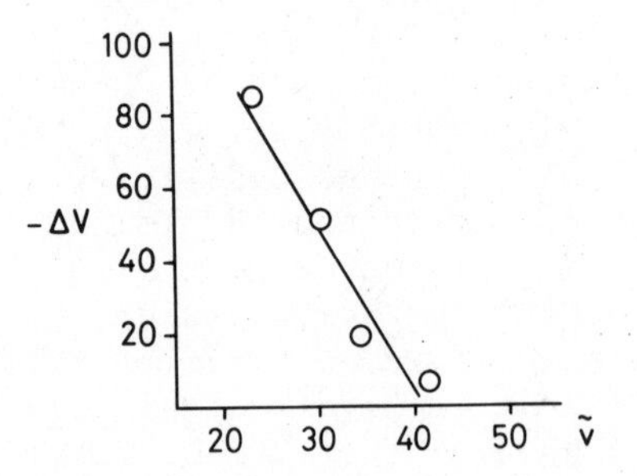

Figure 4. Partial molal volumes for the anions Cl^-, Br^-, NO_3^- and I^- versus the value of ΔV^* for K^+ uptake in their presence. Redrawn from Hall, Ellory and Klein[22].

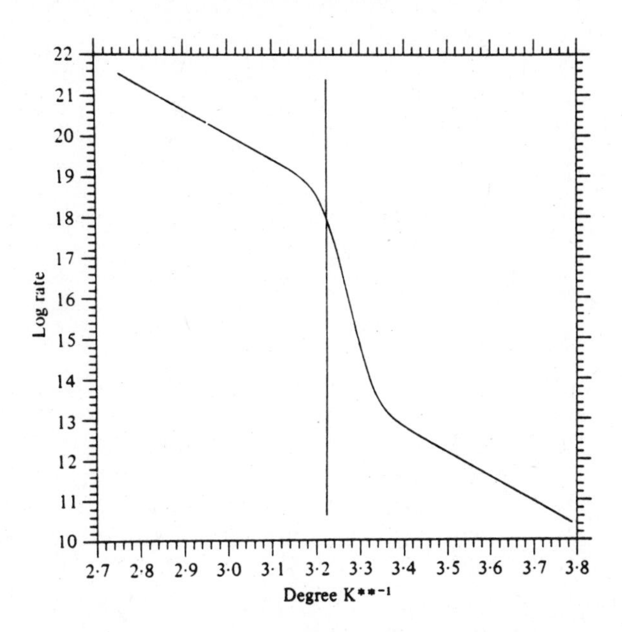

Figure 5. Arrhenius plots for the computer-simulated two-state gel-liquid crystalline partition model with parameters: $\Delta H_1^* = \Delta H_2^* = 48$ kJ mol^{-1}; $\Delta S_1^* = 10$ J mol^{-1} $K^{\circ -1}$; $\Delta S_2^* = -30$ J mol^{-1} $K^{\circ -1}$; $T_m = 310.15°K$; $T_m^{\frac{1}{2}} = 2°K$; partition coefficient, P = 1. From Klein[2].

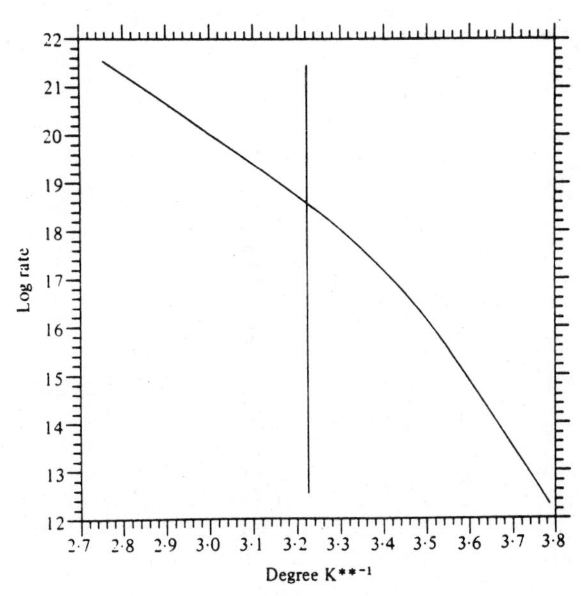

Figure 6. Parameters similar to those in Fig. 5, but with $T_m^{\frac{1}{2}} = 10°K$ and P = 10. From Klein[2].

from cosphere I, and perhaps also removal of solvation water from the region of the membrane channel, then anions with increasing structure-breaking ability would be expected to reduce the overall pressure dependence of the forward reaction as observed experimentally.

It is interesting to note that ΔV^* for passive K^+ diffusion in phospholipid vesicles is of opposite sign and somewhat smaller, around $+20\ cm^3.mol^{-1}$, suggesting a marked difference in mechanism for passive diffusion of K^+ between erythrocytes and lipid vesicles[24].

4. TEMPERATURE DEPENDENCE - a critical look at the interpretation of discontinuities in Arrhenius plots

Svante Arrhenius recognized in 1889, whilst working on the inversion of cane sugar, that reaction rates were proportional to an exponential term involving the reciprocal of the absolute temperature and the heat content μ of the reaction[26]

$$\text{Rate} = A\ \exp(-\mu/_{2T}) \tag{14}$$

The pre-exponential term is now equated with the entropic term, $\exp\ \Delta S^*/_R$, of Eyring transition state theory for rate processes[1]. The experimental activation energy E_a, equivalent to μ in equn. (14) when measured in calories, and obtained experimentally from the relationship

$$\frac{\delta \ln R}{\delta(1/_T)} = \frac{-E_a}{R} \tag{15}$$

is related to the true activation enthalpy

$$\Delta H^* = E_a - RT \tag{16}$$

for reactions in solution.

When data are analyzed in the form of equn. (15), the graphical representation is known as an Arrhenius plot. One particular aspect of Arrhenius plots that has attracted considerable attention over the years has been the interpretation of apparent discontinuities[21]. The postulates put forward by Blackman, Pütter and Crozier[26] were intended to rationalize the early attempts at applying Arrhenius plots to the interpretation of biological rate processes of varying complexity.

In recent years discontinuities or 'breaks' in Arrhenius plots for biological processes have been associated with thermotropic lipid phase transitions[27] often based on meagre evidence. There is, however, ample spectroscopic evidence for a variety of conformational changes in membranes within the physiological temperature range[2]. These should be distinguished from the lipid phase transitions. Moreover, in biological membranes with relatively complex lipid compositions, the phase transition temperature for the gel-liquid crystalline interconversion often lies outside the range of temperature in which biological activity takes place. The actual transition may also take place over a broad range of temperature[28].

What significance can be attributed to a discontinuity in an Arrhenius plot?

One should be able to take for granted the adequacy of the data analysis. Unfortunately this is not always a justifiable assumption, and there are many examples in the literature of the distinction between a straight line or series of straight lines and a curve being based not on an objective statistical treatment but on wishful thinking. The problems of data analysis as applied to Arrhenius plots have been dealt with recently[2,29], and many of the comments made by Ben Chu in this volume on the analysis of photon correlation functions are directly applicable in the general mathematical sense.

Leaving aside any doubts associated with inadequate data analysis or imprecise data, two general special cases are apparent for Arrhenius plots which deviate from single straight line behaviour. Occasionally one may obtain bizarre behaviour as in the example of a flux minimum reported by us[30] with a possible thermodynamic condition

$$\frac{\delta\ \Delta S^*(1/T)}{\delta\ 1/T} = \Delta H^* + RT \qquad (17)$$

The first special case is that of a smooth curve over the range of temperature examined. The trivial explanation involves a number of rate determining steps which cannot be distinguished. The simplest explanation for curvature involves a significant value for C_p^*, the specific heat of activation at constant pressure, since

$$\left(\frac{\delta H}{\delta T}\right)_p = C_p^* \qquad (18)$$

Values of C_p^* which are large enough to produce curvature over the limited physiological range have been reported for enzymatic reactions[31].

In order to distinguish two or more straight line segments the activation enthalpies for the reactions must be very different and the data very precise. Recently I have proposed a thermodynamic two-state model for membrane processes based on previous ideas[2,32], which is capable of generating discontinuities or breaks of slope in the physiological temperature range, without the need for changes in ΔH^* which are unreasonable. The essence of the model is that the process, carrier-enzyme, channel or probe molecule, is thought of as partitioning between the gel and liquid crystalline phase of the membrane hydrocarbon core with a given partition coefficient. Considerable thermodynamic information is available for the lipid phase transition itself[28]. Using Eyring rate theory to construct the formal rate equations for this two-state model, it is possible to generate Arrhenius plots with either two or three straight line segments. Triphasic as well as biphasic plots have been reported in the literature[2].

Two very important features emerge from this model. First that the gradient of the middle, or lower temperature segment in a biphasic plot, is determined predominantly by the thermodynamics of the phase transition and not by the process being studied. And secondly that the positions of the discontinuities and the point of maximum slope are not obviously related to the transition temperature T_m unless the partitioning behaviour of the process is known. Even in the case of a partition coefficient of unity, the transition temperature does not coincide with the point of maximum slope because of the logarithmic transformation used.

In the simplest case in which the difference in rate above and below the transition is determined by entropic factors, i.e., $\Delta H_1^* \simeq \Delta H_2^*$ and $\Delta S_1^* > \Delta S_2^*$, the point detected as the <u>apparent</u> transition temperature (point of maximum slope) is shifted from the <u>true</u> transition temperature by an amount

$$\delta = \left(\frac{1}{T_{app}} - \frac{1}{T_m}\right) \cong \frac{R\ln P}{\Delta H_{VH}} + \frac{(\Delta S_1^* - \Delta S_2^*)}{2\Delta H_{VH}} - \frac{R\ln 2}{\Delta H_{VH}} \tag{19}$$

If one uses membrane probes, such as the fluorescent parinaric acids, to determine the position of the phase transition then the result will depend upon their partition coefficients liquid:solid, and the <u>difference</u> between the reported T_{app} for two probes with partition coefficients P_1 and P_2 will be

$$\left(\frac{1}{T_1} - \frac{1}{T_2}\right) = \frac{R}{\Delta H_{VH}} \cdot \ln \left(P_1/P_2\right) \tag{20}$$

Using literature values for ΔH_{VH} and the partition coefficients for <u>cis</u> and <u>trans</u>-parinaric acids, a relative shift of around

1.0 - 1.5°K would be predicted. This corresponds very closely to the experimental difference observed[33].

Equation (20) may be rewritten in a slightly different form

$$\delta T = (T_1 - T_2) = \frac{T_1 . T_2 . R . \ln (P_1/P_2)}{\Delta H_{VH}} \simeq \frac{T_m^{\frac{1}{2}}}{4} . \ln (P_1/P_2) \qquad (21)$$

to include the width of the transition at half peak height $T_m^{\frac{1}{2}}$. This important approximation indicates that for biological membranes with broad transitions the discrepancy between the reported values for T_m with different probes or membrane processes may be quite large. The data quoted by Thulborn[34] for different probes is of interest here.

As pointed out in my detailed analysis[2], lateral phase separation in the membrane lipid may further complicate an already complex picture[35]. Figures 5 and 6 illustrate the non-coincidence of T_m, shown by the vertical line, with the point of maximum slope for a partition coefficient of unity, and the production of curvature in an Arrhenius plot for a transition with $T_m^{\frac{1}{2}}$ = 10°K, a not unreasonable figure for a biological membrane.

REFERENCES

1. S. Glasstone, K.J. Laidler and H. Eyring, "The Theory of Rate Processes", (McGraw-Hill, New York & London, 1941).
2. R.A. Klein, Quart. Rev. Biophys. xxx 15 (1982).
3. J.F. Danielli, in: "Surface phenomena in Chemistry and Biology", J.F. Danielli, K.G.A. Pankhurst and A.C. Riddiford, eds., (Pergamon Press, London & New York, 1958), pp. 246-265; W.D. Stein, "The movement of molecules across cell membranes", (Academic Press, New York & London, 1967), pp. 70-75.
4. H.P. Burchfield and E.E. Storrs, "Biochemical Applications of Gas Chromatography", (Academic Press, New York & London, 1962).
5. C. Hansch and T. Fujita, J. Amer. Chem. Soc. 1616 86 (1964); C. Hansch, J. Med. Chem. 920 11 (1968).
6. O. Sinanoglu, in: "Molecular Associations in Biology", B. Pullman, ed. (Academic Press, New York & London, 1968), pp. 427-445; O. Sinanoglu and S. Abdulnur, Fedn. Proc., 512 24 (1965); T. Halicioglu and O. Sinanoglu, Ann. N.Y. Acad. Sci. 308 158 (1969); C. Horvath, W. Melander and I. Molnar, Analyt. Chem., 142 49 (1977).
7. F. Franks, "Water", (Plenum Press, New York & London, 1973), pp. 1-54.
8. C. Horvath and W. Melander, Amer. Lab., 17 10 (1978).
9. C. Tanford, Proc. natn. Acad. Sci. U.S.A., 4175 76 (1979).

10. R.B. Hermann, J. Phys. Chem. 2754 76 (1972); Proc. natn. Acad. Sci. U.S.A., 4144 74 (1977).
11. R.A. Klein and J.C. Ellory, J. Membr. Biol. 123 55 (1980).
12. J.A. Reynolds, D.B. Gilbert and C. Tanford, Proc. natn. Acad. Sci. U.S.A., 2925 71 (1974).
13. C. Tanford, "The Hydrophobic Effect: Formation of Micelles and Biological Membranes", (John Wiley, New York & London, 1973).
14. V.L. Sallee, Fedn. Proc. 310 34 (1975); Y. Katz and J.M. Diamond, J. Membr. Biol. 101 17 (1974); J.M. Diamond and Y. Katz, J. Membr. Biol. 121 17 (1974); R.F. Rekker, "The Hydrophobic Fragmental Constant", (Elsevier, Amsterdam, Oxford & New York, 1977).
15. V.L. Sallee, J. Membr. Biol. 187 43 (1978).
16. C. Tanford, Proc. natn. Acad. Sci. U.S.A. 4175 76 (1979); J.H. Hildebrand, J. Phys. Chem. 1841 72 (1968), Proc. natn. Acad. Sci. U.S.A. 194 76 (1979); G. Nemethy, H. Scheraga and W. Kauzmann, J. Phys. Chem. 1842 72 (1968); S.J. Gill and I. Wadsö, Proc. natn. Acad. Sci. U.S.A. 2955 73 (1976).
17. P.J. Flory, "Statistical Mechanics of Chain Molecules", (John Wiley, New York & London, 1969).
18. V.L. Sallee, Fedn. Proc. 310 34 (1975).
19. A.G. MacDonald, "Physiological Aspects of Deep Sea Biology", (Cambridge University Press, London and Cambridge, 1975); P.W. Hochachka, Comp. Biochem. Physiol. 1 52B (1975).
20. A.G. MacDonald, Biochim. biophys. Acta 26 507 (1978); A.G. MacDonald and W. MacNaughtan, J. Physiol. 105P 296 (1979); F. Ceuterick, J. Peeters, K. Heremans, H. De Smedt and H. Olbrechts, Eur. J. Biochem. 401 87 (1978); H. De Smedt, R. Borghgraef, F. Ceuterick and K. Heremans, Biochim. biophys. Acta 479 556 (1979).
21. F.H. Johnson, H. Eyring and B.J. Stover, "The Theory of Rate Processes in Biology and Medicine", (John Wiley, New York & London, 1974).
22. A.C. Hall, J.C. Ellory and R.A. Klein, J. Membr. Biol. 47 68 (1982).
23. F. Hofmeister, Arch. Exp. Pathol. Pharmakol. 247 24 (1888), 1 25 (1889), 395 27 (1890) and 210 28 (1891); H. Davson, Biochem. J. 917 34 (1940); J.L. Kavanan, "Structure and Function in Biological Membranes", (Holden-Day Inc., San Francisco, 1965), Vol I, p. 229.
24. S.M. Johnson and K.W. Miller, Biochim. biophys. Acta 286 375 (1975).
25. S.A. Arrhenius, Z. phys. Chem. 226 IV (1889).
26. F.F. Blackman, Ann. Bot. 281 19 (1905); A. Pütter, Z. allg. Physiol. 574 16 (1914); W.J. Crozier, J. gen. Physiol. 531 9 (1926).
27. A.G. Lee, N.J.M. Birdsall, J.C. Metcalfe, P.A. Toon and G.B. Warren, Biochemistry 3699 13 (1974); P.J. Quinn, Prog. Biophys. molec. Biol. 1 38 (1981); J. Wolfe and

D.J. Bagnall, Ann. Bot. 485 45 (1980).
28. R.N. McElhaney, Chem. Phys. Lipids 229 30 (1982).
29. J. Wolfe and D. Bagnall, in: "Low temperature Stress in Crop Plants: the Rôle of the Membrane", J.M. Lyons, D. Graham and J.K. Raison, eds., (Academic Press, New York, 1980), pp. 527-533.
30. G.W. Stewart, J.C. Ellory and R.A. Klein, Nature (London) 403 286 (1980).
31. H.J. Hinz, D.D.F. Shiao and J.M. Sturtevant, Biochemistry 1347 10 (1971); J.M. Sturtevant and P.L. Mateo, Proc. natn. Acad. Sci. U.S.A. 2584 75 (1978); T.Y. Tsong, R.P. Hearn, D.P. Wratnall and J..M. Sturtevant, Biochemistry 2666 9 (1970); H.F. Fisher, A.H. Colen and R.T. Medary, Nature (London) 271 292 (1981).
32. L. Thilo, H. Träuble and P. Overath, Biochemistry 1283 16 (1977).
33. L.A. Sklar, B.S. Hudson and R.D. Simoni, Biochemistry 819 16 (1977); L.A. Sklar, G.P. Miljanich and E.A. Dratz, Biochemistry 1707 18 (1979).
34. K.R. Thulborn, in: "Fluorescent Probes", G.S. Beddard and M.A. West, eds., (Academic Press, New York & London, 1981), ch. 6, pp. 113-141.
35. E.J. Shimsick and H.M. McConnell, Biochemistry 2351 12 (1973); J.C. Owicki and H.M. McConnell, Biophys. J. 383 30 (1980); B. Snyder and E. Freire, Proc. natn. Acad. Sci. U.S.A. 4055 77 (1980); D. Marsh, A. Watts and P.F. Knowles, Biochemistry 3570 15 (1976).

LIGHT SCATTERING AND PHASE TRANSITIONS IN GMO BILAYER MEMBRANES

G.E. Crawford and J.C. Earnshaw

Department of Pure and Applied Physics
The Queen's University of Belfast
Belfast, BT7 1NN

INTRODUCTION

Bilayer lipid membranes (BLM) are structurally similar to biomembranes but are better characterized physically and chemically. They are therefore particularly appropriate systems for thermotropic phase transition studies. Laser light scattering is a sensitive and reliable technique for measuring the mechanical properties of the BLM, which causes minimum perturbation to the equilibrium system.

The properties of the membrane and its bathing solution are related approximately to the peak frequency ω and the width Γ of the spectrum of the scattered light[1] by

$$\omega = (\gamma_0 q^3/2\rho)^{\frac{1}{2}} \qquad (1)$$

and

$$\Gamma = \eta q^2/\rho \qquad (2)$$

where γ_0 is the interfacial tension of the membrane, η and ρ are the viscosity and density respectively of the aqueous medium, and q is the fluctuation wave-vector.

At the phase transition, the co-operative transisomerization of the acyl chains of the lipid molecules induces changes in the membrane viscoelasticity, affecting both ω and Γ significantly.

EXPERIMENTAL

The optical set-up (Fig 1) used in this experiment is described in detail elsewhere.[2] Due to the low reflectance of BLM a high incident laser power (180 mw) was used. The detector system included a cylindrical lens, giving vertical focussing of the reflected beam, and a weak diffraction grating providing a controllable local oscillator for optical heterodyning. The K7025 Malvern correlator was used for time-domain analysis, the data being transferred to a PDP11/34A for subsequent on-line analysis.

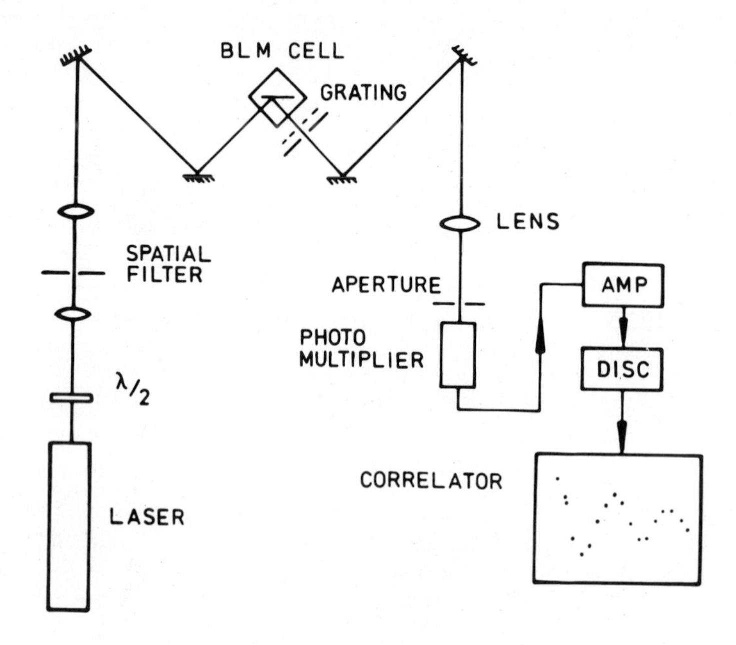

Figure 1 : A schematic representation of the optical system used in photon correlation spectroscopy of bilayer lipid membranes

The cell in which the membrane was formed consisted of an optically polished glass spectrophotometer cuvette. A PTFE septum was lowered through a GMO/cholesterol/n-decane solution spread on the surface of the aqueous medium.[3] The thick film was allowed to drain for several hours before any measurements were taken upon the final black film.

A brass heat-sink surrounding the cell was used to heat and cool the BLM. A PTFE housing gave thermal and acoustic isolation. The temperature was regulated to within 0.1°C by a dual system consisting of circulating thermostatted water and electronically-controlled thermofoil heaters contacting the heat-sink. Slow heating/cooling rates ($0.1^{o}C$/min) were necessary to ensure quasi-equilibrium conditions and to avoid possible premature rupture of the BLM due to thermal gradients. The experiment duration was limited by the relatively short lifetime of the membrane.

The observed correlation data were fitted to the functional form:

$$C(t) = C(0) \cos(\omega t + \phi) e^{-\Gamma t} e^{-\beta^2 t^2/4} \qquad (3)$$

where ϕ is a phase term which accounts for possible deviation of the experimental spectra from Lorentzian form, and β represents the instrumental effect.

The observables ω and Γ obtained can be interpreted in terms of the interfacial tension γ_0 and the membrane viscosity γ' by numerical solution of the dispersion equation:

$$(S^2 + \tau'S + Y)\sqrt{1 + 2S} - (Y + \tau'S) = 0$$

where $S = \dfrac{-i\omega\rho}{2\eta q^2}$, $Y = \dfrac{\gamma_0\rho}{8\eta^2 q}$, $\tau' = \dfrac{\gamma' q}{4\eta}$

The accuracy of the ω and Γ values were essentially limited by the statistical nature of photon correlation spectroscopy. For one set of fifty correlation functions taken consecutively, the standard deviations of ω and Γ were 1.7% and 3.6% respectively. This rather high precision permitted the detection of the small changes which occurred with temperature.

RESULTS AND DISCUSSION

The data obtained upon cooling and subsequent heating of a GMO/cholesterol/n-decane membrane are shown in Fig 2. The evident changes in the variation of both ω and Γ with temperature in the vicinity of $17^{o}C$ are indicative of a phase transition. The peak frequency shows the trend expected from eq 1; γ_0 must decrease as the temperature is lowered, because of the higher packing density of the lipid molecules in the all-trans state. The variation of the damping coefficient on the other hand is complicated and not explicable in terms of the first order approximation. It may be that eq 2 is rendered invalid in the transition region by the effect of the γ' parameter.

While the temperature dependence of γ' is as yet unknown, calculations have shown that plausible variations may cause significant changes in both ω and Γ. Also, the broadening of the transition

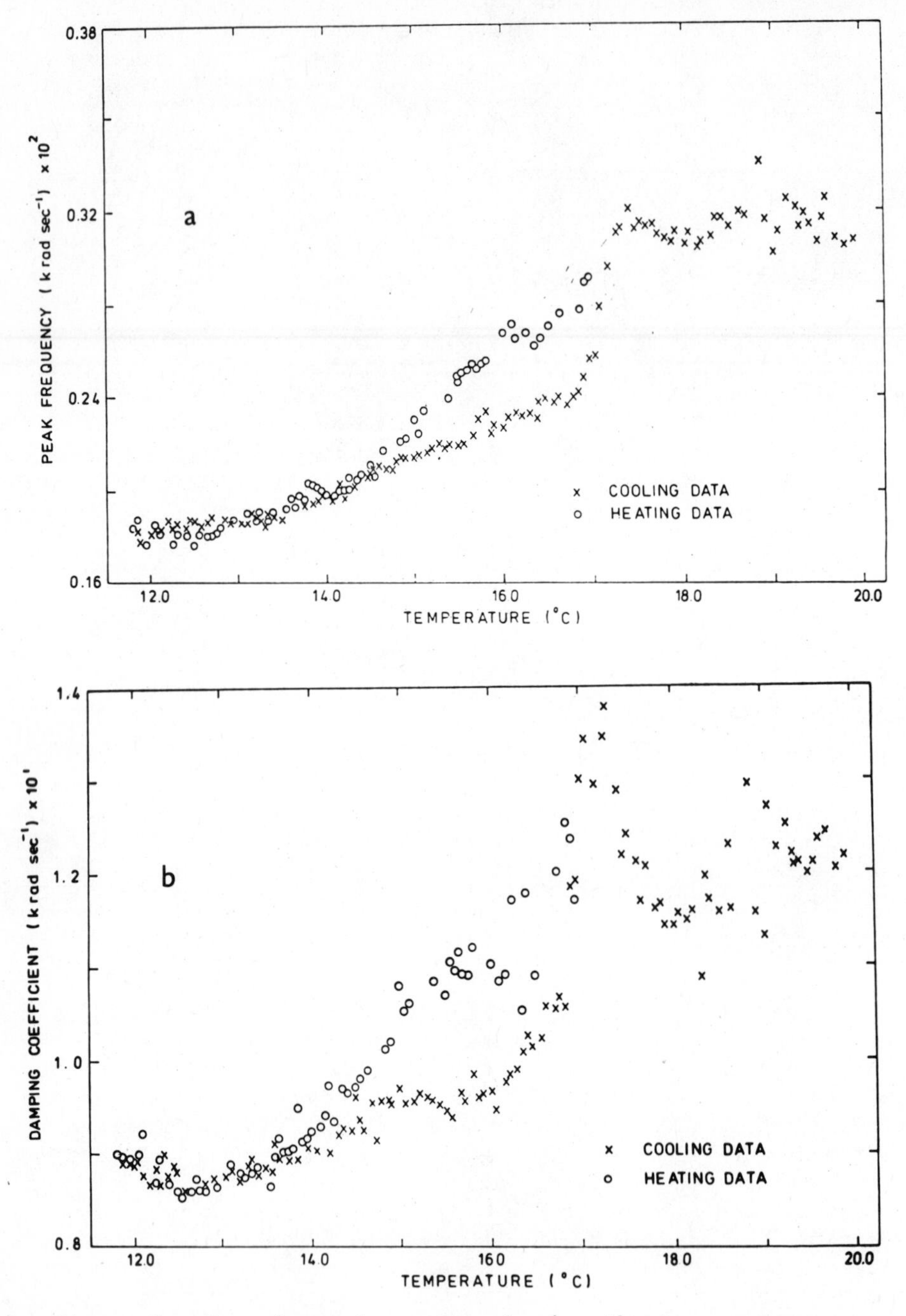

Figure 2 : Experimental results showing the temperature variation of:
(a) Peak frequency ω
(b) Width Γ of the optical spectrum

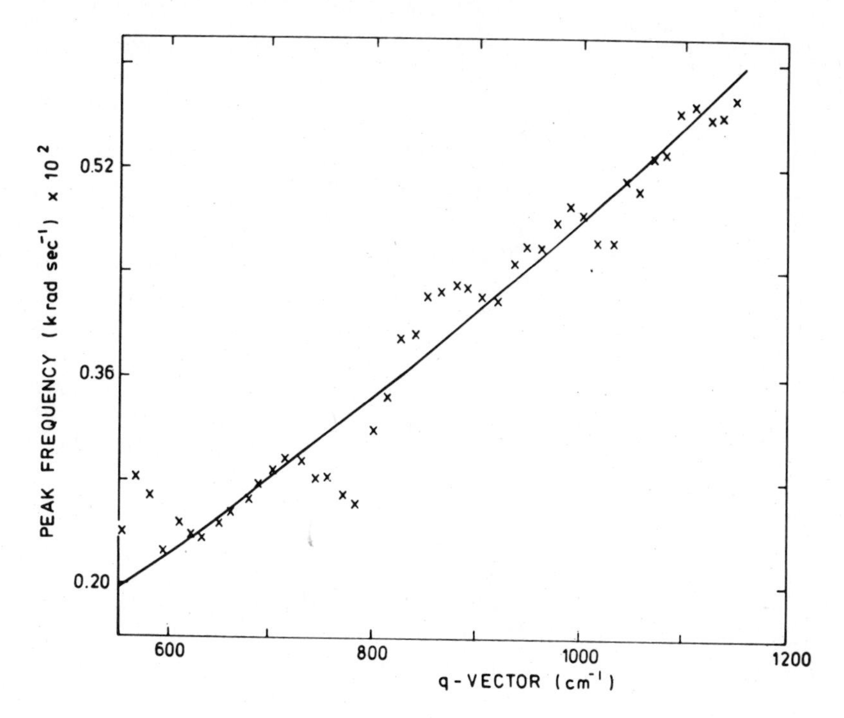

Figure 3 : The dispersion plot of ω versus q. The solid line represents the best-fit to the $q^{3/2}$ variation

due to the presence of cholesterol in the BLM[4] may contribute to the anomalous behaviour of Γ. Examination of a wider temperature range might be helpful in elucidating these points.

The interpretation of the observed data is hindered by experimental difficulties. It has been observed that some membranes change their orientation as they expand and contract with changing temperature, altering the q value under consideration. This obscures the thermal effects and precludes accurate calculation of the viscoelastic parameters. The measurements are further complicated by the very slow oscillation of the membrane about the vertical axis, again altering the q value. This is evident from the deviations of the data in the dispersion plot of ω versus q (Fig 3). The apparatus is presently being modified to ensure that a single q-vector is observed throughout the experiment.

Despite the experimental difficulties, it has been demonstrated that laser light scattering can be used to detect thermotropic phase transitions in BLM. With improved technique, the temperature variation of the viscoelastic parameters could be extracted. From such a set of results it would be possible to investigate the entropic and enthalpic changes associated with the phase transition.[5]

Several other systems, including solvent-free membranes[6] and bilayers formed from lipid mixtures[7] could also be examined.

ACKNOWLEDGEMENTS

This work was supported by a research grant from the SERC. GEC wishes to acknowledge financial support by the Department of Education for Northern Ireland; also assistance with travel funds from the Institute of Physics.

REFERENCES

1. Crilly J.F. and J.C. Earnshaw, in "Fourth International Conference on Photon Correlation Techniques in Fluid Mechanics", W.T. Mayo, Jr, and A.E. Smart, eds. (Stanford University, Stanford) p 21.1 (1980).
2. Crilly J.F., PhD thesis, Queen's University of Belfast (1981). Crilly J.F. and J.C. Earnshaw, to be published.
3. van den Berg J.J., J. Mol. Biol., 12:290 (1965).
4. Hinz H.J. and J.M. Sturtevant, J. Biol. Chem., 247:3697 (1972).
5. Tien H.Ti., "Bilayer Lipid Membranes (BLM)", (Marcel Dekker Inc., New York 1974).
6. White S.H., Biophys. J., 23:337 (1978).
7. Pagano R.E., R.J. Cherry and D. Chapman, Science, 181:557 (1973).

PHOTON CORRELATION STUDIES OF PHASE TRANSITIONS IN LIPID MONOLAYERS

J. F. Crilly and J. C. Earnshaw

Department of Pure and Applied Physics
The Queen's University of Belfast
Belfast BT7 1NN, Northern Ireland

INTRODUCTION

Monolayers of biologically important lipids spread at the air/water interface are used as simple models for certain biomembranes. Study of these systems can yield information on the molecular interactions within their natural counterparts. Measurements of the temperature variation of the viscoelastic properties provide useful thermodynamic data on the molecular basis of monolayer structure. Phase transitions are of considerable importance in the biological milieu since many vital processes depend upon these phenomena. Photon correlation spectroscopy is a sensitive, non-perturbative technique for monitoring changes in monolayer visco-elasticity associated with thermotropic phase transitions. This technique has the advantage that it permits the simultaneous measurement of both static and dynamic monolayer properties.

The theory of light scattering from fluid interfaces is well established.[1] Spectral changes in the scattered light reflect the dynamics of the interface. For thermal excitations the scattered light spectrum is in the form of a Lorentzian for which the corresponding correlation function is a damped cosine wave. Analytic formulae relating the spectral parameters to the interfacial visco-elastic properties can be written as:-

$$\omega_o = (\gamma_0 q^3/\rho)^{\frac{1}{2}} \quad (1)$$

$$\Gamma = \tfrac{1}{2}\eta q^2/\rho \quad (2)$$

where ω_o and Γ are respectively the frequency and width of the spectrum, γ_o is the interfacial tension, ρ the density, η the fluid

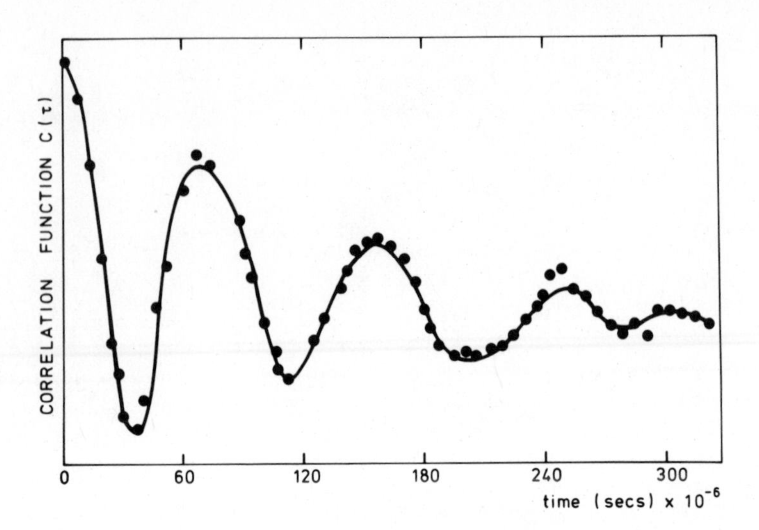

Figure 1. Experimental autocorrelation function (•) at 18°C; the curve is the fitted function (eg. 3).

viscosity and q is the spatial frequency of the interface fluctuation. These equations are only approximate and valid for a limited hydrodynamic regime. It is possible to obtain more exact formulae.[2] However eqs 1 and 2 are useful since they indicate the approximate dependence of ω_0 and Γ on the angle of scattering.

EXPERIMENTAL

Details of the experimental apparatus have been given elsewhere.[3,4] Monolayers of glycerol monoöleate were spread from solution on the surface of aqueous 0.1 M NaCl. Before spreading, the surface of the subphase was swept continuously for several hours to collect surfactant contaminants and dust which were removed by suction. The trough was mounted within a polythene enclosure for acoustic isolation and to prevent excess evaporation from the subphase. Temperature control of the system to within ± 0.1°C was achieved by pumping thermostatted water through a glass coil in the base of the trough. The monolayer was compressed quasi-statically to the fully condensed state and allowed to equilibrate before measurements were taken. Correlation functions were taken at 0.1°C intervals over the range of interest, viz 12°C - 18°C.

The analysis process was simplified considerably by taking data at a large angle of scattering to ensure that instrumental effects

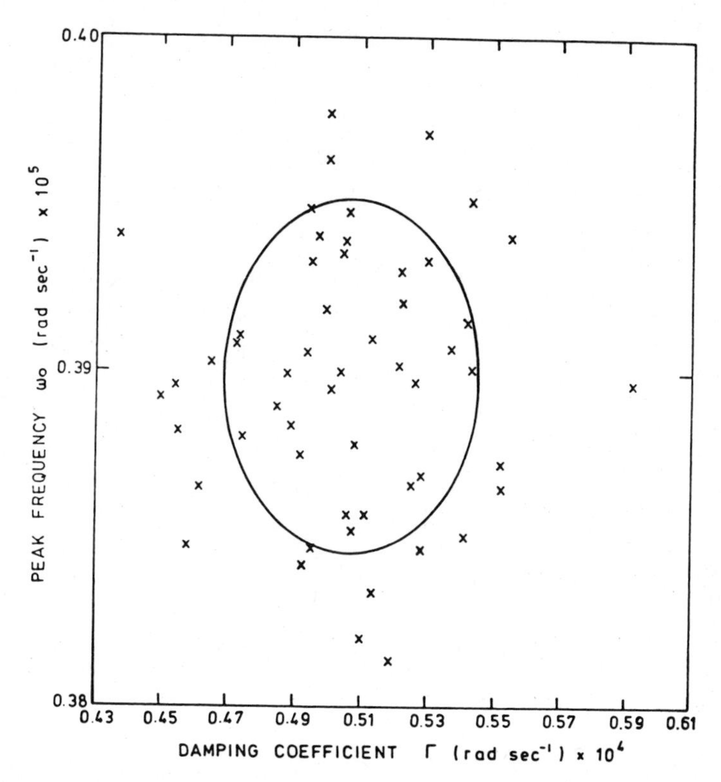

Figure 2. Scatter of measured (ω_o, Γ) values about their statistical average.

were negligible[5]. Experimental correlation functions (as shown in Fig 1) were fitted to an appropriate analytic form:

$$C(\tau) = C(0) \cos(\omega_o \tau + \phi)e^{-\Gamma\tau} \qquad 3$$

The phase term ϕ has been included to account for the deviation of the experimental spectra from an exact Lorentzian form. The values of ω_o and Γ were used to extract the physical properties of the monolayer.

The detection of phase transitions relating to the lipid acyl chain configuration require very precise measurements. It is therefore necessary to consider the limitations imposed by the inherent statistical error of the technique. The measurement accuracy is limited by the statistical scatter of the individual measured values of ω_o and Γ. This is illustrated in the scatter plot of Fig 2 in which the solid line represents the one-standard deviation confidence region. Typical errors on the ω_o values are 0.6%, those on Γ values being considerably larger

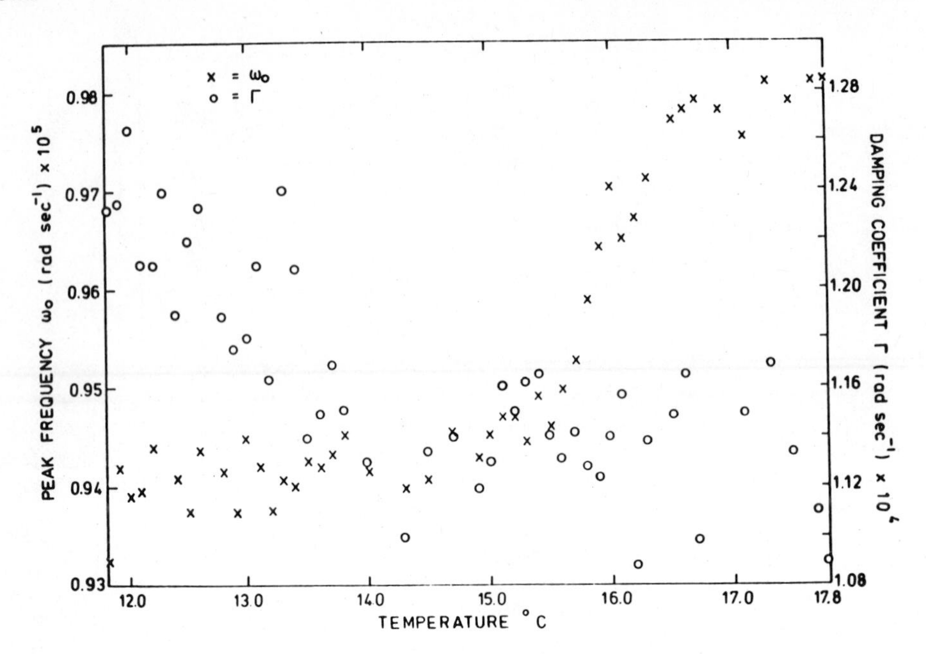

Figure 3. Variation of ω_o (x) and Γ (o) with temperature.

(3.5%). Such precise measurements permit very small changes in the viscoelastic parameters to be discerned.

For a fully condensed monolayer the elasticity and its associated surface dilational viscosity can apparently be neglected[6]. Under this condition the dispersion equation[7] reduces to:-

$$S^2 \left[(1 + S)^2 + Y - \sqrt{1 + 2S} \right] + \beta S^3 = 0 \qquad 4$$

where $Y = \gamma_o \rho / 4\eta^2 q, \quad \beta = \gamma' q / 2\eta, \quad S = - i\omega\rho / 2\eta q^2,$

γ' being the transverse shear viscosity of the surface. The surface pressure of the monolayer π is related to the interfacial tension γ_o by $\pi = \gamma_o - \gamma_c$ where γ_c is the tension of a clean interface at the same temperature. Unique values of π and γ' were deduced by solving eq. 4 with the observed values of ω_o and Γ as the real and imaginary parts of the frequency ω.

RESULTS AND DISCUSSION

Fig 3 shows ω_o and Γ values measured over the range 11.8°C - 17.8°C. The behaviour of ω_o between 15.5°C and 16.5°C indicates an abrupt change in γ_o. This effect can be associated with the main hydrocarbon chain melting transition of GMO in this region.

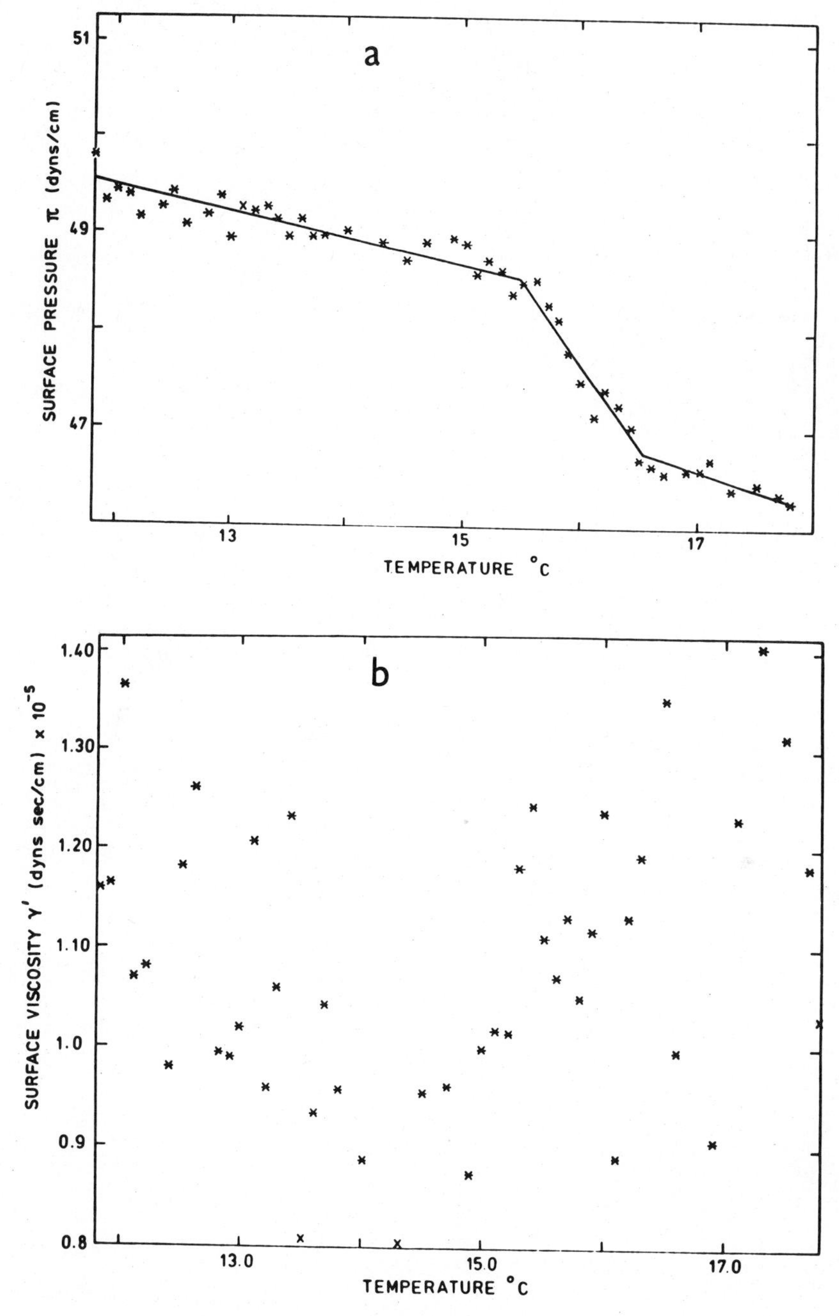

Figure 4. (a) Estimated π values as a function of temperature
(b) Corresponding values for γ'; overall trend is difficult to assess because of the large absolute errors on the viscosity.

The variation of γ_o with temperature acquires a maximum slope at 15.8°C which was therefore taken to be the transition temperature T_c. Physically the transition is not particularly sharp but extends over a temperature range of about one degree. The correspond ing behaviour of Γ in this range is complicated and, as yet, not full understood.

The values of π and γ' obtained by the solution of the dispersion equation and corrected for the temperature variation of the subphase tension and viscosity, are shown in Figs 4 and 5 respectively. Outside the transition region the surface pressure is only weakly dependent on temperature. Therefore the π data were fitted to straight lines above, below and within the transition region. The temperature coefficient dπ/dT was always negative and did not change sign at the transition. Above and below the transitio the temperature coefficients of π were respectively -0.27 dyne/cm/°C and -0.34 dyne/cm/°C. In the transition region the slope of the π-T curve was found to be -1.52 dyne/cm/°C. The change in dπ/dT reflects the energy evolved in the chain melting process which depends upon th particular polymorphic form of the lipid crystal[8]. Quantitative information of this type will be useful in theoretical modelling of the lipid molecular configuration in monolayers and in the further understanding of the energetics of some membrane processes.

The results reported here demonstrate that photon correlation spectroscopy provides a sensitive method for obtaining accurate thermodynamic data which can be related to the structure of the monolayer lipid and to the process of spreading[8]. Clearly this technique is capable of detecting thermotropic phase transitions. The exact nature of the chain melting transition in GMO cannot be determined by light scattering alone, so it must be regarded as complementary to other techniques. The technique has the potential for investigating mixed lipid monolayers or monolayers incorporating interesting chemical modifiers. Some application may indeed prove to be useful in determining configurational entropy effects and also in the study of the structure of water adjacent to the monolayer[9].

ACKNOWLEDGEMENTS

This work was supported by a grant from the SERC.

REFERENCES

1. L. Kramer, J. Chem. Phys. 55: 2097 (1971).
2. D. McQueen and I. Lundstrom, J. Chem. Soc. Faraday Trans I 69: 694 (1973).
3. G.E. Crawford, J.F. Crilly and J.C. Earnshaw, Faraday Symp of the Chem. Soc. No. 16, to be published (1982).
4. J.C. Earnshaw this volume.
5. D. Byrne and J.C. Earnshaw, J. Phys. D. 12: 1145 (1979).

6. J.F. Crilly and J.C. Earnshaw, in: "Biomedical Applications of Laser Light Scattering", D. Sattelle, B. Ware and W. Lee, eds, (Elsevier/North Holland, Amsterdam in the press).
7. D. Langevin and C. Griesmar, J. Phys. D. 13: 1189 (1980).
8. M.C. Phillips and H. Hauser, J. Coll. Interf. Sci. 49: 31 (1974).
9. R.A. Demel, L.L.M. Van Deenen and B.A. Pethica, Biochem. Biophys. Acta. 135: 11 (1967).

LIGHT SCATTERING FROM MICELLAR SOLUTIONS – PROPOSAL FOR A LIGHT SCATTERING STANDARD

Mario Corti and Claudio Minero

CISE S.p.A.
P.O.Box 12081, 20134 Milano, Italy

and

Vittorio Degiorgio

Istituto di Fisica Applicata
Università di Pavia
Via Bassi 6, 27100 Pavia, Italy

CONTENTS

INTRODUCTION

We describe in this paper light-scattering measurements performed on micellar solutions. The emphasis is put here on the experimental problems whereas a detailed account of the physico-chemical information which can be extracted from the data is presented elsewhere. In particular, we discuss a simple and accurate absolute calibration method of the light-scattering apparatus. Such a method is based upon turbidity measurements near the critical consolution point of a non-ionic amphiphile solution.

Laser-light scattering is presently widely utilized for the characterization of macromolecules, micelles, and inorganic colloids in solution[1]. As it is well-known, from a measurement of the average intensity of scattered light it is possible to obtain the molecular weight M of the suspended particles. An intensity correlation measurement may give the translational diffusion coefficient D, and therefore the hydrodynamic radius R_H, of the particles. Whereas the dynamic measurement yields directly D, the molecular weight M is connected with the average scattered intensity, through a calibration constant which has to be derived for each specific light-scattering apparatus. One could be tempted to say that M can be calculated from R_H (assuming a spherical shape) and from the known density of the particles. This is generally not true, because the hydrodynamic volume includes both the volume occupied by the particle and the volume of the solvent which moves with the particle in solution.

The absolute calibration of the light-scattering apparatus usually requires a rather complicated and tedious procedure which involves a careful control of the geometrical and optical parameters of the setup. A common practice is to perform a relative calibration by using a scatterer of known properties, either a pure liquid having a known (and possibly large) scattering cross-section, like benzene and toluene, or a macromolecular solution. There are several biological macromolecules of known size and molecular weight which could be used for such a purpose, but they are not always readily available with good purity and they have often a limited stability in solution. The results reported in this paper suggest that micellar solutions represent a very convenient light-scattering standard for both static and dynamic experiments.

LIGHT SCATTERING FORMULAS

We recall that the average scattering intensity I_{ST} measured at the scattering angle θ and at the distance r from the scattering volume, is given by

$$I_{ST} = I_o\left(\frac{K_o}{n}\right)^4 \frac{V^2\sin^2\gamma}{16\pi^2 r^2}\left\langle|\delta\varepsilon(\vec{K},t)|^2\right\rangle \tag{1}$$

where I_o is the incident intensity, K_o is the incident wave vector, $\vec{K}=\vec{K}_s-\vec{K}_o$, $\vec{K}_s$ is the scattered wave vector, γ is the angle between the scattering direction and the incident electric field, V is the scattering volume, and $\delta\varepsilon(\vec{K},t)$ is the spatial Fourier transform of the dielectric constant fluctuation. In the case of macromolecular solution, I_{ST} can be written as the sum of two contributions, arising one from density fluctuations, I_w, and the other from concentration fluctuations, I_s. For dilute solutions, I_w coincides with the intensity scattered by the solvent alone. The excess intensity I_s which contains information about the solute particles can be written as

$$I_s = I_o\left(\frac{K_o}{n}\right)^4 \frac{V^2\sin^2\gamma}{16\pi^2 r^2}\left(2n\frac{dn}{dc}\right)^2\left\langle|\delta c(\vec{K},t)|^2\right\rangle \tag{2}$$

where $n=\sqrt{\varepsilon}$ is the index of refraction of the solution. The mean square concentration fluctuation $\langle|\delta c(\vec{K},t)|^2\rangle$ is usefully expressed as

$$(|\delta c(\vec{K},t)|^2) = \langle|\delta c(0,0)|^2\rangle\, S(K) \tag{3}$$

where $\langle|\delta c(0,0)|^2\rangle$ is a purely thermodynamic quantity and S(K) is a structure factor.
It is a well-known result that

$$\langle|\delta c(0,0)|^2\rangle = \frac{K_B\, T\, \bar{V}_1\, c}{V\,(\frac{\partial\mu_1}{\partial c})_{T,p}} \tag{4}$$

where K_B is the Boltzmann constant, $\bar{V}_1$ is the partial molal volume of the solvent, and μ_1 is the molal chemical potential of the solvent. When dealing with dilute solutions it is often useful to express $(\delta\mu_1/\delta c)_{T,p}$ by a series expansion in powers of the solute concentration c,

$$-\frac{1}{\bar{V}_1 K_B T}\left(\frac{\partial\mu_1}{\partial c}\right)_{T,p} = N_{AV}\left(\frac{1}{M} + 2Bc + \ldots\right) \tag{5}$$

where N_{AV} is Avogadro number and B is called the second virial coefficient.

The structure factor S(K) is defined here in such a way as to become equal to one as $K\to 0$. It shows an appreciable K-dependence only when the particle size a or the correlation range of concentration fluctuation ξ are comparable to 1/K. Near a critical consolution point ξ diverges, and S(K) is given by the Ornstein-Zernike formula

$$S(K) = \frac{1}{1+K^2\xi^2} \tag{6}$$

The molecular weight of the suspended particles is obtained from the scattered intensity, extrapolated to zero scattering angle ($K\to 0$) and in the limit of negligible interparticle interactions, as given by the relation

$$I_{SO} = I_0\left(\frac{K_o}{n}\right)^4 \frac{\sin^2\gamma}{16\pi^2 r^2}\left(2n\frac{dn}{dc}\right)^2 \frac{cMV}{N_{AV}} \tag{7}$$

In order to be able to compare directly the intensity of light scattered by various liquids and macromolecular solutions, it is convenient to define the Rayleigh ratio R as follows

$$R = \frac{I_S}{I_o} \frac{r^2}{2V\sin^2\gamma} \tag{8}$$

Note that R is K-dependent if I_S is K-dependent.

By writing the scattering volume V as the product of a cross-sectional area and a length L, the power P_S scattered into a solid angle $\Delta\Omega$ is

$$P_S = 2P_o RL\Delta\Omega\sin^2\gamma \tag{9}$$

Therefore R represents the fraction of the incident power scattered into a unit solid angle per unit length of the scattering volume. R has units of reciprocal length, cm^{-1}. The total scattered power, obtained by integration over $\Delta\Omega$ in the case in which P_S is not K-dependent, is

$$P_{S,tot} = P_o(8\pi/3)RL \tag{10}$$

The total attenuation coefficient $\alpha=(8\pi/3)R$ is called the turbidity of the solution.

Besides the average intensity I_S, the other quantity of interest in a light scattering experiment from a macromolecular solution is the intensity correlation function $G_2(\tau)$. The time-dependent part of $G_2(\tau)$ is proportional to $\exp(-2DK^2\tau)$, therefore the mass diffusion coefficient D can be derived directly from the experimental results once K^2 is known. In the limit of non-interacting particles D coincides with the translational diffusion coefficient of the particle, D_o For a dilute solution, D can be expressed, analogously to $(\delta\mu/\delta c)_{T,p}$ by the first terms of a series expansion as

$$D = D_o(1+K_Dc+\ldots) \tag{11}$$

where the coefficient K_D plays here the same role as the virial coefficient B in Eq.(5).

MICELLAR SOLUTIONS

Micelles are aggregates spontaneously and reversibly formed in aqueous solution by amphiphile molecules or ions[2]. Micelle formation is not strictly equivalent to a phase separation. However, if the aggregation number is reasonably large the transition from the predominantly unassociated amphiphile to the micellar state does occur over a narrow critical range of concentration. It is usual to define a single concentration within this transition zone as the critical micelle concentration (cmc). An important feature of micelle formation is that the free amphiphile concentration c_o in equilibrium with micelles changes very slowly with the concentration of micelles, so that it is usually a good approximation to assume that c_o is constant and equal to the cmc when $c \geqslant$ cmc. Near the cmc micelles have a globular (not necessarily spherical) shape, with a typical aggregation number $m \simeq 100$. An equilibrium mixture should contain micelles with a range of different aggregation states. However, near the cmc the relative variance of the m-distribution is small ($V<0.1$).

The measurement of micellar parameters, either by light scattering or by other techniques as centrifugation, osmometry and viscosimetry, is usually performed at amphiphile concentrations well above the cmc. As we have discussed in several papers[3-5], intermicellar interactions may have a considerable effect on solution properties, so that the interpretation of the experimental data in terms of individual micelle parameters is not always straightforward. In this paper we will focus our discussion on aqueous solutions of nonionic amphiphiles. Such systems often present a consolution curve (called cloud curve in the micellar literature) with a minimum (critical point) at low amphiphile concentrations (1-2%). The existence of a phase separation process implies strong intermicellar interactions, so that the micellar solution is, in the critical region, highly nonideal. The light scattering experiment may yield true micellar parameters only if performed sufficiently far away from the critical point, as discussed in Ref.5 and shown later on in this paper. The available results show that critical concentration fluctuations play an important role for the description of solution properties in a large temperature region around the critical temperature T_c. The critical behaviour of micellar solutions presents, however, some peculiarities not observed in usual critical binary mixtures. This indicates that the usual models of critical phenomena may not be appropriate for describing critical micelle solutions[6]. A very important step toward the understanding of the problem could be represented by a direct measurement of the temperature-dependence of micelle size. Although there are neutron scattering experiments[7] indicating that such a dependence is small for some systems, there appears to be no general agreement about this point, and it is still unclear whether the results obtained with a specific system can be extrapolated to all the homologues.

APPARATUS AND MATERIALS

A scheme of the light scattering apparatus is presented in Fig.1. The light source is an Argon ion laser (Spectra Physics Model 165) operating on the TEMoo mode at the wavelength λ=514.5 nm. To avoid sample heating the laser output power is kept typically at 50 mW and

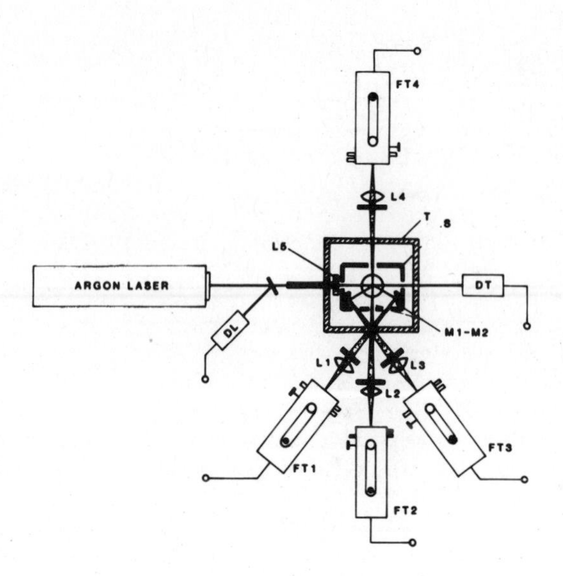

Figure 1. Light scattering setup.

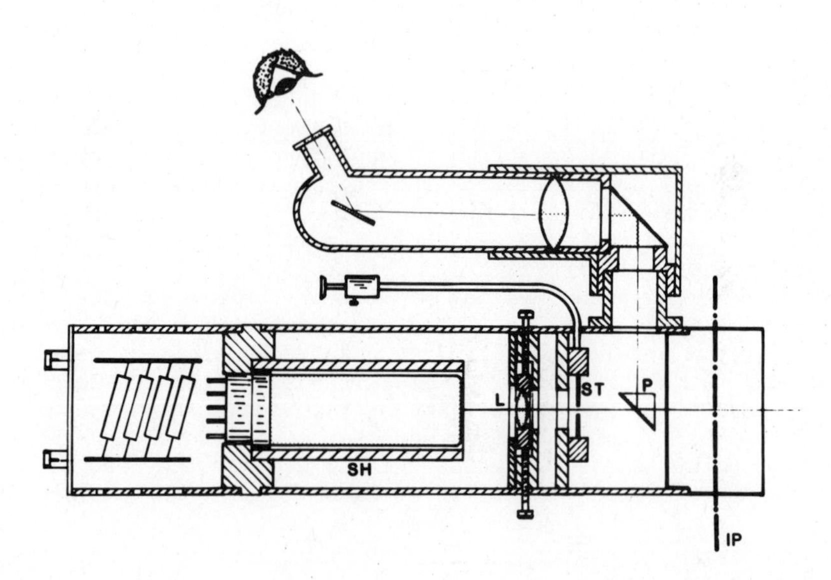

Figure 2. Collecting optics for the light scattering measurement.

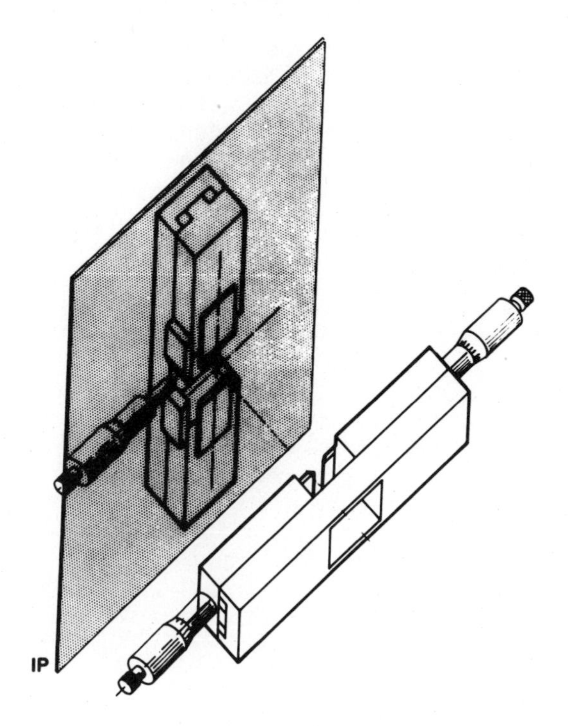

Figure 3. Detail of Fig.2.

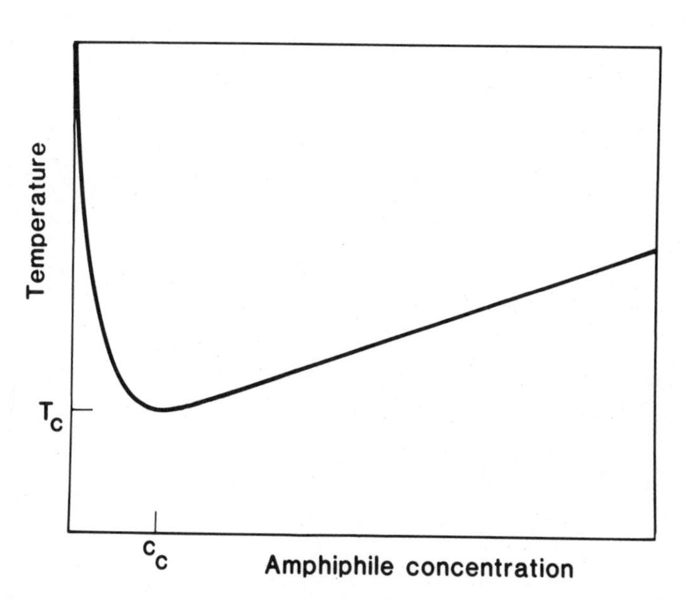

Figure 4. Phase diagram of a non-ionic amphiphile in water showing the consolution curve.

never above 200 mW. The laser intensity has a long term stability of 1% and a short term stability of 0.1%. We found unnecessary to perform spatial filtering of the laser beam. Incidentally it must be said that spatial filtering may itself introduce some intensity noise, because of mechanical vibrations and thermal drifts. The overall mechanical stability is assured by rigid mountings blocked to a granite table 10 cm thick.

The laser beam is mildly focused in the center of the scattering cell by a single lens, L5 in Fig.1, with a focal length of 8 cm. The resulting angular spread is $\pm 1^{\circ}$. The experiments described in this paper have been performed with a square 1x1 cm cell, 0.8 cm high, specially built by Hellma in fused silica of optical quality. The cell is temperature controlled with a long term stability of 0.001°K. We use a double stage thermostat. The metal box T is stabilized withi $\pm 0.01^{\circ}$K by means of a circulating water system. The internal box S is isolated from T and thermally controlled with a Peltier reversible heat pump. Temperature is sensed by a fast response thermistor which is set into a resistive bridge. The amplified error signal provides the driving current to the Peltier element. Absolute temperature is measured with an accuracy of 10 mK° by means of a quarz thermometer. Temperature variations are evaluated with much better precision, that is 0.0002°K.

The scattered light is collected at fixed angles with four detectors, two at 90° and the others at 20° and 150°. Mirrors M_1 and M_2 are used to direct light through a single hole in the outer shield T for the 20° and 150° photomultipliers. This design allows good thermal and mechanical stability, still with reasonable flexibility for asymmetry measurements.

Photomultiplier FT2, at 90°, is used for intensity measurements only. It is therefore operated with many coherence areas in the collecting optics, with stable high voltage kept on day and night and with direct analogue output from anode without amplification. This photomultiplier allows measurements of the light intensity scattered by pure water with quite good reproducibility from day to day. This feature can be profitably used in monitoring dust decontamination of the scattering cell during filtration.
Photomultiplier FT4 is used only for the dynamic measurements at 90°. The collecting optic apertures can be optimized during the experiment for best performance in dynamic measurements without worrying about intensity calibrations. The 90° detectors are ITT FW130, while the other two are EMI 9863 KB. All of them have an effective photocathode area of about 2.5 mm. The overall long term stability of the photomultiplier gain is about 1%.

The collecting optics is the same for the four angles. A lens images the scattering volume on the slit plane IP at the entrance of the photomultiplier housing, see Fig.2. A prism P can be inserted to view the scattering volume through the slit set, in the plane IP, and the collecting optics. The viewing system is focused on the plane IP. Therefore, the operator can align the collecting optics easily and then choose the scattering volume size by acting on the micrometricallycontrolled slit set, see Fig.3. Once the prism P is removed the scattered light is collected by the photomultiplier. The light spot is centered on the small photocathode area by means of the lens L.

The scattering cell is filled through a microporous filter with 0.2 μm pore size. We found that an accurate cleaning of the cell plus the microporous filtering could eliminate almost completely the dust. To avoid sample contamination, the connecting tubes and the filter holder are made with Teflon type material. The minimum amount of sample which is needed to fill cell and dead volumes is 1.3 cm^3. For a very accurate sample filtering a large amount is required, typically 5-8 cm^3.

The average scattered intensity is obtained from a measurement of the analogue average photocurrent performed with a stable and sensitive digital voltmeter.
The intensity correlation function is measured by a digital 108-channel real-time correlator which operates on standardised photoelectron pulses. The Koppel fit to the autocorrelation function is performed by an HP 9825 desk computer which is directly connected to the correlator.

The results reported in this paper have been obtained with the polyoxyethylene nonionic surfactant $C_{12}E_8$ which shows a 12 carbon alkylic chain and an hydrophilic part made of 8 ethylene oxide groups. Other measurements not discussed here have been performed with $C_{12}E_6$ and C_8E_4. High-purity monodisperse $C_{12}E_8$ was obtained from Nikko Chemicals, Tokyo. The nonionic surfactant in crystalline form was dissolved without further purification in doubly distilled and degassed water at a temperature of about 40°C. It is absolutely important to avoid contact with oxygen because the polyoxyethylene may undergo an oxidation process which changes its physicochemical properties. We have also found that prolonged contact of the nonionic surfactant solution with metals (Al,Fe) should be avoided because the critical temperature may be changed by some degrees by metal ions. A typical phase diagram is shown in Fig.4. The consolution curve separates the single phase micellar region (below) from the region (above) in which two isotropic solutions coexist. In the case of $C_{12}E_8$ the critical temperature, as determined by visual observation of the meniscus, is 76.1 °C and the critical concentration c_c is about 15 mg/cm^3.

EXPERIMENTAL RESULTS

We report in Fig.5 the average intensity of scattered light for a $C_{12}E_8$ aqueous solution at the critical concentration c=15 mg/cm^3 in the temperature range 15-75 °C. The scattered intensity depends

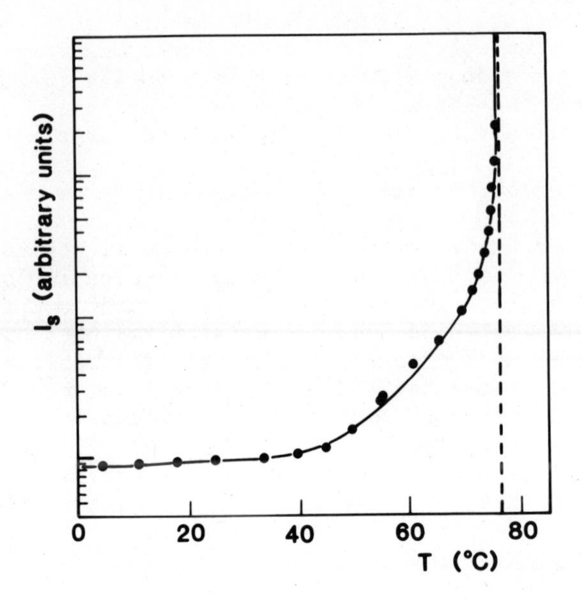

Figure 5. Scattered light intensity measured as function of the temperature for a $C_{12}E_8$-H_2O solution at the amphiphile concentration 15 mg/cm^3.

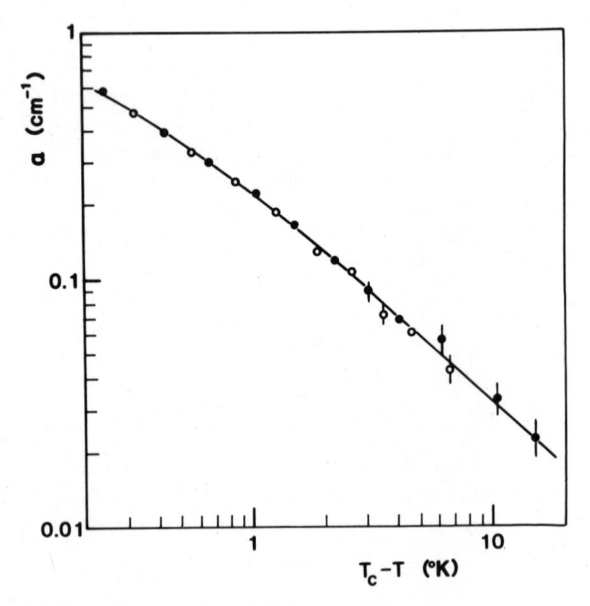

Figure 6. The turbidity α measured as function of the temperature for the same solution as in Fig.5.

weakly on the temperature in the region 15-40 °C, whereas a strong enhancement is found above 40 ° C, and I_S seems to diverge as T approaches the critical temperature T_c=75 °C. The divergence follows a simple power law behaviour as a function of the reduced temperature T_c-T, as shown in Ref.6.

The scattered intensity does not show any dependence on the scattering angle up to temperatures around 70 °C. Above this temperature, the intensity scattered at θ=20° is larger than that scattered at 90°. For the data above 70 °C, we have reported in Fig.5 the extrapolation at $K \to 0$ of the scattered intensity obtained by postulating a Ornstein-Zernike dependence of I_S on K (see Eq.6), namely $I_S = I_{S0}(1+K^2\xi^2)^{-1}$, where I_{S0} is the extrapolated intensity and ξ is the correlation length of the concentration fluctuations.

We have measured the turbidity of the $C_{12}E_8$ solution at the critical concentration at various temperatures. The transmission of the scattering cell becomes appreciably smaller than 1 only at temperatures above 65 °C. The results are reported in the logarithmic plot of Fig.6 as a function of the temperature distance from the critical point.

In order to perform the absolute calibration of the apparatus it is necessary to know the temperature dependence of the refractive index increment dn/dc. We have measured this dependence for the $C_{12}E_8$ solution and found that it is well described by the linear law

$$dn/dc = 0.135 - 0.00024(T-293°)$$

where T is the absolute temperature and dn/dc is expressed in cm^3/g.

We have measured the concentration dependence of the scattered intensity at several temperatures in the range 20-50 °C. We report, as an example, in Fig.7 the results obtained at 20 °C.

Fig.7 contains also the diffusion-coefficient data obtained from the intensity-correlation measurements performed along the isotherm at 20 °C.

ABSOLUTE SCATTERED INTENSITY CALIBRATION

A full interpretation of the light-scattering data is presented elsewhere[5,6]. We limit here the discussion to a specific problem, the absolute calibration of the scattered intensity, which is of general interest for micellar and macromolecular solutions.

The simplest way to calibrate the apparatus is by comparison with a scatterer of known Rayleigh ratio. In our first experiments on micellar solutions[8] we have used benzene as the known scatterer.

Later on we have simply used pure water. Although water is a rather weak scatterer, there are some advantages in using water as a standard scatterer when working with dilute aqueous solutions. In fact, because of the negligibly small change of the index of refraction all refraction effects in the collected scattered light remain substantially constant. Furthermore, with our fixed cell system, it is convenient in any case to fill the cell with pure water before and after the measurement with the micellar solution in order to flush out the previous solution and to check whether any dust is present in the cell. We have computed the Rayleigh ratio R_w of pure water at λ=514.5 nm for linearly polarized light illumination by interpolation of the data of Kratohvil et al.[9]. We find R_w=1.42·10^{-6} cm^{-1}. The uncertainty of this value is at least 10%. Our data show, for instance that a 10 mg/cm^3 solution of $C_{12}E_8$ scatters 65 times the intensity of pure water. This allows to compute immediately the Rayleigh ratio of the $C_{12}E_8$ solution as R=65×1.42·10^{-6} cm^{-1}=0.92·10^{-4} cm^{-1}.

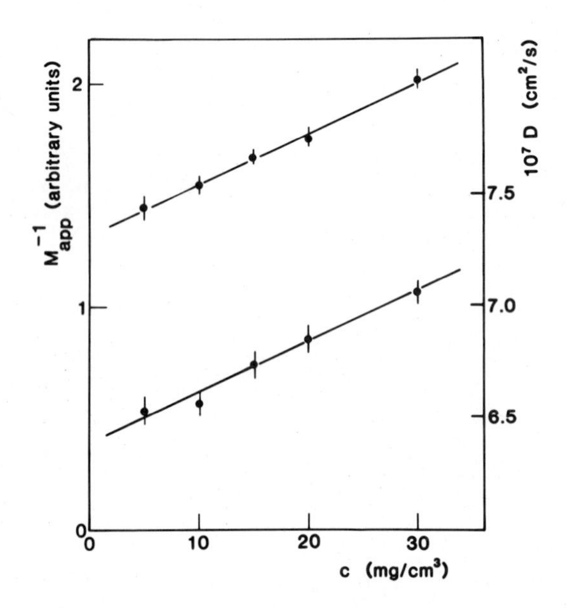

Figure 7. The apparent micellar weight and the mass diffusion coefficient of $C_{12}E_8$-H_2O solutions at 20 °C as a function of the amphiphile concentration.

What is usually more relevant is the knowledge of the molecular weight of the micelle M. By using the expressions reported above, it is easy to show that the apparent molecular weight M_{app} is given by

$$M_{app} = A\left(\frac{dn}{dc}\right)^{-2} \frac{I_r - 1}{c - c_o} \qquad (12)$$

where I_r is relative scattered intensity as compared to pure water, and A is a calibration constant defined as $A=R_w/K$, with $K=2\pi^2n^2/(\lambda_o^4 N_A)$, λ_o being the wavelength of incident light and N_A Avogadro number. By taking n=1.335, we obtain, at λ=514.5 nm, A=0.1705 cm^3/g. By using this value, we find that the value of M_{app} extrapolated for $c \to c_o$ is M=65,000. The extrapolated value of the diffusion coefficient is $D_o=6.4 \cdot 10^{-7}$ cm^2/s, which corresponds to a hydrodynamic radius R_H=3.4 nm.

The concentration dependence of M_{app} (and of D) shown in Fig.7 cannot be attributed to an actual change in the aggregation number because it is unreasonable to find a decrease of M with concentration. We can therefore assume that M does not depend on c and that the c-dependence of M_{app} (and of D) is due to interactions. The positive values of the slopes shown in Fig.7 is indicative of repulsive interactions[4]. Since the micelles are not charged, the observed slopes should only be due to excluded volume effects.

The turbidity data of Fig.5 can be used to obtain the Rayleigh ratio of the $C_{12}E_8$-H_2O solution at high temperatures. Since I_S is K-dependent, R is also K-dependent, and the turbidity α represents an integral over the Ornstein-Zernike intensity distribution. The relation between the Rayleigh ratio at $\theta \to 0°$, R(0), and α is

$$R(0) = \frac{\alpha}{\pi f(\beta)} \tag{13}$$

where $\beta=2K_o$ and $f(\beta)$ is a known function[10]. For our calibration purposes, it is sufficient to consider the data at a single temperature. Take, for instance, the point at T=76.175 °C, 0.325 °C away from T_c. The turbidity α is (0.19±0.01) cm^{-1}. The correlation range ξ, as derived by the intensity asymmetry and by dynamic measurements, is 35 nm. The function f is 1.86, and therefore $R(0)=3.1 \cdot 10^{-2}$ cm^{-1}. In principle, it is possible to use the $C_{12}E_8$-H_2O solution at the concentration 15 mg/cm3 and temperature T=76.175 °C as an absolute light scattering standard. From a practical point of view, however, since light scattering experiments on micellar and macromolecular solutions are usually performed at room temperature, it is not convenient to use a high-temperature standard. There are two other reasons why such a standard is not appropriate. First, the turbidity, and hence the Rayleigh ratio, is strongly temperature-dependent in the critical region, so that a very high temperature-stability of the scattering cell would be required. Secondly, the properties of the micellar solution in the critical region are very sensitive to the presence of impurities, mainly because impurities shift the critical temperature. We can easily derive the Rayleigh ratio at any temperature, however, by using the intensity data of Fig.5. We find in this way $R=(1.01 \pm 0.05) \cdot 10^{-4}$ cm^{-1} at 20 °C. This value is in very good agreement with the value derived above starting from a known value of R_w.

ACKNOWLEDGMENTS

This work was supported by the Progetto Finalizzato Chimica Fine e Secondaria of the Italian National Council of Research, Rome, Contracts Nos. 81.001735.95 and 81.01910.95.

REFERENCES

1. See Refs 1-7 in V. Degiorgio, this volume
2. C. Tanford "The Hydrophobic Effect. Formation of Micelles and Biological Membranes", Wiley, New York (1980)
3. M. Corti and V. Degiorgio, Laser-Light-Scattering Investigation on the Size, Shape and Polydispersity of Ionic Micelles, Ann. Physique (Paris) 3:303 (1978)
4. M. Corti and V. Degiorgio, Quasielastic Light Scattering Study of Intermicellar Interactions in Aqueous Sodium Dodecyl Sulphate Solutions, J. Phys. Chem., 85:711 (1981)
5. M. Corti and V. Degiorgio, Micellar Properties and Critical Fluctuations in Aqueous Solutions of Nonionic Amphiphiles, J. Phys. Chem., 85:1442 (1981)
6. M. Corti, V. Degiorgio and M. Zulauf, Nonuniversal Critical Behavior of Micellar Solutions, Phys. Rev. Letters, 48:1617 (1982)
7. M. Zulauf and J. P. Rosenbusch, J. Phys. Chem. (in press); R. Triolo, L. J. Magid, J. S. Johnson and H. R. Child, J. Phys. Chem. (in press)
8. M. Corti and V. Degiorgio, Light Scattering Study on the Micellar Properties of a Nonionic Surfactant, Opt. Commun., 14:358 (1975)
9. J. P. Kratohvil, M. Kerker and L. E. Oppenheimer, Light Scattering by Pure Water, J. Chem. Phys., 43:914 (1965)
10. V. G. Puglielli and N. C. Ford, Turbidity Measurements in SF_6 Near Its Critical Point, Phys. Rev. Letters, 25:143 (1970)

LASER LIGHT SCATTERING STUDY OF THE FRACTIONATION OF CASEIN MICELLES IN SKIM MILK BY CONTROLLED PORE GLASS CHROMATOGRAPHY

M.C.A. Griffin and M. Anderson

Physical Sciences Department
National Institute for Research in Dairying
Shinfield, Reading RG2 9AT, Berkshire, England

Casein is a major protein component of milk. It consists of several different polypeptides: α_{s1}, α_{s2}, β and κ casein, as well as varying amounts of proteolytic fragments of these polypeptides. The polypeptides are bound together by calcium phosphate and citrate to form aggregates of a wide range of sizes (Holt et al, 1978) up to 680 nm in diameter (McGann et al, 1980). The size distribution of these casein aggregates, known as 'micelles', in milk may change with different conditions such as variation in Ca^{2+} concentration, pH and history of heat treatment. The nature of the size distribution may affect functional properties of the milk such as enzymic coagulation on addition of rennet (Ekstrand et al, 1980; Dalgleish et al, 1981). It has proved desirable to characterize the size distribution of casein micelles in milk so as to investigate related properties further.

In the work described here, controlled pore glass (CPG), with a nominal pore size of 300 nm, was used as a medium for exclusion chromatography. The buffer with which the column was equilibrated, and which was also used for elution, was a simulated milk ultrafiltrate (SMUF), pH 6.6, described by Jenness and Koops (1962). Friesian milk, taken from a bulk storage tank, was centrifuged for 20 min at 1000g and the cream removed. To prevent dissociation of the micelles on dilution the skim milk was fixed by treatment with glutaraldehyde (.5%) and 1 ml samples were loaded onto the column at 4°C. 1 ml (approx.) fractions were collected. UV absorbance measurements at ca 340 nm and at 280 nm were made on each fraction, the former being a crude measure of elastically scattered light, while the latter resulted in part from the absorption at 280 nm by the peptide backbone. In Figure 1 the numbering starts at the first fraction to have any appreciable absorbance. The large

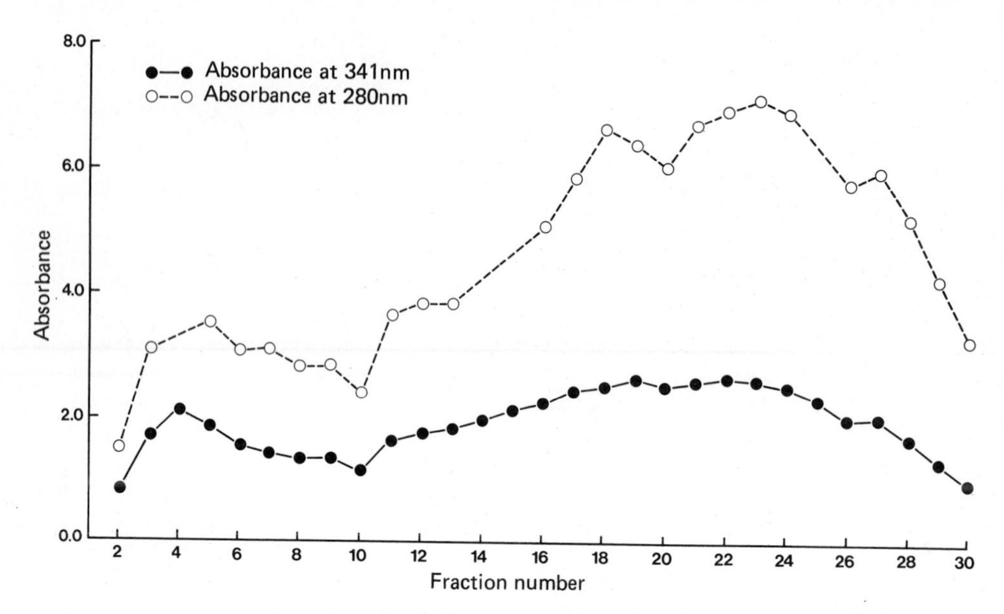

Figure 1. Absorbance profiles of fractions eluted from the CPG column after application of 1 ml skim milk.

casein micelles were eluted first, followed by soluble casein and whey proteins.

The fractions were then examined by photon correlation spectroscopy using a Malvern multibit correlator type K7025 and spectrometer with a Spectra-Physics HeNe 15mW laser. Samples from the fractions were diluted into SMUF, to a concentration at which there were no observable effects of multiple scattering, and the temperature of the cuvettes was kept at 20°C. Auto-correlation functions were obtained at 90° scattering angle, and the normalised correlation data, $g^{(2)}(\tau)$, were analysed by a least squares fit of $\ln(g^{(2)}(\tau)-1)$ to a third order polynomial:

$$\ln \left(g^{(2)}(\tau)-1\right)^{\frac{1}{2}} = A_o - \bar{\Gamma}\tau + \frac{\mu_2\tau^2}{2!} - \frac{\mu_3\tau^3}{3!}$$

omitting higher order terms, and weighting the data as $(g^{(2)}(\tau)-1)^2$. From the value of $\bar{\Gamma}$ the diffusion coefficient was calculated from $D = \bar{\Gamma}/K^2$ where $K = \frac{4\pi n}{\lambda} \cdot \sin \theta/2$, the length of the scattering vector. An 'average' diffusion coefficient, $\bar{D}$, of particles in each fraction was obtained using a version of the method described by Brown and Pusey (1974) in which the values of D were linearly extrapolated to $\bar{\Gamma}\tau_{max} = 0$.

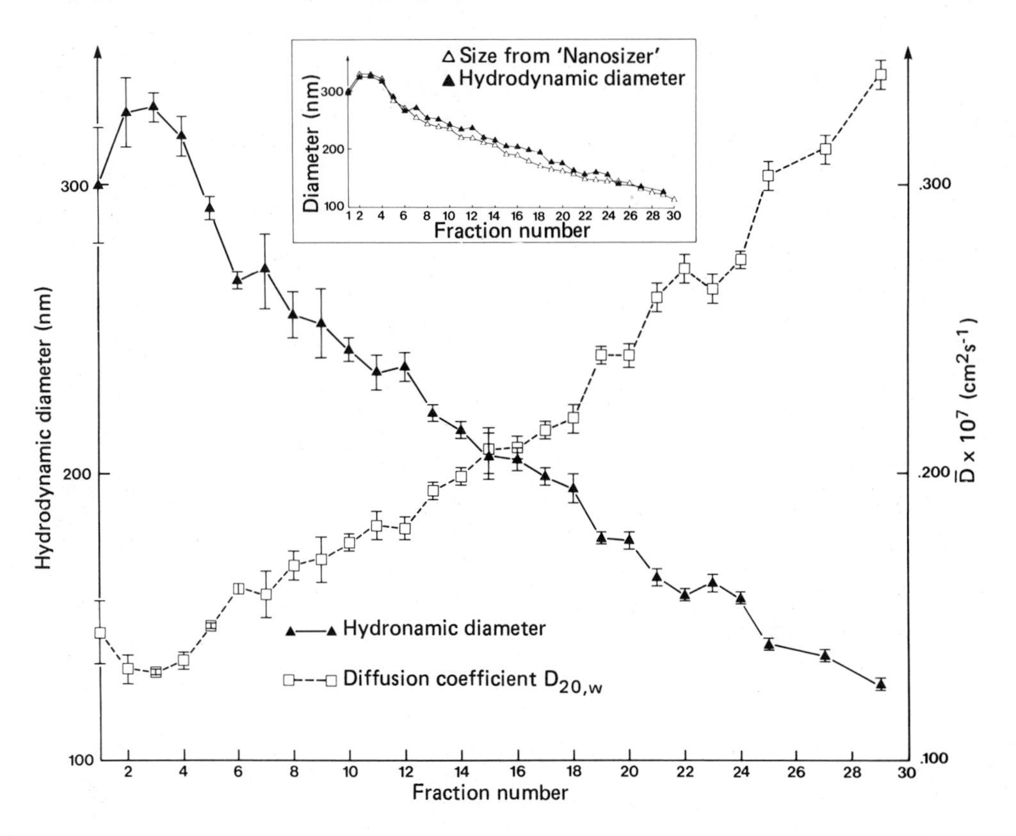

Figure 2. Profile of diffusion coefficients and hydrodynamic diameters obtained for each fraction from the same column-run described in Figure 1.

These average values of the diffusion coefficients of the casein particles in each fraction are shown in Figure 2 which also shows the hydrodynamic diameters (obtained assuming that the Stoke's relationship holds: $d = \frac{kT}{3\pi\eta\bar{\bar{D}}}$).

For the purpose of comparison, Figure 2 shows results for the hydrodynamic diameters obtained using an automatic photon correlation spectrometer (a Coulter 'Nano-Sizer', kindly made available to us by Express Dairy Foods Ltd) which operates on the same principles as the Malvern spectrometer used in the present study, but which produces size and polydispersity data without the need for software, according to unrevealed cabalistic formulae! (The Coulter Nano-Sizer, 1980).

A study was made of the angular variation of the diffusion coefficient (calculated from data obtained using the Malvern correlator, as described above) of a few individual fractions from a different column-run to that shown in Figures 1 and 2. Figure 3 shows some of these measurements, together with those obtained from diluted glutaraldehyde-fixed skim milk. Figure 4 shows electron micrographs of thin sections, prepared by the method described by Holt et al, of skim milk and of two casein fractions from a CPG column-run.

The size range of casein micelles presents a difficulty in interpretation of the average diffusion coefficient obtained, as the form factors at large angles are not unity, and vary with individual particle size and shape. In the skim milk the wide range of particle sizes is reflected in an average diffusion coefficient that varies significantly with K^2 (Figure 3). Moreover, while it should be possible, in theory, to construct a 'Zimm-plot' to yield the z-average diffusion coefficient as the intercept at $\theta = 0$ (Brown et al, 1975; Brehm and Bloomfield, 1975), in practice, with skim milk, the standard deviations of the measured decay constants increase markedly with decreasing scattering angle. The much smaller variation in the average diffusion coefficient with scattering angle, and the lower percentage standard deviations in the average diffusion coefficients, of samples prepared from single casein fractions from the CPG column give us confidence in the use of the value measured at 90° scattering angle to characterize the size of particles in the sample.

In Figure 4 the distribution of sizes in the individual fractions is clearly much narrower than that in skim milk (sectioning of a monodisperse solution of spheres in any case produces a distribution of sizes because the plane of sectioning does not generally pass through the centre of the spheres), which is consistent with the light scattering results shown in Figure 3.

Although we have taken care to collect correlation data for a sufficient time to have 10^6-10^7 counts per channel we have found with this system that $\frac{\mu_2}{\overline{\Gamma}^2}$ was not a good measure of polydispersity, as, even though values were in general lower for casein fractions than for skim milk, the size of the error was very large, the standard deviations often being as great as the value of $\frac{\mu_2}{\overline{\Gamma}^2}$.

Hitherto, characterization of the size of casein particles in the fractions from CPG chromatography of skim milk has been carried out by electron microscopy of several pools of bulked fractions and absorbance measurements (McGann et al, 1980; Larsson-Raznikiewicz et al, 1979). Photon correlation spectroscopy provides a rapid method for obtaining information on the size of the particles in each fraction. The elution of micelles smaller than the 300 nm

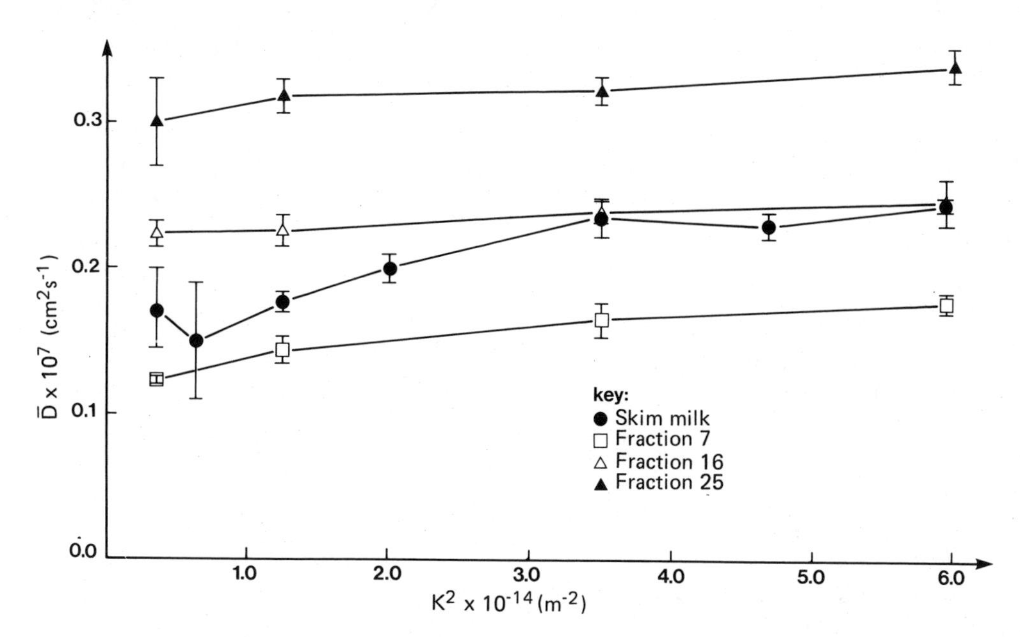

Figure 3. The variation of measured diffusion coefficient with K^2 of skim milk and of three fractions from a column-run.

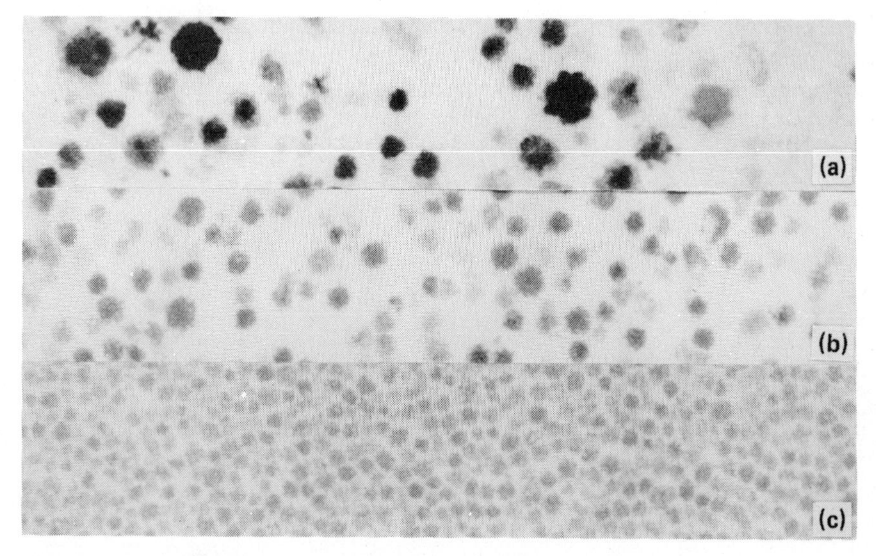

Figure 4. Electron micrographs of thin sections of skim milk and of two casein fractions. Magnification is x 24,300.

(a) Skim milk, (b) Fraction 10, (c) Fraction 15.

pore size of the CPG has been shown to be according to size, and a good separation is effected.

This separation of CPG chromatography will be useful in providing a description of the milks to be examined, and in addition the availability of casein micelles of a narrow size range will enable more precise light scattering studies of casein properties to be carried out.

Acknowledgements

We thank Mr B.E. Brooker for the electron micrographs, Miss C. Moore for technical assistance and Dr W.G. Griffin for advice on the analysis of data.

References

Brehm, G.A. & Bloomfield, V.A. (1975) Macromolecules 8: 663-5.

Brown, J.C. & Pusey, P.N. (1974) J. Phys. D 7: L31-L35.

Brown, J.C., Pusey, P.N. & Dietz, R. (1975) J. Chem. Phys. 62: 1136-1144.

Dalgleish, D.G., Brinkhuis, J. & Payens, T.A.J. (1981) Eur. J. Biochem. 119: 257-261.

Ekstrand, B., Larsson-Raznikiewicz, M. & Perlmann, C. (1980) Biochim. Biophys. Acta 630: 361-366.

Holt, C., Kimber, A.M., Brooker, B. & Prentice, J.H. (1978) J. Coll. Int. Sci. 65: 555-565.

Jenness, R. & Koops, J. (1962) Neth. Milk & Dairy J. 16: 153-164.

Larsson-Raznikiewicz, M., Almlöf, E. & Ekstrand, B. (1979) J. Dairy Res. 46: 313-316.

McGann, T.C.A., Donnelly, W.J., Kearney, R.D. & Buchheim, W. (1980) Biochim. Biophys. Acta 630: 261-270.

'The Coulter Nano-Sizer' by the Fine Particle Group, Coulter Electronics Ltd, Luton, UK. January 1980 pp. 10.

STRUCTURAL STUDIES ON BOVINE CASEIN MICELLES BY LASER LIGHT SCATTERING

C. Holt

Hannah Research Institute
Ayr
Scotland. KA6 5HL

Bovine casein micelles are more-or-less spherical particles of colloidal dimensions, composed of 4 phosphoprotein types. The weight average molecular weight is about 5×10^8 and average radii by light scattering methods are generally 80-150nm, although there is a broad range of sizes[1]. Besides the phosphoproteins, casein micelles contain Ca, Mg, P_i and citrate, but the principal salt constituents are Ca and P_i in the form of a calcium phosphate salt and calcium bound directly to sites on the casein. It appears that the phosphate groups of the phosphoproteins are largely incorporated in the lattice of the calcium phosphate[2] and hence that the calcium phosphate could be responsible, in part, for holding together casein micelles. On the other hand, caseins can associate in the presence of mM concentrations of Ca^{2+} to give aggregates of the same size as natural casein micelles. In many physico-chemical studies this has been assumed to be the mechanism of formation of micelles and the possible role of calcium phosphate has been neglected.

Recently it has been shown[3] that the concentrations of free ions such as Ca^{2+}, Mg^{2+} and HPO_4^{2-} and complexes such as $CaCit^-$ and $MgH_2PO_4^+$ can be calculated accurately. Also, the solubility product governing the phase separation of milk calcium phosphate has been defined so that various synthetic buffers, saturated with respect to calcium phosphate but with different free ion concentrations, can be prepared. In this work, the effect of varying ion concentrations on the stability of bovine casein micelles is described. Two types of experiments were performed. In dilution experiments, skim-milk (0.5ml) was diluted by the buffer of interest (40ml) and turbidity measured as a function of time. In dialysis experiments, skim-milk was dialysed against a large volume of the buffer for several days and aliquots taken at various times for physical and chemical measurements. In a given buffer the rate of dissociation in dilution experiments was found to be an order of magnitude faster than in dialysis experiments.

The dissociation of micelles following dilution by simple $CaCl_2$ buffers containing 20mM imidazole and 60mM NaCl, pH 6.7, occurs rapidly at [Ca^{2+}] close to or below the 2-3mM which is characteristic of the natural ionic environment of micelles. Some factor other than [Ca^{2+}] is required to maintain micelle stability. Micelles are completely unstable if Mg^{2+} is used in place of Ca^{2+} and the addition of Mg^{2+} to a buffer containing Ca^{2+} slows down the rate of dissociation only slightly, so Mg^{2+} ions are not the stabilising factor.

When phosphate is added to the dilution buffer, rates of dissociation decrease, at a given free [Ca^{2+}], as the buffer approaches saturation. However, if free [Ca^{2+}] is about 0.5mM or less, micelles dissociate even if the buffer is supersaturated in calcium phosphate, whereas at a higher [Ca^{2+}] of 1.6mM, dissociation is much reduced and casein micelles are virtually indefinitely stable at saturation. By including Mg^{2+} and citrate, buffers can be devised which resemble closely the natural ionic environment of casein micelles. As was found with simple $CaCl_2$ buffers, inclusion of Mg^{2+} appears to slow down only slightly the rate of dissociation of micelles at a given [Ca^{2+}] and degree of saturation, and other ions (Na^+, K^+, Cl^-, Citrate) are of lesser importance. These experiments suggest that necessary conditions for native micelle stability at pH 6.7 are a [Ca^{2+}] above about 1.4mM and a solution saturated with respect to calcium phosphate.

To investigate further the importance of calcium phosphate for micelle integrity, a dialysis experiment was carried out using a dialysis buffer containing no phosphate but having [Ca^{2+}] = 3mM; a concentration slightly higher than that in the skim-milk. During dialysis, calcium phosphate passed into solution and turbidity decreased as the micelles dissociated although calcium ion binding, as distinct from calcium phosphate binding, to casein is hardly changed. However, diffusion coefficient measurements showed that the hydrodynamic radius was not significantly reduced during the first 40h of measurement, during which the molecular weight of the micelles decreased by over 30% (Fig. 1). Thus it appears that in this initial phase there was a loss of mass from the micelle while the framework of the particle remained intact and it was only later, when calcium phosphate content had been reduced further, that the framework was disrupted.

This result is essentially the same as that observed by Lin et al.[4], on addition of EDTA to casein micelle fractions, though in their work it is not possible to separate the effect of reducing [Ca^2] from that of dissolving the colloidal calcium phosphate.

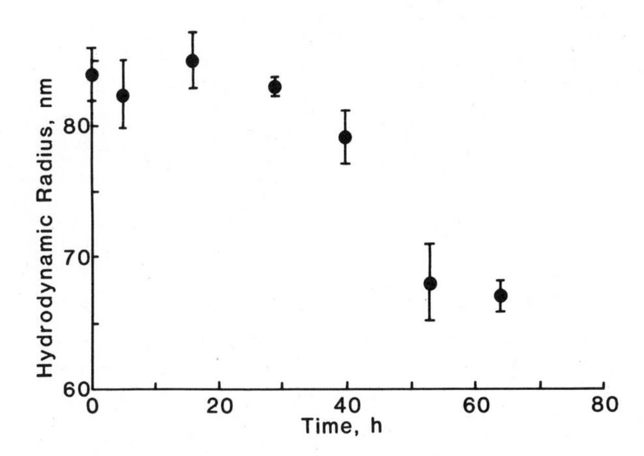

Fig. 1. Dissociation of bovine casein micelles by dialysis against 3 mM calcium chloride buffer.

Other dissociation experiments have been carried out in which the calcium phosphate content of micelles is held constant while $[Ca^{2+}]$ is varied. Calcium ion binding to the casein is dependent on $[Ca^{2+}]$ in the dialysing buffer following a binding isotherm, and micellar dissociation occurs at low $[Ca^{2+}]$. Dynamic light scattering measurements have not yet been made under these circumstances but the intention is to compare the patterns of dissociation observed when $[Ca^{2+}]$ and calcium phosphate are varied independently and obtain information on how these two factors separately affect micelle structure and stability.

REFERENCES

1. C. Holt, A. M. Kimber, B. E. Brooker, and J. H. Prentice, Measurement of the size of bovine casein micelles by means of electron microscopy and light scattering, J. Coll. Interf.

Sci. 65:555 (1978).
2. C. Holt, S. S. Hasnain and D. W. L. Hukins, Structure of bovine milk calcium phosphate determined by x-ray absorption spectroscopy, Biochim. Biophys. Acta (in the press).
3. C. Holt, D. G. Dalgleish and R. Jenness, Calculation of the ion equilibria in milk diffusate and comparison with experiment, Anal. Biochem. 113:154 (1981).
4. S. H. C. Lin, S. L. Leong, R. K. Dewan, V. A. Bloomfield, and C. V. Morr, Effect of calcium ions on the structure of native bovine casein micelles, Biochemistry 11:1818 (1972).

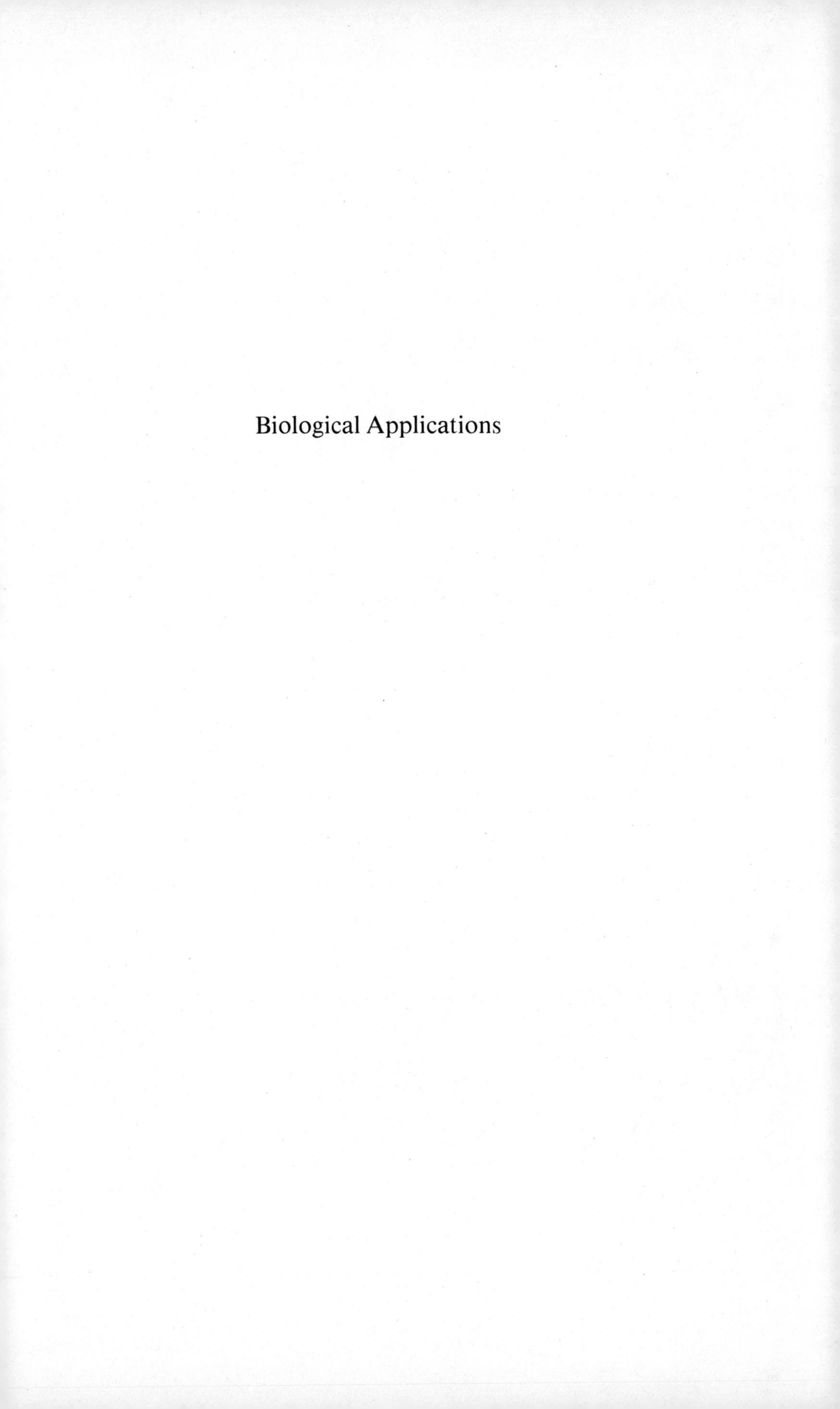

Biological Applications

VESICLES

Martin W. Steer

Botany Department
The Queen's University of Belfast
Northern Ireland BT7 1NN

CONTENTS

1. INTRODUCTION

Vesicles are important cellular components that are too small (50-400 nm diameter) to be resolved by light microscopy. They are formed inside cells by the Golgi apparatus, or in plant cells by equivalent structures, the dictyosomes[1]. Evidence from labelling experiments and from static electron micrographs is consistent with the view that vesicles move from their sites of formation to the plasma membrane[2]. The vesicle membrane then fuses with the plasma membrane and releases the contents of the vesicle to the extra-cellular environment. The released product is called a secretion, and contains proteins (frequently enzymes), carbohydrates, lipids or lower molecular weight materials. After release these are suspended or dissolved in an aqueous medium that is partly derived from the movement of water out of the secretory cell. Secretions may, as in the case of digestive enzymes for example[3], flow along ducts away from the cell surface, or may, as in the case of collagen[4] and plant cell walls[5], contribute to an extracellular matrix that remains in contact with the cell that secreted it.

It is not possible to follow the movements and behaviour of vesicles in living cells by light microscopy, while electron microscopy only yields static images. It is possible to detect vesicle migration _in vitro_ by laser light scattering[6,7] and, since vesicles constitute a fairly homogeneous group of particles in the cytoplasm, it should be possible to record their movements in living cells. This seminar will provide the background information on the formation, transport and fusion of vesicles necessary for such studies.

2. SECRETORY TISSUES AND CELLS

Although most, if not all, animal and plant cells show some secretory activity, only a very few have been used for detailed studies; mostly because of problems of access to individual cells in the surrounding tissues and of collection and quantification of the secreted product.

In animals by far the most popular tissue for study is the pancreas[3] (usually from mouse, rat or rabbit), which secretes a variety of digestive enzymes into the gut (exocrine, Fig.1) or a number of regulatory hormones into the bloodstream (endocrine). Studies have been performed on the whole tissue and on living, functional cells released into a suitable medium[8]. For light scattering work it is essential that functional cells can be isolated in this way; although this requirement severely limits the number of cellular systems that are accessible to the technique. Other secretory cells that can be isolated include hepatocytes[9] (liver cells), parotid glands[10], lacrimal glands[10], thyroid[11] and pituitary[12] cells. Isolates of free secretory cells, such as mast cells, have also been examined[13]. An important secretory system that cannot be separated as isolated functional cells is the production and release of acetylcholine from synaptic vesicles in nerve cells[14].

Some unicellular animals, such as _Tetrahymena_[15] and _Paramecium_[16] have been used for the study of the secretory system responsible for the placement of specialised surface structures in these organisms.

In plants vesicle production and release is an essential step in the formation of the plasma membrane and of the carbohydrate wall that surrounds each cell. As this activity is dispersed throughout the thin layer of cytoplasm that lies between the plasma membrane and the tonoplast, it is not easily studied. Various plant and fungal cells, however, exhibit tip growth during which a tube-like extension is formed. In these cells the secretory processes, that is vesicle formation, transport and fusion, are concentrated at the growing tips of the tubes[17]. Examples include pollen tubes[18], root hairs[19], algal rhizoids[19] and fungal hyphae[20]; all are accessible for laser light scattering experiments. As in animals, plants

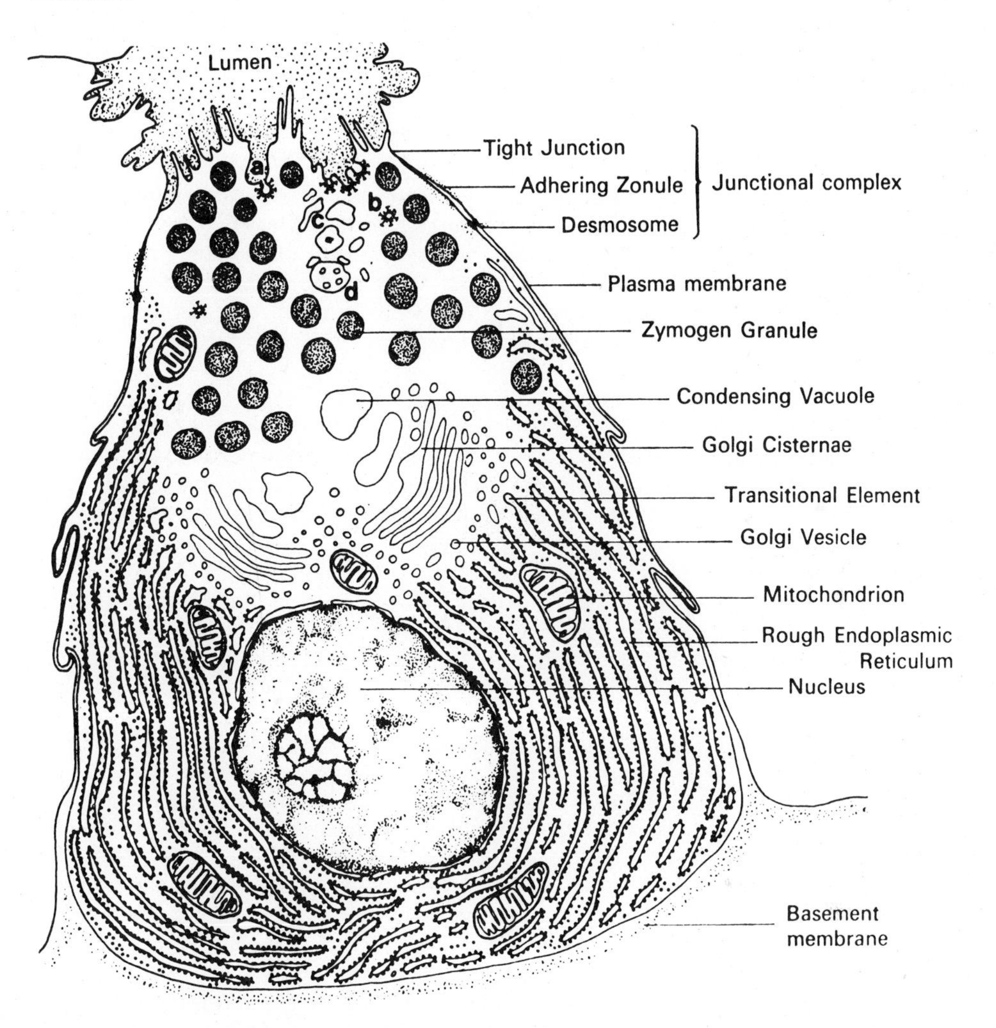

Fig.1 Pancreatic acinar cell. The diagram shows the main cell components with the secretory vesicles (zymogen granules) concentrated at the apical end of the cell. A possible membrane retrieval sequence is also illustrated with (a) coated vesicles forming from the plasma membrane surface; (b) moving through the cytoplasm; (c) coalescing to form endocytotic vesicles and (d) multivesicular bodies. Reproduced from ref.3.

produce various secretions including nectar, salt (sodium chloride), digestive enzymes and mucilage[21]. Most of these involve cells that are inaccessible and difficult to isolate in a functional state from the surrounding tissues and which lack a polarised cytoplasmic structure that concentrates secretion in one particular part of the cell.

3. VESICLE FORMATION

Vesicles arise from the edges and faces of the Golgi apparatus in animals, or the dictyosomes in plants. The bounding membrane of the cisterna forms a bleb that swells and is pinched off, forming a vesicle. Virtually nothing is known of the structural events or forces involved in this process. The vesicle membrane is clearly derived from that of the cisternae lumen.

These cisternal components originate from various sites within the cell. The proteins of the membranes and secretory proteins (enzymes) are synthesised on the rough endoplasmic reticulum and transferred to the Golgi via small transitional vesicles (Fig.1). Some proteins may be subsequently added to the membranes of the cisternae and the vesicles from the cytoplasmic pool. Many of the secretory proteins are modified in the cisternae by the addition of single, di- or poly-saccharide units. Lipid membrane components are synthesised initially on the endoplasmic reticulum and then transferred as bounding membranes of the transitional vesicles. This membrane is modified in the Golgi by the selective removal and addition of individual lipid molecules[22]. Polysaccharides are synthesised inside the cisternal lumen, and their synthesis may continue within the vesicles after their release from the Golgi apparatus[23]. It is probable that cisternal components of plant dictyosomes also originate from the same sites in the cell, although there is some confusion over the importance of the role of the endoplasmic reticulum in the supply of lipids and proteins[24].

4. VESICLE TRANSPORT

Released vesicles move away from the Golgi towards the cell surface. They may accumulate in the cytoplasm of these apical regions until the cell receives a signal which initiates their mass release, or they may be released continuously. In either case control is frequently seen to be exerted over the exact site of vesicle delivery and release. This suggests that transport to the surface is not a random diffusion process. Evidence for the involvement of microfilaments in this transport process comes from the inhibition of vesicle movement by cytochalasins[25]. This involvement presumably requires a method of attaching vesicles to the microfilaments which guide the vesicles to the correct discharge point on the cell surface.

Little is known about the way in which a force could be imparted from the microfilaments to the vesicles. Two types of system have been proposed. One, based on muscle actin/myosin systems, would involve a myosin-type molecule, embedded in the vesicle membrane, sequentially binding to and releasing from the microfilament fibres[26]. The other envisages links between the vesicles and specific subunits (G-actin) on the microfilament fibres. Movement of these links, and hence the attached vesicles, would be accomplished by removal of subunits from the fibre near the cell surface and addition of subunits to the microfilament below the vesicle. This process, termed treadmilling, has been demonstrated in both microtubules and microfilaments, and could result in the movement of a link along the chain[27]. Membrane proteins that bind to microfilaments have been characterised, so there is some support for the view that direct connections are established between vesicles and microfilaments[28].

5. VESICLE FUSION

On arrival at the cell surface a complex series of events ensures that the lipid bilayer of the vesicle membrane is successfully integrated into the plasma membrane. Specific 'docking' sites seem to exist on the plasma membrane[29]. These are characterised by the general absence of membrane proteins from the lipid bilayer[14]. Changes may also occur in the lipids themselves, of both the plasma membrane and the approaching vesicle. Large head groups of some lipids seem to present structural barriers to fusion that are removed either by displacement of the lipid out of the fusion zone or by enzymatic attack on the head groups to remove them[30]. Calcium ions play a crucial role in the fusion of the apposed bilayers, either by creating relatively hydrophobic conditions, or by neutralising surface charges[31]. This promotion of membrane fusion can be demonstrated in both _in vitro_[32] and _in vivo_[33] systems. The culmination of these events is the flowing together of the two membranes to form a single continuous bilayer[34].

6. VESICLE DYNAMICS

Clearly the synthesis of vesicles must be geared to the secretory requirement of the cell. Little is known about the regulation of vesicle synthesis, although it is clear that controls do exist. For example, in cells that only release vesicles for a short period on stimulation, vesicle formation is limited to the replacement of the discharged vesicles until the next round of stimulation[35]. In some cells stimulation is required to initiate vesicle synthesis before secretion can commence[36] and in others production appears to be unrelated to the requirement for the secreted product[37]. It is clear therefore that information is required on the rate of vesicle production, and on changes in this rate that occur under different conditions.

Two different approaches to this problem have been developed to date. The first relies on equating some parameter of the secretion with the size of vesicles in the cell. Thus both the rate of volume increase of secretion[38] and the rate of area increase of the plasma membrane[39] have been related to a nominal size of vesicles, determined by electron microscopy, to give the rate of vesicle formation required to maintain the observed flow rates. The major drawback of this approach is that part of the volume of secreted fluid is derived from direct water flow across the plasma membrane and that much of the additional surface area of membrane contibuted by the vesicles is recycled back into the cytoplasm (Fig.1).

The second, more direct, approach relies on inhibiting the vesicle transport system with cytochalasin, so that vesicles accumulate in the cytoplasm as they are formed[25]. The cells are then fixed for electron microscopy at known time intervals after inhibition. Quantitative analysis of the electron micrographs[40] is used to estimate the vesicle population at the time of inhibition and at each succeeding time interval thereafter. While this approach is more accurate and direct than the one discussed above it still suffers from the drawback that changes in activity can only be detected over a relatively long time interval (5-10 minutes) in the presence of a drug that may have undesirable side effects. In addition this method entails a considerable amount of work, so that it may take several weeks to make a single determination.

7. STUDY OF VESICLES BY LASER LIGHT SCATTERING

A new method of following changes in vesicle numbers and rates of movement in living cells is required. Laser light scattering techniques provide non-invasive estimates of particle motion and number which might be of value in the study of vesicle dynamics. We have already seen that motions of vesicles _in vitro_ can be followed by such techniques[6,7], however application to _in vivo_ studies would depend on a number of additional factors. These include the size, number and motions of other cell components in the scattering volume relative to those of the vesicles. Quantitative electron microscopy reveals that vesicles usually form a unique size class with a numerical density of 10-20 μm^{-3}. Few estimates are available for their velocity of migration through the cytoplasm; this may be of the order of 10 $\mu m\ min^{-1}$.

It seems probable that laser light scattering techniques would be able to provide information on relative changes in the number of vesicles in a cell, and possibly also on their rate of movement through the cytoplasm. The form of such motions may provide information on the mechanisms generating the motive force. Thus vesicles may move smoothly and continuously at a given velocity, or move in short discrete steps, or longer bursts.

Vesicle movement involves their interaction with other structures in the cell, probably the microfilaments. These may undergo some characteristic motion during vesicle movement which could be detectable by laser light scattering. Such motions may be curtailed or modified when vesicle movement is inhibited or stimulated.

Laser light scattering techniques could make significant contributions to the study of secretory processes, providing information on the regulation of vesicle production and on the cellular processes involved in movement and discharge of the vesicles.

REFERENCES

1. M. W. Steer, this volume: 000-000 (1983).
2. D. J. Morré and H. H. Mollenhauer, in: "Dynamic Aspects of Plant Ultrastructure", A. W. Robards ed, McGraw-Hill, London, 84-137 (1974).
3. R. M. Case, Biol. Rev. 53: 211-354 (1978).
4. B. R. Olsen and R. A. Berg, Symp. Soc. exp. Biol. 33: 57-78 (1979).
5. P. M. Ray, W. R. Eisinger and D. G. Robinson, Ber. Deutsch. Bot. Ges. 89: 121-146 (1976).
6. D. P. Siegel, B. R. Ware, D. J. Green and E. W. Westhead, Biophys. J. 22: 341-346 (1978).
7. B. R. Ware, this volume: 000-000 (1983).
8. J. A. Williams, Am. J. Physiol. 235: E517-E524 (1978).
9. M. Prentki, C. Chapponnier, B. Jeanrenaud and G. Gabbiani, J. Cell Biol. 81: 592-607 (1979).
10. V. Herzog and F. Miller, Sym. Soc. exp. Biol. 33: 101-116 (1979).
11. V. Herzog and F. Miller, Cytobiologie, 18: 207 (1978).
12. M. G. Farquar, E. H. Skutelsky and C. R. Hopkins, in: "The Anterior Pituitory Gland", A. Tixier-Vidal and M.G. Farquar ed, Academic Press, New York and London, 83-135 (1975).
13. D. Lawson and M. C. Raff, Symp. Soc. exp. Biol. 33: 337-348 (1979).
14. R. Marchbanks, Symp. Soc. exp. Biol. 33: 251-276 (1979).
15. B. Satir, C. Schooley and P. Satir, J. Cell Biol. 56: 133-178 (1973).
16. H. Plattner, C. Westphal and R. Tiggemann, J. Cell Biol. 92: 368-377 (1982).
17. A. Sievers and E. Schnepf, in: "Cell Biology Monographs Vol. 8. Cytomorphogenesis in plants", O. Kiermayer ed., Springer-Verlag, Wien, N.Y., 265-299 (1981).
18. J. M. Picton, J. C. Earnshaw and M. W. Steer, this volume: 000-000 (1983).
19. H. D. Reiss and W. Herth, Planta, 146: 615-621 (1979).

20. R. J. Howard, J. Cell Sci. 48: 89-103 (1981)
21. A. Fahn, Secretory Tissue in Plants, Academic Press, London, New York and San Francisco (1979)
22. D. J. Morré, in: "Cell Surface Reviews, Volume 4, The Synthesis Assembly and Turnover of Cell Surface Components", G. Poste and G. L. Nicolson eds, North-Holland Publishing Co. Amsterdam, 1-83 (1977)
23. P. J. Harris and D. H. Northcote, Biochim. biophys. Acta 237: 56-64 (1971)
24. T. M. Shannon, Y. Henry, J. M. Picton and M. W. Steer, Protoplasma 112:189-195 (1982)
25. J. M. Picton and M. W. Steer, J. Cell Sci. 49: 261-272 (1981)
26. R. L. Murray and M. W. Dubin, J. Cell Biol. 64: 705-710 (1975)
27. T. L. Hill and M. W. Kirschner, Proc. Natl. Acad. Sci. USA 79: 490-494 (1982)
28. J. A. Wilkins and S. Lin, Biochim. biophys. Acta 642: 55-66 (1981)
29. G. Dahl, R. Ekerdt and M. Gratzl, Symp. Soc. exp. Biol. 33: 349-368 (1979)
30. D. Allan and R. H. Michell, ibid: 323-336 (1979)
31. R. Fraley, J. Wilschut, N. Duzgunes, C. Smith and D. Papahadjopoulos, Biochemistry 19: 6021-6029 (1980)
32. G. Dahl and M. Gratzl, Cytobiologie 12: 344-355 (1976)
33. I. Schulz and H. H. Stolze, Ann. Rev. Physiol. 42: 127-156 (1980)
34. P. Pinto da Silva and M. L. Nogueira, J. Cell Biol. 73: 161-181 (1977)
35. M. Kaliner and K. F. Austen, J. Immunol. 112: 664-674 (1974)
36. D. W. Schwab, E. Simmons and J. Scala, Amer. J. Bot. 56: 88-100 (1969)
37. A. Tartakoff and P. Vassalli, J. Cell Biol. 79: 694-707 (1978)
38. E. Schnepf, Z. Naturforschung 16b: 605-610 (1961)
39. W. J. Vanderwoude, D. J. Morré and C. E. Bracker, J. Cell Sci. 8: 331-351 (1971)
40. M. W. Steer, "Understanding Cell Structure", Cambridge University Press (1981)

STRUCTURE AND DYNAMICS OF DISK MEMBRANE VESICLES

Hyuk Yu

Department of Chemistry
University of Wisconsin
Madison, Wisconsin 53706, U.S.A.

CONTENTS

1. INTRODUCTION

This paper has a simple objective. It is to show that the laser light scattering techniques constitute the crucial parts of a battery of physical methods employed to probe an interesting and important problem, i.e., the molecular mechanism of vertebrate photoreceptor function. In order to do so, I will first describe briefly what is known about the photoreceptor function of rod cells of the vertebrate retina. That will be followed by the description and justification for focusing on a particular organelle of a rod cell, i.e., disk membrane, and how we isolate them into vesicles *in vitro* and characterize their size and shape. Subsequently four specific studies with these vesicles that are relevant to the molecular mechanism of photoreception will be chosen to illustrate the point.

Visual receptor cells in vertebrates are of two types, i.e., rod and cone, and the rod is responsible for scotopic (night) vision while the cone is for photopic (color) vision. The gross morphology of rod and cone cells differs only slightly from each other in a given species and among various vertebrate species.

Farthest from the eye's external surface are the protoplasmic extensions that join to the neurons carrying impulses to the optic nerve. The nucleus is located above the synaptic junction which is in turn surmounted by the metabolically active inner segment. The inner segment is densely packed with mitochondria and the photoreceptor biosyntheses occur in this region.[1] The outer segment is attached to the inner segment by a single cilium. Both rod and cone outer segments are constituted of stacks of the order of a thousand disk membranes which are the photoactive surfaces on which are imbedded visual photopigments, rhodopsin, which acts as the locus of the primary event in the photoreception mechanism. Rhodopsin in rod outer segment (ROS) absorbs a photon in the appropriate frequency range and is "bleached" to an apoprotein, opsin, and all-*trans*-retinal which is known to take place in picoseconds.[2] An hypothetical enzyme, retinal isomerase, then converts all-trans-retinal to 11-*cis*-retinal which is subsequently bound back to opsin to close the visual cycle.[3] While little is known about the enzymatic conversion of all-*trans*-isomer to 11-*cis*-isomer, the bleaching of rhodopsin has been studied in great detail.[4]

The photochemical change of photopigment is connected to the photoreceptor mechanism of ROS disk membranes according to Hagins and Yoshikami as follows. There exists a substantial dark Na^+ current inward through the plasma membrane of ROS. Upon receiving photons, the conformational change of the photopigment molecules causes the Ca^{2+} efflux to take place from the intradiskal space to cytoplasm and these Ca^{2+} will diffuse to the circumference edge of disks and react with the binding sites on the interior of plasma membranes to block the dark Na^+ current. This is then somehow transmitted to the synaptic junction of rod cells as the receptor potential which is known to occur within milliseconds of photoreception.[5] There exists however a set of conflicting experimental evidence with respect to this calcium transmitter hypothesis,[6,7,8] commonly known as the Hagins' model.[9] In particular, a number of attempts to determine the photoreleased Ca^{2+} from ROS or isolated disks have failed to find the requisite released amount even though the total content of Ca^{2+} is consistent with the model.

2. SYSTEM: DISK MEMBRANE VESICLE

From the foregoing brief discussion, it is obvious that the disks of ROS are the organelle that requires a detailed study of their structure and properties in order to focus on the photoreceptor function. It is possible to separate the disks from other organelles of a rod cell, swell them in spherical vesicles in hypotonic media,[10] suspend them into dilute solutions, and study them by various physical methods. Thus our system of study is the isolated swollen disks which we call the disk membrane vesicles (DMV). Since the characterization of DMV relative to its size and shape by elastic and quasielastic light scattering has been detailed elsewhere,[11,12] I will only summarize here the results. Before doing so, it should be noted that the crucial element in our successful attempt to deduce the size and shape by light scattering is the relatively narrow size distribution. Since this was found to be the case with bovine disks,[11] we have dealt exclusively with the those of bovine retinae. In hypotonic media of about 1 mM osmolarity or less, DMV may be modelled as spherical shells and their radius is about 450-480nm. This is deduced by both the light scattering techniques; the elastic scattering method gives rise to a scattering profile with appropriate features that are consistent with the model of spherical shell with a 450 nm radius and the quasielastic light scattering method provides a Stokes radius of 480 nm from the observed translational diffusion coefficient. In addition, the breadth of the size distribution as expressed by the Schulz-Zimm distribution function[13,14] is given by h = 100 or more; such a value of the distribution parameter h amounts to about 70-80% of the sample being found within the 10% range of either side of the mean radius.

3. SPECIFIC TOPICS AND TECHNIQUES

We have chosen to study the following four topics, and in each case is presented a brief indication of the relevance in reference to what has been outlined in the INTRODUCTION. They are:

(i) Osmotic deformation -- here, we can use the osmotic gradient contributed by different ions to gauge the membrane permeability. Thus, one may sensibly raise the question of how special or unique is Ca^{2+} relative to its passive transport across the disk membrane in the bleached state.

(ii) Surface charge density -- disk membrane surface should be charged as with any other membranes at physiological pH but there has been no attempt to characterize the surface charges and to identify their sources. With the use of DMV in dilute suspensions, we have been successful in determining the titration valency and the electrokinetic charge of DMV. Again, these are possible because we can unambiguously characterize the size and shape of DMV in specific suspending media.

(iii) Calcium ion binding -- instead of the photoinduced efflux of calcium ion from the inside of disks to cytoplasm, one can devise a competitive model which supposes photoinduced changes of calcium ion binding on the cytoplasmic side of disks, based on the same set of experimental observations. In fact, it is not possible to distinguish the calcium efflux from the calcium release off the disk surface. Using the electrophoretic light scattering technique, we can address this question by delineating the Ca^{2+} binding isotherm with respect to the photochemical state of DMV.

(iv) Photopigment lateral diffusion -- by virtue of having an optically anisotropic equatorial plane within DMV (see below) and photopigment movement on the DMV surface, a weak electric field induces birefringence in a suspension of DMV. Upon taking advantage of this phenomenon, we can deduce the lateral diffusion coefficient of photopigments from dynamic electric birefringence measurements. One of the proposals for photopigments functioning as the photo-induced Ca^{2+} pore on disk membranes entails light-triggered aggregation of photopigments. If this is true and such aggregates moves about on the membrane surface as single entities, then there should exist a difference in the lateral diffusion of photopigment with respect to its photochemical state. This technique should be made to probe such an event in the context of the calcium efflux hypothesis.

This concludes the brief discussion of the topics and how they were chosen. It should be emphasized that we must know how well we prepare the system before we try to change the conditions and see how it responds, and this is done by the two light scattering techniques. We now turn to discuss each topic in detail.

A. Osmotic Deformation

The disk membranes are known to respond to an osmotic gradient in native ROS[15,16,17] as well as DMV[18,19] hence the membrane permeability to individual ions can be probed by observing how sensitively it deforms by varying concentration of a given non-electrolyte or ion in the suspending medium. If a given solute is completely permeative to DMV, there should be no chemical potential gradient across the bilayer and DMV should remain undeformed, independent of the solute concentration. On the other hand, if the membrane is totally impermeable to a solute, an addition of that solute to the suspending medium will induce the transport of permeative components, principally water, until an equilibrium is established such that the residual chemical potential gradients are counterbalanced by the membrane stiffness.

A crucial experiment here is the effect of the DMV permeability to Ca^{2+} relative to the photochemical state. In terms of the Hagins' model, there should exist a large difference in the Ca^{2+} permeability between the bleached and unbleached states, at least in the efflux direction, i.e., from the disk interior to cytoplasmic side. To the extent that the membrane permeability, unlike active transport, is not known to be asymmetric, the influx and efflux directions should not be different. Thus the permeability probed through the osmotic deformation of DMV could provide us with a clue relative to the calcium transmitter model at the vesicular level, i.e., at hypotonic state. Equally plausible is the possibility that any difference in the permeability of the native state might have been all obliterated upon suspending the disks in hypotonic media to swell them into balloon shape and to remove a variety of peripheral components that were present in the native state.

The scheme is to perform two kinds of light scattering experiments whereby each kind provides one of the two parameters characterizing the shape and size of the deformed DMV. The osmotic deformation has been shown to proceed via oblate ellipsoidal shell to native disk shape with its semi-major axis dimension conserved.[19] Since the elastic light scattering profile is most sensitive to the semi-major axis b of an oblate ellipsoidal shell, its structural features remain practically unchanged as a DMV undergoes osmotic deformation. This is shown in Fig. 1 below where the scattering profile represented by $x^2P(x)$ is plotted for different axial ratios ρ; x is the product of the scattering wave vector κ and the semi-major axis of oblate ellipsoidal shell b, $x \equiv \kappa b$, and P(x) is the isotropic part of the intraparticle scattering form factor. Hence, $x^2P(x)$ should be proportional to the product of the polarized scattering intensity at angle θ, $I_{VV}(\theta)$, and $\sin^2(\theta/2)$ in dilute suspensions where interparticle interference contribution to the scattering is negligible, i.e., $x^2P(x) \propto \sin^2(\theta/2) I_{VV}(\theta)$.

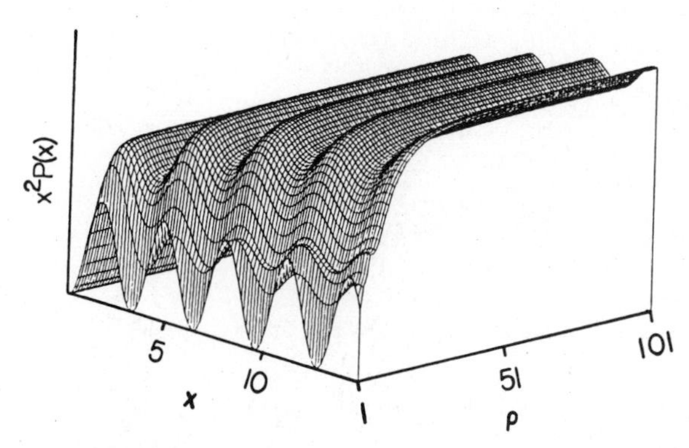

Fig. 1. A three-dimensional representation of the theoretical scattering profiles of oblate ellipsoidal shell model for $0 \le x \le 15$ and $1 \le \rho \le 101$.

More specifically, the extremum positions of a modulation profile are insensitive to the axial ratio ρ and are an inverse linear function of the semi-major axis b within narrow windows for each order extremum. This is displayed in Fig. 2.

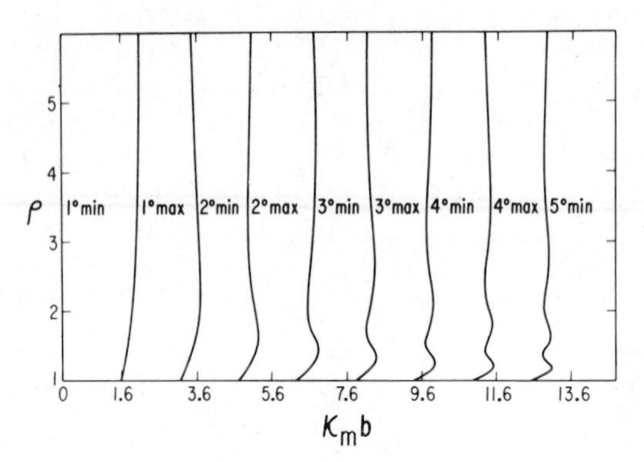

Fig. 2. Theoretical extremum positions of the modulation profile of elastic light scattering of oblate ellipsoid shell are displayed at different axial ratios, where K_m stands for the magnitude of scattering wave vector at each order of the extrema.

Thus having shown that the extremum positions of the elastic light scattering profile are sensitive only to the semi-major axis b, we note an additional feature of the 3-dimensional plot of Fig. 1. As ρ is increased the amplitude of modulation profile is progressively decreased while the mean level of the profile increases with ρ up to about 10 and then practically levels off to the asymptotic limit. We use this feature of the scattering profile as a qualitative measure of the deformation. We now turn to the second experimental method which is sensitive to the axial ratio ρ. This is accomplished by performing quasielastic light scattering. The translational diffusion coefficient D of oblate ellipsoids as given by Perrin[20] is

$$D = (kT/6\pi\eta b)\rho G(\rho) \tag{1}$$

with

$$G(\rho) = \tan^{-1}(\rho^2-1)^{\frac{1}{2}} / (\rho^2-1)^{\frac{1}{2}} \tag{2}$$

where η is the viscosity of suspension and kT has the usual meaning. The term ρG(ρ) in Equation (1) is a monotonically increasing function of ρ, hence D should increase monotonically up to π/2 times its value at ρ = 1 as long as b remains the same for the entire range of ρ.

Combining these two methods, we first have shown[19] that there is a difference in the osmotic deformation profiles of DMV suspended in sucrose in the "unbleached" and bleached state as probed by a low power (~1 mW) He/Ne laser which has only a slight photobleaching effect.[21] The profiles of elastic light scattering are shown in the inset of Fig. 3; the point of this figure is to demonstrate experimentally that the modulation amplitude decreases with ρ, which is separately deduced from the translational diffusion coefficient through Equations (1) and (2).

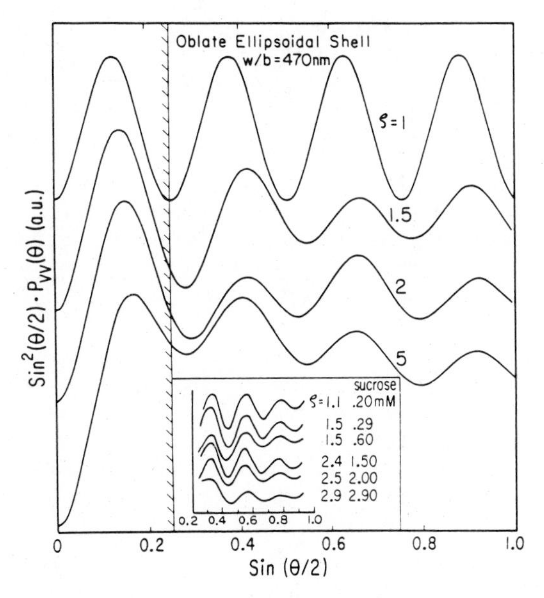

Fig. 3. The theoretical scattering profiles of oblate ellipsoidal shells at indicated values of ρ with the refractive index of scattering medium of water (1.333), the incident blue light (λ_o = 546 nm) and the semi-major axis b = 470 nm; $P_{VV}(\theta)$ is the same as $I_{VV}(\theta)$. The vertical shaded bar indicates the instrumental scattering angle limit below which data are not taken. The inset shows the experimental profiles of bleached DMV whose amplitude decreases with increasing ρ.

Thus elastic light scattering alone gives rise to the observation that b remains the same and ρ increases with concentration of sucrose in the suspending medium; the former is established quantitatively while the latter is given only qualitatively.

We then examine the membrane permeability to Na^+, K^+, Ca^{2+} and Mg^{2+} with chloride as the common anion.[22] The elastic scattering profiles are displayed in Fig. 4 and the corresponding results of quasielastic scattering in NaCl are displayed in Fig. 5.

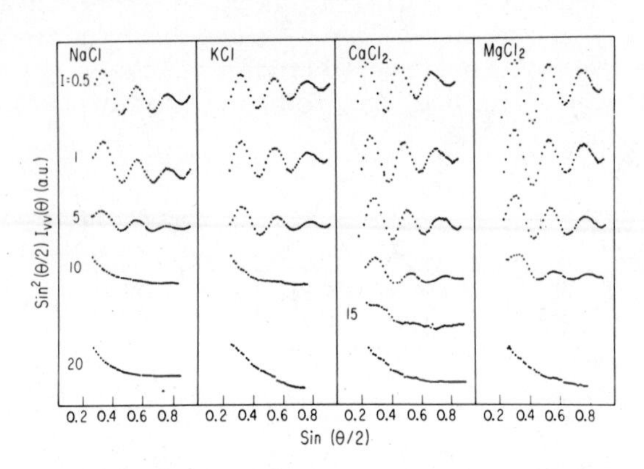

Fig. 4. The scattering profiles of chlorides of Na, K, Ca and Mg in 1 mM imidazole at total ionic strength of 0.5, 1, 5, 10, 15 (Ca^{2+} only) and 20 mM.

Combining these results, we reach the conclusion that the DMV permeability to Na^+, K^+, Ca^{2+} and Mg^{2+} is about the same in the bleached state and there appears to be a little difference in the "unbleached" state as examined by a low power He/Ne laser. We thus come to an important observation: Ca^{2+} is not unique and DMV shows no difference in the permeability to Ca^{2+} relative to the photochemical state. This is not in accord with the expectations of Hagins' model.

B. Surface Charge Densities

Upon exchanging every cation with H^+ and every anion with OH^-, one can titrate a given suspension of DMV with base or acid under appropriate ionic strength. The results can be combined with those of the light scattering experiment to deduce the titration valency and the corresponding surface charge density because the DMV number density in the suspension and the surface area per DMV are known. We have very recently performed such experiments.[23] Under the assumption that a mixed bed of ion exchange resins was completely efficacious in exchanging all cations and anions with H^+ and OH^- respectively, we deduce the titration valency of 1.1×10^6 electronic charge per DMV at pH 7 if the isoelectric point is at pH 4.3 (see below) and the surface charge density of about 0.4 electronic charge per $100\mathring{A}^2$ at the same pH. In addition, the main component of negative charges in the pH range of 5-7 is suspected to be carboxylate because we found pK_a value of the dissociating species to

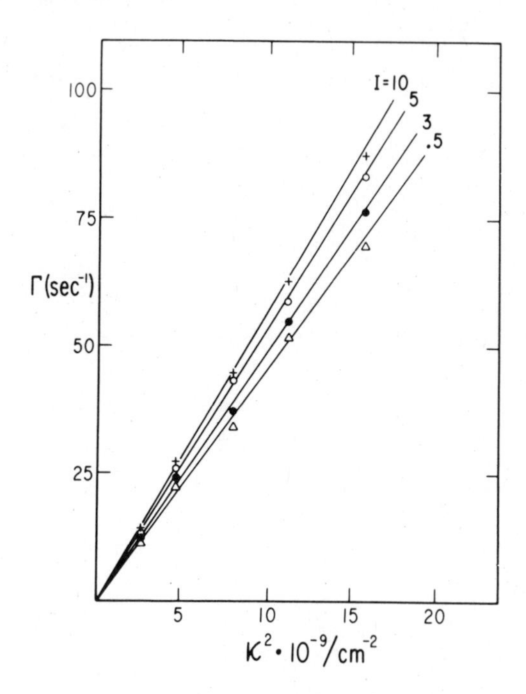

Fig. 5. The decay constant Γ of the single exponential autocorrelation function of photocount of the quasielastically scattered light is plotted against κ. The number on each straight line indicates the total ionic strength in mM in 1 mM imidazole, pH 7 in various NaCl concentrations, and the slope of each line gives the diffusion coefficient D.

be at 5.0. In the process of characterizing the size and shape of DMV over the entire range of pH, 3.5-9.5, we have also found the following: The elastic light scattering profile remains the same while the translational diffusion coefficient develops a dip at around pH 4.3. This is displayed in Fig. 6. Since the decrease and recovery of D in the pH range of 3-5 is reproducible and found in two different suspending media, we attribute the dip to a reversible aggregation of DMV in the vicinity of its isoelectric point which is separately confirmed by the pH dependence of electrophoretic mobility (see below).

The scattering cell, the modulation field and the analysis scheme of electrophoretic light scattering have been given elsewhere.[24] The electric field induced Doppler shifted power spectra $S(\nu)$

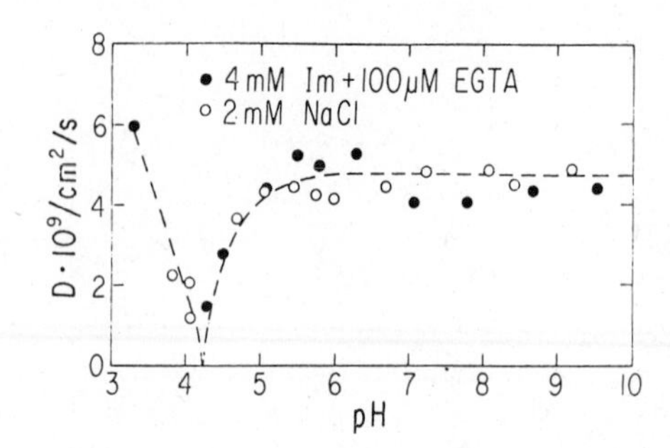

Fig. 6. The translational diffusion coefficient, corrected to 20°C with the viscosity of water at the temperature, is plotted against pH in two suspending media.

of DMV scattering are displayed in Fig. 7 where (A) shows the pH dependence of $S(\nu)$ and (B) shows the calcium ion dependence of $S(\nu)$. Focusing on the results shown in (A) for the moment, it is clear that the shifted peak increases in frequency as pH is raised. The power spectra at a given pH are acquired at different scattering angles and field strengths, and the frequency shifts $\Delta\nu_s$ at these conditions are collected to determine the electrophoretic mobility μ at that pH. The results so obtained are assembled as a function of pH and displayed in Fig. 8. It is obvious that the isoelectric point is at around pH 4.3, which is in complete accord with the expected value from Fig. 6, supporting our contention that the dip in D is indeed due to the DMV aggregation in the vicinity of isoelectric point. From the value of 3.5×10^{-4} cm^2/V·s for the electrophoretic mobility at pH 7, we deduce the electrokinetic charge of 8×10^4 per DMV with the aid of Henry's law[25] to correct for the Debye-Hückel screening effect. We must therefore conclude that only about 7% of the titration valency is electrokinetically active and the rest of the surface charges are well screened away by counter-ions. In terms of the electrokinetic surface charge density, there are only about 3 electronic charges per 10^4 Å^2. This, we must note, is a very small value. Hence the disk membrane surface should be regarded as a very weakly charged plane.

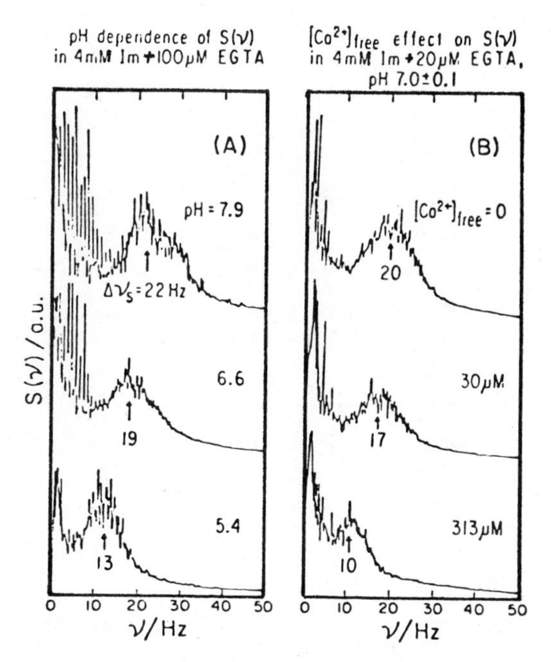

Fig. 7. (A) Doppler shifted power spectra in different pH conditions and (B) in different free calcium ion concentrations at pH 7 which are adjusted by calcium buffer.

C. Calcium Ion Binding

In conjunction with the study of surface chargè densities detailed in the previous section, we examine[23] the electrophoretic mobility dependence on the free calcium ion concentration. As shown in the right column of Fig. 7, the peak position of Doppler shifted power spectrum decreases in frequency as free Ca^{2+} concentration in the suspending medium is increased. This would be possible only if the original electrokinetic charge on DMV surface is neutralized by Ca^{2+} but its total binding could be far greater than the decrement of electrokinetic charge because any ion exchange between Ca^{2+} and other cation would not result in lowering of the mobility but would account for the Ca^{2+} binding. Hence, whatever we deduce from the observed decrement of electrophoretic mobility should set the lower bound of Ca^{2+} binding. Keeping such a stipulation in mind, the binding isotherm at 20°C has been determined from the mobility dependence on free Ca^{2+} concentration. Fig. 9 shows the results, where the solid curve represents a model of two binding sites deduced from the best fit to the data points given by open circles, which

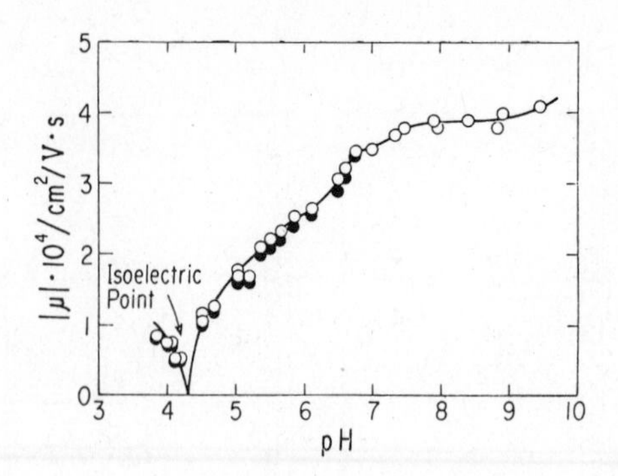

Fig 8. Magnitude of electrophoretic mobility as a function of pH in 2 mM NaCl. Filled circles are the experimental points and they are distinguished from the ionic strength reduced points (open circles) when they are different enough.

are all reduced to the ionic strength of 2 mM. The model parameters are determined as:

$$K_1 = 4.5 \times 10^4 \ (M^{-1}), \ n_1 = 1.5 \times 10^4$$

$$K_2 = (4 \pm 2) \times 10^2 \ (M^{-1}), \ n_2 = 0.7 \times 10^4$$

Comparing our K_1 and n_1 to those reported by Hendricks,[26] it is remarkable indeed that the two sets of data are in agreement despite entirely different methods employed in each set. We would take this agreement as supporting evidence for no cation exchange, at least in the high affinity binding regime.

D. Photopigment Lateral Diffusion

Recalling the osmotic deformation study presented in A., it is quite natural to suppose that there must exist in DMV a structural feature that resists the osmotic deformation, hence is different from the rest of the membrane constitution. We call this the equatorial rim which may be characterized as a structurally distinct entity dividing at the equator the two hemispheres of DMV. The origins of anisotropy of optical and electrical polarizabilities have been attributed to this rim plane.[27] Thus the electric field

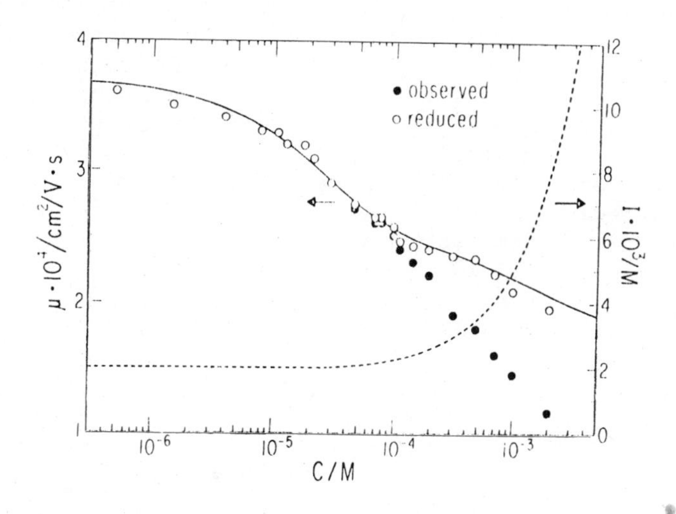

Fig. 9. Electrophoretic mobility is plotted against free calcium ion concentration adjusted by a calcium buffer ($CaCl_2$/EGTA). The solid curve is the binding isotherm deduced with a two-binding site model and the chained curve is for the total ionic strength of suspending medium. The observed mobilities (filled circles) are all reduced to those in total ionic strength of 2 mM.

induced optical birefringence of DMV suspensions comes from the rim plane within DMV despite its spherical shape; photopigment molecules execute field-induced displacement within each hemisphere of a DMV but cannot cross the equatorial rim because it acts as a displacement barrier. Based on two sets of static and dynamic birefringence results,[27,28] the following birefringence mechanism has finally been proposed,as schematically depicted in Fig. 10. (a) Prior to application of an external electric field, we have the uniform distribution of photopigments on both hemispheres and the random orientation of the rim plane relative to the field direction. (b) With the field on, the photopigment distribution is perturbed away from the uniform one. Since the rim acts as the displacement barrier, the induced dipole moment appears in the equatorial plane. At the same time, the induced dipole moment forces DMV to orient preferentially with the rim plane parallel to the field direction. (c) The steady state in the orientational distribution is attained after a sufficient duration with the field on although the degree of orientation depends on the field strength and the temperature. (d) When the field is off, the photopigment distribution starts to return to the unperturbed state, and at the same time, the rim plane orientation starts to recover back to the random distribution. (e) Both the photopigment distribution and rim plane orientation recover back to complete random distribution. In the light of the proposed mechanism, it is quite reasonable to ascribe the birefringence transient to the slowly induced dipole (SID) moment caused by the field-induced displacement of photopigments on the DMV surface. Hence with use of a rapidly reversing bipolar electric field, we can deduce the photopigment displacement rate from the rise and dip-recovery regions if the global rotatory motion is accounted for from the field free decay region.[29] In the limit of vanishingly small electric field, the displacement rate is no more than the lateral diffusion rate. An example of such an experiment is shown in Fig. 11. The solid curve is the theoretical prediction; the rise region is drawn with the SID relaxation time of 300 ms and the dip-recovery region is drawn with the corresponding time of 500 ms, both with the global rotatory relaxation time of 96 ms for the field free decay region. Finally the lateral diffusion coefficient is estimated as $3.3 \pm 1.2 \times 10^{-9}$ cm^2/s which compares favorably with $3.5 \pm 1.5 \times 10^{-9}$ cm^2/s obtained by Poo and Cone.[30]

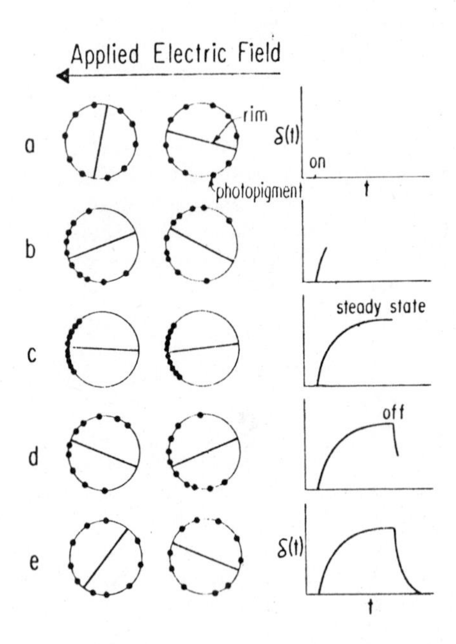

Fig. 10. Schematic illustration of the proposed birefringence mechanism.

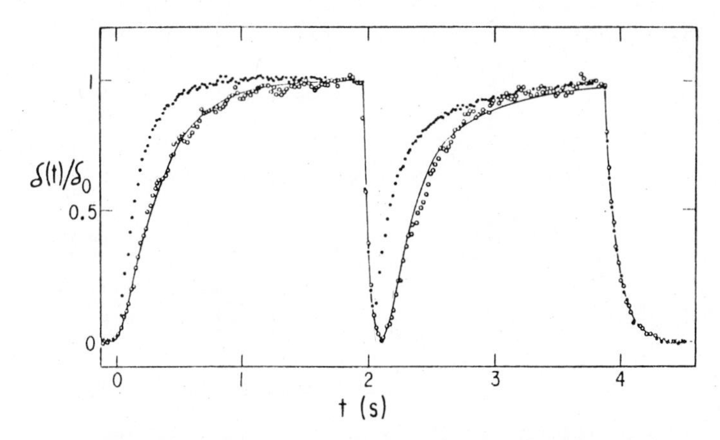

Fig. 11. Normalized electric field induced birefringence transient, $\delta(t)/\delta_o$, by a rapidly reversing electric field. The open and filled circles refer to the results for 3 and 10 V/cm field strength, respectively. The solid curve is the theoretical predictions.

ACKNOWLEDGEMENTS

I am most grateful for continuing support by an NIH grant EY01483 and the Biomedical Sciences Support Grant of NIH administered through the Graduate School of the University of Wisconsin-Madison. All of the studies presented here are the works of many of my past and current associates and students to whom I am greatly indebted. They are Takashi Norisuye, W. F. Hoffman, G. B. Caflisch, E. J. Amis, D. M. Piatt, Hideo Takezoe, Taihyun Chang, Toshiaki Kitano, D. A. Davenport, D. J. Wendt and E. D. Erickson.

REFERENCES

1. R. Rodieck, The Vertebrate Retina, (W. H. Freeman, San Francisco, 1973).
2. G. E. Busch, M. L. Applebury, A. A. Lamola and P. M. Rentzepis, Proc. Nat. Acad. Sci. U.S.A. 69: 7802 (1972).
3. G. Walds, Science 162: 230 (1968).
4. Accounts Chem. Res. 8: (1975) (Special Issue on the Chemistry of Vision, Ed., E. L. Menger).
5. J. K. Wong and S. E. Ostroy, Arch. Biochem. Biophys. 154: 1 (1973).
6. P. A. Liebman, Invest. Opthalmol. 13: 700 (1974).

7. E.Z. Szutz and R. A. Cone, Biochim. Biophys. Acta 458: 194 (1977).
8. P. P. M. Schnetkamp, F. J. M. Daemen and S. L. Bonting, Biochem. Biophys. Acta 458: 259 (1977).
9. W. A. Hagins, Annu. Rev. Biophys. Bioeng. 1: 131 (1972).
10. H. G. Smith, Jr., G. W. Stubbs and B. J. Litman, Exp. Eye Res. 20: 211 (1975).
11. T. Norisuye, W. F. Hoffman and H. Yu, Biochemistry 15: 5678 (1976).
12. H. Yu, Methods in Enzymology, Vol. 81, p. 616 (1982).
13. G. V. Schulz, Z. Phys. Chem. B43: 25 (1939).
14. B. H. Zimm, J. Chem. Phys. 16: 1099 (1948).
15. J. Heller, T. J. Ostwald and D. Bok, J. Cell Biol. 48: 633 (1971).
16. D. G. McConnell, J. Biol. Chem. 250: 1898 (1975).
17. J. I. Kerenbrot, D. T. Brown and R. A. Cone, J. Cell Biol. 56: 389 (1973).
18. R. A. Raubach, P. P. Nemes and E. A. Dratz, Exp. Eye Res. 18: 1 (1974).
19. T. Noyisuye and H. Yu, Biochim. Biophys. Acta 471: 436 (1977).
20. F. Perrin, J. Phys. Radium 7: 1 (1936).
21. W. F. Hoffman, T. Norisuye and H. Yu, Biochemistry 16: 1273 (1977).
22. E. J. Amis, D. J. Wendt, E. D. Erickson and H. Yu, Biochim. Biophys. Acta 664: 201 (1981).
23. T. Kitano, T. Chang, D. M. Piatt, G. B. Caflisch and H. Yu, to be published.
24. G. B. Caflisch, T. Norisuye and H. Yu, J. Colloid Interface Sci. 76: 174 (1980).
25. D. C. Henry, Proc. Roy. Soc. (London) Ser. A 133: 106 (1931).
26. Th. Hendricks, P. M. M. van Haard, F. J. M. Daemen, and S. L. Bonting, Biochim. Biophys. Acta 467: 175 (1977).
27. H. Takezoe and H. Yu, Biophys. Chem. 14: 205 (1982).
28. H. Takezoe and H. Yu, ibid. 13: 49 (1981).
29. H. Takezoe and H. Yu, Biochemistry 20: 5275 (1981).
30. M.-M. Poo and R. A. Cone, Nature (London) 247: 438 (1974).

VESICLE DYNAMICS IN POLLEN TUBES

Jill M. Picton and Martin W. Steer

Department of Botany

John C. Earnshaw

Department of Physics

The Queen's University of Belfast
N. Ireland, BT7 1NN

CONTENTS

1. INTRODUCTION

Pollen tubes are apically growing plant cells in which extension is accompanied by the fusion of vesicles with the tube tip. The vesicles arise from numerous dictyosomes, elements of the Golgi apparatus, present within the cytoplasm (Fig. 1). At the tip, the vesicles provide both the membrane and the cell wall carbohydrates necessary for the extension of the tube.

We are able to quantify the numbers of vesicles present within a pollen tube and the rate with which they move to, and fuse with the tip. Changes in these numbers can be brought about by altering the growth conditions or by treatment with drugs. We are attempting

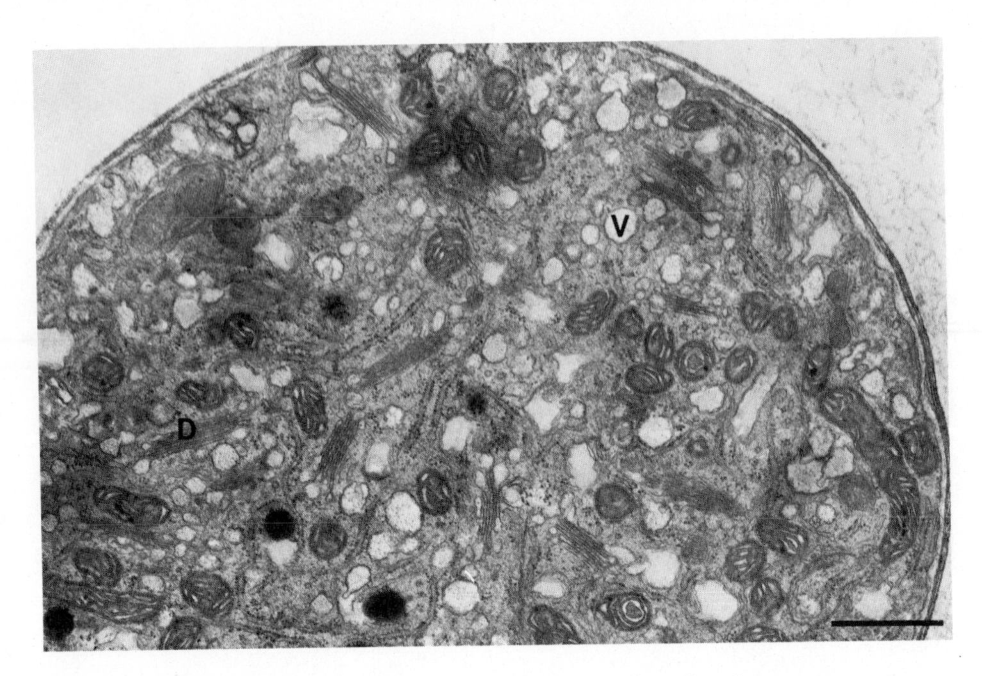

Fig. 1. Transverse section on a pollen tube showing numerous dictyosomes (D) and their secretory vesicles (V). Bar = 1 μm.

to correlate such changes in tube structure and growth pattern with changes observed in the correlation functions obtained by laser Doppler microscopy.

2. STEREOLOGICAL ANALYSIS OF VESICLE TRANSPORT

The number and velocity of vesicles travelling to the tube tip depends on the vesicle production rate, the vesicle fusion rate and the vesicle residence time in the cytoplasm. Determination of these parameters requires initially that the number of vesicles in the cytoplasm be determined.

A. Vesicle size distributions

The vesicle profiles present in a known area of cytoplasm are measured from electronmicrographs and assigned to size classes based on their diameter using a Kontron Image Analyser. As the vesicles are not necessarily sectioned at their maximum diameter, the size distribution of profiles is not equal to the size distribution of vesicles present within the cytoplasm. The conversion is made using the method of Rose[1] which also corrects for section thickness effects.

B. Vesicle production rates

We have developed a method to determine vesicle production rates by the dictyosomes.[2] Essentially it involves treatment of pollen tubes with cytochalasin D which interferes with microfilament action. This prevents vesicle migration and causes them to accumulate around the dictyosomes.[3] Assuming that cytochalasin D does not affect vesicle production rates over the short treatment time (10 minutes), the rate of vesicle accumulation will equal the rate of vesicle production. Accumulation rates obtained by this method are of the order of 3 min^{-1} μm^{-3}. This represents a total vesicle production rate per tube of approximately 8,000 vesicles per minute.[4]

C. Vesicle migration velocity

The relationships between the various parameters which can be determined during pollen tube growth are shown in Fig. 2.

The residence time for vesicles in the cytoplasm is typically 5-10 minutes.[4] The vesicle transport velocity is equal to the fusion rate (F) divided by the standing vesicle population (S) and the cross-sectional area of the tube (A). Typical values for these parameters are $F = 8{,}000\ min^{-1}$, $S = 20\ \mu m^{-3}$ and $A = 50\ \mu m^2$, giving a vesicle transport velocity of 8 $\mu m\ min^{-1}$. This is slower than the streaming rate observed in these tubes but faster than the *in vitro* results obtained for treadmilling of microfilaments.[5]

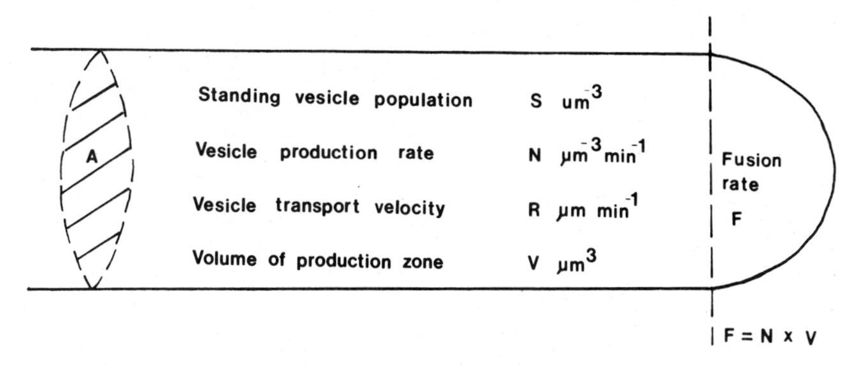

$$\text{Residence time in production zone} = \frac{S}{N}$$

$$\text{Vesicle transport velocity} = \frac{F}{S \times A}$$

Fig. 2. Diagram illustrating the relationships between the various parameters of pollen tube growth.

3. LASER DOPPLER STUDIES OF POLLEN TUBES

Vesicle movements can be studied in living pollen tubes by laser Doppler microscopy. Interpretation of the correlation functions obtained depends on parallel electronmicroscope studies to establish the origins of each signal within the cell. This information will then allow subsequent studies to be carried out directly, avoiding the poor time resolution of the quantitative electron microscopy approach.

As pollen tube tips do not stream and lack all organelles except for the accumulating vesicles, we have begun our experiments in this region.

After treatment with cytochalasin D, which inhibits vesicle transport, the tip vesicle population is reduced and becomes completely absent after 10 minutes.[2] Comparison of correlation functions obtained from a normally growing tube tip (Fig. 3.) with those obtained from a cytochalasin D treated tip (Fig. 4.) shows that the signal levels obtained after cytochalasin D treatment are much lower. We conclude that the reduction in signal strength represents either the loss of the contribution of the vesicles or the inactivation of microfilaments. These two possibilities cannot be distinguished at present.

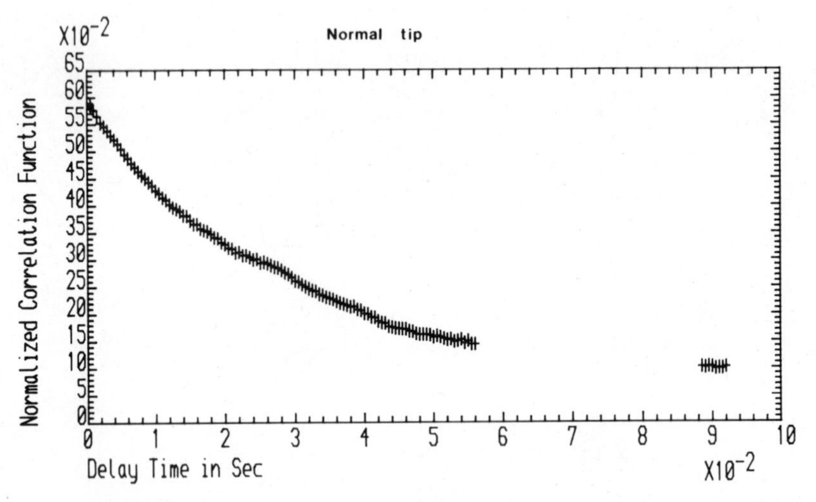

Fig. 3. Correlation function obtained from normally growing pollen tube tip.

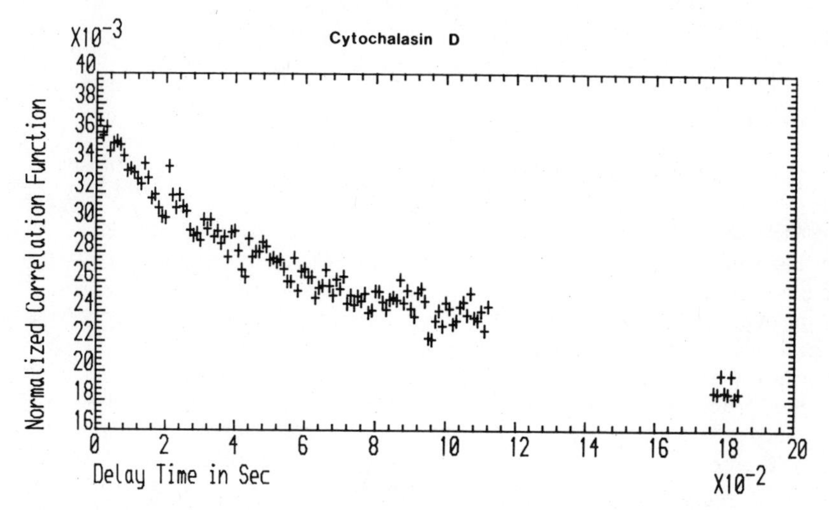

Fig. 4. Correlation function obtained from a pollen tube tip following treatment with cytochalasin D.

Pollen tube growth requires the correct calcium ion balance to be maintained at the tube tip[6,7]; the presence of both suboptimal and supraoptimal Ca^{2+} levels in the growth medium inhibit tube growth.[7]

A correlation function obtained from a high Ca^{2+} inhibited pollen tube tip is shown in Fig. 5. In this case the reduction in

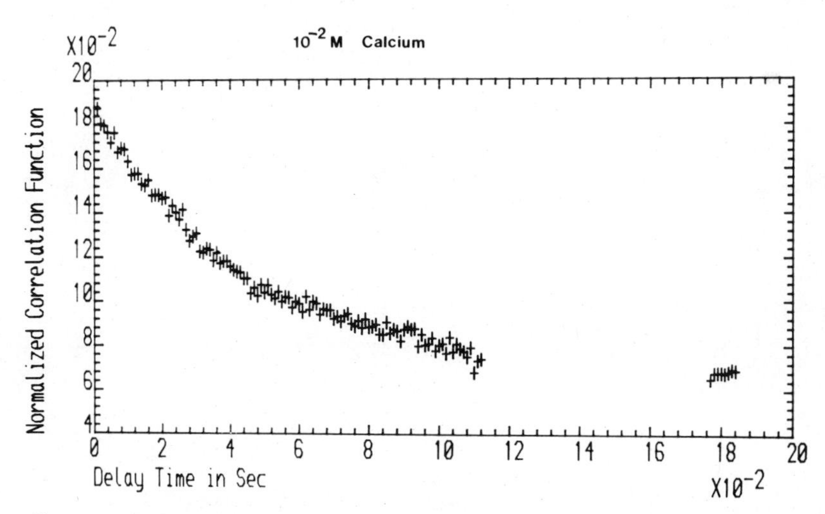

Fig. 5. Correlation function obtained from a pollen tube tip following inhibition by high Ca^{2+} levels.

signal level is not so substantial. Ultrastructurally, such tubes still contain vesicles at their tips but the tip wall becomes abnormally thickened preventing extension.[7] The vesicles continue to move towards the tip as in normal tubes.

4. INTERPRETATION OF CORRELATION FUNCTIONS

A major problem with this study is the interpretation of the correlation functions. We have attempted to extract the various diffusion components present using the computer programs by Provencher known as DISCRETE[8] and CONTIN.[9] These usually reveal the presence of two components, one with Γ in the range 1-30 and the other with Γ in the range 50-800. In normal tubes the slower component has a signal level about 10 times that of the faster component, while in cytochalasin D treated tubes the signal levels for each are similar. The interpretation of these correlation functions is complicated by the possibility that the reference beam level is substantially increased by static scatterers in the non-growing tubes.

5. CONCLUSIONS

We have developed a method for the real time *in vivo* detection of changes in pollen tube tips following certain drug treatments. Further work will concentrate on determining the nature of the changes that can be detected by LDM and on quantification of signal levels with numerical density of cellular structures.

ACKNOWLEDGEMENTS

We wish to thank Mr. G. W. McCartney and Mr. R. Reed for providing excellent electronmicroscopy facilities and Professor L. T. Threadgold for the use of the Kontron Videoplan. This work was supported by grants from the U. K. Science and Engineering Research Council and from the Company of Biologists.

REFERENCES

1. P. E. Rose, *J. Microscopy*, 118: 135-141 (1980).
2. J. M. Picton and M. W. Steer, *J. Cell Sci.*, 49: 261-272 (1981).
3. H. H. Mollenhauer and D. J. Morré, *Protoplasma*, 87: 39-48 (1976).
4. J. M. Picton, Ph.D. thesis, Queen's University, Belfast (1981).
5. T. L. Hill and M. W. Kirschner, *Proc. Natl. Acad. Sci. USA*, 79: 490-494 (1982).
6. J. M. Picton and M. W. Steer, *J. theor. Biol.*, (in press) (1982).
7. J. M. Picton and M. W. Steer, *Protoplasma*, (in press) (1982).
8. S. W. Provencher, *J. Chem. Phys.*, 64: 2772-2777 (1976).
9. S. W. Provencher, *Chem.*, 180: 201-209 (1979).

A PRELIMINARY RHEOLOGICAL INVESTIGATION OF LIVING PHYSARUM ENDOPLASM

Masahiko Sato, Terence Z. Wong*, and Robert D. Allen

Department of Biology, *Thayer School of Engineering
Dartmouth College
Hanover, N.H. 03755

ABSTRACT

An improved magnetic viscoelastometer has been developed which is applicable to large syncytial systems. In combination with the double chamber method (16) and videomicroscopy (1), characterization of the endoplasm of *Physarum polycephalum* has been carried out using the parameters - viscosity, elasticity, and yield point - similar to the experiments of Yagi (28) and Hiramoto (14) but with much improved quantitation.

The yield strength ($\tau_{.x}$) was probed in the direction perpendicular to the plasmodial vein with the magnetic particle technique. The mean value determined is 0.58±0.13 dyne/cm^2. Trajectories of these particles showed the endoplasm reproducibly manifests a yield point and elastic recoil on termination of the magnetic force (F_m). The form of the trajectory was reminiscent of a Maxwell model. Thus, the endoplasm behaves as a viscoelastic material beyond the yield point.

1. INTRODUCTION

Despite the remarkable advances in the development of analytical techniques in the past thirty years, the physical state of living cytoplasm remains crudely characterized. Biorheological investigations *in vivo* have generally been either quantitatively inadequate (6,11,14,27,28) or were disruptive to the cell (3,4,10, 12).

In this paper we present some preliminary results of the application of an improved magnetic particle technique, capable of direct quantitative rheological analysis of large cells or syncytial systems. The immediate objective was to characterize the cytoplasm (endoplasm) of the acellular slime mold Physarum polycephalum using the parameters of viscosity, elasticity, and yield strength. Cytoplasm is a complex non-Newtonian material (11,13,14,27,28) which can not be adequately characterized by one parameter, as has been attempted with viscosity for actin solutions (5,22,25). Thus, the magnetic particle technique presented here combined with videomicroscopy (1) and the double-chamber method (16) has allowed the determination of several new rheological phenomena inherent in Physarum endoplasm.

2. MATERIALS

Magnetic Particles

The nickel, iron, molybdenum alloy Hy Mu 800 was provided for us by Carpenter Technology (Reading, Pa.). The alloy had been argon gas atomized to produce spheroidal particles, and then fractionated. We received the -400 sieve mesh fraction (36μm diameter and less).

A method was developed to anneal these particles in H_2 gas at 899°C for two hours and cooled at 93°C/hr. to maximize magnetic permeability (26).

Cultures

The albino strain CH 975/943 was the generous gift of the late C.E. Holt (MIT). Macroplasmodia were cultured on peptone yeast extract (PYE) nutrient agar dishes (2) on a four day cycle. Suitable veins and plasmodial masses from fourth day cultures were excised and manipulated with flame drawn glass needles.

3. METHODS

Double Chamber Specimens

Modified Kamiya (16) chambers were made from an acrylic block into which two 5/16 inch wells were drilled 7/16 inch apart. Smaller vent holes (1/16 inch) were drilled into the sides to which brass tubes (3/32 inch O.D.) were affixed (Fig. 1). One of the two sides is connected with tubing to the manometer which allows the modulation of endoplasmic shuttle streaming with balance pressure (16).

Physarum macroplasmodia will phagocytize magnetic particles (Hy Mu 800), especially when "sweetened" with culture medium. Cultures were grown on PYE agar dishes (2) layered with dialysis membrane (12k, Spectrum Medical Industries, Inc., Los Angeles, CA) to facilitate handling of plasmodial veins. Plasmodial masses were sprinkled with PYE coated Hy Mu 800 at least two hours previous to

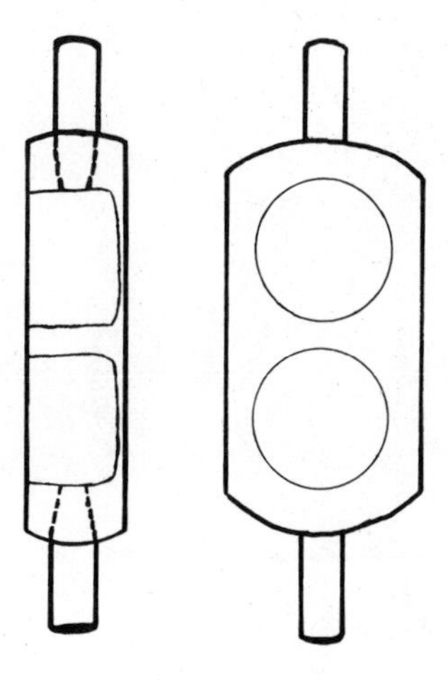

Fig. 1. Modified double chambers were made from acrylic blocks into which two 5/16" wells were drilled 7/16" apart. Vent holes (1/16") were drilled in the sides to which 3/32" O.D. brass tubing was affixed.

experimentation. Double chamber preparations were assembled by the following procedure:

1. One side was half filled with 2% agar (Difco Laboratories, Detroit, MI). The chamber was slanted to ensure that the vent hole remained free of agar. This step was repeated with the other side.

2. A plasmodial vein (about 0.5mm diameter) from fourth day cultures was carefully placed on the double chamber with each end resting on the agar of opposite wells. Plasmodial masses were added to the ends and allowed to fuse. The preparation was incubated in a moist chamber for two hours.

3. Particle impregnated masses were added to one end and the preparation was further incubated for 40 minutes.

4. When the vein had attained suitable clarity with particles visibly flowing in the endoplasm, the top of the preparation was layered with 37°C, 2% Low Gel Temperature Agarose (Sigma Chemical Co., St. Louis, Mo.). A no. 2 coverslip was placed on top and pressed gently to flatten the vein slightly between the acrylic and the slide.

5. The chamber was then mounted onto the modified mechanical stage of a 90° tilted WL microscope (Carl Zeiss, Inc., NY) Fig. 2.

6. Tubing from the manometer was connected to one side of the double chamber. When a suitable particle came into view in the microscope field, the streaming was stopped with balance pressure. The endoplasm was held motionless at least 30 sec.

Fig. 2. Notice the 90° tilted WL (Carl Zeiss) microscope with the matched set of objectives (UD 40/0.65, 6.9mm W.D.). The origin (X=Y=Z=0) is located at the specimen point surrounde on both sides by magnetic coils.

before measurements were taken to minimize effects from possibl shock due to initial application of pressure.

Videomicroscopy

The tilted WL microscope (Carl Zeiss) was fitted with matched long working distance objectives (Zeiss UD 40/0.65,6.9 W.D.) to maximize accessibility to the specimen. 12W tungsten illumination with H_2O infrared and 546nm narrow band interference filters were used.

Brightfield microscopy was used principally with the MK II 65 camera (Dage-MTI, Inc., Michigan City, IN) with auto/manual black, auto gamma, and external sync. drive capabilities. This camera was driven by a second camera (SC-15A, Sylvania Electric Products, Inc., Bedford, MA) through a video screen splitter/inserter (model V270SP, Vicon Industries, Inc., Plainview, NY). This enabled the simultaneous observation of the microscope field, manometer, and ammeter to the magnet coils on the monitor screen.

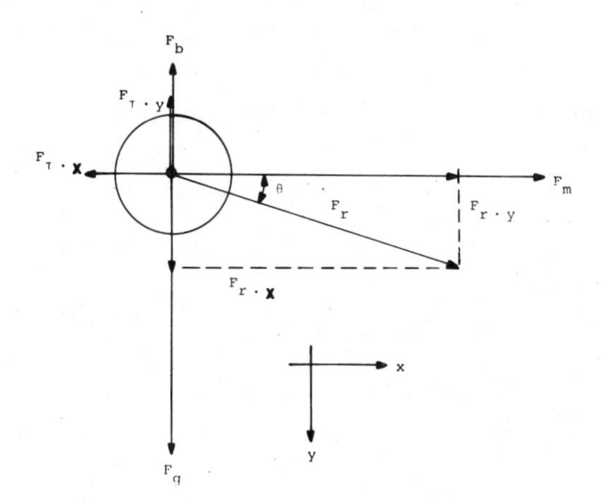

Fig. 3. Force balance on a sphere. F_g refers to the gravitational force, F_b the buoyant force, and F_m the magnetic force. $F_{r.x}$ and $F_{r.y}$ refers to the static yield force which exist only in the organism. F_r is the resultant force used in the empirical determination of F_m for the magnet. F_r equals the force due to drag (F_f).

4. CALCULATIONS

Coordinate System

The origin (X=Y=Z=0) is located at the specimen point as defined by the microscope. The X-axis coincides with the central axis of the magnetic pole pieces on either side of the specimen (Fig. 2). X increases toward the left magnet. The Y-axis is defined by the vertical axis and increases in the direction of the force due to gravity. The Z-axis is defined by the optic axis of the microscope and increases in the direction towards the examiner, or camera (Fig. 2).

Force Balance on a Sphere

The major forces to be considered on a sphere are shown in Fig. 3:

$$F_g = \frac{4\pi r^3}{3} \rho_b g \ , \qquad \text{gravitational force}$$

$$F_b = \frac{4\pi r^3}{3} \rho_m g \ , \qquad \text{buoyant force}$$

$$F_{\tau.x,y} = \tau_{.x,y}\pi r^2 C \;, \quad \text{static yield force}$$

$$F_f = 6\pi r\eta u \;, \quad \text{drag force on a sphere}$$

$$F_m = 1.185\text{x}10^2(2.2045I^2+23.490I+24.541)r^3 \;, \quad \text{magnetic force}$$

F_g is defined as the force due to gravity and is a function of the radius (r) in cm, density of Hy Mu 800 (ρ_b = 8.74g/cm^3), and the acceleration due to gravity (g = 980cm/s^2). F_b is defined as the buoyant force acting on the sphere which is a function of the radius (r) in cm, density of the surrounding medium (ρ_m = 0.8739g/cm^3, 25°C) and gravitational acceleration (g).

$F_{\tau.}$ is defined as the static yield force in the X ($F_{\tau.x}$) and Y ($F_{\tau.y}$) direction. $\tau.$ is the static yield strength (dyne/cm^2) and C is a geometrical correction for a sphere roughly equal to 1.75 (22). These static yield forces do not exist in Newtonian fluids, but do inside the organism. F_f is defined as the force due to drag and is proportional to viscosity η in poise (P) and velocity u (cm/s). The Reynolds number for spheres in 40cP viscosity standards (Cannon Instruments Co., State College, Pa.) is < 10^{-5} so inertial contributions can be neglected. F_m refers to the magnetic force (in dynes) which is a function of magnet current I (in amperes) and particle radius r (in cm.). This force was empirically determined for the left magnet (26).

Calibration

The magnetic viscoelastometer was calibrated in ASTM certified viscosity standards (Cannon Instruments Co.). Less than 0.002% w/w Hy Mu 800 particles were sonicated for 40 min. (model B-12, Branson Cleaning Equip. Co., Shelton, CT) in calibration chambers (Fig. 4) while connected to a temperature controlled circulatory regulator (model K2, Messgerate-Werk Lauda, Brinkman Instruments, Westbury, NY) adjusted to 25±0.1°C. Cooling coils similarly connected (25±0.1°C) were placed around the chambers during sonication as a precaution.

Force Analysis

A sphere falling with respect to gravity in a Newtonian fluid is governed by the following force relationship (Fig. 3):

$$F_f = F_g - F_b$$

or:

$$\eta = \frac{2r^2 g(\rho_b - \rho_m)}{9u}$$

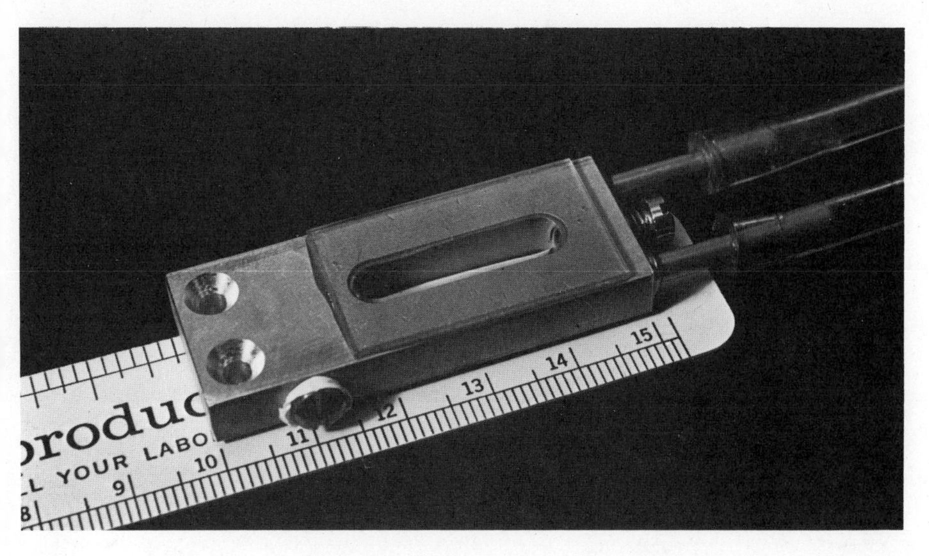

Fig. 4. Calibration chambers made from aluminum blocks were connected to a temperature regulated circulator (25°±0.1 C) (Messgerate-Werk Lauda).

In the organism, a sphere falling parallel to the long axis of the vein in the Y direction experiences this force balance:

$$F_f = F_g - F_b - F_{\tau.y}$$

or:

$$\eta = \frac{2r^2g(\rho_b-\rho_m)}{9u} - \frac{\tau.rC}{6u}$$

The density of the endoplasm has been estimated to be approximately 1.12 g/cm^3 (Belcher unpublished, (21)). The static yield strength in the Y direction $\tau_{.y}$ has been estimated to be approximately 1.05 dyne/cm^2 (26).

Faxen's correction (8) was used to account for ectoplasmic wall effects.

Thus:

$$\eta_y = \frac{\eta}{1-2.108\xi + 2.09\xi^3 - 0.95\xi^5}$$

η_y is defined as the Faxen corrected viscosity in the Y direction where r and R_c are the radii of the sphere and vein, respectively, and $\xi=r/R_c$.

Magnetic Force Determination

The force balance on a particle with the magnet on, in viscosity standard (Cannon Instruments Co.) is (Fig. 3):

$$F_m = (F_g - F_b)\cot\theta$$

Magnetic force is also equal to:

$$F_m = K'VB\frac{dB}{dx}$$

with V = particle volume (cm^3), B = magnetic field strength (gauss), and dB/dx = the gradient of the magnetic field strength (gauss/cm). K' is a constant representative of the magnetic properties of the nickel alloys.

Thus: $$(F_g - F_b)\cot\theta = K'VB\frac{dB}{dx}$$

This equality was used to derive the empirical formula characteristi for the left magnetic coil (26):

$$F_m = 1.185\text{x}10^2(2.2045I^2 + 23.490I + 24.541)r^3$$

Magnetic Determination of Yield Strength ($\tau_{.x}$)

In the magnetic *in vivo* experiments, the following force relationship occurs on a magnetic sphere (Fig. 3):

$$F_f = F_m - F_{\tau.x}$$

Particles less than 3.7μm in diameter are usually stationary with respect to a reference granule, in which case F_f=0. Thus:

$$F_m = F_{\tau.x}$$

or: $$\tau_{.x} = \frac{1.185\text{x}10^2(2.2045I^2 + 23.490I + 24.541)r}{\pi C}$$

5. RESULTS

Magnetic Particles

Annealing of the -400 sieve mesh fraction of the Hy Mu 800 powder yielded spheroidal particles with increased magnetic permeability. Measurements *in vivo* were taken with nearly perfect spheres only, as examined through the microscope.

Calibration

Hy Mu 800 particles suspended in ASTM certified viscosity standards (Cannon Instruments Co.) in temperature regulated calibration chambers (Fig. 4) were used to characterize the response of the magnet to current input. The magnets were driven with a DC regulated power supply (model 6274B, Hewlett Packard, Lexington, Mass.) in the current regulated mode. The magnetic field strength (B) and gradient dB/dx were measured as a function of current and distance with a Bell 640 gaussmeter (Bell Inc., Columbus, OH) and found to be constant 2mm off the X-axis at the specimen point.

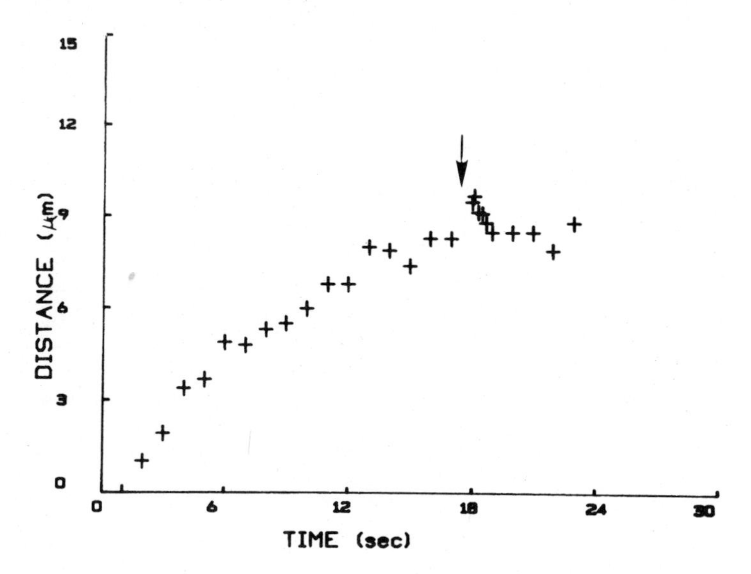

Fig. 5. The trajectory of a 3.9μm diameter sphere pulled in the X direction with $F_m = 9.235 \times 10^{-8}$ dyne is characteristic of a Maxwell model. The arrow points to the time at which F_m was cut.

This is primarily a consequence of the large pole piece to specimen separation distance of 1.2cm. Through the use of the calibration chamber, the magnetic particle technique showed a ±10% variation between particles in a Newtonian fluid. This variability is probably the result of heterogeneities in composition between particles. For the same particle in a Newtonian fluid pulled with a given magnetic force (F_m), the variability of the technique is of the order of ±1%.

Viscoelasticity In Vivo

A 3.9μm diameter particle stationary in both X and Y directions with respect to an endoplasmic reference granule was observed to move towards the left magnet when a magnetic force (F_m) of 9.235×10^{-8} dyne was applied. The trajectory was plotted versus time in Fig.5. The particle was seen to exhibit an elastic recoil on termination of the magnetic force. The form of this curve is reminiscent of a coiled spring in series with a dashpot: a Maxwell element. Therefore, the endoplasm appears to have viscoelastic properties beyond the yield point.

Structural Anisotropy

Table I lists 8 different particles, including the particle of Fig. 5, which were stationary in both X and Y directions until pulled in the X direction with the corresponding F_m. The F_y column lists the corresponding force in the Y direction. In every case but one, $F_y > F_m$; yet, the particle was stationary with respect to a reference granule until the application of the lesser F_m.

The F_m applied was used to calculate the static yield strength in the X direction (τ_x). For the eight particles, τ_x was calculated to be 0.58 ± 0.13 dyne/cm^2. τ_y has been estimated to be 1.05 dyne/cm^2 based on the observation that particles of 3.7μm diameter do not fall parallel to the long axis of the plasmodial vein with respect to a reference granule (26). This difference in static yield strength, together with the force relationship of Table I is evidence for stuctural anisotropy in the endoplasm.

Viscosity

Falling ball viscometry was performed _in vivo_ on balance pressure stabilized endoplasm. The particle of Fig. 6 was 17.4μm in diameter and followed from 18(min):39.53(sec) to 29: 04.99. The particle was manoeuvered to the center of the vein (70μm diameter) with the magnets and allowed to fall in the Y direction, parallel to the long axis of the vein. The velocity was derived off the monitor and used to calculate Faxen's corrected viscosity (η_y) as detailed in the METHODS.

The viscosity values (η_y) were found to fluctuate more than three fold (1.3-4.2P) in magnitude over a ten minute span. The sphere was centered with respect to the vein with the magnets previous to each run. The fluctuations were reproducible (26) but were not regular enough to deduce a pattern or warrant the calculation of a period.

A dynamoplasmogram was initiated at 29:00.00 for the same specimen and followed for ten minutes. It was not possible to generate viscosity measurements and dynamoplasmographs simultaneously. Both Fig. 6a and 6b were generated by fitting the data points (denoted by "+") to a curve using a cubic-spline interpolation algorithm (15). The dynamoplasmogram appeared to have a period of 78.75 sec. as calculated for 4 shuttle streaming cycles. The pressure and viscosity changes with time were compared, under the assumption that the frequency content of the dynamoplasmogram remained relatively unchanged throughout the experiment. Cautious attempts at frequency correlation between these two curves were made, but no obvious relationships were discernable. Our initial impression from Figures 6a and 6b is that the viscosity and shuttle streaming cycles seem to occur

Table 1
Static Yield Strength *In Vivo*

diameter (μm)	$\tau_{.x}$ (dyne/cm^2)	F_m(dyne)	F_y(dyne)
2.0	0.532	2.925×10^{-8}	3.13×10^{-8}
2.5	0.654	5.62×10^{-8}	6.11×10^{-8}
2.5	0.7799	6.7×10^{-8}	6.11×10^{-8}
2.7	0.394	3.95×10^{-8}	7.7×10^{-8}
2.8	0.639	6.89×10^{-8}	8.58×10^{-8}
2.9	0.537	6.209×10^{-8}	9.54×10^{-8}
3.3	0.675	1.01×10^{-7}	1.405×10^{-7}
3.9	0.444	9.235×10^{-8}	2.32×10^{-7}
	mean: 0.58 ± 0.13		

independently from one another. However, it was not feasible to generate the simultaneous data required to substantiate this.

6. DISCUSSION

The magnetic particle technique as applied to *Physarum* allows both falling ball viscometry (7,9,22) in the Y direction and a similar magnetic procedure in the X direction (26). The microscopic probe size allows for the analysis of minute sample volumes in the microliter range, and is applicable to large syncytial systems like *Physarum polycephalum*. The endoplasm has been shown to reproducibly manifest a static yield strength ($\tau_{.x}$) on the order of 0.58 ± 0.13 dyne/cm^2. The observation that these particles did not move with respect to an endoplasmic reference granule before the application of F_m even though $F_y > F_m$ is evidence of structural anisotropy. Nakajima and Allen (24) have observed a puzzling negative birefringence in the endoplasm of plasmodial veins which may correlate with these results.

The trajectory of a 3.9μm particle (Fig. 5) is reminiscent of the behavior of a Maxwell model in creep experiments. Thus, the endoplasm appears to be a viscoelastic material beyond its yield point.

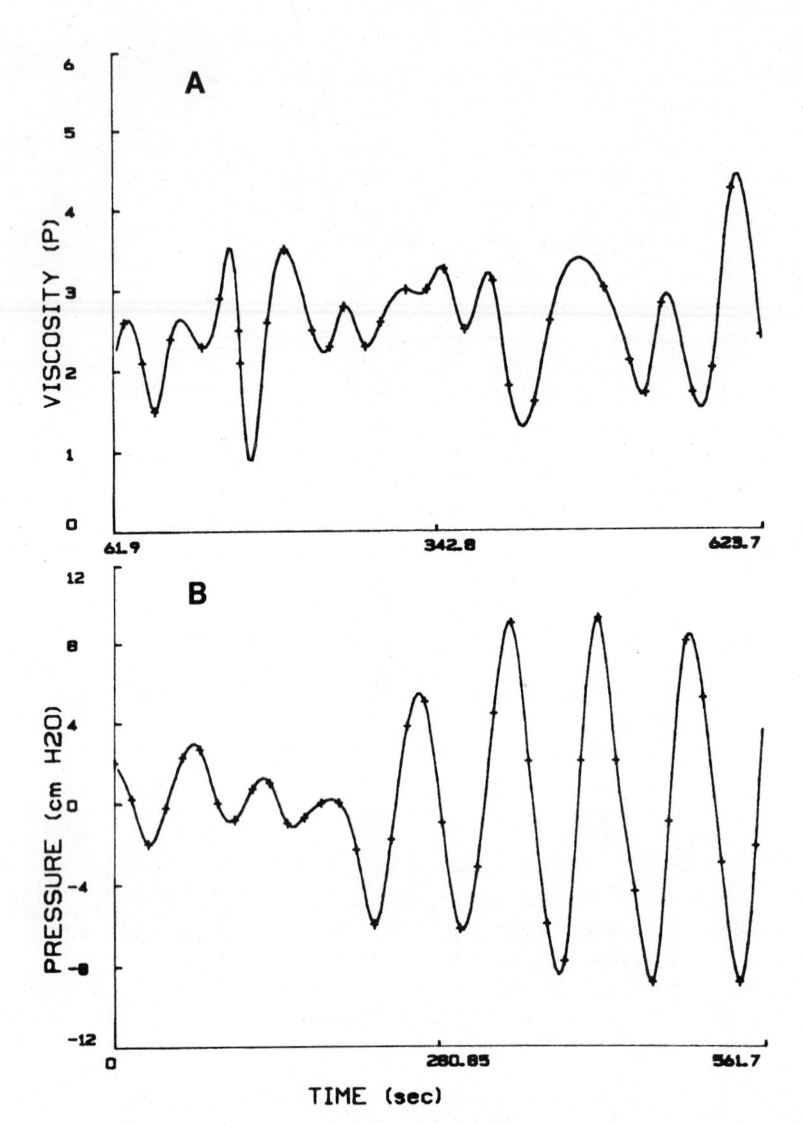

Fig. 6. (A) shows the variation of Faxen corrected viscosity (η_y) in the Y direction as a function of time. The abscissa is from 61.92 to 623.7 sec. (B) is the dynamoplasmogram for the specimen in (A). The ordinate refers to the cm of H_2O of balance pressure required from the manometer to stabilize streaming. The abscissa is from 0 to 561.75 sec. Note that the abscissas in (A) and (B) do not correspond in real time.

Viscosity in the Y direction was probed by the falling ball method (7,9,22). Faxen's correction (8) was used to account for the effect of the ectoplasm on particle velocity. The value of η_y was observed to fluctuate from 1.3 - 4.2P over a ten minute span.

Faxen's correction (8) and the geometrical correction C may not be applicable to the non-Newtonian Physarum endoplasm; therefore, the viscosity η_y and yield strength τ_x values may be subject to further correction. In viscoelastic materials, viscosity is often a complex function of shear stress; thus, a shear stress versus rate of shear curve should be generated for better characterization of the endoplasm. Such curves have been derived in both X and Y directions, as detailed in (26).

The η_y fluctuations were observed in the middle region of the plasmodial vein, using the same specimen with the same particle. The endoplasm was maintained in approximately the same state during the measurements, which occurred at approximately 30 sec. intervals for 10 minutes. Thus, the rheology of the endoplasm would appear to be capable of spontaneous changes and may not be wholly passive in relation to the ectoplasm as is generally believed (17 - 20).

The value of the static yield strength and viscosity measurements are based on the assumption that the endoplasmic granules do not behave differently from the endoplasmic (cytoplasmic) ground substance. This may not be the case as there is evidence to the contrary (23). Laser doppler spectroscopy may have the potential to elucidate the behavior of the true endoplasmic ground substance. This information is important in clarifying the results presented here.

Our purpose has been to present some preliminary results of an attempt to characterize Physarum endoplasm rheology in vivo with a calibrated direct measurement technique. The variability of the magnetic particle technique is significantly below that exhibited by the organism. The ±10% variability of our technique is primarily attributable to differences among the magnetic particles rather than the magnetic particle technique itself; measurements in a Newtonian fluid using a given particle were quite reproducible (±1%). The technique is also minimally disruptive to the cell, as phagocytized particles have been found in the endoplasm two days later with no adverse effects on viability.

The rheology of the endoplasm of Physarum polycephalum has been shown to be more complex than had previously been assumed (11,13,27). The endoplasm has been shown to be a complex viscoelastic material with a yield point capable of changing its properties several fold within tens of seconds.

REFERENCES

1. R.D. Allen and N.S. Allen, Protoplasma 109:209 (1981).
2. E.N. Brewer and A. Prior, Phy. Newsletter 8:45 (1976).
3. D.E.S. Brown and D.A. Marsland, J. Cell Comp. Physiol. 8:159 (1936).
4. R. Chambers and E.L. Chambers, "Explorations into the Nature of the Living Cell", (Harvard Univ. Press, Cambridge, MA, 1961).
5. J.A. Cooper and T.D. Pollard, Methods in Enzymology, in press (1982).
6. F.H.C. Crick and A.F.W. Hughes, Exp. Cell Res. 1:37 (1950).
7. D.A. Cygan and B. Caswell, Trans. Soc. Rheology 15:663 (1971).
8. H. Faxen, Arkiv. Mat. Astron. Fyzik 17:1 (1922).
9. R.H. Geils and R.C. Keezer, Rev. Sci. Instrum. 48:783 (1977).
10. E.N. Harvey and D.A. Marsland, J. Cell Comp. Physiol. 2:75 (1932).
11. A. Heilbronn, Jahrb. f. Wissensch. Botan. 61:284 (1922).
12. L.V. Heilbrunn, "The Dynamics of Living Protoplasm", (Academic Press, New York, N.Y., 1956).
13. L.V. Heilbrunn, Protoplasmatologia II(C1):2 (1958).
14. Y. Hiramoto, Exper. Cell Res. 56:201 (1969).
15. R.W. Hornbeck, "Numerical Methods", (Quantum Press, New York, N.Y., 1975).
16. N. Kamiya, Science 92:462 (1940).
17. N. Kamiya, Protoplasmatologia 8(3a):1 (1959).
18. N. Kamiya and K. Kuroda, Protoplasma 49:1 (1958).
19. N. Kamiya, Y. Yoshimoto, and F. Matsumura, "International Cell Biology ('80-'81)", H.G. Schweiger Ed., (Springer-Verlag, Berlin, 1981).
20. H. Komnick, W. Stockem, and K.E. Wohlfarth-Bottermann, Int. Rev. Cytol. 34:169 (1973).
21. H. Leontjew, Zeitschr. f. vergleich Physiol. 7:195 (1928).
22. S.D. MacLean-Fletcher and T.D. Pollard, J. Cell Biol. 85:414 (1980).
23. R.V. Mustacich and B.R. Ware, Protoplasma 91:351 (1977).
24. H. Nakajima and R.D. Allen, J. Cell Biol. 25:361 (1965).
25. T.D. Pollard and J.A. Cooper, Methods in Enzymology, in press (1982).
26. M. Sato, T.Z. Wong, and R.D. Allen, in preparation.
27. W. Seifriz, Brit. J. Exper. Biol. 3:1 (1924).
28. K. Yagi, Comp. Biochem. Physiol. 3:73 (1961).

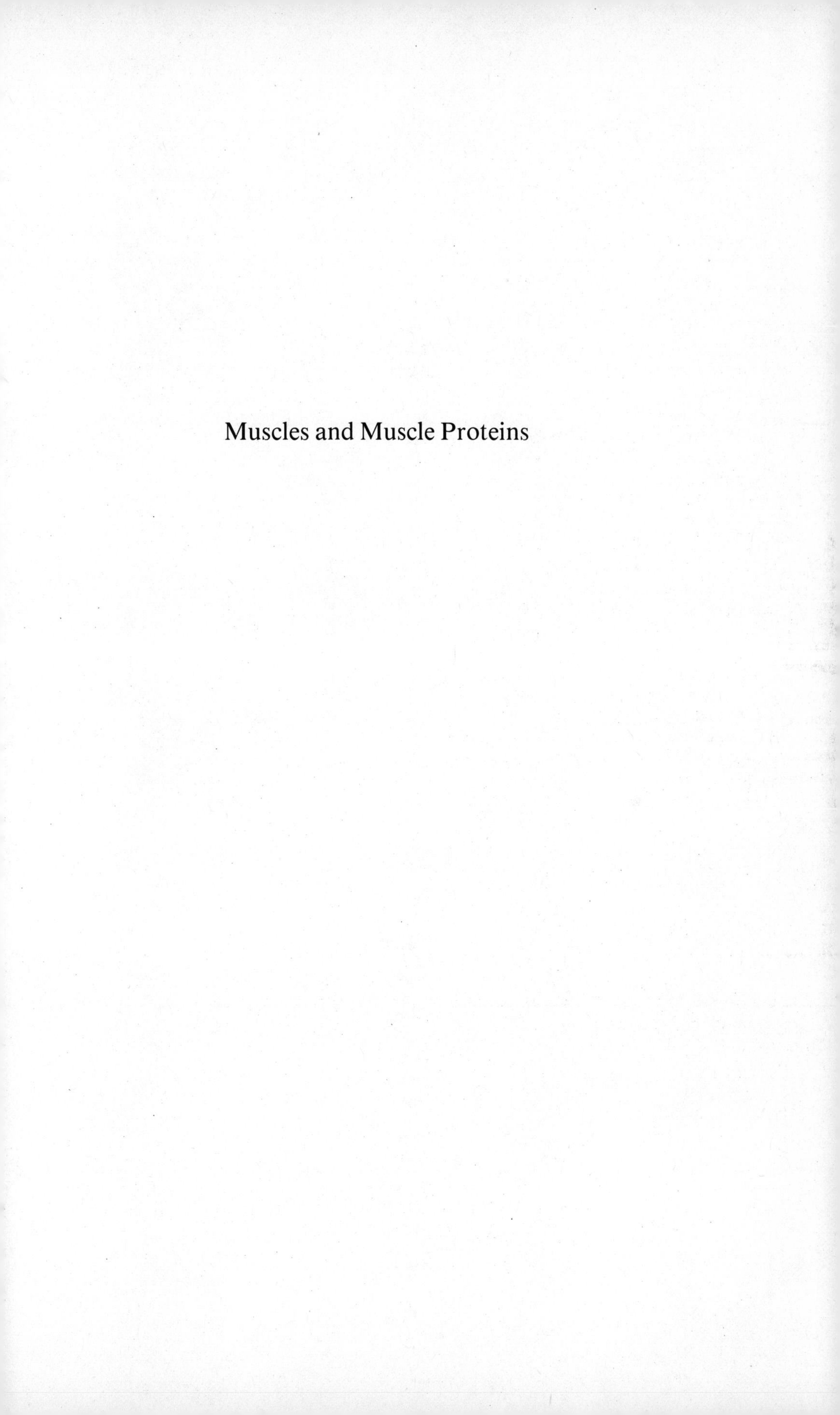

Muscles and Muscle Proteins

THE APPLICATION OF QUASI-ELASTIC LIGHT SCATTERING TO THE STUDY OF MUSCULAR CONTRACTION

Francis D. Carlson

Dept. of Biophysics
The Johns Hopkins University
Baltimore, Maryland 21218

CONTENTS

ABSTRACT

In these lectures, our primary focus will be on the phenomena of biological contractility and on how photon correlation spectroscopy can be used to study the basic molecular processes from which these phenomena arise. Photon correlation spectroscopy (PCS) permits the measurement of macromolecular diffusional relaxation times ranging from 10^{-6} to 10^{2} seconds and these measurements can be made with great accuracy on small quantities of material in solutions, in gels, and in intact cells. Much has been learned about the structural dynamics of contractile proteins and organelles from photon correlation and transient electric birefringence (TEB) studies. The internal structural dynamics of contractile cells have also been successfully studied with PCS.

However, the condensed state of the contractile proteins within the cell and the complexity of their interactions makes the interpretation of PCS data singularly difficult. PCS studies by themselves are not sufficient to lead to a solution of the problem of muscular contraction. On the other hand, when combined with time resolved X-ray diffraction studies and nuclear magnetic resonance studies, PCS studies can be highly useful in narrowing the field of possible models for the contractile mechanism.

INTRODUCTION

We shall begin this review in section 2 with a detailed description of the structure, physiology and biochemistry of striated muscle as it is understood today. Our current view of the structure of muscle is derived from optical studies on intact muscle, optical and electron microscopic studies on fixed muscle cells, and small angle and wide angle X-ray diffraction studies on intact muscle cells in various physiological states. The structure of the muscle cell's organelles and their constituent molecules, as revealed by both electron microscopy and standard physical chemical techniques, will be reviewed. The physiological properties of striated muscle, the characteristics of its ability to develop tension and shorten, will be covered and related to the structural characteristics of muscle. A survey of the biochemistry and chemical energetics of contraction and a presentation of the Huxley Simmons model of the cross bridge cycle completes Section 2. In Section 3, the results of PCS and related studies on isolated contractile proteins and organelles are reviewed and discussed in detail. Finally, Section 4 treats exclusively the results of recent PCS studies on single striated muscle fibers. Discussions of experimental technique, data analysis, and interpretation of results are presented throughout the review.

STRUCTURE, PHYSIOLOGY, AND BIOCHEMISTRY OF STRIATED MUSCLE

Structure

The contractile proteins of striated muscle cell are organized in highly crystalline arrays. This has made possible the use of optical, electron microscopical and X-ray diffraction techniques to determine in detail the structure of muscle and to characterize the structural changes which muscle undergoes when it passes from one physiological state to another. Muscle can exist in three physiological states: the resting state, the contracting or active state, and the rigor state. Along with the different physiological properties, implicit in the very names of these states, there are structural and biochemical differences.[1,2,3,4,5,6,7]

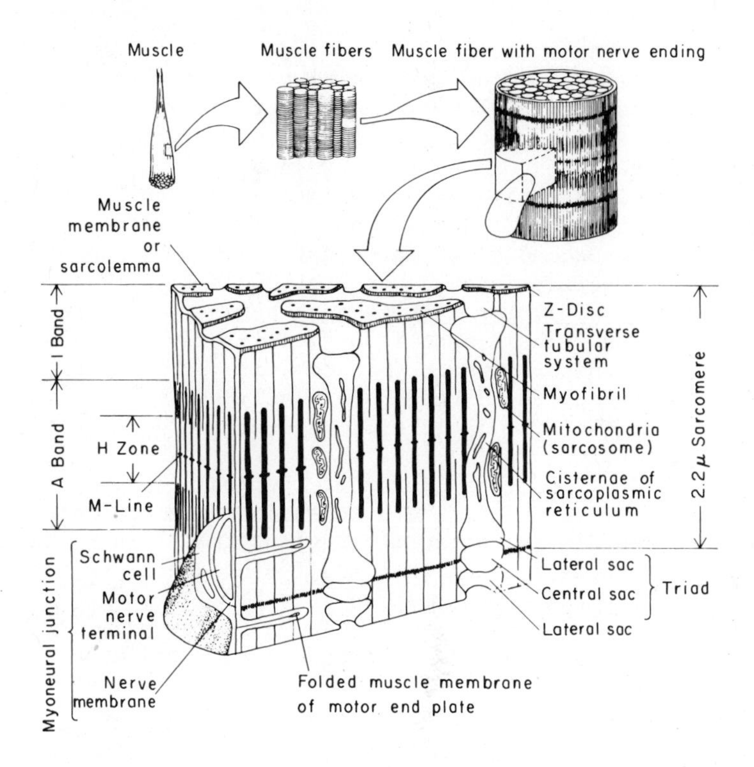

Figure 1. Enlarged section of a single striated muscle fiber showing the myofibrillar system, the transverse tubular system, the sarcoplasmic reticulum, and the myoneural junction.

A single muscle fiber is the basic cellular unit of the voluntary mucle. Under the light microscope the muscle fiber shows a characteristic banded or striated appearance. The repeating unit in the pattern of striations is called the sarcomere. The muscle fiber is also subdivided longitudinally into myofibrils. The fiber itself has a diameter of 50-100 um (um designatates micron throughout text) and the myofibrils have diameters of 1 um. A muscle fiber contains several thousand myofibrils. The striations of muscle are due to refractive index variations along the myofibrils which are aligned in register as shown in Figure 1. The muscle fibers are surrounded by a membrane (sarcolemma), contain soluble salts and proteins, are multinucleated and contain large numbers of mitocondria spaced in rows between the myofibrils. The sarcoplasmic reticulum, a lace-like membraneous network, surrounds the myofibrils and plays an important role in the coupling of excitation to the contractile process of the muscle fiber, a subject which will not be of concern to us. The striated appearance of the muscle fiber is not due to a periodic variation in absorbancy but rather it arises from a periodic axial variation in the refractive index of the fiber. In a whole muscle fiber preparation this refractive index variation is strong enough to refract light out of the objective aperture of a microscope and give the muscle a banded or striated appearance. The alternating high and low refractive index bands are referred to as the A and I bands (discs) respectively. A thin dark band, the Z-line (disc), divides the I-band. The A-band is optically anisotropic and shows positive uniaxial birefringence that contains both a form and intrinsic component. The I-bands are weakly optically anisotropic. The repeating unit of the myofibril, the sarcomere, is the structural unit between one Z-line and another. The Z-line is highly refractile. Near the center of the A-band there is a region of reduced refractive index, the H-zone, in the center of which there is a narrow highly refractile region, the M-line. The A-band has a higher refractive index than the I-band due to the presence of a higher concentration of protein.

Electron microscopy of thin sections of striated muscle show that the sarcomeres of the myofibrils consist of two sets of filaments. The A-band consists of a set of thick filaments and the I-band consists of a set of thin filaments as shown in Figure 2.

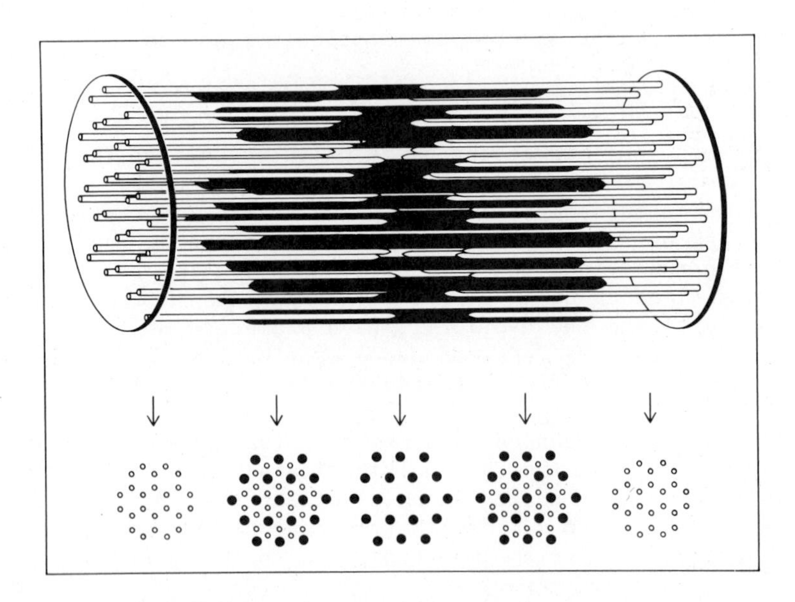

Figure 2. A schematic reconstruction showing the interdigitating thick (shaded) and thin filaments of vertebrate striated muscle.

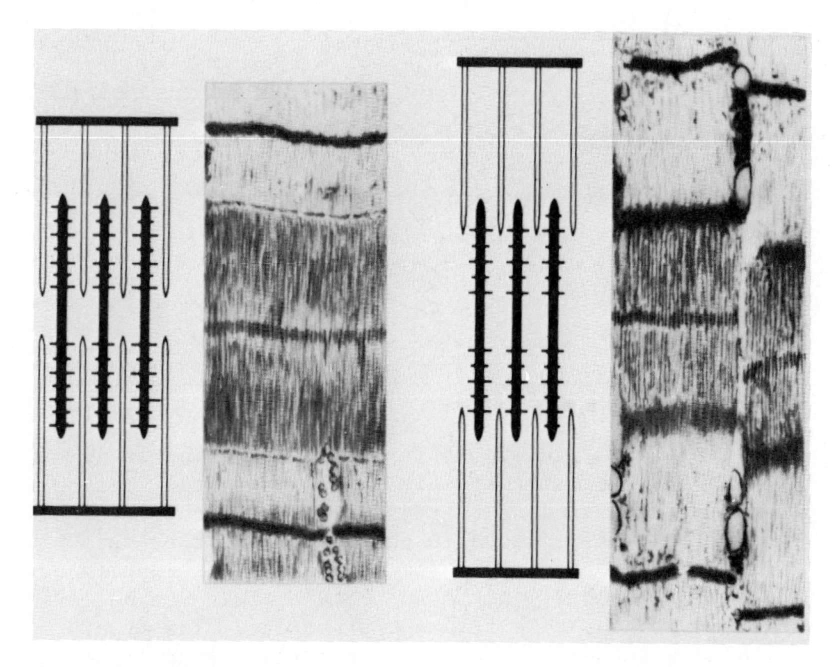

Figure 3. Diagram illustrating the sliding of thick and thin filaments.

The thick filaments are organized on a close packed hexagonal lattice into which the thin filaments interdigitate at the trigonal points (in vertebrate muscle) as shown in Figure 3. If the muscle is stretched appreciably and fixed before viewing in the electron microscope, the thin filaments appear withdrawn from the thick filament array. During this stretch the length of the A-band remains constant while that of the I-band is increased. X-ray diffraction studies on intact muscle fibers have confirmed the existence of the interdigitating thick and thin filaments in live muscle. The ability of the thick and thin filaments to slide past one another is an essential component of the sliding filament theory or model of muscular contraction.

TABLE 1

Relative Proportions of Myofibrillar Proteins in Rabbit Skeletal Muscle

Protein	% of total structural protein
Myosin	55
Actin	20
Tropomyosin	7
Troponin	2
C protein	2
M proteins	<2
α-Actinin	10
β-Actinin	2

An analysis of the protein content of the myofibrils is given in Table 1. Myosin, the major protein in the myofibril, is located exclusively in the A-band. Its physicochemical properties are listed in Table 2. Also found in the A-band are the M protein (in the M-line) and the C protein. The thin filaments contain actin, tropomyosin and troponin. Alpha and beta actinin are believed to be associated with the thin filament structure. The myosin molecule has six subunits, two heavy chains and four light chains as shown in Figure 4. The molecule can be cleaved proteolytically to give subfragments called light meromyosin, heavy meromyosin, and the S-1 and S-2 subfragments as indicated in Figure 4. A single

TABLE 2

Structural Features of Rabbit Myosin

Molecular weight	460,000±10,000 daltons
Radius of gyration	470±30Å
Total length	1400–1500Å
Length of tail	1300–1400Å
Diameter of tail	20±5Å
Diameter of single lobe (S-1 fragment) of head	70–100Å
Number of lobes (S-1 fragments) in head	2
Number of heavy chains	2
Number of light chains (2 DTNB + A-1 + A-2)	4
Molecular weight of A-1 light chain	21,000 daltons
Molecular weight of A-2 light chain	17,000 daltons
Molecular weight of DTNB light chain	18,000 daltons

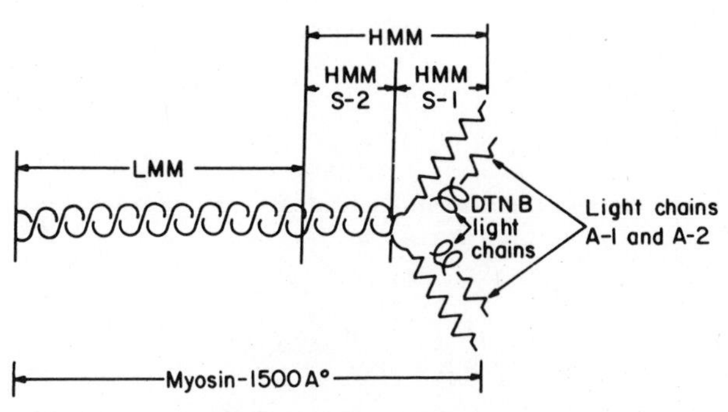

Figure 4a. Schematic drawing of myosin molecule showing subunit structure consisting of two heavy chains and four light chains. The heavy chains are shown as two intertwined coils separated at one end. Also indicated are the subfragments of myosin.

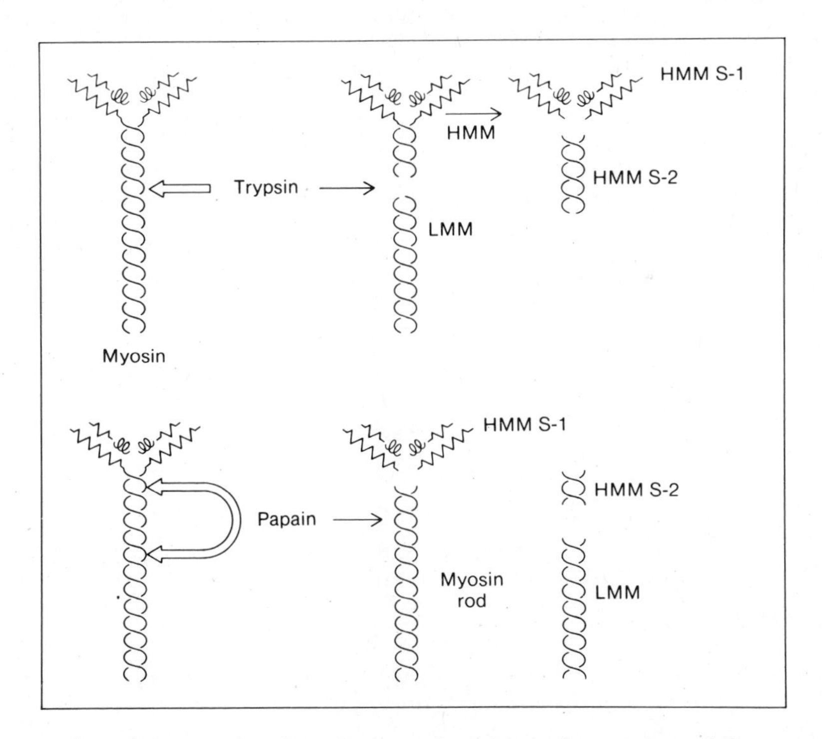

Figure 4b. Scheme showing the production of myosin subfragments by treatment with the proteolytic enzymes trypsin and papain.

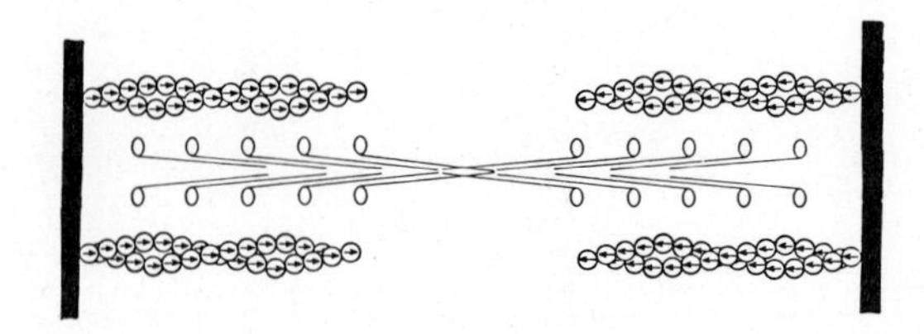

Figure 5. Diagram of thick and thin filament structure showing polarity of structure.

thick filament contains 300 to 400 myosin molecules packed side by side in an as yet unknown way. There is a reversal of the head-to-tail sense of the myosin molecules in the two halves of a thick filament about the midpoint as shown in Figure 5. The heavy meromyosin parts of the myosin molecules give rise to a rough surface along the thick filament except in the central region where there is a bare central zone free of the S-1 heads of myosin. The protruding S-1 elements of the heavy meromyosin subfragment are called cross-bridges. M proteins and the C protein are distributed in narrow bands along the thick filament and are believed to connect single myosin molecules together to form the close packed hexagonal array that makes up the thick filament. Synthetic thick filaments can be made from pure myosin.

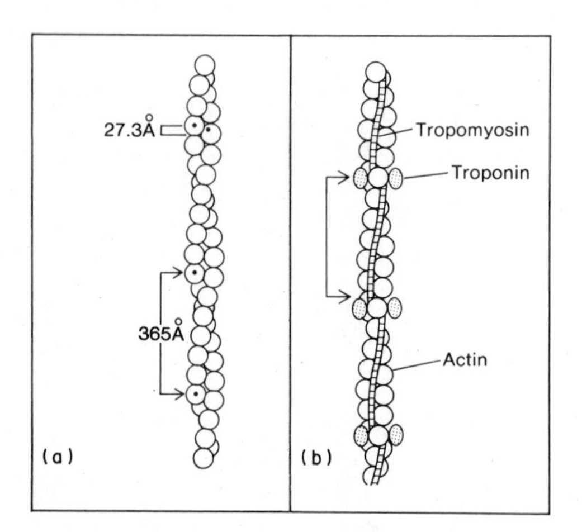

Figure 6a. F-actin structure showing arrangement of G-actin monomers as deduced from X-ray diffraction and electron microscopy studies.

Figure 6b. Highly schematic model of thin filament showing bound tropomyosin and troponin. There is one tropomyosin and one troponin for every seven G-actin monomers.

The arrangement of tropomyosin and troponin along the F-actin strand of the thin filament is shown in Figure 6. The F-actin itself consists of subunits called G-actin. When isolated in solutions of low ionic strength, G-actin has a molecular weight of 41,700 daltons and contains a single adenosine triphosphate (ATP) molecule bound to the protein. At high ionic strengths G-actin polymerizes to produce F-actin with hydrolysis of ATP to ADP. The thin filaments have a sense or polarity which is reversed on either side of the Z-line as shown in Figure 5. The myosin molecule reverses its polarity about the midpoint of the bare zone of the thick filament. Thus, the stereochemical relationships between single myosin molecules in the thick filament and single actin molecules in the thin filament are preserved while their polarities are reversed about the midpoint of the sarcomere. This reversal of structural organization of the myosin and actin molecules orients these molecules about the midpoint of a sarcomere in such a way that the force produced by their stereospecific interactions results in a net contractile component along the fiber axis.

The Physiological States of Striated Muscle

In the resting or relaxed state striated muscle exhibits a very low rate of oxygen utilization and ATP turnover. At sarcomere lengths near 2.0 to 2.4 um, it is readily extensible but becomes less so for longer sarcomere lengths as shown in Figure 7. This high degree of extensibility about the muscle's normal operating length is functionally important. Muscles can only contract so they operate in antagonistic pairs about a joint to produce limb movement. The fact that one pair relaxes while the other contracts means that the relaxed muscle offers little resistance to rotation of the limb. The coordination of antagonistic pairs of muscles to produce purposeful motion of the trunk and limbs is one of the major functions of the brain.

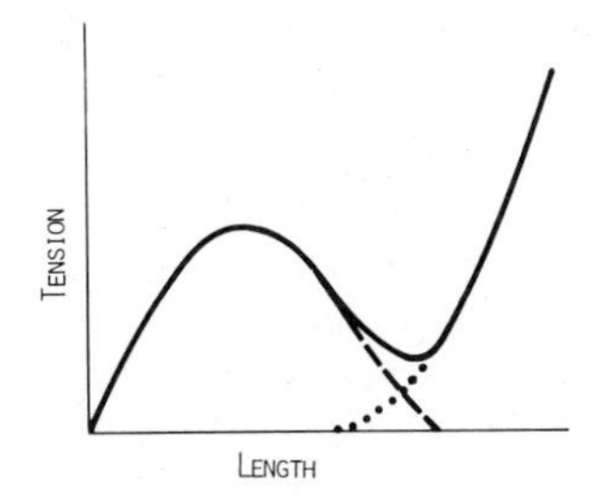

Figure 7. Schematic representation of tension on sarcomere length in vertebrate striated muscle. Total tension, solid curve. Relaxed tension, dotted curve. Tension developed, dashed curve.

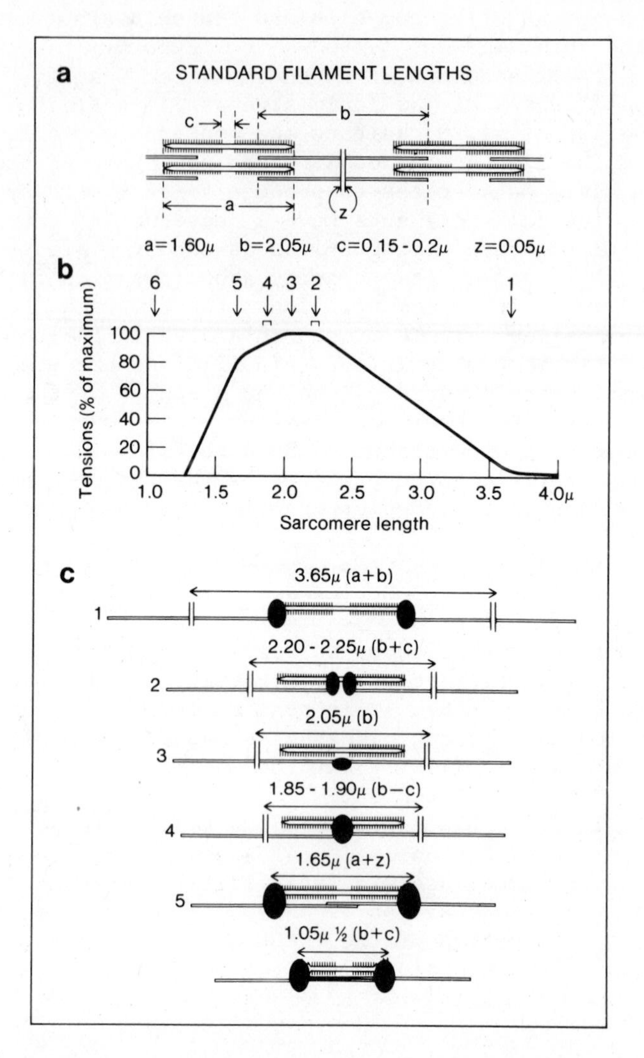

Figure 8. (a) Standard filament lengths. (b) Tension-length curve from part of a single muscle fiber (schematic summary of results). The vertical arrows along the top show the various critical stages of overlap that are portrayed in (c). (c) Critical stages in the increase of overlap between thick and thin filaments as a sarcomere shortens. The numbers on the left refer to the various discontinuities in (b) above. Tinted oval areas indicate regions of critical behavior of the thick and thin filaments. [From A.M. Gordon, A.F. Huxley, and F.J. Julian (1966), J. Physiol. 184, 170-192.]

When constrained at its ends, a stimulated muscle will contract isometrically and develop tension. If not constrained it will shorten and do work. A brief stimulus produces an impulse of force called a twitch. Repetitive stimulation at sufficiently high frequencies produces a smooth sustained contraction called a tetanus. In both the twitch and the tetanus there is a substantial increase in the tension in the muscle, the amount of which depends on the sarcomere length, or more precisely, on the amount of overlap between the thick and thin filaments. For sarcomere lengths above 2.05 um the active tension developed during contraction falls off linearly with thick and thin filament overlap as shown in Figure 8. The linear decrease in active tension with the decrease in the length of the region of overlap between thick and thin filaments suggests that the contractile force is produced only in the region of overlap of these filaments. This is the region in which the myosin cross bridges are in sufficiently close proximity to the thin filaments of actin for an interaction to occur.

In the rigor state, muscle is stiff, highly inextensible, and completely devoid of ATP and produces no heat. If, as can be done with excised muscle, an ATP generating system is provided, the rigor state can be reversed and the muscle can be made to relax or contract by controlling the amount of calcium ions present. In the rigor state virtually all of the cross bridges are tightly and rigidly, but not covalently, bound to the thin filaments. This tight cross linking of the thick and thin filaments to one another endows the muscle with a high degree of stiffness and inextensibility that is characteristic of the rigor state.

Biochemistry and Energetics of Muscular Contraction

Biochemical and thermochemical studies of intact muscle and isolated contractile proteins have given us a detailed but not yet complete picture of the chemistry of contraction. The contractile proteins, myosin and actin, convert the energy derived from the hydrolysis of ATP to adenosine diphosphate (ADP) into work through a complex sequence of chemical events shown in a simplified form in Figure 9. Although the myosin molecule binds ATP and is capable of hydrolyzing this molecule in solution, the hydrolysis rate of ATP is extremely low under the conditions found in relaxed muscle. However, upon stimulation of the muscle, calcium ions are released from the sarcoplasmic reticulum and bind to the troponin molecules that are attached to the thin filament. The positions of troponin and tropomyosin then change and expose actin elements of the thin filament to the ATP-activated S-1 head of the myosin molecule (cross-bridge). This in turn allows the attachment of an active cross-bridge to an available actin site in the thin filament. The already hydrolyzed ATP, still bound to the attached cross-bridge, is then rapidly released as ADP and phosphate ion and tension is developed by the attached cross-bridge in a manner not yet fully understood. It has been suggested that tension development is the result of a rotation of the S-1 elements about their points of attachment to actin.[8,9] After the cross-bridge produces its tension impulse and releases ADP and phosphate ion it remains attached to actin in the rigor configuration. The ADP released from the myosin is converted into ATP by the phosphorylating system found in the sarcoplasm. ATP from the sarcoplasm rapidly diffuses to the rigor type actomyosin cross-bridge where it is bound and causes the dissociation of myosin from the actin filament. The dissociated myosin molecule then hydrolyzes the bound ATP and, undergoes a conformational change while retaining ADP and phosphate ion in a bound or activated state. The activated myosin cross-bridge can then rebind to a free actin site and repeat the process cyclically, delivering a brief impulse of tension to the thin filament on each cycle. One ATP molecule is hydrolyzed for each cross-bridge in a cycle and the asynchronous cycling of the many cross-bridges during a tetanus gives rise to a steady tension or shortening depending upon the constraints applied to the muscle.

The biochemical scheme outlined here has been developed for the most part from studies of biochemical reaction kinetics conducted on myosin that has been made soluble by enzymatic fragmentation. As we have already seen, myosin in the intact muscle exists in a highly condensed, nearly crystalline state. It is not at all clear that conclusions drawn from studies of reaction kinetics in solution on fragmented myosin molecules apply to the intact muscle cell. Indeed, the highest ATP hydrolysis rates that have been obtained on solubilized myosin are an order of magnitude or less than the rates of ATP splitting that have been observed in intact muscle during a tetanus.

Models of Contraction

Several models have been postulated for the origin of the contractile force. The earliest[9] requires that when attached to the thin filament the S-1 subfragment produces a couple that causes rotation about the S-1 - S-2 junction. To keep the tension of a cross-bridge independent of interfilament spacing it was assumed that the HMM and LMM subfragments were joined by a hinge as shown in Figure 9b. Numerous modifications of this model have been proposed but all retain rotational freedom for S-1 about its junction to S-2. More recently,[10] a more elaborate model endows the S-2 element with the elastic properties of a spring and the S-1 element several sites for attachment to the actin filament. As the S-1 element changes its attachment site it stretches the S-2 spring thereby transmitting tension to the body of the thick filament. This model includes features that explain some of the transient mechanical behaviour of the attached cross-bridge. Yet another proposal[11] locates the tension developing mechanism in the S-2 element itself through a helix-coil transition.

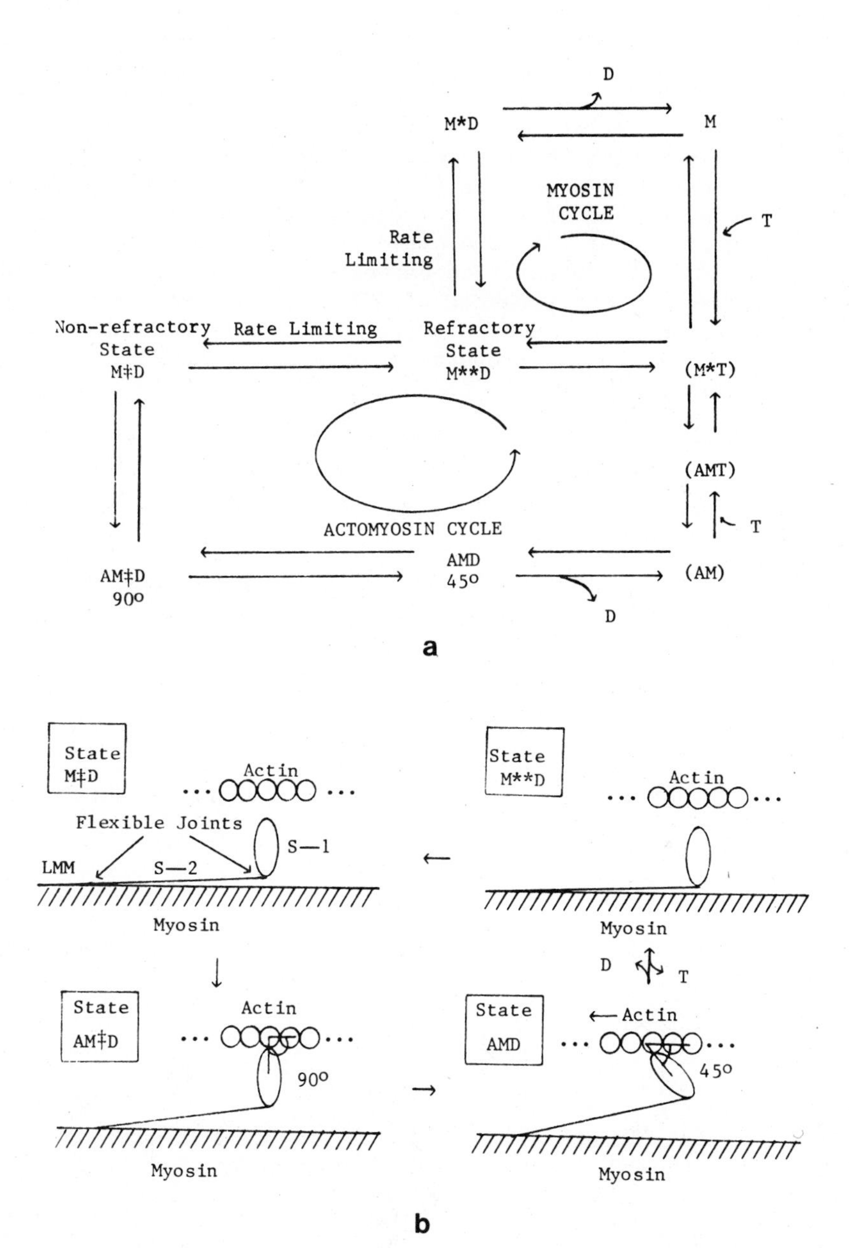

Figure 9a. Highly schematic diagram of sequence of reaction steps in cross-bridge cycle.

Figure 9b. Schematic diagram of cross-bridge cycle.

For our present purposes we can summarize the present state of our knowledge of muscle as follows. The two interdigitating sets of filaments in muscle serve to provide the spatial configurations and mechanical linkages that are necessary to constrain the forces of contraction to the long axis of the muscle fiber. Along with the sarcoplasmic reticulum, the troponin and tropomyosin, bound to the thin filament, comprise the control mechanism that enables the rapid switching, on and off, of the contractile mechanism in response to a nerve impulse. The basic tension developing reaction, however, is due to an interaction between two contractile proteins, actin and myosin, in the presence of ATP, the chemical source of energy which the muscle converts to work. In what follows we shall concentrate on the dynamic properties of the contractile proteins and the structural dynamics of the muscle fiber itself.

STRUCTURAL DYNAMICS OF THE CONTRACTILE PROTEINS

We have discussed the chemical aspect of the mechanochemistry of contraction and found them fairly well understood while the mechanical aspects remain far from clear. We turn now to the mechanical characteristics of the contractile proteins, myosin, actin, and their complex actomyosin and the thick and thin filaments, with the expectation that a knowledge of the structural dynamics of these molecules and organelles is essential to a full understanding of the mechanochemistry of contraction.

According to existing theories, the myosin molecule is not rigid but possesses two hinges, or joint-like regions that endow it with considerable segmental flexibility. Indeed, it is generally believed that the flexibility properties of myosin are an essential part of the mechanism for the conversion of chemical energy into work when myosin interacts with actin. Although the thin filaments show flexibility under some conditions, it is not clear to what extent this mechanical characteristic contributes to the contractile process in muscle, although a mechanism has been suggested.[12] However, F-actin definitely plays a major role in many forms of cellular motility, and so its structural dynamics are of general interest.[13] Here, a brief review of non-hydrodynamic studies of the flexibility of the contractile proteins is presented, followed by a detailed summary of the results of PCS studies on the hydrodynamic properties of these molecules.

FLEXIBILITY OF MYOSIN

All of the models proposed for the mechanism of force development by the myosin molecule endow this molecule with a flexible joint between the S-1 subfragment and the S-2 subfragment and a joint of limited flexibility at the LMM-HMM junction. A variety of studies of myosin and its subfragments have been directed toward obtaining evidence for flexibility at these junctions in the myosin molecule. Generally speaking, these studies are of two types: non-hydrodynamic and hydrodynamic.

Non-hydrodynamic studies of myosin's flexibility have involved the use of such techniques as: Time-Resolved Fluorescence Polarization Anisotropy, Saturation Transfer Electron Paramagnetic Resonance, Electron Microscopy, Chemical Cross-Linking Analysis, and Fluorescence Fluctuation Spectroscopy. All of these techniques require that the particle being studied be chemically modified by the rigid attachment of a suitable label to the S-1 head. The label in the case of the fluorescence studies, called a fluorophore, endows the S-1 head with fluorescence. In the electron paramagnetic studies a spin-label is used. Electron microscopy requires staining or shadowing of the molecule and observation in a high vacuum. Chemical cross-linking introduces chemical cross-links between the myosin molecules in the thick filament. The cross-linked structure is then enzymatically fragmented and the fragment distribution is determined by SDS gel analysis. An extensive review of these non-hydrodynamic studies has appeared recently.[14,15] They can be summarized as follows. In dilute solutions, the S-1 subfragment has an axial ratio between 3.5 and 5 if it is modelled as a prolate ellipsoid. Under the electron microscope S-1 appears as a prolate ellipsoid with axes 15.0 nm by 5.0 nm and having a poorly resolved thin region at the end that joins the rod. S-1 rotational relaxation times increase slightly when S-1 is joined to the rod and electron micrographs of myosin show the occurence of all orientations of S-1 with respect to the S-2 axis. Thus, the S-1 - S-2 junction appears to have the properties of a freely rotatable universal joint.

In myosin filaments, Electron Paramagnetic Resonance studies of the rapid rotational motion of S-1 show that it is considerably slowed on binding to F-actin in the absence of ATP, thus establishing that actin and the S-1 head form a rigid unit in the absence of ATP. Further, the motion of S-1 is coupled to ATPase activity and occurs in intact myofibrils. In the absence of ATP the S-1 head is completely immobilized (rigor state).[16,17,18] These studies also showed that the S-1 angular orientational distribution is broad in the relaxed state and very narrow in the rigor state. However, glycerinated fibers with the S-1 heads labelled with a fluorophore showed no fluorescence fluctuations in S-1 orientation, either in the resting or the rigor state.[19,20] In the contracted state a $2s^{-1}$ rate of fluorescence fluctuation was observed. This decay rate was attributed to the cycle-time for ATP hydrolysis in this preparation.

Early X-ray diffraction studies on live muscle indicated that there was a net movement of X-ray scattering material from the thick filament to the thin filament. This could have been the result of a net change in the average S-1 orientation during contraction.[21,22,23,24] Recent time-resolved X-ray diffraction studies on contracting muscle[25] indicated that in a tetanus the S-1 heads change their tilt (orientation) following a quick stretch or a quick release and recover from these transients with a 50-100 ms time constant. As discussed below, this long recovery time matches that of PCS measurements on contracting single muscle fibers.

Full flexibility in the rod at the HMM-LMM junction is supported by electron microscopy studies.[26,27] Viscometric studies[28] have provided evidence for a temperature sensitive region at the LMM-HMM junction that controls the viscometric properties of rod presumably by melting the α-helical part of the rod in the HMM-LMM region resulting in an increased flexibility.

This observation has formed the basis of a model for the origin of the contractile force in muscle in which the α-helix-to-coil transition is assumed to be the essential process.[29] However, evidence obtained from some of the fluorescence studies suggests that only a limited flexibility exists at the HMM-LMM junction. Chemical cross-linking studies also support the conclusion that flexibility at the LMM-HMM junction is limited.

In summary, the non-hydrodynamic studies indicate that the S-1 - S-2 junction is a freely flexible universal joint. The LMM-HMM junction has a limited flexibility which viscometric studies show to be temperature dependent in a way that is compatible with observed changes in the α-helical content of the HMM subfragment. Since these non-hydrodynamic techniques involve the attachment of fluorophores or spin labels to the S-1 head of the myosin molecule, or other chemical modifications of the myosin molecule, the possibility exists that the presence of the labels may strongly perturb the structural dynamics of the myosin molecule. Since these labels must be rigidly attached to the molecule in order to be reliable indicators of rotational motion, they introduce a new chemical group into the S-1 part of the myosin molecule that may change its internal dynamics.

PCS AND TEB STUDIES OF MYOSIN FLEXIBILITY

We turn now to the results of PCS and TEB studies of the structural dynamics of the myosin molecule . The PCS technique measures fluctuations and is therefore free of the objections that apply to perturbaton techniques. The TEB studies, recently improved, introduce weak perturbations and require no chemical modification of the molecule. In this section we present a detailed review of the PCS and TEB studies on the translational and rotational diffusion constants of myosin, its subfragments, G- and F-actin, and isolated thin filaments. For rigid rods, rotational relaxation times, measured by TEB, vary inversely as the cube of the length of the rod. Furthermore, the rotational relaxation time is strongly dependent upon the flexibility of the rod while the translational diffusion coefficient, measured by PCS, is not. Recent improvements in TEB methodology have made it an extremely powerful technique for studying rotational diffusion.[30]

Sample Preparation For PCS Studies

Reliable PCS measurements can only be made in the absence of large, unwanted contaminants, henceforth referred to as "dust". The introduction of dust into a sample may be the result of accidental contamination of the cuvette, or the solvents used, or as is frequently the case with proteins, denaturation and irreversible aggregation due to any of a variety of causes.

Regardless of origin, the presence of dust with a high light scattering power seriously limits the precision of PCS measurements. There are several procedures available for reducing or eliminating dust from preparations of contractile proteins. There are also ways of collecting and editing data that will reduce the errors that arise in solutions containing small amounts of dust. Usually the dust in a sample diffuses much more slowly than does the substance of interest. In this case $g^{(2)}(\tau)$, can be written as:[31]

$$g^{(2)}(\tau) = 1 + \frac{I_S^2}{\langle I \rangle^2} |g^{(1)}(\tau)|^2 + \frac{2 I_S I_D}{\langle I \rangle^2} |g^{(1)}(\tau)| + \frac{I_D^2}{\langle I \rangle^2} X$$

where I_S is the intensity of the light scattered by the sample particle and I_D is the intensity of light scattered by the dust. The total intensity, $I = I_S + I_D$, and $g^{(1)}(\tau)$ is the normalized field autocorrelation function associated with the scatterer of interest. X is given by: $X = (\langle|h|^4\rangle/\langle|h|^2\rangle^2)-1$ where $h \equiv h(\tau)$ is the complex amplitude of the fluctuating component of the light scattered by the dust alone. There are four cases of interest when the dust diffuses <u>much</u> <u>more</u> <u>slowly</u> than the particle of interest. (1) No dust, $I_D = 0$, and there is no error in $g^{(2)}(\tau)$ due to dust. (2) Scattering from the walls of the container or from a rigid unknown gel phase present in the sample itself, so-called "stationary dust". In this case there is no direct way of determining the intensity of the light scattered from the dust or gel phase and that scattered from the sample particles of interest. (3) If at all times there is a substantial amount of dust present in the scattering volume, then $h(\tau)$ is a Gaussian variable, and X = 1. In this case:

$$g^{(2)}(\tau)-1 = \left[\frac{I_S}{\langle I \rangle} |g^{(1)}(\tau)| + \frac{I_D}{\langle I \rangle}\right]^2$$

and $g^{(1)}(\tau)$ can be recovered from a measurement of $G^{(2)}(\tau)$ and the <u>background</u>. To be sure, this is an unlikely situation, but it can occur in the study of the diffusion of monomers in the presence of a monodisperse polymeric form. (4) When only one or two dust particles occasionally pass through the scattering volume, and the fraction of time that a dust particle is in the scattering volume is small, then the "more-than-Gaussian-dust" situation is obtained. If X >> 1 and if $X \cdot I_D \gg I_S$ then:

$$g^{(2)}(\tau)-1 = \frac{I_S^2}{\langle I \rangle^2} |g^{(1)}(\tau)|^2 + \frac{I_D^2}{\langle I \rangle^2} X$$

In the above expression, the second term on the right is a constant and so $g^{(1)}(\tau)$ can be obtained readily from a determination of $g^{(2)}(\tau)$ and the measured background. In this situation, I_D is much greater than I_S, and the presence of the dust is readily detected as a transient increase in the count rate. Consequently, it is possible to switch-off the data collection when dust is in the scattering volume.[32] This high count rate makes it possible to edit the data and correct for the effects of dust as described below.

All aqueous solutions used in PCS studies should initially be deionized and filtered to produce the equivalent of that produced by the Milli-Q-reagent grade water system (Millipore Corporation, Bedford, MA, U.S.A.). This system produces water which contains particulate matter less than 0.22 um in diameter and has a conductivity equal to that of pure water. This water can be further purified by filtration through a 0.05 um filter (Millipore Corp., Bedford, MA, U.S.A.). Super clean water prepared in this way, and contained in a clean cuvette, exhibits the light scattering properties of pure water, with only an occasional flash due to dust.

The presence of dust in a purified protein sample can be ascertained by viewing the light scattered from the sample at small angles. The presence of bright transient flashes of light indicates that the sample contains dust.

There are several ways of removing dust from the original preparation prior to filling the cuvette. G-actin preparations can be filtered through a 0.08 um filter or centrifuged at 35,000 g for two hours using an angle head rotor. This centrifugation procedure also works well for native thin filaments and F-actin. Myosin must be centrifuged at 65,000 g for two hours before transference to a cuvette. Transference of any protein solution to a clean light scattering cuvette (see below) should be done in a dust-free box. To avoid the formation of large irreversible aggregates of the protein, the sample should be introduced along a side of the cuvette to minimize drop or bubble formation. Before introducing the sample the cuvette should be filled with filtered solvent first, then emptied before filling with the protein solution. At low concentrations, the binding of myosin to glass can appreciably reduce the protein concentration. This difficulty may be avoided by filling the cuvette with myosin solution and allowing it to stand for several minutes so that binding can occur. The cuvette is then emptied and refilled with fresh myosin solution. Binding of a protein to the walls of the cuvette can be monitored simply by measuring the intensity of the light scattered at a fixed angle. If binding to the glass occurs, the intensity of the scattered light will gradually drop to a constant level. Whatever the situation, the final concentration of the protein in the scattering cuvette must be determined either spectrophotometrically or by some other means. It is absolutely essential that the cuvette itself be clean. In our experience cuvettes can be cleaned best by the following sequence of operations: (1) washing in warm detergent (2) rinsing several times with warm distilled water and then with N-propanol (3) inverting and allowing to dry at 50°C and then refluxing in acetone.[33] Micro-pipettes that are to be used in transfering protein solutions should also be cleaned in this way to avoid contaminating the sample with dust. Cleaned cuvettes should be capped immediately and stored under an inverted beaker or other dust shield.

Once the cuvette has been filled as described above, the sample can be checked again for dust by examining the light scattered near the forward direction as described previously. If significant quantities of dust are present, as indicated by this simple test, or by a widely fluctuating count rate when the sample is in the spectrometer, there is one further procedure that may be used. The filled cuvette can be placed in a swinging bucket rotor and centrifuged at 500 g. The stoppered cuvette must be placed in a centrifuge tube adapter that has been fitted with a rubber cushion in the bottom to provide a flat base. In addition, water should be added to the adapter to a level that approximately matches the level of the solution in the cuvette. If this is done correctly, the cuvette is likely to survive several hours of centrifugation and the dust content of the sample will be substantially reduced.

Finally, because both the structural properties of the contractile proteins and their diffusion coefficients depend critically on temperature, it is essential that the scattering cuvette be placed in a well regulated thermal bath ($\pm$ 0.05° C) that allows access to scattered light over at least an angular range from 0 to 150°.[33] Our bath and sample holder are shown in Fig. 10. In conducting PCS studies on contractile proteins it is important to insure that scattering from back reflected beams (beams reflected back along the incident beam at interfaces) are kept at a minimum. For Rayleigh scatterers, the back reflected beam can introduce systematic errors at all scattering angles and for non-Rayleigh scatterers large systematic errors can occur for scattering angles greater than 90°, while at smaller angles the error is less. Of course, at 90° scattering angle, such errors do not arise. However, for most cases of interest, restricting measurements to a scattering angle of 90° would seriously limit the usefulness of PCS studies. For further discussion of these errors see below. Errors due to back reflections can be eliminated by means of well designed beam stops or by tilting the sample cuvette.[33,34] Data collection and editing procedures designed to reduce errors due to dust are effective once the dust has been reduced to fairly low levels, that is, a dust particle occasionally

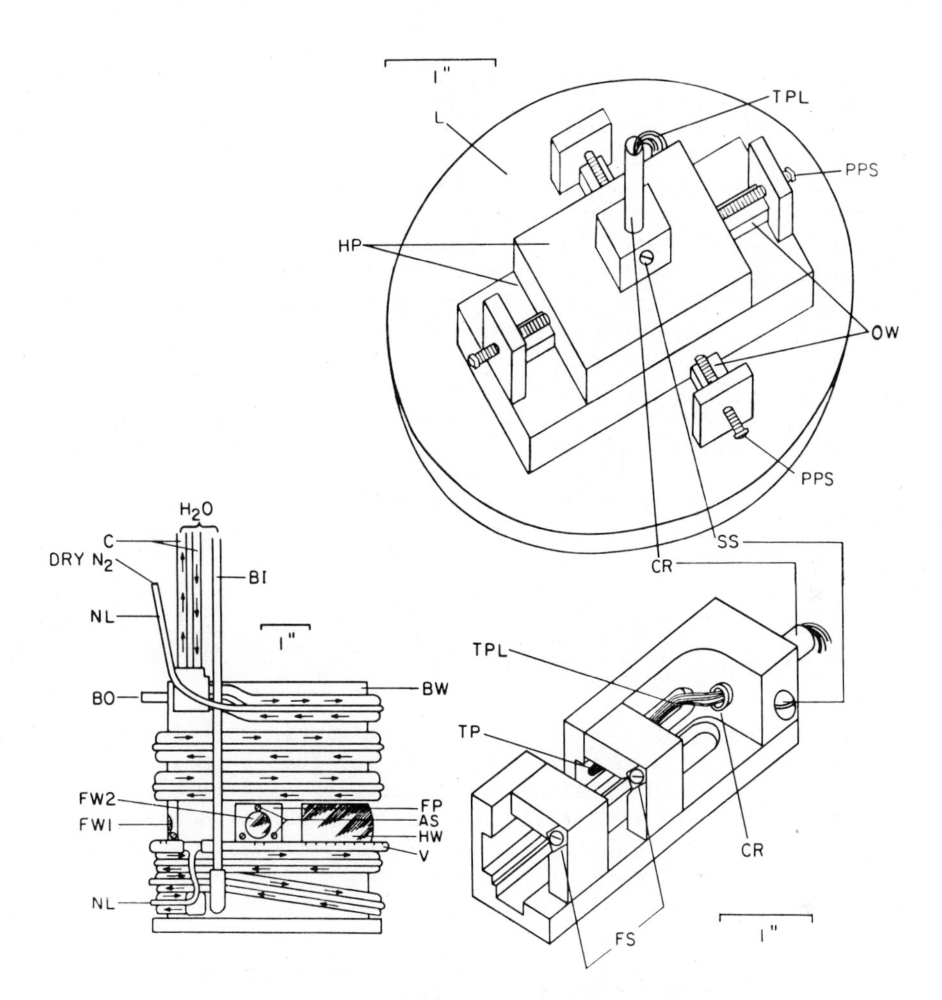

Figure 10 The cuvette is immersed in a water bath that is temperature regulated by circulation of water through coils (C) surrounding the bath wall (BW). Water can also be circulated through the bath itself through an inlet (BI) and an overflow (BO), permitting rapid temperature equilibration and filtering of the bath water to remove dust. The entrance window for the laser is one of two optically flat windows (FW). Scattered light is detected through a hemispherical equatorial zone window (HW). The cuvette is held in the cuvette holder by two phosphor-bronze flat springs (FS). The holder is attached to a translator (HP,OW) on the bath lid (L) by a rod (CR) through which leads of a temperature probe (TP) pass. For further details see reference 33.

drifts through the scattering volume. Data collected in a series of short experiments may be examined by recording the total number of counts in each short experiment. This will reveal those experiments with inordinately high counts. It is presumed that the high counts are due to dust, and samples showing high counts can be eliminated from further data analysis. Editing the data in this way requires that each short experiment be sufficiently long to give valid estimators of the statistical parameters of the short experiment.[35] Collecting data as a batch of several short experiments has the important advantage of enabling each short experiment to be treated as a separate data point and allows an estimation of the standard deviation in $g^{(2)}(\tau)$ for each value of τ. The effects of dust can also be reduced by fitting $g^{(2)}(\tau)-1$ to a single exponential plus a constant, or equivalently, by extrapolating the first cumulant to a zero quality factor, Q, where $Q = 4C/R^2$, C = second cumulant, and R = first cumulant.[36] In our experience, data-editing techniques must be used advisedly, and they are only useful if the sample has been rendered nearly free of dust.

PCS Measurements on protein solutions should always include an independent measurement of the background count and data fitting procedures should include the measured background as a determined quantity and not a parameter to be fit. If a measured $g^{(2)}(\tau)-1$ does not approach the measured background within the limits of experimental error then it must be presumed that there are components of $g^{(2)}(\tau)$ present in the PCS that have time constants much larger than the time-scale of the experimental measurements. Failure to reach the background level can be the case if dust is present, if high polymeric forms are present, if a gel phase is present, or if accidental transient heterodyning occurs. Whatever the cause the knowledge that $g^{(2)}(\tau)$ does or does not approach the measured background is very useful. Fitting the background level should be avoided because it obscures important properties of the system and can produce a systematic error in the observed coherence time.

Hydrodynamic Models from PCS Data

The reason for collecting PCS and TEB data on the contractile proteins is to provide the data needed to develop hydrodynamic models whose translational and rotational coefficients of diffusion in dilute or semi-dilute solutions are equivalent to that of the particle under study. As indicated, the size and general shape of myosin and its subfragments is approximately known from electron microscopy and other physical chemical studies. What is required is a formalism to employ in calculating diffusion coefficients of appropriate models so that a set of shape parameters for the model can be developed that best agree with the PCS data. In its current form, hydrodynamic modeling theory permits the calculation of the translational and rotational frictional coefficients of rigid macromolecular structures by modeling them with an assemblage of small spherical subunits. This technique takes into account hydrodynamic interactions within the same particle. For a rigid particle with a complex shape, the particle can be modeled as an aggregate of rigid spheres rigidly attached to one another to give a resultant particle with an overall size and shape that closely approximates the particle of interest. Such a model has been made for myosin.[37] The model leads to a set of simultaneous hydrodynamic interaction equations which can be solved by standard numerical computational techniques to yield the requisite frictional factors. Using this technique, one can arrive at a shape and a set of dimensional constants for myosin or its subunits and calculate the expected translational and rotational diffusion coefficients (and intrinsic viscosity as well) for comparison with observed results. Although this technique takes into account intra-particle hydrodynamic interactions, it does not lend itself readily to making calculations on particles that are flexible, or segmentally jointed. However, one can calculate the translational and rotational diffusion coefficients for various configurations of the particle, each of which is assumed to be rigid, and then average over all configurations.[37]

Hydrodynamic theory allows the direct calculation of the diffusion coefficients in dilute solutions for such simple shapes as ellipsoids, spheres, and cylinders.[38,39,40,41,42] Recently, hydrodynamic theory has been extended to include the calculation of diffusion coefficients for segmentally flexible macromolecules in which rigid segments are connected by freely flexible universal joints.[43,44] In addition, an approximate theory has been developed for calculating the PCS of light scattered from a dilute solution of two rigid rods joined by a universal joint, once-broken rods, in which it is assumed that intramolecular optical interference effects are negligible and hydrodynamic interactions between segments are negligible.[45] The theory of the PCS of the once- broken rod was developed with specific reference to the myosin molecule. When the segments of a once broken rod are joined by a free universal joint, the two half-rods execute rotational diffusion independently of one another with an altered rotational relaxation time. If, for example, the two segments are approximately the same length, the rotational diffusion coefficient of each, D_R, can be approximated by the average, that is $D_R = (D_R' + D_R'')/2$ and the rotational decay rate is $6D_R$, where D_R' and D_R'' indicate the two separate rotational diffusion coefficients of the segments. Also, when there is an upper limit to the allowed values of the angle between the two segments of a broken rod a single rotational diffusion constant can be obtained on the assumption that the two rods execute their limited rotational diffusion independently of one another.[46] For a very long, freely diffusing, once-broken rod, neglecting hydrodynamic interactions, the end over end rotational diffusion coefficient for a rod-half is 4.67 times as great as that for the full unbroken rigid rod. For short rods, this factor is smaller. If the once-broken rod is translationally constrained at its break point, the rotational diffusion coefficient of the half-rod drops by a factor of 2.[43]

Diffusion Constants for Myosin and Its Subfragments

Table 3 lists the coefficients of diffusion for myosin and its subfragments, both experimental and theoretical. With regard to the translational coefficient of diffusion, the tabulated data show that removal of the S-1 heads from the myosin molecule, to yield the myosin rod, results in only a 20% increase in D_T. On the other hand, the removal of the LMM subfragment (a much smaller total mass) to yield HMM, results in an approximately 80% increase in D_T. Thus, the translational diffusion properties of the myosin molecule are primarily determined by the rod portion and the S-1 heads produce a change in D_T which is in the nature of an end effect. Further, the observed D_T for myosin rod is in good agreement with the theoretical value calculated from hydrodynamic theory[39,40] for a rigid rod or for a once-broken rod.[44] This result is in agreement with the hydrodynamic theory which predicts only a small difference between the translational diffusion coefficient for a rigid rod and for a once-broken rod with the universal joint at its center.[44]

The tabulated data on the recent depolarized light scattering measurements on LMM and rod are worthy of note for they give a clear indication of the flexibility at the LMM-HMM junction. These results indicate that LMM is a rigid rod and that the myosin rod is best described as a once-broken rod with almost free rotational diffusion in a conical volume that subtends an angle of approximately 128°. The depolarized PCS data collected on rods could not be fit with a single exponential and yielded rotational coefficients of diffusion much higher than that calculated for a rigid rod with the length of the myosin rod.[47] In this study, however, polarized PCS measurements yielded a value of 4.8 ± 0.8 ksec^{-1} for q^2D_T of myosin rod, under the conditions of their experiments. This value is significantly less than the 8.0 ksec^{-1} expected from hydrodynamic theory for the rod. The depolarized PCS for myosin rod were better fit by a sum of two exponentials of the form:

$$g^{(2)}(\tau) = A + (B_1 \exp(-\tau/\tau_1) + B_2 \exp(-\tau/\tau_2))^2$$

where A, B_1, B_2, τ_1 and τ_2 are constants. This kind of double exponential form corresponds to that expected theoretically for the case where intraparticle optical interference effects are negligible.[45] Using the observed value for B_1/B_2 of

TABLE 3

MYOSIN AND MYOSIN FRAGMENTS

	$D_{20,w} \times 10^7$ cm^2/s (Theory)*	Rotational Relaxation Time us (20°C) (Theory)*	Reference
Monomer	1.15	38	64,65
	1.24		66
	1.11		67
	(1.17,1.19)	(52.9,20.2)	68
			69,69
Dimer	.84		66
HMM	1.9	5	49,70
	1.93		68
	(1.68)	(9.6)	69,69
LMM	1.89	7.6	68,71
		7.7	47
S1	11.3	.25	49,70
	4.8		50
	(5.76)	.13-.15	69,72
		.16,.22	16,16
S2	3.98	3.5	68,71
		2.0	70
Rod	1.24	24.0	68,71
		21.7 or 3.2 and 43.5	47
	(1.41,1.43)	(33.4,15.6)	69,69

* First value calculated with bend angle of 180°, second value calculated with bend angle of 90° (see reference 69.)

1.2 ± 0.4 and the theory for light scattering by the once-broken rod with limited segmental rotation, the average half-angle through which rotational diffusion occurred was found to be 128° and corresponded to a rotational diffusion constant D_R of 24 ± 6 krad/sec. This quantity should be compared with the average of the measured values of D_R for LMM and S-2 free in solution, which is 35 krad/sec. Thus, the observed rotational diffusion constant, D_R, for the once-broken rod is 69% of this average. Hydrodynamic theory predicts that for a once-broken rod D_R should be 53% to 78% of the value for a free segment, or 16-28 krad/sec, depending upon the extent to which hydrodynamic interactions along the length of the particle are taken into account.[43,48] Thus, the observed value for D_R appears to be in good agreement with theoretical expectations if one assumes that myosin rod is a once-broken rod with the break near the center and rotation restricted to a half-angle of 128° about the break point. The double exponential fit gave a value of 3.7 $\pm$ 0.8 krad/sec for the second exponential, which according to the theory corresponds to the rotational diffusion constant for the unbroken, rigid rod. For such a rigid rod a value of 5.3 krad/sec is expected from hydrodynamic theory. A value of 4.8 $\pm$ 0.8 krad/sec was obtained when the first 8 us of data were disregarded and a single exponential fit made to the remaining data. Although these recent depolarized PCS studies support the existence of a high degree of flexibility at the LMM-HMM junction in rod, the discrepancies reported in the observed value of D_T and D_R for the rigid myosin rod are perplexing. The rigidity of LMM seems well established .

The different values listed for the translational diffusion coefficient of S-1 require some clarification. Using the shape of S-1 as obtained from electron microscopy studies,[26] a translational diffusion coefficient of 5.76 x 10^{-7} cm^2/s was obtained.[37] This compares unfavorably with the reported value of 11.3 x 10^{-7} cm^2/s[49,50] and it differs significantly from other values that have been reported. Thus, while there are some troublesome discrepancies to be sorted out, the hydrodynamic studies lead to a picture of myosin as a molecule with a high degree of flexibility at the S-1 - S-2 junction and at the HMM - LMM junction, although in the latter case the flexibility seems to be somewhat limited.

Table 4 summarizes the diffusion coefficients for actin in its various forms and Figures 11a and 11b are plots of the data on G-actin obtained from PCS studies.[51] The PCS measurements on G-actin had to be corrected for the back-reflected beam even though a refractive index matching beamstop was used. For monodisperse Rayleigh scatterers it can be shown that the intensity weighted $\overline{\Gamma}$, (the first cumulant), for the sum of the light scattered from the main and the back-reflected beam is:

$$\overline{\Gamma} = \left[\left(\frac{1-\alpha}{1+\alpha}\right) q^2 + \left(\frac{4\pi}{\lambda}\right)^2 \frac{\alpha}{(1+\alpha)} \right] D_T'$$

where: α = the reflection coefficient and D_T' = the true translational diffusion coefficient. Thus, a plot of $\overline{\Gamma}$ is linear in q^2, passes through the true Γ (90°), but has a positive intercept for $q^2 = 0$. For a back-reflected beam that is 1.3% of the main beam, the experimental slope is 3% less than the true

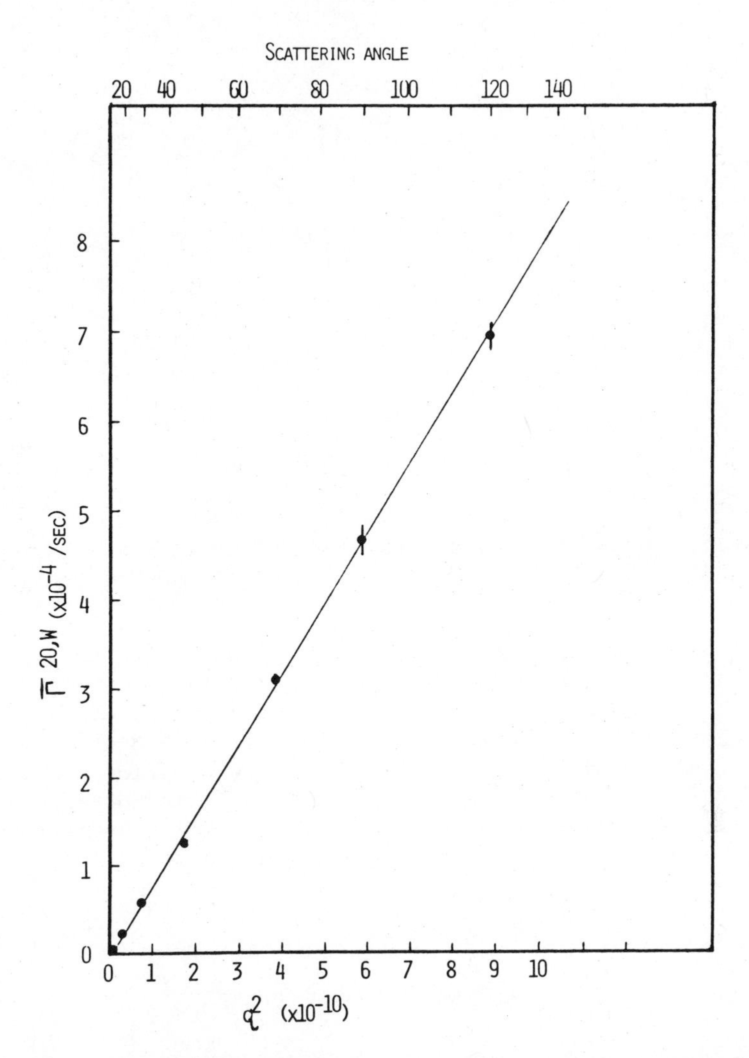

Figure 11a. Plot of the decay rate of the field autocorrelation function at 20°C in water vs. the square of the scattering vector. Data were collected at 15°C from a 0.64 mg/ml solution of G-actin. The measured $\overline{\Gamma}$ was corrected for temperature and viscosity to get $\overline{\Gamma}_{20,w}$. Error bars represent the standard deviation for repeated experiments (4-20 repetitions). Error bars not shown are smaller than the symbols.

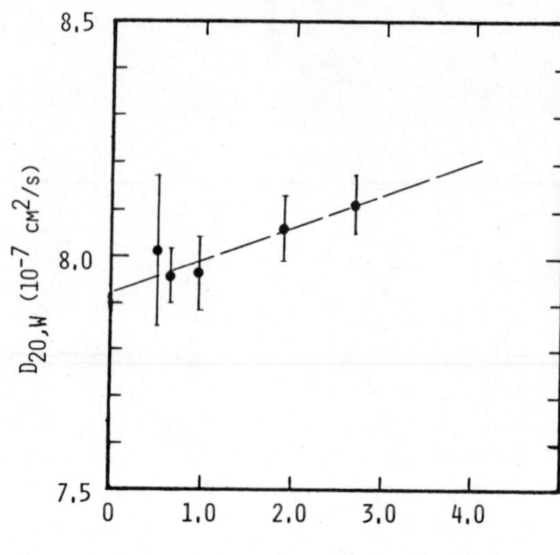

Figure 11b. Translational diffusion constant at 20°C in water vs. concentration of actin. $D_{T}20,w$ was obtained from the slope of $\overline{\Gamma}^{20,w}$ vs. q^2 for each concentration of actin. Error bars represent the estimated error based on the fit of $\overline{\Gamma}^{20,w}$ vs. q^2 plots.

slope. When this correction was made a value of 8.3×10^{-7}cm/^{2}s was obtained for the translational diffusion coefficient of G-actin at 20°C in the limit of zero concentration. This corresponds to a sphere with a radius of 2.6 nm and a 50% hydration or to a prolate elipsoid with axes 2.0 nm and 4.0 nm and a 30% hydration, a shape

TABLE 4

ACTIN AND ACTIN COMPLEXES

	$D_{20,w} \times 10^7$ cm^2/s	Rotational Relaxation Time us	References
G-actin	8.13	.024	51,86
	6.1		58
	5.28		59
F-actin	.199		54
Native Thin Filaments	.198		60

that is in good agreement with X-ray diffraction results on actin-DNase 1 complex and with electron microscopy studies on crystalline actin sheets.[52,53]

F-actin and complexes of F-actin with myosin subfragments have been extensively studied in recent years in an effort to determine the flexibility characteristics of F-actin and the way in which these characteristics are influenced by complexing with myosin subfragments.[54,55,56,57,58,59] As it is usually prepared, F-actin is highly polydisperse, and as a result its $g^{(2)}(\tau)$'s cannot be fit by a single exponential, nor does it give plots of Γ versus q^2 that are straight lines passing through the origin. The Γ versus q^2 plot is concave upward and as $q^2 \rightarrow 0$ it extrapolates through the origin as would be expected for a highly polydisperse system.

F-actin, as we have already discussed, activates the ATPase activity of myosin, HMM and S-1 subfragment. If F-actin is mixed with a solution containing HMM and ATP the F-actin activates the HMM ATPase activity and it is possible to follow the PCS of the mixture as the ATP is gradually hydrolysed to ADP. As long as there is ATP present, the intensity autocorrelation function of the mixture is only slightly different from that of F-actin alone. The HMM fragments are weaker scatterers than F-actin and they cycle through the cross-bridge cycle attaching and detaching from an actin filament splitting an ATP molecule for each cycle. Because of the great length of the F-actin molecule, and because the HMM does not remain permanently attached to it in the presence of ATP, HMM does not alter the PCS of the reacting mixture very much. However, when the ATP is all hydrolyzed it is no longer possible for the HMM that is bound to the F-actin to be dissociated, and as a result, a rigor cross-bridge is formed with the F-actin. Since there are two S-1 heads on each HMM, one HMM fragment can form a rigor link with two different actin filaments and act as a cross-linking agent. When this occurs, the F-actin filaments in solution become cross-linked and a gel is formed. The PCS characteristics of the acto-HMM gel are strikingly different from those of F-actin and HMM in the presence of ATP. The amplitude of $g^{(2)}(\tau)$ is greatly reduced and the correlation time is altered as the F-actin molecules are immobilized to form the gel network.[54,49] When S-1 is used instead of HMM, no gel state develops after the initial ATP is fully hydrolyzed, even though the S-1 subfragments are attached (rigor link) to the F-actin. This is to be expected since S-1 can bind only to one F-actin molecule and is unable to cross-link two F-actins together to form a gel. These PCS studies on the interaction of HMM and S-1 in the presence and absence of ATP indicate that both HMM and S-1 under conditions of maximum ATPase activity can exist in a refractory state in which they are temporarily unable to bind to F-actin and remain free in solution. Only after the S-1 converts from the refractory to the activated state (see Fig. 9), can it form an active cross-bridge with F-actin.

The availability of a nearly monodisperse preparation of isolated thin filaments from the scallop has made possible a study of the flexibility of these filaments using PCS.[60] For these thin filaments the observed value of D_T at 5°C was $1.24 \pm 0.06 \times 10^{-8}$ cm^2/s. Assuming a thin filament diameter of 90Å, this value for D_T corresponds to a thin filament having a length of 1.06 ± 0.06 um, in good agreement with the value of 1.01 ± 0.09 um obtained by electron microscopy.[61] Since the translational diffusion coefficient of a rigid particle depends on the ratio of the absolute temperature to the viscosity of the solvent at that temperature, $g^{(2)}(\tau)$ for such a particle should scale as the absolute temperature over the viscosity on the τ-axis (T/η-scaling). Accordingly, if $g^{(2)}(\tau)$ is measured at several different temperatures, and the values are scaled to a standard temperature by the T/η-scaling rule, the observed $g^{(2)}(\tau)$ at different temperatures will superimpose if T/η-scaling holds. While T/η-scaling was observed at low temperatures and for forward scattering it was not observed in the temperature range of 30 - 40°C at higher scattering angles. In this temperature range, it was found that while the tropomyosin dissociated from the thin filament, there was no appreciable change in the length distribution of the tropomyosin-free filaments. This failure of T/η-scaling associated with the loss of tropomyosin can be interpreted to mean that the flexibility of the thin filament is increased when the tropomyosin is removed. Further, using the distribution of filament lengths obtained from electron microscopy and the method of splines to approximate this distribution, computer-generated Γ vs. q^2 plots were obtained. For any reasonable distribution, these plots could not be made to fit the observed Γ vs. q^2 behavior for thin filaments. However, Γ vs. q^2 data for fd virus particles could be fit by this procedure. These results indicate that polydispersity alone cannot account for the Γ vs. q^2 behavior of thin filaments. The amplitude of the flexural motion can be estimated. For such a motion to be discernible by PCS the displacement of one point on a thin filament relative to another must be greater than $1/q$ which at the high scattering angles used in this experiment was several hundred Å. T/η-scaling is not unambiguous evidence for particle rigidity, for there are conditions under which a semi-flexible particle may show T/η-scaling and there are also conditions under which a particle may fail to show T/η-scaling when it is actually rigid.[62]

SUMMARY

PCS studies on the contractile proteins and organelles have provided strong evidence for nearly complete flexibility at the

LMM - HMM junction, rigidity of LMM, and support for the size and shape of the S-1 fragment that has been determined from electron microscopy. The diffusion coefficient measurements on G-actin are in excellent agreement with data obtained by other techniques and support the suggestion that G-actin is a prolate ellipsoid with an axial ratio of about 2. The use of PCS to monitor the interactions of F-actin with HMM and S-1 in the presence of ATP has provided strong unambiguous evidence for the existence of a cross-bridge cycling that includes a refractory state of the S-1 head of myosin. These studies have established the existence of an acto-HMM gel in the absence of ATP where the HMM acts as a cross-linker between two actin filaments. This gel state corresponds to the rigor state in intact muscle. Finally, they have provided evidence to support the view that F-actin is flexible and that this flexibility is appreciably diminished by the binding of tropomyosin, in agreement with other results.[63] Although most of these results are reasonably firm and consistent, some gaps and troublesome discrepancies remain for further studies.

PHOTON CORRELATION SPECTROSCOPIC STUDIES ON SINGLE SKELETAL MUSCLE FIBERS

We have summarized the evidence for the cycling cross-bridge, sliding filament model of muscular contraction. This evidence indicates that the contractile force originates from the independent interaction of the activated S-1 heads of a myosin molecule in the thick filament with an actin molecule in the thin filament. Each activated S-1 head (cross-bridge) cycles independently through (1) attachment to actin filaments, (2) tension generation, (3) detachment from actin, and (4) reactivation by ATP. Here we shall review the results of various experimental efforts to demonstrate independent cycling of cross-bridges.[73,19]

If it is assumed that cross-bridges are attached with a probability, p, or detached with a probability (1-p) and that each develops tension in the attached state independently of the state of all other cross-bridges then the number of attached bridges in each half of a thick filament should be given by the binomial distribution. As the number of attached cross-bridges fluctuates so will the net force acting on the thick filament, and with it, the tension developed by the interacting thick and thin filaments in a sarcomere.[74] Accordingly, fluctuations in (1) the tension developed by a muscle, or (2) the polarized fluorescence of labeled S-1 heads in muscle cells, or (3) the light quasi-elastically scattered from contracting muscle might be detectable under certain conditions.

Fluctuations in Tension and Polarized Fluorescence

It has been shown that during contraction tension fluctuations do occur in small bundles of myofibrils obtained from glycerinated preparations of rabbit psoas muscle.[73] Furthermore, fluctuations in polarized fluorescence have been observed in the same preparation after labeling the S-1 heads with a fluorophore.[19] The observed rate constant in this case was 2 s^{-1}, a value compatible with the rate of ATP hydrolysis in this glycerinated preparation, but far less than the ATP hydrolysis rates found in intact muscle, thus leaving the significance of these results unclear.

When excited with polarized ultraviolet light of the proper wavelength, live muscle fibers emit polarized fluorescent light which arises from the polarization of the intrinsic fluorescence of the amino acid tryptophan. Since tryptophan is present in substantial amounts in the S-1 head of the myosin molecule it was thought that it could serve as a label for following S-1 orientation. Initial experiments indicated that the polarization of the intrinsic tryptophan fluorescence changed when a muscle passed from the resting state to the contracting state or to the rigor state.[75,76] However, more recent studies have shown that the polarization of the intrinsic tryptophan fluorescence does not change with the physiological state of the muscle, and that the changes originally observed were due to the limitations of the amount of ATP available to the muscle under the conditions of the experiments.[77] For the present, at least, we must conclude that in the intact muscle fiber tryptophan fluorescence is not sensitive to changes in the orientation of the cross-bridge, or the cross-bridge does not rotate, or both.

PSC Studies on Whole Muscles and Fiber Bundles

Early PCS measurements on whole muscles yielded values of 2-3 ms for the decay time of $g^{(2)}(\tau)$ during an isometric tetanus at full overlap.[78] This short decay time observed on whole muscle is to be compared with the much longer, 70 ms, decay time observed on single fibers[79] and discussed in detail below. Other PCS studies on thin bundles of live sartorious muscle fibers, on both glycerinated and skinned rabbit psoas muscle fiber preparations under conditions of low tension development, also showed 2-4 ms delay times.[80] However, for the vertically oriented small bundles, these studies showed that the amplitude of the normalized intensity autocorrelation function was zero for light in the zeroth order diffraction maximum in disagreement with the early studies on whole sartorius muscle.[78] Interpretation of these PCS studies on whole muscles and small bundles of fibers were flawed because of mechanical and optical heterogeneities that could neither be controlled nor evaluated. The isolated single muscle fiber, though extremely difficult to prepare, is the preparation of choice for PCS studies because its use finesses most of the problems that are inherent in heterogeneous multifiber preparation.

PCS Studies on Isolated Single Muscle Fibers

The organization of the contractile proteins in a muscle fiber endows the thick and thin filaments with a light scattering power far greater than that of the soluble sarcoplasmic proteins. Consequently, the myofibrillar structures dominate the light scattering properties of the fiber. Although myosin cross-bridges make up 35% of the myofibilar protein, it is likely that the amplitude of their translational motion is too small to be directly detected by dynamic light scattering. On the other hand, independent and asynchronous cycling of the cross-bridges could produce larger rms displacements ($\sim$ 100 nm) of individual thick filaments or even entire myofibillar A-bands. These large scale displacements, if present, might be measurable by PCS.[74]

As established earlier on in this book, the straightforward interpretation of PCS data requires that certain conditions be met. This is particularly so when the object of study is as structurally complex as is the interior of a single muscle fiber. Specifically, it is necessary to show that (1) dynamic light scattering by a muscle fiber is quasi-stationary, (2) that the scattered field is Gaussian with zero mean, and (3) that when biological conditions restrict the duration of a PCS measurement to a small number of coherence times, appropriate corrections for the resultant bias in the normalized intensity autocorrelation function, $g^{(2)}(\tau)$, must be made.[79,81] Fortunately for PCS measurements, single muscle fibers are strong scatterers and the number of scattered photon counts detected in a coherence time is always considerably larger than 1, and the statistical accuracy of $g^{(2)}(\tau)$ measurements is limited primarily by the number of coherence times contained in the time span of a stable tension plateau of a tetanus.[79] Finally, since it is possible to repeat the measurements of the PCS during a 1-2 s tetanus on the same fiber it is possible to combine these repeated measurements on a single fiber in a way that reduces the effects of bias and biological variability.[79] The measurements of $g^{(2)}(\tau)$ reported below were conducted in a way that verified the conditions enumerated above and provided justification for the method of data analysis used.[79]

Preparation, Apparatus, and Procedure

Single muscle fibers, dissected from the dorsal head of the semitendinosus muscle of Rana pipiens in the standard way,[82] were used. These fibers were approximately 11 mm long at rest length where they had a sarcomere length of 2.1 um and diameters of 50 to 120 um. The single fibers were attached to hooks made of fine platinum wire by means of thin aluminium clips.[79,82]

The chamber used for dissection also served as a cell for light scattering studies. Made of plexiglass, with a volume of 40 ml, its bottom was fitted with a glass window and its lid, similar to the bottom window, was fitted to the chamber with an O-ring seal. Small glass tubes passed through the walls of the chamber making it possible to circulate filtered and thermostated Ringer's solution through the interior of the chamber. A platinum hook at the tip of a rotatable shaft slipped through the hole in the aluminium foil clip at one end of the single fiber. The shaft was rotatable and was used to eliminate twists in the fiber. The aluminium foil clip at the other end of the fiber was hooked to a platinum wire that passed through a grease vaseline seal in the chamber wall and connected to the tension transducer mounted on the outside of the chamber. This arrangement made it possible to orient the single fiber vertically or horizontally and keep it continuously bathed in a thermostated, oxygenated Ringer's solution under conditions where it could be electrically stimulated and its tension recorded, while it was illuminated by a laser beam and PCS measurements were being made. Direct electrical stimulation with pulses of 0.2 ms duration and alternating polarity was applied across the fiber through two parallel platinum plate electrodes (3 mm by 15 mm by 1 mm) spaced 4 mm apart. With the glass cover removed, it was possible to examine the muscle fiber with water immersion optics during a 2 s tetanus. The rate of axial drift of the fiber in the 300 um field of view was always less than 10 um/s, and motion of one region of the fiber relative to another was, if discernable, less than 3 um/s. The sealed chamber filled with Ringer's solution was immersed in a thermostated bath ($\pm$ 0.02°C). The bath was fitted with an entrance and exit window for the incident beam and a spherical zone window (0-150°) through which the scattered light exited.

Figure 12 is a diagram of the PCS used in these studies. An argon ion laser provided an incident beam with a diameter of 2.5 mm and a single TEM 00 axial mode at 488 nm. The power incident upon the muscle fiber was less than 1 mw and the fiber axis was always perpendicular to the laser beam in either the vertical or, as was usually the case, the horizontal orientation. The incident beam was vertically polarized with respect to the horizontal scattering plane unless otherwise stated. Detection optics consisted of an aperture (250 um diameter), a lens that imaged the fiber on a rear slit, a prism analyzer set to select vertically polarized light, a rear slit (usually 240 um x 1 mm) with its long axis perpendicular to the fiber axis, and a photomultiplier (F4085 ITT). The fiber-front aperture distance and the front aperture-rear slit distance

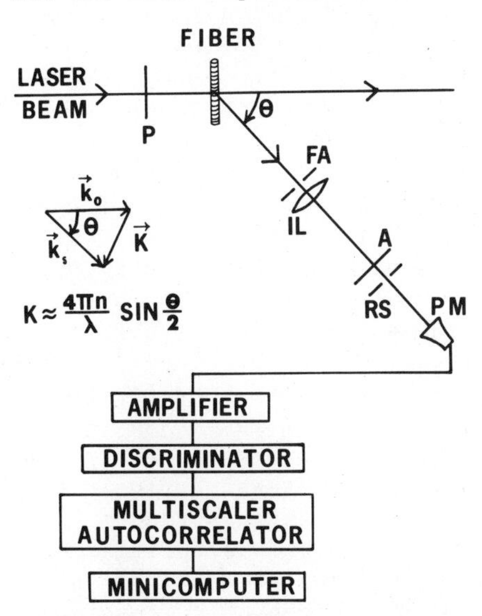

Figure 12 IFS spectrometer and data accumulation electronics: (P) polarizer, (FA) front aperture, (IL) imagine lens, (A) analyzer, (RS) rear slit, and (PM) photomultiplier. (Reproduced from the Biophysical Journal, 33, 1981, pp 39-62 by copyright permission of the Biophysical Society.)

were each 20 cm. With these optics, the photocathode viewed the full diameter of the muscle fiber and its surrounds for a length of approximately 200 um with a numerical aperture that subtended a coherence area or so.

After amplification and discrimination, photomultiplier pulses were counted by a multiscaler-photoncorrelator, and the collected counts were periodically transferred to a minicomputer for further processing. In this way, the numbers of counts arriving in each of 40 contiguous 2 ms time intervals was obtained by the multi-scaler, then during the next 20 ms the multiscaler transferred these data to the minicomputer, then another record of 40 contiguous 2 ms long samples was obtained, and so on for the duration of the tetanus. The normalized intensity autocorrelation function, $g^{(2)}(\tau)$, and, when necessary, the photocounting probability density, P(n), were calculated. Here P(n) is the probability of observing n counts in a 2 ms sample time. For sample times less than 2 ms , the correlator calculated $g^{(2)}(\tau) = \langle n_s(t)n(t+\tau)\rangle$, $\tau \neq 0$ directly. The scaled counts, $n_s(t)$, were obtained by generating one count for every s^{th} photoncount.

Stimulation of the fiber for a period of 2.0 s produced a contraction which plateaued within 0.3 s and allowed data collection from 0.4 s to 2.0 s. Data were collected at scattering angles, θ, of 7.5 to 45°. The detection optics were centered on the minimum of a diffraction pattern except when a diffraction maximum itself was being studied. The length of the fiber seen by the detection optics was (180/cos θ) um normally, but was varied from (48/cos θ) to (360/cos θ) um. Sarcomere lengths were between 2.0 to 2.1 um except when the effects of sarcomere length were being studied.

Tension plateaus were flat to within 5% and values from 1.4 to 3.3 $Mdyn/cm^2$ were obtained. For each set of parameters in an experiment, 3 to 6 successive 2.0 s tetani were observed. Data in a group of repeat experiments were combined by averaging the individual $G^{(2)}(\tau)$ values and normalizing by the group average counts per sample time. This procedure yielded a statistical accuracy for a group of three 1.6 s tetani equivalent to that of one 4.8 s tetanus with the bias for the group average amplitude and half time slightly less than the bias associated with a single continuous 4.8 s measurement, namely 7-9%. Combining data in this way assumes that the intensity fluctuations in successive tetani arise from a single stationary random process with the same mean intensity. This assumption was verified experimentally. The total number of counts, n_p, in a tetanic plateau fluctuates from one tetanus to another in a group of successive tetani on the same single fiber. The observed ratio, r, of δ_p, the standard deviation of n_p, to the mean of n_p, $\langle n_p\rangle$, ($r = \delta_p/\langle n_p\rangle$) was determined experimentally and compared to the theoretical value of r for a stationary random process using data obtained from 18 groups of tetani on nine fibers, with θ ranging from 25° to 45°. A value of $r = 0.25$ was observed. The theoretical value for a single stationary random process was $r = 0.28$. Accordingly, data obtained from a group of tetani were combined by averaging the individual $G^{(2)}(\tau)$ and then normalizing by the group average counts per sample time to obtain $g^{(2)}(\tau)$.

RESULTS

A typical measurement of $(g^{(2)}(\tau) -1)$ made during the plateau of four successive tetani of a horizontally mounted single fiber and a scattering angle of 45° is shown in Figure 13. $g^{(2)}(\tau)$ decays from its initial value, $(1 + A)$ to 1 for long τ. We define, $\tau_{1/2}$ as the time required for $(g^{(2)}(\tau)-1)$ to decay to $A/2$, where A is the value of $[g^{(2)}(0)-1]$. The observed value, $\tau_{1/2} = 92$ ms, was typical. Eighteen different measurements on nine fibers for $\theta = 25$ to 45° yielded a mean decay rate, $\langle 1/\tau_{1/2} \rangle$, of 13.7 s^{-1}. This corresponds to an effective $\tau_{1/2}$ of 73 ms. Table 5 summarizes these results. The variation in $\tau_{1/2}$ is due primarily to variation among individual fibers and to a lesser extent to a correlation of $\tau_{1/2}$ with scattering angle. These values for $\tau_{1/2}$ are about 30 times those previously reported.[78,80]

Measurements of $g^{(2)}(\tau)$ with the autocorrelator revealed that this quantity was flat on a time scale shorter than a few milliseconds. This made it possible to collect data in the multiscale mode and thus calculate the P(n) and check the validity of the assumption of a Gaussian field with zero mean. During the tension plateau the probability of obtaining n counts in a 2 ms. sample time was measured in four successive tetani and the result is plotted in Figure 14. The upper curve in Figure 13 is the plot of $(g^{(2)}(\tau)-1)$ obtained from this same experiment.

Under ideal conditions P(n) should have Bose-Einstein statistics and the amplitude A should be approximatley 1.0. However, P(n) can be altered significantly by such non-ideal conditions as a detector with finite area and experimental durations of only 20 to 30 times the field correlation time. To circumvent these complications, P(n) was measured on scatterers known to produce a Gaussian field with zero mean, namely a suspension of polystyrene latex spheres in ethylene glycol. This suspension was substituted for the muscle, but the same detection optics remained in place. Further, measurements on the suspension were made using experimental durations of 20 to 40 coherence times, to include the effects of bias. Gaussian statistics for light scattered by the polystryene latex sphere suspension was verified by measuring the amplitude of $(g^{(2)}(\tau)-1)$ in the limit of very small detector areas and showing that it approached one. The suspension yielded a

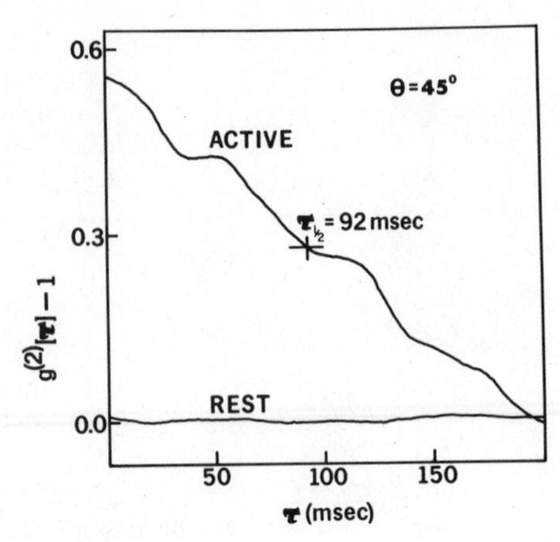

Figure 13. Active: $g^{(2)}(\tau) - 1$ measured during the tension plateau, 0.4-2.0 s after first stimulus, of four successive tetani. Rest: $g^{(2)}(\tau)-1$ measured during rest (three successive data batches, each of 1.7 s duration). (Reproduced from the Biophysical Journal 33, 1981, 39-62 by copyright permission of the Biophysical Society.)

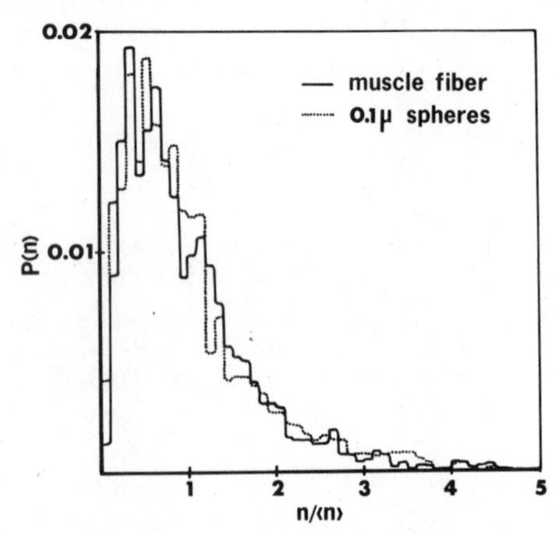

Figure 14. Solid line: P(n) measured during the tension plateau, 0.4-2.0 s after the first stimulus, of four successive tetani (group average counts per sample time was 47.9 and A = 0.55). Dotted line: P(n) measured using solution of 0.1 um diam spheres and focused beam (four successive data batches, each of 1.6 s duration; group average counts per sample time was 45.8 and A = 0.59). (Reproduced from the Biophysical Journal 33, 1981, 39-62 by copyright permission of the Biophysical Society.)

half-time of 75 ms for θ = 16° and the values of P(n) obtained in four successive 1.6-second experiments at θ = 16° are plotted as the dotted line in Figure 14. While the two curves shown in

Figure 14 for P(n) superimpose closely, the average value for the amplitude in the muscle measurements, A = 0.58, is lower than both the theoretical value and the experimentally measured value for the polystyrene latex spheres, A = 0.66. This difference is probably significant and may indicate the presence in the muscle fiber measurements of a slightly greater statistical bias than estimated, or possibly a small amount of the total light scattered by the muscle fiber is temporarily incoherent. The possibility that the muscle fiber elastically scatters a small amount of light is excluded because the observed deviation of P(n) from that expected for a Gaussian field was inconsistent with this possibility. There is, therefore, a good experimental basis for believing that under the conditions of our experiments, light quasi-elastically scattered by a contracting muscle fiber gives rise to a Gaussian field with zero mean, and thus the Seigert relationship is applicable.

To test the quasi-stationarity of light scattered from a fiber during a tetanic contraction, the mean intensity of the scattered light, $\langle I \rangle$, the amplitude, A, and the half time, $\tau_{1/2}$, were examined for systematic variation during the tension plateau. Values of these quantities were calculated from data obtained separately in the first and second halves of 28 groups of plateaus on 10 fibers for θ ranging from 7.5° to 45°. The average values for the differences between the intensities in the two halves, $\langle D_I \rangle$, the amplitudes, $\langle D_A \rangle$, and the halftime, $\langle D_\tau \rangle$, are shown in Table 5. The mean intensity is very nearly the same in the two halves of the plateau but the amplitude decreased by 23%, and $\tau_{1/2}$ increased by 46%. This result indicates that the scattering process is only quasi-stationary and strict care must be taken to restrict comparisons to equivalent portions of the tension plateau when values of a particular experimental parameter are changed.

TABLE 5

STATISTICAL ANALYSIS OF MEASURED PARAMETERS

Parameter	Mean	Standard deviation	Standard deviation of the mean
(A)	0.58	±0.11	±0.03
$(1/\tau_{1/2})$	13.7 s^{-1}	±7.2 s^{-1}	±1.7s^{-1}
(D_1)	0.08	±0.35	±0.07
(D_A)	0.23	±0.57	±0.11
(D_τ)	-0.46	±0.64	±0.12
(r)	0.25	±0.14	±0.03
r_{theor}	0.28		

Relaxed Muscle

Light scattered from relaxed (resting) muscle had a constant intensity for periods of several minutes. The value of $(g^{(2)}(\tau)-1)$, obtained at $\theta = 45°$ on relaxed muscle, is shown in Figure 13. An average amplitude, $A < 0.01$, was obtained from a series of 100 measurements each of 1.7 s duration. This low amplitude fluctuation could have been due to a small amount of dust in the Ringer's solution (background) that surrounded the muscle fiber even though background scattering from the solution was only 0.5% of the total scattered intensity.

Over periods of 15 to 30 minutes significant slow fluctuations in the intensity of the scattered light did occur and they appeared to be correlated with changes in the visible fine structure of the maxima of the diffraction pattern produced by the sarcomere periodicity in the fiber. These slow fluctuations in intensity were more rapid immediately following a tetanus.

$g^{(2)}(\tau)$ was measured at various scattering angles for a single fiber and one for a small bundle of eight fibers. A plot of $1/\tau_{1/2}$ against the projection of the scattering vector on the fiber axis, $K_F = (2\pi n/\lambda)\sin\theta$, for horizontal orientation of the fiber, is shown in Figure 15. $1/\tau_{1/2}$ varied roughly linearly with K_F for a given preparation, although there was considerable

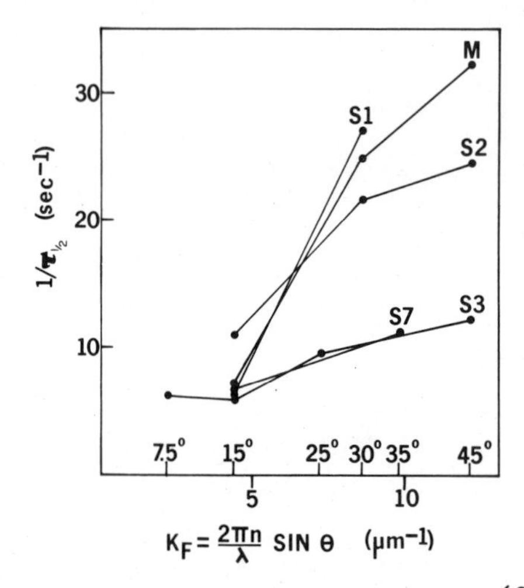

Figure 15. The reciprocal of the half time of $g^{(2)}(\tau)$ is plotted versus the projection of the scattering vector on the fiber axis. The fiber axis was oriented horizontally. Solid lines connect data points obtained from the same preparation. S1, S2, S3, and S7 were single fiber preparations, and M was a bundle of eight fibers. (Reproduced from the Biophysical Journal 33, 1981, 39-62 by copyright permission of the Biophysical Society.)

variation among preparations. For fiber S2, $g^{(2)}(\tau)$ measurements were made with its axis oriented a few degrees off the vertical so that the detection optics sampled scattered light from a region just off the zeroth order maximum ($K_F \leq 1.5$ um^{-1}). The values of $\tau_{1/2}$ for both the horizontal and vertical orientation of the fiber are shown in Figure 16. With the fiber axis oriented vertically the small observed values of $1/\tau_{1/2}$ mean that the motion of the scattering material in the fiber is primarily along the fiber axis.

The linear relation between $1/\tau_{1/2}$ and K_F indicates that the scattering material in the fiber maintains nearly constant relative axial velocities for time intervals longer than the field coherence time. Further, polarizability fluctuations which are independent of K_F do not seem to be present because $1/\tau_{1/2}$ drops sharply with K_F, with little or no indication of a positive intercept.

If the muscle is modeled as a cylindrical volume element containing randomly positioned Rayleigh scatterers, each moving independently with a constant velocity along the cylinder axis, the scattered light from the volume element will give PCS with the characteristics shown in Figures 15 and 16, and its sarcomeres will show regions of tilt as shown in Figure 17. Further, if the rms relative axial velocities of the scatterers is given by $\delta_v = 0.81/K_F\tau_{1/2}$, δ_v will be 1-2 um/s for $\tau_{1/2} = 70$ ms and $\theta = 30°$.[79]

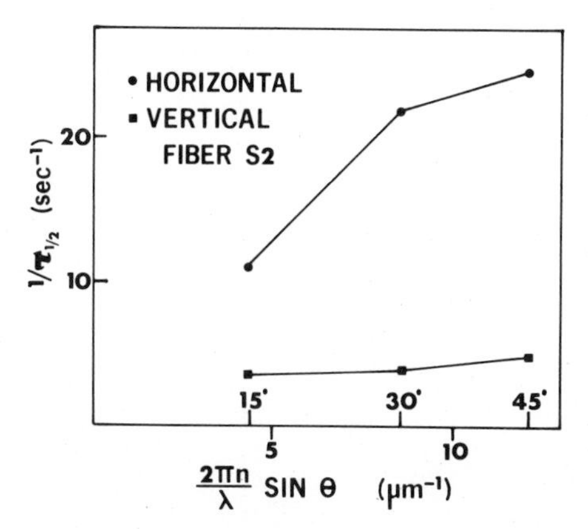

Figure 16. The reciprocal of the half time of $g^{(2)}(\tau)$ is plotted versus $(2\pi n/\lambda)\sin\theta$ for the horizontal (circles) and vertical (squares) orientation of the fiber axis of the same preparation (S2). For the horizontal orientation, $K_F = (2\pi n/\lambda)\sin\theta$, whereas for the vertical orientation, $K_F \leq 1.5$ um^{-1}. (Reproduced from the Biophysical Journal 33, 1981, 39-62 by copyright permission of the Biophysical Society.)

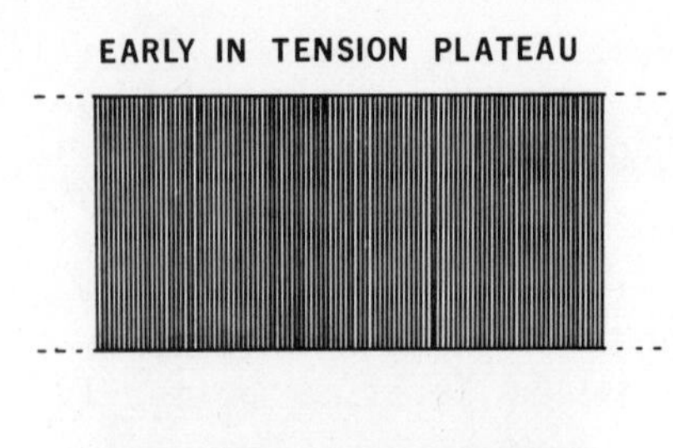

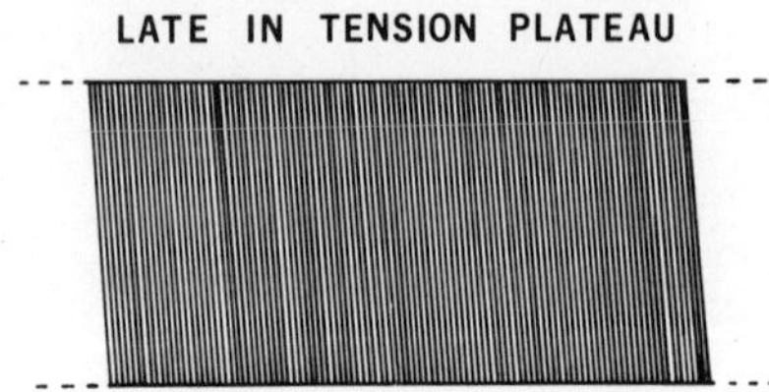

Figure 17. Sketches of a muscle fiber early and late in the tension plateau of a 2 s tetanus. The scale of the sketches is appropriate for a 100-um diam fiber and a sarcomere length of 2.0 um. When the depicted situations are assumed to occur at a time interval of 1.6 s, the rms relative velocity of scattering material is 2.5 um/s, and the resulting half time of $g^{(2)}(\tau)$ is 38 ms. (Reproduced from the Biophysical Journal 33, 1981, 39-62 by copyright permission of the Biophysical Society.)

Dependence of $\tau_{1/2}$ on Length Viewed

During contraction a slow elongation (less than 1%/sec) could be detected as a slow movement of the first order diffraction maximum of the fiber sarcomere repeat. Generally, this diffraction maximum showed movement only after 20-30 two-second tetani and the movement was associated with lower values of $\tau_{1/2}$ than those obtained when no such movement occurred. To study this movement, measurements of $g^{(2)}(\tau)$ were made on a fiber for several values of the lengths of the fiber actually viewed by the detection optics. The length viewed was varied by changing the dimensions of the front aperture and rear slit in the detection optics. A change in the width of the rear slit by a factor of 7.5 produced an increase in $1/\tau_{1/2}$ by a factor of only 2.7. Using the randomly-positioned-scatterer model and assuming that the relative motion of the scatterers arose solely from the slow elongation of the fiber, the dependence of $1/\tau_{1/2}$ on the length of the fiber viewed was calculated. If the only relative motion of the scatterers during the tension plateau had been that due to a uniform elongation too small to yield a visible movement of the diffraction maximum, then the measured $1/\tau_{1/2}$ should have increased by a factor of 9.6 for a

7.5-fold increase in the length of the rear slit. The observed increase was only by a factor of 2.7. We can conclude, therefore, that while some very slow elongation of the length of the fiber viewed may have occurred, this lengthening was not the dominant component of $g^{(2)}(\tau)$ when no visible movement (<0.5%/s) of the diffraction maximum was observed.

Angular Dependence of the Amplitude

The amplitude of $g^{(2)}(\tau)$ for scatterers with axial displacements constrained to have amplitudes less than $1/K_F$ is significantly less than that for free scatterers. A model consisting of independent, identical, one dimensional, harmonically-bound, Rayleigh scatterers, randomly distributed throughout the scattering volume and having rms displacements of δ_D, has an amplitude of $g^{(2)}(\tau)$, $A = g^{(2)}[0]-1$, given by $(1-\exp[-2K_F^2\delta_D^2])$.[55] A comparison of the observed dependence of A on θ with the values calculated for the harmonically-bound Brownian particle indicated that in fibers exhibiting no elongation, the possibility that the motion responsible for the decay of $g^{(2)}(\tau)$ arose from particles constrained to rms relative axial displacements as small as 0.3 um could not be rigorously excluded (see Table 6).[54,83]

$g^{(2)}(\tau)$ on the Diffraction Maxima

Low values for the amplitude of $g^{(2)}(\tau)$ are expected on diffraction maxima, for these maxima arise from the interference of elastically scattered light. Plots of $g^{(2)}(\tau)$ versus τ from measurements made in a diffraction minimum and on a first order

TABLE 6

AMPLITUDE OF $g^{(2)}(\tau)$ RESULTING FROM MOTION CONSTRAINED TO AN RMS DISPLACEMENT OF δ_D FOR VARIOUS SCATTERING ANGLES

RMS Displacement	$A=g^{(2)}(0)-1= 0.66\,(1-\exp[-2K_F^2\delta_D^2])$		
δ_D	θ=7.5°	θ=15°	θ=30°
(micrometers)			
0.01	0.001	0.003	0.01
0.05	0.02	0.06	0.20
0.10	0.06	0.22	0.51
0.20	0.22	0.52	0.66
0.30	0.39	0.64	0.66
observed	>0.25	>0.30	0.58

(Reproduced from the Biophysical Journal 33, 1981, 39-62, by copyright permission of the Biophysical Society.)

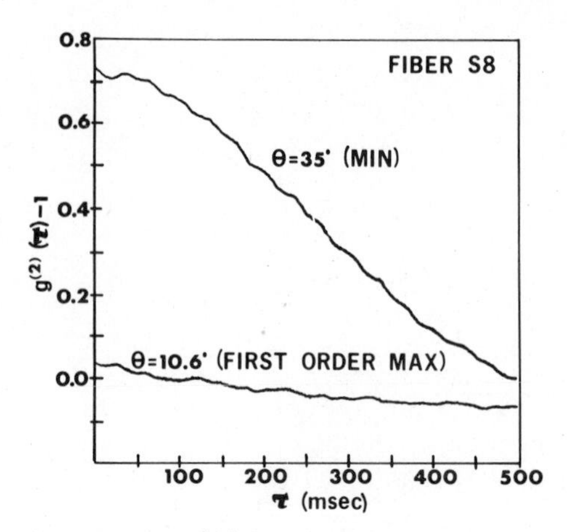

Figure 18. Measurements of $g^{(2)}(\tau)$ were made on the first order diffraction maximum (θ = 10.6°; data taken during the tension plateau, 0.5-2.0 s after first stimulus, of four successive tetani) and in a diffraction minimum (θ=35°; data taken during the tension plateau, 0.5-2.0 s after first stimulus, of four successive tetani) for fiber S8. (Reproduced from the Biophysical Journal 33, 1981, 39-62 by copyright permission of the Biophysical Society.)

maximum are shown in Figure 18. The value of the amplitude on the maximum was 0.03, while the amplitude in the minimum was 0.73, in agreement with expectations. This result further reinforces the evidence for the constancy of the sarcomere length during the 1.6 s duration of the tetanus plateau. Had the sarcomere length changed, the diffraction maximum would have swept across the front aperture of the detection optics and the result shown in Figure 18 would not have been obtained. The sarcomere length had to be constant to better than 0.5%/s during the time that $g^{(2)}(\tau)$ was measured on the diffraction maximum.

Dependence of $\tau_{1/2}$ on Sarcomere Length

Measurements of $g^{(2)}(\tau)$ were obtained sequentially at sarcomere length of 2.05, 2.80, 3.40, 4.00, and 2.3 um. Plots of $1/\tau_{1/2}$ versus sarcomere length are given in Figure 19 together with the tension records. For sarcomere lengths above 2.05 um the tension rose rapidly at first and then slowly crept to its plateau level. During the creep phase the sarcomeres in the region under observation slowly elongated as movement of the first order maximum indicated. At sarcomere lengths of 4.00 um there was still significant tension development, but no plateau, even though at this sarcomere length there is no overlap of the filaments in the region of the fiber that was viewed by the detection optics. These results are in good agreement with earlier findings.[82] Decay rates for

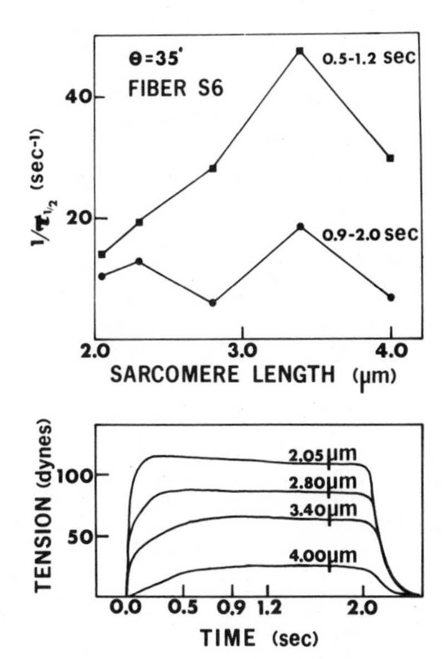

Figure 19. Lower half: tension traces from fiber S6 at four different sarcomere lengths. Upper half: the reciprocal of the half times during the tension plateau (circles) and during the tension creep phase (squares). (Reproduced from the Biophysical Journal 33, 1981, 39-62 by copyright permission of the Biophysical Society.)

the creep phase (0.5 to 1.2 s) and the plateau phase (0.9 to 2.0 s) are plotted as a function of sarcomere length in the upper panel of Figure 19. During the plateau, both the amplitude and decay rate of $g^{(2)}(\tau)$ are essentially independent of sarcomere length. However, during the creep phase when the viewed region of the fiber was being elongated the decay rate was always increased.

Birefringence Changes and Fluctuations

Using the known structures of the thick and thin filaments, their volume fraction in muscle, and the requisite refractive indices, we have calculated the changes in birefringence, B, of a muscle fiber that would be expected to occur with changes in cross-bridge orientation.[84] Calculated values are given in Table 7 together with observed values from the literature. The calculated values of B are interesting because they clearly show the strong dependence of B on the assumed distribution of cross-bridge orientations. The tabulated results clearly indicate the need for reliable measurements of both the changes in B associated with changes in the physiological state and the polarizability of the S-1 head of myosin. Our own efforts to measure B and its changes with physiological state have revealed that major sources of error in such measurement arises from thickness non-uniformities in the single muscle fiber and a slow variable twisting of the fiber with

TABLE 7

Physiol. State	For θ distribution shown	B (theory)* ($x10^{+3}$)	B (observed) ($x10^{+3}$)	
Rest/Relaxed	$\theta=0$	1.83	1.92 ± .03	(87)
	$0^\circ \leq \theta \leq 45^\circ$	1.69		
	$0^\circ \leq \theta \leq 90^\circ$	1.45	1.67 ± .05	(88)
Contracting	$45^\circ \leq \theta \leq 90^\circ$	1.21	1.76 ± .02	(87)
Rigor	$\theta=90^\circ$	1.07	1.46 ± .08	(88)
	$60^\circ \leq \theta \leq 76^\circ$	1.17		
	$\theta=45^\circ$	1.45		

*$B=n_{\parallel} - n_{\perp}$ where $n_{\parallel}$ and $n_{\perp}$ are refractive indices for E-field parallel and perpendicular to the fiber axis respectively. The theoretical value of B, B (theory), was calculated using the relation given below (Haskell, Blank, and Carlson, 1979, unpublished).

$B=n_{\parallel} - n_{\perp}$ = average value of birefringence,
where:

$$n_{\parallel}^2 = \frac{(n_1^2-1)\left[f_1+f_H(B_H+(A_H-B_H)\langle\cos^2\theta\rangle_\theta)\right]+(n_2^2-1)f_2}{f_1+f_H(B_H+(A_H-B_H)\langle\cos^2\theta\rangle_\theta)+f_2} + 1$$

$$n_{\perp}^2 = \frac{(n_1^2-1)\left[f_1B_1+f_H(B_H+(A_H-B_H)\langle\sin^2\theta\rangle_\theta/2)\right]+f_2(n_2^2-1)}{f_1B_1+f_H(B_H+(A_H-B_H)\langle\sin^2\theta\rangle_\theta/2)+f_2} + 1$$

$$A_H = \left[1+\frac{(n_1^2-n_2^2)A'}{n_2^2}\right]^{-1} \qquad B_H = \left[1+\frac{(n_1^2-n_2^2)A}{n_2^2}\right]^{-1} \qquad B_1 = \left[1+\frac{(n_1^2-n_2^2)}{2n_2^2}\right]^{-1}$$

f_1= volume fraction of filaments exclusive of S-1 heads = 0.061.

f_H= volume fraction of S-1 heads = 0.03.

f_2= volume fraction of sarcoplasm = $1-f_1-f_H$.

n_1= refractive index of proteins = 1.53

n_2= refractive index of sarcoplasm = 1.35

A and A' are constants that depend on the axial ratio of the ellipsoid used to model the S-1 cross-bridge. Axes assumed, 18.5 and 4.5 nm.

θ = angle between the long axis of the cross-bridge and the axis of muscle fiber. $\langle \ldots \rangle$ denotes the average over the distribution of θ.

changes in its physiological state. Experimental techniques designed to circumvent these difficulties are now being developed.

Preliminary birefringence fluctuation correlation function measurements of single fibers in the resting and contracting state have been made.[84] Our photon correlation spectrometer was adapted to measure correlations in birefringence fluctuations simply by including a high quality rotatable $\lambda/4$-plate (mica) in the detection optics. The $\lambda/4$ plate was inserted in the detection optics between the sample (muscle fiber) and the rotatable analyzer as is regularly done in conventional birefringence measurements. The E-field of the incident laser beam was oriented at 45° to the axis of the muscle fiber (oriented vertically) and the fast axis of the $\lambda/4$-plate was oriented parallel to the incident E-field. In this arrangement, the scattering plane is normal to the fiber axis. Upon passing through the birefringent muscle fiber, the laser light becomes elliptically polarized. It is then rendered plane polarized upon passing through the $\lambda/4$-plate and rotated by an angle of $\delta/2 = 2\pi d\,(n_{||}-n_{\perp})/\lambda$, where d is the thickness of the preparation (not usually uniform for a muscle fiber), $n_{||}$ and $n_{\perp}$ are the refractive indices of the fiber with the E-field parallel and perpendicular to the fiber axis (fast axis) respectively, and λ is the wavelength of the incident light. Rotation of the analyzer through an appropriate angle restores the extinction condition and results in a minimum in the transmitted intensity and the photomultiplier count rate. If this compensating adjustment is made during a trial tetanus, then in subsequent tetani one can measure the PCS of the intensity fluctuations in the transmitted (or scattered) light that are due to birefringence fluctuations. Such birefringence fluctuation correlation function measurements should reflect the fluctuating behavior of the birefringence of the muscle fiber that occurs during contraction due to the presumed asynchronous cycling of cross-bridges through different orientations.

When care is taken to use strain free, clean and properly mounted optics, the experimental arrangement described above will given an overall extinction ratio less than 1×10^{-6}. Relaxed muscle fibers gave extinction ratios of 2×10^{-5} when the incident E-field was parallel to the fiber axis. During contraction the extinction ratio increased to 10^{-3} and complete and precise compensation for this change was not possible simply by presetting the analyzer. Nevertheless, birefringence autocorrelation function measurements could be made on the zeroth order diffraction maximum that gave a correlation function with an amplitude of 0.005 at most and an indeterminate $\tau_{1/2}$. This result means that if birefringence fluctuations arising from cycling of the S-1 heads occur at all, their rms amplitudes are less than 2% of the measured birefringence of resting muscle. Had the birefringence fluctations been greater, the amplitude of the observed autocorrelation function would have been larger.[84]

DISCUSSION

These results on single muscle fibers differ in two important respects from other results reported on whole muscle and on small bundles of fibers.[78,80] In all other earlier PCS studies on muscle, the observed values of $\tau_{1/2}$ were 5 ms or less at a scattering angle of 45°. This is substantially smaller than the average value of 73 ms. reported here. In addition, the results obtained on single fibers indicate that the measured amplitude of $g^{(2)}(\tau)$ for vertically oriented fibers was essentially zero, a result in agreement with the recent studies on small bundles of fibers and skinned fibers[80] but in disagreement with earlier results on whole muscles of frog.[78] While there is some indication that values of $\tau_{1/2}$ obtained from measurements on bundles of fibers are lower than those obtained on a single fiber, it does not appear likely that this is the sole source of the discrepancy in $\tau_{1/2}$. The fact that the whole sartorious muscle which has a thickness of 1 mm showed a large amplitude when oriented vertically while small bundles and single fibers showed virtually zero amplitude for the same orientation suggests that the thicker preparation behaves as a deep phase screen. Furthermore the whole muscle cannot be reliably viewed under the microscope whereas a fiber can. The possibility that interfiber motion occured during an isometric contraction cannot be rigorously excluded in the case of whole muscle. Such a motion, would produce random phase modulations of the transmitted and scattered light that could produce a fast, angle-independent decay time for $g^{(2)}(\tau)$. Twisting of the whole muscle preparation about its long axis would also lead to a non-zero amplitude in the zeroth order maximum, and this possibility was not excluded in the earlier studies. The intact single fiber preparation was also monitored through the microscope during contraction, and thus it is possible to verify its structural stability and mechanical uniformity.

The observations reported here cannot be attributed to a drift of part of the single fiber preparation through the laser beam. The maximum drift rate of the entire 200 um illuminated section of a fiber was less than 10 um/s. Hence only about 10% of the 200 um long scattering volume would be replaced by a new portion of the fiber in the 2 s observation time. This time scale is 20 to 40 times longer than that actually observed. Estimates of the period of the longitudinal and transverse oscillations in the fiber during an isometric tetanus yield values in the range of 0.1 ms and 1 ms , respectively, and such oscillations should be rapidly damped.[79]

If the S-1 cross-bridge assumes different orientations of its long axis in the different physiological states, then the measured birefringence, B, in these physiological states should differ. Although early studies on the polarization of intrinsic tryptophan fluorescence in muscle indicated that the S-1 head did indeed take on different orientations in the different physiological states,[75,76] recent results indicate that this quantity shows no difference between the resting and contracting muscle, indicating that the distribution of cross-bridge orientations does not change appreciably in these different physiological states.[77] Further, recent polarized fluorescence studies of actin-bound S-1 demonstrated that if rotation of actin-bound S-1 occurs, the lifetime of the bound state at a fixed angle is long or the rotation is through a small angle or both.[85] Of particular interest are the very recent millisecond time-resolved x-ray diffraction studies[25] indicating that in a tetanus the S-1 heads change their tilt on both a quick release and a quick stretch transient, and for both of these transients there is a slow recovery of the orientation of the S-1 heads following the length perturbation over a period of 50-100 ms. The quick release perturbation shows a very fast component (5 ms) of recovery in addition to the slow recovery phase. The PCS studies reported above showed relative axial velocities of 1-2 um/s for myofibrillar structures during the plateau of a tetanus are compatible with these recent x-ray findings. However, interpretation of the x-ray results is hampered by a lack of information about the shape of the S-1 head and its capabilities for changes in shape.

SUMMARY

We conclude that the 200 um central segment of the muscle fiber observed during an isometric tetanus contains no scattering material possessing relative axial velocities greater than 3 um/s that are accompanied by displacements greater than 0.05 um. That is to say, if all the scattering material in a fiber had relative velocities with an rms value of 10 um/s and rms displacements of 0.05 um, the observed decay rate of $g^{(2)}(\tau)$ at $\theta = 30°$ would have contained a component having a $\tau_{1/2} = 10$ ms and an amplitude of 0.2. Had such a component been present it would have been detected in our experiments, but none was observed. The dependence of $1/\tau_{1/2}$ on scattering angle and on fiber orientation indicates that the motion of the scatterers is primarily along the fiber axis and relative velocities are 1-2 um/s. The large observed amplitudes of $g^{(2)}(\tau)$ at small scattering angles excludes the possibility that the motion of the scatterers was constrained to relative displacements having an rms amplitude $\leq$ 0.1 um. On the other hand, these results alone are not sufficiently precise to exclude the possibility that the scatterers actually had relative rms displacements as small as 0.3 um. This possibility is not inconsistent with the observed results, for a measured $\tau_{1/2}$ of 70 ms at $\theta = 30°$ corresponds to an rms relative velocity of 1.3 um/sec and a coherence time of approximately 200 ms. Thus, velocities constant over a coherence time would result in displacements of 0.3 um.

Structural fluctuations due to imbalance in independent cross-bridge cycling at the level of the myofibrillar sarcomere are limited to relative velocities of thick or thin filaments of 3 um/s having axial displacements greater than 0.05 um. However, if such structural fluctuations had made a major contribution to the observed decay rate of $g^{(2)}(\tau)$, a large change in the amplitude and decay rate of $g^{(2)}(\tau)$ would be expected for conditions of little or no overlap of the thick and thin filaments when only a very few cross-bridges are present in the region of overlap. However, neither the amplitude nor the decay rate of $g^{(2)}(\tau)$ showed any dependence on sarcomere length as would be expected if the origin of the structural fluctuations was entirely due to independent, asynchronous cycling of cross-bridges at the myofibrillar level.

Direct microscopic observation using a 500x water immersion lens revealed axial translations of one region of a fiber relative to another during an isometric contracton.[79] Although difficult to measure by eye, these observed relative motions had velocities in the range of 1-2 um/s and were consistent with the existence of the relative motion of groups of myofibrillar sarcomeres moving axially relative to other groups of myofibrillar sarcomeres.

Possible explanations for the absence of evidence in these studies for structural fluctuations at the level of the myofibrillar sarcomere are: (1) the present model of asynchronous cross-bridge cycling may not be correct, (2) the structural rigidity imparted to the thick and thin filament arrays by the M- and Z-lines may constrain their relative displacements to less than 0.05 um, (3) cross-bridges may be attached to the thin filaments for most of their cycle time, thus resulting in a structure that is too rigid to allow resolvable relative displacements, and (4) the upper limit of a 3 um/s relative velocity for structural fluctuations set by our results is approximately the same as the maximum relative velocity of thick and thin filaments during an unloaded shortening of a muscle fiber, and relative velocities of thick and thin filaments less than 1-2 um/s would be obscured by the larger shortening motions.

REFERENCES

FOR LECTURE #1:

1. F. D. Carlson and D. R. Wilkie, "Muscle Physiology," Prentice Hall, Inc., Englewood Cliffs, NJ (1974).
2. A. Weber and J. M. Murray, Physiol. Rev. 53:612 (1973).
3. J. M. Squire, Ann. Rev. Bioeng. 4:137 (1975).
4. H. G. Mannherz and R. S. Goode, Ann. Rev. Biochem. 45: 427 (1976).
5. R. A. Murphey, Ann. Rev. Physiol. 41:737 (1979).
6. S. B. Marston, R. T. Tregear, C. D. Rodger and M. L. Clarke, J. Mol. Biol. 128:11 (1979).
7. E. Taylor, CRC Crit. Rev. Biochem. 6:103 (1979).
8. A. F. Huxley and R. M. Simmons, Nature (Lond.) 233:533 (1971).
9. H. E. Huxley, Science 164:1356 (1969).
10. A. F. Huxley and R. M. Simmons, C.S.H.S.Q.B. 37:669 (1972).
11. W. F. Harrington, Proc. Natl. Acad. Sci. U.S.A. 68:685 (1971).

FOR LECTURE #2:

12. F. Oosawa et al., C.S.H.S.Q.B. 37:277 (1972).
13. "Organization of the Cytoplasm", C.S.H.S.Q.B. XLVI (1982).
14. R. H. Mendelson, in: "Cell and Muscle Motility," R. Dowben and J. Shea, eds., Plenum Publ. Corp. (1982) pg 257.
15. S. C. Harvey and H. C. Cheung, in: "Cell and Muscle Motility," R. Dowben and J. Shea, eds., Plenum Publ. Corp. (1982) pg. 279.
16. D. D. Thomas et al., Proc. Natl. Acad. Sci. U.S.A. 72:1729 (1975).
17. D. D. Thomas and R. Cooke, Biophys. J. 25:19a (1979).
18. D. D. Thomas et al., Biophys J. 32:873 (1980).
19. J. Borejdo et al., Proc. Natl. Acad. Sci. U.S.A. 76:6346 (1979).
20. J. Borejdo and S. Putnam, Biochem. Biophys. Acta 459:578 (1977).
21. H. E. Huxley and W. Brown, J. Mol. Biol. 30:383 (1967).
22. J. C. Hazelgrove and H. E. Huxley, J. Mol. Biol. 77:549 (1973).
23. J. C. Hazelgrove et al., Nature (Lond.) 261:606 (1976).
24. R. W. Lymn and G. H. Cohen, Nature (Lond.) 258:770 (1975).
25. H. E. Huxley et al., Proc. Natl. Acad. Sci. U.S.A. 78:2297 (1981).
26. A. Elliott and G. Offer, J. Mol. Biol. 123:505 (1978).
27. K. Takahashi, J. Biochem. 83:905 (1978).
28. M. Burke et al., Biochem. 12:701 (1973).
29. W. F. Harrington, Proc. Natl. Acad. Sci. U.S.A. 76:5066 (1979).
30. J. Newman and H. Swinney, Biopolymers 15:301 (1976).

31. H. Z. Cummins and P. N. Pusey, in: "Photon Correlation Spectroscopy and Velocimetry," H. Z. Cummins and E.R. Pike, eds., Plenum Publ. Corp. (1977) pg. 164.
32. Y. Allen and A. Hochberg, Rev. Sci. Inst. 46:381 (1975).
33. C. Montague and F. D. Carlson, in: "Advances in Enzymology," S. Colowick and N. Kaplan, eds. (in press).
34. T. J. Racey, R. Hallett, and B. Nickel, Biophys. J. 35:557 (1981).
35. C. J. Oliver, Adv. Phys. 27:387 (1978).
36. P. Nieuwenhuysen, Macromolecules 11:832 (1978).
37. J. G. de la Torre and V. A. Bloomfield, Biochem. 19:5118 (1980).
38. S. J. Broersma, J. Chem. Phys. 32:1626 (1960).
39. S. J. Broersma, J. Chem. Phys. 32:1632 (1960).
40. J. Newman et al., J. Mol. Biol. 116:593 (1977).
41. H. Yamakawa, "Modern Theory of Polymer Solutions," Harper and Row, NY (1971).
42. P. J. de Gennes, "Scaling Concepts in Polymer Physics," Cornell Univ. Press, Ithaca, NY (1979).
43. W. A. Wegener et al., J. Chem. Phys. 73:4086 (1980).
44. W. A. Wegener, Biopolymers 19:1899 (1980).
45. K. Zero and R. Pecora, Macromolecules, (in press).
46. C. C. Yang and R. Pecora, J. Chem. Phys. 72:5333 (1980).
47. S. Highsmith et al., Biochem. 21:1192 (1982).
48. M. Fujiwara et al., Reports on Progress in Polymer Physics in Japan 23:531 (1980).
49. A. B. Fraser et al., Biochem. 14:2207 (1975)
50. J. T. Yang and C. C. Yoo, Biochem. 16:578 (1977).
51. C. Montague, K. W. Rhee, and F. D. Carlson, J. Cell Motility and Muscle Res. (in press).
52. U. Aebi et al., Nature 288:296 (1980).
53. D. Suck, W. Kabsch, and H. G. Mannherz, Proc. Natl. Acad. Sci. U.S.A. 78:4319 (1981).
54. F. D. Carlson and A. B. Fraser, J. Mol. Biol. 89:273 (1974).
F. D. Carlson, J. Mol. Biol. 95:139 (1975).
55. S. Fujimi, J. Phys. Soc. Japan 29:751 (1970).
S. Ishiwata and S. Fujimi, J. Phys. Soc. Japan 31:1601 (1971).
56. S. Fujimi and S. Ishiwata, J. Mol. Biol. 62:251 (1971).
57. S. Ishiwata and S. Fujimi, J. Mol. Biol. 68:511 (1972).
58. K. Mihashi, Arch. Biochem. Biophys. 107:441 (1964).
59. F. Lanni, D. L. Taylor, and B. R. Ware, Biophys. J. 35:351 (1981).
60. J. Newman and F.D. Carlson, Biophys. J. 29:37 (1981).
61. B. M. Millman and P. M. Bennett, J. Mol. Biol. 103:439 (1976).
62. T. Maeda and S. Fujimi, J. Phys. Soc. Japan 42:1983 (1977).
T. Maeda and S. Fujimi, Macromolecules 14:809 (1981).
63. A. Wegner, J. Mol. Biol. 131:839 (1979).
64. C. Montague, unpublished data.

65. R. W. Rosser et al., Macromolecules 11:1239 (1978).
66. T. J. Herbert and F. D. Carlson, Biopolymers 10:2231 (1971).
67. A. D'Albis and W. Gratzer, J. Biochem. 251:2825 (1976).
68. S. Lowey et al., J. Mol. Biol. 42:1 (1969).
69. J. G. de la Torre and V. A. Bloomfield, Biochem. 19:5118 (1980).
70. S. Kobayasi and T. Totsuka, Biochem. et Biophys. Acta 376:375 (1975).
71. S. Highsmith et al., Proc. Natl. Acad. Sci. U.S.A. 74:4986 (1977).
72. R. Mendelson et al., Biochem. 12:2250 (1973).

FOR LECTURE #3:

73. J. Borejdo and M. F. Morales, Biophys. J. 20:315 (1977).
74. F. D. Carlson, Biophys. J. 15:633 (1975).
75. C. G. Dos Remedios et al., J. Gen. Physiol. 59:103 (1972).
76. C. G. Dos Remedios et al., Proc. Natl. Acad. Sci. U.S.A. 69:2542 (1972).
77. K. Guth, Biophys. Struct. Mech. 6:81 (1980).
78. R. F. Bonner and F. D. Carlson, J. Gen. Physiol. 64:555 (1975).
79. R. C. Haskell and F. D. Carlson, Biophys. J. 33:39 (1981).
80. S. Fujimi, in: Proc. NATO Adv. Study Inst. on Scattering Techniques Applied to Supramolecular and Nonequilibrium Systems, S. H. Chen, B. Chu, and R. Nossal, eds., Plenum Press (1981) pg 725.
81. B. Saleh, "Photoelectron Statistics," Springer-Verlag, Inc. NY (1978).
82. A. M. Gordon et al., J. Physiol. (Lond) 184:170 (1966).
83. F. D. Carlson, J. Mol. Biol. 95:139 (1975).
84. R. C. Haskell and F. D. Carlson, unpublished.
85. T. Yanagida, J. Mol. Biol. 146:539 (1981).
86. K. Mihashi and P. Wahl, FEBS Lett. 52:8 (1975).
87. A. Eberstein and A. Rosenfalck, Acta Physiol. Scand. 57:144 (1963).
88. D. L. Taylor, J. Cell Biol. 68:497 (1976).

DYNAMIC LIGHT SCATTERING STUDY OF MUSCLE F-ACTIN IN SOLUTION*

Satoru Fujime, Shin'ichi Ishiwata and Tadakazu Maeda

Mitsubishi-Kasei Institute of Life Sciences, Tokyo 194

CONTENTS

INTRODUCTION

Structural basis of F-actin

G-actin is globular in shape (Fig. 1a). Its molecular weight is about 42k daltons. G-actin polymerizes into F-actin under physiological salt concentrations (Fig. 1b). Based on observations by electron microscopy, a "pearl-and-necklace" model is proposed for the ultrastructure of F-actin. F-actin is a two-stranded helical polymer. The half pitch of the helix is 35 nm and within this length, there are 13 G-actins. The total length of F-actin varies according to polymerization conditions and, roughly speaking, is longer than 1 μm. As might be supposed from its structure, F-actin is rather stiff. Electron micrographs show the images of gradually curved F-actin. Tropomyosin is a rodlike protein (Fig. 1c). When tropomyosin molecules are added to the solution of F-actin, they bind to F-actin and settle in the grooves of F-actin helix forming tropomyosin strands (Fig. 1d). Myosin has two heads called subfragment-1 (S-1) and binds to F-actin in the absence of ATP. Partial digestion by some kind of proteases produces heavy meromyosin (HMM) and also S-1 (Fig. 1e).

Brief review of previous light-scattering studies of F-actin

Using an ac-coupled spectrum analyzer, the power spectra of

---

* S. Ishiwata (visiting scientist from Waseda University, Tokyo 160) contributed to section 3 and T. Maeda to section 4.

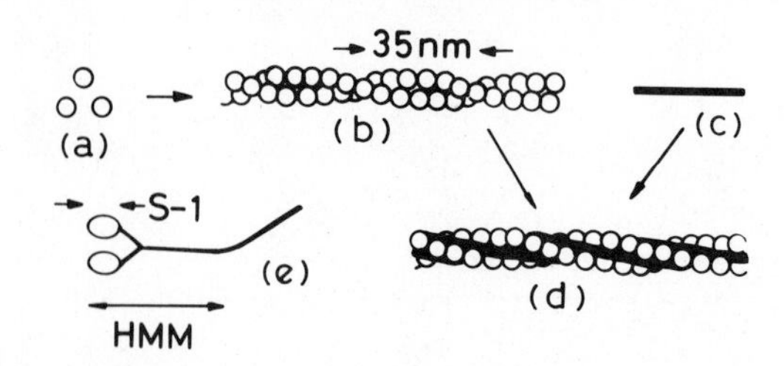

Figure 1. Illustration of the structures of muscle proteins.

light scattered from solution of F-actin were first obtained by Fujime (1970) and were approximated using a single Lorentzian having a half width Γ. The Γ vs K^2 relation thus obtained has a form of $\Gamma = AK^2 + B$ for $K^2 \geq 2 \times 10^{10}$ cm^{-2}, where A and B were non-zero constants and K is the length of the scattering vector. The extensive studies by Ishiwata and Fujime (1971, 1972) showed that both A and B were delicately dependent on solvent conditions and the state of F-actin interacting with other muscle proteins such as tropomyosin and HMM.

Using a digital correlator, the correlation functions of light scattered from solutions of F-actin and its complexes with HMM or S-1 were studied by Carlson and his associates (1974, 1975). The correlation functions for F-actin are highly non-exponential and the initial decay rate deduced from a cumulant expansion method has a form of $\overline{\Gamma} = A(K^2)K^2$ where $A(K^2)$ is an increasing function of K^2. Maeda and Fujime (1977) reconfirmed this result, and further reported the observation of a very long tail (which decays in a range of tens of seconds) in the correlation function.

Since the long tail behavior of correlation functions is expected to come from entanglements of long filaments such as F-actin in semi-dilute solutions, Oplatka and his associates (1977) studied F-actin and its complexes with HMM at relatively low actin concentrations. In order to avoid complexities coming from sample polydispersity of *in vitro* reconstituted F-actin, Newman and Carlson (1980) used intact thin filaments of scallop adductor muscle (F-actin/tropomyosin complex in Fig. 1d). Based on a theoretical model (Fujime, 1970; Fujime and Maruyama, 1973), Maeda and Fujime (1981) succeeded in deducing the flexibility of F-actin from the experimental results of Newman and Carlson.

In this article, we will present mainly the experimental results on *in vitro* reconstituted F-actin and its complexes with other muscle proteins in a relatively wide range of actin concentrations, studied by digital autocorrelation and fast Fourier methods. Then we will briefly discuss the experimental results based on our theoretical model outlined in Fujime, Maeda and Ishiwata (1982).

METHODS

The time correlation function G(t) and the power spectrum S(f) of a stochastic signal i(t) are connected with each other by the Fourier integral. Let us consider a model situation where the signal after digital-to-analog (D/A) conversion is first fed to a low-cut filter (LCF) and then to a correlator after analog-to-digital (A/D) conversion (Fig. 2a). Then we will have

$$\bar{G}(t) = \int_{-\infty}^{\infty} [1 - F(f)]S(f)e^{2\pi i f t}df = G(t) - \tilde{F}(t)*G(t) \tag{1a}$$

$$\bar{S}(f) = [1 - F(f)]S(f) \tag{1b}$$

where the asterisk means convolution and $\tilde{F}(t)$ is the Fourier transform of F(f). The filter function [1 - F(f)] is assumed, for convenience of algebraic manipulation, to be given by

$$F(f) = \frac{\gamma}{2}\frac{2\gamma}{(2\pi f)^2 + \gamma^2} \quad \text{or} \quad \tilde{F}(t) = \frac{\gamma}{2}e^{-\gamma t} \tag{2}$$

where $\gamma/2\pi$ is the critical frequency in Hz units of the low-cut filter. The [1 - F(f)] in this particular choice of F(f) just corresponds to the frequency characteristics of a low-cut filter consisting of a two-stage cascade RC-network. Then we have for G(t) = $\exp(-\Gamma t)$, for example,

$$\tilde{F}(t)*G(t) = \gamma\frac{\Gamma e^{-\gamma t} - \gamma e^{-\Gamma t}}{\Gamma^2 - \gamma^2} \tag{3}$$

Experimental verification of the validity of eq.(1a) with eq.(3) was made by measuring the intensity of light scattered from solution of polystyrene latex spheres (Fig. 2b).

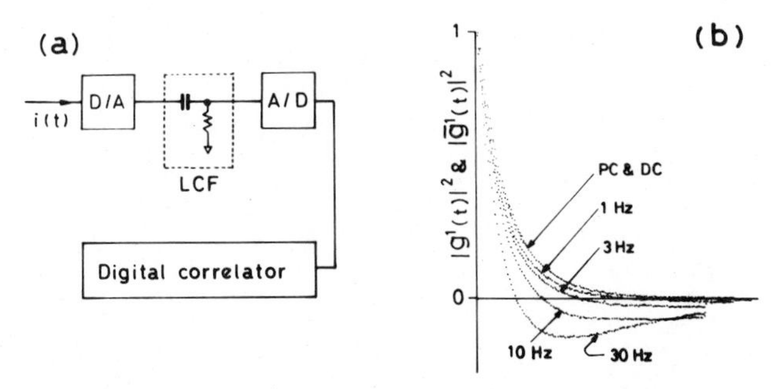

Figure 2. (a) Block diagram of the electronics arrangement.
(b) Correlation functions measured at varied values of the critical frequency of the low-cut filter. (See text)

Let us consider another model situation, where $i(t)$, or $n(t)$ in a photon-counting experiment, is recorded with a sampling time T [s]. Then we have a set of sampled values of $i(t)$ such as $i(t,T)$, $i(t+T,T)$, ... , $i(t+mT,T)$, ... , which we simply write as $\{i(m)\}$. We divide this big set into many subsets $\{i(m)\}_0$, $\{i(m)\}_1$, ... , $\{i(m)\}_k$, ... , where $\{i(m)\}_k$ is such a subset as it contains components $i(N_k)$, $i(N_k+1)$, ... , $i(N_k+N-1)$ and $N_k = kN$. Now we have the correlation function $G_k(t)$ and the power spectrum $S_k(f)$ for each of the above subsets. According to Rice (1944), the correlation function $G(t)$ and the power spectrum $S(f)$ for the big set are given by the (ensemble) average of $G_k(t)$ and $S_k(f)$ over all subsets k, respectively. Here we have to take account of the Nyquist sampling theorem. For a sampling time T, we do not have any information on frequency components higher than 1/2T (Hz). In addition to this, for data points N of each subset, say 1024, we do not have any information on frequency components lower than 1/2NT (Hz). When we compute $S_k(f)$ by use of a fast Fourier transform (FFT) algorithm, we subtract the mean value of $i(m)$ from each of the sampled values, i.e., $\{i(m) - \langle i\rangle\}_k$. The mean value $\langle i\rangle = \langle i(m)\rangle_k$ for any subset k does not correspond to the dc component of $i(t)$ during the time interval for the subset k, but it contains frequency components lower than 1/2NT (Hz). Figure 3 depicts the frequency characteristics of an "FFT spectrum analyzer." Likewise, when we adopt the usual definition of the baseline level of a correlation function, the baseline level B_k of $G_k(t)$ is given by $N\langle i(m)\rangle_k{}^2$. The summation of B_k over all k is not equal to the baseline level B of $G(t)$. This comes from the fact that $\langle i(m)i(m+N)\rangle \neq \langle i(m)i(m+\infty)\rangle$ for a finite value of N, say 1024. The correlation function $G_k(t) - B_k$ has no frequency components higher than 1/2T and lower than 1/2NT (Hz). Thus we have

$$\bar{G}(t) = \text{Ensemble average of } G_k(t) - B_k \tag{4a}$$

$$\bar{S}(f) = \text{Ensemble average of } S_k(f) \tag{4b}$$

where the critical frequency of a low-cut filter is close to 1/2NT. The consideration given above is important in what follows.

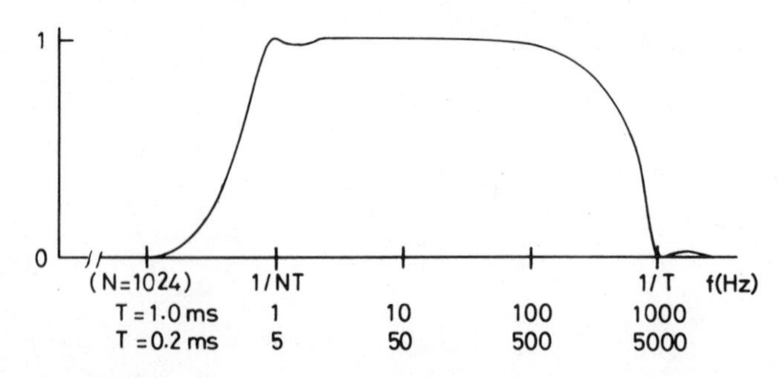

Figure 3. The frequency characteristics of an FFT analyzer.

The time correlation function $G^2(t)$ of the intensity of light scattered from solutions of F-actin and F-actin complexed with other muscle proteins were measured by use of a 128-channel digital correlator. Our correlator has 16 extra channels whose delay times start from $128T(1 + M)$, where T is the channel width (sampling time) and M = 1, 2 or 3. For discussion that follows, we define the following three types of normalized correlation functions:

$$[g^1(t)]^2 = [G^2(t) - G^2(\infty)]/[G^2(T) - G^2(\infty)] \tag{5}$$

$$[g^1_*(t)]^2 = [G^2(t) - G^2(\overline{512T})]/[G^2(T) - G^2(\overline{512T})] \tag{6}$$

$$[\bar{g}^1(t)]^2 = [\bar{G}^2(t) - \bar{G}^2(\infty)]/[\bar{G}^2(T) - \bar{G}^2(\infty)] \tag{7}$$

where $t = (1 + n)T$; $n = 0, 1, \ldots, 127$, and $G^2(\overline{512T})$ is the average of the above-mentioned 16 extra channels for M = 3.

We also recorded sampled values of photon-counts, $\{n(t+mT,T)\} = \{n(m)\}$, on magnetic tapes. From these data, we computed both correlation functions and power spectra defined in eqs(4a and b) for N = 1024. For our hardware system, see APPENDIX.

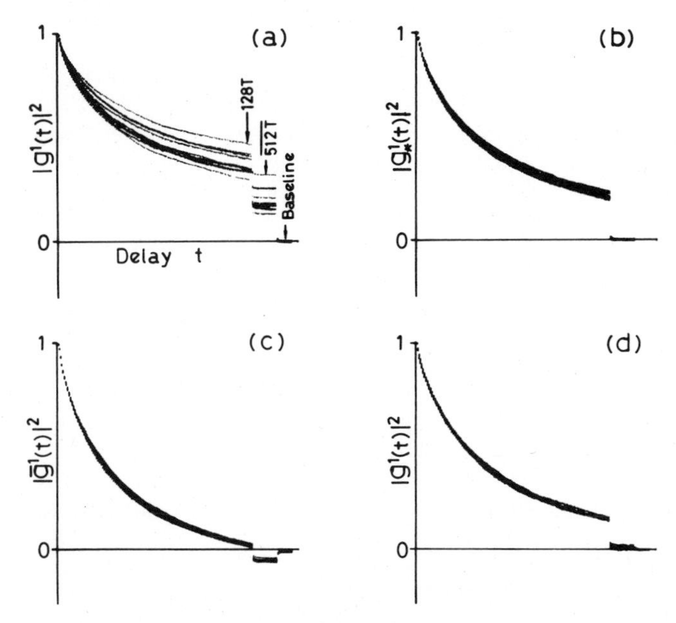

Figure 4. Experimental correlation functions (10 runs each) of the intensity of light scattered from solution of F-actin. Data accumulation period = 300 s/run, temperature = 3 °C, T = 0.2 ms and $K^2 = 0.9 \times 10^{10}$ cm^{-2}. (a-c): at 0.5 mg/mL F-actin, (b) is a replot of (a) and, in (c), $\gamma/2\pi = 3$ Hz. (d): at 0.05 mg/mL F-actin. For details, see text.

EXPERIMENTAL

Figure 4a shows examples of $[g^1(t)]^2$ at 0.5 mg/mL F-actin. Strong non-reproducibility of the profiles of correlation functions comes from the contribution of very slowly decaying components, because $[g_*^1(t)]^2$ in Fig. 4b and $[\bar{g}^1(t)]^2$ in Fig. 4c show nice reproducibility. Figure 4d shows examples of $[g^1(t)]^2$ at 0.05 mg/mL F-actin. Here observed profiles of correlation functions have nice reproducibility and no long tail can be seen. As a measure of the existence of slowly decaying components, let us define

$$R = [g^1(\overline{512}T)/g^1(128T)]^2 \tag{8}$$

Figure 5 shows the R vs concentrations of F-actin. The R-values drastically changed at F-actin concentrations between 0.2 and 0.5 mg/mL. However, the decay characteristics of $[g_*^1(t)]^2$ was rather similar irrespective of the concentrations of F-actin as the values of $[g_*^1(128T)]^2$ in Fig. 5 show. Figure 6 shows another example of correlation functions. A poor reproducibility of the profiles of $[g^1(t)]^2$ measured for 300 s (Fig. 6a) was improved very little even when we measured them for 3600 s (Fig. 6b). On the other hand, when we cut-off the frequency components lower than 0.3 Hz, a rather nice reproducibility of the profiles of $[\bar{g}^1(t)]^2$ was observed even for a data accumulation period of 300 s (Fig. 6c), and a very nice reproducibility was observed for a data accumulation period of 4500 s (Fig. 6d). A measuring time (data accumulation period) of 3600 s is not long enough to observe such slowly decaying components with a high statistical accuracy.

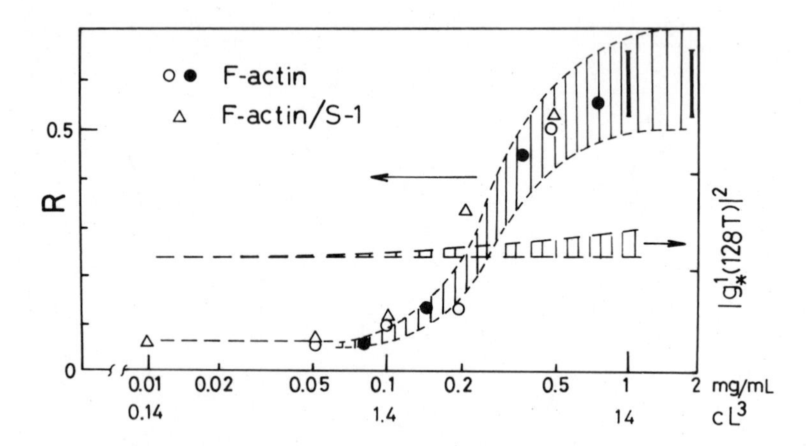

Figure 5. R and $[g_*^1(128T)]^2$ vs concentrations of F-actin (average of 30-90 runs per point). Experimental conditions were the same as those in Fig. 4. $[g_*^1(128T)]^2$ values of F-actin/S-1 were a bit larger than those for pure F-actin for all concentrations studied. For details, see text.

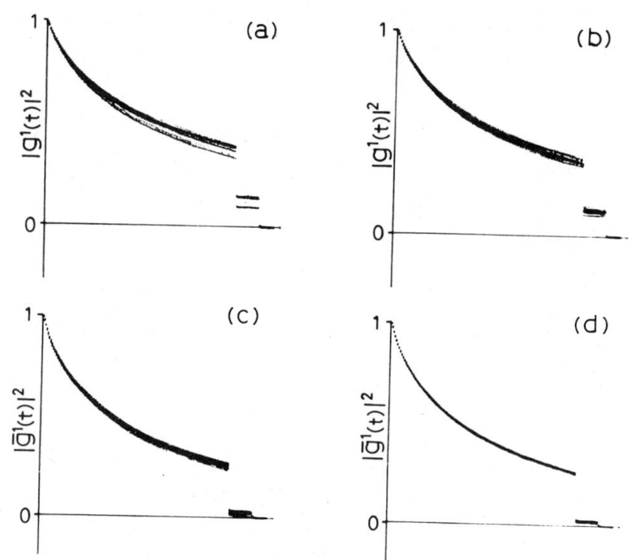

Figure 6. Correlation functions (10 runs each) of the intensity of light scattered from solution of F-actin/S-1 at 0.2 mg/mL F-actin and 1.25 mg/mL S-1. (a): $[g^1(t)]^2$ at the measuring time of 300 s/run and (b): 3600 s/run. (c): $[\bar{g}^1(t)]^2$ at $\gamma/2\pi$ = 0.3 Hz and the measuring time of 300 s/run and (d): 4500 s/run. Other conditions are the same as those in Fig. 4.

Figure 7 shows examples of correlation functions $[\bar{g}^1(t)]^2$ defined in eq.(4a). Even for solution at 1 mg/mL F-actin, a relatively nice reproducibility of the profiles of correlation functions was observed for an effective measuring time of 80 s (i.e., the average of 400 $[G_k(t) - B_k]$'s). This is a natural consequence of the fact that the

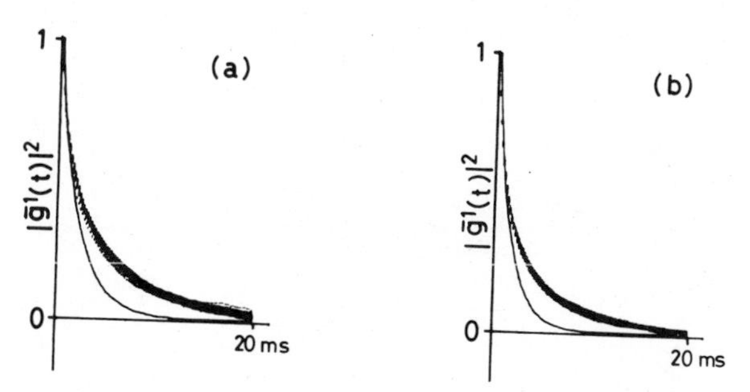

Figure 7. Correlation functions (12 runs each) of the intensity of light scattered from solution of F-actin at 1 mg/mL. T = 0.2 ms, N = 1024, temperature = 15 °C, effective measuring time = 80 s/run and the scattering angles of 30° for (a) and 50° for (b). (—): single exponentials.

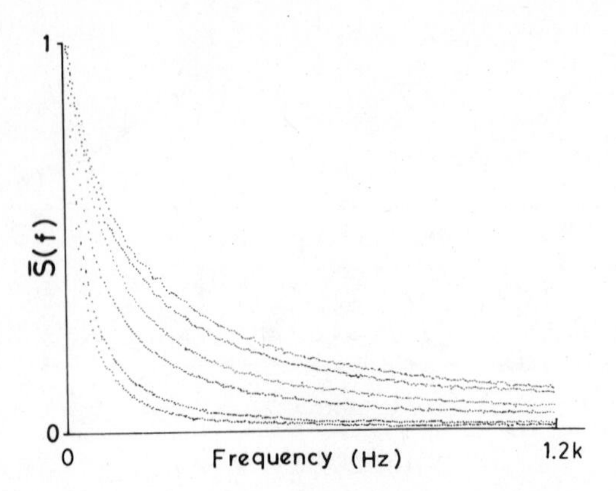

Figure 8. Power spectra (an average of 12 runs) of the intensity of light scattered from solution of F-actin. The data used for this computation are the same as those used for computation of correlation functions in Fig. 7. Scattering angles are 30°, 50°, 70°, 90°, 110° and 130°, respectively, for narrow to wide spectra.

frequency components lower than 2.5 Hz (T = 0.2 ms and N = 1024) were filtered out. The solid lines in Fig. 7 represent the single exponentials. Corresponding to a large deviation between observed correlation functions and single exponentials, the power spectra defined in eq.(4b) have large components at low frequency regions (Fig. 8). Note that the delay time of 10 ms in a correlation function corresponds to the frequency of $1/(10 \text{ ms} \times 2\pi) = 17$ Hz in a power spectra.

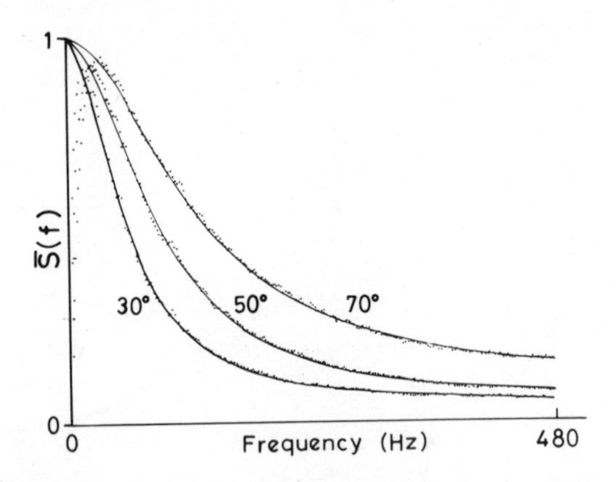

Figure 9. Power spectra (an average of 10 runs) of the intensity of light scattered from solution of F-actin at 1 mg/mL. T = 1 ms, N = 1024, data accumulation period = 400 s/run and temperature = 15 °C. The FFT spectra were modified using eq.(1b) with $\gamma/2\pi = 30$ Hz. For details, see text.

We have to take account of the fact that the initial decay rates of correlation functions have larger values whereas the half-height widths of corresponding power spectra have smaller values.

As mentioned in Introduction, our earlier results were obtained using the ac-coupled spectrum analyzer. To compare the present results with the previous ones, the FFT spectra were further modified using the relation given in eq.(1b), where $\gamma/2\pi$ was chosen to be 30 Hz, the equivalent over-all frequency characteristics of our old spectrum analyzer. Figure 9 shows some examples of the spectra thus obtained. — See also the second paragraph in APPENDIX. — The solid lines represent the least-squares fitted curves to

$$\overline{S}(f) = P\frac{\Gamma}{f^2 + \Gamma^2} + Q \qquad (P/\Gamma + Q = 1) \tag{9}$$

where P and Q are constants. The Q contains contributions from high frequency components as well as the frequency-independent shot noise. The magnitude of Q was less than 0.05 at the scattering angles lower than 70°, but it became 0.1 to 0.15 at the higher scattering angles.

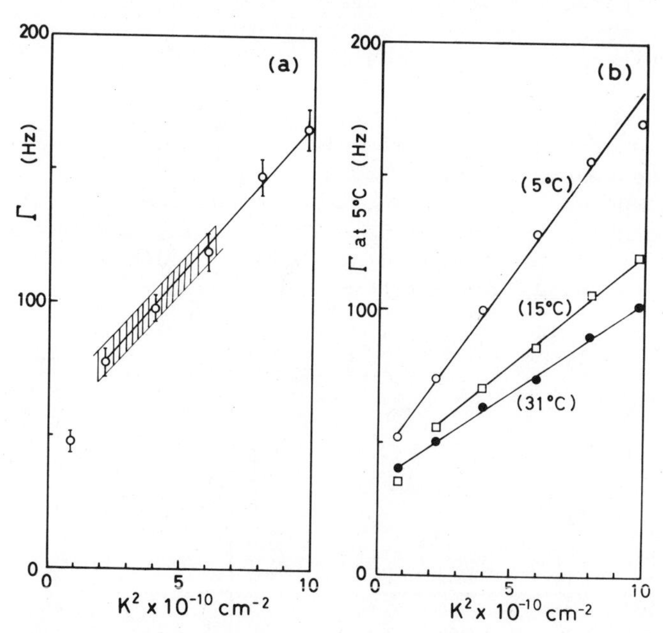

Figure 10. The Γ vs K^2 relations of the power spectra for solutions of F-actin at 1 mg/mL. The relations, eqs(1b, 4b and 9), were used. (a): T = 1 ms, N = 1024, temperature = 15 °C and $\gamma/2\pi$ = 30 Hz. The shaded area indicates the region where the previous results are included. (b): the Γ vs K^2 relations at different temperatures after T/η scaling to 5 °C. For details, see text.

Figure 10a shows the Γ vs K^2 relation of the spectra $\bar{S}(f)$ modified with eq.(1b). This result is in good agreement with our old results. In the present approximation of eq.(9), the contribution to $\bar{S}(f)$ from the fast components of fluctuations results in an increase of the Q-value, because very wide Lorentzians with small amplitudes are practically approximated as the frequency-independent component. In order to see how this Γ vs K^2 relation changes with variation of environmental conditions, we studied the temperature dependence of power spectra and correlation functions. Figure 10b shows Γ vs K^2 relations at three different temperatures, where Γ values have been corrected with the ratio of the absolute temperature T to the solvent viscosity η. The higher the measured temperature, the smaller the Γ values corrected to 5 °C. Since it has been known that F-actin becomes more flexible as temperature rises, this result means that the Γ values become small when the filament becomes flexible. This is opposite to the change in $\bar{\Gamma}$ of correlation functions, that is, the initial decay rates $\bar{\Gamma}$ become larger as the filament flexibility increases (Maeda and Fujime, 1981). Figure 11 shows examples of correlation functions $[\bar{g}^1(t)]^2$ at different temperatures. The correlation function at 31 °C decays faster than that at 5 °C, but after T/η scaling the former (the solid line) decays more slowly than the latter *at longer time regions*. Reflecting this situation, Γ measured at 31 °C and corrected to 5 °C is smaller than Γ measured at 5 °C. The same trend was also observed for F-actin/tropomyosin complexes.

Because of the reciprocal relation between the correlation function and the power spectrum, the correlation function is sensitive to the fast components of fluctuations whereas the power spectrum to the slow components of fluctuations. The apparent discrepancy between $\bar{\Gamma}$ and Γ comes from a very wide distribution of decay rates in the case of F-actin in semidilute solution.

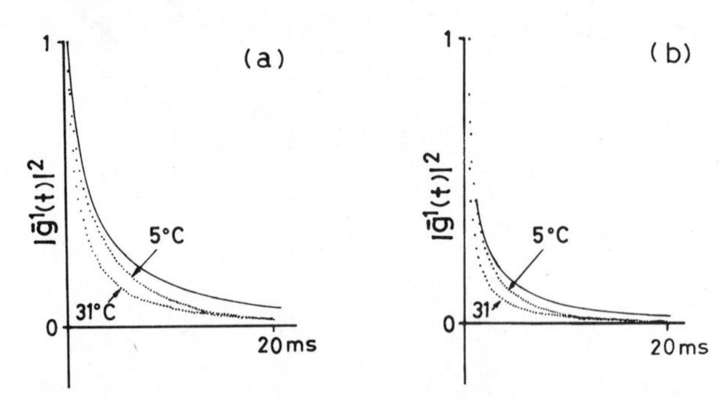

Figure 11. Correlation functions $[\bar{g}^1(t)]^2$ at 5 °C and 31 °C. At 1 mg/mL F-actin, T = 0.2 ms, N = 1024 and effective data accumulation period = 80 s × 10 runs. The solid line represents the correlation function at 31 °C after T/η scaling to 5 °C. The scattering angles were 50° in (a) and 90° in (b). For T/η scaling, see Discussion.

THEORETICAL

Now we consider a theoretical model which may account for the experimental facts at least qualitatively.* Let us define the conformation of a long, thin and semiflexible filament by a space curve $\vec{r}(s,t)$, where s is the coordinate of a segment and t is time:

$$\vec{r}(s,t) = \vec{R}(t) + s\vec{t}(t) + \sum_{n\geq 2} \vec{q}(n,t)Q(n,s) \tag{10}$$

Here $\vec{R}(t)$ is the position vector of the center of resistance of, and $\vec{t}(t)$ is the unit vector parallel to the tangent at s of the filament. The terms under the summation sign represent the bending motions of the semiflexible filament. The field correlation function $g^1(t)$ of light scattered from solution of semiflexible filaments in this model is given by (L being the length of the filament)

$$g^1(t) = \iint_{-L/2}^{L/2} J(s,s',t)dsds' \Big/ \iint_{-L/2}^{L/2} J(s,s',0)dsds' \tag{11}$$

$$J(s,s',t) = \langle \exp[i\vec{K}\cdot\{\vec{r}(s,t) - \vec{r}(s',0)\}]\rangle$$

$$= \langle(0th \text{ and } 1st \text{ modes})\rangle \times \prod_{n\geq 2} \langle nth \text{ mode})\rangle \tag{12}$$

At first, we assume that the filament is a stiff rod. Let us denote by D_1 (= D_2) and D_3 the sideways and lengthways translational diffusion constants respectively of, and by Θ the rotational diffusion constant of the long rod. The distribution function $F(\vec{R},\vec{t};t)$ for the configuration of rods is known to satisfy

$$[\partial/\partial t - D_3(\vec{t}\cdot\partial/\partial\vec{R})^2 - D_1\{\partial^2/\partial\vec{R}^2 - (\vec{t}\cdot\partial/\partial\vec{R})^2\} - \Theta\nabla^2]F = 0 \tag{13}$$

By putting F_k to be the space Fourier transform of F, we have

$$[\partial/\partial t - \Theta\{\nabla^2 - \mu^2\cos^2\theta\}]f_K = 0 \tag{14}$$

$$\mu^2 = (D_3 - D_1)K^2/\Theta = (KL)^2/12 \quad \text{and} \quad F_K = \exp(-D_1K^2t)f_K \tag{15}$$

where θ is the angle between $\vec{t}$ and $\vec{K}$, and $\Theta = 12D_1/L^2$ was assumed. For $\mu^2 \gg 1$, i.e., $KL \gg 1$, the Green function to eq.(14) has been known (Doi and Edwards, 1978). Using this Green function, we have

$$\langle(0th \text{ and } 1st \text{ modes})\rangle$$

$$= \exp(-D_1K^2t)\cosh^{-1}(\mu\Theta t)\int_0^1 \cos[K(s-s')\eta]\exp[-\mu\eta^2\tanh(\mu\Theta t)]d\eta \tag{16}$$

Integration of eq.(16) over s and s' gives $G^1(t)$ for a rod at $KL \gg 1$. For short times $t \ll 1/(\mu\Theta)$, we have ($k = KL/2$ and $j_o(z) = (\sin z)/z$)

*A more detailed descriptions on the model will be found in Fujime, Maeda and Ishiwata (1982).

$$G^1(t) = \exp(-D_1K^2t)\int_0^1 [j_\circ(k\eta)]^2\exp[-(D_3 - D_1)K^2t\eta^2]d\eta \tag{17}$$

Equation (17) includes *continuous relaxation rates from D_1K^2 to D_3K^2 and hence has a bit long tail.*

In the case of semidilute solutions, we have another problem. Let c be the number of rods in unit volume. For high concentrations $c >> 1/L^3$ (but $c << 1/(dL^2)$ where d is the diameter of the rod), the rotational motion of each rod is severely restricted as well as the sideway translation. On the other hand, the lengthway translation is almost free. According to Doi (1975), the rotational diffusion constant of the rod under such a condition is given by

$$D_r = (D_3/L^2)/(cL^3)^2 = \Theta/(cL^3)^2 \tag{18}$$

Except for very small values of KL, we have

$$\tilde{\mu}^2 = D_3K^2/D_r = (KL)^2(cL^3)^2 >> 1 \tag{15'}$$

Then eq.(16) with $\mu\Theta \longrightarrow \tilde{\mu}D_r$ and $D_1 \longrightarrow 0$ results in

$$<(0th \text{ and } 1st \text{ modes})> = \cosh^{-1}(\tilde{\mu}D_rt)\int_0^1 \cos[K(s-s')\eta]\exp[-\tilde{\mu}\eta^2\tanh(\tilde{\mu}D_rt)]d\eta \tag{16'}$$

Integration of eq.(16') over s and s' gives $G^1(t)$ for rods at $c >> 1/L^3$, which has *a very long tail.* Indeed, for short times $t << 1/(\tilde{\mu}D_r)$ we have (Doi and Edwards, 1978)

$$G^1(t) = \int_0^1 [j_\circ(k\eta)]^2\exp(-D_3K^2t\eta^2)d\eta \tag{17'}$$

which includes *continuous relaxation rates from 0 to D_3K^2.*

When we take into account the internal bending motions of a *slightly bendable* filament, we have, to a good approximation,

$$J(s,s',t) = \exp(-D_1K^2t)\cosh^{-1}(\mu\Theta t)\int_0^1 d\eta\ \cos[K(s-s')\eta]$$

$$\times \exp[-\mu\eta^2\tanh(\mu\Theta t)]\exp[-(K^2/6){\sum}''\Phi(n,s,s',t)] \tag{19}$$

$$J(s,s',t) = \cosh^{-1}(\tilde{\mu}D_rt)\int_0^1 d\eta\ \cos[K(s-s')\eta]\exp[-\tilde{\mu}\eta^2\tanh(\tilde{\mu}D_rt)]$$

$$\times \exp[-(K^2/2)(1 - \eta^2){\sum}''\Phi(n,s,s',t)] \tag{19'}$$

$$\Phi(n,s,s',t) = \langle q_n^2\rangle[Q(n,s)^2 + Q(n,s')^2 - 2Q(n,s)Q(n,s')\exp(-t/\tau_n)] \quad (20)$$

where $\langle q_n^2\rangle$ is the expectation value of $\vec{q}(n,t)$, and τ_n is the relaxation time of the n*th* mode (Maeda and Fujime, 1981). Equation (19) is suitable for dilute solutions at $KL >> 1$, and eq.(19') for semidilute ones. The present model is a hybrid one of Doi and Edwards (1978) and Maeda and Fujime (1981).

To visualize the characteristic feature of the model, some examples of the profiles of $g^1(t)$'s based on eq.(19') are shown in Fig. 12. In order to show the long-time behavior of $g^1(t)$'s, the time scale is intentionally chosen to be too large so that the initial decay of $g^1(t)$'s is not clear in this figure. Comparing the simulated $g^1(t)$'s for a semiflexible filament with those for a rigid rod, it is evident that the rapid decay of the former comes from the contribution from the internal bending motions of the semiflexible filament. It is a characteristic feature of the present hybrid model that the initial decay rate of $g^1(t)$ does not depend on cL^3, whereas the long tail behavior of $g^1(t)$ does. This comes from the fact that for short times, both $\mu\eta^2\tanh(\mu\Theta t)$ in eq.(19) and $\tilde{\mu}\eta^2\tanh(\tilde{\mu}D_r t)$ in eq.(19') become $D_3K^2t\eta^2$.

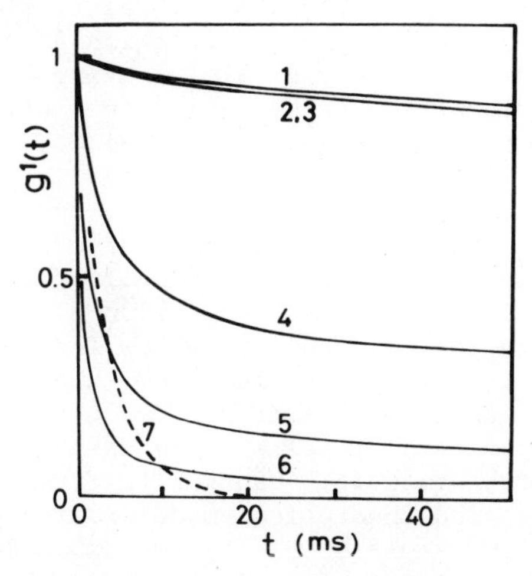

Figure 12. Examples of simulated correlation functions for semidilute solution of very long, semiflexible filaments. $L = 2$ µm, $D_o = (2D_1+D_3)/3 = 0.8 \times 10^{-8}$ cm^2/s at 5 °C and $cL^3 = 160$. (1-3): for $\gamma L = 0$ (rigid rod) at $K^2 = 2$, 5 and 10×10^{10} cm^{-2} respectively. (4-6): for $\gamma L = 0.5$ at $K^2 = 2$, 5 and 10×10^{10} cm^{-2} respectively. (7): $\exp(-D_1K^2t)$ at $K^2 = 5 \times 10^{10}$ cm^{-2}. The parameter γL is a measure of the filament flexibility (Maeda and Fujime, 1981).

DISCUSSION

Non-reproducibility of the profiles of correlation functions of F-actin at a relatively high concentration (Carlson and Fraser, 1974; Fig. 4a in this article) originates from the existence of very slowly decaying components (Maeda and Fujime, 1977). Very large far point values of correlation functions in Fig. 4a suggest a possibility that there is an occupation number fluctuation of very small number of very large aggregates at this relatively high concentration of F-actin. At the moment, however, we do not believe this possibility. We have several pieces of reasonable evidence to this. First of all, correlation functions measured at different scattering volumes keeping the coherence condition unaltered, showed the same decay behavior. Single clipped and full or scaled correlation functions showed the same behavior. The experimental amplitude of normalized intensity correlation functions was about 1.7 at a very short channel width. It did not exceed 2. (Its value for dilute solution of polystyrene latex spheres was about 1.8 at the same machine condition.) So, we believe that strong non-reproducibility of the profiles of, and large far point values of correlation functions at high actin concentrations come from very slow Gaussian fluctuations. We are considering that these very slow fluctuations come from constraint due to entanglement of very long filaments. Actually, at relatively low concentrations of F-actin, these extremely slow components disappeared (Hochberg et al, 1977; Fig. 4d). Even at relatively high concentrations of F-actin, the profiles of correlation functions were fairly reproducible provided that very slowly decaying components were eliminated by assuming that the baseline level of the correlation function was equal to the far point value, i.e., $G^2(\overline{512T})$, and/or by filtering (Figs. 4, 6 and 7).

The drastic and big change in the R-value in Fig. 5 seems to be due to the transition from dilute to semidilute regime as the cL^3 values suggested. There may be no objection if we study the F-actin solution at a dilute regime. But, very low intensities of scattered light at higher scattering angles make the experiment very difficult. In order to extract the physical properties of F-actin, for example, its flexibility parameter, it seems to be useful to study correlation functions in the forms of $[g^1_*(t)]^2$ and/or $[\bar{g}^1(t)]^2$ of semidilute solution. If we have a good theoretical model for computation of correlation functions for semiflexible filaments in semidilute regime, a comparison between theoretical and experimental $[g^1_*(t)]^2$ and/or $[\bar{g}^1(t)]^2$ will give us information on dynamics of individual filaments in the same way as we have discussed for dilute regime (Maeda and Fujime, 1981).

As shown in Fig. 5, the decay characteristics of $[g^1_*(t)]^2$ is rather insensitive to the concentration of F-actin. Figure 13 shows examples of the third order cumulant fitting of correlation functions. These examples again suggest a possibility to extract the dynamic

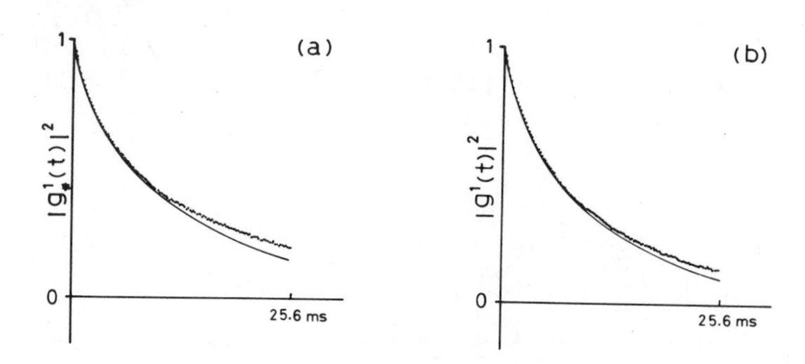

Figure 13. The third order cumulant fitting.
(a): $[g^1_*(t)]^2$ at 0.5 mg/mL F-actin, the simple average of 10 runs in Fig. 4b. (b): $[g^1(t)]^2$ at 0.05 mg/mL F-actin, the simple average of 10 runs in Fig. 4d.

properties of individual filaments in semidilute solution. A study along this line is underway.

Apparent discrepancy between the initial decay rate $\bar{\Gamma} = A(K^2)K^2$ of correlation functions and the half-height width $\Gamma = AK^2 + B$ of power spectra is clarified to a great extent. The $\bar{\Gamma}$ has a clear meaning, but its estimation is sometimes difficult as in the present case. On the other hand, the physical meaning of Γ is not clear yet, but reproducibility of its value is fairly nice and it is a good indicator of the dynamic properties of filaments in semidilute solution such as F-actin in the present case. It is noted that in some cases, experimental correlation functions are not necessarily related *strictly* to experimental power spectra because of the limited range in time and frequency domains where experimental quantities are measured.

Quantitative and critical testing of the theoretical model mentioned above has not been made yet. Qualitatively speaking, however, the model seems to be consistent with experimental results. The following statement is qualitative but intuitive. For dilute regime, the correlation function has a form (from eq.(19))

$$[g^1(t)]^2 = \exp(-2D_1K^2t)[f(t)]^2 \qquad (f(0) = 1) \qquad (21)$$

where the factor f(t) includes the effect of the lengthways translational (D_3), rotational and bending motions of filaments. Thus, the correlation function decays with an initial decay rate larger than or equal to $2D_1K^2$ (see Curve 7 in Fig. 12). On the other hand, for semidilute regimes, the decay behavior of the correlation function is quite different from that for the dilute regime. First, we put $\langle q_n^2 \rangle = 0$ in eq.(20). Then, eq.(19') or eq.(16') gives us very slow decay of the correlation function (Curves 1-3 in Fig. 12). Experimental correlation functions, however, decay to a great extent with a much larger decay rate and then with a very small decay rate (Fig. 4a). The model including the effect of the filament flexibility, i.e., $\langle q_n^2 \rangle \neq 0$, is compatible with this situation (Curves 4-6 in Fig. 12).

For long times $t \gg \tau_n$, $\Phi(n,s,s',t)$ in eq.(20) becomes $\langle q_n^2 \rangle \times [Q(n,s)^2 + Q(n,s')^2]$. Since $\tau_n \ll 1/(\mu D_r)$, for $t \gg \tau_n$ the correlation function derived from eq.(19') decays in the same way as that for a rigid rod derived from eq.(16'). Let us write

$$G(t) = G_{t,r}(t) + [G(t) - G_{t,r}(t)] \tag{22}$$

where $G(t)$ is derived from eq.(19') and $G_{t,r}(t)$ from eq.(19') with $t/\tau_n = \infty$ in $\Phi(n,s,s',t)$. The first term in the right-hand-side of eq.(22) decays with a very small decay rate and is eliminated by the low-cut filter. The second term includes mostly the contribution from bending motions of filaments and decays in the time range of the order of τ_2. This situation is compatible with experimental findings (Figs. 4c,6c and 7).

Finally we would like to mention about T/η scaling of correlation functions. Since both D and Θ are proportional to T/η, the ratio of the absolute temperature to the solvent viscosity, the correlation functions for a stiff rod measured at different temperatures can scale as $(T/\eta)t$, or in other words, if T/η scaling is absent, the scatterers will be non-rigid (Newman and Carlson, 1980). We have studied this T/η scaling theoretically, and concluded that if flexibility of the scatterer is independent of temperature and/or solvent conditions, the correlation functions can scale as $(T/\eta)t$ irrespective of the flexibility value, or in other words, if T/η scaling is absent, the flexibility parameter of the scatterer is temperature-dependent (Fujime and Maeda, 1982). As shown in Fig. 11, $[\bar{g}^1(t)]^2$ for F-actin could not scale as $(T/\eta)t$. This suggests that F-actin is flexible and its flexibility parameter depends on temperature.

REFERENCES

Doi, M., 1975, *J. Physique*, 36:607.
Doi, M. and Edwards, S. F., 1978, *J. Chem. Soc. Faraday II*, 74:560.
Carlson, F. D. and Fraser, A. B., 1974, *J. Mol. Biol.*, 89:273.
Fraser, A. B., Eisenberg, E., Kielley, W. W. and Carlson, F. D., 1975, *Biochemistry*, 14:2207.
Fujime, S., 1970, *J. Phys. Soc. Jpn.*, 29:751.
Fujime, S. and Ishiwata, S., 1971, *J. Mol. Biol.*, 62:251.
Fujime, S. and Maruyama, M., 1973, *Macromolecules*, 6:237.
Fujime, S. and Maeda, T., 1982, *Biophys. J.*, 38:213.
Fujime, S., Maeda, T. and Ishiwata, S., 1982, *in* "Biomedical Applications of Laser Light Scattering," eds D. Sattelle et al, Elsevier/North Holland Biomedical Press, in press.
Hochberg, A., Low, W., Tirosh, R., Borejdo, J. and Oplatka, A., 1977, *Biochim. Biophys. Acta*, 460:308.

Ishiwata, S. and Fujime, S., 1972, *J. Mol. Biol.*, 68:511.
Maeda, T. and Fujime, S., 1977, *J. Phys. Soc. Jpn.*, 42:1983.
Maeda, T. and Fujime, S., 1981, *Macromolecules*, 14:809.
Newman, J. and Carlson, F. D., 1980, *Biophys. J.*, 29:37.
Rice, S. O., 1944, see "Elementary Statistical Physics," C. Kittel, John Wiley, New York (1958), p. 117.

APPENDIX

To record sampled values of $\{n(t+mT,T)\}$ and compute correlation functions $G_k(t)$ and power spectra $S_k(f)$ in eq.(4), we used a hardware system shown in Fig. 14. Sixteen kilowords of memory were used for data accumulation area. Every time this area was filled up, data on memory were transferred to a magnetic tape in unit of four kilowords by a free format mode in a machine level. Our minicomputer (Nippon Data General, 02/30) possessed hardware multiply/divide and floating point units. For N = 1024, computation times were 3 sec for $G_k(t)$ with 100 delay points and 1 sec for $S_k(f)$. We also used an Eclipse S-140 minicomputer (Nippon Data General) for off-line computation.

We sometimes placed the *D/A-LCF-A/D circuit* in Fig. 2a between DISCRIM and COUNTER in Fig. 14. By setting $\gamma/2\pi$ = 30 Hz, we could obtain the FFT power spectra very close to those in Fig. 9 without any modification with $[1 - F(f)]$ in eq.(1b). The power spectra in this case just correspond to our old ones (Fujime, 1970).

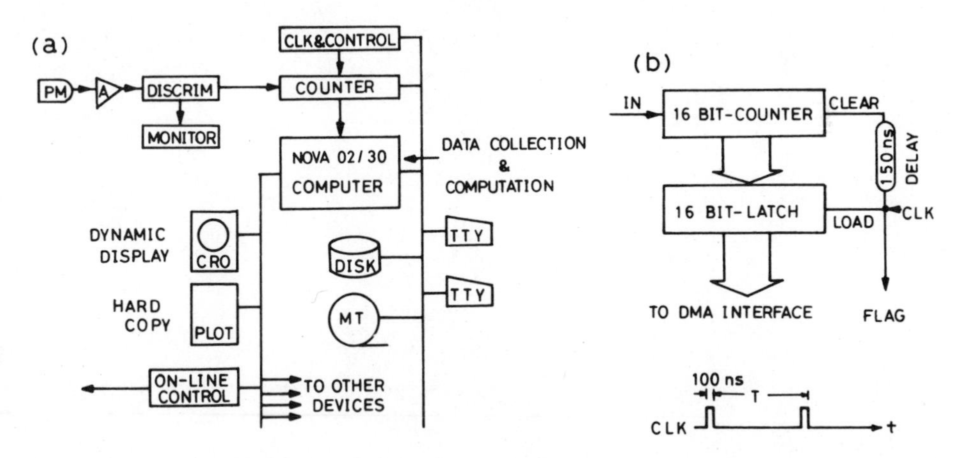

Figure 14. (a) Block diagram of the electronics system. PM: photomultiplier, A: preamplifier, DISCRIM: discriminator, and CLK & CONTROL: clock signal generator and controller for COUNTER.
(b) A 16 bit-counter/latch circuit and a waveform of clock signal, which enable registering of $n(t+mT,T)$ in a direct memory access (DMA) mode in a contiguous manner.

FURTHER EVIDENCE OF CROSS-BRIDGE MOTIONS IN LIMULUS THICK MYOFILAMENT SUSPENSIONS

S.-F. Fan, M. M. Dewey and D. Colflesh

Department of Anatomical Sciences
Health Sciences Center
State University of New York at Stony Brook
Stony Brook, New York 11794

B. Chu

Department of Chemistry
State University of New York at Stony Brook
Stony Brook, New York 11794

1. INTRODUCTION

The contraction of striated muscle is known to be caused by an active relative sliding of the thick and thin myofilaments.[1,2] In addition, there is an increasing amount of evidence suggesting that active sliding between protein filaments occurs not only between the thin actin-containing filaments and the thick myosin-containing filaments, involved in the motility of a wide variety of cells,[3] but also in the cilia motion involving dynein and tubulin rather than myosin and actin.[4] The most widely held view for the molecular mechanism of the sliding process in striated muscle is that the cross-bridges which are projected outward from the thick myofilament moves cyclically upon activation. The cross-bridge motions in turn pull the thin myofilament. So far, this scheme has been supported only by indirect evidence.[5,6]

Quasi-elastic light scattering has been used successfully to investigate the dynamics of submicroscopic structure in solution,[7,8] including the thin myofilament suspensions[9,10] and the in situ myofibrils.[11] Recently, we have studied the dynamic light scattering of Limulus thick myofilament suspensions.[12] Since the Limulus muscle is activated through both thin and thick myofilaments by calcium ions,[13] it might be expected that calcium ions may even activate the isolated thick myofilament. This expectation has been supported by the fact that the average linewidth $\bar{\Gamma}$ increases tremendously at high K values after the calcium ion concentration of the solution is increased, where K is the magnitude of the momentum transfer vector. Although we have attributed the additional high frequency components to cross-bridge motions, we had not ruled out conclusively that the observed high frequency components could be the result of an increase in flexibility for the activated thick myofilaments. In this article we present further evidence relevant to this point.

2. EXPERIMENTAL METHODS

2.1. Isolation of Thick Myofilaments

Muscle bundles of levator of the telson of Limulus (Tachypleus polyphenus), isolated and fixed in length with the telson in the down position, were soaked in relaxing solution (100mMKCl, 5mM EGTA, 1mM $MgCl_2$, 5mM Tris, 5mM ATP, pH 7.4) or in an equal volume mixture of glycerol and relaxing solution at 4°C for 24-48 hours. After homogenizing, the thick myofilaments were separated by gradient centrifugation at 1.2×10^5G for about 15 minutes. Before light scattering measurements, the filament suspensions were dialyzed at 4°C for 2 to 4 days against the solution containing the required composition which had 0.1% glycerol added.

All light scattering measurements were performed immediately after centrifuging the suspensions at 5×10^3 G for one hour at 4°C in order to remove the dust particles and aggregates, if any.

2.2. Light Scattering Measurement

The light scattering spectrometer has been described previously.[12] The measured single-clipped time correlation function $G_k^{(2)}(\tau)$ has the form

$$G_k^{(2)}(\tau) = \langle n_k(t)n(t+\tau)\rangle = A(1+b|g^{(1)}(\tau)|^2) \qquad (1)$$

where n(t) is the number of photoelectron counts per sample time at time t, k is the clipping level, A is a background, b is a fitting parameter, $g^{(1)}(\tau)$ is the first-order electric field correlation function and τ is the delay time.

For monodisperse long semiflexible filaments of length L and $KL \geqslant 3$, the contribution of high frequency internal motions, such as bending and rotational motions, will increase the $\bar{\Gamma}$ value where

$$\overline{\Gamma} = \int G(\Gamma)\Gamma \, d\Gamma \qquad (2)$$

with $G(\Gamma)$ being the normalized linewidth distribution function. We used the cumulants method[14] to determine $\overline{\Gamma}$ and μ_2 defined as

$$\mu_2 = \int G(\Gamma)(\Gamma-\overline{\Gamma})^2 \, d\Gamma \qquad (3)$$

with $\mu_2/\overline{\Gamma}^2$ being the variance of the linewidth distribution function. In studying the thick myofilament suspensions, we found that $\mu_2/\overline{\Gamma} \lesssim 0.3$ over all accessible ranges of K. Thus, our linewidth measurements were carried out at $\overline{\Gamma}\tau_{max} \simeq 3$, where τ_{max} denotes the maximum delay time range. We used the second-order cumulants fit for most fittings. On occasion, a third-order cumulants fit was performed to check the goodness of the second-order analysis.

3. RESULTS AND DISCUSSION

For Limulus striated muscle, in addition to the change in the extent of the overlap between thick and thin myofilaments, thick myofilaments have been observed to shorten during contraction.[15,16] Even isolated, thick myofilaments shorten after an increase in the calcium ion concentration, as has been shown by electronmicrographs with L decreasing from 4.0 to 3.0 μm and the diameter d increasing from 24 to 30 nm. In our earlier experiments, the relaxed (long) thick myofilaments were activated by dialyzing against an activating solution (100 mM KCl, 5mM $CaCl_2$, 1 mM $MgCl_2$, 5 mM Tris, 5 mM ATP, pH 7.4). After activation, the values of $\overline{\Gamma}$ at large values of K become much larger than those from the relaxed (long) state. This experimental observation suggests either the existence of additional high-frequency internal motions caused by the cross-bridges or an unlikely increase in flexibility in the shortened thick myofilaments because of an increase in the calcium ion concentration. We shall recapitulate one of our earlier results[12] as shown in Figure 1. The two possibilities can be inferred by comparing the K-dependence of $\overline{\Gamma}$ between curve (b) and curve (c). Curve (b) was obtained after the filaments were dialyzed against an activating solution. It represents the shortened state with active cross-bridge motions. Curve (c) was obtained from filaments dialyzed against a relaxing solution after obtaining curve (b). The length of the filaments was not changed. It represents the shortened state in the absence of active cross-bridge motions. However, if we take curve (b) as direct experimental evidence of high frequency cross-bridge motions, we must show that (1) other forms of activation of the cross-bridge motions should produce similar effects as those exhibited in curve (b), and (2) the removal of cross-bridge motions by other means should also remove the high-frequency motions. Figure 2 shows

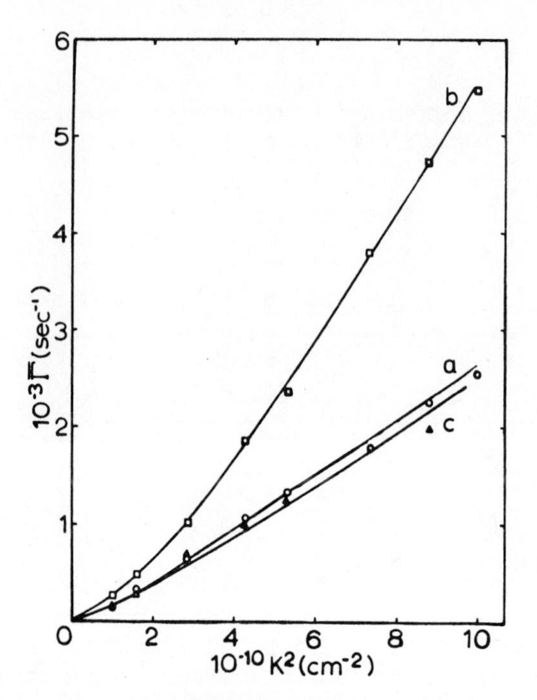

Fig. 1. Plots of $\bar{\Gamma}$ versus K^2 for Limulus thick myofilament suspensions. Most fittings were carried out to a second-order cumulants fit. (a) Filaments dialyzedagainst a relaxing solution. (b) Filaments dialyzed against an activating solution. (c) Filaments dialyzed against a relaxing solution after measurements of curve (b).

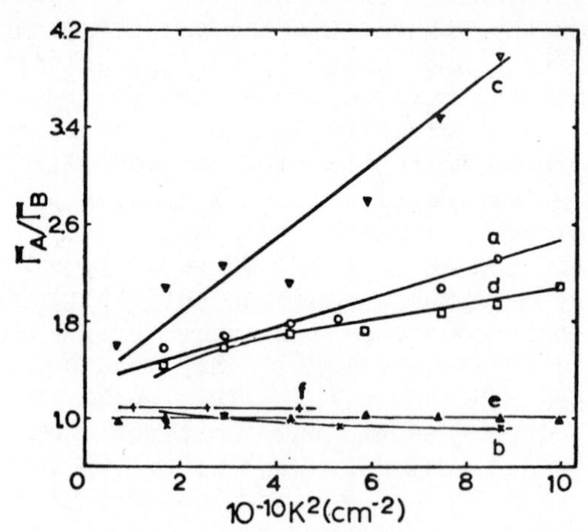

Fig. 2. Plots of $\bar{\Gamma}_A/\bar{\Gamma}_B$ versus K^2 under various treatments. See text for detailed conditions.

plots of $\overline{\Gamma}_A/\overline{\Gamma}_B$ versus K^2 after the isolated thick myofilaments have been treated, as follows:

Curve a: Subscript A denotes that the myofilament suspension, isolated in a relaxing solution, was dialyzed against an activating solution, while subscript B denotes the same myofilament suspension dialyzed against a relaxing solution.

Curve b: Subscript A denotes that the myofilament suspension of A in curve a was then dialyzed against a relaxing solution while subscript B is the same as B in curve a.

Curve c: As isolated myofibrils can be activated by Congo Red,[17] $\overline{\Gamma}_A$ and $\overline{\Gamma}_B$ are the values obtained after and before adding 0.01% Congo Red (final concentration) into the thick myofilament suspensions in a relaxing solution. The values of $\overline{\Gamma}_A/\overline{\Gamma}_B$ at high K ranges are even greater than those in curve a suggesting the possibility that addition of Congo Red may activate the cross-bridge motions and increase the flexibility of the thick myofilaments.

Curve d: Heating the muscle fiber to 44°C for ten minutes was believed to denature the cross-bridges.[18,19] $\overline{\Gamma}_A$ and $\overline{\Gamma}_B$ are the values obtained before and after heating the filament suspended in an activating solution.

Curve e: $\overline{\Gamma}_A$ and $\overline{\Gamma}_B$ are the values obtained before and after heating the filament suspension in a relaxing solution.

Curve f: Vanadate was reported to form a complex with myosin and ADP[20,21] which might be a stable analogue of the myosin-ADP-phosphate complex. Since the latter is believed to be a key intermediate in the myosin ATPase mechanism, vanadate is expected to inhibit the cross-bridge motion. $\overline{\Gamma}_A$ and $\overline{\Gamma}_B$ are those values obtained with filaments suspended, respectively, in an activating solution and a relaxing solution, each containing $\sim$10 mM vanadate.

The results support our earlier assertion that, after raising the calcium ion concentration of the solution, the increase in the values of $\overline{\Gamma}$ at large KL ranges is due to the presence of cross-bridge motions and not due to an increase in the flexibility of the shortened thick myofilament. Our reasoning is as follows.

Curve a: State A corresponds to activated short filaments while state B corresponds to relaxed long filaments. As $K \to 0$, $\overline{\Gamma}_A/\overline{\Gamma}_B$ corresponds to essentially the ratio of the reciprocal of the length of the filaments ($\overline{\Gamma}_A/\overline{\Gamma}_B \sim L_B/L_A \sim 1.2$).

At higher K values, the increase in $\overline{\Gamma}_A/\overline{\Gamma}_B$ can be attributed to cross-bridge motions and/or the activated short filament is more flexible than the long relaxed filament.

Curve b: State A corresponds to rerelaxed short filaments, while state B is the same B (relaxed long) as in curve a. As $K \to 0$, $\overline{\Gamma}_A/\overline{\Gamma}_B$ is approximately equal to L_B/L_A. However, in the rerelaxed state, the slightly more rigid short filament (with $\overline{\Gamma}_A/\overline{\Gamma}_B < 1$ at finite K values) has no high frequency components as that in the activated state as shown in curve a.

Curve c: State A: activation by Congo Red instead of calcium ions; activated short filaments in relaxing solution differing from those in state A of curve a because now the activated short filaments are suspended in a solution containing virtually no calcium ions. State B is the same B (relaxed long) as in curve a. As $K \to 0$, $\overline{\Gamma}_A/\overline{\Gamma}_B$ is again approximately equal to L_B/L_A. Curve c is above curve a indicating the more dramatic effect of Congo Red when compared with activation by calcium ions. By merely comparing between curves a and c, we still cannot distinguish the origin of the high-frequency components.

Curve d: State A is the same as A (activated short) in curve a. State B corresponds to the short relaxed state suspended in activating solution. Therefore, at high K values, we have shown that the high frequency components (with $\overline{\Gamma}_A/\overline{\Gamma}_B$ comparable to those in curve a) remain whether B corresponds to the long relaxed state with virtually no calcium ions or the short relaxed state in the presence of calcium ions.

Curve e: Curve e is a control for curve d. State A is the same as B in curve a. State B corresponds to heating the long-relaxed myofilaments in the same relaxing solution.

Curve f: State A: inhibition of cross-bridge motions of long myofilaments by vanadate in the presence of calcium ions. State B: relaxed long myofilaments in the presence of vanadate. Both states result in long myofilaments. However, as $\overline{\Gamma}_A/\overline{\Gamma}_B \sim 1$, calcium ions have not changed the flexibility of the myofilaments in the presence of vanadate.

If the calcium ions, in the absence of vanadate, have not changed the flexibility of myofilaments, curve a represents an observation of cross-bridge motions. If heat, which is expected to denature the cross bridge, does not reduce the flexibility of myofilaments, we may again draw the same conclusion. However, curve e shows that heat has no effect on the filament flexibility. Consequently, we have gotten further support for our assertion on the observation of cross-bridge motions by photon correlation spectroscopy.

REFERENCES

1. H. E. Huxley, Science, 164:1356 (1969).
2. A. F. Huxley, J. Physiol. 243:1 (1974).
3. T. D. Pollard and R. R. Weihing, C. R. Crit. Rev. Biochem. 2:1 (1974).
4. K. E. Summen and I. R. Gibbons, PNAS U.S.A. 68:3092 (1971).
5. H. E. Huxley and W. Brown, J. Mol. Biol. 30:383 (1967).
6. J. Borjedo, S. Outman and M. F. Moral, PNAS U.S.A. 76:6346 (1979).
7. B. Chu, Phys. Scripta, 19:458 (1979).
8. S. H. Chen, B. Chu and R. Nossal, ed., Scattering Techniques Applied to Supramolecular and Non-equilibrium Systems, pp. 928, Plenum Press, New York (1981).
9. T. Maeda and S. Fujime, Macromolecules, 14:809 (1981).
10. J. Neuman and F. D. Carlson, Biophys. J., 29:37 (1980).
11. R. C. Harkell and F. D. Carlson, Biophys. J., 33:39 (1981).
12. K. Kubota, B. Chu, S.-F. Fan, M. M. Dewey, P. Brink and D. Colflesh, submitted to J. Mol. Biol.
13. W. Lehman and A. G. Szent-Gyorgyi, J. gen. Physiol., 66:1 (1975).
14. D. T. Koppel, J. Chem. Phys., 57:4814 (1972).
15. M. M. Dewey, R. J. C. Levine, D. Colflesh, B. Walcott, L. Braun, A. Baldwin and P. Brink in Cross-Bridge Mechanism in Muscle Contraction (Sugi, H. and Pollack, G. H., ed.) 3-22, University of Tokyo Press, Tokyo (1979).
16. M. M. Dewey, D. Colflesh, P. Brink, S.-F. Fan, B. Gaylein and N. Gural in Basic Biology of Muscle: A Comparative Approach. (Twarog, B., Levine, R. J. C. and Dewey, M. M., ed.) Raven Press, New York (1982).
17. S.-F. Fan, Scientia Sinica, 13:692 (1964).
18. H. E. Huxley, Cold Spring Harbor Symp. Quant. Biol., 37:361 (1972).
19. S.-F. Fan and Y.-S. Wen, Acta Physiol. Sinica, 31:227 (1979).
20. C. C. Goodno, PNAS U.S.A., 76:2620 (1979).
21. C. C. Goodno and E. W. Taylor, ibid., 79:21 (1982).

STRETCH-INDUCED TRANSPARENCY CHANGE ASSOCIATED WITH CROSS-BRIDGE DEFORMATION IN ACTIVE FROG'S MUSCLE

F.W. Flitney and J.C. Eastwood

Department of Physiology & Pharmacology
Bute Medical Building
University of St Andrews
St Andrews, Scotland

INTRODUCTION

In a recent study of the transparency changes produced by stretching active frog's muscle (Flitney & Eastwood, 1982), we identified a component of the optical signal which was associated with the extra tension generated. This was referred to as the tension-dependent transparency change, or TDC, to distinguish it from another component which appeared to be related to the change in muscle length. The amplitude of the TDC was found to vary linearly with the tension increment, ΔP, when this was made to change by altering the rest length of the muscle prior to stimulation. Since muscle stiffness (and hence, ΔP) depends upon the extent of filament overlap (see Discussion), this observation suggested to us that the TDC is generated by strain in the cross-bridges. However, for various reasons we could not exclude the possibility that the filaments themselves might be partially (or even wholly) responsible.

The experiments we now report are concerned with the effect of altering the angle ($\emptyset$) between the plane of polarisation of the incident laser beam and the long axis of the muscle. The aim was to try and obtain information on the spatial orientation of the structures responsible for the transparency change.

MATERIALS & METHODS

Experiments were performed using the sartorius muscle of the frog, <u>Rana temporaria</u>. The isolated muscle was mounted vertically in a glass-walled chamber and a small area ($\sim$1mm diam., 5-10 mm from the pelvic end) illuminated with a He-Ne laser (15 mW). The resting sarcomere length, s, was calculated from the separation of the 0th and 1st order spectra in the diffraction pattern generated by the muscle, using the relation $s = \lambda/\sin\theta$ ($\lambda = 0.6328\ \mu m$; θ = angle subtended at the muscle by the 0th and 1st orders). The intensity of the transmitted (undeviated) light was recorded using a photodiode, located 10 cm behind the muscle. Muscles were stimulated under isometric conditions (square pulses, 7-12V, 0.2 ms; frequency: 15-25 Hz; for 1s) and a stretch applied 500 ms after the start of the contraction (1.8 mm; 50 ms duration). Experiments were performed at 5-7°C.

The apparatus is similar to that used previously (Flitney & Eastwood, 1982), with some modification. First, a $\lambda/2$ mica plate was interposed between the laser and muscle. This was mounted on a rotating turret, so that the angle ($\emptyset$) between the plane of polarisation of the laser beam and the long axis of the muscle could be changed; recordings of the transparency change were made at $\emptyset = 0^{\circ}$ and $\emptyset = 90^{\circ}$. Second, the timing of events and the collection and processing of data were performed with the aid of a microcomputer (Hewlett Packard Ltd., type HP 85). The outputs from the tension recorder and zero order photocell were captured using transient recorders (Datalab Ltd., type DL 902) and then transferred to the computer for storage and subsequent analysis.

RESULTS

Fig. 1 illustrates the method used to analyse the transparency change, described in more detail elsewhere (Flitney & Eastwood, 1982). It depends upon two assumptions: first, that the optical response is made up of a tension-dependent component (TDC), superimposed on a length-dependent one (LDC); and second that the delayed <u>increase</u> in transparency (phase 3; Fig 1A, seen after the stretch has ended) represents a reversal of the <u>decrease</u> (phase 1) which parallels the initial (steep) tension increment (ΔP, Fig 1B). The H P 85 is programmed to eliminate the baseline shift due to the LDC. The amplitude of this component is the difference between the steady-state light intensity ultimately reached (esimated by computing the mean of the 50 data points preceding the last stimulus) and the light intensity immediately before commencing stretch. The 'corrected' transparency change is obtained by subtracting the recorded light intensity values from those indicated by the dashed line.

The amplitude of the TDC is defined as $\Delta I/I$ (Fig. 1 C). The tension increment ΔP is the extra force per unit area of muscle ($kg.\ cm^{-2}$) generated by the stretch.

Fig. 2 shows paired optical and tension recordings from two

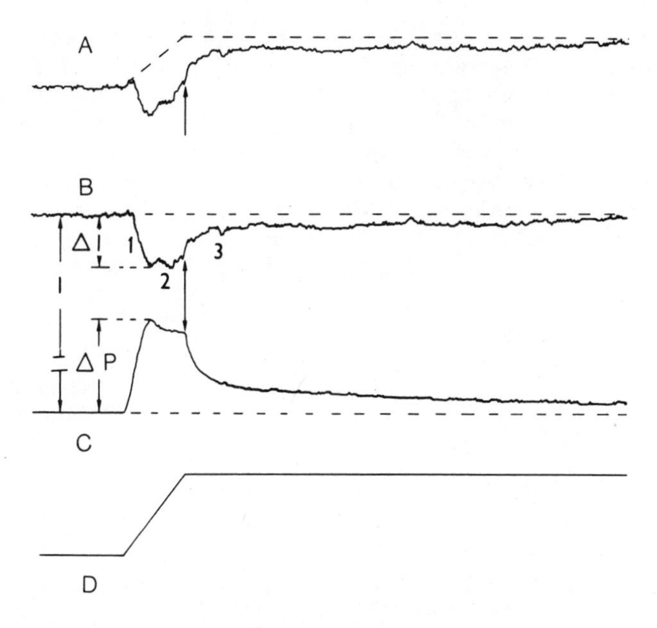

Fig. 1. Analysis of optical recordings. A, original trace. B, 'corrected' recording. C, tension increment. D, muscle length.

different muscles (A,B). $\emptyset$ was changed from 0° to 90° between each tetanus. The changes in light intensity have been normalised, so that they are directly comparable, and all recordings have been corrected to remove the LDC of the transparency change (see Methods).

Two important effects of changing $\emptyset$ were seen.

First, it was found that $\Delta I/I$ is greater when $\emptyset = 0^{\circ}$ than when $\emptyset = 90^{\circ}$. This was consistently seen at all sarcomere lengths studied. The values obtained in experiments with 6 muscles where the starting sarcomere length was varied from 2.0 to 3.2 µm, are shown in Fig. 3. Both sets of results confirm our previous finding (Flitney & Eastwood, 1982), that $\Delta I/I$ varies linearly with ΔP. However, the slopes of the two lines ($= (\Delta I/I).(kg.cm^{-2})^{-1}$, denoted by $k_{0^{\circ}}$ and $k_{90^{\circ}}$) are clearly different. Linear regression analysis of the data gave the following results: $k_{0^{\circ}} = 0.148$ (n = 34 recordings) and $k_{90^{\circ}} = 0.063$ (n = 44 recordings).

Secondly, the transparency change recorded with $\emptyset$ at 90° is not simply a scaled down version of the response seen with $\emptyset$ at 0°. This is an important observation, because if the actin and/or myosin filaments make a major contribution to the transparency change, then we could expect the two kinds of recording to show quantitative differences, but to be qualitatively similar. A careful study of the differences which exist (see Fig. 2) has yet to be made. One approach currently being explored is to subtract recordings made at $\emptyset = 90^{\circ}$ from those obtained with $\emptyset = 0^{\circ}$. The 'difference' traces (Fig. 4) show several distinct maxima and minima. One of these (denoted by α) is consistently seen, and occurs just after the abrupt decrease in muscle slope stiffness.

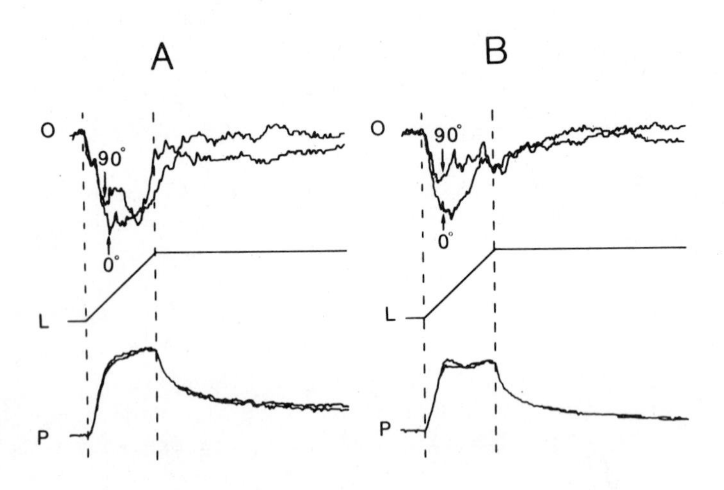

Fig. 2. Transparency recordings (corrected) made at $\emptyset = 0^{\circ}$ and 90°. L, muscle length. P, tension increment.

DISCUSSION

The TDC of the transparency change appears to originate in elements of the contractile system which become strained as the muscle is stretched. It now seems probable that the cross-bridges themselves make a major contribution to the optical response. The evidence on which this conclusion rests in necessarily indirect.The argument hinges on the results of a previous study (Flitney & Hirst, 1978) in which sarcomere length changes were monitored throughout stretches similar to those used here. It was found that the sudden decrease in muscle slope stiffness which occurs during a stretch coincides with an abrupt lengthening of the sarcomeres, termed

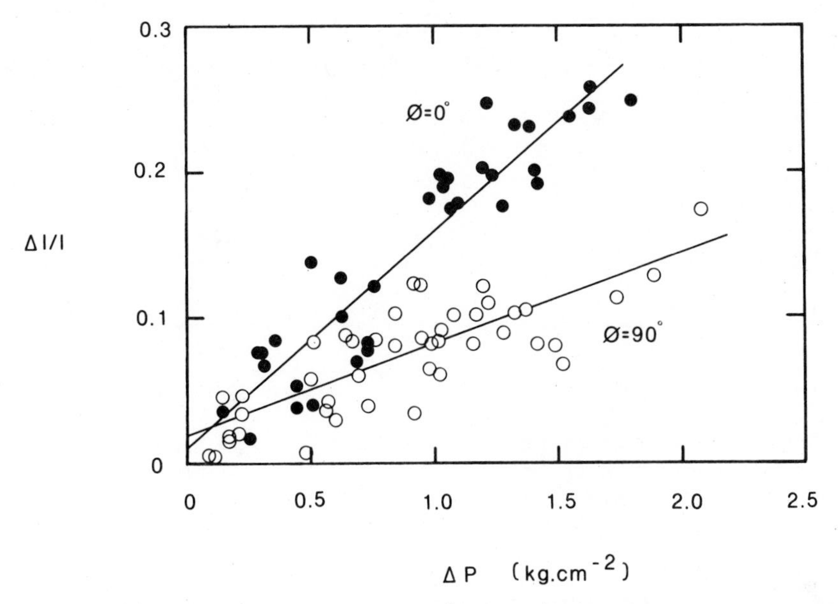

Fig. 3. Relationship between $\Delta I/I$ and ΔP, recorded at different muscle lengths. Filled circles, $\varnothing = 0^\circ$; open circles, $\varnothing = 90^\circ$.

sarcomere 'give'. This was attributed to forcible detachment of cross- bridges and it happened when the actin and myosin filaments had been displaced by 11-12 nm from their steady-state (isometric) position. The same experiments established that the stiffness of the sarcomeres varies in direct proportion to the extent of filament overlap.

These observations show that the cross-bridges provide the major source of compliance in a contracting muscle, and that the filaments are comparatively inextensible structures. The same conclusion was reached by A.F. Huxley & Simmons (1973), based upon their analysis of the tension transients generated by subjecting isometrically-contracting muscle fibres to rapid ($<$1ms) length changes.

The ability of a particle to scatter light is a function of its charge distribution, and this will be changed if it becomes strained. This is evidently what happens to the cross- bridges in a muscle during stretch. The fraction of light deviated in this way (D) will be some function of the strain in each element, s, and the number, n, involved ($D = f(s,n)$) and the same two parameters will also determine the tension increment ($\Delta P = f(s,n)$). We can

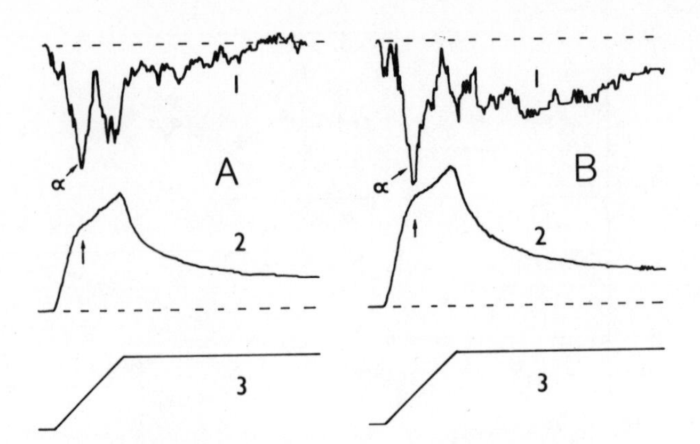

Fig. 4. 'Difference' traces (1) made by subtracting records obtained with $\phi = 90°$ from those recorded using $\phi = 0°$. Results from 2 muscles (A,B). Note that difference reaches a maximum just after the decrease in muscle stiffness.

therefore expect the form of the tension increment and that of the transparency change to show certain similarities, and this is so: the initial decrease in transmitted light intensity parallels the early (steep) rise of tension, and the forcible breaking of cross-bridges, indicated by the reduction in muscle stiffness, coincides with an abrupt decrease (or even reversal) in the rate at which the light intensity changes. The same line of reasoning predicts that $\Delta I/I$ and ΔP should vary in a similar way as the rest length of a muscle is changed prior to stimulation. This results in a proportional change in the extent of filament overlap and, because the cross-projections on the myosin filaments are evenly spaced, every 14.3 nm, a corresponding change in the number of sites potentially available for cross-bridge formation. The results of Fig.3 (and those of Fig. 5 in Flitney & Eastwood, 1982) confirm that $\Delta I/I$ changes linearly with ΔP.

These observations are consistent with the view that the cross-bridges are involved in generating the transparency change. However, they do not rule out the possibility that the filaments themselves contribute to the response, because although relatively inextensible, they are comparatively large structures and a small deformation might contribute substantially to the transparency change. Moreover, they would be subject to stresses which would vary with filament overlap in the same way as those acting on the cross-bridges.

The present experiments were undertaken in an attempt to resolve this uncertainty. The stress applied to the muscle is in the direction of its long axis, and the component of strain in an asymmetric scattering element will be maximal if its long axis is aligned in this direction, but less if inclined at any other angle (Θ) to it. In general, then, the scattering effect will be maximal when $\varnothing = \Theta$. The results of Fig. 3 show that $\Delta I/I$ is considerably greater (by 2.5 x) when $\varnothing = 0^{\circ}$ than when $\varnothing = 90^{\circ}$. Recordings were not made at intermediate values of $\varnothing$, and so all that can be said about this observation is that the scattering elements are probably more closely aligned with the long axis of the muscle than at right angles to it.

The difference in the form of the transparency change recorded at 0° and 90° appears to be potentially more informative. A.F. Huxley & Simmon's (1973) experiments led them to conclude that each cross-bridge is a 'jointed' structure, made up of two elements in series: a force-generating component (the myosin head) and an 'instantaneous' elastic element. They postulated that the force required to propel the filaments is generated by a stepwise movement of the attached heads along the actin filament. The stepping motion is permitted by rotation about the junction between the head and elastic element. The study of sarcomere 'give', referred to earlier, showed that cross-bridge detachment occured when the filaments had been displaced by 11-12 nm. It was inferred (see Flitney & Hirst, 1978b; p 477 for details) that most of this sliding movement is actually taken up by backward rotation of the attached head and only a small amount by lengthening of the elastic element.

These considerations lead us to conclude that several factors may be involved in generating the transparency change: (i) longitudinal strain in the elastic elements; (ii) shearing strain in the attached cross-bridge heads; and (iii) backward rotation of the heads along the actin filaments. It is reasonable to assume that the contribution made by each will depend upon $\varnothing$. The change in the angle of attachment of the heads during stretch is likely to be important, since it would be expected to have opposite effects on light polarised at 0° and 90°. Indeed, we are currently exploring the idea that the onset of the α-wave (Fig. 4) marks the point at which the extra force in the elastic elements exceeds that anchoring the heads in position on the actin filament, causing them to rotate backwards and eventually detach.

We are grateful to the Wellcome Trust for supporting this work, and to the Physiological Society and Company of Biologists for travel grants.

REFERENCES

Flitney, F.W. & Hirst, D.G., 1978. Cross-bridge detachment and sarcomere 'give' during stretch of active frog's muscle. J. Physiol., 276: 449-465.

Flitney, F.W. & Hirst, D.G., 1978a. Filament sliding and energy stored by the cross-bridges in active muscle subjected to cyclical length changes. J. Physiol. 276: 467-479.

Flitney, F.W. & Eastwood, J.C., 1982. Transparency changes associated with force enhancement during stretch of active frogs' muscle. In: Biomedical applications of laser light scattering. Elsevier/North - Holland Biomedical Press.

Huxley, A.F. & Simmons, R.M. 1973. Mechanical transients and origin of muscular force. Cold Spring Harb. Symp. Quant. Biol., 37: 669-680.

ACTIN POLYMERIZATION IN CELL CYTOPLASM

E. Del Giudice[+], S. Doglia[o] and M. Milani

Istituto di Fisica dell' Università, Via Celoria, 16,20133 Milano (Italy); [+]Istituto Nazionale di Fisica Nucleare, Sez. di Milano; [o]Gruppo Nazionale di Struttura della Materia del C.N.R., Milano

Recent theoretical[(1)] and experimental[(2)] research strongly supports the existence of coherent electric vibrations inside living matter. Biomolecules, schematized as sets of polar oscillators open to an external energy flow, can oscillate coherently in a particular vibrational mode when appropriate conditions are fulfilled, namely the external energy supply exceeds a certain threshold and is completely dissipated outside. The energy supply in a living cell can be provided by metabolic chemical reactions. Giant coherent oscillations, with typical frequencies of the order of magnitude of 10^{11}-10^{13}Hz, can then be produced and propagate inside cells.

Moreover, among different systems, each oscillating à la Fröhlich, long (up to 1 μm) range forces arise, which become strongly attractive when the interacting systems oscillate on the same frequency.

Raman and microwave spectroscopies in living cells have been developed by which Fröhlich's coherent oscillations may be detected, should they exist, and studied. The results have been presented in a recent review [(2)] and we summarise here the main findings. First, some microwave frequencies between 30 and 140 GHz appear to be absorbed, by active cells, more strongly than others; second, the absorption of such frequencies alters the vital sequences of metabolic activities and may increase or decrease the rate of cell proliferation; third, Raman shift lines from 5 to 3000 cm^{-1} have been observed in the spectra of active cells which do not appear in the spectra of inactive ones. The line intensities and frequencies are time dependent and, in particular, frequency values exhibit a variation with the age of the cell; fourth the Stokes Raman lines have Antistokes counterparts of an intensity greater than that expected if a normal thermal distribution existed in living cells

Experimental difficulties have to be overcome in order to obtain these spectra[3]. First of all, one has to guarantee that cells remain alive all through the experiment. Moreover, it must be kept in mind that biological processes are not stationary ones, but are strongly time-dependent and are influenced, too, by many factors such as nutrients, concentrations, temperature, external fields and so on; laser power also is a critical factor. Then it is more important, at least in a first approach, to concentrate on the regularities shown by the data rather than on their numerical values.

Actually individual spectra exhibit rather variable features, but it is possible [4,5] to recognize a fully reproducible regularity law among the different spectra. All the lines between 300 and 3000 cm^{-1} are the integer multiples or sums or differences of few fundamental frequencies. The fundamentals are the first to appear in the cell life cycle, while generated frequencies appear later; the higher the frequency, the greater the appearance time. Lower frequencies disappear when higher frequencies are switched on. This regular pattern among variable individual spectra suggests that we are not facing an artifact, but an objective law in living matter. We have proposed[6] a model, based on Fröhlich's ideas, where we try to get an understanding of the origin of these laws.

Here we wish to discuss an important consequence of the above mentioned frequency mixing, the so-called self-focussing[7,8]. Actually, if the observed vibrations were just the coherent ones predicted by Fröhlich, a nonlinear relationship between the polarization $\vec{P}$ of the medium and the incoming electric field $\vec{E}$ should exist. Neglecting the vectorial notations and calling, as usual ε_0 the vacuum dielectric constant and χ the electric susceptibility:

$$(1)\quad P = \varepsilon_0 \, \chi(E)E$$

Consequently, by using the Clausius-Mossotti equation, the refractive index n of the medium (supposed isotropic) would depend on E and be expressed as a sum of the even powers of E:

$$(2)\quad n = n_0 + n_2 \, |E|^2 + ...$$

We can disregard higher powers of E in the range of values we are interested in.

It is interesting to note that the sign of n_2 is the same as the sign of the angle δ which appears in the electric birefringence in the Kerr effect.

A positive birefringence in a Kerr medium would then imply that the nonlinear refractive index is higher than the linear one. Consequently, as the incoming beam goes through such a medium, it undergoes a lens effect and after a distance, because of the counteracting diffraction spreading, becomes trapped in a small cross-section waveguide, whose length R increases with the beam power. For greater distances the diffraction dominates and the beam spreads out.

In the framework of the cell dynamics, as revealed by Raman spectroscopy above, one could search for direct biological evidence for filamentation processes, which could be connected with the Fröhlich waves and their self-focussing. Actually modern cytology[9] has detected microstructures inside the cytoplasm, the "cytoskeleton", which are strongly time-dependent and vary with the cell activity. They are made up mainly of polar proteins, such as actin.

We try now to show that an aqueous solution of a protein - for instance the G-actin monomer which gives rise under polymerization to the filamentous F-actin - becomes, if a Fröhlich activity is switched on, a two-phase medium formed by water and a filamentous protein network.

First of all, we remark that aqueous actin solutions - both in the monomer (G-actin) and polymer (F-actin) form - exhibit positive birefringence[10,11] when the electric field E has a strength such as in living cells (up to 10^7 V/cm). Then self-focussing is expected, when Fröhlich waves are brought in[12].

The self-focussed Fröhlich beam defines a region, where the refractive index $n = n_o + n_2 |E|^2$ $(n_2 > 0)$ is greater than the refractive index $n = n_o$ outside. Consider now the net volume force acting on a charge distribution ρ within a medium of susceptibility χ. Disregarding pure gradient terms, which could account for local charge fluctuations we get[13].

$$(3) \quad \vec{E}_V = \rho\vec{E} - \tfrac{1}{2}\,\varepsilon_o\, E^2 \nabla\chi$$

For high values of E the second term dominates. This term defines a force normal to the beam axis and oriented inward the filament.

So the actin monomers in the solutions will be sucked inside the Fröhlich waveguides, producing their alignment along the axis, while their dipole moments will be oriented transversally. Under proper conditions this self-association could lead to the production of the F-actin polymer.

In conclusion the propagation of Fröhlich coherent electric waves in a solution of polar molecules, characterized by positive birefringence, can provide a mechanism for self-association and polymerization.

Additional evidence for the polymerizing effect of the Fröhlich waves is provided by a recent paper of Rowlands et al.[14]. Erythrocytes are shown to be tied together forming chains when metabolic activity is going on and when some macromolecules are present in the solution. The absence of macromolecules inhibits the effect.

When examined under the electron microscope, neighbouring erythrocytes are connected by fibrils which can be long up to 100 μm. These fibrils exhibit always small gaps at both ends, where they attach themselves to the cell membranes.

The effect is strongly pH-dependent and reaches the maximum size at pH = 7.4. The macromolecules able to transmit the interaction have extended chain structures, like fibrinogen, while compact molecules like albumin are unable to do the job[15].

It is interesting to note that the molecules able to form fibrils are characterized by a Kerr behaviour similar to the actin one. In particular the positive birefringence of fibrinogen[16] is strongly pH-dependent, having a peak just between pH = 7 and 8.

REFERENCES

1. H. Fröhlich in "Advances in Electronics and Electron Physics" L. Marton and C. Marton eds., Academic Press, New York (1980) vol. 53, p. 85.
2. S.J. Webb, Physics Reports 60: 201 (1980).
3. S.J. Webb, Collective Phenomena 3: 313 (1981).
4. E. Del Giudice, S. Doglia and M. Milani, Phys. Lett. 85A: 402 (1981).
5. E. Del Giudice, S. Doglia and M. Milani and S.J. Webb, to appear in Phys. Lett. A.
6. E. Del Giudice, S. Doglia and M. Milani, to appear in Physica Scripta.
7. S.A. Akhmanov, R.P. Sukhorukov and R.V. Khokhlov, in "Laser Handbook", F.T. Arecchi and E.O. Schulz-DuBois, North-Holland, Amsterdam (1972) Ch. E3.
8. Y.R. Shen, Progr. Quant. Electr. 4: 1 (1975).
9. I.I. Wolosewick and K.R. Porter, J. Cell Biol. 82: 114 (1979); A. Hoglund, R. Karlsson, E. Arro, B. Fredriksson and U. Lindberg, J. Muscle Res. Cell Motility 1: 127 (1980). J.S. Clegg, Collect. Phenom. 3: 289 (1981). R.L. Margolis and L. Wilson, Nature 293: 705 (1981).

10. S. Kobayasi, H. Asai and F. Oosawa, Biochim. et Biophys. Acta 88: 528 (1964).
11. S. Kobayasi, Biochim. et Biophys. Acta 88: 540 (1964).
12. E. Del Giudice, S. Doglia and M. Milani, Phys. Lett. 90A: 104 (1982).
13. W.H. Panofsky and M. Phillips in "Classical Electricity and Magnetism", Addison-Wesley, London (1955).
14. S. Rowlands, C.P. Eisenberg and L.S. Sewchand, to appear in J. Biol. Physics.
15. L.S. Sewchand, D. Roberts and S. Rowlands, to appear in Cell Biophys.
16. I. Tinoco - J. Am. Chem. Soc. 77: 3476 (1955).

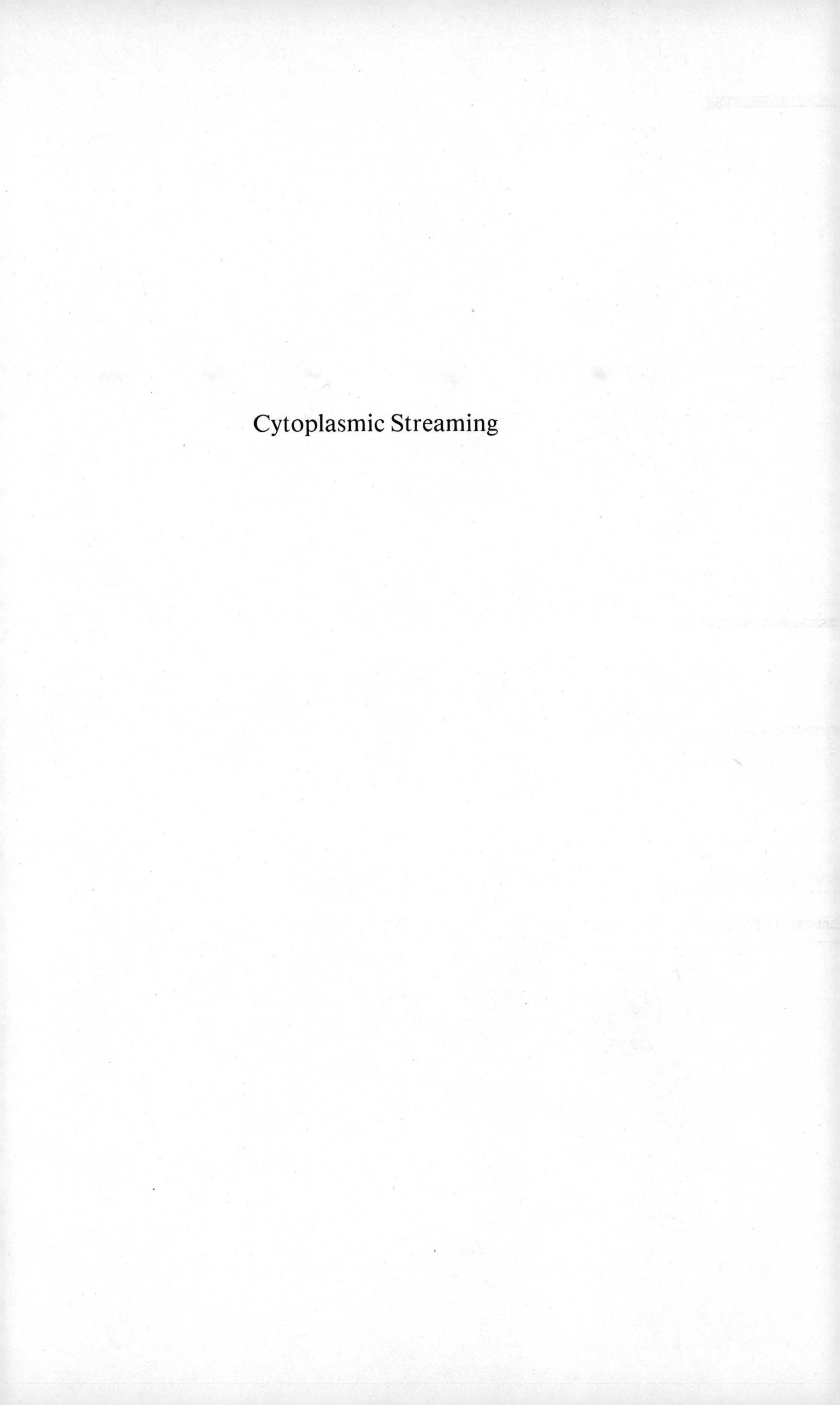

Cytoplasmic Streaming

DYNAMIC CELLULAR PHENOMENA IN PHYSARUM

POSSIBLY ACCESSIBLE TO LASER TECHNIQUES

K. E. Wohlfarth-Bottermann

Institute for Cytology
University of Bonn, FRG

CONTENTS

ABSTRACT

Contractile and motile phenomena in plasmodia of Physarum are considered with respect to possibilities of applying laser techniques for registration and for an analysis of their molecular basis. In addition to protoplasmic strands, the advantages of two special models are discussed, protoplasmic drops and endoplasmic veins. The presentation is focussed on the following phenomena: endoplasmic shuttle streaming, contractile activities of the force-generating cytoplasmic actomyosin, dynamics of actin transformations (sol $\rightleftharpoons$ gel transitions) and actin-myosin interactions, synchronisation and wave phenomena of contractile activities, and membrane flow.

1. INTRODUCTION

The plasmodial phase of acellular slime molds can be easily cultured under laboratory conditions (axenic as well as non-sterile) and is used mainly in two areas of experimental biology, namely developmental biology and cell motility research. Some investigations applying lasers for an analysis of cellular motility phenomena were reviewed by Earnshaw and Steer.[1] The aim of the present contribution is to summarize those dynamic cellular phenomena that presumably can be analysed by laser techniques.

The evaluation of the possibilities of laser application and the interpretation of laser data depend on an intimate knowledge of the object, from the molecular level up to the macroscopic dimension. In the last few decades, considerable knowledge was accumulated due to the fact that Physarum polycephalum became a preferred object for the analysis of cytoplasmic actomyosin.[2] Morphological, biochemical and physiological data are available; however it is not possible to review details here. The main dynamic phenomena interesting for laser techniques will be characterized in form of diagrams; the cited references are far from being complete, literature from our laboratory is given preference.

Fig. 1 shows the integration of the plasmodial phase (1) into the life cycle. The plasmodium represents a giant cell with a size ranging between microscopic dimensions and several square dcm, a multinuclear mass of cytoplasm with prominent motile phenomena (intracellular mass transport and locomotion). Migrating plasmodia[3] (Fig. 1 (1)) are differentiated into a frontal (compact) protoplasmic sheet and a rear region differentiated into single but interconnected plasmodial strands with a diameter of > 50 μm and < 1 mm.

2. PLASMODIAL STRANDS

A. Shuttle streaming

Protoplasmic strands (Fig. 2) show a complicated architecture. Principally, they are differentiated into an ectoplasmic tube and an endoplasmic channel. The ectoplasmic tube as well as the endoplasmic channel contain all cell organelles; the differences can be characterized as follows: whereas the ectoplasmic tube contains a complcated system of plasmalemma invaginations and light-microscopically visible cytoplasmic actomyosin fibrils, these differentiations are lacking in the more fluid endoplasm.[4] Within the endoplasmic channel the endoplasm flows rhythmically back and forth (period approximately 1.3 min) with a high velocity (up to 1.3 mm/sec). The velocity profile indicates a pressure flow mechanism,[5] because the velocity decreases from the mid stream to the border, namely the ectoplasmic tube with its innermost circular plasmalemma invaginations[6] (Fig. 2) Cell organelles, including the numerous nuclei (diameter 2.5 - 4 μm)

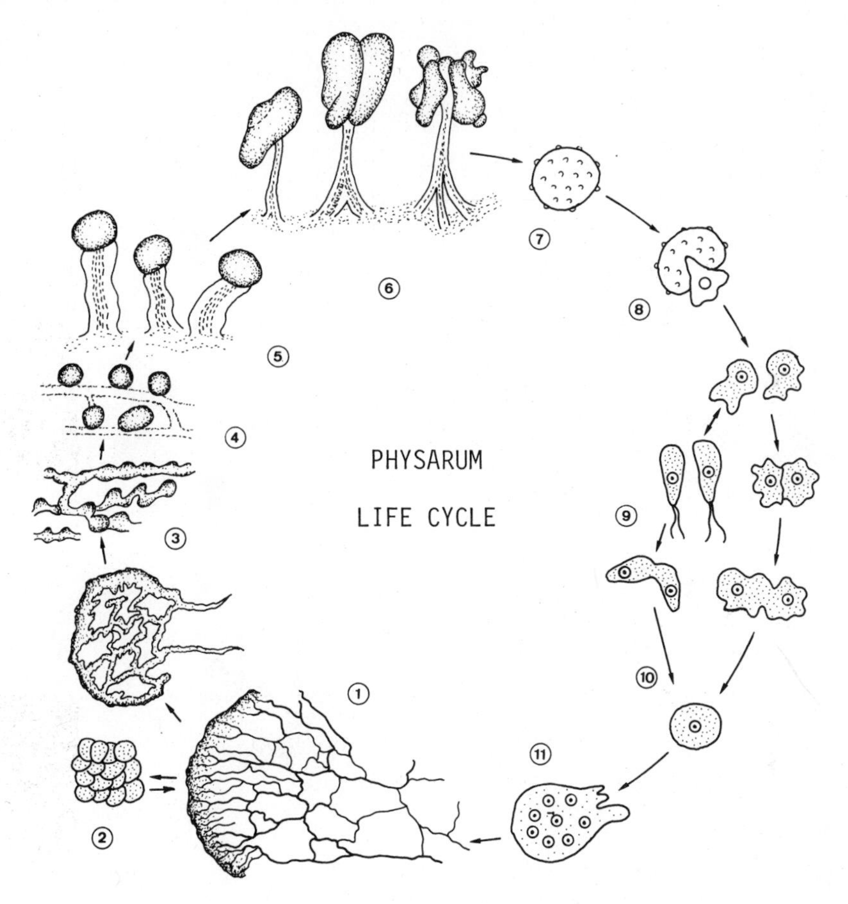

Fig. 1 The life cycle of the acellular slime mold Physarum polycephalum.
Predominantly the plasmodial vegetative (nutritional) phase (1) represents an object for laser studies. The other phases of the life cycle can be characterized as follows: (2) = Sclerotia formation, occurring under unfavourable environmental conditions; (3) - (6) = Sporangia formation and differentiation of haploid spores; (7) - (8) = Spore germination resulting in myxamebae; (9) = fusion of two haploid cells; (10) = diploid cell; (11) = division of nuclei without cytokinesis resulting in a multinuclear plasmodium (1) (Shraideh, unpublished).

and mitochondria (diameter 0.5 - 1.5 µm), are transported in this way throughout the plasmodium.

The endoplasm passively follows hydrostatic pressure differences generated by the rhythmic contractile activity[7] of cytoplasmic

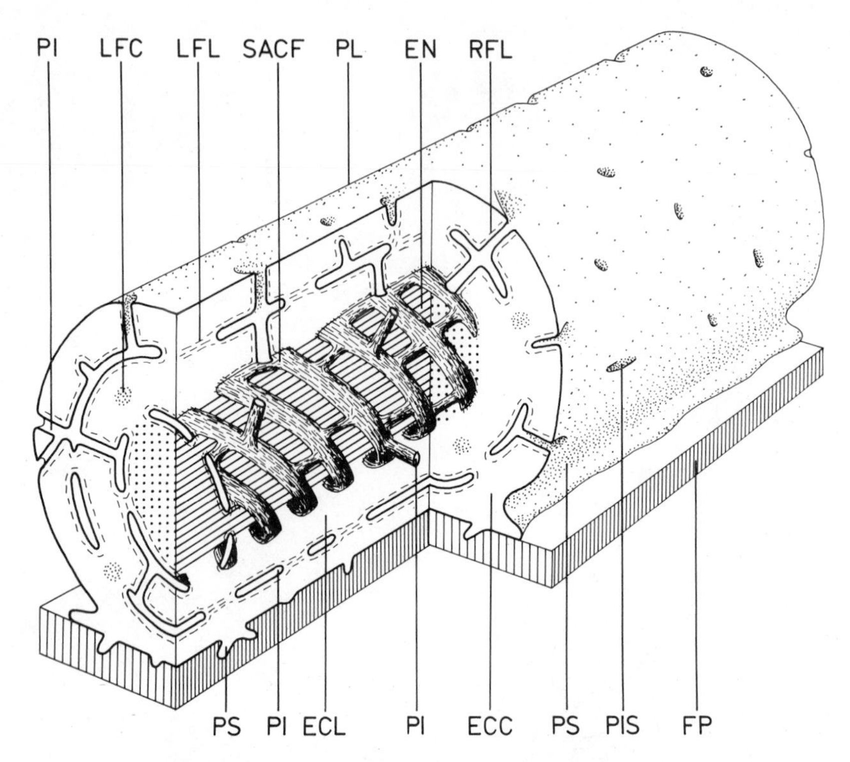

Fig. 2 Architecture of a plasmodial strand.
PI = plasmalemma invagination; LFC = longitudinally running actomyosin fibril in cross section, LFL = in longitudinal section; SACF = surface view onto actomyosin sheets covering circularly and diagonally running plasmalemma invaginations which surround the endoplasmic channel (dotted and hatched, respectively, in cross and longitudinal section); PL = plasmalemma; EN = endoplasm; RFL = radial actomyosin fibrils in longitudinal section; PS = pseudopodia; ECL = ectoplasmic tube in longitudinal section, ECC in cross section; PIS = surface view onto openings of the plasmalemma invagination system; FP = filter paper (substrate under laboratory conditions).

actomyosin fibrils and sheets inserted onto the outer plasmalemma and its invaginations (Fig. 2). The plasmalemma invaginations contain an extracellular space system which has a structural continuity and the form of a complicated labyrinth[6] (compare also Fig. 3). The width of the invaginations varies between 0.1 and 20 µm, depending on the culture method.[8]

Whereas the main endoplasmic stream runs in the direction of the long axis of the strand, laser studies should consider that

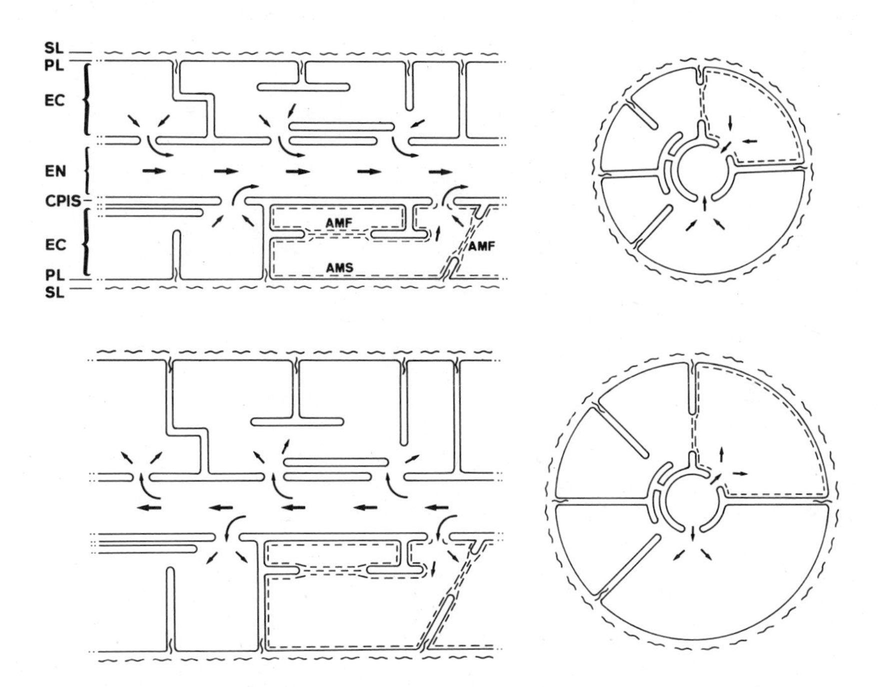

Fig. 3 Details of streaming phenomena in a plasmodial strand with shuttle streaming.
Left side: longitudinal sections; right side: cross sections. Upper pictures: contraction phase (constriction) of the strand; lower pictures: relaxation phase (dilation). Thick arrows: shuttle streaming direction within the endoplasmic channel (EN); small arrows: sideward directed smaller endoplasmic streamlets, leaving or entering the ectoplasmic tube (EC) via apertures of the circular plasmalemma invagination system (CPIS); SL = outer slime layer; PL = outer plasmalemma; AMF = actomyosin fibrils connecting plasmalemma invaginations; AMS = actomyosin sheets sandwiching plasmalemma and invaginations.

there are sideward-directed streamlets of endoplasm, i. e., perpendicular to the direction of the main stream[1, 9-11] (Fig. 3). These endoplasmic streamlets flow through apertures of the circular plasmalemma invagination system at the borderline between ectoplasm and endoplasm.[12] The direction of the streamlets changes with the direction of the endoplasmic main stream, and there is a definite, but not constant correlation between the endoplasmic streaming cycles and the contraction cycles of the strand.[13]

The permanent exchange of ectoplasm and endoplasm at all sites of a plasmodial strand requires that permanent ecto ⇌ endoplasm-interconversions (sol ⇌ gel transitions) have to be considered throughout the ectoplasmic tube. These interconversions are intimately and obligatorily interwoven with the contraction-relaxation cycle of the force generating machinery, the cytoplasmic actomyosin.[14] Details of the contraction-relaxation cycle[2] of cytoplasmic actomyosin cannot be discussed here. Their analysis is much more complicated than in muscle actomyosin due to the dynamics of cytoplasmic actin,[15,16] the lower polymer status of cytoplasmic myosin and the simultaneous existence of mass transport phenomena (cytoplasmic streaming), as compared with muscle contraction. This situation demands an application of laser methods.

Due to the streamlets running between the endoplasmic channel and the ectoplasmic tube, there is a permanent change in the volumes of the endoplasmic channel and the ectoplasmic tube in the course of each contraction cycle in a distinct profile of a plasmodial strand.[9-11] These volume changes are correlated with the contraction-relaxation cycle of the strand such that during its contraction phase the ectoplasmic volume decreases, whereas it increases during relaxation, i. e., dilation of the strand (Fig. 3). Furthermore, for the interpretation of laser data it has to be taken into account that the volume of the extracellular space within the plasmalemma invaginations (width 0.1 to 20 µm) varies with the contraction-relaxation activity of the strand.[17] Thus, there is a certain rhythmical exchange of a small part of the extracellular medium within the invagination system in phase dependence on the contraction cycle. In any case, the contraction cycle-related dynamics of the plasmalemma invagination system must be considered when evaluating laser data. A quantitative reduction of the invagination system and its delayed de novo formation is possible by applying the drug caffeine.[18,19]

It is of major significance for laser studies that the endoplasm can be substituted by different inert media (endoplasm substitution technique,[20] Fig. 4, E). This offers the possibility to analyse the activity of the ectoplasmic tube in a specimen without endoplasmic streaming and thus to differentiate between effects originating from ectoplasm and endoplasm, respectively.

B. Contractile activities

Protoplasmic strands can be easily isolated from the plasmodium and their contractile activities can be measured by tensiometry[21,22] (Fig. 4, C - D, Fig. 8). The period of the radial contraction activity (Fig. 4, C) is on an average 1.3 min, i. e., the same value as the period of shuttle streaming.[23] The average period of longitudinal contraction activity is 2.1 min (Fig. 4, D). After a few minutes, motile phenomena such as cytoplasmic streaming are reassumed under tensiometric conditions. Simultaneous tensiometric and laser registrations should offer valuable information for the interpretation of laser data. In the last few years such combined studies in connection with computer data analysis are performed at the Moscow

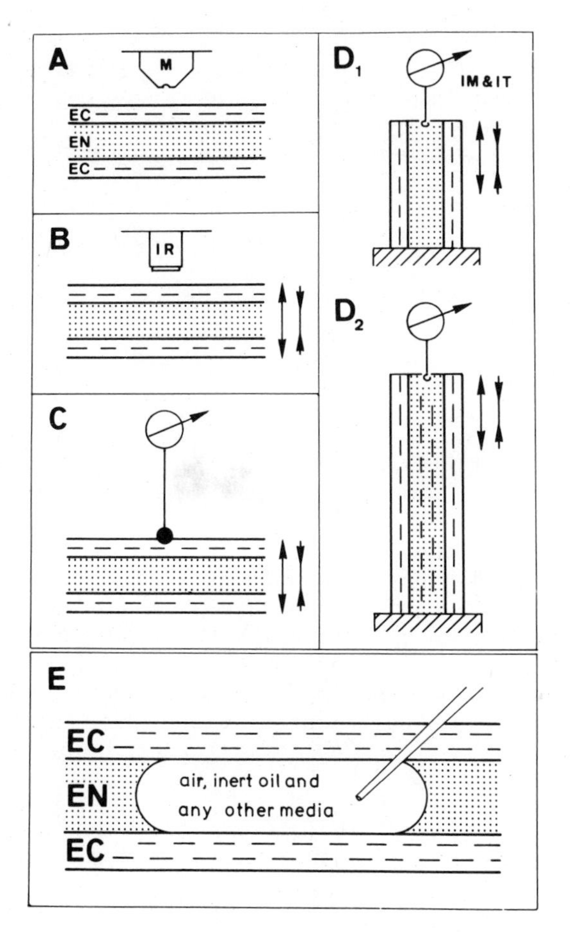

Fig. 4 Diagrammatic representation of methods for the analysis of shuttle streaming (A), radial contraction and wave phenomena (B - C), force measurements (D), effects of stretching ($D_1 \rightarrow D_2$) and the preparation of endoplasm-substituted protoplasmic strands (E). M = microscope; EC = ectoplasmic tube; EN = endoplasmic channel; IR = infrared sensor; IM and IT = isometric and isotonic, respectively; double arrows; changing directions of the contractile phenomena of the strands.

State University (Prof. Y. M. Romanovsky).

There are further possibilities of combining a registration of streaming phenomena, contraction activities and mass nett transport[7,9] with suitable laser techniques. The experimental advantage of the large plasmodial mass is that the object can easily be adapted to varying conditions without decreasing its vitality, e. g.,

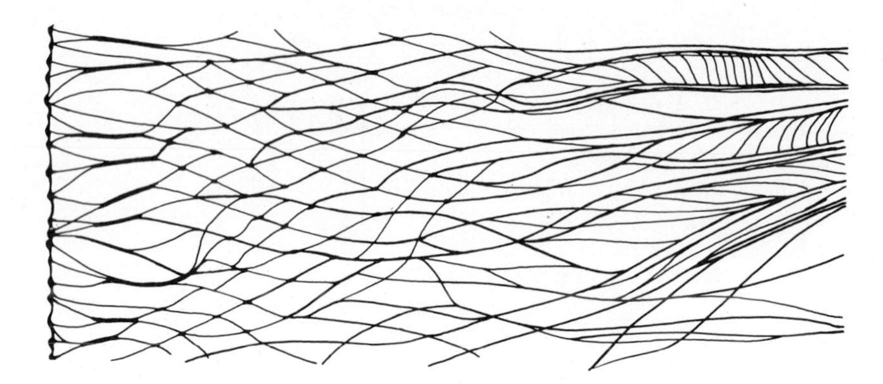

Fig. 5 An extremely delicate pattern of actin as revealed in the front-area (left side) of very thin migrating plasmodia (thickness $< 10\ \mu m$) by immunofluorescence cytochemistry (staining with antibodies to actin). The right side of the picture is in the direction of the posterior region in which endoplasmic pathways (functionally comparable with plasmodial strands) begin to differentiate. Note the diagonally running pattern of actin fibrils which surround these primary endoplasmic channels. Closer to the front region (left side), the endoplasmic channels diverge into very small pathways up to the tip of the plasmodium and show shuttle streaming activity[28].

after more than 12 hours of tensiometric registration (as shown in Fig. 4 (D)), the amputated protoplasmic strand develops to a new plasmodium under appropriate nutrional conditions, but also for a short time span without them, because the cell contains a high reservoir of glycogen.

C. Actin transformations and actin-myosin interactions

These molecular phenomena occurring permanently at all sites of the ectoplasmic tube of plasmodial strands are of actual interest for the function of cytoplasmic actomyosin. Changes in the polymer status of cytoplasmic actin are an obligatory part of the contraction-relaxation cycle.[14] Because our present knowledge results predominantly from biochemical studies, one should anticipate that information gained from in situ-conditions would contribute to an analysis of the molecular contraction mechanism of cytoplasmic actomyosin. Whereas under normal conditions actomyosin fibrils are found only in the ectoplasmic tube[4,24,25] (compare Fig. 2), stretching the strand to about 50 % of its original length (Fig. 4, $D_1 \rightarrow D_2$) leads to a formation of light-microscopically visible fibrils also in the endoplasmic channel.[26,27] This leads to an increased longitudinal force output and possibly favours attempts in using laser techniques.

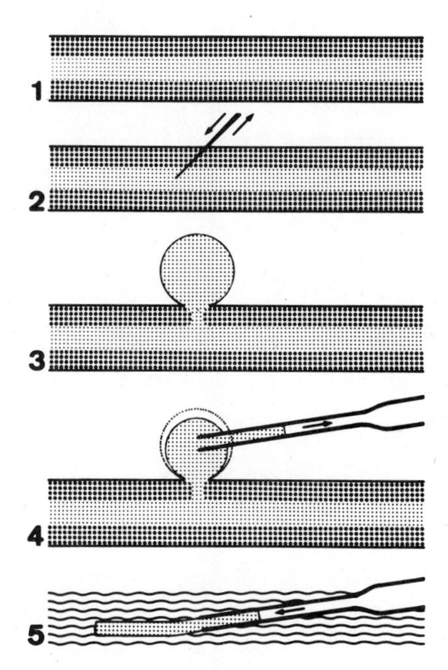

Fig. 6 Procedure for the preparation of endoplasmic veins. A protoplasmic strand (1) is punctured with the aid of a glass needle (2). The protruding endoplasmic drop (3) is punctured by a pipette and immediately thereafter a part of the protruding endoplasm is sucked up (4) and ejected into a physiological solution (5) [29].

3. THIN PLASMODIAL SHEETS

In the case that laser techniques require much thinner objects as plasmodial strands, it is possible to prepare protoplasmic sheets between agar sandwiches with a thickness of only 5 to 10 μm (Fig. 5). Such stages reach the dimension of normal cell layers in tissue culture, and the plasmodia show, in principal, all vital phenomena of the thicker plasmodial stages.[28]

4. ENDOPLASMIC DROPS AND VEINS

Protoplasmic strands isolated from the plasmodial mass show a rather broad morphological variation depending on different factors such as topographical location within the plasmodium, culture or starving conditions, substrate etc. A much more regular, i. e., more reproducible appearance of specimens is available when using endoplasmic drops[19,24] and endoplasmic veins.[29] Further advantages of these models are that they show strictly time-dependent differentiation processes allowing the study of additional dynamic cellular processes, e. g., membrane flow mechanisms. Fig. 6 demonstrates

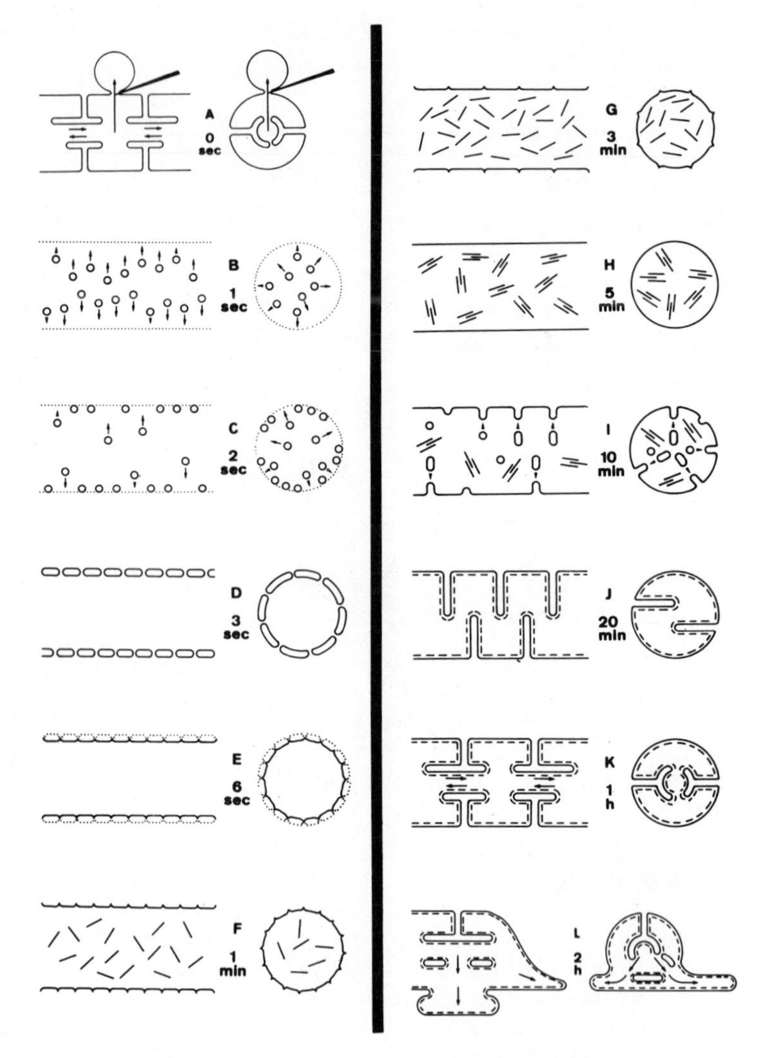

Fig. 7 Sequence (A - L) of time-dependent differentiation processes in endoplasmic veins (pictures to the left = longitudinal sections) and endoplasmic drops (pictures to the right). Also, the pictures to the right indicate the cross sectional aspects of endoplasmic veins.
(A) = Generation of an endoplasmic drop by puncturing a protoplasmic strand (compare Fig. 6). (B) - (E) = plasmalemma regeneration in endoplasmic veins and drops. (F) - (G) = actin polymerisation. (H) = Actomyosin fibrillogenesis. (I) = Regeneration of plasmalemma invaginations. (J) = Formation of actomyosin sandwiches adjacent to the plasmalemma invagination system; de novo uptake of oscillating contraction rhythmicity. (K) = Uptake of endoplasmic shuttle streaming.
(L) = Begin of locomotory activity by formation of pseudopods.

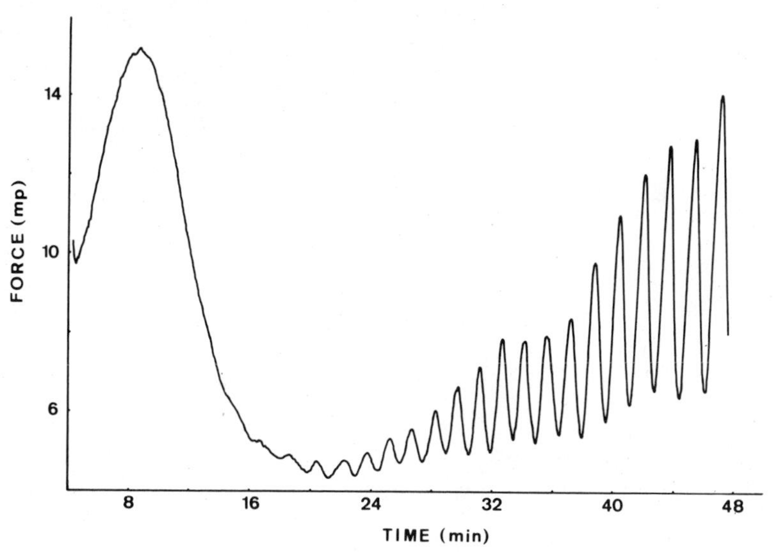

Fig. 8 Tensiometric registration of the de novo uptake of rhythmical contraction activity of endoplasmic veins[29].

the generation of endoplasmic drops and endoplasmic veins. Endoplasmic drops with a volume between 0.04 and 4.0 cubic mm (favourable size 0.4) can be generated by puncturing a plasmodial strand with a glass needle. Immediately after the protrusion of large drops, the endoplasm can be sucked up into pipettes and pushed out into a fluid resulting in a long protoplasmic cylinder designated as endoplasmic vein (diameter 0.2 - 0.3 mm, length up to 3 cm).

A. Membrane flow

In endoplasmic drops and veins, a new plasmalemma is formed within 6 sec after the generation of these models (Fig. 7, A - E). This process is performed by a very rapid vesicle transport to the periphery of the naked protoplasmic mass (Fig. 7, B - C), followed by an alignment of the vesicles at the cell periphery (Fig. 7, D) and a lateral fusion of these vesicles.[30] The peripheral part of the vesicle membrane is discarded (Fig. 7, E) and the centripetal part forms the new plasmalemma (Fig. 7, F).

Further membrane flow mechanisms, however with a lower velocity, are responsible for the regeneration of the plasmalemma invagination system (Fig. 7, I - J). The plasmalemma invaginations grow to the interior side of the vein via fusion with intracellular vesicles.[19,31] In endoplasmic veins, the membrane surface of the invaginations can reach 75 % of the total plasmalemma area, the volume of the invaginations 10 - 15 % of the total protoplasmic volume.[29]

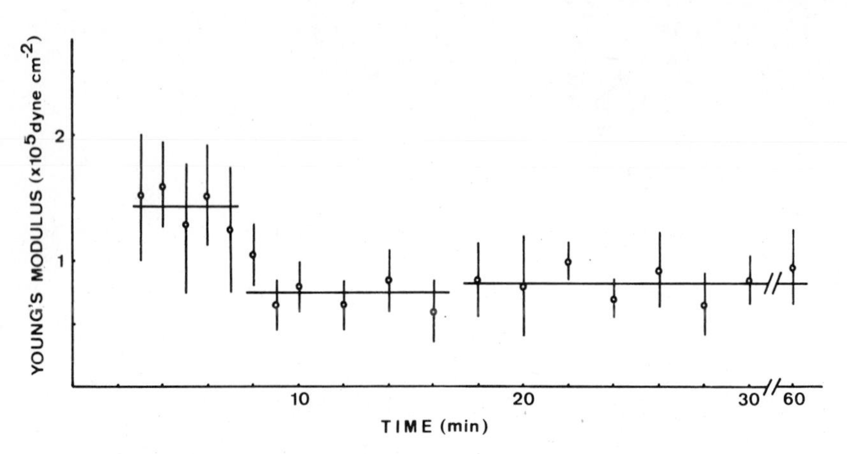

Fig. 9 Changes of Young's modulus in the time course of the morphogenetic processes in endoplasmic veins (compare Fig. 7)[29].

B. Actin transformations and actin-myosin interactions

Endoplasmic veins and drops primarily represent the pure endoplasm with relative low viscosity and contain a considerable amount of actin with a low polymer status, i. e., a short chain length.[15,16] A strong increase in viscosity of the models at an age between 1 and 3 min (Fig. 7, F - G) is based on actin polymerisation followed within several min by an actomyosin fibrillogenesis (Fig. 7, H).[19,24,29,31] Both strictly time-dependent processes should be readily available to laser techniques. Both processes can also be registered and evaluated by tensiometry[21,29] (Figs. 8 and 9), thus, it is possible to apply complementary techniques simultaneously during the recording of laser data.

C. Wave phenomena

Intact plasmodia as well as isolated protoplasmic strands and endoplasmic veins (Fig. 10) show wave phenomena[32,33] in their contraction rhythms. In general the plasmodia seem to represent a more or less "monorhythmic system"[34] such that the contraction periods do not differ greatly within various plasmodial regions. Synchronisation phenomena have been studied with tensiometric methods,[35,36] indicating that the endoplasmic stream is responsible for a synchronisation of contractile activities.[13] An exact registration of the phase behaviour of radial contractile rhythmicity reveals metachronous phenomena (Fig. 10), which have to be interpreted as propagated waves.[29,32,33] Also these phase relations should be accessible to laser analysis in combination with other methods, e. g., infrared registration[37] of radial rhythms or photometrical techniques.[38]

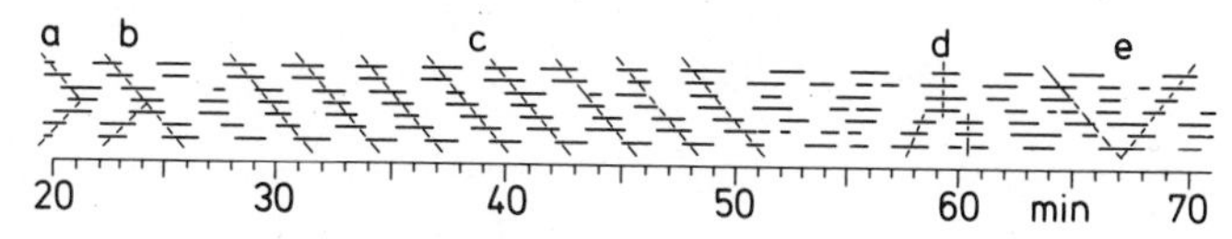

Fig. 10 Pattern of metachronous radial contraction activities of an endoplasmic vein demonstrating wave phenomena along the longitudinal axis of an endoplasmic vein.
The pattern represents a phase evaluation of radial contraction behaviour in the age range between 20 and 70 min. Infrared registration at seven points of a vein with a total length of 2.5 cm. The original, oscillating contraction curves were transposed to this diagram by plotting the duration of successive relaxation periods (black bars) during the course of the experiment (abscissa). The vertical sequence of bars represents measuring points 1 to 7 according to their sequence along the longitudinal axis of the vein. The resulting patterns (a - e), indicated by dotted lines, reveal phase shifts, i. e., metachronic radial contractions which can be interpreted as propagated waves of radial contraction rhythmicity [29].

Recently the plasmodial structure and motility of Physarum has been comprehensively reviewed by Kessler[39] in a monograph comprising the cognate genus Didymium and a taxonomic survey.

5. CONCLUSIONS

The principal possibilities of using laser techniques on this object were demonstrated by Mustacich and Ware and by Sattelle (see Earnshaw and Steer [1]) by revealing streaming directions transverse to the longitudinal axis of the strand, and a correlation with the radial contraction rhythmicity. Laser Doppler techniques should deliver a more detailed knowledge of flow rates and flow profiles than that resulting from light microscopy.

Time-related cellular reactions, such as sol $\rightleftharpoons$ gel transformation based on actin polymerisation and actin-myosin interactions, i. e., movements of particles below the resolution of the light microscope, should be accessible to laser analysis as well as movements of membrane compartments, e. g., during plasmalemma regeneration.

A more specific evaluation of the advantages and possible difficulties of this object for laser techniques is the aim of this symposium and a matter of discussion during the meeting. As general advantages of Physarum the following can be nominated.

1.) Availability of different vital plasmodial stages in a range between 10 µm thickness and 2 mm thickness and an object size between normal cell size and several square dcm. This means: optional adjustment of the scattering volume.

2.) Relative simple experimental handling and adaptation to the requirements of the techniques without danger of a decreased vitality.

3.) The possibility to study cellular dynamic phenomena of different nature ranging between the molecular and the microscopic level in combination with established techniques (light microscopy, tensiometry and other registration techniques).

4.) A strictly time-dependent behaviour, e. g., of the streaming and contraction activities in plasmodial strands, flow of membrane compartments, actin transformations, actin-myosin interactions and wave phenomena in models such as endoplasmic veins.

5.) The possibility to substitute the endoplasm in plasmodial strands by inert media reduces the sample dispersity and allows a discrimination of scattering phenomena resulting from different regions.

ACKNOWLEDGEMENTS: The author thanks Dr. R. L. Snipes (Giessen) for reading the manuscript, Dr. R. Stiemerling and Mr. S. Schumacher for preparing the diagrams.

REFERENCES

1. J. C. Earnshaw and M. W. Steer, Studies of cellular dynamics by laser Doppler microscopy, Pestic. Sci. 10:358 (1979).
2. K. E. Wohlfarth-Bottermann, Ursachen von Zellbewegungen, Cytoplasmatische Actomyosine und ihre Bedeutung für Protoplasmaströmungen und Zellmotilität, Leopoldina 21;85 (1975).
3. H. Komnick, W. Stockem and K. E. Wohlfarth-Bottermann, Cell Motility: Mechanisms in protoplasmic streaming and ameboid movement, Int. Rev. Cytol. 34:169 (1973).
4. K. E. Wohlfarth-Bottermann, Weitreichende fibrilläre Protoplasmadifferenzierungen und ihre Bedeutung für die Protoplasmaströmung. X. Die Anordnung der Actomyosinfibrillen in experimentell unbeeinflußten Protoplasmaadern von Physarum in situ, Protistologica XI:19 (1975).

5. N. Kamiya, Protoplasmic streaming, in: "Protoplasmatologia VIII. 3a", L. V. Heilbrunn and F. Weber, eds., Springer, Wien (1959).
6. K. E. Wohlfarth-Bottermann, Plasmalemma invaginations as characteristic constituents of plasmodia of Physarum polycephalum, J. Cell Sci. 16:23 (1974).
7. K. E. Wohlfarth-Bottermann, Oscillatory contraction activity in Physarum, J. Exp. Biol. 81:15 (1979).
8. F. Achenbach, W. Naib-Majani and K. E. Wohlfarth-Bottermann, Plasmalemma invaginations of Physarum dependent on the nutritional content of the plasmodial environment, J. Cell Sci. 36:355 (1979).
9. N. Hülsmann and K. E. Wohlfarth-Bottermann, Spatio-temporal relationships between protoplasmic streaming and contraction activities in plasmodial veins of Physarum polycephalum, Cytobiologie 17:317 (1978).
10. N. Hülsmann and K. E. Wohlfarth-Bottermann, Räumliche und zeitliche Analyse von kontraktionsabhängigen Oberflächenbewegungen bei Physarum polycephalum, Cytobiologie 17:23 (1978).
11. A. Grebecki and M. Cieslaswka, Dynamics of the ectoplasmic walls during pulsation of plasmodial veins of Physarum polycephalum, Protoplasma 97:365 (1978).
12. K. E. Wohlfarth-Bottermann and F. Achenbach, Lateral apertures as passage-ways between ectoplasm and endoplasm in plasmodial strands of Physarum, Cell Biol. Int. Rpts. 6:57 (1982).
13. U. Achenbach and K. E. Wohlfarth-Bottermann, Synchronization and signal transmission in protoplasmic strands of Physarum. The endoplasmic streaming as a pacemaker and the importance of phase deviations for the control of streaming reversal, Planta 151:584 (1981).
14. K. Götz v. Olenhusen and K. E. Wohlfarth-Bottermann, Evidence for actin transformations during the contraction-relaxation cycle of cytoplasmic actomyosin: Cycle blockade by phalloidin-injection, in: "Cell Motility, Molecules and Organization", S. Hatano, H. Ishikawa and H. Sato, eds., University of Tokyo Press, Tokyo (1979).
15. G. Isenberg and K. E. Wohlfarth-Bottermann, Transformation of cytoplasmic actin. Importance for the organization of the contractile gel reticulum and the contraction-relaxation cycle of cytoplasmic actomyosin, Cell Tiss. Res. 173:495 (1976).
16. K. E. Wohlfarth-Bottermann and G. Isenberg, Dynamics and molecular basis of the contractile system of Physarum. in: "Contractile Systems in Non-Muscle Tissues", S. V. Perry, A. Margreth and R.S. Adelstein, eds., North/Holland Publ. Comp., Amsterdam (1976).

17. W. Gawlitta, H. U. Hoffmann and W. Stockem, Morphology and dynamic activity of the cell surface in different types of microplasmodia of the acellular slime mold Physarum polycephalum, Publ. Univ. of Innsbruck 120:176 (1979).
18. K. Götz v. Olenhusen, H. Jücker and K. E. Wohlfarth-Bottermann, Induction of a plasmodial stage of Physarum without plasmalemma invaginations, Cell Tiss. Res. 197:463 (1979).
19. F. Achenbach, U. Achenbach and K. E. Wohlfarth-Bottermann, Plasmalemma invaginations, contraction and locomotion in normal and caffeine-treated protoplasmic drops of Physarum, Eur. J. Cell Biol. 20:12 (1979).
20. T. Ueda, K. Götz v. Olenhusen and K. E. Wohlfarth-Bottermann, Reaction of the contractile apparatus in Physarum to injected Ca^{++}, ATP, ADP and 5' AMP, Cytobiologie 18:76 (1978).
21. K. E. Wohlfarth-Bottermann, Tensiometric demonstration of endogenous oscillating contractions in plasmodia of Physarum polycephalum, Z. Pflanzenphysiol. 76:14 (1975).
22. N. Kamiya, Physical and chemical basis of cytoplasmic streaming, Ann. Rev. Plant Physiol. 32:205 (1981).
23. K. E. Wohlfarth-Bottermann, Oscillating contractions in protoplasmic strands of Physarum: Simultaneous tensiometry of longitudinal and radial rhythms, periodicity analysis and temperature dependence, J. Exp. Biol. 67:49 (1977).
24. K. E. Wohlfarth-Bottermann, Weitreichende, fibrilläre Protoplasmadifferenzierungen und ihre Bedeutung für die Protoplasmaströmung. II. Lichtmikroskopische Darstellung, Protoplasma 57:747 (1963).
25. K. E. Wohlfarth-Bottermann, Weitreichende, fibrilläre Protoplasmadifferenzierungen und ihre Bedeutung für die Protoplasmaströmung. III. Entstehung und experimentell induzierbare Musterbildungen, Roux'Archiv Entwicklungsmech. 156:371 (1965).
26. K. E. Wohlfarth-Bottermann and M. Fleischer, Cycling aggregation pattern of cytoplasmic F-actin coordinated with oscillating tension force generation, Cell Tiss. Res. 165:327 (1976).
27. M. Fleischer and K. E. Wohlfarth-Bottermann, Correlation between tension force generation, fibrillogenesis and ultrastructure of cytoplasmic actomyosin during isometric and isotonic contractions of protoplasmic strands, Cytobiologie 10:339 (1975).
28. W. Naib-Majani, W. Stockem, K. E. Wohlfarth-Bottermann, M. Osborn and K. Weber, Immunocytochemistry of the acellular slime mold Physarum polycephalum. Spatial organization of cytoplasmic actin, Eur. J. Cell Biol. (in press).
29. Z. Baranowski and K. E. Wohlfarth-Bottermann, Endoplasmic veins from plasmodia of Physarum polycephalum: a new strand model with defined age, structure and behaviour, Eur. J. Cell Biol. 27:1 (1982).

30. K. E. Wohlfarth-Bottermann und W. Stockem, Die Regeneration des Plasmalemmas von Physarum polycephalum, Roux'Archiv Entwicklungsmech. 164:321 (1970).
31. F. Achenbach and K. E. Wohlfarth-Bottermann, Morphogenesis and disassembly of the circular plasmalemma invagination system in Physarum polycephalum, Differentiation 19:179 (1981).
32. Z. Baranowski, The contraction-relaxation waves in Physarum polycephalum plasmodia, Acta Protozoologica 17:377 (1978).
33. Z. Hejnowicz and K. E. Wohlfarth-Bottermann, Propagated waves induced by gradients of physiological factors within plasmodia of Physarum polycephalum, Planta 150:144 (1980).
34. A. Grebecki and M. Cieslawska, Plasmodium of Physarum polycephalum as a synchronous contractile system, Cytobiologie 17:335 (1978).
35. Y. Yoshimoto and N. Kamiya, Studies on contraction rhythm of the plasmodial strand. III. Role on endoplasmic streaming in synchronization of local rhythms, Protoplasma 95:111 (1978).
36. Y. Takeuchi and M. Yoneda, Synchrony in the rhythm of the contraction-relaxation cycle in two plasmodial strands of Physarum polycephalum, J. Cell Sci. 26:151 (1977).
37. K. E. Samans, K. Götz v. Olenhusen and K. E. Wohlfarth-Bottermann, Oscillating contractions in protoplasmic strands of Physarum: Infrared reflexion as a non-invasive registration technique, Cell Biol. Int. Rpts. 2:271 (1978).
38. F. Achenbach, U. Achenbach, K. E. Samans and K. E. Wohlfarth-Bottermann, An inexpensive "silicon photo device" for transmicroscopic registration of rhythmical movement phenomena, Microscopica Acta 84:43 (1981).
39. D. Kessler, Plasmodial structure and motility, in: "Cell Biology of Physarum and Didymium", H. C. Aldrich and J. W. Daniel, eds., Academic Press, New York (1982).

AMOEBOID MOVEMENT IN *CHAOS CAROLINENSIS*

Robert D. Allen

Department of Biological Sciences
Dartmouth College
Hanover, N.H. 03755

1. SCOPE

Amoeboid movement is often regarded as one of the simplest forms of cell motility. However, it consists of even simpler phenomena that either constitute, or are encountered in, other forms of motility, namely cytoplasmic streaming, changes in cell shape, and the formation of temporary locomotor appendages, pseudopodia.

It is a truism that amoeboid movement is the way(s) amoebae move, but the details of the process are so diverse among the hundreds of known species of free-living and parasitic amoebae and among the hundreds of amoeboid tissue cells that we cannot be certain at present that what is learned about one cell type is broadly applicable to others. In fact, there is evidence of diversity in mechanisms of amoeboid movement (Allen, 1968).

This paper is intended to encourage and provide a background for the use of laser-Doppler spectroscopy to study amoeboid movement. This and other biophysical approaches are best applied to the giant, carnivorous, multinucleate amoeba, *Chaos carolinensis*. This cell has been the material of choice for the study of cell morphology, patterns of streaming, cytoplasmic consistency and ultrastructure of amoebae. Some promising beginnings have been made in understanding the molecular basis of movement in this species. However, it is not easy to grow *Chaos* to gram quantities, and when this is done, one must recognize the disadvantage that *Chaos* (and other large amoebae) contain cytoplasmic symbionts. Therefore, the proteins extracted from them are not all expressions of the *Chaos* genome.

Biochemical studies of the contractile and cytoskeletal proteins have progressed more rapidly with smaller cells, such as the soil amoeba, Acanthamoeba and the amoeboid stage of the cellular slime mold, Dictyostelium. While the results of these studies are important, they may or may not be applicable to Chaos. However, they may reveal important principles governing the control of contractility, consistency and cell shape, whether or not the responsible cytoskeletal components are identical.

The scope of this paper will be limited to general background: what the laser-Doppler-spectroscopist should know about cell morphology, the complex pattern of streaming, cytoplasmic consistency, and the mechanisms of movement in this well studied model cell, Chaos carolinensis.

2. MORPHOLOGY

Chaos carolinensis is protected from its external environment by its plasmalemma, which consists of three parts: an outer glycocalax of ca. 100 nm long surface filaments resembling a shag-rug in scanning electron micrographs, the plasma membrane permeability barrier (represented in electron micrographs of thin sections as "unit membrane") and a supportive cytoskeletal infrastructure.

When tweaked with a micromanipulator needle, the plasmalemma appears to be a tough, elastic coat, but it becomes fragile enough to rupture under the pressure of cytoplasmic streaming at submicromolar external concentrations of calcium ions (Taylor, Condeelis, Moore and Allen, 1973). Thus its mechanical properties depend on the external calcium concentration.

During cell locomotion, the entire cell surface is believed to be displaced forward over unattached, cylindrical pseudopodia. This is shown by the behavior of particles attached to the glycocalyx. Particles adhering to the ventral surface advance around points of attachment to the substratum, indicating that the surface is fluid (Griffin and Allen, 1960). This is shown even more dramatically when an amoeba is impaled by a needle. Cell movement is not in any way impeded; the membrane merely flows around the needle.

Immediately beneath the plasma membrane is a clear layer, the hyaline ectoplasm. Its physical consistency is not known with certainty, but there are some indications that cytosol (the interstitial fluid of the cytoplasm) may move from the front to the rear in this layer under some conditions (Allen, 1961a, 1973).

The remainder of the cytoplasm contains numerous organelles of various sizes and shapes, including nuclei, mitochondria, lysosomes, Golgi bodies, ER, contractile vacuoles, etc., and inclusions such as refractile bodies, crystals of truiret, and oil droplets (Allen,1961a)

The granular cytoplasm is divisible into two layers, the outer ectoplasm, which normally remains relatively stationary in relation to the substratum, while the inner layer of endoplasm streams into extending pseudopodia. As the endoplasm reaches the tip of each pseudopodium, it everts to form the ectoplasmic tube.

The ectoplasmic tube of an extending pseudopodium may be as simple in shape as cylindrical in monopodial species, or as complex as several cylinders extending from a shapeless body of cytoplasm in polypodial specimens. In monopodial specimens there tends to be a greater diameter near the middle of the cell, and the tail region becomes progressively thinner.

It has recently been shown by Bynum and Allen (1980) that the more typical polypodial cells can often be caused to become monopodial by centrifugation. Under these conditions, the tails of these cells develop a helical form, and the tail ectoplasm undergoes torsional movements that may proceed in either sense.

Having seen helical form and torsional rotation of the tail in monopodial _Chaos_, Bynum and Allen showed that there was a lesser but nevertheless demonstrable tendency toward helical organization on the part of retracting pseudopodia in normal polypodial amoebae.

Pseudopodia of _Chaos_ may be classified as either simple or compound. Simple pseudopodia are initiated by the formation of a single blister of clear fluid called the hyaline cap. The endoplasm either partially or completely invades the hyaline cap region as it becomes everted to form the cylindrical ecoplasmic tube. The hyaline cap reappears periodically at the tips of pseudopodia, especially those in which the streaming is sporadic.

Compound pseudopodia are usually initiated by the appearance of two or more hyaline caps at different loci on the cell surface. Each hyaline cap is invaded by a stream of endoplasm which flows independently of other streams, but sharing the same non-cylindrical ectoplasmic tube. Thus compound pseudopodia may have asynchronous bursts of streaming and hyaline cap formation in two or more endoplasmic streams (Allen, 1973).

3. CONSISTENCY

It has been believed for over a century that there are local variations in cytoplasmic consistency in amoebae (for reviews, see de Bruyn, 1947; Allen, 1961a). Mast (1926) first suggested that the apparent sol gel changes in amoeboid movement were a fundamental aspect of the mechanism. He proposed a very simple model for amoeboid movement in which a pressure gradient generated by contraction, coupled with a reversible sol⇌gel transition in the cytoplasm accounted for the most general aspects of amoeboid movement.

The discovery that cytoplasm could stream when removed from the cell (Allen, Cooledge and Hall, 1960) led not only to the abandonment of Mast's theory but introduced doubt that the cytoplasm existed in true sol and gel states with a transformation as envisaged by theories of the time (Allen, 1961a).

The rule of thumb that "if something flows it must be a sol and if not, it must be a gel" seemed to be invalidated by the comparison of flow patterns in attached and unattached amoebae (see Section IV). In unattached monopodial amoebae, the cytoplasm "flows" forward as endoplasm, then turns and "flows" backward as a sleeve of ectoplasm. In attached cells, only the endoplasmic flow is apparent, therefore it appears to be "fluid" (Allen, 1961a).

In thinking about the significance of flow for the consistency of cytoplasm, one must distinguish between the displacement of the fluid and shear developed within it. Restating the situation in those terms, one can say that in either attached or unattached cells the endoplasm is sheared as it is displaced, while the ectoplasm is hardly sheared at all, whether it is displaced or not. (Actually, the situation is more complex, because the tail ectoplasm does experience some shear, while the axial region of the endoplasm experiences very little shear).

Thus while there is some justification for believing that a qualitative change (sol→gel) in consistency may occur in the frontal region, and that a reciprocal (gel→sol) change occurs in the tail region, we lack an adequate quantitative assessment of what this change might entail in terms of rheological parameters.

Several efforts have been made to define both qualitatively and quantitatively the consistencies of different regions of the cell. Analysis of velocity profiles (Allen and Roslansky, 1959), analysis of centrifuge microscope records (Allen, 1960), magnetic sphere experiments (Yagi, 1961) and strain birefringence measurements (Francis and Allen, 1971) have been undertaken. The results have been reviewed elsewhere (Allen, 1973). One can summarize by stating that the endoplasm, despite its appearance of fluidity, is a weak gel with complex viscoelastic properties. The explanation for these complex rheological properties was found in the interactions of macromolecules that comprise the contractile cytoskeleton of the cell (Taylor, Condeelis, Moore and Allen, 1973; Allen and Taylor, 1975; Taylor and Condeelis, 1979).

To comprehend the mechanism of amoeboid movement fully, it will be necessary to determine the rheological properties of the cytoplasm locally in order to understand precisely where the contractile events and changes in rheological properties occur.

A beginning has been made in this direction in the work to be reported at this conference by Masahiko Sato. The magnetic sphere method has been refined in his hands to the point that it can begin to obtain the data that are required.

4. PATTERN OF FLOW

There are two ways to observe and quantify the pattern of streaming in Chaos. The first to be attempted was the measurement of transverse velocity profiles (in the fluid dynamic sense: v=f(r), as opposed to the statistical sense used in laser-Doppler spectroscopy). The velocity profile of a Newtonian fluid in a capillary is a paraboloid of revolution. The question asked in measuring the velocity profile was: "if the motive force for amoeboid movement is assumed to be a pressure gradient, is the velocity profile consistent with Newtonian properties?" The answer was negative (Allen and Roslansky, 1959). Although the velocity profile in the uroid was found to be more pointed than parabaloid and passed quickly through a parabaloid shape, it became a truncated parabaloid in the anterior half of the cell. This observation found an explanation in the consistency information described earlier. However, the results indicated nothing about the validity of the pressure hypothesis that was then being tested.

The second way of quantifying the pattern of streaming is to measure longitudinal or sagittal velocity profiles. Here one finds that along a body of endoplasm, streaming always begins first at a new pseudopod tip and dies out last there (Allen, 1973). It is also seen that the greatest fluctuations in velocity occur near the front of a pseudopodium. In a compound pseudopodium the fluctuations of velocity in two streams sharing the lumen of the same ectoplasmic tube are often asynchronous and may even have different frequencies. Each stream thus has its own signature and dynamics. For this reason, it will be absolutely essential in laser-Doppler spectroscopy experiments to record simultaneously the behavior and morphology of the cell in order that the spectroscopic data can be interpretable.

5. MOLECULAR MECHANISMS

Theories of amoeboid movement have been the subject of many lively debates over the past 150 years (see de Bruyn, 1947; Allen, 1961a; Taylor and Condeelis, 1979 for reviews). It is remarkable that some of the very earliest ideas on the subject that were expressed by Felix Dujardin (1835) are consistent with those found in the recent literature based on sound experimental evidence. Dujardin called the substance of amoebae and other protists "le sarcode," implying by the term (and stating explicitly) that the living substance had the property of contractility.

The literature of the present century has stressed the involvement of two processes, contractility and "sol⇌gel transformations" or consistency changes in the mechanisms of amoeboid movement. Differences in the relative importance of these processes were expressed in the first half of the century (reviewed by de Bruyn, 1947). All of these hypotheses assumed that flow resulted from a pressure gradient generated by contraction of the ectoplasmic tube. Doubt was cast on this assumption by the observation that cytoplasm released from amoebae confined in glass or quartz capillaries could stream, not only in the characteristic "fountain" pattern, but cyclotically and in apparently contractile streamlets (Allen, Coolege and Hall, 1960). Although this observation led to the proposal of several theories of amoeboid movement (see Allen and Allen, 1978 for review), the pressure theory was not refuted experimentally until a decade later, when it was shown that the creation of a "pressure sink" by suction applied to one pseudopodium could not reverse streaming into other pseudopodia (Allen, Francis and Zeh, 1971). This result indicated that a pressure difference cannot be the primary mechanism, but it does not eliminate the possibility that pressure differences might result from another mechanism, such as one involving directed forces.

The frontal contraction theory (Allen, 1961b) proposed that a contractile force acted at the tip of each advancing pseudopodium to draw the viscoelastic endoplasm toward the contracting region. Each portion of cytoplasm, according to this concept, advances toward the region of contraction where it simultaneously contracts, gelates and becomes everted to form the ectoplasmic tube. This theory rationalizes the horizontal and vertical velocity profiles, the consistency changes, and certain behavioral responses that suggest their control by contractile events at the front of the cell.

The molecular basis for cytoplasmic contractility in *Chaos* was first demonstrated by the application of specially designed solutions analogous to contraction-, relaxation-, and rigor-solutions for muscle myofibrils applied to the naked cytoplasm of amoebae (Taylor, Condeelis, Moore and Allen, 1976). In a rigor-solution, amoeba cytoplasm showed photoelastic behavior. When 0.5 mM ATP was added, this behavior was lost (relaxation). On addition of micromolar calcium, contraction occurred. At the correct balance of calcium and magnesium ions and ATP, the isolated cytoplasm streamed actively for many minutes, showing muscle-like contractions and relaxations.

The control of cytoplasmic contractility by the same physiological factors that control contractility in muscle myofibrils turned out not to be coincidental. Negatively stained endoplasm in relaxing solution was shown to contain numerous structures that were identifiable morphologically as f-actin filaments and myosin aggregates. Both actin and myosin, which had been isolated and purified earlier from smaller amoebae were also found in *Chaos* extracts (Condeelis, 1977). However, the heavy chain of myosin of *Chaos* was found to be of somewhat higher molecular weight than that of the myosin of rabbit striated muscle, in contrast to the "minimyosin" that had been found earlier in Acanthamoeba (Pollard and Korn, 1973).

Although much compelling evidence supports the view that the actin-myosin contractile system of the amoeba does its work in the frontal region of the cell, it is also true that not all of the evidence points in that direction. It has been suggested, for example, that contractions might occur in the tail as well as the front (Stockem, Hoffman and Gawlitta, 1982). In support of that idea are the results of aequorin luminescence studies showing a higher overall concentration of free calcium ions in the tail as well as the expected bursts of calcium entry at the front (Taylor, Blinks and Reynolds, 1980). The meaning of these results is not entirely clear. Certainly the cell consists of many calcium "compartments" (the plasmalemma, the mitochondria, the calcium binding proteins of the cytoskeleton and of the cytosol, ER and other membranous structures, etc). Since the fluxes among these and their binding constants within the bounds of physiological pH are largely unknown, it is difficult to relate the free calcium concentration directly to any one function such as the control of contractility.

The question of whether a contraction of the tail ectoplasm might assist the frontal contraction mechanism was posed recently by applying blasts of a laser microbeam to the frontal and caudal portions of amoebae (Cullen and Allen, 1980). It was found that sublytic doses of irradiation had an immediate and measurable effect only when applied to the frontal portion, not the caudal portion. In addition, a lytic dose applied to the caudal region of an amoeba osmotically equilibrated with its environment opened the caudal end without any immediate effect on the velocity of forward streaming. These results are consistent with the frontal contraction theory and would appear to eliminate the need to postulate two localized motive forces in the cell in order to account for amoeboid movement.

Much remains to be learned about contractility and its control in the giant amoeba. The role of calcium ions may be subordinate to control over both contractility and consistency by actin-binding and other cytoskeletal proteins. So far, research on these aspects of amoeboid movement has progressed most rapidly on small amoebae and various tissue cells (see Taylor and Condeelis, 1979 for review). How applicable these results are to the giant amoebae remains to be determined.

REFERENCES

Allen, R.D., 1960, The consistency of amoeba cytoplasm and its bearing on the mechanism of amoeboid movement. II. The effects of centrifugal acceleration observed in the centrifuge microscope, J. Biophys. and Biochem. Cytol., 8:379.

Allen, R.D., 1961a, Amoeboid movement, in: "The Cell", vol. II, 135-216, A.E. Mirsky and J. Brachet, eds., Academic Press, New York.

Allen, R.D., 1961b, A New theory of amoeboid movement and protoplasmic streaming, Exp. Cell Res., Suppl., 8:17.

Allen, R.D., 1968, Differences of a fundamental nature among several types of amoeboid movement, S.E.B. Symposia, 22:151.

Allen, R.D, 1973, Biophysical aspects of pseudopod extension and retraction, in: "The Biology of Amoeba", K.W. Jeon, ed., Academic Press, New York.

Allen, R.D. and Allen, N.S., 1978, Cytoplasmic streaming in amoeboid movement, Ann. Rev. Biophys. Bioeng., 7:497.

Allen, R.D., Cooledge, J.W., and P.J. Hall, 1960, Streaming in cytoplasm dissociated from the giant amoeba, Chaos chaos, Nature, 187:896.

Allen, R.D., Francis, D.W., and Zeh, R., 1971, Direct test of the positive pressure gradient theory of pseudopod extension and retraction in amoebae, Science, 174:1237.

Allen, R.D. and Roslansky, J.D., 1959, The consistency of amoeba cytoplasm and its bearing on the mechanism of amoeboid movement. I. Analysis of velocity profiles of Chaos chaos, J. Biophys. Biochem. Cytol., 6:437.

Allen, R.D. and Taylor, D.L., 1975, The molecular basis of amoeboid movement, in: "Molecules and Movement", R. Stephen and S. Inoué, eds., Raven Press, New York.

Bynum, R.D., and Allen, R.D., 1980, Torsional movements in the amoeba, Chaos carolinensis, suggest a helical cytoskeletal organization, J. Protozoology, 27:420.

Condeelis, J.S., 1977, The isolation of microquantitites of myosin from Amoeba proteus and Chaos carolinensis, Anal. Biochem., 78:374.

Cullen, K, and Allen, R.D., 1980, A laser microbeam study of amoeboid movement, Exp. Cell. Res., 125:1.

De Bruyn, P.P.H., 1947, Theories of amoeboid movement, Quart. Rev. Biol., 22:1.

Dujardin, F., 1838, Reserches sur les organisms inférieurs, Ann. Sci. Nat. Zool., 4:343.

Francis, D.W. and Allen, R.D., 1971, Induced birefringence as evidence of endoplasmic viscoelasticity in Chaos carolinensis, J. of Mechanochemistry and Cell Motility, A. Oplatka, ed.,

Griffin, J.L., and Allen, R.D., 1960, The movement of particles attached to the surface of amoebae in relation to current theories of amoeboid movement, Exp. Cell Res., 20:619.

Mast, S.O., 1926, Structure, movement, locomotion and stimulation in amoeba, J. Morph., Physiol., 41:347.

Pollard, .D. and E. Korn, 1973, Isolation from Acanthamoeba castellanii of an enzyme similar to muscle myosin. J. Biol. Chem., 248:4682.

Stockem, W., Hoffman, A.-V., and W. Gawlitta, 1982, Spatial organization and fine structure of the cortical filament layer in normal locomoting Amoeba proteus, Cell and Tissue Res., 221:505.

Taylor, D.L., J.R. Blinks and G. Reynolds, 1980, The contractile basis of amoeboid movement. VIII. Aequorin luminescence during amoeboid movement, endocytosis and capping. J. Cell Biol., 86:599.

Taylor, D.L. and Condeelis, J.S., 1979, Cytoplasmic structure and contractility in amoeboid cells. Int. Rev. Cytol., 56:57.

Taylor, D.L., Condeelis, J.S., Moore, P.L., and Allen, R.D., 1973, The contractile basis of amoeboid movement: I. The chemical control of motility in isolated cytoplasm. J. Cell. Biol., 59:378.

Yagi, K., 1961, The mechanical and colloidal properties of Amoeba protoplasm and their relations to the mechanism of amoeboid movement. Comp. Biochem. Physiol., 3:73.

CYTOPLASMIC STREAMING IN PLANT CELLS AND ITS RELATION TO THE CYTOSKELETON

Nina Strömgren Allen

Department of Biological Sciences
Dartmouth College
Hanover, New Hampshire 03755 U.S.A.

CONTENTS

ABSTRACT

The known patterns and rates of cytoplasmic streaming in plant cells as well as the investigations on plant cells with laser Doppler spectroscopy will be reviewed briefly. Our current knowledge of the molecules involved in the movement of cytoplasm and/or saltation of particles in cytoplasm and their possible location in cells will be described.

The classical systems often used in the study of streaming were selected as particularly suitable for light-beating laser spectroscopic investigations. These are (1) the giant fresh-water alga, *Nitella*, (2) the giant marine alga, *Acetabularia*, and (3) germinating pollen grains of *Amaryllis*. All of these cells are large, easily cultured and exhibit rapid transport of various sized particles in their cytoplasm. Furthermore, a body of knowledge has been accumulated describing the morphology, physiology, and biochemistry as well as the motile behavior of these cells.

Several theories that have been proposed to account for the rotational streaming observed in Nitella will be discussed. Some challenging experiments will be proposed to study these cells employing the laser Doppler spectroscopic techniques.

1. INTRODUCTION

The dynamic process of cytoplasmic streaming can be appreciated only in living cells. Since 1774, when streaming in Chara was first described by the Italian, Corti, observations have been carried out with increasingly sophisticated light microscopes. Streaming, which occurs during some or all parts of the life cycle of organisms, is an irreversible, continuous deformation of cytoplasm at the expense of endogenous energy. In most plant cells, the cell wall confines the living portion of the cells, although the interior areas of cytoplasm stream and shift within these walls. Most plant cells also contain one or several membrane bound vacuoles through and around which cytoplasm may move.

The rheological properties of the cytoplasm are not well characterized. This information is needed in order to understand movements occurring within the cells. Are there areas of higher or lower viscosity in which the same motive force would cause slower or faster streaming rates to occur?

The literature relevant to cytoplasmic streaming in plants has been reviewed by Kamiya[1,2] and N. and R. Allen[3]. Laser Doppler spectroscopic investigations of particle displacements in plant cells have been reported by two groups, Mustacich and Ware[4] and Langley, Sattelle, Piddington and Ross[5]. This work is briefly discussed in reference 3.

2. OBSERVATIONS ON STREAMING

A. The observed patterns and rates of streaming in plant cells

Plant cells exhibit many and varied patterns of streaming at various rates ranging from barely measurable movements to the intermediate streaming rates of from 2-10 μm sec^{-1} as observed in epidermal cells of onion and barley coleoptiles or Acetabularia cells to the very rapid rates (40-90 μm sec^{-1}) observed in characean internodal cells. These movements can be constant or intermittent, and uni-, bi- or multidirectional and they can occur in various patterns, from seemingly random to highly directional movements. Underlying the various patterns to be described are fibrillar elements near or in contact with which particles and cytoplasm are seen to move. These fibrillar elements have been difficult to observe and record, but whenever better methods of observation, staining or fixation are found, more fibrillar elements are seen.

At least three types of linear elements, microfilaments, microtubules and intermediate filaments are found in cells. The 7 nm f-actin microfilaments occur singly and in bundles. The f-actin, probably in association with myosin can cause movement and appears to be the common denominator in plant streaming. Various cytochalasins reversibly inhibit streaming and are known to disrupt actin-myosin based movements. Microtubules (25 nm wide) are found in plant cells, but movements of the cytoplasm have been found to be inhibted by the microtubule-inhibitor colchicine only in a few plant cells. One case is *Acetabularia*[7], where the faster migration of the nuclei from the rhizoid to the cap seems to be microtubule-dependent, while the slower movements are microfilament-related. Another linear element may be observed in the light microscope and be mistaken for one of the above filaments: thin membranous tubules of the endoplasmic reticulum. Fibrillar elements often show what appear to be independent movements. One can in some cell types observe the underlying filaments shifting their positions resulting in a shifting pattern of particle movements.

Although some of the parts of the plant cell cytoskeleton have been revealed and are similar to those of animal cells, it is not yet clear exactly which components of the cytoskeleton are directly involved in cell motility. Even with the best light microscopic techniques one cannot observe all the fibrils present in the cytoplasm. The type of cells, like *Nitella*, *Acetabularia* and epidermal cells of higher plants which are among the most favorable for studying cytoplasmic streaming are unfortunately also the least well preserved by electron microscopic fixation techniques. For example, in epidermal cells of oat coleoptiles many more filaments can be observed with sensitive techniques of optical microscopy than by transmission electron microscopy (personal observation with N. Parthasarathy, Cornell University).

When a fixative is introduced to any of these cells, the cells do not stop streaming immediately, but the entire cytoplasm often contracts. This is the reason that rapidly frozen specimens yield valuable data, because the cytoplasm is fixed relatively unaltered in the outermost cell areas. Because of the difficulty of fixation particularly of the filaments and since we need to know more about their behavior in living cells, the light microscopic studies and laser Doppler studies are particularly valuable for the study of these cells.

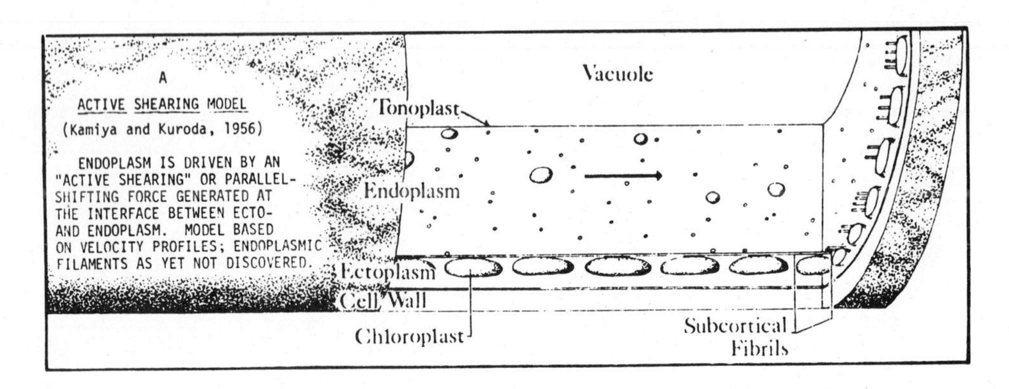

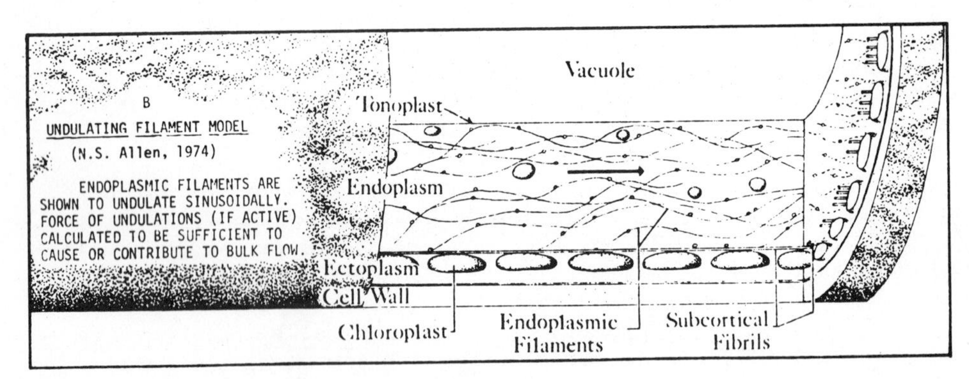

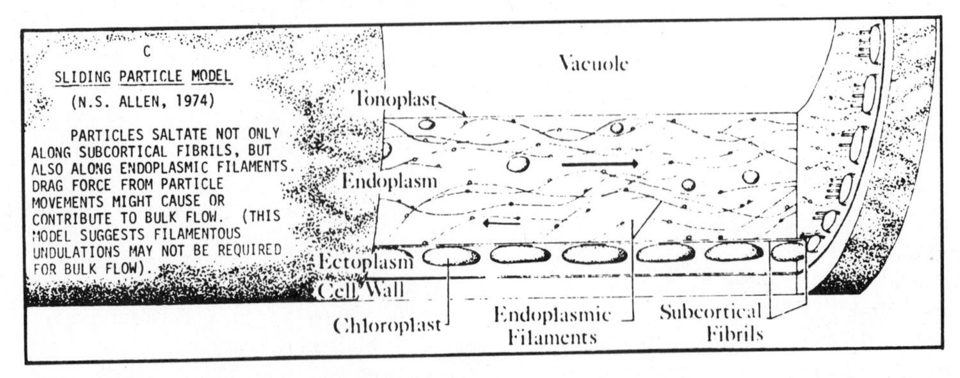

Fig. 1. Theories that may account for the observed cytoplasmic streaming in *Nitella*.

The five basic types of streaming have been described[1,2] for plant cells. 1. Saltation (agitation) is characterized by discontinous rectilinear or curvalinear excursions of particles which exceed the distance the particle would be expected to move by Brownian motion. 2. Circulation (e.g. in Tradescantia stamen hairs) exhibits fairly stable streaming along the cortex and inside transvacuolar strands. 3. Fountain and reverse-fountain streaming occurs in pollen tubes of lilies (Lilium) and Amaryllis respectively. In reverse-fountain streaming the outer sleeve of endoplasm streams towards the tip of the pollen tube and when it reaches the end the stream bends and moves as a rod up the center of the pollen tube toward the pollen grain. This type of streaming is more ordered than circulation, but less ordered than Type 4. 4. Rotational streaming found in Elodea and Nitella refers to the well-organized movement of the entire cytoplasm as a belt around the periphery of the cells. 5. Multistriate streaming is observed in Acetabularia and is very complex and interesting. The stiff cortex has ridges along which cytoplasm with small and large particles moves at rates varying from 0.1 $\mu m\ s^{-1}$ to 10 $\mu m\ s^{-1}$. The streaming is bidirectional and generally parallel to the cell wall with contiguous streams moving in opposite directions at different rates.

Patterns of streaming may change over shorter or longer time periods during the development of a plant cell. Often different, more complex patterns will occur during the development of a cell.

B. Methods of rate measurement

Various methods are used to measure the rates at which particles move in cells. They range from measuring distances traversed over time elapsed with an ocular micrometer and an observer controlled stopwatch, to cinématographic analysis of various complexity to the more recently developed computer-based analysis of video - or ciné images. All of these methods are laborious and time-consuming as well as expensive. Laser Doppler spectroscopy is capable of more rapid, less observer-biased measurements of movements based on the light scattering of objects in cells. In order to use this method to its best advantage, it is important to have a clear understanding of the geometry, morphology and ultrastructure of the cell under examination. It is important to combine good light microscopic observations with the data obtained with this method because laser Doppler spectroscopy fails to record the spatial pattern of velocities and instead records a statistical profile of velocities irrespective of position in the cell. Thus microscopic and laser Doppler spectroscopic data are complementary; each important to the interpretation of the other.

3. SYSTEMS SELECTED FOR THE STUDY OF CYTOPLASMIC STREAMING

I. NITELLA

As already mentioned particle movements in Nitella internodal cells have been measured using laser Doppler spectroscopy[4,5]. Nitella internodal cells were selected since 1) they stream at a constant, continuous and rapid rate, 2) their morphology was fairly well known and 3) their geometry was simple. They are particularly interesting because they possess rates of streaming which are almost an order of magnitude faster than that observed in most plant cells.

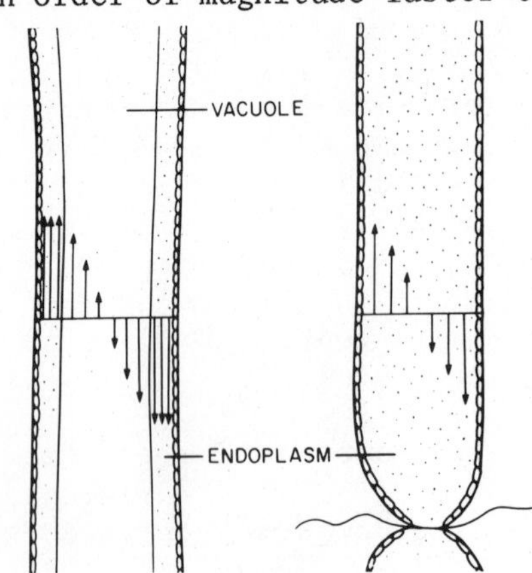

Fig. 2. a. Velocity profiles of internodal cells of Nitella. The length of the arrows indicate the relative velocities of streaming and are the same throughout the endoplasm and then decreasing to zero at the center of the vacuole.
b. Velocity profile of vacuoleless internodal cell of Nitella. In contrast to the intact cell, the shear occurs throughout the endoplasm with the highest velocities observed in contact with the cortex[8].

1. Cell structure and theories about the mechanisms of streaming in Nitella

Large, 1-5 cm long and ca. 1 mm wide internodal cells of Nitella are isolated from the long algal strands obtained from bodies of fresh water. The overall cell architecture is seen in Fig 1 and 2. The plasma membrane lines the inside of the highly birefringent cell wall and surrounds the cytoplasm. The outermost thin layer of cytoplasm is the cortex or ectoplasm in which the stationary chloroplasts are anchored. Subcortical fibrils (SF) known to be composed of f-actin, stretch for long distances along the inner surfaces of chloroplasts (see Fig. 3 and 4). A belt of endoplasm, in which streaming rates from 40-90 μm sec^{-1} are observed, moves spirally in one direction along the cell until it comes to the end where it turns and streams back also spirally occupying the opposite half of the cell's crossection. The two oppositely moving streams are separated by "indifferent zones" on either the top or bottom side of the cell. With ordinary light microscopy of internodal cells the endoplasm appears to be a clear fluid in which a variety of particles (spherosomes, mitochondria and nucleii to name a few) move at the same velocity from the subcortical region to the tonoplast surrounding the large central vacuole.

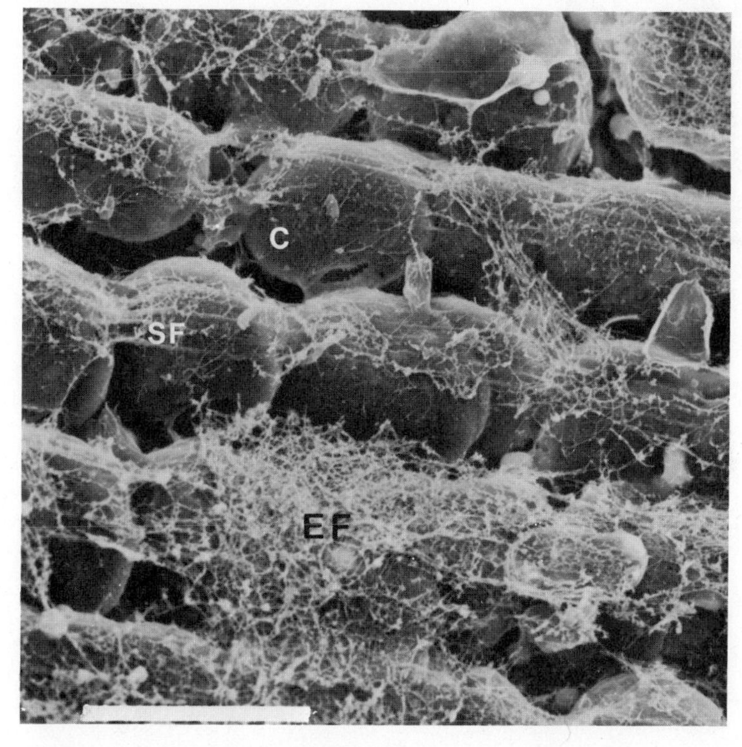

Fig. 3. Scanning electronmicrograph of the cytoplasm of _Nitella furcata_ (_var_. _megacarpa_) viewed from the inside, revealing rows of chloroplasts (C) connected by subcortical fibrils (SF) in turn connected to small endoplasmic filaments (EF). Bar equals 7 μm.

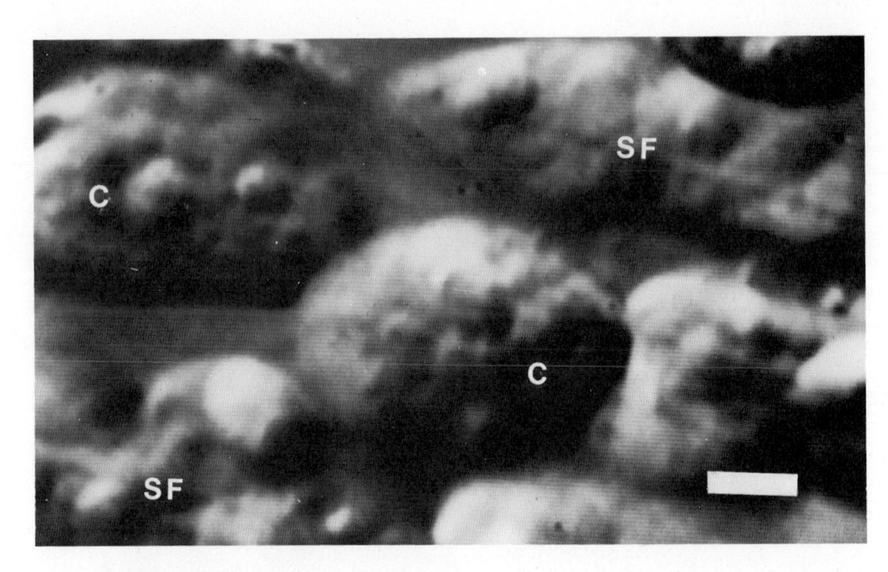

Fig. 4. Videomicrograph of the subcortical region of a 2 cm Nitella cell. The large structures are chloroplasts (C) containing light-scattering starch granules. Subcortical fibrils (SF) stretch between and just above the chloroplasts. Bar equals 2 µm.

The best images one can obtain using differential interference microscopy (DIC) or video-enhanced contrast (AVEC-DIC) microscopy reveal that particles always seem to move on or very close to either SF's or endoplasmic filaments (EF's). Fig. 4 demonstrates the presence of SF's in intact internodal cells. The presence of EF's in optical sections several µm from the subcortical region can be seen on the videoscreen. This cell had no experimental treatments.

Figure 2 shows the velocity profiles reported by Kamiya and Kuroda[8] in a) untreated cells and b) in a cell in which the vacuole has been eliminated. In a, the particles move at the same rate throughout the entire endoplasm. Kamiya and Kuroda proposed the active shearing theory (Fig. 1a) from the observation of an apparant plug-flow velocity profile with the steep velocity gradient adjacent to the cortex. According to this theory, the motive force for streaming is generated at the interface between the endoplasm and ectoplasm where the SF's are found. SF's, bundles of actin ca. 0.1 - 0.2 µm in diameter are postulated to interact with myosins located in the endoplasm to create the observed flow of the entire endoplasm[2].

In 1974 I observed that the endoplasm contained numerous, often undulating endoplasmic filaments[6] along or upon which particles moved. (Fig. 1b). The endoplasmic filaments appeared in light,

Fig. 5. Transmission Electronmicrograph of freeze-fractured, deep-etched *Nitella* endoplasm demonstrate the extensive fibrillar networks present in the endoplasm. The knife did not graze the cell parallel to the cell wall, but at an angle. Thus more 2 and 4 nm filaments and fewer long 7 nm filaments are seen. (Work done with G.C. Ruben). Bar equals 1 μm.

scanning electron and high voltage electron microscopy very much like SF's except that their diameters (ranging from 40-80 nm) usually are less than that of the SF's. The EF's branch from the SF's and are also composed of actin. The early observations of EF's were made in "window cells"[6] and critics claimed that the EF's might be an artifact of the window preparation. The recently developed AVEC-DIC method[9] permitted the recording of the SF's (Fig. 4) as well as EF's (see video tape) in untreated, highly light-scattering cells. This method allows cheaper, faster and clearer recording of the particle movements and one frequently observes somewhat variable rates of streaming in closely adjacent areas of endoplasm. Small areas or strands of endoplasm are observed fairly frequently to stream in the direction opposite to that of the rest of the endoplasm. Neither of these two events could occur if we assume that the active shearing theory is correct.

Based on the available data I have proposed (Fig. 1b and c) a theory to explain the observed saltations of particles as well as their sometimes undulatory motions. When one observes the videotapes* of streaming in Nitella, there can be no doubt 1) the endoplasmic filaments exist, 2) that they sometimes undulate and possibly propel the cytoplasm[6] and 3) that saltations of particles at somewhat varying rates occur along both SF's and EF's (whether undulating or not). In fact, when cells that have stopped streaming due to application of cytochalasin B or electrical or mechanical stimulation are observed in the thin (0.2 μm) optical sections characteristic of DIC images, the spherosomes and other cytoplasmic particles begin their saltatory motions along EF's throughout the endoplasm at the same time that they begin in the SF's. In scanning electron micrographs one can observe the presence of large numbers of EF's. In freeze-fractured, deep-etched Nitella endoplasm[10,11] we find extensive arrays of intermeshing filaments (Fig. 5) as networks which are composed of 2, 4, and 7 nm wide filaments. We conclude from the size and structure of the 7 nm filaments that they are f-actin. Particles of the same size as the ubiquitous, refractile spherosomes (ca. 0.1 μm diameter) are observed in both SEMicrographs and TEMicrographs to be closely attached to the f-actin filaments by ca. 2 nm wide filaments.

In contrast to the views of Kamiya[2], it seems far more likely to conclude that the interactions between f-actin (EF's and SF's) and myosin attached to the membranes of the moving particles occurs throughout the endoplasm as well as in the subcortical region. The force that generates the observed movements is applied locally to

*A videotape showing streaming in plant cells is available at cost from N.S. Allen, Dartmouth College, Hanover, N.H. 03755.

the endoplasm as well as at the subcortical layer. The observed undulations of endoplasmic filaments seem to be real. Although we have no evidence that these undulations drive the endoplasm around the cell, it has been calculated that the observed undulations could generate sufficient force to cause movements at the observed rates[6]. An alternative (but less attractive) explanation for these observed undulations might be that they are a passive reflection of a force on the endoplasmic filament network applied in the subcortical region.

II. ACETABULARIA

I will briefly describe the streaming in *Acetabularia*, because I think it might be an interesting system to study with laser Doppler spectroscopy.

Both *Nitella* and *Acetabularia* are large algal cells. The irregular streaming rate in *Acetabularia* ranges from ca. 1-10 μm sec^{-1}, which is much slower than that in *Nitella*. There are two types of streaming, one occurs in the "fast" strands and then there is an underlying "slower" movement. *Acetabularia* is a single cell with a large nucleus located in its rhizoid. Contrary to the situation in *Nitella*, the nuclei move through the cytoplasm only once in the lifecycle, and the chloroplasts are not stationary. Inside the cell membrane lies a thin layer of cytoplasm which streams rather slowly on cortical ridges (Fig. 6). There is a large central vacuole, but it is not as clearly defined as the vacuole in *Nitella*. Thin strands of cytoplasm traverse it and streaming also occurs within these.

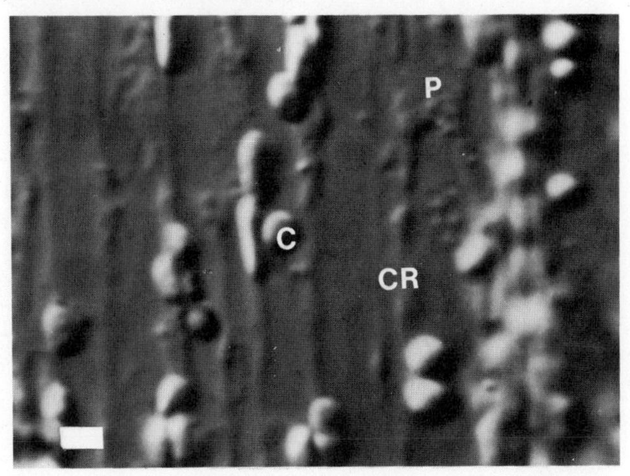

Fig. 6. Videomicrograph of a 1 cm long *Acetabularia* cell, which exhibits multistriate streaming of various-sized particles (P) and chloroplasts (C) on linear cortical ridges (CR). Bar equals 2 μm.

I have recently observed using AVEC-DIC microscopy (Fig. 7) a thin membranous polygonal network (PN) lying against the plasma membrane. These networks shift and attach themselves to membranes associated with the cortical ridges and thus with the streaming. I have seen this type of polygonal network in quite a few other plant cell types and although its significance is not clear now, they should be investigated.

These cells can be manipulated readily and variously. They can form cytoplasmic droplets, they can be ligated. The streaming rates exhibit a circadian rhythm and can be manipulated by both the quantity and quality of light the cells receives. The cells, like Nitella, have a measurable action potential and a considerable amount of knowledge exists about their electrophysiological properties.

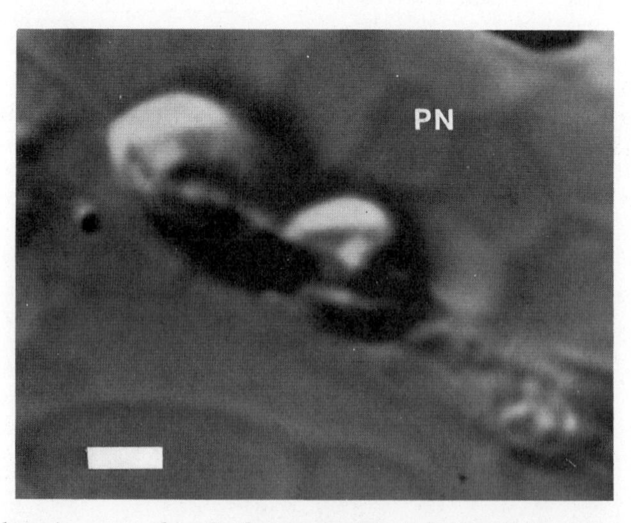

Fig. 7. Videomicrograph of the same Acetabularia cell as in Fig. 6. A membranous, polygonal network (PN) which is ER-like is present closely appressed to the plasmamembrane. This anastomosis and fuses with the cortical ridges and would seem to be involved in the streaming process. Bar equals 2 μm.

III. GERMINATING AMARYLLIS POLLEN GRAINS

Much of what is known about pollen grain and pollen tube streaming appears in Kamiya's review[1]. The pollen tube[12] is a very rapidly expanding system. The pollen tube grows down the stigma and style of the flower and new cell wall is constantly being made at the tip of the pollen tube, and materials must be moved constantly to the tip. The reverse-fountain streaming observed in the pollen tube (Fig. 8) of *Amaryllis* occurs at a rate of 4-5 μm sec^{-1} away from the pollen grain in the outer sleeve of endoplasm and at a rate of 6-7 μm sec^{-1} towards the pollen grain in the center section of the endoplasm. AVEC-DIC microscopy enabled us to easily record the small moving particles and occasionally to record fibrillar elements along which these particles moved.

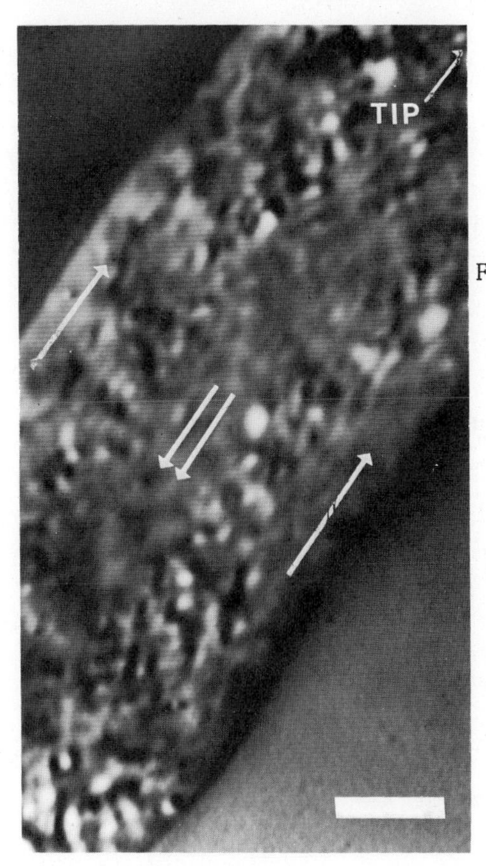

Fig. 8. Videomicrograph of an *Amaryllis* pollen tube. Cytoplasmic streaming near the tip of this pollen tube is in a reverse-fountain pattern. Particles move rapidly in both the center and outer segments in opposite directions. The movement is faster in the center zone. Arrows indicate direction of flow. Bar equals 2 μm.

It is now possible to better observe and record the particle movements in these pollen tubes. It would be interesting to try to detect if there are particular size-classes of particles which move at different rates and to characterize the streaming rates better by the use of correlations between microscopic and laser Doppler spectroscopic methods.

4. CHALLENGES TO LASER DOPPLER SPECTROSCOPY

I have chosen to discuss three systems in which rapid particle motion analysis should be carried out. I will briefly suggest a few experiments and I hope that the brief descriptions of particle motions in plant cells will suggest other and possibly better studies to the reader.

The reported velocity distributions measured in *Nitella* by laser Doppler spectroscopy do not correspond to the plug flow velocity profiles measured by Kamiya and Kuroda[8]. The former is an intermediate between a parabolic distribution and plug flow[4]. Furthermore, as mentioned before, the spatial information obtained so well by light microscopy is not obtained using laser Doppler spectroscopy. It is therefore important to try to reconcile the information obtained by these two methods[13,14,15].

There have been several reports of very rapid (up to 1000 μm sec^{-1}) movements in *Nitella*. These have not been observed with the light microscope, and it certainly would be hard to record such rapid movements, but again these data need to be confirmed and their underlying basis needs to be understood.

Using the AVEC-POL microscopic technique, I have recently found that the SF's and EF's are weakly birefringent. Is it possible to detect movements of birefringent objects such as these?

Neither *Acetabularia* nor *Amaryllis* pollen tube movements have been analyzed using laser light scattering methods and I suggest that these systems should be investigated with videomicroscopic recordings and laser light scattering techniques.

ACKNOWLEDGEMENTS

I am grateful to D. T. Brown for his videomicroscopy of germinating *Amaryllis* pollen tube. I also wish to thank K. Orndorff, D. Brown and J. Kenealy for expert technical assistance and S. Stearns for the excellent typing of this manuscript.

REFERENCES

1. N. Kamiya, Encycl. Plant. Phys. XVII/2: 979 (1962).
2. N. Kamiya, Ann. Rev. Plant Phys. 32: 205 (1981).
3. N. S. Allen and R.D. Allen, Ann. Rev. Biophys. Bioeng. 7: 497 (1978).
4. R. V. Mustacich and B.R. Ware, Biophysical Jour. 16: 373 (1976).
5. K. H. Langley, R. W. Piddington, D. Ross and D. B. Satelle, Biochim. Biophys. Acta 444 (3): 893 (1976).
6. N. S. Allen, J. Cell Biol. 63: 270 (1974).
7. H. -U. Koop and O. Kiermayer, Protoplasma 102: 147 (1980).
8. N. Kamiya and K. Kuroda, Bot Mag. (Tokyo) 69: 544 (1956).
9. R. D. Allen, N. S. Allen and J. L. Travis, Cell Motility 1: 291 (1981).
10. N. S. Allen, Can. Jour. Bot. 58: 786 (1980).
11. N. S. Allen and G. C. Ruben, J. Cell Biol. (Abstract) 83: 328a (1979).
12. J.M. Picton, M.W. Steer and J.C. Earnshaw, this volume.
13. D.B. Sattelle, D.J. Green and K.H. Langley, Physica Scripta 19: 471 (1979).
14. R.V. Mustacich and B.R. Ware, Phys. Rev. Lett. 33: 617 (1974).
15. R.V. Mustacich and B.R. Ware, Biophys. J. 17: 229 (1977).

THE ROTATION MODEL FOR FILAMENT SLIDING AS APPLIED TO THE CYTOPLASMIC STREAMING *

Robert Jarosch and Ilse Foissner

Institut für Botanik der Universität
Lasserstraße 39
A-5020 Salzburg, Austria

CONTENTS

ABSTRACT

Many details of cytoplasmic streaming and related motions performed by cellular filaments and microtubules are imitated by working models with rotating helices (film), indicating the validity of these models. The rotations are obviously connected with the winding and unwinding motions of cytoskeleton-associated helical filaments (filament-associated proteins or MAPs). These winding motions cause sliding as shown by the working models: the same rotational direction of two connected elements means parallel sliding; the opposite direction, antiparallel sliding. The latter system shows three velocities related by the formula

$$v_W = \frac{v_{F_2} - v_{F_1}}{2},$$

*Cordially dedicated to Professor N. Kamiya who devoted his life to the study of protoplasmic streaming.

where v_W is the velocity of the helical waves which are unidirectional on both rotating elements. v_{F2} is the sliding velocity of the quicker element. v_{F1} is the sliding velocity of the slower element in the opposite direction.

This model is applied to the actin filaments of a plasmodial strand, where $v_{F1} \sim v_{F2}$ and $v_W \sim$ zero. It explains important dynamic features of the strand, especially its contraction-relaxation cycle, as the molecular basis of the shuttle streaming. Applied to the actin filaments of a characean cell, it explains the two velocities of the protoplasmic streaming, if v_{F2} means the rapid sliding velocity of the endoplasm (about 50 μm/s at 20° C), and v_W the slower motion of particles closely attached to the subcortical fibrils (about 20 μm/s). The unknown v_{F1} results: $v_{F2} - 2\ v_W =$ 10 μm/s. From this it can be expected that the subcortical filaments slide temporarily with a velocity of about 10 μm/s in the opposite direction of the streaming.

1. INTRODUCTORY REMARKS ON THE HISTORY OF CYTOPLASMIC STREAMING AS A PHYSICAL PHENOMENON

Bonaventura Corti who discovered the cytoplasmic streaming in characean algae in the autumn of 1773 was a professor of physics in Lucca. He was unable to find a physical explanation for the two streams that slide in opposite directions in a capillary tube. In his first publication on that point (1774) Corti[1] thus postulated the existence of a septum between the two streams. He corrected this hypothesis in a second paper in 1776.[2] After the rediscovery of cytoplasmic streaming in 1807 by Treviranus in Bremen (published in 1810/1811),[3] the German botanist Martius (1815)[4] also declared that this movement would be hydrodynamically impossible. Corti's pupil Gozzi (1818)[5] later produced two or more circuits by tying off the cells which definitely abolished the idea of a septum.

What is more important, however, is the question concerning the physical basis of the "motive force". In the course of time nearly all known physical forces have been used as an explanation for cytoplasmic streaming without success (cf. ref 6). The motive force which in characean cells originates in the chloroplast-containing cortical cytoplasmic layer was described surprisingly early; first in 1818 by Amici[7] in Modena (who supposed an electrical force), then in 1838 by Donné[8] and in 1837, 1838 by Dutrochet[9,10] in Paris, and was recognized again by subsequent scientists[11-15,17] (see also ref. 16, p. 167). In his last paper (1846) Dutrochet[18] designated the motive force as a specific vital force. He is absolutely correct inasmuch as this "sliding-" or "shearing-force" which depends on cellular filaments does not exist in the non-vital range of the chemical-physical phenomena.

2. THE DYNAMICAL PHENOMENA OF THE CHARACEAN CYTOPLASMIC FIBRILS

The characean cells contain thick bundles of actin filaments, the subcortical fibrils which cause cytoplasmic streaming by producing the "active shearing force". They can be studied light-microscopically in the living cell by dark-field observation,[19] or by interference contrast,[20,21] but even better in squeezed out cytoplasmic droplets[14,22] where they show a series of remarkable dynamic phenomena: 1) particle displacement (=cytoplasmic streaming), 2) ring formation, 3) undulation, 4) rolling motion and 5) net-dynamics all of which can be imitated in model experiments with rotating helices (film).[60] The details of these processes can be most easily explained by assuming quick rotations of the filaments (about 200/s) which thereby cause the "active shearing force" as a result of the wave displacement. The laws expected to occur here can be described by a special branch of mechanics, the so-called screw mechanics.[23,24]

3. CYTOPLASMIC STREAMING AND THERMODYNAMICS

Particles which are carried along with the cytoplasmic mass-streaming may exhibit Brownian motion. On the other hand, the shift of particles along the filaments is completely smooth. Obviously the dynamical filaments as well as the adhering particles are not subject to molecular movement. Some authors (e. g. A. Meyer[25]) therefore interpreted the cytoplasmic streaming as a "directed molecular movement". This phenomenon can possibly be explained by the self-stabilization of the filaments during quick rotation. If a helix made of wire rotates quickly in a viscous fluid (e. g. glycerol or honey) it stabilizes rectilinearily owing to the strong hydrodynamical forces and moves perfectly smoothly. The general structural destabilization and the domination of surface tension forces (e. g. the pearl-necklace-like transformation of cytoplasmic strands and filopods, the rounding of chloroplasts) following damage and death of cells is probably due to the cessation of filament rotations.

Although the single thermodynamical fluctuations do not influence filament rotation, the sum of all thermal motions will determine the resistance in the environment of the rotating filament (temperature dependence of life motions): the higher the temperature, the lower the viscosity, the lower the resistance, the faster the rotations.

4. THE MODEL OF PERIODICAL WINDING AND UNWINDING OF CYTOSKELETON-ASSOCIATED PROTEIN HELICES

Contrary to the opinion that the "active shearing force" originates in the subcortical fibrils of the characean cells, seemingly different findings have been made in recent years. They

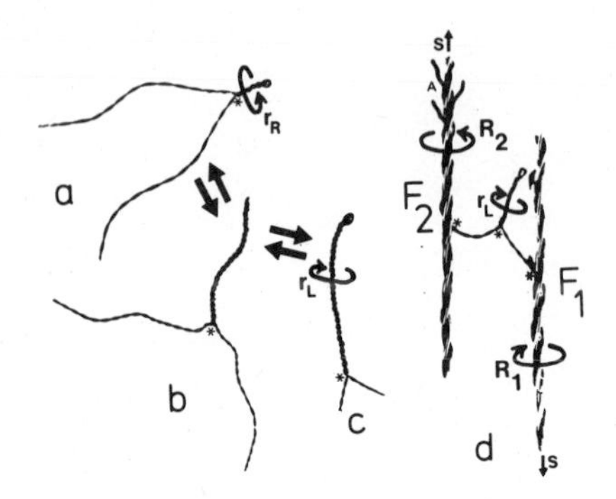

Fig. 1. a,b,c. Structure model of a myosin-like protein filament which performs torsional oscillations with self-intertwining. The filament is shown in three different phases which - as pointed out by the arrow pairs - may readily change into each other. The arrows r_R and r_L indicate clockwise or counter-clockwise rotation of the double filament.
d: Attachment of the filament ends to the actin filaments F_1 and F_2 brings about the rotation of the latter (arrows R_1 and R_2). The sites of winding and unwinding are marked by asterisks. Here we expect enzymatic effects like phosphate transfer. In the case of the rotational direction R_2 A indicates the unwinding of polarily attached associated proteins.

have stressed the importance of Ca^{2+}-sensitive endoplasmic factors such as myosin for cytoplasmic streaming[26,27] (cf. ref. 28). Especially the electronmicroscopical demonstration of fine helical filaments (diameter 30-40 µm) which are attached not only to the membranes of the endoplasmic organelles but also to the subcortical fibrils[29-31] has stimulated the discussion and has accounted for the elaboration of a new sliding model.[32] Besides myosin and tropomyosin, a number of other actin-binding proteins have been demonstrated in other cells all of them with extreme high molecular weights (e. g. "actin-binding protein", filamin, spectrin[33]). These proteins aggregate to oligomers by head-to-tail-association thereby forming longer filaments (cf. for example 34). Like tropomyosin and myosin they probably consist of perfectly elastic coiled-coil α-helices which may rotate and self-intertwine ("telephone-cable-phenomenon") due to torsions which result from charge modifications at the polar side-chains.[6] Because of the self-intertwining we have to expect a periodical displacement of adhering cations leading again to a change in the charge distribution and to counter-rotation. Thus the filament oscillates between two different states as shown by the models in Fig. 1a,b,c. This behaviour is only possible when the ends of the filaments are not free but fixed or when they combine to form a closed ring which allows the torsional energy to be stored. Changes in ion concentration (Ca^{2+} and $MgATP^{2-}$ as antagonists) will strongly influence the extent of winding; for instance even the addition of mM ATP^{2-} can lower the number of bound cations! Fig. 1d depicts a filament the ends of which are bound to actin filaments of parallel polarity (F_1 and F_2). It is to be noticed that the filament ends point downwards at F_1 and upwards at F_2 according to the polarity. The active torsion-caused rotation (arrow r_L) results in opposite rotations of F_1 and F_2 (arrows R_1 and R_2). Because of winding, F_1 slides downwards, F_2 upwards (arrows S, "antiparallel sliding" in spite of the same polarity). According to this model there exist actin filaments more or less wrapped with actin-associated proteins. Fig. 2 shows 4 phases of a periodic oscillation in a working model. Owing to the opposite rotations (arrows r_L and r_R) and the ensuing winding and unwinding of the helix the rods periodically move apart ("relaxing phase") and together ("contracting phase").

The sites where molecules which are bound to the filaments will meet or separate are indicated by an asterisk (Fig. 1). At these sites we expect enzymatic rearrangements ("proximity-effect", "rupture-effect"), expecially phosphate transfer and ATP hydrolysis.

5. APPLICATION TO THE CYTOPLASMIC STREAMING IN PLASMODIA

The sliding model in Fig. 2 is especially suited as a basis of periodical movements in plasmodia although it is a simplification to deduce from the behaviour of a single pair of filaments the behaviour of the plasmodial strands. In the latter the sol endoplasm flows back and forth due to periodical contraction of the ectoplasmic gel (compare ref. 28). Wohlfarth-Bottermann[35-37] was the first to demonstrate that filaments in the ectoplasmic gel are the structural basis of these movements. Kamiya concluded as early as 1968:[38] "The unit filaments, represented probably by F-actin, may possibly slide with one another being intermediated by myosin molecules in such a way that the bundles themselves shorten and exert pressure on the endoplasm". The presence of an actomyosin protein-complex in the plasmodium was shown by Loewy[39] as early as 1952 and was later studied mainly by Hatano and co-workers (for references - also about other actin-associated proteins - see 28). Kamiya[38] found moreover "that a glycerol-treated plasmodial strand can oscillate sponaneously" and "that the rhythm of the movement is based on the intrinsic character of the protein molecules themselves. It

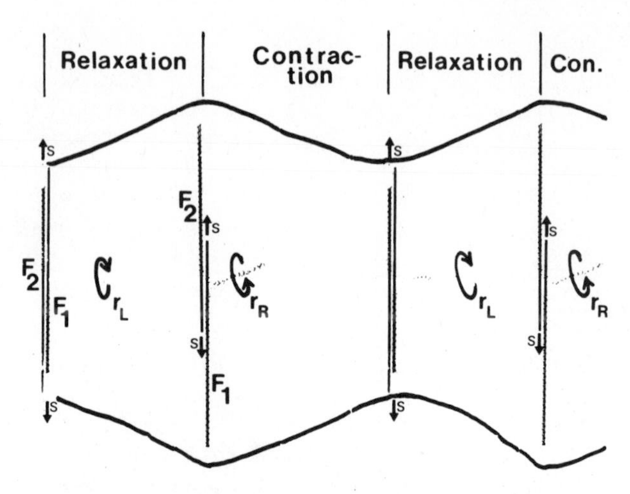

Fig. 2. The model of Fig. 1d as a working model to explain the periodicity in plasmodial strands. One helix-branch winds along the rod F_1 downwards, the other upwards along F_2. The rotation r_L causes "relaxation" by "sliding" apart of the rods (arrows S), the opposite rotation r_R causes "contraction". There are no helical waves: v_W = zero, $v_{F1} = v_{F2}$. For further explanation see text.

shows that a simple system composed of contractile protein, ATP and some simple ions can make a feed-back loop which produces the oscillatory motion." In the electron micrographs made by Nagai and Kato[40] side-arms or cross-bridges between the actin filaments of plasmodia are clearly visible. The periodical transformation of actin filaments to a network corresponding to the contraction cycle[41,42] is possibly due to the periodical unwinding of filaments which reduces the stability and stiffness of the actin filaments. This may be connected with F-actin depolymerization.[43] Hatano's flexible Mg-polymer of F-actin[44] may be identical with the unwound actin filaments - devoid of tropomyosin. The actin filaments are aligned largely parallel and show strongest birefringence in the phase of maximal dislocation and in the incipient contracting phase during which the endoplasm streams away backward from the front region.[28,45-47] The period lasts approximately 1.3 min under natural conditions at room temperature[48-50] indicating very slow winding and unwinding (about one rotation/s). The tension curve shows that the contracting phase lasts longer than the relaxation phase[51] probably because the tension force is produced against a resistance, especially under isometric conditions. Thus the higher the resistance caused by the system for measuring the tension the longer the period (up to 3 min). The overcoming of this resistance does not take place continuously but step-wise, possibly because of the elastic properties of the force producing filaments. This behaviour is indicated as a shoulder in the contracting phase of the tension curve.[51,52] Plasmodia can be activated by stretching, just as muscle.[51,53,54] Stretching increases tension as shown by the higher amplitude. We suppose that stretching (= tensile stress in the direction of the arrows S, Fig. 1) causes passive rotations of the filaments F_1 and F_2 in the direction of the arrows R_1 and R_2. This means further unwinding of the double helix (arrow r_L) and an increase in its torsional tension.

The rhythmical rotation or torsion of a free hanging plasmodial strand was described as early as 1954 by Kamiya and Seifriz.[55] The free end of this strand rotates in the same direction as the lower end of the models in Fig. 1 and 2, namely counter-clockwise (as seen from above) in the contracting phase.[51] This implies a rotation of right-handed helices as represented by the actin filaments. The phenomenon is explained by a twisted arrangement of filaments in the wall of the tubular strand.[48] Since the twisting motion of one direction predominates and therefore may cause an infinite rotation in this direction, we may suppose that the filament sliding in one direction is favoured (as in characean cells, see below). This can also be concluded from the forward movement of the whole plasmodium. An additional presupposition for the directed movement of the plasmodium is the synchronization of all filament displacements to a uniform rhythm as described for example by Grebecki and Cieslawska.[56]

The period of oscillation in ATP concentration is the same as that of the contraction-relaxation cycle of the plasmodial strand.[57] Accordingly ATP could be synthesized and released during winding and split during unwinding (see Fig. 1d, asterisks). The counter-oscillating Ca^{2+} concentration could be achieved by the displacement of attached Ca^{2+} ions during winding along the F-actin and by the re-accumulation during unwinding.

6. APPLICATION TO THE CHARACEAN CYTOPLASMIC STREAMING

The unidirectional streaming of the endoplasm in characean cells originates in the shearing force produced by the subcortical fibrils. The problem of how a cyclic filament motion may lead to this continuous movement was first referred to and mathematically treated by Donaldson.[58] In the case of F-actin the favouring of one direction of movement could be achieved by polarily binding proteins like heavy meromyosin (HMM). The "arrow heads" in subcortical filament bundles obtained after treatment with HMM point in the opposite direction of rotational streaming.[59] Such proteins will therefore unwind in the case of the rotational direction R_2 (at A in Fig. 1d) but wind up with the opposite rotational direction R_1.[60] The increased frictional resistance after unwinding suppresses the displacement of the waves. Hence only a screwing motion opposite to

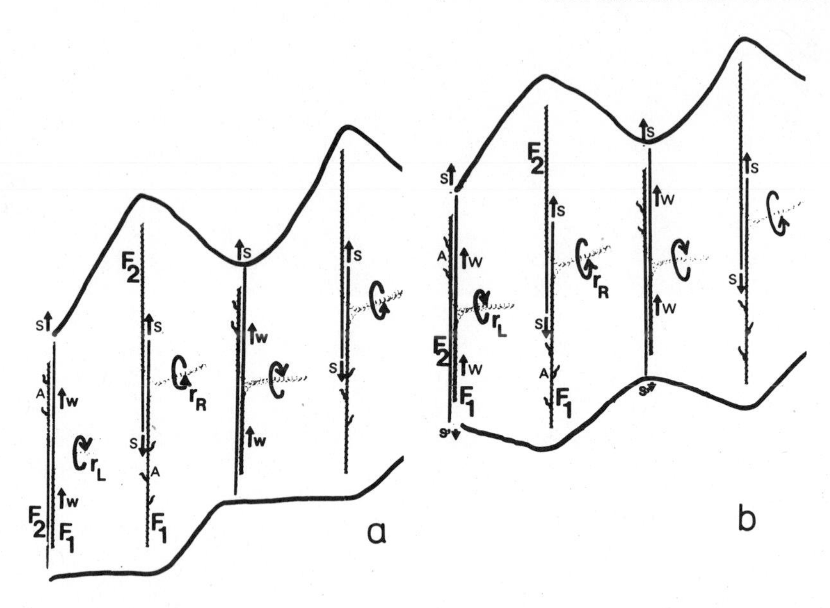

Fig. 3. Working models of periodical filament sliding to explain the start of cytoplasmic streaming in a characean cell. In a the subcortical filament F_1 is completely prevented from sliding downwards ($v_{F1} = 0$, $v_{F2} = 2$, $v_W = 1$), in b only partly ($v_{F1} = 0.5$, $v_{F2} = 2.5$, $v_W = 1$). F_2 represents an endoplasmic filament. Helical waves (arrows W) occur only in every second phase. For further explanation see text.

the waves is possible. The models in Fig. 3 and 4 show the behaviour of an actin filament which - according to this conception - "sets up its bristles". In the models of Fig. 2 both elements (F_1 and F_2) are equivalent and the shifting proceeds upwards and downwards without hindrance (arrows S). However, this is not the case with the subcortical filaments of the characean cell (represented by the F_1-element) because of their binding to the stationary chloroplast files: The rotation (r_L) of a helical endoplasmic filament which is attached to F_1 will cause the rotation R_1 of F_1 (Fig. 1). If F_1 is prevented from sliding downwards, the waves and the side arm move upwards (arrows W in Fig. 3a). If the direction of rotation changes $r_L \rightarrow r_R$, $R_1 \rightarrow R_2$, the postulated polarily attached proteins unwind, thereby preventing the dislocation of the waves and of the side arm backwards. Now, F_1 performs a screwing motion upwards (arrow S). This mechanism favours the upward movement though a short downward dislocation during the phase with rotation r_L may remain (short arrows S' in Fig. 3b). Periodical winding and unwinding along an element implies a step-wise movement upwards (compare the curves

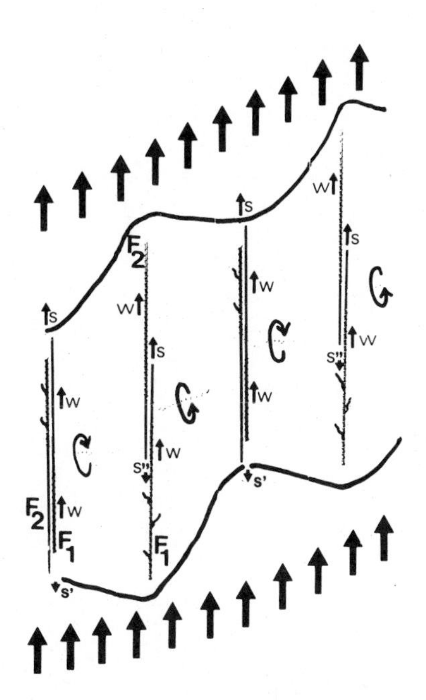

Fig. 4. Working model to explain steady cytoplasmic streaming in a characean cell. The mass-streaming of the endoplasm, as indicated by the large arrows, prevents the sliding of F_2 in an opposite direction of streaming (arrows S"). Helical waves (arrows W) therefore occur in both phases. For further details see text.

in Fig. 3 above). Now let us look at the element F_2 with a helix adhering as depicted. If F_1 is completely prevented from downward sliding (Fig. 3a), F_2 slides upward with double velocity as compared to the wave displacement along F_1 and F_2 (arrows W). When F_1 is only incompletely hindered from sliding (Fig. 3b) we can distinguish three velocities: v_{F1} = sliding velocity of F_1 downwards, v_{F2} = sliding velocity of F_2 upwards, v_W = velocity of wave flow upwards, connected by the formula

$$v_W = \frac{v_{F2} - v_{F1}}{2}$$

After irritation of a vital internodal cell and cessation of streaming one can find random particle motions which exceed the extent of Brownian movement. The particles obviously move only by means of the endoplasmic filaments to which they are attached. At the beginning of endoplasmic streaming single particles which are closely attached to the subcortical filaments perform saltatory motions in the streaming direction. The distance travelled does not exceed several µm. The saltatory motions probably correspond to the step-wise forward movement of side arms and waves as described in the model of Fig. 3. The continuous movement of the endoplasm then starts and increases gradually until it reaches its maximal velocity. Now, particles which closely adhere to the subcortical fibrils do not attain velocities higher than 15-20 µm/s (at 20° C) while others move beyond them with 50-60 µm/s. This faster movement corresponds to the maximal velocity of the endoplasm under non-disturbed conditions. It exceeds the velocity of the particles adhering to the subcortical fibrils by twice or three times the amount.[19,21] The slower movement (10-15 µm/s) is also observed at the "single subcortical fibrils" of Kamitsubo[61] in the absence of mass streaming. In the model of Fig. 3b the dislocation of waves (arrows W) along F_1 probably corresponds to the slower movement, and the sliding motion of F_2 (arrows S upwards) to the fast movement of the endoplasm. If we take the known values from undisturbed streaming cells, i. e. $v_W = 20$ being the slow motion and $v_{F_2} = 50$ being the fast motion and put them in the equation above, we obtain $v_{F_1} = 10$ for the sliding motion of F_1 (arrows S' downwards). Accordingly, the subcortical filaments are expected to move back and forth periodically in the direction of streaming. This behaviour is expressed by the curves in Fig. 3b and 4 below. If the retrograde motion (10 µm/s) did not exist, the fast movement of the endoplasm would be twice as fast as the slow one. The generation of the motive force for cytoplasmic streaming proceeds obviously via many small steps which can be recognized only at the beginning of streaming in the saltatory motion of particles as described by the models in Fig. 3. All these steps together produce the continuous movement of the endoplasm. This mass streaming brings about changes in the resistance against sliding: F_2 in Fig. 4 is pushed forwards (arrows S) without resistance but scarcely backwards (short arrows S"). Thereby F_1 moves faster in the forward direction (arrows S). This causes the side arm which showed step-wise motion at the beginning of streaming (cf. Fig. 3) now to move continuously forwards. Under undisturbed conditions F_1 and F_2 may behave in the same way in their oscillating displacement.

Finally it must be noted that the behaviour of the two filaments described here cannot be applied to a filament bundle without restriction since further possibilities of filament winding and unwinding exist. Additionally, we may expect new hydrodynamical effects which have not been considered in our model.

ACKNOWLEDGEMENTS: The authors thank Prof. Dr. K. E. Wohlfarth-Bottermann for valuable discussions. This work was supported by the Österreichische Nationalbank, Jubiläumsfonds Projekt Nr. 1927.

REFERENCES

1. B. Corti, Osservazioni microscopiche sulla tremella e sulla circolatione del fluido in una pianta acquajuola, Lucca (1774).
2. B. Corti, Journal de Physique (Rosier) 8:232 (1776).
3. L. C. Treviranus, Beiträge zur Pflanzenphysiologie, Göttingen (1810/11).
4. K. F. P. v. Martius, Über den Bau und die Natur der Charen, Nova Acta Leop. 9 (1815)
5. Gozzi, Giorn. di Fis., Brugnatelli, (1818).
6. R. Jarosch, The torsional movement of tropomyosin and the molecular mechanism of the thin filament motion, in: "Cell Motility: Molecules and organization", S. Hatano, H. Ishikawa and H. Sato, eds., University of Tokyo Press, Tokyo, 291-319 (1979).
7. G. B. Amici, Osservazioni sulla circolazione del succhio nella Chara, Mem. mat. fis. Soc. Italiana, Modena 18:183-202 (1820).
8. A. Donné, Note sur la circulation du Chara, Ann. Sci. Nat., 2. sér. 10:346-348 (1838).
9. H. J. Dutrochet, Observations sur le Chara flexilis, C. R. Acad. Sci. Paris 5:775-784.
10. H. J. Dutrochet, Observations sur la circulation des fluides chez Chara fragilis Desvaux, Ann. Sci. Nat., 2. sér. 9:5-38, 65-79 (1838).
11. G. Hörmann, Studien über die Protoplasmaströmung bei den Characeen, Jena (1898).
12. K. Linsbauer, Untersuchungen über Plasma und Plasmaströmung an Chara-Zellen. I. Beobachtungen an mechanisch und operativ beeinflußten Zellen, Protoplasma 5:563-621 (1929).
13. N. Kamiya and K. Kuroda, Velocity distribution of the protoplasmic streaming in Nitella cells, Bot. Mag. Tokyo 69:544-554 (1956)
14. R. Jarosch, Plasmaströmung und Chloroplastenrotation bei Characeen, Phyton (Argentina) 6:87-107 (1956).
15. R. Jarosch, Die Protoplasmafibrillen der Characeen, Protoplasma 50:93-108 (1958).
16. N. Kamiya, Protoplasmic streaming, in: "Protoplasmatologia VIII. 3a", L. V. Heilbrunn and F. Weber, eds., Springer, Wien (1959).
17. T. Hayashi, Experimental studies on protoplasmic streaming in Characeae, Scientific Papers of the College of Gen. Education, Univ. of Tokyo 10:245-282 (1960).
18. H. J. Dutrochet, Le magnétisme peut-il exercer de l'influence sur la circulation du Chara? C. R. Acad. Sci. Paris 22:619-622 (1846).
19. R. Jarosch, Die Dynamik im Characeen-Protoplasma, Phyton (Argentina) 15:43-66 (1960).
20. E. Kamitsubo, Motile protoplasmic fibrils in cells of Characeae, Proc. Jap. Acad. 42:507-511, 640-643 (1966).
21. E. Kamitsubo, Motile protoplasmic fibrils in cells of Characeae, Protoplasma 74:53-70 (1972).

22. K. Kuroda, Behaviour of naked cytoplasmic drops isolated from plant cells, in: "Primitive motile systems in cell biology", R. D. Allen and N. Kamiya, eds., Academic Press, New York and London (1964).
23. R. Jarosch, Grundlagen einer Schraubenmechanik des Protoplasmas, Protoplasma 57:448-500 (1963).
24. R. Jarosch, Screw-mechanical basis of protoplasmic movement, in: "Primitive motile systems in cell biology", R. D. Allen and N. Kamiya, eds., Academic Press, New York and London (1964).
25. A. Meyer, Die Plasmabewegung, verursacht durch eine geordnete Wärmebewegung von Molekülen, Ber. dtsch. bot. Ges. 38:36-43 (1920).
26. T. Hayama and M. Tazawa, Ca^{2+} reversibly inhibits active rotation of chloroplasts is isolated cytoplasmic droplets of Chara, Protoplasma 102:1-9 (1980).
27. Y. Tominaga and M. Tazawa, Reversible inhibition of cytoplasmic streaming by intracellular Ca^{2+} in tonoplast-free cells of Chara australis, Protoplasma 109:103-111 (1981).
28. N. Kamiya, Physical and chemical basis of cytoplasmic streaming, Ann. Rev. Plant Physiol. 32:205-236 (1981).
29. R. Nagai and T. Hayama, Ultrastructure of the endoplasmic factor responsible for cytoplasmic streaming in Chara internodal cells, J. Cell Sci. 36:121-136 (1979).
30. R. E. Williamson, Filaments associated with the endoplasmic reticulum in the streaming cytoplasm of Chara corallina, Eur. J. Cell Biol. 20:177-183 (1979).
31. N. S. Allen, Cytoplasmic streaming and transport in the characean alga Nitella, Can. J. Bot. 58:786-796 (1980).
32. R. Jarosch and I. Foissner, A rotation model for microtubule and filament sliding, Eur. J. Cell Biol. 26:295-302 (1982).
33. J. H. Hartwig and Th. P. Stossel, Structure of macrophage actin-binding protein molecules in solution and interacting with actin filaments, J. Mol. Biol. 145:563-581 (1981).
34. J. S. Morrow and V. T. Marchesi, Self-assembly of spectrin oligomers in vitro: A basis for a dynamic cytoskeleton, J. Cell Biol. 88:463-468 (1981).
35. K. E. Wohlfarth-Bottermann, Weitreichende, fibrilläre Protoplasmadifferenzierungen und ihre Bedeutung für die Protoplasmaströmung. I., Protoplasma 54:514-539 (1962).
36. K. E. Wohlfarth-Bottermann, Weitreichende, fibrilläre Protoplasmadifferenzierungen und ihre Bedeutung für die Protoplasmaströmung. II., Protoplasma 57:747-761 (1963).
37. K. E. Wohlfarth-Bottermann, Differentiations of ground cytoplasm and their significance for the generation of the motive force of amoeboid movement, in: " Primitive motile systems in cell biology", R. D. Allen and N. Kamiya, eds., Academic Press, New York and London (1964).

38. N. Kamiya, The mechanism of cytoplasmic movement in a myxomycete plasmodium, in: "Aspects of cell motility", Symp. Soc. Exp. Biol. Cambridge 22:199-214 (1968).
39. A. Loewy, An actomyosin-like substance from the plasmodium of a myxomycete, J. Cell. Comp. Physiol. 40:127-156 (1952).
40. R. Nagai and T. Kato, Cytoplasmic filaments and their assembly into bundles in Physarum plasmodium, Protoplasma 86:141-158 (1975).
41. R. Nagai, Y. Yoshimoto and N. Kamiya, Changes in fibrillar structures in the plasmodial strand in relation to the phase of contraction-relaxation cycle, Proc. Jap. Acad. 51:38-43 (1975).
42. R. Nagai, Y. Yoshimoto and N. Kamiya, Cyclic production of tension force in the plasmodial strand of Physarum polycephalum and its relation to microfilament morphology, J. Cell Sci. 33:205-225 (1978).
43. G. Isenbert and K. E. Wohlfarth-Bottermann, Transformation of cytoplasmic actin, Cell Tiss. Res. 173:495-528 (1976).
44. H. Tanaka and S. Hatano, Conformational changes induced in plasmodium actin polymer by Ca^{2+} in the presence of muscle native tropomyosin, J. Mechanochemistry, Cell Motility 3: 195-200 (1976).
45. H. Nakajima and R. D. Allen, The changing pattern of birefringence in plasmodia of the slime mold Physarum polycephalum, J. Cell Biol. 25:361-374 (1965).
46. M. Ishigami, R. Nagai and K. Kuroda, A polarized light and electron microscopic study of the birefringent fibrils in Physarum plasmodia, Protoplasma 109:91-102 (1981).
47. H. Sato, S. Hatano and Y. Sato, Contractility and protoplasmic streaming preserved in artificially induced plasmodial fragments, the "caffeine drops", Protoplasma 109:187-208 (1981).
48. N. Hülsmann and K. E. Wohlfarth-Bottermann, Räumliche und zeitliche Analyse von kontraktionsabhängigen Oberflächenbewegungen bei Physarum polycephalum, Cytobiologie 17:23-41 (1978).
49. N. Hülsmann and K. E. Wohlfarth-Bottermann, Spatio-temporal relations between protoplasmic streaming and contraction activities in plasmodial veins of Physarum polycephalum, Cytobiologie 17:317-334 (1978).
50. K. E. Wohlfarth-Bottermann, Contraction phenomena in Physarum: New results, Acta Protozool. 18:59-73 (1979).
51. N. Kamiya and Y. Yoshimoto, Dynamic characteristics of the cytoplasm. A study on the plasmodial strand of a myxomycete, in: "Aspects of cellular and molecular physiology", K. Hamaguchi, ed., Tokyo University Press (1972).
52. N. Kamiya, R. D. Allen and R. Zeh, Contractile properties of the slime mold strand, Acta Protozool. 11:113-124 (1972).
53. N. Kamiya, Contractile properties of the plasmodial strand, Proc. Jpn. Acad. 46:1026-1031 (1970).

54. K. E. Wohlfarth-Bottermann and M. Fleischer, Cycling aggregation patterns of cytoplasmic F-actin coordinated with oscillating tension force generation, Cell Tiss. Res. 165:327-344 (1976).
55. N. Kamiya and W. Seifriz, Torsion in a protoplasmic thread, Exp. Cell Res. 6:1-16 (1954).
56. A. Grebecki and M. Ciéslawska, Plasmodium of Physarum polycephalum as a synchronous contractile system, Cytobiologie 17: 335-342 (1978).
57. Y. Yoshimoto, T. Sakai and N. Kamiya, ATP oscillation in Physarum plasmodium, Protoplasma 109:159-168 (1981).
58. I. G. Donaldson, Cyclic longitudinal fibrillar motion as a basis for steady rotational protoplasmic streaming, J. theor. Biol. 37:75-91 (1972).
59. Y. M. Kersey, P. K. Hepler, B. A. Palevitz and N. K. Wessells, Polarity of actin filaments in characean algae, Proc. Natl. Acad. Sci. USA 73:165-167 (1976).
60. I. Foissner and R. Jarosch, The motion mechanics of Nitella filaments (cytoplasmic streaming): Their imitation in detail by screw-mechanical models, Cell Motility 1:371-385 (1981).
61. E. Kamitsubo, Effect of supraoptimal temperatures on the function of the subcortical fibrils and an endoplasmic factor in Nitella internodes, Protoplasma 109:3-12 (1981).

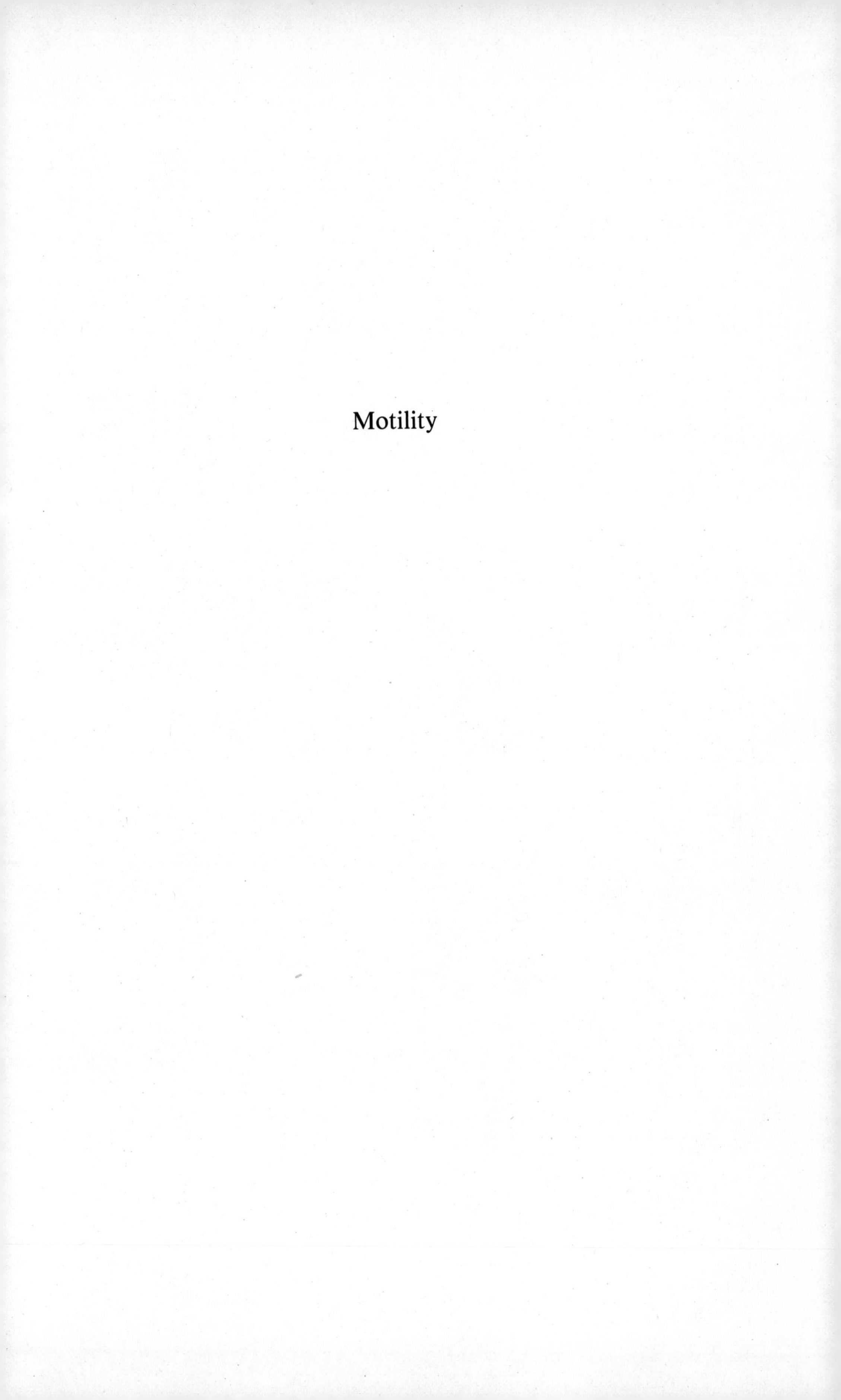

Motility

MOTILITY OF LIVING CELLS AND MICRO-ORGANISMS*

Jean-Pierre Boon
Université Libre de Bruxelles
Faculté des Sciences, C.P. 231
1050 Bruxelles, Belgium

CONTENTS

---

*Work supported by the Fonds National de la Recherche Scientifique (F.N.R.S., Belgium).

1. MOTILITY AND SMALL-SCALE HYDRODYNAMICS

Motility is what characterizes the dynamics of living cells and organisms with autonomous motion. Such are flagellated bacteria and spermatozoa which thus can be assigned a velocity. Objects without locomotion apparatus or cells whose apparatus machinery has been inhibited merely undergo Brownian motion. The mobility of Brownian systems is characterized by their diffusion coefficient.

The dynamical behavior of a solid body moving in a fluid medium is governed by inertia and by viscosity (or friction). The ratio of inertial forces to viscous forces defines the Reynolds number, $R = \ell v/\nu$, where ℓ is the size of the object, v its velocity and ν the kinematic viscosity of the medium. The value of R characterizes the hydrodynamic regime. Motile cells and living organisms move in fluids with virtually no change in kinematic viscosity (10^{-1}-10^{-2} stokes) whereas their sizes typically range from 10^{-4} to 10^{2} cm and their corresponding velocities cover 6 to 7 decades from 10^{-4} cm sec^{-1}. Thus Reynolds numbers are of the order of 10^{6} for large size animals which thus live in an essentially turbulent regime where inertial forces dominate. Such forces are totally negligible for micro-organisms like bacteria and flagellated cells like spermatozoa. So small-scale hydrodynamics is characterized by very low Reynolds numbers ($\sim 10^{-6}$) and the efficiency of motion of living cells in this regime is far from obvious. For instance, there is no simple answer to the question "why do flagellated bacteria swim" ? This problem was discussed in three fascinating papers by Carlson (1962), by Purcell (1976), and by Berg and Purcell (1977) (for a review see also Roberts, 1981). However before we turn to the question "Why ?", we shall consider the complementary question "How ?", i.e. we start with a brief description of the flagellar

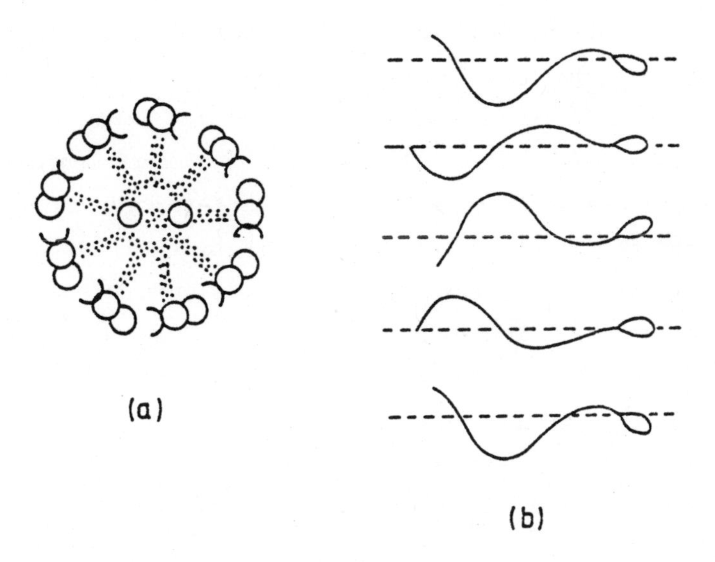

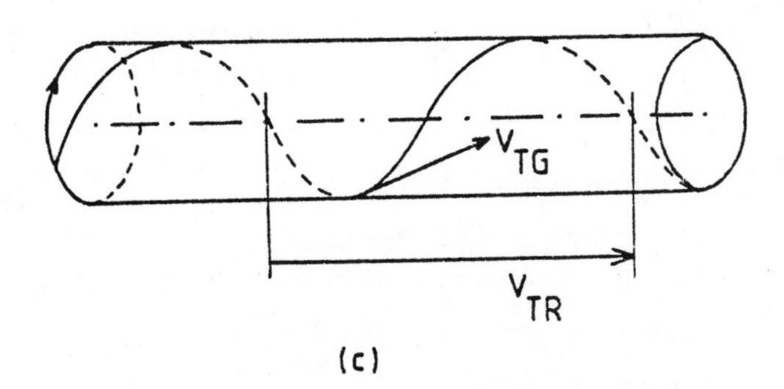

Fig. 1 (a) 9 + 2 axonemal structure of micro-tubules in flagellum (Brokaw and Gibbons, 1975) ; (b) Wavy motion of bull spermatozoon (Gray, 1958) ; (c) Helical trajectory, V_{TR} = translational velocity, V_{TG} = tangential velocity.

apparatus of bacteria and spermatozoa and of the type of motion that results from two structurally different locomotion devices.

There exists a wide variety of shapes of spermatozoa in the living world. The shape may even vary within the same species ; however the head size is nearly always of the same order of magnitude, i.e. a few microns. Typically the morphological components of the spermatozoon are the head ($\sim$ 5-10 μm), the neck ($\sim$ 10 μm), the middle piece ($\sim$ 10 μm), and the flagellum ($\sim$50 μm) with a thin end piece. There are species in nature whose spermatozoa have a different morphology ; those considered here have a structure that conforms to the above general description. The flagellum comprises a set of axially aligned microtubules - the so-called 9 + 2 axoneme surrounded by an extension of the cytoplasmic membrane - whereby bending and wave motion is generated (see Fig. 1.a,b). The movement of the spermatozoon results in a trajectory that can be modeled on the basis of helical motion; then the displacement can be characterized by a translational velocity and a tangential (or instantaneous) velocity (see Fig. 1.c).

From the viewpoint of dynamical behavior, the most studied bacteria are Escherichia coli and Salmonella typhimurium. These micro-organisms have a cellular body with a size of 2.5 μm and about half-a-dozen helical flagella with a diameter $\sim$ 150 Å and a length $\sim$ 10 μm. Contrary to eukaryotic flagella, bacterial flagella have a single naked, filament. The flagellar base comprises the hook and the basal body that consists of inner and outer rings mounted on a shaftlike structure inserted in the cell membrane (see Fig. 2.a). The motion of the bacterium is generated by rotation of the filaments by means of the mechanical arrangement of the rod and rings. Thus the basal body of the flagellum plays the role of a motor apparatus such that all fila-

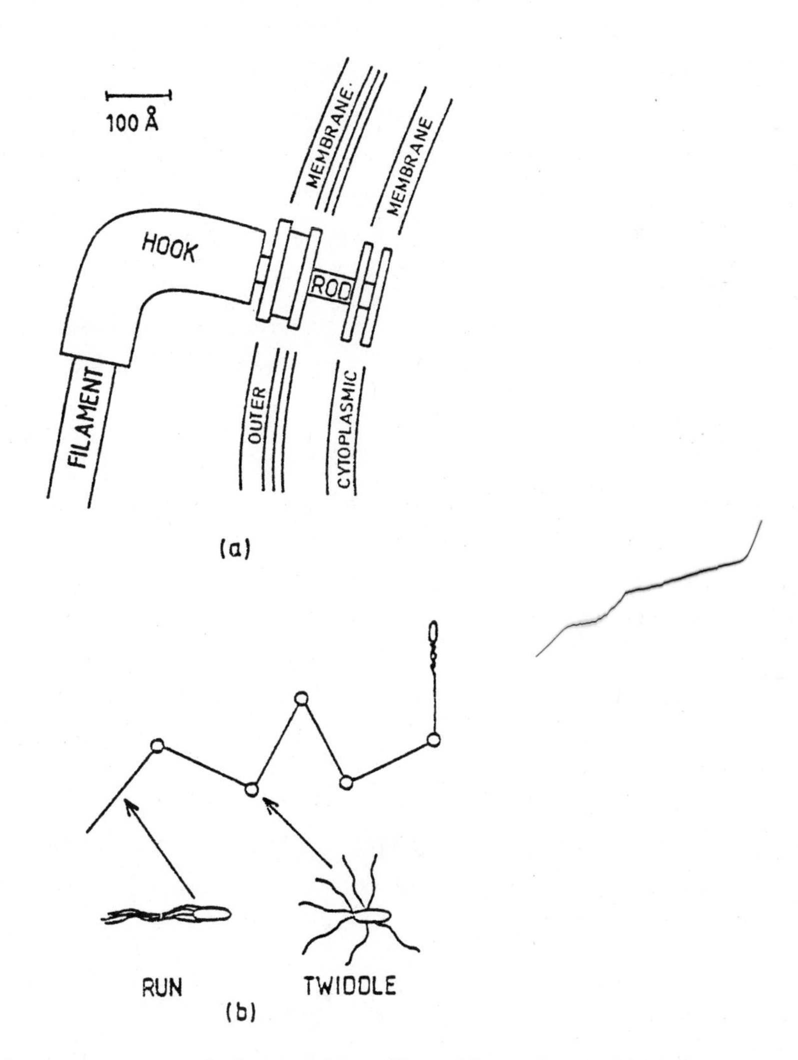

Fig. 2 (a) Basal structure of bacterium flagellum (DePamphilis and Adler, 1971) ; (b) Two phase motion of bacterium.

ments can rotate in synchronism. There are two phases in bacterial motion, the run during which the bacterium undergoes translational displacement over a distance of the order of ten times its size, and the twiddle where it undergoes erratic rotational movement with no net displacement (see Fig. 2.b). During the run, the rotary motors operate such that the filaments form a coordinated bundle acting as a propellor ; upon reversal of rotation sense, the filaments disperse and the cell enters the twiddle phase. In normal strains, the two phases alternate in regular sequence : the bacterium runs along a more or less linear trajectory, then stops and twiddles ; the twiddling phase is followed by a new run in a direction different from that of the previous one. In a homogeneous medium angles between successive runs are randomly distributed, run lengths are constant on the average, and so are the durations of each phase.

In conclusion, there is some degree of similarity between the motion of the spermatozoon and of the bacterium : when they move in rather smooth straight lines, their net displacement is along the axis of the helical trajectory of the cell body. As far as the basic mechanism of cell dynamics is concerned, they have as much in common as a fish with a submarine.

Whereas its ultimate goal gives the spermatozoon obviously good reasons to swim, the motivation for bacterial motility is all the less evident in that there exist non-flagellated bacteria which are equally capable of living. It may be reasonably assumed that reasons why flagellated bacteria move should be directly related to nutrient uptake (Carlson, 1962). Substrates that are used as sources of energy are sugars and amino-acids, in addition to which oxygen is essential to aerobic metabolism. Since in aqueous media these substances diffuse rather slowly (with a difusion coefficient D $\sim 10^{-5}$ cm^2 sec^{-1}) the question arises as to what is the efficiency of motility versus diffusion; in particular two effects of motility are considered in this respect : (i) stirring, and (ii) swimming (Purcell, 1976).

In hydrodynamics, the Sherwood number is defined as $S = \ell v/D$ (compare to the Reynolds number) ; S may be interpreted as the ratio of the characteristic diffusion time : $t_D \simeq \ell^2/D$, to the characteristic time needed to move a molecule over a distance ℓ by stirring the medium with velocity v : $t_S \simeq \ell/v$. When the Sherwood number is large, t_S is small compared to t_D, and something is to be gained by stirring the medium (see the discussion on sperm chemotaxis; section 3.3). For bacteria with $\ell \sim 10^{-4}$ cm and $v \sim 10^{-3}$ cm sec^{-1}, and substances with $D \sim 10^{-5} cm^2 sec^{-1}$, S is of the order of 10^{-2}, indicating in this case the inefficiency of stirring. The second effect of motility to be considered is the relative increase in current of material collected by a swimming cell with respect to the current of diffusing material. The latter follows straightforwardly from integration of Fick's law over a sphere surrounding a non-motile cell, i.e.

$$J_0 = 4\pi \ell D (c_\infty - c_0) \qquad (1.1)$$

where c_∞ and c_0 are the substrate concentrations at a distance from the cell $\gg \ell$ and at the cell surface respectively. (It is assumed that the concentration inside the cell is negligibly small). Considering the membrane permeability $k_m = (J_0/4\pi\ell^2)\, c_0^{-1}$, one has

$$J_0 = (4\pi \ell D c_\infty)(1 + D/\ell k_m)^{-1} \quad (1.2)$$

wherefrom two cases follow :

(i) $k_m \ll D/\ell$; then $J_0 \simeq 4\pi \ell^2 k_m c_\infty$. The limiting factor is the membrane permeability so that whether the cell is moving or not makes no difference in nutrient collection.
(ii) $k_m \gg D/\ell$; then $J_0 \simeq 4\pi \ell D c_\infty$. Since here the limiting factor is diffusion, it is appropriate to evaluate the relative increase in current for collecting material by swimming, J_s , with respect to diffusion current, J_0 .

The relevant quantity $\Delta \equiv (J_s - J_0)/J_0$ is a function of S (Berg and Purcell, 1977) with limiting behavior

$$\Delta \equiv (J_s - J_0)/J_0 = const \times (\ell v/D)^n \quad (1.3)$$

where n = 2 for $S < 10^{-1}$ and $n = 1/3$ for $S > 1$. It follows from (1.3) that $\Delta \propto (\ell \eta a)^n$ which indicates that for large cells $(\ell > 10\,\mu m)$ and/or media with high viscosity ($\eta > 10^{-1}$ poise) and/or large substrate molecules (a > 100 Å), the increase factor Δ may take an appreciable value ($\Delta \geq 1$) and swimming becomes efficient (Carlson, 1962). However in aqueous medium with sugars and amino-acids as nutrients ($D \sim 10^{-5}$ cm^2 sec^{-1}), bacteria ($\ell \sim 10^{-4}$ cm) would have to move with a velocity of several mm/sec to outswim diffusion. With the realistic figure $\ell v/D \simeq 3 \times 10^{-2}$, Δ is at best of the order of a couple of percents and swimming is uneconomical for the bacterium.

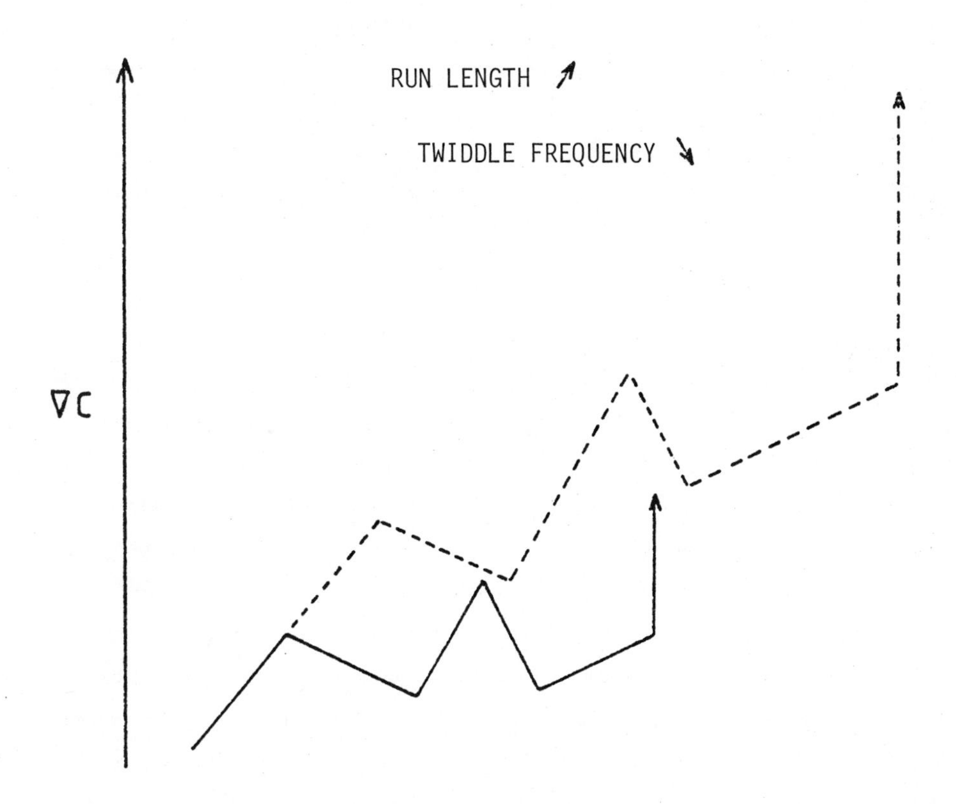

Fig. 3 Trajectory of bacterial motion in the presence of an attractant concentration gradient.

Eppur si muove !

So far we have examined possible reasons for bacterial motility and they turned out to be unjustified for low Reynolds number and low Sherwood number homogeneous media. However the situation becomes different for inhomogeneous media ; then motivation for bacterial motility is to be found in the way bacteria optimize their physiological environment by swimming to "better" regions. A striking case is chemotaxis, the phenomenon by which motile cells (and higher organisms) respond to external chemical stimulations, in particular substrate concentration gradients. Independently of their specific properties as nutrients, substances act as attractants, that is bacteria tend to swim to regions with higher concentration of such substances. Repellents are chemicals that have the opposite effect : bacteria swim away from regions rich in repellent. When moving towards-or away from - such regions, the bacterium now executes a biased random walk, as illustrated in Fig. 3. The solid lines describe the paths of the motile cell in a homogeneous medium (same as Fig. 2b) whereas the dashed lines show the trajectory of the cell in the presence of an attractant gradient indicated by the vertical arrow. When the bacterium swims with a velocity vector that has a positive component along the gradient direction, the run length is elongated (roughly twice its value in the homogeneous case) and as a result the frequency of twiddling stages decreases : what is now the efficiency of "chemotactic motion" ?

Consider the characteristic length $\ell_c = D/v$ where v is the velocity of the bacterium and D the diffusion coefficient of the attractant molecule ; ℓ_c measures the distance over which motion with velocity v balances diffusion. So for lengths $\ell^* < \ell_c$, diffusion dominates, whereas if $\ell^* > \ell_c$, the velocity is such that bacteria outswim diffusion. For E. coli with $v \sim 30\ \mu m/sec$ and e.g. glucose with $D \sim 10^{-5} cm^2 sec^{-1}$, $\ell_c \sim 30\ \mu m$; so if the length of the run is not larger than this value, glucose uptake is governed by diffusion. A bacterium runs over distances of about ten times its size $(2\text{-}3\ \mu m)$ in homogeneous media, but the run length roughly doubles in chemotactic motion oriented along the direction of an attractant concentration gradient. Then ℓ^* becomes larger than ℓ_c ; so motility is efficient in inhomogeneous media. Because of their biassed random walk, bacteria will exhibit a net displacement towards regions with higher concentration of attractant. As a result, coordinated motion leads to cooperative effects at the level of the bacterial population, a macroscopic manifestation of which is the formation and migration of chemotactic bands (see section 3).

2. LIGHT SCATTERING STUDY OF MOTILITY

The analysis of the dynamical behavior of biological systems is in general not simple. Complications arise from the usually complex structure of cells and accordingly because of the complexity of the trajectories of their motion. In addition there is often heterogeneity and structural dispersity, which are difficult to evaluate quantitatively but cannot be ignored except in ideal cases. So when studying the dynamical behavior of biological cells one seeks to set up a program that establishes criteria for motility characterization, such as :

- distribution of translational velocities ;
- distribution of non-translational components of motion (rotation,...) ;
- percentage of motile cells ;
- ratio of normal swimmers to defective swimmers ;
- for spermatozoa studies, it is also important to know the total number of cells in the collected sample.

It may be appropriate in simple cases and for "good" samples, to restrict the analysis to the evaluation of characteristic parameters like the mean translational velocity (e.g. for monodisperse, highly motile, spherical cells), the pitch in helical trajectories,...

It is important to obtain a reliable characterization of motility factors because when these are known it becomes possible to investigate the influence of physiological factors (physical, chemical, and biological) by quantitative measurements of changes and variations in the motility factors. The analysis is to be complemented by theory (or theoretical models) for physical interpretation. Such research programs devoted to the study of motility by laser light scattering spectroscopy started fifteen years ago at Saclay (Bergé, Dubois, Volochine and co-workers, 1967) for spermatozoa studies and about ten years ago at MIT (Nossal, Chen and co-workers, 1971) for bacterial motion. Since these pioneering studies, considerable developments have been accomplished and several groups have been actively contributing to the study of motility by light scattering. In this chapter, we restrict ourselves to a survey of those studies that concern spermatozoa and bacteria motion and chemotaxis. Light scattering spectroscopy has also been applied successfully to other motile organisms like algae which are discussed elsewhere in the present volume (Craig, Racey and Hallett, 1982 ; Ascoli and Frediani, 1982). The motile behavior of prokaryotic and eukaryotic cells as studied by quasi-elastic light scattering has been reviewed recently by Chen and Hallett (1982); see also the earlier review by Cummins (1977).

2.1. Time domain analysis

The quantity accessible to laser light scattering spectroscopy measurements is the intensity correlation function (see the chapter by Degiorgio in this volume) or the square of the normalized scattered electric field autocorrelation function

$$\left| \langle \underline{E}^*(0) . \underline{E}(t) \rangle / \langle |E|^2 \rangle \right|^2 = |g^{(1)}(t)|^2 \quad (2.1)$$

where $g^{(1)}(t)$ can be identified here with the Van Hove intermediate scattering function

$$g^{(1)}(t) = F_s(k,t) = \int_{-\infty}^{+\infty} d^3r \, e^{i\underline{k}.\underline{r}} \, G_s(\underline{r},t) \quad (2.2)$$

$G_s(\underline{r},t)$ is the self correlation function

$$G_s(\underline{r},t) = \langle n_1(0,0) \, n_1(\underline{r},t) \rangle / \langle n \rangle \quad (2.3)$$

with n_1 the number density and $\langle n \rangle$ its average value. It follows from space Fourier transformation of (2.3) that

$$F_s(k,t) = \langle \exp[i \, \underline{k} . \underline{r}(t)] \rangle \quad (2.4)$$

with

$$\underline{r}(t) = \int_0^t dt' \, \underline{v}(t') \quad (2.5)$$

where $\underline{v}$ is the velocity of the scattering particle. When particles move with constant velocity through the scattering volume, (2.4) reads

$$F_s(k,t) = \int_0^\infty d^3v \, P(\underline{v}) \exp(i\underline{k}.\underline{v}t) \qquad (2.6)$$

where $P(\underline{v})$ is the velocity distribution function of the scatterers' population. For isotropic velocity distribution, integration over the angles yields

$$F_s(k,t) = 4\pi \int_0^\infty dv \, v^2 P(v) \, \frac{\sin(kvt)}{kvt} \qquad (2.7)$$

To arrive at this result, the assumption was made that the scatterers are identical, independent, and structureless (point scatterers with spherical optical symmetry). Note that if v were not constant, the full dynamical structure factor $G_s(\underline{r},t)$ should be considered. Fortunately, the assumption v = const is usually satisfied for bacteria and spermatozoa. Indeed considering the time scale for changes in the spectral function $F_s(k,t)$, i.e. $\tau \simeq (kv)^{-1}$, with v typically of the order of $20\,\mu m/sec$ and $k \sim 10^4\,cm^{-1}$, one has $\tau \sim 50\,msec$. Since motile cells usually move on smooth trajectories for about one second, their velocity can safely be assumed to remain constant for times $\sim$ 50 msec. (Chen and Hallett, 1982).

It follows from (2.7) that the scattering function exhibits scaling versus x =kt. So non-scaling behavior will appear as indication of deviation from the above model. As far as size and shape are concerned, the criterion for scaling is $kd \ll 1$ where d is the cell size which in general is of the order of (or larger than) the wavelength of the incident light. Therefore spectra from motile cells like bacteria and spermatozoa will show scaling only at sufficiently small scattering angles. An example is given in fig. 4. for the bacterium Salmonella typhimu-

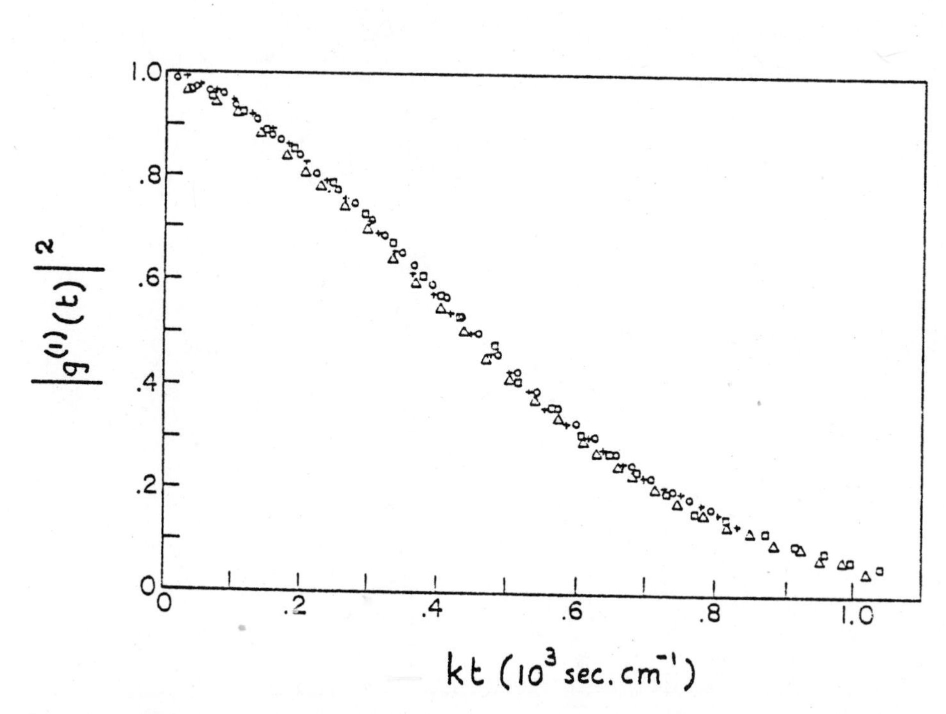

Fig. 4 Scaling of scattering function at low k (Θ = 3°, 5°, 6°, 7,5°) for spectra from S. Typhimurium (Stock, 1978).

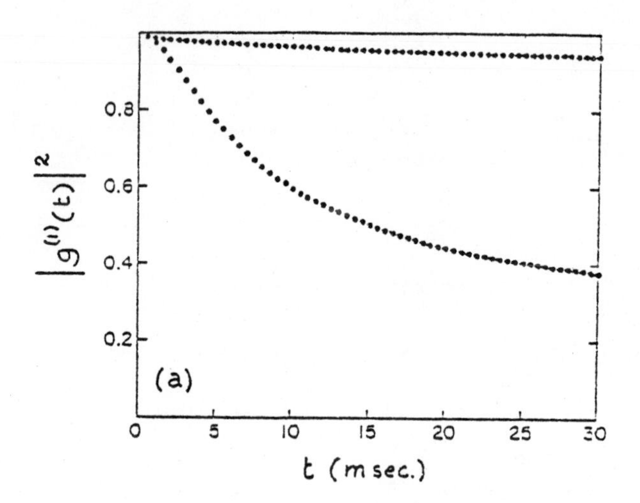

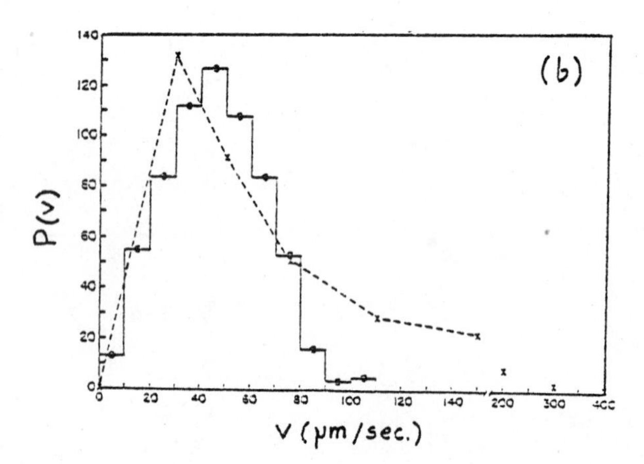

Fig. 5 (a) Intensity correlation function of human spermatozoa in seminal plasma (θ = 12°) (upper curve = dead spermatozoa); (b) Swimming speed distribution ; crosses : splines analysis of data from (a), circles : videomicroscopic assay (Frost and Cummins, 1981).

rium. When there is scaling, data inversion can in principle be performed to obtain the velocity distribution function (Nossal, Chen and Lai, 1971). Indeed from (2.7), one obtains

$$P(v) = \frac{2v}{\pi} \int_0^{\infty} dx \; x \, F_s(x) \sin(xv) \quad ; \quad x = kt \tag{2.8}$$

Fig. 5 illustrates the inversion procedure for the case of human spermatozoa. Various techniques have been developed to extract velocity distribution from light scattering spectra (Stock, 1976; Frost, 1979 ; Earnshaw, 1982). The evaluation of the velocity distribution is of importance in that it provides a "good signature" of the dynamical behavior of motile species and consequently constitutes a valuable probe of the effect of changes in physiological conditions. However experience has shown that -in particular for biological samples-such techniques must be applied with great cautiousness ; this problem is discussed at length elsewhere in the present volume (Chu, 1982). In particularly simple cases (e.g. highly motile samples of spherical cells) a simplified procedure may be adopted to characterize the species merely by an average velocity (see section 2.3)

2.2. Frequency domain analysis.

In frequency domain analysis, the quantity that is measured experimentally is the power spectrum of the density-density correlation function, i.e. the space and time Fourier transform

$$\begin{aligned} S_s(k,\omega) &= (2\pi)^{-1} \int_{-\infty}^{+\infty} dt \, e^{-i\omega t} \int_{-\infty}^{+\infty} d^3r \, e^{i\underline{k}.\underline{r}} G_s(\underline{r},t) \\ &= (2\pi)^{-1} \int_{-\infty}^{+\infty} dt \, e^{-i\omega t} F_s(k,t) \end{aligned} \tag{2.9}$$

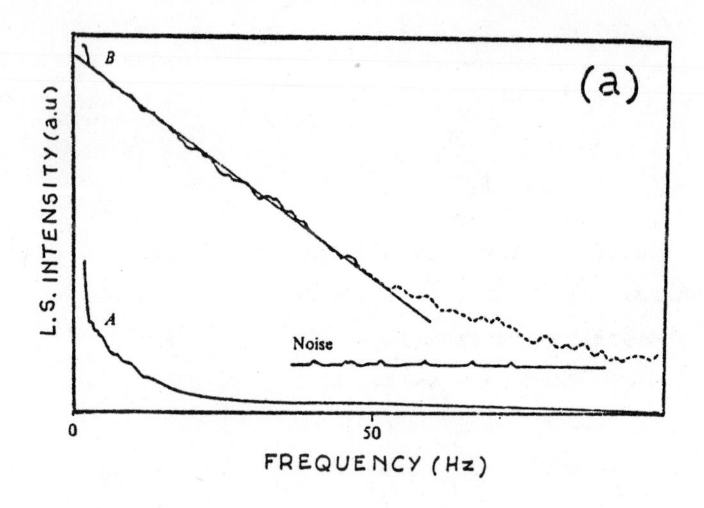

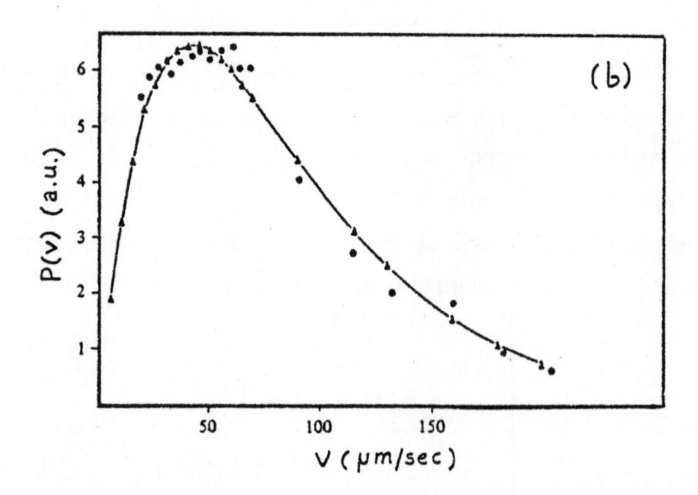

Fig. 6 (a) Light scattering spectrum from human spermatozoa, A = $S_s(k,\omega)$; B = ln A ; (b) Swimming speed distribution obtained from point to point derivative of A, Eq. (2.11) (dots) and from calculation using the value of the slope of B (triangles) (Jouannet, Volochine, Deguent and David, 1976).

For the model system considered in the previous section, (2.9) combined with (2.7) yields the following expression for the dynamical structure factor

$$S_s(k,\omega) = \frac{1}{2k}\int_{k/\omega}^{\infty} dv \, \frac{P(v)}{v} \qquad (2.10)$$

Inversion can be performed by differentiation to obtain the velocity distribution function, that is

$$-2k^2 \frac{\partial}{\partial\omega} S_s(k,\omega) = \frac{P(v)}{v} \qquad (2.11)$$

This procedure (Jouannet et al, 1976) shows that the velocity distribution can be obtained experimentally by a point to point differentiation of the frequency spectrum as illustrated in Fig. 6. Experimental observation shows that human spermatozoa samples exhibit an exponentially decaying power spectrum of the f rm $S_s(k,\omega) \propto \exp - \omega/k v_c$ where v_c is the characteristic velocity. It then follows from (2.11) that $P(v)$ can be identified with the Gamma function

$$P(v) = \frac{4v}{v_c^2} \exp\left(-\frac{2v}{v_c}\right) \qquad (2.12)$$

in which case, a measure of the characteristic velocity is obtained directly from the slope of the log of $S_s(k,\omega)$ (see fig. 6). Within the limits of applicability of this model, one can evaluate additional characteristic factors like the ratio of motile to non-motile cells and the total number of spermatozoa in the sample (see the paper by Volochine in the present volume).

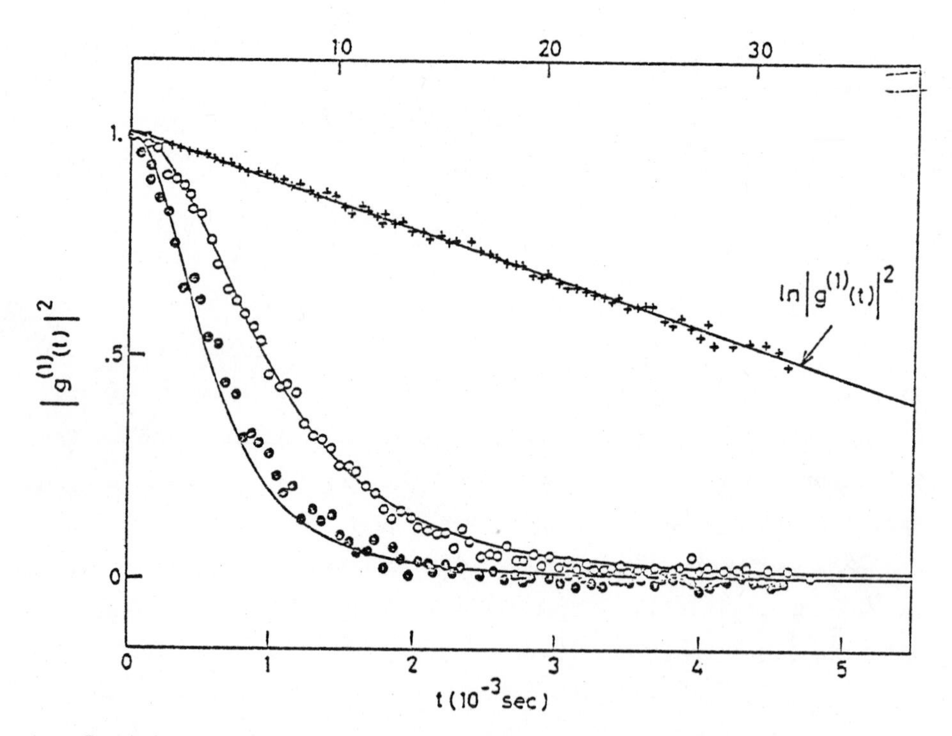

Fig. 7 Light scattering spectra from Asterias spermatozoa in sea water at $\Theta = 90°$ (dots) and $\Theta = 50°$ (circles) ; crosses = dead cells (Herpigny and Boon, 1979).

2.3. Application and limitation of the simple model.

Asterias rubens, a red starfish of the North Sea, has spermatozoa with spherical head (diameter $\sim 3\,\mu m$) and excellent motility (very few defective swimmers). For model scatterers the analysis of section 2.2 yields

$$F_s(k,t) = \left[1 + \left(\frac{k v_c t}{2}\right)^2\right]^{-1} \tag{2.13}$$

Note however that since not all cells are motile and because motile cells are also mobile, the complete scattering function reads

$$g^{(1)}(t) = \left[\alpha F_s^m(k,t) + (1-\alpha)\right] \exp\left(-D_{eff} k^2 t\right) \tag{2.14}$$

where α denotes the fraction of motile spermatozoa and D_{eff} is the effective diffusion coefficient of non-motile cells. D_{eff} is easily determined from experimental spectra obtained by light scattering from dead spermatozoa. The two quantities of interest in (2.14), with $F_s^m(k,t)$ given by (2.13), are α and v_c ; they are obtained by the method of "integrated spectra" (Herpigny and Boon, 1979) which amounts to solve nummerically the set of two equations obtained by time integration of two spectra taken at two different scattering angles; see Fig. 7. In addition to simplicity, the method has the advantage of being applicable to a wide class of velocity distribution functions. The species that served as a test for this method is an echinoderm, that is a species with external fecondation. Organims with external fecondation are good candidates for spermatozoa motility studies by light scattering because sperm samples are diluted in aqueous suspension under experimental conditions that are convenient for spectroscopic measurements and that can be made equivalent to those of natural environment in actual fecondation. (see also the discussion on sperm chemotaxis in section 3).

Such conditions are usually not met for sperm motility studies of mammalians and in particular of human spermatozoa and bull spermatozoa which have been most actively studied. Therefore further dificulties arise in the interpretation of light scattering results in addition to those related to factors mentioned earlier : non-uniformity in size and shape of head, morphological differences inducing different swimming properties, complexity of trajectories of motion, heterogeneity of spermatozoa population. To a lesser extend, similar considerations apply to bacteria as well. It is then to be expected that non-scaling in spectra will be the common rule rather than the exception.

2.4. Structural and dynamical effects.

Non-scaling arises because of static structural effects and because of complexity in dynamical structure of scatterers. Static effects build up when the scattering cells are large and lack spherical symmetry. Then the static structure factor $S_s(k) \equiv F_s(k, t=0) \neq 1$ and $F_s(k,t)$ depends on orientation of the symmetry axis of the scatterer with respect to the direction of the scattering vector $\underline{k}$. This is the case for bacteria which exhibit twiddling and wiggling superimposed on translational displacement and for spermatozoa whose trajectories are close to helical. As a result there is deviation from a linear trajectory over distances that may be shorter than k^{-1} and additional time correlations must be included in the computation of $F_s(k,t)$ which has the general form

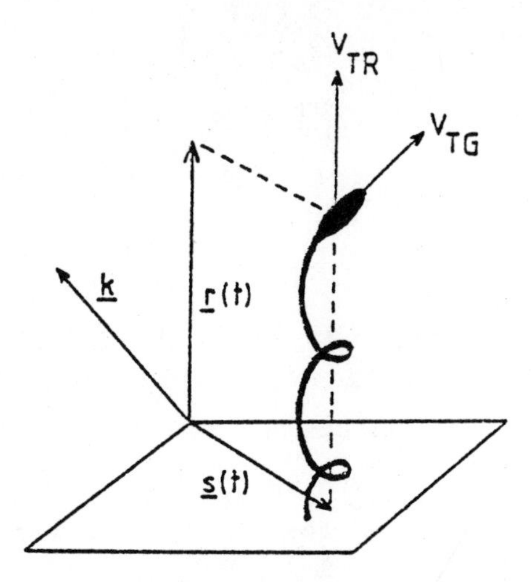

Fig. 8 Geometry for non-rectilinear trajectory of cell motion.

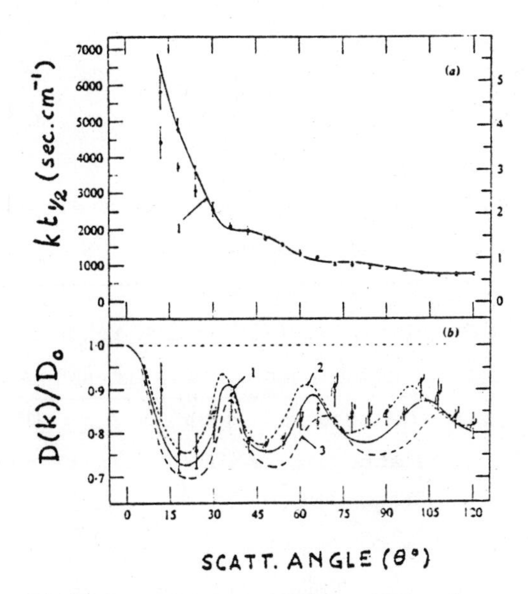

Fig. 9 (a) Halfwidth of scattering function and (b) scaled effective diffusion coefficient from light scattering spectra of E. coli. Angular dependence of experimental data (dots and circles) and of model calculations for coated ellipsoids (lines)(Holz and Chen, 1978).

$$F_s(k,t) = \langle b_k^*(0)\, b_k(t) \times \exp i\underline{k}.[\underline{r}(t)-\underline{r}(0)] \exp i\underline{k}.[\underline{s}(t)-\underline{s}(0)] \rangle \tag{2.15}$$

where $b_k(t)$ is the dynamic form factor and $\underline{r}(t)$, $\underline{s}(t)$ are the components of the scatterer's position along and perpendicular to the average displacement direction ; see Fig. 8.

The evaluation of the form factor from total intensity measurements as a function of the scattering angle, i.e.

$$I(\theta) \propto F_s(k,t=0) = \langle |b_k(0)|^2 \rangle \tag{2.16}$$

requires model calculation. The ellipsoidal scatterer provides a good representation for bacteria (Chen, Holz, and Tartaglia, 1977) ; in particular the coated ellipsoid model has been used successfully to analyze the experimental data as illustrated in the top graph of Fig. 9. Spectra obtained from non-motile bacteria decay approximately exponentially in time ; non-scaling is observed because of structural effects which renders the effective diffusion coefficient k-dependent. This is shown in the bottom graph of Fig. 9 where the measured quantity $D_{eff}(k)$ is plotted as a function of the scattering angle. Comparison with model calculation is quite satisfactory.

Similarly dynamical structure factors that describe non-translational components of motion are evaluated by means of theoretical models. One of the first models that was proposed describes spermatozoa motion as the helical trajectory of a point scatterer (Combescot, 1970). Off-axis motion was accounted for in a less simplified way by Schaefer, Banks and Alpert (1974) who

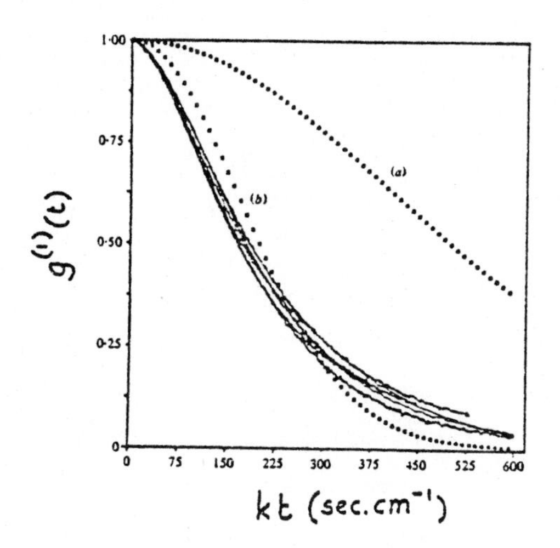

Fig.10 Scattering functions from E. coli ; experimental data at θ = 20°, 35°, 50°, 70°, 90° (curves) compared to helical motion model (b) ; (a) = computed spectrum for pure translation (Schaefer , Banks, and Alpert, 1974).

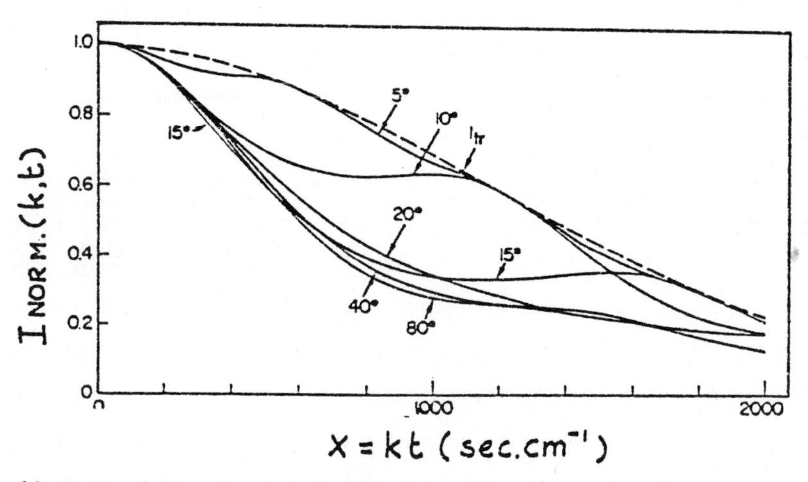

Fig.11 Scattering functions computed from dumb-bell model ; Itr = computed spectrum for pure translation (Boon, Nossal, and Chen, 1974).

modeled the dynamical behavior of the bacterium as a rod of point scatterers with helical motion. The corresponding spectrum can be calculated explicitly provided the function $F_s(k,t)$ is factored so that translation and rotation of the cell are uncoupled. Comparison of the computed spectra with those obtained experimentally from E. coli suspensions is shown in Fig. 10.

The same decoupling assumption was made in the "dumb-bell" model of Boon, Nossal and Chen (1974) where wiggling motion of E. coli is described by the oscillations of a 2-point scatterer cell moving along a linear trajectory. The spectral functions were calculated for a variety of scattering angles and translational velocities. As shown in Fig. 11, the results clearly indicate the importance of the contribution of non-translational components of motion ; the resulting non-scaling effects due to wobbling motion were also confirmed by light scattering studies of S. typhimurium by Stock and Carlson (1975).

The complex structure of the bull spermatozoon and of the human spermatozoon renders light scattering spectral analysis of such systems even more complicated. Studies by Craig, Hallett and Nickel (1979) and by Harvey and Woolford (1980, 1982) on bull spermatozoa motility have demonstrated that rotational motion of the head gives a major contributon to the light scattering spectrum. Based on the observation that the flat shape of the head of the bull spermatozoon could be represented by an oblate ellipsoid, a model was developped by Craig et al (1979) who described the spermatozoon dynamical behavior by an oblate ellipsoid with center of mass moving along a helical trajectory. They also included velocity and rotational distribution functions in their analysis. The result of the comparison between theory and experiment is shown in Fig. 12. The helical model was recently extended by Craig, Hallett, and Chen (1982) ; they investi-

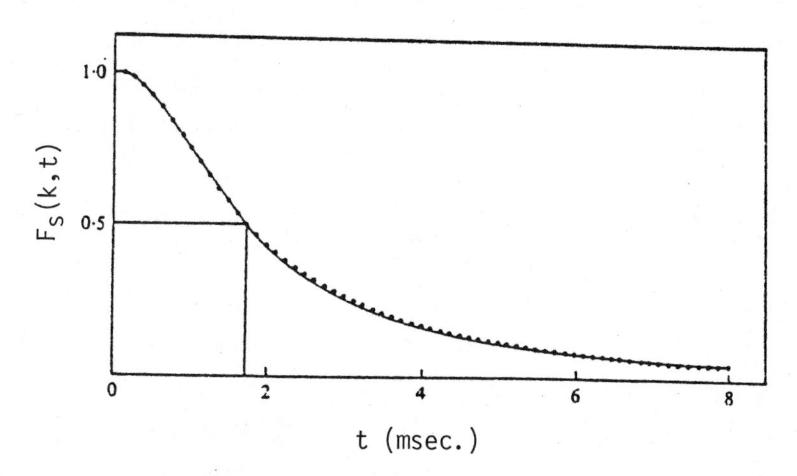

Fig. 12 Comparison of experimental data from bull spermatozoa (dots) with model calculation for ellipsoid on helical trajectory (curve) (Craig, Hallet and Nickel, 1979).

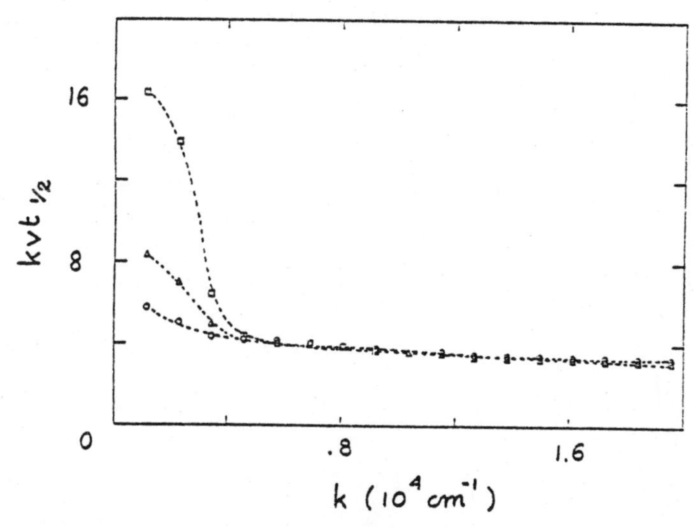

Fig. 13 Scaled halfwidth as a function of wavenumber for an ellipsoid moving on helical trajectories with different pitches (squares = 3 μm ; triangles = 12 μm ; crosses = 18 μm (Craig, Hallet, and Chen, 1982).

gated the behavior of the spectral function over a wide range of scattering angles and they analyzed the effects of parametric changes in the ellipsoid semi-axes and in the helical trajectory (pitch, radius, frequency). With these degrees of freedom, the model acquires sufficient generality for applications to the study of a variety of motile cells and micro-organisms. It is also a virtue of this model analysis to provide clear indication of which dynamical components can be probed under given experimental conditions. An illustration is given in Fig. 13 where the spectral halfwidth is plotted versus scattering wavenumber for three different helical pitches, all other parameters being kept constant. The graph shows that deviations from scaling increase when the pitch becomes small, that is when the curvature of the trajectory is larger ; such deviations become observable when k is sufficiently small, that is at small scattering angles. When the latter have larger value, the wavelength of the probe becomes short and the detected dynamical component is the instantaneous velocity which is independent of the pitch.

The experimental and theoretical investigations of motility by laser light scattering spectroscopy provide us with considerable quantitative information on the dynamical behavior of living cells and microorganisms. On the other hand the variety and complexity of biological motions renders it difficult to draw general rules that could be used as simple recipes for the interpretation of spectroscopic results. However guided by experimental observation and theoretical analysis, one can arrive at least at some conclusions : light scattering spectra contain in general important contributions from off-axis motion ; contribution from rotational and oscillatory components are dominant features of spectra taken at large scattering angles (typically $> 20°$) ; in the high-k range, where scaling may be found for some species, velocity measurements yield information on the instantaneous (or

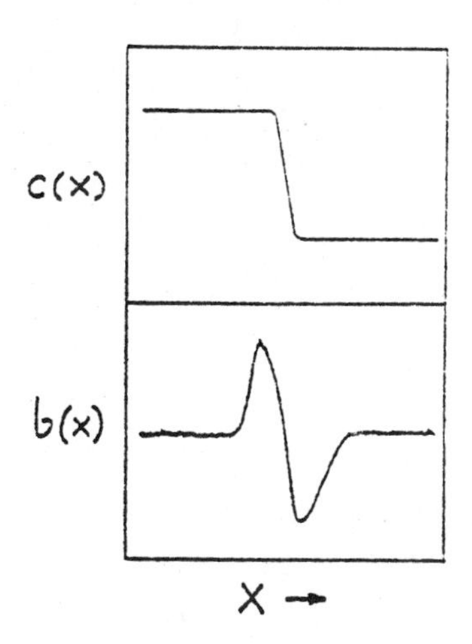

Fig. 14 Chemotactic response of S. typhimurium to a spatial attractant concentration gradient. (a) Initial time serine concentration profile ; (b) Bacterial distribution after 8 minutes (Tsang, McNab, and Koshland, 1973).

Fig. 15 Chemotactic band migration of E. coli in response to glucose. (a) At initial time, bacteria are in the lower part of cell ; (b) Bacterial band after 2 hours. (Herpigny, 1982)

tangential) speed ; contribution from translational motion is more important at small scattering angles (typically $<$ 10°) ; progressive (or displacement) velocities may be measured in the low-k spectral range.

3. MOTILITY AND CHEMOTAXIS

The dynamical behavior of bacteria is characterized by two phases of motion, the run and the twiddle (see Fig. 2.b). In an inhomogeneous medium, bacteria tend to swim from regions poor in attractant to those that are richer in attractant ; they undergo a biased random walk (see Fig.3) which results into a net displacement towards higher concentration zones. As a result of the cooperative effect of the microscopic biased motion of the individuals, there is formation of bacterial bands, a macroscopic phenomenon visible with naked eye. An example is given in Fig. 14. An initially uniform bacterial suspension of S. typhimurium is subjected to an imposed concentration gradient of serine ($c(x)$ line). After about ten minutes, bacteria have moved away from the low concentration region nearest to the gradient zone where they accumulate as shown by the peak in the profile of bacteria concentration measured by light scattering intensity (Dalquist, Lovely, and Koshland, 1972). Another example is shown in the photograph of Fig. 15. which illustrates the chemotactic response of E. coli population to glucose. At initial time, the bacterial suspension is located in the lower part of a test tube (Fig. 15a) filled with buffer solution containing glucose. Bacteria consume glucose thereby creating a local concentration gradient. They respond to the attractant gradient by swimming towards the higher concentration zone where they deplete the medium. As a result the gradient shifts continously upwards along with a slab of bacteria that forms a migrating band (Fig. 15b).

3.1. Dynamic light scattering as a probe of chemotactic effects

Small scattering angles may be chosen such that $kd \ll 1$, d being the size of the scatterer (see section 2.1). Then the form factor is unity and dynamic light scattering probes the translational motion of the center of mass of the scattering object. For E. coli and S. typhimurium, such conditions should be realized with a scattering angle of $\sim$ 3-4°. In the run phase, the bacterium moves with a velocity $\sim$ 20-30 $\mu m\ sec^{-1}$; since the duration of the run is of the order of one second, the run length is larger than k^{-1} and it is expected that $F_s(k,t)$ exhibit scaling versus $x = kt$. However it is found experimentally that the scaled half-widths of spectra taken at the center of a migration band of E. coli responding chemotactically to oxygen deviate considerably from scaling at low scattering angles (Holz and Chen, 1978). This observation suggests that there is an additional component of bacterial motion that should be taken into account. Since twiddling appears as jitter motion of the center of mass, it can be modelled as a small step random walk process with an effective diffusion coefficient D_t. Then combining the contributions from runs and twiddles one has for a fully motile population (Holz and Chen, 1978).

$$F_s(k,t) = \alpha_t \exp(-D_t k^2 t) + \alpha_r F_{run}(k,t) + F_{run \rightleftarrows tw} \tag{3.1}$$

where α_t and α_r are the weighted fractions of the bacterial population in the twiddle and run phases respectively. The third term in (3.1) accounts for the transition from one phase to the other. Scaling versus $x = kt$ is obtained for the contribution from runs. On the other hand, the half-width of the scattering

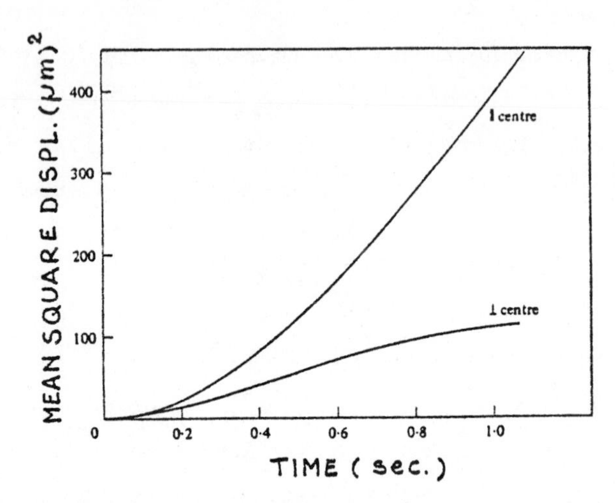

Fig. 16 Mean square displacement of E. coli in center of migrating band in the direction of (∥) and perpendicular to (⊥) the oxygen gradient (Wang and Chen, 1981).

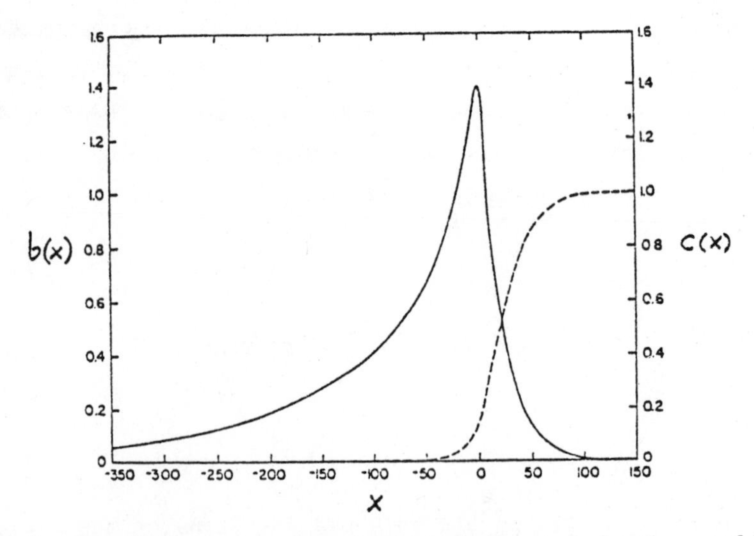

Fig. 17 Model calculation of chemotactic bacterial band : solid line = bacterial density distribution, dashed curve = substrate concentration profile (Lauffenburger, 1979).

function in the first term is $kt_{1/2} = (\ln 2)/(D_t k)$ which behaves like k^{-1} and so increases dramatically at low scattering angles. Thus light scattering can be used as a probe of twiddle motion as well as of run motion. Since their respective contributions are indicative of the chemotactic biased random walk of the bacterium (see Fig. 3), the method provides a measure of the effects of chemotaxis.

The spectral values of light scattered from a chemotactic migration band were further exploited by Wang and Chen (1981). They measured light scattering spectra in three zones of the band, front, center and tail, in each case with wave vector oriented parallel and perpendicular to the direction of the oxygen gradient which is also the direction of migration. The behavior of the scattering function $F_s(k,t)$ is Gaussian in the short time domain, then becomes exponential, and finally decays with a long time tail. Characteristic parameters of bacterial motion are then extracted from numerical fits to the experimental data. Most important are the following results : (i) the average velocity component along the gradient is larger in the center of the band than in its wings ; (ii) the mean square displacement grows in time much faster in the direction of the gradient than perpendicular to it (Fig. 16). Wherefrom the following conclusions can be drawn : (i) the bacterial velocity is larger where the gradient is steeper so that bacteria do not spread out downstream, but form a band which persists since v_C (in the center) $> v_F$ (in the front) ; on the other hand since v_C is also larger than v_T (in the tail), bacteria are left behind (upstream) as a result of which the migration band is a trailing band (see section 3.2) ; (ii) the finding that $\langle \Delta z^2(t) \rangle$ (along the gradient direction) becomes much larger than $\langle \Delta X^2(t) \rangle$ (in the plane perpendicular to the gradient) provides clear indication that biased random motion produces a net displacement towards the high concentration region. Such conclusions are important in that they establish a connection between microscopic motility and macroscopic chemotaxis.

3.2. Phenomenological theory of bacterial chemotactic bands

The discovery of chemotaxis goes back to the late nineteen century (Pfeffer, 1881 ; Engelman, 1883) but it was not until the midsixties that the phenomenon began to be investigated in a more systematic fashion. It was the pioneer work of the groups of Adler (1966), of Berg (1972), and of Koshland (1972) that initiated the major developments in bacterial chemotaxis. A considerable activity in this domain of research followed : a recent literature survey lists nearly 400 papers published between 1958 and 1981 (Eisenbach, 1981) ; for an up to date status of the art, see the review papers by Berg (1975), Adler (1978), and Koshland (1980). During the last fifteen years, the biophysical aspects of bacterial motility and chemotaxis have been studied intensively from both the microscopic point of view (individual cell motion ; see section 1 and section 3.1) and from the macroscopic point of view (band formation and migration) which is discussed in this section. Simultaneous with the development of experimental work, Keller and Segel (1971) proposed a theoretical analysis of bacterial band migration ; their model sets the grounds for a phenomenological theory of bacterial chemotactic band formation and migration. Reviews of theoretical model studies are available in the recent literature (Boon, 1975 ; Segel, 1978 ; Nossal, 1980 ; Lapidus and Levandowsky, 1981).

The basic set of equations is given by (i) the conservation equation for bacterial density

$$\partial_t b = -\nabla . \underline{J} + G \quad (3.2)$$

and (ii) the evolution equation for the substrate (attractant) concentration

$$\partial_t c = D \nabla^2 c - K b \quad (3.3)$$

In Eq. (3.2) $b \equiv b(\underline{r},t)$ is the bacterial density ; G, the bacterial growth term (set here equal to zero as chemicals essential for bacterial growth are not present in the experiments considered) ; and $\underline{J} \equiv \underline{J}(\underline{r},t)$ the current density

$$\underline{J} = \underline{J}_\mu + \underline{J}_c \quad (3.4)$$

with

$$\underline{J}_\mu = -\mu \nabla b \; ; \quad \underline{J}_c = \underline{v}_c b \quad (3.5)$$

where $\underline{J}_\mu$ denotes the current due to motility with μ the motility coefficient, and $\underline{J}_c$ denotes the chemotactic current with $\underline{v}_c = f(c)$, the chemotactic velocity, a function of the attractant concentration. In Eq. (3.3), $c \equiv c(\underline{r},t)$ is the substrate concentration ; on the r.h.s. the first term is the diffusion term with D, the substrate diffusion coefficient, and the second term describes substrate consumption with rate constant K. When more than one substrate is efficient, (as e.g. in aerobic metabolism of sugars and amino-acids) additional equations of type (3.3) are required to complete the description of the system. Because of the functional form of the chemotactic term and of the consumption term, these equations, subject to the appropriate initial conditions, can become highly non-linear and may therefore have different branches of solutions that characterize various types of system behavior. Here we shall restrict our discussion to a couple of typical situations that have been studied experimentally by static light scattering (total integrated intensity). Detailed analyses are given in the literature (Dahlquist, Lovely and Koshland, 1972; Tsang, Mcnab and Koshland, 1973; Lauffenburger, 1979; Wang, 1982; Herpigny, 1982.).

When a bacterial population responds to a chemical stimulation, the resulting effect is a function of the concentration of the attractant. The response depends on the concentration variation in the medium as - for an attractant - bacteria swim from low concentration regions to high concentration regions. Thus the response should be expressed as a function of the concentration gradient. However since stimuli are not sensed below some threshold concentration and because inhibition occurs when the concentration exceeds a saturation value, the response function also depends on the absolute value of the attractant concentration. The simplest analytic function introduced initially to model the chemotactic term in Eq. (3.3) was obtained by setting $\underline{v}_c = \chi \nabla c$ with $\chi = \delta c^{-1}$, where δ denotes the chemotactic coefficient (Keller and Segel, 1971). Then with the further assumption that the consumption rate factor K = const, the set of Eqs. (3.2) and (3.3) can be solved analytically to describe steady state band migration. Fig. 17 shows the bacterial density profile of the migrating band centered around the point of steepest change in attractant concentration.

Bacterial density profiles can be measured by static light

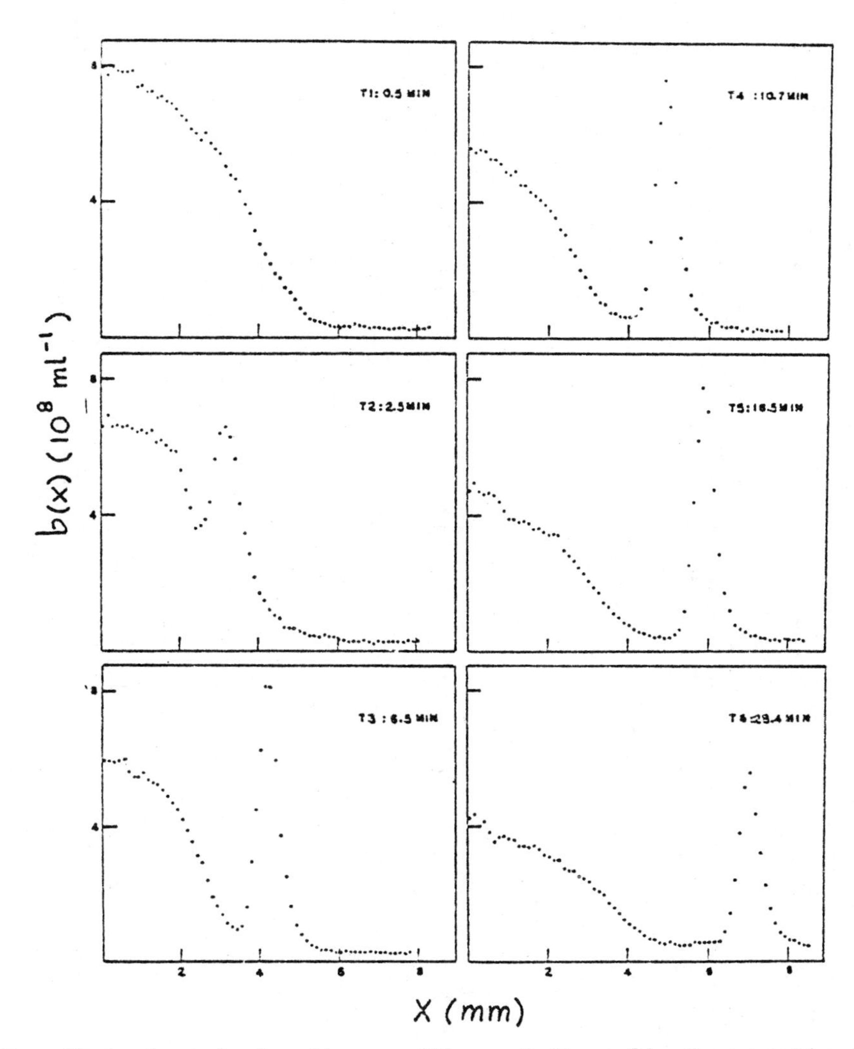

Fig. 18 Bacterial density profiles of E. coli chemotactic band responding to oxygen : time evolution of bacterial distribution as measured by light scattering intensity (Wang, 1982).

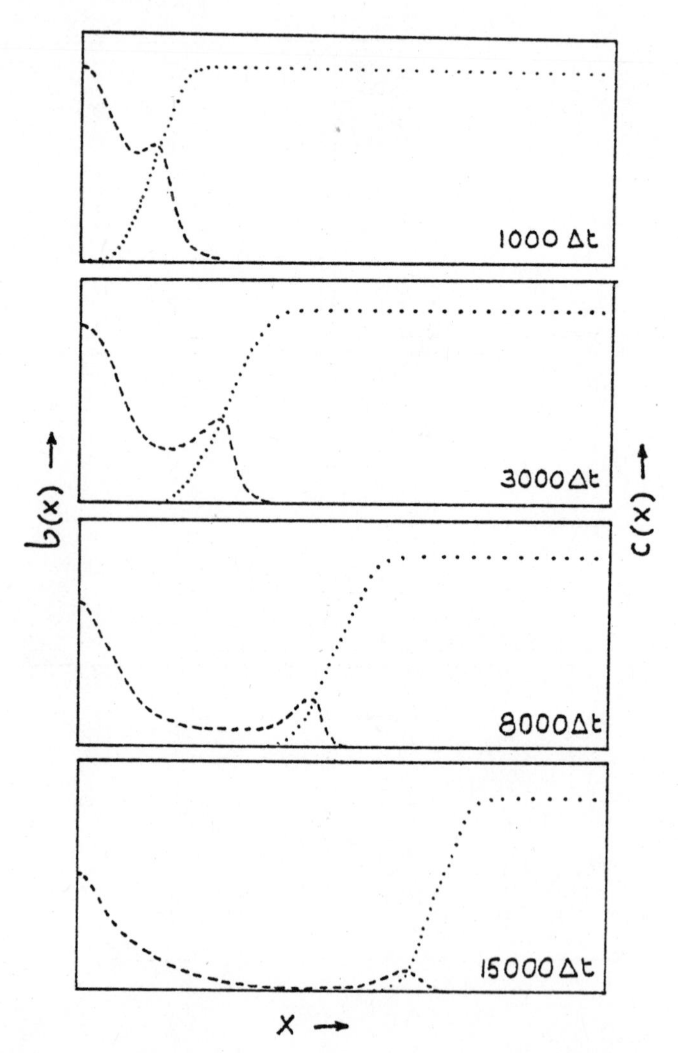

Fig. 19 Model calculation of the evolution of bacterial density, b(x) (dashed curves), in response to spontaneous substrate concentration gradient, c(x) (dotted curves) (Herpigny and Boon, 1982).

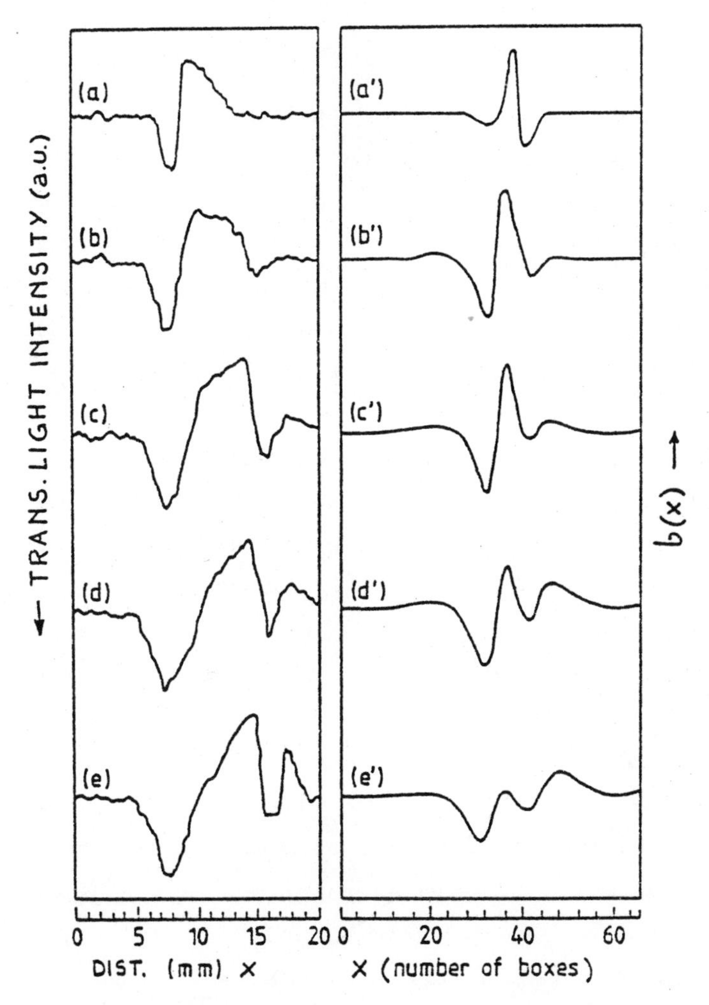

Fig. 20 Chemotactic response of E. coli to simultaneous effects of two attractants (glucose and oxygen). Experimental results obtained by absorption photometry (left frame ; after 10 min (a), 20 min (b), 30 min (c), 40 min (d), 70 min (e)) ; theoretical model results (right frame) (Boon and Herpigny, 1982).

scattering by monitoring the scattered light intensity through the sample cell as a function of distance. So when a band forms from an initial inoculum and migrates as described in the introduction to section 3 (see Fig. 15), the evolution of the bacterial density distribution can be followed as shown in Fig. 18 which illustrates the response of E. coli to oxygen : the progressive formation of the band followed by its migration is clearly seen (Wang, 1982). Theoretical analysis was performed on the basis of a generalization of Eqs. (3.2) and (3.3) (Boon and Herpigny, 1982) ; the results are shown in Fig. 19. Qualitative agreement is obtained between theory and experiment. The quantitative differences are significant : they are related mainly to the value of the ratio δ/μ which measures the efficiency of chemotaxis (via the systematic driving term $\underline{J}_c$) versus motility (via the diffusion term $\underline{J}_\mu$) . Band trailing as obtained here in the theoretical computation and as observed in various experimental cases appears as the macroscopic manifestation of bacterial motility variations in the band as discussed in section 3.1.

The second example describes the response of an initially uniform bacterial population to an imposed attractant gradient. In the present case bacteria are suspended in an oxygen saturated medium with an initial glucose step gradient at mid-height of the cell. It would be expected from logical intuition that bacteria would accumulate towards the high glucose concentration zone (here the bottom of the cell, i.e. 0 mm $< X <$ 10 mm in Fig. 20). The evolution of the profile of the bacterial distribution as measured by static light scattering (Fig. 20, left frame) shows (i) that bacteria migrate towards the region initially poor in glucose, where they form a broad band, (ii) that a second narrower band forms beyond the broad band. The basic reason for the appearance of a bi-modal distribution is to be found in the response of E. coli to the simultaneous effects of two attrac-

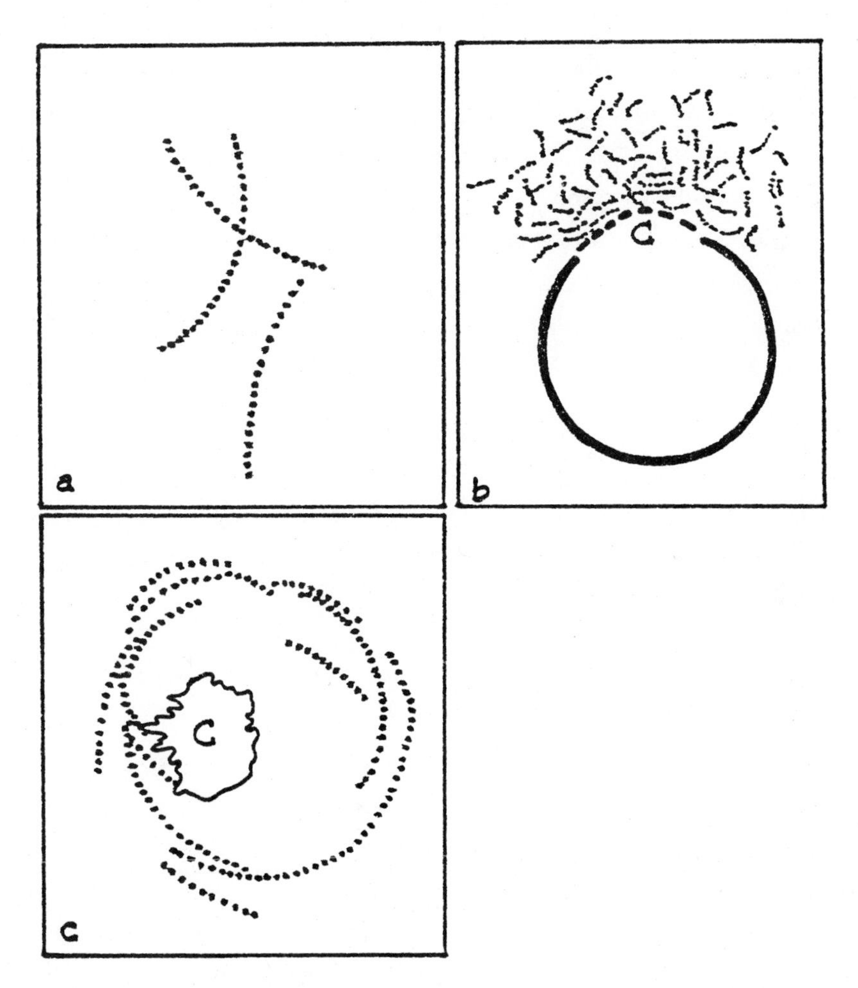

Fig. 21 Sperm chemotaxis in Muggiae kochi. (a) Trajectories of spermatozoa motion, (b) Spermatozoa clouding around the egg's cupule, C ; (c) Trajectories of spermatozoa around isolated cupule. (Carré and Sardet, 1981).

tants : glucose and oxygen. Bacteria first consume glucose aerobically thereby creating an oxygen gradient ; they are then subject to oppositely directed gradients and therefore form two bands. To describe this situation analytically, a generalized theoretical model is constructed to include evolution equations and chemotactic currents for both substrates . The results presented in Fig. 20 (right frame) are in good agreement with experimental observation. Further analysis provides the evolution of the concentration profiles and of the chemotactic velocity wherefrom physical interpretation follows. Most important is the conclusion that light scattering experiments and theoretical analysis can be combined to investigate aspects of bacterial motility and chemotaxis that provide valuable information complementary to microscopic observation and biological studies.

3.3. Sperm chemotaxis

Chemotaxis has also been observed for spermatozoa in species with external fecondation. (Miller 1973, 1977). In particular it was shown recently that chemotactic effects exist and are clearly important in *Muggiaea kochi*, a species of siphonophores (planktonic cnidarians) of the Mediterranean sea (Carré and Sardet, 1981). Microscopic observation shows that *M. kochi* spermatozoa swim in sea water with slightly curved trajectories (Fig. 21.a). When an egg is present in the medium, spermatozoa that wander in the vicinity of the cupule, an extracellular organelle located at the egg's maturation pole, tend to remain in that zone, where they execute a "circular dance", while more and more spermatozoa are attracted in this region (Fig. 21.b). The phenomenon is observed with isolated cupules (Fig. 21.c) whereas cupule deprived eggs do not attract spermatozoa.Quantitative measurements of the spermatozoa motion show that the curvature of their trajectory in the vicinity of the cupule increases by a factor of 4 with respect to the free swimming value (compare Figs. 21.a and 21.c) whereas the average speed and flagellar beat frequency remain essentially unchanged. The presence of the cupule is not necessary for fertilization but is indispensable for attraction, which persists when the cupule is dissolved. The chemical nature of the specific substance responsible for attraction has not been determined yet, but analysis by microelectrophoresis in polyacrylamide gradient gel has characterized the the attractant as a substance that migrates electrophoretically with a mobility close to that of bovine serum albumin (BSA). So they should be of similar molecular weight and the attractant would have a diffusion coefficient of the order of 8×10^{-7} $cm^2 sec^{-1}$ in aqueous medium and with higher value in BSA.

Let us take these findings into consideration and go back to the analysis of the value of motility as discussed in section 1. Remember that the Sherwood number defined as $S = \ell v / D$ measures the efficiency of stirring with respect to diffusion. M. kochi spermatozoa have a size $\ell \sim 20\ \mu m$ and move with an average speed $v \sim 40\ \mu m/sec$. Assuming that the attractant has a diffusion coefficient of the order of that of BSA, one finds $S \sim 10$! Then the factor that measures the ratio of relative increase in material current due to stirring with respect to the diffusion current becomes here (see Eq.(1. 3)).

$$\Delta = 0.7 \times (\ell v / D)^{1/3} \simeq 2 \qquad (3.6)$$

that is J(stirring) is three times larger than J (diffusion). If so, spermatozoa motion around the egg's cupule would have the most important purpose of stirring the medium so that attractant molecules spread into a larger zone of space where more spermatozoa can be attracted towards the egg. Newly attracted spermatozoa contribute to the stirring thereby increasing the volume of the attraction zone. As a result a cloud of spermatozoa forms around the maturation pole of the egg (see Fig. 21.b). Within the limits of conjecture, a conclusion may be drawn as to the fertilization value of motility not only as such (that is motion as a mean for the spermatozoon to reach the egg) but also by way of stirring the medium (thereby increasing the probability of fertilization) for species with external fecondation.

REFERENCES

ADLER, J., (1966), Science 153 : 708.

ADLER, J., (1978), in The Harvey Lectures (Series 72), Academic Press, New York, p195.

ASCOLI C., and FREDIANI, C., (1982) in the present volume.

BERG, H.C., (1975), Ann. Rev. Biophys. Bioeng. 4 : 119.

BERG, H.C., and BROWN, D.A., (1972), Nature 239 : 500.

BERG, H.C., and PURCELL, E.M., (1977), Biophys. J., 20 :193.

BERGE, P., VOLOCHINE, B., BILLARD, R., and HAMELIN, A., (1967) C.R. Acad. Sc., Paris 265 : 889.

BERGE, P. and DUBOIS M., (1973) Rev. Phys. Appl. 8 :89.

BOON, J.P., (1975), in Membranes, Dissipative Structures, and Evolution. John Wiley-Interscience, New York, 169.

BOON, J.P., and HERPIGNY, B., (1982), Bacterial Chemotaxis and Band formation (to be published).

BOON, J.P., NOSSAL, R., and CHEN, S.H., (1974), Biophys. J. 14 : 847.

BROKAW C.J., and GIBBONS, I.R., (1975) in Swimming and Flying in Nature, Plenum Publ. Co., New York, 89.

CARLSON, F.D., (1962) in Spermatozoon Motility, AAAS, Washington, 137.

CARRE, D., and SARDET, C., (1981) Sperm Chemotaxis in Siphono mores.

CHEN, S.H. and HALLET, F.R., (1982), Quart.Rev.Biophys. 15: 131.

CHEN, S.H., HOLZ, M., and TARTAGLIA, P., (1977) Appl. Opt. 16: 187.

CHU, B. (1982), in the present volume.

COMBESCOT, R., (1970), J. Phys. 3:767.

CRAIG, T., HALLETT, F.R., and CHEN, S.H., (1982), Appl. Opt.

CRAIG, T., HALLETT, F.R., and NICKEL, B., (1979), Biophys. J. 28 : 457.

CRAIG, T., RACEY T.J., and HALLETT, F.R., (1982), in the present volume.

CUMMINS, H.Z., (1977), in Photon Correlation Spectroscopy and Velocimetry, Plenum Press, New York, 200.

DAHLQUIST, F.W., LOVELY, P., and KOSCHLAND, D.E., (1972) Nature New Biology, 236 : 120.

DE GIORGIO, V., (1982), in the present volume.

DEPAMPHILIS, M.L., and ADLER, J., (1971), J. Bact. 105 : 384.

DEPAMPHILIS, M.L., and ADLER, J., (1971), J. Bact. 105 : 396.

ENGELMAN, T.W., (1881), Pflugers Arch. Ges. Physiol. 25 : 285.

EARNSHAW, J.C., (1982), in the present volume.

EARNSHAW, J.C., and MUNROE, G. (1982), in the present volume.

EISENBACH, A. (1981), A Bibliography on Motility and Behavior of Bacteria.

FROST, J.W., (1977), Ph. D. Thesis, N.Y.U.

FROST, J.W., and CUMMINS, H.Z., (1981), Science, N.Y. 212 : 1520.

GRAY, J. (1958), J. Exp. Biol. 35 : 96

HARVEY, J.D., and WOOLFORD, M.W., (1980) Biophys. J. 31 : 147.

HERPIGNY, B., (1982), Ph. D. Thesis, University of Brussels.

HERPIGNY, B., and J.P. BOON, (1979) J.Phys. 40 : 1085.

HOLZ, M., and CHEN, S.H., (1978) Appl. Opt. 17 : 1930.

JOUANNET, P., VOLOCHINE, B., DEGUENT, P., and G. DAVID, (1976) Progr. Repr. Biol. 1 : 28.

KELLER, E.F., and SEGEL, L.A., (1971), J. Theor. Biol. 30 : 235.

KOSHLAND, D.E., (1980) Bacterial Chemotaxis as a Model Behavioral System, Raven Press, New York.

LAPIDUS, J.R. and LEVANDOWSKY, M., (1981), in Biochemistry and Physiology of Protozoa, Academic Press, New York, 235.

LAUFFENBURGER, D.A., (1979), Ph. D. Thesis, University of Minnesota.

MILLER, R.L., (1973) in Behaviour of Micro-organisms, Plenum Press, New York, 31.

MILLER, R.L., (1977), Adv. Invert. Repr. 1 : 99.

NOSSAL, R., (1980), in Biological Growth and Spread, Springer-Verlag, Berlin, 410.

NOSSAL, R., and CHEN, S.H., (1972), J. Phys. 33 : C1-171.

NOSSAL, R., CHEN, S.H., and LAI C.C., (1971), Opt. Comm. 4 : 35.

PFEFFER, W., (1883) Ber. Dtsch. Bot. Ges., 1 : 524.

PURCELL, E.M., (1976), in Physics and our World, A.I.P., New York, 49.

ROBERTS, A.M., (1981), in Biochemistry and Physiology of Protozoa, Vol. 4, Academic Press, New York, 5.

SCHAEFFER, D.W., BANKS, G., and ALPERT S.S., (1974), Nature, 268: 162.

SEGEL, L.A., (1978), in Studies in Mathematical Biology, Vol. 15, Part I, Mathematical Association of America, Washington, 170.

STOCK, G.B., (1976), Biophys. J., 16 : 535.

STOCK, G.B., (1978), Biophys. J., 16 : 79.

STOCK, G.B., and CARLSON F.D., (1975) in Swimming and Flying in Nature, Plenum Press, New York, 57.

TSANG, N., McNAB, R.M., and KOSHLAND, D.E., (1973) Science 181 : 60.

VOLOCHINE, B. (1982), in the present volume.

WANG, P.C., (1982), Ph. D. Thesis, M.I.T.

WANG, P.C., and CHEN, S.H., (1981), Biophys. J. 36 : 203.

WOOLFORD, M.W., (1980) Ph.D. Thesis, University of Waikato.

WOOLFORD, M.W., and HARVEY, J.D., (1982), Biophys. J.

CHEMOTAXIS AND BAND FORMATION OF ESCHERICHIA COLI STUDIED BY LIGHT SCATTERING

Paul C. Wang and Sow-Hsin Chen

Nuclear Engineering Department
Massachusetts Institute of Technology
Cambridge, Massachusetts 02139

I. INTRODUCTION TO CHEMOTAXIS OF BACTERIA

The phenomenon that certain motile bacteria move toward chemicals that aid their survival and away from chemicals that are harmful, is called "chemotaxis". For instance, bacteria are attracted by nutritious sugars and amino acids, and repelled by phenol and their own excretory products such as acids and alcohols. The macroscopic aspects of bacterial chemotaxis have been known ever since the end of the 19th century, through the work of Engelmann (1) and Pfeffer (2). In 1966, Adler (3) revived the study using modern microbiology techniques. Subsequently, Berg and Brown (4) introduced the tracking microscope, and MacNab and Koshland (5) developed the temporal gradient apparatus. These efforts combined with other genetic analyses provide a detailed microscopic picture of bacterial chemotaxis. Escherichia coli and Salmonella typhimurium were commonly chosen for study because vast knowledge of their biochemistry and genetics exists. We have chosen E. coli (wild type K12) as a model system because its light scattering properties have been extensively investigated by our group during the last ten years (6,7,8).

The motions of E. coli K12 wild type bacteria can be characterized by two states: either a "run" state in which a bacterium moves in a fairly straight path or a "twiddle" state in which a bacterium jigs around locally. After twiddle, a bacterium makes a run in a new randomly chosen direction. This run-twiddle motion repeats sequentially. In a homogeneous medium the twiddle occurs with a constant frequency. However, in a case of chemotaxis the random walk of a bacterium is biased toward the direction of increasing nutrient by lowering frequency of twiddle. A bacterium tends to run longer and twiddle less in a favorable direction (9,10,11). As a result there is a net displacement toward the direction of increasing nutrient.

E. coli cells are about 1 μm in diameter and 2 μm long. They propel themselves at as much as 20 μm per second by means of four to eight flagella. Flagella are thin helical appendages some 7 μm long that protrude at various places on the cell surface. The flagella form a bundle during the process of swimming. They rotate counterclockwise like a propeller. Reversal of flagellar rotation causes the bundle to disperse, separating the flagella, and then causes asymmetric pushing by the dispersed flagella on the bacterial cell body which forces the bacterium to turn or twiddle. Return to smooth swimming occurs on reassembly of the flagellar bundle.

In order to receive a signal from the outside environment, bacteria must have appropriate receptors. The receptors so far identified have all been shown to be protein molecules. Approximately 20 different kinds of attractant receptors and 10 different kinds of repellent receptors have been identified in E. coli. There are more than 5000 receptor molecules of each kind distributed over the bacterial surface. Each receptor is specifically sensitive to only one or two particular chemicals. The receptors all appear to be either in the periplasmic space or on the inner membrane. The bacterium integrates information gathered by receptors over time. That is to say it constantly compares the change of the fraction of receptors that form complexes with the chemoeffective chemicals within a certain interval of time. This information is used to adjust the twiddle frequency through the signal processing system. By suppression of twiddle, the bacterium effectively runs toward the favorable direction.

These microscopic biochemical processes control the run-twiddle motion of every single bacterium. The collective effect of these individual bacterial motions is to form a travelling chemotactic band. In a motility buffer (10^{-2}M KH_2PO_4, 10^{-6}M L-methionie and pH 6.9) saturated with oxygen, E. coli, which are initially confined in a finite region, will consume oxygen locally. After the local oxygen has been depleted, they will move out towards a region of higher oxygen concentration. Due to the oxygen consumption and diffusion of substrate oxygen, a travelling oxygen concentration gradient is formed. On one side oxygen is completely depleted and on the other it is saturated. Bacteria will follow this travelling oxygen concentration gradient collectively and form a chemotactic band. Bacteria located in different portions of the band sense different chemical concentrations in the environment. Study of the average motion of bacteria in different portions of the band can provide a means to test different proposed chemotaxis models.

The microscopic bacterial motion in a chemotactic band can be described by four dynamic parameters: τ_1 the average duration of twiddle motion, τ_2 the average duration of run motion, V_2 the run speed, D the equivalent diffusion constant of twiddle motion. The small angle laser light scattering technique has been successfully used to extract all these four parameters simultaneously using data taken in a very short interval of time (2 min.) in our laboratory (6, 7,8). Bacterial density in a chemotactic band is typically as high as 10^8/ml. At this high density, the traditional stroboscopic photomicrograph (5) and the tracking microscopic technique (4) cannot be applied. Macroscopically the bacterial migration in a chemotactic band can be described by two quantities, bacteria density and nutrient concentration as functions of position and time. These two quantities are coupled through two governing differential equations. The relationship between the macroscopic migration and microscopic motions of the bacteria can be articulated through the present theory. All the parameters in the two theories are determined simultaneously by the laser light scattering experiment. The study of the kinetics of bacterial motions in a chemotactic band is complete.

II. BAND FORMATION

Interaction between bacterial density and substrate concentration gives rise to a macroscopic manifestation of chemotaxis. A phenomenological mathematical model was first proposed by Keller and Segal (KS) (12) to describe this macroscopic aspect of chemotaxis. There are two continuity equations expressing the conservation of bacteria density b (z,t) and substrate concentration C (z,t).

$$\frac{\partial b(z,t)}{\partial t} = -\frac{\partial}{\partial z}\left[-\mu\frac{\partial b(z,t)}{\partial z} + V_c(c)b(z,t)\right] \tag{1}$$

$$\frac{\partial C(z,t)}{\partial t} = -\frac{\partial}{\partial z}\left[-D\frac{\partial C(z,t)}{\partial z}\right] - k(c)b(z,t) \tag{2}$$

For simplicity an initial condition can be chosen in such a way that the model equations are one-dimensional. Bacteria initially are layered on the bottom of a cuvette in which there is a buffer with a constant distribution of substrate concentration. As bacteria start to consume substrate they will create a substrate concentration gradient. Part of the bacteria will follow this self-created substrate concentration gradient and move out.

The first term in the bracket on the right hand side of Eq. (1) expresses a diffusion current due to the random motion. This can be taken as a macroscopic manifestation of the microscopic random walk. μ is the motility coefficient. The second term is the effective chemotactic current. V_c is a net drift velocity. If df/dz is defined as an indicator of the chemotactic signal, then we can postulate a linear relationship:

$$V_c = \delta\frac{df}{dz} \tag{3}$$

δ is the chemotactic coefficient. Based on some empirical observations of sensation, Keller and Segal (12) proposed that f equals $\ln C/C_{th}$, where C_{th} is the threshold concentration in which bacteria start to have chemotactic responses. V_c is then equal to $(\delta/C)\ \partial C/\partial z$. From the considerations of equilibrium binding and dissociation of the protein receptors and the chemoeffectors, Mesibov et al. (13) derived an expression for f, by equating it to the fraction of bound receptors, i.e., $f = C/(C + k_d)$, where k_d is the dissociation constant of receptor - chemoeffector complex. V_c in this model is,

$$V_c = \frac{\delta k_d C}{(C + k_d)^2}\left(\frac{\partial C}{C\partial z}\right) \cdot \tag{4}$$

This model, based upon the concept of dissociation of receptor-chemoeffector complex, is called the "modified" model. These two different models lead to predictions of different initial drift velocities and steady state bacterial density distributions. In the substrate conservation equation, Eq. (2), D is the diffusion coefficient of the substrate in the buffer, the k(C) is the number of substrate molecules consumed per second per bacterium.

Since Keller and Segal first proposed the phenomenological model of bacteria chemotaxis, many researchers (7, 14) have tried to solve these two governing equations (Eq. (1) and (2)) either numerically or analytically under some special assumptions. Most of these efforts are focused on the solutions applicable to a propagating band with a constant migration speed. In order to have a complete picture of the development of a chemotactic band, we also study the initial state of the problem in this paper. We solve Eq. (1) and (2) theoretically for the initial state case and also carry out light scanning experiments to monitor the development of a band. Many interesting features which relate the theory to the experimental results about chemotaxis are obtained.

Solving the initial part of the band formation analytically, we assume that the bacteria density does not change appreciably during the time interval in which we are interested. The density is constant from the bottom of the cuvette to the top of the inoculant. Then we solve the substrate concentration and its derivatives as functions of positions at different times. From the spatial derivatives of substrate concentration, chemotactic velocities can be calculated, provided that the fraction of the bacteria which move out from the original location is known. If we consider the inoculant region and the rest of the cuvette as two separate regions, then this problem can be formulated as follows:

in inoculant region I: $$\frac{\partial C_I}{\partial t} = D\frac{\partial^2 C_I}{\partial z^2} - kb_0 \quad (5)$$

in the buffer region II: $$\frac{\partial C_{II}}{\partial t} = D\frac{\partial^2 C_{II}}{\partial z^2} \quad (6)$$

b_0 is the original bacteria density. Boundary conditions are

at $z = 0$, $$\frac{\partial C_I}{\partial z} = 0 \quad (7)$$

$z = z_0$, $$C_I(z_0,t) = C_{II}(z_0,t) \quad (8)$$

$$\frac{\partial C_I(z,t)}{\partial z} = \frac{\partial C_{II}(z,t)}{\partial z} \quad (9)$$

and $z = \infty$, $$C_{II}(z,t) = C_0 \quad (10)$$

The initial condition is

$$t = 0,\ C(z,t) = C_0 \quad (11)$$

Eq. (7) means the cuvette is closed at the bottom. Eq. (10) means the buffer extends to infinity. The initial substrate concentration is constant, C_0, throughout the cuvette. Eq. (8) and (9) show that substrate concentration and its derivative are continuous at the boundary z_0. Eq. (5) and (6) are solved rigorously by a Laplace transformation method (15). The solutions are

$$\tilde{C}_I = 1 - \tilde{t} + \frac{1}{2}\Big[(\tilde{t} + \frac{(1-\tilde{z})^2}{2\tilde{D}})\mathrm{Erfc}(\frac{(1-\tilde{z})}{2\sqrt{\tilde{D}\tilde{t}}}) + (\tilde{t} + \frac{(1+\tilde{z})^2}{2\tilde{D}})$$

$$\mathrm{Erfc}(\frac{(1+\tilde{z})}{2\sqrt{\tilde{D}\tilde{t}}}) - \sqrt{\frac{\tilde{t}}{\pi\tilde{D}}}(1-\tilde{z})\exp(-\frac{(1-\tilde{z})^2}{4\tilde{D}\tilde{t}}) - \sqrt{\frac{\tilde{t}}{\pi\tilde{D}}}(1+\tilde{z})\exp(-\frac{(1+\tilde{z})^2}{4\tilde{D}\tilde{t}})\Big] \quad (12)$$

$$\tilde{C}_{II} = 1 - \frac{1}{2}\Big[(\tilde{t} + \frac{1}{2}\frac{(\tilde{z}-1)^2}{\tilde{D}}\ \mathrm{Erfc}(\frac{(\tilde{z}-1)}{2\sqrt{\tilde{D}\tilde{t}}}) - (\tilde{t} + \frac{(\tilde{z}+1)^2}{2\tilde{D}})$$

$$\mathrm{Erfc}(\frac{(\tilde{z}+1)}{2\sqrt{\tilde{D}\tilde{t}}}) - \sqrt{\frac{\tilde{t}}{\pi\tilde{D}}}(\tilde{z}-1)\exp(-\frac{(\tilde{z}-1)^2}{4\tilde{D}\tilde{t}}) + \sqrt{\frac{\tilde{t}}{\pi\tilde{D}}}(\tilde{z}+1)\exp(-\frac{(\tilde{z}+1)^2}{4\tilde{D}\tilde{t}})\Big] \quad (13)$$

$$\frac{d\tilde{C}_I}{d\tilde{z}} = \sqrt{\frac{\tilde{t}}{\pi\tilde{D}}}\ \Big[\exp(-\frac{(1-\tilde{z})^2}{4\tilde{D}\tilde{t}}) - \exp(-\frac{(1+\tilde{z})^2}{4\tilde{D}\tilde{t}}) +$$

$$\frac{1}{2\tilde{D}}\left[(1+\tilde{z})\mathrm{Erfc}\left(\frac{1+\tilde{z}}{2\sqrt{\tilde{D}\tilde{t}}}\right) - (1-\tilde{z})\mathrm{Erfc}\left(\frac{1-\tilde{z}}{2\sqrt{\tilde{D}\tilde{t}}}\right)\right] \tag{14}$$

$$\frac{d\tilde{C}_{II}}{d\tilde{z}} = \sqrt{\frac{\tilde{t}}{\pi\tilde{D}}}\left[\exp\left(-\frac{(\tilde{z}-1)^2}{4\tilde{D}\tilde{t}}\right) - \exp\left(-\frac{(\tilde{z}+1)^2}{4\tilde{D}\tilde{t}}\right) + \right.$$

$$\left.\frac{1}{2\tilde{D}}\left[(\tilde{z}+1)\mathrm{Erfc}\left(\frac{\tilde{z}+1}{2\sqrt{\tilde{D}\tilde{t}}}\right) - (\tilde{z}-1)\mathrm{Erfc}\left(\frac{\tilde{z}-1}{2\sqrt{\tilde{D}\tilde{t}}}\right)\right]\right. \tag{15}$$

These solutions are presented in a dimensionless form in order to be compared with any experimental results. These dimensionless variables are

$$\tilde{z} = \frac{z}{z_0}, \quad \tilde{C}_{I,II} = \frac{C_{I,II}}{C_0}, \quad \tilde{t} = \frac{t}{\tau}, \quad \tilde{D} = \frac{D\tau}{z_0^2}, \quad \tau^{-1} = \frac{kb_0}{C_0} \tag{16}$$

z_0 is the height of the inoculant, 0.25 cm is used in this computation. C_0 is the saturated substrate concentration. For oxygen in water at room temperature it is 2×10^{-4}M. D is the oxygen diffusion constant which is taken to be 2×10^{-5}cm^2/sec. Bacteria density, b_0, is 10^8/ml. τ is 1400 sec. These dimensionless solutions of $\tilde{C}$ as functions of $\tilde{z}$ are also presented in graphic form in Fig. 1. Fig. 2 is the chemotactic speed, V_c, derived from the KS model. Fig. 3 is that derived from the modified model. $\overline{\delta}$ is taken to be 2 in calculation. These graphs may be used to find out the theoretical chemotactic speed for a given experimental condition. For example, we generally find that 10% of the bacteria from the inoculant form a band. We can therefore draw a line at $\tilde{z} = 0.9$ and the intersections would give the chemotactic speeds at different time. Suppose we experimentally observe that the band formed 2 min. ($\tilde{t} = 0.09$) after inoculation, then the intersections of the $\tilde{z} = 0.9$ line with the V_c curves for $t < 2$ min. would give us the minimum chemotactic speed, V_c, required for the bacteria to move out and form a band. If t is taken to be 1 min. ($\tilde{t} = 0.045$), the intersections are 0.5 in Fig. 2 and 0.12 in Fig. 3. After conversion, the corresponding chemotactic speeds are 1.0 μm/sec. for the KS model, and 0.24 μm/sec. for the modified model. A comparison between these theoretical calculated values with experimental measured V_c provides a means of testing the credibility of the different models.

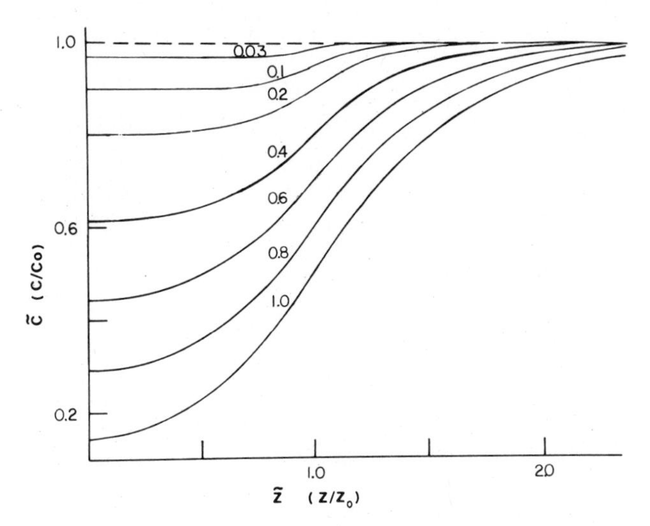

Fig. 1. Theoretical curves of dimensionless substrate concentration, $\tilde{C}$, computed by Eq. 12 and 13, as function of positions at different time. The inoculant height z_0 is 0.25 cm. THe saturated substrate concentration C_0 is 2×10^{-4} M. The numbers on each curve are the dimensionless time. Diffusion constant D is taken to be 2×10^5 cm^2/sec. b_0, the bacteria density, is 10^8/ml.

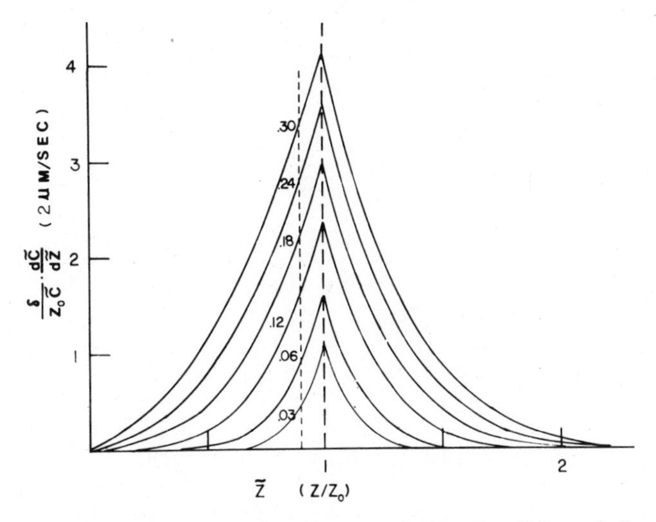

Fig. 2. Chemotactic speed V_c derived from the KS model according to $V_c = \delta df/d\tilde{z}$, where $f = \ln \tilde{C}/\tilde{C}_{th}$. Theoretical values of V_c can be read out directly if the fraction of the bacteria moving out is known. For example, if 10% of bacteria move out to form a band then we draw a $\tilde{z} = 0.9$ dash line, the cross point gives $V_c = 1.0$ μm/sec. at $\tilde{t} = 0.045$.

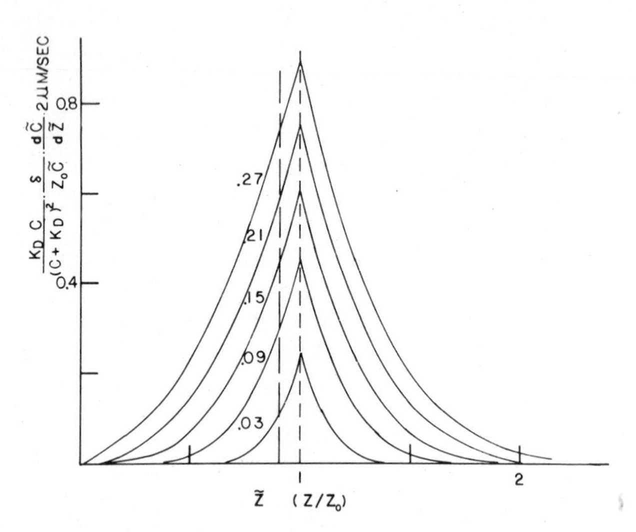

Fig. 3. Chemotactic speed V_c derived from the modified model according to Eq. 4. If 10% of bacteria move out to form a band, then the corresponding V_c is 0.24 μm/sec. at $\tilde{t}$ = 0.045 which is 1 min. after inoculation in real time.

III. BAND PROPAGATION

When L-serine is added to a motility buffer which is saturated with oxygen, two chemotactic bands may be formed. Whether E. coli will form one or two bands depends on the relative concentrations of oxygen and L-serine (see Fig. 4 and 5). E. coli can consume L-serine either aerobically or anaerobically. The saturated oxygen concentration at room temperature is 2×10^{-4}M. If the L-serine concentration is more than 5×10^{-3}M, only one band is formed. This is an oxygen limited band. The bacteria consume all the oxygen and part of the L-serine and follow the gradient up to the top of the solution where the oxygen concentration is high. Since there is too much L-serine in the buffer, with the number of bacteria inoculated, the bacteria can never consume all the L-serine locally. This means that the diffusion rate of L-serine is faster than the bacterial L-serine consumption rate. If the L-serine concentration is between 7×10^{-4}M to 5×10^{-3}M, two bands are formed and the second band (denoted as the L band) is an L-serine limited band. In the first band, bacteria use up all the

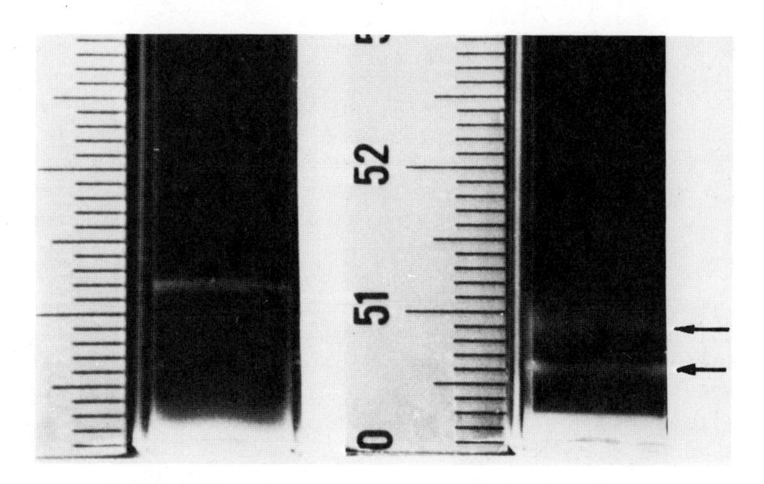

Fig. 4. Two examples of chemotactic bands in serine solution. Figure on the right was taken 1800 seconds after the inoculation in a solution containing 8×10^{-4} M of serine. The top band is a combined serine-oxygen band and the bottom one is a pure serine band. Figure on the left was taken 3300 seconds after the inoculation in a solution containing 3×10^{-4} M serine. There is only one band because the serine concentration is approximately equal to the equilibrium oxygen concentration $C_0 = 2 \times 10^{-4}$ M.

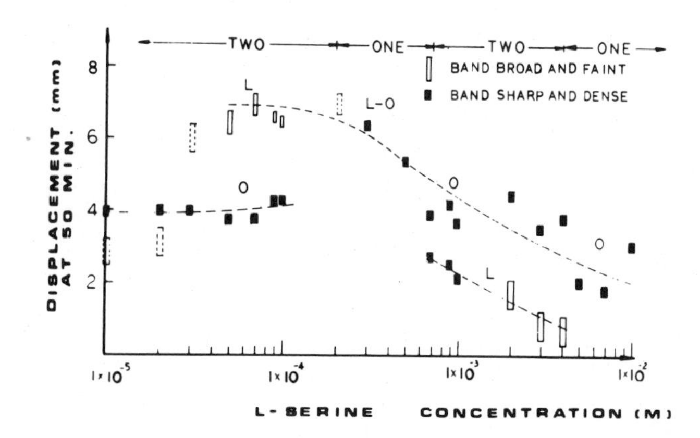

Fig. 5. Positions of chemotactic bands in oxygen saturated serine solutions 50 minutes after inoculation. L stands for a serine band and 0 stands for an oxygen band. It is worth noting that between serine concentrations of 3×10^{-4} - 6×10^{-4} M there is only one serine-oxygen band. The dash line blocks represent the top positions of the bands which had diffused away at the time the data was taken.

oxygen and part of the L-serine. This first band follows the combined oxygen and L-serine gradients upwards. Bacteria behind the first band use L-serine anaerobically and create a L-serine concentration gradient. This second band follows the L-serine concentration gradient. If the L-serine concentration is between 3×10^{-4}M and 6×10^{-4}M, there is only one band. This is bacause bacteria use up all of the L-serine and oxygen simultaneously. For the cases where the L-serine concentration is less than 1×10^{-4}M, there are two bands again. In these cases the first band is L-serine-limited and the second band is oxygen-limited.

A similar light scanning setup as Holz's (7) is used to map out the band profiles as functions of time. When a migrating chemotactic band passes through a stationary laser beam, the intensity of 90° scattered light is proportional to the bacterial density in the scattered volume. The scattered light is detected by a photomultiplier. The output of the photomultiplier is amplified and shaped into logic pulses which are then fed to the input of a multichannel analyzer (MCA). By repeatedly recording the scattered light intensities from the MCA, the chemotactic band profiles as functions of time are then obtained. Positions of the band peaks as functions of time are recorded from a sequence of band profiles. Migration velocity is then computed. From the profiles, we can also find out how many bacteria move out to form a band. The lowest point in a profile is taken to be the boundary of a band. The bacteria density is calculated from the initial inoculant density and its corresponding scattered light intensity. The area under the peak of a profile gives the total number of bacteria in the band. All of our experiments show that about 9 to 13% of bacteria from the inoculant move out to form a band. The number of bacteria in the band is constant for a long time and gradually decreases because of the bacterial ageing effect. At that time, the band migration slows down.

E. coli K12 wild type bacteria are grown in a L-broth to mid-log phase at 37°C shaking scheme. The bacteria density is 5×10^8/ml. A scattering cuvette (cross section 3 x 3 mm^2) is first rinsed and filled with a motility buffer which contains 8×10^{-4}M L-serine. 70 μl of the culture medium is carefully layered at the bottom of a cuvette. The inoculant height is about 8 mm. The starting point of scanning is adjusted to be 3 mm below the top of the inoculant. This assures that in every scan the band as well as some part of the original inoculant will be scanned through. By doing so, the band positions can have a reference point and the bacteria density of the inoculant can be monitored. Fig. 6 shows the band profiles in this above case. Two bands are formed. The first one is oxygen-limited and the second is L-serine-limited. The second band is formed about 11 minutes after inoculation. We can see the bacteria density decreasing as time goes on. This leaving of bacteria behind makes the process of generating a chemoattractant concentration gradient harder simply because total substrate consumption rate is lower for

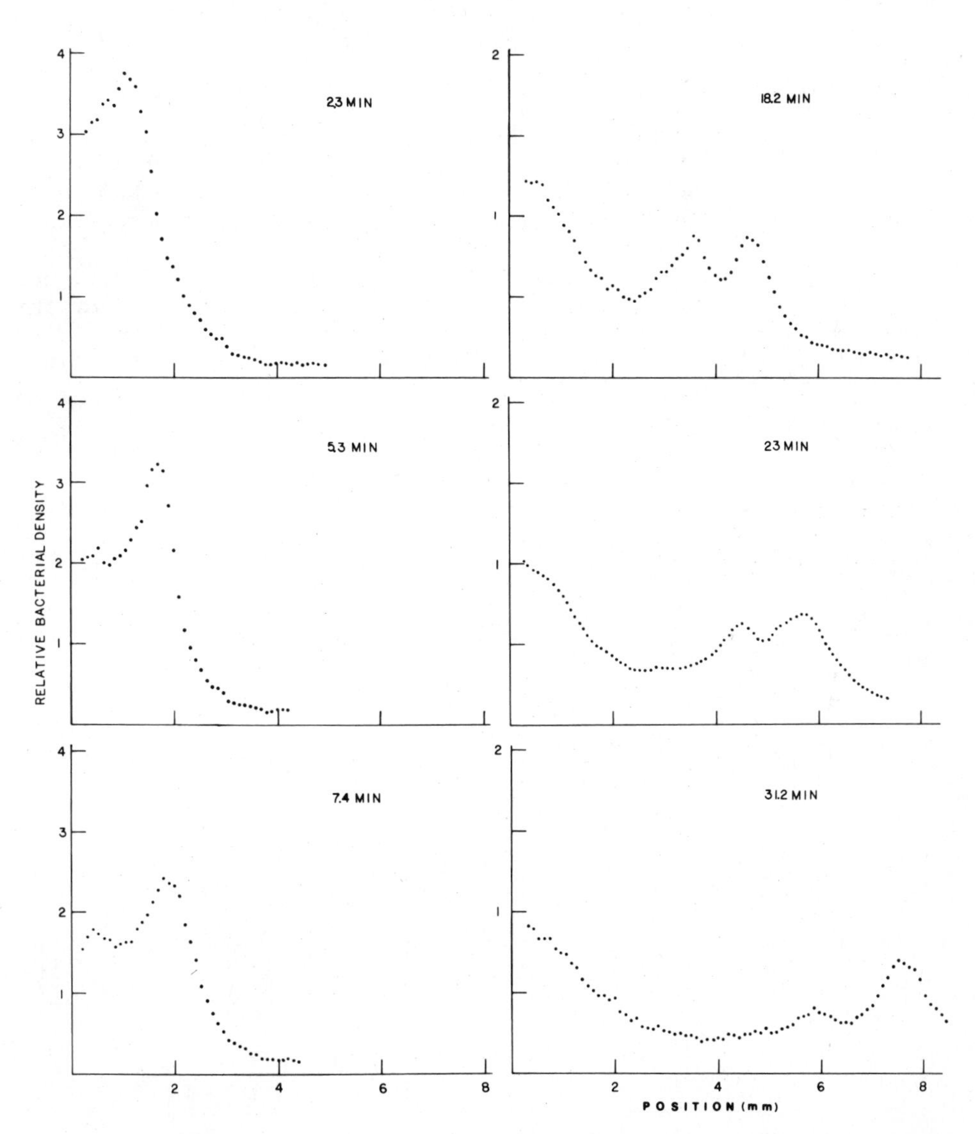

Fig. 6. Formation of two bands. Additional 8×10^{-4} M L-serine is put in the buffer as attractant. The second band, which is serine limited, is formed 10-15 minutes after inoculation. The relative low bacteria densities for these bands at later stage show that the bands become weaker gradually.

fewer bacteria. The shallower concentration gradient makes bacteria less chemotactic; in response they disperse. Fig. 7 is the peak positions of both the oxygen and serine-limited bands. The migration speed of the oxygen-limited band is 3.4 μm/sec. and for the serine-limited band it is 2.9 μm/sec.

The motility coefficient μ and chemotactic coefficient δ are the two most important macroscopic quantities describing chemotaxis. In the oxygen case, these coefficients have been extracted by Holz and Chen (7) using a moment analysis method. Using the same method we carefully analyzed the second band in 8×10^{-4}M L-serine case. This is a pure L-serine-limited band. Thus the δ and μ values for E. coli in serine are obtained. The fifth picture in Fig. 6 is used for this analysis. Holz and Chen first solve Eq. (1) and (2) numerically for different combinations of the $\overline{\delta}$ and γ values. $\overline{\delta}$ is defined as δ/μ and γ is μ/D. They get a theoretical distribution function $b(x)/b_{max}$ for every set of $\overline{\delta}$ and γ. The moments of these theoretical distribution functions are compared with the moments of the experimental bacteria distribution function, and $\overline{\delta}$ and μ are thus extracted. The values we get for this serine case are $\overline{\delta} = 2.5$ and $\mu = 5.5 \times 10^{-6} cm^2/sec$.

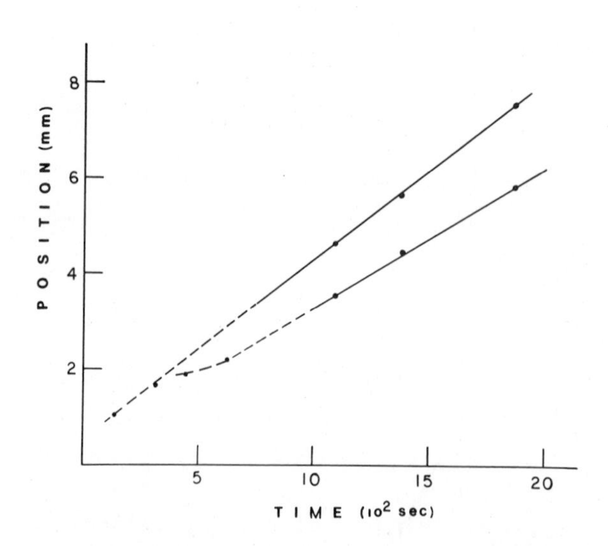

Fig. 7. The peak positions of the oxygen and serine bands from Fig. 6. The top line indicates positions of the oxygen band at various times. The migration speed is 3.4 μm/sec. The second line indicates the positions of the serine band at various times. Its migration speed is 2.9 μm/sec.

IV. PHOTON CORRELATION MEASUREMENTS OF ANISOTROPIC MOTIONS

The dynamic parameters of the microscopic bacterial motion in a chemotactic band can be extracted by a small angle laser light scattering technique (8,16). In this section, we present a modified run-twiddle two-step model. This model describes bacterial preferential motions towards an attractant. Experimentally, an improved chopping measurement method is introduced. This chopping scheme can measure the photon correlation function in both parallel and perpendicular directions simultaneously.

Theory

In the photon correlation technique (17) the measured photo-count correlation function is reducible to the self-correlation function

$$F_s(q,t) = \langle e^{iq\cdot[R(t)-R(0)]}\rangle \tag{17}$$

where R(t) and R(0) are the position vectors of a typical bacterium at time t, and 0, respectively. Time 0 is some arbitrarily chosen beginning time. q is the magnitude of the scattering vector given in terms of the wavelength of laser λ, the index of refraction of medium n, and the scattering angle θ by $(4\pi/\lambda)\ n \sin(\theta/2)$. For a scattering angle $\theta \sim 3°$, $q \sim 0.52 \times 10^4 cm^{-1}$, and the typical dimension of E. coli as $d \sim 1 \times 10^{-4} cm$, we have $qd < 1$. For data taken at this small scattering angle we can safely treat an E. coli as a point particle and model only its center of mass motion.

In a homogeneous medium where motions are isotropic, $F_s(q,t)$ depends only on the magnitude of $\vec{q}$. On the other hand, when motions are not isotropic, such as in the case of bacteria in a chemotactic band, $F_s(q,t)$ depends on the direction of the $\vec{q}$ vector. Let us take the direction of the oxygen gradient (vertical direction) as z-direction and the direction perpendicular to it as x-direction. Then we have

$$F_{\parallel}(q,t) = \langle e^{iq[z(t)-z(0)]}\rangle \tag{18}$$

$$F_{\perp}(q,t) = \langle e^{iq[x(t)-x(0)]}\rangle \tag{19}$$

Since the origin of time is arbitrary, we shall call $Z(t) = z(t) - z(0)$, $X(t) = x(t) - x(0)$.

A. Free Motion

For bacteria in the run state, we can set $Z(t) = v_z t$ for the motion in z-direction and thus

$$F_{\parallel}(q,t) = \langle e^{iqv_z t}\rangle = \langle e^{iqvt\mu}\rangle = \int_0^{\infty}\int_{-1}^{1}\int_0^{2\pi} e^{iqvt\mu}\, p(v)p(\mu)p(\varphi)v^2 dv\, d\mu\, d\varphi \tag{20}$$

Where $\mu = \cos\theta$. It has been shown by a direct optical tracking of E. coli (19) in an isotropic medium that the speed distribution is closely approximated by a Maxwell distribution

$$p(v) = (2\pi V_2^2)^{-3/2}\exp(-v^2/2V_2^2),\ p(\varphi) = p(\mu) = 1 \tag{21}$$

Substituting Eq. (21) into Eq. (20) we get

$$F_{\parallel}(q,t) = \exp\ (-q^2V_2^2\ t^2/2) \tag{22}$$

We can also identify the average speed $<v> = (8/\pi)^{\frac{1}{2}}V_2$, drift velocity $<v_z> = 0$ and $<v_z^2> = V_2^2$. With a simple integration $F_{\perp}(q,t)$ can also be written in the same form as Eq. (22) with the same $<v>$, $<v_x>$ and $<v_x{}^2>$ values.

In a chemotactic situation, we assume that the speed distribution $p(v)$ and the azimuthal distribution $p(\varphi)$ are the same as in the isotropic case. However the angular distribution $P(\mu)$ has to be modified as $p(\mu) = 1 + \Delta\mu$, where Δ is a small dimensionless quantity.

It can easily be shown, using this new velocity distribution, that $<V_x> = <V_y> = 0$, $<V_z> = \Delta<v>/3$ and $<V_x{}^2> = <V_y{}^2> = <V_z{}^2> = V_2^2$. These results show that although bacteria maintain the same rms speed in the isotropic case, they now have a net drift velocity, $\Delta<v>/3$, in the z-direction. One can estimate the dimensionless quantity, Δ, by equating the experimental band migration speed, $\overline{V}$, to $<v_z>$,

$$\overline{V} = \frac{1}{3}\ \Delta\ <v>. \tag{23}$$

Since $\overline{V} \sim 1$ µm/sec., $<v> \sim 20$ µ/sec., Eq. (23) gives $\Delta \sim 3/20 \sim 0.15$. The correlation function can then be worked out to be

$$F_{\parallel}\ (q,t) = \exp\ [-\ q^2V_2^2t^2/2] + i\ \Delta\ <j_1(qvt)> \tag{24}$$

$$F_{\perp}\ (q,t) = \exp\ [-\ q^2V_2^2t^2/2], \tag{25}$$

which show that the correlation function in the perpendicular case is the same as in isotropic case, but the parallel case there is a small correction due to band drifting. $j_1(qvt)$ is a spherical Bessel function of the first kind.

In a homodyne case one measures

$$|F_{\parallel}(q,t)|^2 = \exp\ [-\ q^2V_2^2t^2] + \Delta^2|<j_1qVt>|^2 \tag{26}$$

While in a heterodyne case, one measures

$$\text{Re}F_{\parallel}(q,t) = \exp\ [-\ q^2V_2^2t^2/2] \tag{27}$$

In this model, the effects of the band drift due to chemotaxis is included. This is an improvement compared with our previous model. This modification gives an additional term $\Delta^2|\langle j_1(qvt)\rangle|^2$ in homodyne measurements. However, it does not enter into a heterodyne measurement because in this case only the real part of the self correlation functions counts. In our E. coli aerotaxis band migration experiment, with drift velocity $\langle v_z\rangle$ = 0.8 μm/sec., V_2 = 14 μm/sec., the correction is only about 1%. Fig. 8 shows the relative importance of this correction term.

This free motion approximation is good when the mean free path of the free motion L_2 is larger than q^{-1}. For example, a typical run L_2 may be 20 μm and this criterion is amply satisfied for q values at all attainable scattering angles.

B. Short Step Motion: Gaussian Approximation (6)

E. coli in the twiddle state can be modeled as a series of small step motions, as far as the center of mass motion is concerned. When the change of direction is so rapid that $qL_1 < 1$ then the motion can be modeled as a random walk of step length L_1. This is a Gaussian

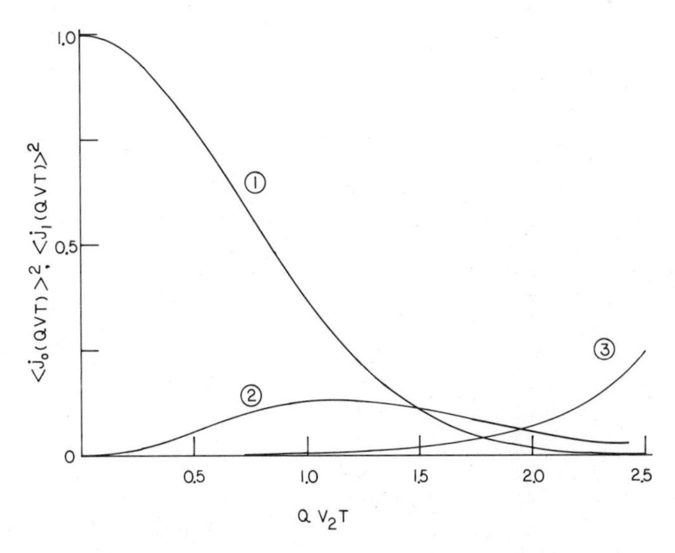

Fig. 8. Two contributions to a measured homodyne photon correlation function from the real and imaginary parts of a self-correlation function based on a modified run-twiddle model. qV_2t is in a dimensionless scale. Curve 1 is from the real part of Eq. (24). Curve 2 is $\langle j_1(qvt)\rangle^2$. Curve 3 is the relative importance of these two terms. α is taken to be 0.15.

random process and a well-known theorem (20) states

$$F(q,t) = \langle e^{iqZ(t)} \rangle = \exp(-\, q^2 \langle Z^2(t) \rangle /2) = \exp(-\, q^2 D_E t) \quad (28)$$

where D_E is the equivalent diffusion constant of E. coli in a motility buffer.

The total self-correlation function can be written as (8)

$$F_s(q,t) = \frac{\tau_1}{\tau_1+\tau_2} P_{run}(t)\, F_{run}(q,t) + \frac{\tau_2}{\tau_1+\tau_2} P_{tw}(t) F_{tw}(q,t)$$

$$+ \frac{2}{\tau_1+\tau_2} \int_0^t dt_1 P_{run}(t-t_1) P_{tw}(t_1) F_{run}(q,t-t_1) F_{tw}(q,t_1)$$

$$+ \text{higher order terms} \quad (29)$$

where F_{run} and F_{tw} are the correlation functions of Free and Short Step Motion. τ_1 and τ_2 are the average run and twiddle times. The third term is the contribution from the bacteria which make one transition, either from run to twiddle or from twiddle to run, during the period of observation. If the observation time t (the time scale of the correlation function measured) is small compared with both τ_1 and τ_2, then during observation the majority of bacteria can make at most only one transition. We therefore safely neglect the higher order terms.

Experiment

The basic experiment technique is the same as in our previous studies (8) except that a chopping apparatus is introduced. Fig. 9 shows a modified small angle spectrometer. The scattered light passes through a pinhole on a mask, and is focussed upon a photocathode in the photomultiplier. The scattering vector $\vec{q}$ is determined by the pinhole position d and distance L_1. With the geometric configuration indicated in Fig. 9 the scattering angle can be chosen to be 2.5°, 3.0° and 3.5° depending on which pinhole is used. The scattering angle is adjustable by changing L_1 and L_2. A mask is driven to move up and down by a motor. It opens up two pinholes, one of which is in a parallel, or z, direction, and the other of which is in perpendicular, or x, direction, alternatively. These up and down motions are synchronized with the start function of MCA to cause the measured photon correlation functions in these two directions to be addressed to different quarters on the MCA. Each quarter has 256 channels. This synchronization insures that the measured scattered light is from the same scattered volume at the same time in both directions. This process removes the ambiguity which arose from previous study of whether the scattered light is from the same region of the chemotactic band. The methods of signal processing, computer control, data storing on cassette tape and

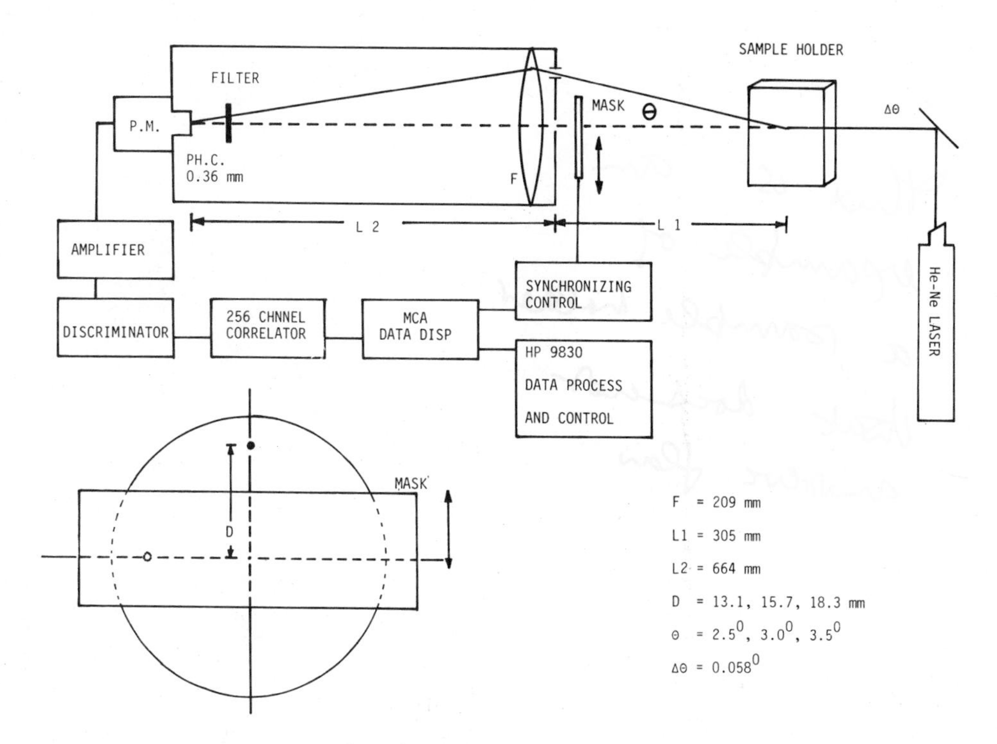

Fig. 9. Horizontal-Vertical small angle spectrometer. Laser light (6328Å) is scattered by the bacteria in the scattering cuvette, through a pinhole on the mask, then it is focused upon a photocathode in the photomultiplier. The scattering angle can be chosen as 2.5°, 3.0° and 3.5° in this configuration. The movements of the motor-driven mask are synchronized with MCA recording so that the correlation functions in both directions will be recorded into two different quarters in MCA.

extracting self correlation functions are all the same as in our previous studies (8).

The photon correlation functions of both oxygen and serine bands in perpendicular and parallel directions are measured in the 8×10^{-4}M L-serine experiment. The experiment conditions are described in section III. Two correlation functions (see Fig. 10) for the center of the serine band in both directions are carefully analyzed.

In this series of experiments there is always some heterodyne signals mixed with homodyne signals being measured. The normalized

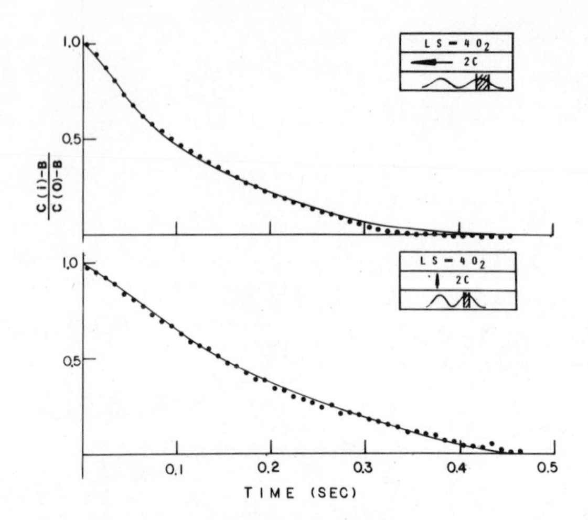

Fig. 10. Normalized scattered light intensity correlation functions taken from the center of a serine band. The L-serine concentration 8×10^{-4} M is four times that of the saturated oxygen concentration. The solid lines are the fitted curves according to Eq. (30). The fitted results are in table I.

measured photon correlation functions can be written as (21)

$$\frac{\langle n(0)n(t)\rangle - B}{\langle n(0)n(0)\rangle - B} = \alpha f(A)\mathrm{Re}F_s + (1-\alpha)f'(A)|F_s|^2 \tag{30}$$

where $\alpha \equiv \dfrac{2\langle n_L\rangle\langle n_s\rangle}{2\langle n_L\rangle\langle n_s\rangle + \langle n_s\rangle^2}$.

n(t) is the photon counts at time t. $\langle n_s\rangle$ and $\langle n_L\rangle$ stand for scattered and unscattered light, respectively. f(A) and f'(A) are coherent factors which depend on the optical geometry of the set up. F_s is the total correlation function as shown in Eq. (29), which contains both real and imaginary parts. The microscopic parameters of bacterial motions extracted by this method are listed in Table I. This table shows that in a direction parallel to that of the travelling band, bacteria tend to run for a longer time, that is relatively less bacteria are in the twiddle state (smaller β). The equivalent diffusion constants, D_E, for twiddle motions are the same in these two directions. V_2 is the width of the Gaussian distribution of

Table I

Motional Parameters of E. coli (K12, Wild type) in a L-serine gradient obtained by the run-twiddle analysis.

Position	τ_1(s)	τ_2(s)	$\beta(\tau_1/(\tau_1+\tau_2))$	$D_E(\mu m^2/s)$	$V_2(\mu m/s)$
Parallel Center	0.43	0.60	0.42	18	10
Perpendicular Center	0.43	0.17	0.72	18	14

velocity. Smaller V_2 in the parallel case means it has a narrow range of velocity distribution and $\langle v \rangle$ is smaller also. However, because of this smaller β value there is a chemotactic drift in the parallel direction.

V. RELATION BETWEEN MICROSCOPIC AND MACROSCOPIC PARAMETERS

The macroscopic description of chemotaxis, δ and μ are the two key parameters indicating bacterial motility and chemotactic ability. The most important parameters in microscopic description are τ_1, τ_2 and V_2. They tell us the details of motion of each bacterium. Theoretically, these two sets of parameters are linked together. Experimentally, we can perform scanning and small angle scattering experiments to measure these parameters. The linkage of these parameters makes the whole study complete. We can write a generalized model of chemotaxis as follows:

$$\text{response} = \int_{\text{sensing time}} (\text{signal})\,dt \tag{31}$$

By tracking microscope technique, Berg and Brown (4) have shown that there is a run distance increase, ΔL, in the favorable direction. Therefore, the response can be written as $\Delta L/L$, where L is the run distance in a homogeneous medium. Adler (3) and coworkers argued that the chemotactic response is a function of the change in receptor occupancy over a period of time, df/dt. The sensing period is of the order of run time, τ_2. Eq. (31) then becomes

$$\frac{\Delta L}{L} = \int_0^{\tau_2} \frac{df}{dt}\,dt. \tag{32}$$

ΔL and L can be written in terms of the drift velocity V_c and run speed $\langle v \rangle$ as follows: $V_c\tau_2 = \Delta L$ and $\langle v \rangle \tau_2 = L$. From Eq. (32) we get

$$V_c = \tau_2 \langle v \rangle \overline{\frac{df}{dt}} = \tau_2 \langle v \rangle^2 \overline{\frac{df}{dz}} = \delta \overline{\frac{df}{dz}} \quad (33)$$

where the bar is a time average and $\delta = \tau_2 \langle v \rangle^2$. τ_2 and $\langle v \rangle$ are values derived from measurements taken in the chemoattractant gradient direction. In our case it is the parallel direction. The motility coefficient μ can be obtained from a simple kinetic theory argument,

$$\mu = \frac{1}{3} \langle V \rangle^2 \quad (\tau_1 + \tau_2)/(1 - \overline{\cos\theta}). \quad (34)$$

θ is the average direction change between two runs. $\overline{\cos\theta}$ is taken to be 0.5. Microscopic parameters in Eq. (34) should be taken from an experiment where there is no chemotaxis. We can take data from bacteria in a homogeneous medium or from bacteria in the front part of a band where the gradient is small.

For example, from the data of the serine case, we have $\tau_1 = 0.43$ sec., $\tau_2 = 0.17$ sec. and $\langle v \rangle = (8/\pi)^{\frac{1}{2}} V_2 = 22.3$ μm/sec. in the perpendicular direction. From Eq. (34), we get $\mu = 2.0 \times 10^{-6}$ cm^2/sec. In the parallel direction, $\tau_2 = 0.6$ sec. and $V_2 = 16$ μm/sec., therefore, δ equals 1.5×10^{-6} cm^2/sec. and $\overline{\delta} = 1.33$. From the macroscopic measurement we got $\overline{\delta} = 2.5$ and $\mu = 5.5 \times 10^{-6}$ cm^2/sec. The agreement between the two sets of parameters is fairly good.

In this paper theoretical results have been presented which are the initial-state solutions of KS phenomenological equations for the formation of chemotactic bands. A modified theoretical model for bacterial self-correlation function based on two-state motions has been developed to extract the mean square speed of run motion and relative probability of twiddle vs. run at the centers of the chemotactic bands. This modified model includes the effects of anisotropic bacterial chemotactic motions. Experimentally, a small angle light scattering technique with chopping ability has been successfully used to study detailed bacteria motions in a L-serine-limited band. We have used light scanning and moment analysis to study L-serine-limited band profiles at the steady-state stage. The chemotactic coefficient δ and the motility coefficient μ of bacteria in serine are extracted. The macroscopic and microscopic pictures of chemotaxis are successfully unified by the linkage of δ and μ with τ_1, τ_2 and V_2.

This work is supported by a National Science Foundation grant no. 8111534-PCM.

References

1. T. W. Engelmann, Neue Methode zur Untersuchung der Sauerstoffausscheidung pflauzlicher und thierischer Organismen, Pflugers Arch. Ges. Physiol. 25:285-292 (1881).
2. W. Pfeffer, Uber chemotaktische Bewegungen von Bakterien. Flagellation und Volvocineen, Unters. Botan. Inst. Tubingen. 2:582-663 (1888).
3. J. Adler, Chemotaxis in bacteria, Science (Wash., D.C.) 153:708-716 (1966).
4. H. C. Berg and D. A. Brown, Chemotaxis in Escherischia coli analyzed by three-dimensional tracking, Nature, 239:500-504 (1972).
5. R. M. MacNab and D. E. Koshland, Jr., The gradient-sensing mechanism in bacterial chemotaxis, Proc. Natl. Acad. Sci. USA, 69: 2509-2512 (1972).
6. M. Holz and S. H. Chen, Quasi-elastic light scattering from migration chemotactic bands of Escherichia coli, Biophys. J. 23:15-31 (1978).
7. M. Holz and S. H. Chen, Spatio-Temporal structure of migrating chemotactic band of Escherichia coli. I. Traveling band profile, Biophys. J. 26:243-261 (1979).
8. P. Wang and S. H. Chen, Quasi-elastic light scattering from migrating chemotactic bands of Escherichia coli. II. Analysis of anisotropic bacterial motions, Biophys. J. 36:203-219 (1981).
9. J. Adler, Chemotaxis in bacteria, Annu. Rev. Biochem., 44: 341-356 (1975).
10. H. C. Berg, Chemotaxis in bacteria, Annu. Rev. Biophys. Bioeng., 4:119-136 (1975).
11. D. E. Koshland, Jr. "Bacterial Chemotaxis as a Model Behavioral System " Raven Press, N.Y., N.Y. (1980).
12. G. F. Keller and L. A. Segal, Traveling Bands of Chemotactic Bacteria: A Theoretical Analysis, J. Theor. Biol. 30:235-248 (1971).
13. R. Mesibov, G. W. Ordal, and J. Adler, The range of attractant concentrations bacterial chemotaxis and the threshold and size of response over this range - Weber law and related phenomena, J. Gen. Physiol. 62:203-223 (1973).
14. T. L. Scribner, L. A. Segel, and E. H. Rogers, A numerical study of the formation and propagation of traveling bands of chemotactic bacteria, J. Theor. Biol. 46:189 (1974).
15. H. Bateman, "Tables of integral transforma". Vol. 1, McGraw-Hill, New York,(1954).
16. S. H. Chen and P. C. Wang, Light Scattering Measurement of the Two-State Motional Parameters of Escherichia Coli in Chemotactic Bands, "Biomedical Applications of Laser Light Scattering", Sattelle, D. et al. eds., North Holland, (1981).
17. S. H. Chen, W. B. Veldkamp, and C. C. Lai, Simple digital clipped correlator for photon correlation spectroscopy. Rev. Sci. Instrum. 46:1356 (1975).

18. R. Nossal, S. H. Chen, and C. C. Lai, Use of laser scattering for quantitative determinations of bacterial motility, Opt. Comm. 4:35, (1977).
19. M. Holz, and S. Chen, Tracking bacterial movements using a one-domensional fringe system, Opt. Lett. 2:109, (1978).
20. S. Chandrasekhar, Stochastic problems in physics and astronomy, Rev. Mod. Phys. 15:1, (1943).
21. J. C. Oliver, Correlation techniques, "Photon Correlation and Light Beating Spectroscopy", H. Z. Cummins, and E. R. Pike, ed., Plenum Press, New York (1974).

A COMPARISON OF MODELS USED IN THE ANALYSIS OF QUASI-ELASTIC LIGHT SCATTERING DATA FROM TWO MOTILE SYSTEMS: SPERMATOZOA AND *CHLAMYDOMONAS REINHARDTII*

T. Craig, T. J. Racey and F. R. Hallett

Physics Department
University of Guelph
Guelph, Ontario, Canada

INTRODUCTION

It has been over ten years since it was first indicated that the motility of microorganisms might be quantitatively studied using quasi-elastic light scattering.[1] Two important assumptions were used in those studies; first, that the scatterer was a point particle or at least spherically symmetrical and second, that it moved in straight lines for times long compared with $(qv)^{-1}$. It appears that both of these restrictions are not upheld at the same time in any real motile systems.

We now present some new low angle scattering data from the spherical system *Chlamydomonas Reinhardtii* to look more closely at trajectory effects. We also will reexamine some previous bull spermatozoa scaling data[2] and extend the theoretical predictions of the models used by these authors to the low scattering angle region.

THEORETICAL CALCULATIONS

The low angle scattering calculations on *Chlamydomonas* will be based on the so-called "parallel oscillation" model,[4] a model which certainly violates the straight line motion assumption for low q. In this model the trajectory of the cell is described by

$$r(\tau) = v\tau + A(\sin(\omega\tau + \phi) - \sin\phi) \qquad (1)$$

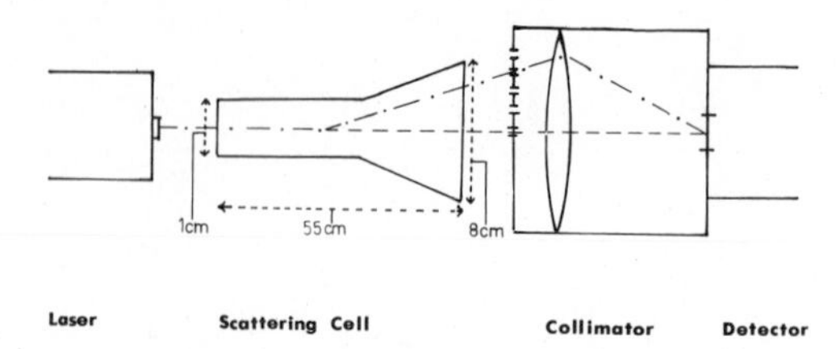

Figure 1. Schematic drawing of the low angle scattering apparatus.

which is essentialy a sine wave superimposed on a ramp function. This trajectory after being averaged over an appropriate speed distribution, yields the following electric field autocorrelation function[4]

$$g^{(1)}(\tau) = X_m \int_1^{-1} d\nu \frac{\left[\left(\frac{4}{v}\right)^4 + (q\nu\tau)^4 - 6\left(\frac{4}{v}\right)^4 (q\nu\tau)2\right]}{\left[\left(\frac{4}{v}\right)^2 + (q\nu\tau)^2\right]^4} J_o\left(2\nu q 4 \sin\left(\frac{\omega}{2}\tau\right)\right), \tag{2}$$

$+ (1 - X_m)$ non-motile function.

The numerical calculations were performed as in that paper.

In the case of the bull spermatozoa the correlation functions were calculated using earlier models[2] and similar numerical methods. Scaling curves generated by this approach show sizeable oscillations at high q due to spermatozoan structure. For this case it is expected that in the low q region deviations from this average high q scaling level would be seen.[5]

EXPERIMENTS

The experimental apparatus and methods for bull spermatozoa studies have been described elsewhere.[2] A new experimental arrangement will be described for the acquisition of the low angle scattering data for the Chlamydomonas.

The low angle scattering spectrometer (Fig. 1) consisted of a horn shaped plastic scattering cell of length 55 cm consisting of a 1 cm diameter optical, glass entrance window and a 8 cm diameter exit window. The light source was a 15 mW Optikon laser (C. W. Radiation, Mountainview CA). The light collecting apparatus

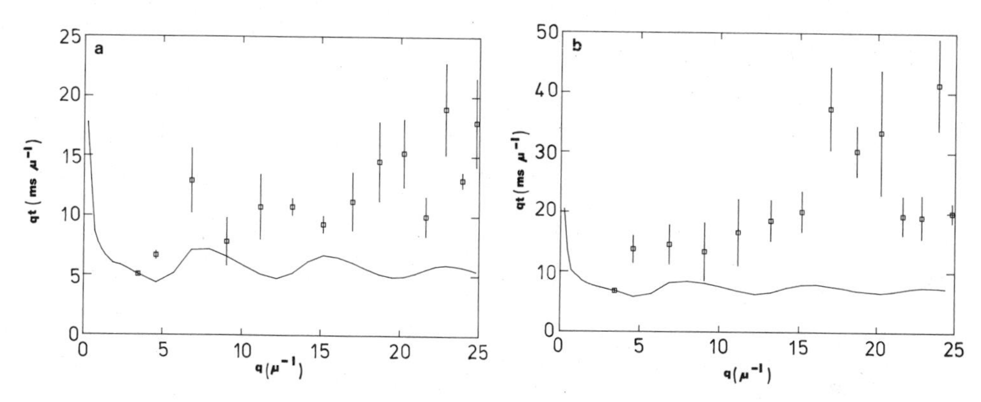

Figure 2. Exptl (□) and theoretical (–) scaling curves for autocorrelation functions from a) normal b) defective bull spermatozoa. The vertical bars indicate the standard error of the mean for the averages shown.

consisted of a series of 400 μm pinholes located on either side of a lens. The lens was used to focus the light emerging from the front pinhole on the face of the photodetector. The second pinhole placed immediately in front of the detector, allowed only the light from the scattering volume to strike the photoactive surfaces, i.e. it defined the depth of field for the system. The scattering angle, θ, after making refractive corrections, depended on the distance from the scattering volume to the front pinhole and the distance from the optical axis to this pinhole. With this arrangement scattering angles of 3.3°, 2.6° and 1.7° were obtainable. These angles were verified using 0.6 μm latex spheres.

RESULTS

A routine sample of bull spermatozoa contains three different types of cells; normally swimming cells which swim along helices, defectively swimming cells which exhibit a planar sinusoidal motion and non-motile cells which sink and rotate slowly as they settle. Figure 2 shows the halfwidths of experimental and theoretical autocorrelation functions obtained from samples of two of these types of cells scaled by the magnitude of the scattering vector q and plotted versus q. Although the curves have average values over q which are quite different (due to different trajectory speeds[3]) the peaks at values of q of 5.0 μm^{-1} or more lie in the same locations. These peaks are due to static structural effects caused by the asymmetry of the cell. They have a spacing of ~7 μm^{-1}. This

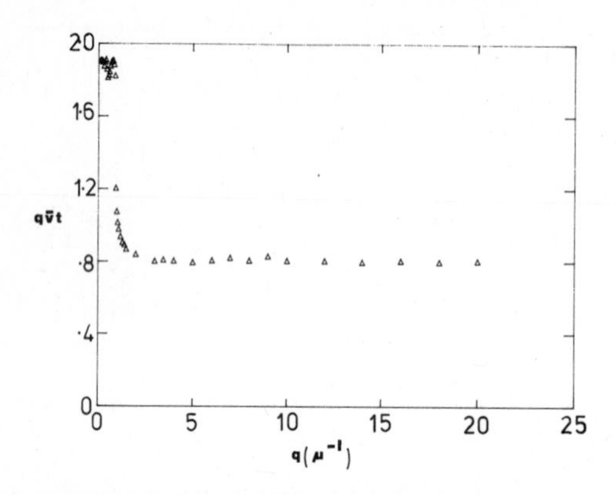

Figure 3. Large q theoretical scaling curve for autocorrelation functions for Chlamydomonas.

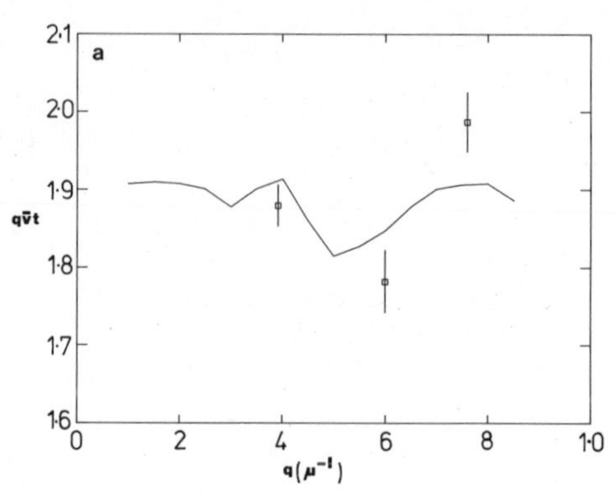

Figure 4. Low q theoretical scaling curves (–) for autocorrelation functions from Chlamydomonas as well as low angle data (□).

indicates that they arise due to a feature of size $(2\pi/q) \sim 1$ μm. Only the coat thickness of the model particle (0.15 μm) or the overall thickness of the cell (1.0 μm) lie in this range. Since these peaks persist if the coat is removed[2] it is likely they are caused primarily by the cell thickness. All other morphological size parameters of the cell are too large to produce such widely spaced peaks. This is consistent with other observations[3] where theoretical scaling curves similar to figure 2 were plotted and ellipsoids with smaller dimensions showed much larger "diffraction" type effects.

At q values less than 5.0 μm^{-1} one sees an overall increase in the so called scaling level due to trajectory effects, i.e. q^{-1} is now of such a magnitude that it samples large pieces of the helical path[6] over which the trajectory exhibits considerable curvature. At very low q the light scattering sees an overall progressive speed which is much smaller than the trajectory speed thereby increasing the halfwidth dramatically.

The situation in the case of Chlamydomonas is much different. Figure 3 shows a theoretical scaling curve for the halfwidths determined using the parallel oscillation model.[4] Here, since the average speed of the cells is well known, the dimensionless parameter $q\bar{v}t$ is plotted versus q. In the higher q range no features are seen in the scaling curves. This would be predicted due to the spherical nature of the cell. This figure was generated using the parameters which best fit the individual autocorrelation functions taken. However, in the lower q range we see as before a

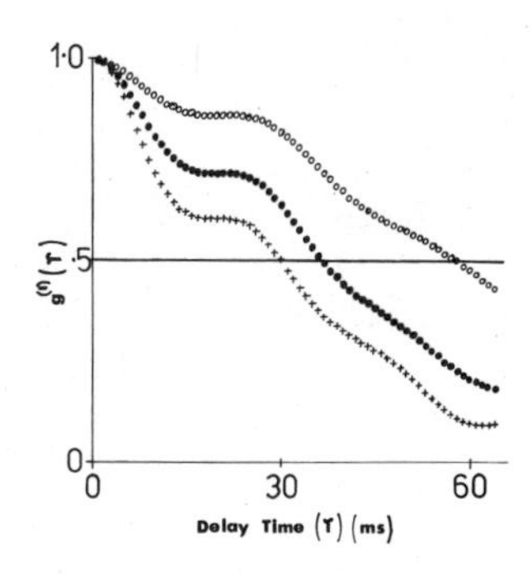

Figure 5. Exptl autocorrelation functions of Chlamydomonas at a) 1.7° (o) b) 2.6° (●) and c) 3.3° (+)

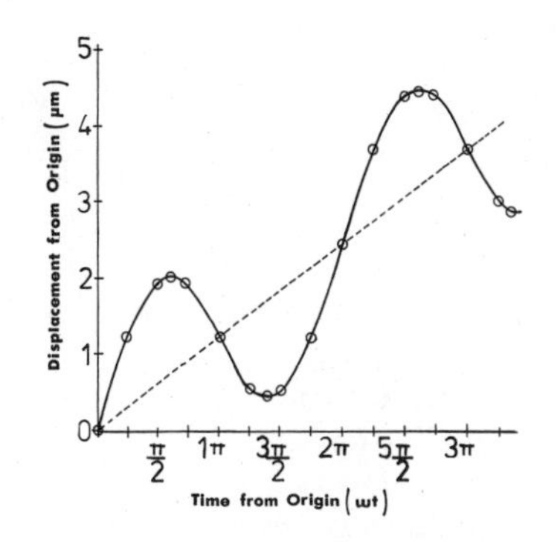

Figure 6. Schematic representation of about two oscillations of the trajectory of a Chlamydomonas cell.

large increase in the scaling level. This is due again to the fact that the cells are no longer travelling in straight lines for times on the order of $(qv)^{-1}$ because of the oscillation in their movement. Figure 4 narrows in on this low q area of the scaling curve. The data points scaled by the appropriate $\bar{v}$(84 μm/s) are shown. These were taken at 1.7°, 2.6° and 3.3° using the low angle scattering apparatus shown previously. The features that should be noticed here are the oscillations in the theoretical and experimental scaling curves.

The individual correlation functions are interesting as well. Figure 5 shows the correlation functions taken at angles of 1.7°, 2.6° and 3.3°. All these functions contain a "plateau" or peak region of non-zero delay times. The plateaus occur in all of the correlation functions at about 23 ms. To understand this plateau figure 6 showing the trajectory of the cell as a function of time must be examined. The cells position should be most highly correlated with the t=0 value when it is again closest to the origin. This occurs at ~11/16 of an oscillation. At the average frequency of 34 Hz this agrees very well with the position of the plateau. This occurs again later in the trajectory and so causes the repetitive pattern seen.

Figure 5 also shows the reason for the oscillations in the scaling curves and also their sizes. The large oscillation in the scaling curve is caused by the fact that at angles of 1.7°, 2.6° and 3.3° one brackets the area in q space from the large plateau to the next in the correlation function, that is, it is examining an area of the cellular oscillation between 11/16 of the first oscillation and 11/16 of the second oscillation. This is also confirmed by the fact that q^{-1} for 1.7^o, 2.6^o and 3.3^o yields values of 1.90, 1.25 and 0.98 fractions of cellular oscillations. This results in the cell being close to the origin, far away and then close again all in relative terms. The smaller oscillation at $q \sim 0.3\ \mu m^{-1}$ results from looking at the next plateau or oscillation in the trajectory of the cell. This however is too far away to be highly correlated i.e. the difference in being "close" to the origin and "far" from the origin is small compared to the absolute distance from the origin. As well, the distance to the first plateau represents the final drop from the progressive speed detection to the trajectory speed detection as noted earlier.

DISCUSSION

Chlamydomonas and bull spermatozoa are two contrasting systems from the point of view of light scattering. *Chlamydomonas* have quite simple morphological characteristics whereas bull spermatozoa are highly asymmetrical leading to the featureless scaling curve in the former case or to a highly structured scaling curve in the latter in the large q regime. On the other hand in the low q regime we are able in the case of the *Chlamydomonas* to see structure in the scaling curve due to the details of its more complicated trajectory. In the intermediate regime ($q \sim 3\ \mu m^{-1}$) we see in both cases the area of q-space of some concern, in that here we no longer have pure trajectory speed or pure translational speed[3] but a mixture of both.

REFERENCES

1. R. Nossal, Spectral analysis of laser light scattered from motile organisms, *Biophys. J.* 11:341-354 (1971).
2. T. Craig and F. R. Hallett, Halfwidth scaling of electric field autocorrelation functions of light scattered from bull spermatozoa, *Biophys. J.* 38:71-78 (1982).
3. T. Craig, F. R. Hallett and S.-H. Chen, Scaling properties of light scattering spectra for particles moving with helical trajectories, *Applied Optics* 21:648-653 (1982).
4. T. J. Racey, F. R. Hallett and B. G. Nickel, A quasi-elastic light scattering and cinematographical investigation of motile *Chlamydomonas Reinhardtii*, *Biophys. J.* 35:557-571 (1981).
5. S.-H. Chen and F. R. Hallett, Determination of motile behavior of prokaryotic and eukaryotic cells by quasi-elastic light scattering, *Quart. Rev. Biophys.* 15:131-222 (1982).

LIGHT-SCATTERING STUDIES OF BIOLOGICAL POPULATIONS AND BIOLOGICAL STRUCTURES

B. Volochine

DPh-G/PSRM, Orme des Merisiers, CEN-SACLAY, 91191 Gif-sur-Yvette Cedex,
Département de Physique, Université Paris V, 45 rue des Sts Pères, 75270 Paris Cedex 06, France

I°) LIGHT-SCATTERING MEASUREMENT OF THE MOTILITY OF HUMAN SPERMATOZOA
- a - Introduction
- b - Method
- c - Dynamical parameters of a given population
- d - Application

II°) HUMAN CERVICAL MUCUS : ENTANGLED MACROMOLECULES OR HYDROGEL ?
- a - Introduction
- b - Theoretical analysis of the two models
- c - Results
- d - Conclusion and prospectives

I°) MOTILITY OF HUMAN SPERMATOZOA

a - Introduction

Among the parameters which are usually studied in human sperm, motility is one which is the most correlated with fertility.

Motility is mostly determined by microscopic observation on several fields of a sperm drop, placed between slide and coverslip at 37°C. The observer makes a global estimation between progressive and reduced motility (i.e. no progressing spermatozoa or poorly motile spermatozoa). The evident subjectivity, the lack of precision and the variability of the results being major disadvantages, many methods of objective sperm motility estimation have been investigated.

Unfortunately most of the proposed techniques cannot be easily applied in practice. The microphotography (Janick and Mac Leod 1970) and the microcinematography (Rothschild 1953 ; van Duijn, van Voorst and Freund 1971) or the "Photokymography" (Castenholz 1974) which bring the best information, require a far too long analysis and in practice allow only the study of a small number of spermatozoa. Electronic analysis devices such as the one proposed by Jecht and Russo in 1973 using 3 TV cameras coupled to a huge computer (the NASA'S) are certainly the best methods which could be imagined, but unfortunately most laboratories cannot afford them.

The method based on optical density changes, caused by sperm set in movement and kept motionless, gives only a motility index (Timourian and Watchmaker 1970) but neither the motility quality nor the velocity distribution are estimated by this method.

In order to discuss the validity and the objectivity of a given method of motility measurement, one must first define the essential requirements to be satisfied by any ideal motility measurement technique.

Thus, in 1973, Jecht and Russo have defined six basic objectives which must be satisfied by any method.

1 - Objective measurement of sperm concentration

2 - Determination of the distribution of total number of moving spermatozoa as a function of their velocities

3 - To find the mean velocity of a given population of spermatozoa

4 - To determine the percentage of spermatozoa with zero velocity

5 - To determine the distribution of total number of moving spermatozoa as a function of the direction of their pathways *and* type of swimming

6 - To avoid altering the environment or including artificial

factors while determining 1 through 5.
To these 6 requirements, we added three more conditions which we consider as essential.

7 - Possibility to perform measurements in a wide range of concentration [from (0.5 to 200).10^6 spermatozoa/ml]

8 - Measurements must be performed in a short time; spermatozoa motility is not a stable parameter

9 - The manipulation of the device must be simple and rapid.

Several years ago, a scattered light technique was proposed by several physicists (Bergé, Volochine, Billard and Hamelin 1967; Boon, Nossal and Chen 1974) in order to evaluate the motility of living micro-organisms and of spermatazoa. We decided to adapt this technique to the specific problem of human sperm.

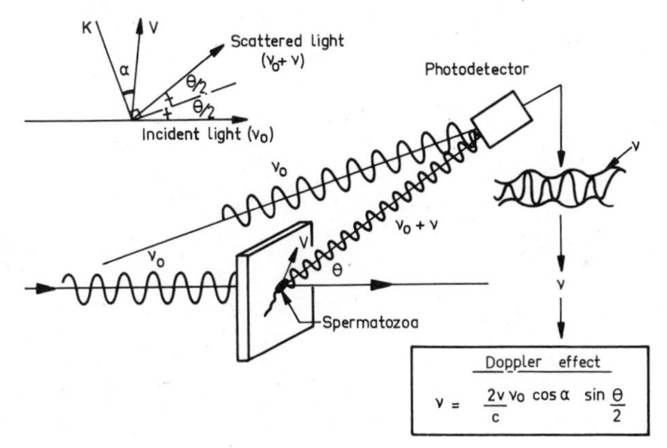

Fig. 1 - Doppler effect : principle of detection.

b - Method

Fig.1 gives the principle of this method. If a beam of monochromatic light falls on a population of moving spermatozoa, by Doppler effect, the intensity of the scattered light is spread in frequencies giving a continuous spectrum depending on dynamic parameters characterizing the motility of the population.

In order to detect and analyse this scattered light, the heterodyne photon beating technique is used (Forrester 1961). The principle of this technique consists of mixing on the photocathode of the photodetector the light to be analysed with a great quantity of the incident light at frequency ν_0. Thus, there appear in the photocurrent of the photodetector beats, giving for each velocity V and for each direction α of the velocity a modulation of the photocurrent at the corresponding frequency ν_s. The spectrum of the scattered light is the sum of all these components. The photocurrent is analysed by a real-time wave analyzer which gives directly the spectrum on a X-Y recorder, either in a semi-log system of coordinates or in a linear system of coordinates.

Fig.2 shows the block diagram of the experimental device which was adapted in its definitive form to the study of spermatozoa by Bergé and Dubois (1973).

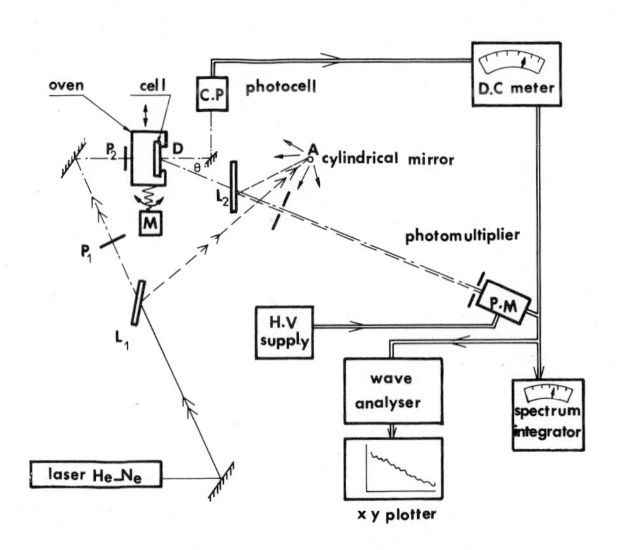

Fig. 2 - Block diagram of experimental apparatus.

Fig.3 shows an example of a real spectrum obtained by this method, curve (a) is the spectrum of scattered light in a semi-log plot ; curve (b) represents the same spectrum in a linear plot. The shot noise represented here in a semi-log plot is the random photon noise of the laser. In order to use the spectrum (a) one must first subtract from it this noise.

c - Dynamical parameters of a given sample

1 - Concentration : The scattered intensity is given by the equation :

$$I_s = C \cdot N_T + I_P$$

where N_T represents the concentration of spermatozoa, I_P is the scattered intensity (background) due to all other elements which are in the sperm (seminal plasma, parasite cells, crystals, etc..) and the coefficient C is related to the scattering cross section of a spermatozoon by the relation :

$$C = C_1 \ f(\theta) \ P_o$$

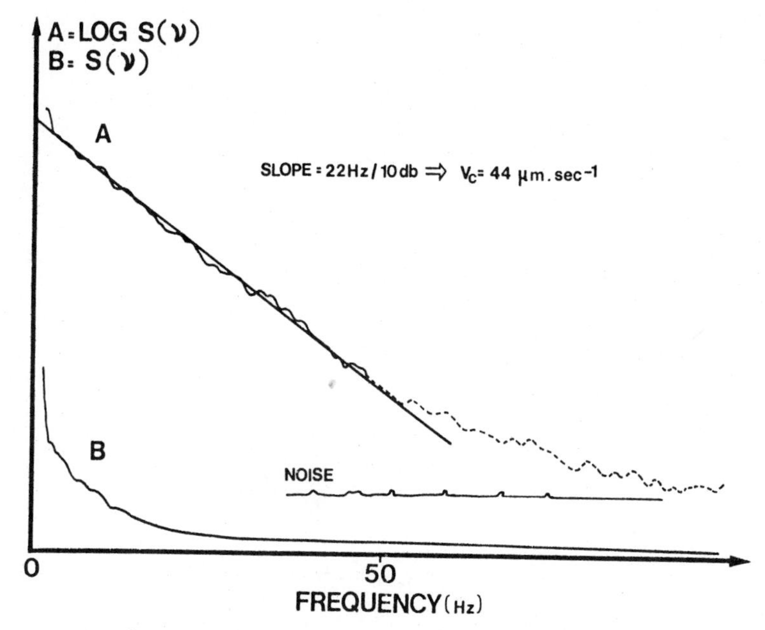

Fig. 3 - Example of a frequency spectrum.

where C_1 is a parameter taking in account the scattering cross-section and geometrical parameters of the experimental device, θ is the scattering angle, P_o is the input power. C is determined experimentally. If P_o and θ are fixed, $I_s - I_p$ is then proportional to N_T and the slope gives the value of C.

2 - Velocity

The spectrum is given by the equation :

$$S(\nu) = N_m \int_{v=\frac{\nu}{k}}^{\infty} \frac{p(v)}{kv} \, dv \qquad (1)$$

where p (v) is the probability density of the velocity v in the studied population, N_m is the number of motile spermatozoa and k the wave vector :

$$k = \frac{2\nu_o}{C} \sin \frac{\theta}{2}$$

The normalization of p(v) implies :

$$\int_{v=o}^{v=\infty} p(v)\,dv = 1$$

From (1) we have : $\frac{dS(\nu)}{d\nu} \div \frac{p(v)}{kv} \Rightarrow p(v) \div \nu \frac{dS(\nu)}{d\nu}$ with $v \leftrightarrows \frac{\nu}{k}$.

The experimental spectra, in semi-log plot give straight lines; this implies that the velocity distribution is given by the relation:

$$p(v) = \frac{1}{V_c^2} v \exp\left(-\frac{v}{V_c}\right)$$

where V_c is the modal velocity of the population (the most represented velocity in the population).

V_c is obtained from the slope β of the spectrum in semi-log plot:

$$V_c = \frac{1}{\beta k} .$$

3 - Number of motile spermatozoa : N_m

The integral of the spectrum corresponding to motile spermatozoa, obtained by a pass-band frequency filter, is given by :

$$S_m = \int_{\nu_{min}}^{\nu_{max}} S(\nu)d\nu \approx S_m(\nu)\Delta\nu \div N_m$$

hence : $N_m = \frac{S_m}{hI_{Lo}C_1 f(\theta)P_o}$ where h is the heterodyne efficiency and I_{Lo} is the intensity of the local oscillator.

4 - Percentage of motile spermatozoa (Fig.4)

This parameter is determined by the ratio between the integral S_m of the spectrum due to motile spermatozoa and S_T the integral of the total spectrum which is obtained, when the sample is set in motion perpendicular to the incident beam. So the same Doppler effect is induced in the unmotile spermatozoa and to the motile ones, and the whole spectrum is shifted inside the pass-band frequency filter.

$$\% = \frac{S_m}{S_T} = \frac{N_m}{N_T} .$$

d - Application to immunology of human reproduction

One of the numerous applications of this method is the study of the kinetics of immobilization of antisperm antibodies on human spermatozoa.

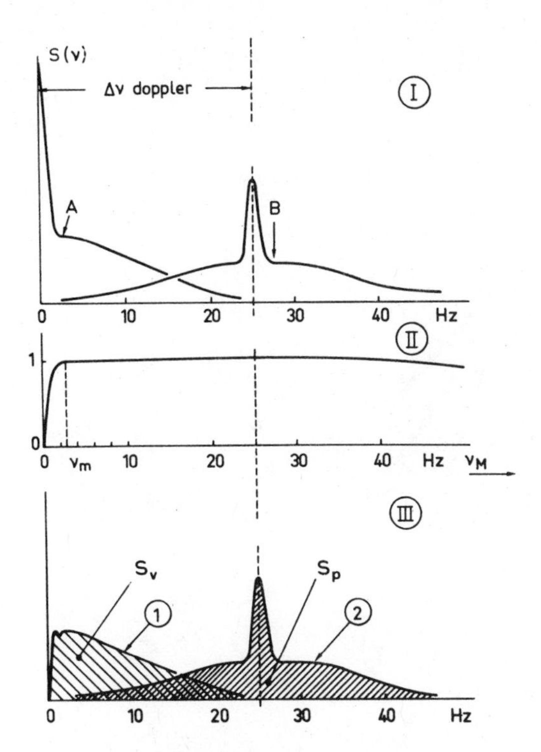

Fig. 4 - Principle of the measurement of the percentage of motile spermatozoa. I. A spectrum : the cell being motionless ; B spectrum : the cell being in motion. II. Response of the pass-band filter. III. Measured integrals.

Fig.5 shows in a linear plot the time evolution of the characteristic velocity of a given population in presence of antibodies and complement. We plot here the ratio between the characteristic velocity measured at time t to the characteristic velocity of the check sample (the same population without antibodies).

Fig.6 gives the same results in a semi-log plot. The obtained straight lines show that the normalized characteristic velocity decreases exponentially with time. This result gives the possibility to characterize objectively the *activity* of a given serum at a given dilution by the characteristic decay-time τ of the characteristic velocity.

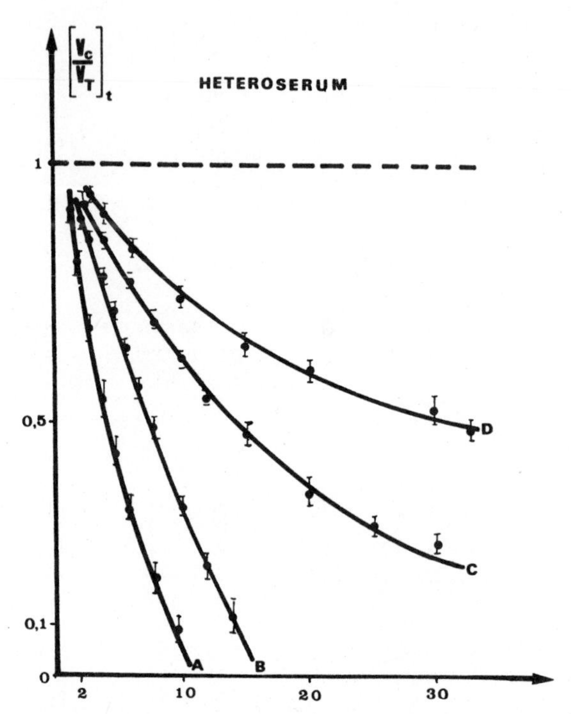

Fig. 5 - Time dependence of the ratio $(V_C/V_T)_t$. Each experimental point is the mean value of 7 experiments performed on 7 different semen. The curves A,B,C,D correspond respectively to different dilutions of the heteroserum : 1/1, 1/2, 1/4, 1/8 at room temperature (25°C).

Fig.7 shows, for a dilution of 1/32, the time evolution of the velocity distribution of a given sample.

Fig.8 shows (same experimental conditions) the time evolution of the velocity distribution for the same spermatazoa after washing (i.e. separated from their seminal plasma).

These results give new possibilities for describing the mechanism of action of antibodies on human spermatozoa, fundamental to biology of reproduction.

The development of this technique was supported by Grant n° 74-7-0846 from DGRST.

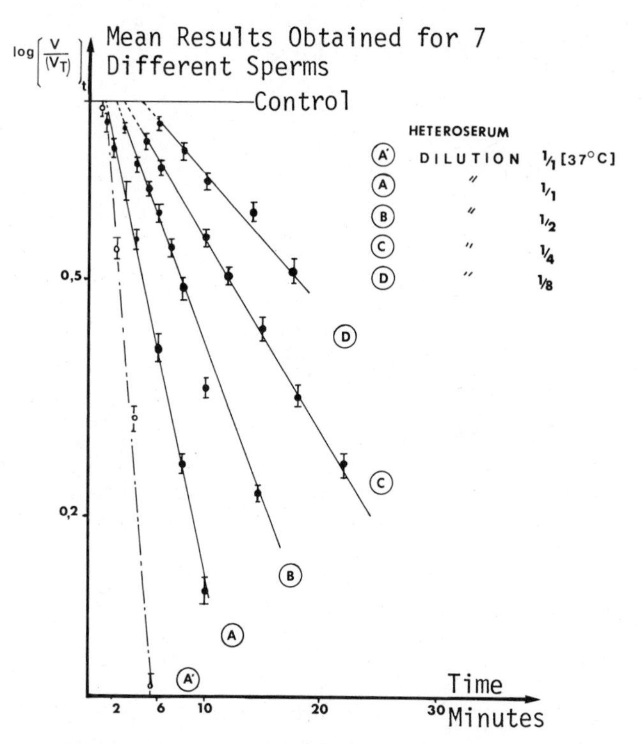

Fig. 6 - Time dependence of the ratio $(V_c/V_T)_t$ in a semi-Log plot. The dotted line A' corresponds to a dilution of 1/1, but at 37°C instead of 25°C (line A).

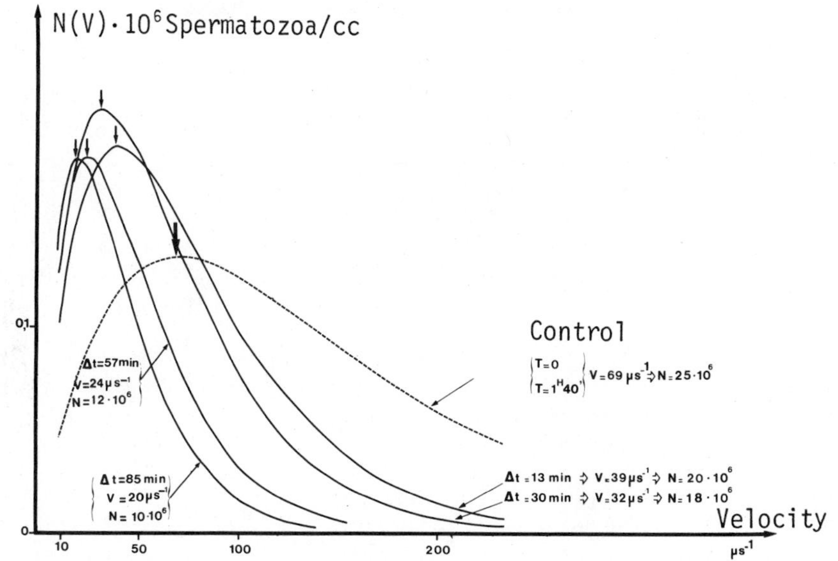

Fig. 7 - Time dependence of the velocity distribution for washed spermatozoa (dilution 1/32).

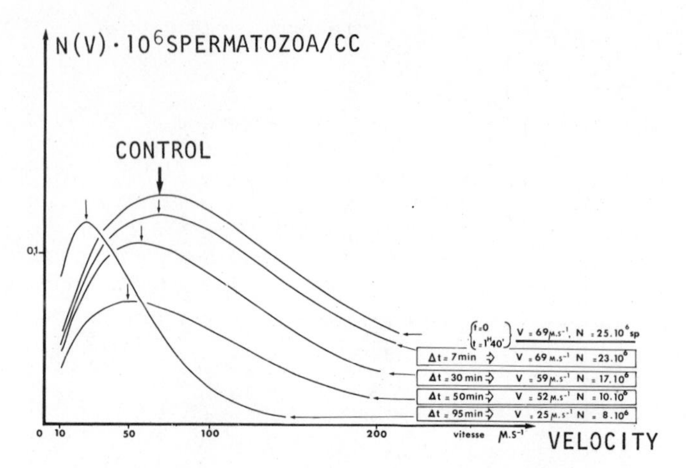

Fig. 8 - Time dependence of the velocity distribution for the same unwashed population (dilution 1/32). Compare this figure with the last one.

II°) HUMAN CERVICAL MUCUS (ENTANGLED MACROMOLECULES OR HYDROGEL ?)

a - Introduction

The secretion produced in the upper part of the cervix by the columnar secretory cells constitutes not only a channel for the male gametes but it also plays, amongst other functions, the role of a cyclic barrier against the spermatozoa.

It has long been known that the so-called "cervical mucus" is a favorable environment for the migration of sperm only around the time of ovulation, when conception is possible.

In addition to many biochemical variations, drastic physical changes in rheological properties which are known to occur cyclically in the mucus under hormonal influence appear to be responsible for the hostility of the mucus at other times. Indeed, the parallelism between the penetrability of cervical mucus by spermatozoa and its 'spinnability' suggests that its obstructive effect may be based on a mechanism that modifies it macromolecular structure.

A theoretical spatial organization of the cervical mucus infrastructure has been proposed by Odeblad (1968) on the basis of nuclear magnetic resonance (NMR) studies : according to the data obtained by this technique, the glycoprotein solid phase is organized as a network of fibrils linked together by oblique or transverse bonds, in a "tricot-like macromolecular arrangement".

At mid cycle, the glycoprotein chains, which constitute the framework of the mucoïd hydrogel, organize themselves into fibrils, thus forming a meshwork large enough to allow spermatozoa to pass through.

Outside the ovulatory period, however, the glycoprotein chains would, according to Odeblad's theory, have a greater autonomy and would rarely be associated as fibrillar micelles ; the number of cross-linkages would be sufficient to ensure the arrangement of a much denser network, thus opposing a most effective barrier to sperm cells.

Odeblad also suggested that these "micelles" can be looked upon as harmonic oscillators and that sperm cells probably swim into the intermicellar spaces, on average 3 μm apart, being assisted in their forward progress by "thermal modulations of the medium" which cause the cavities to expand and contract more or less rhythmically (Odeblad 1962).

The use of freeze-drying and critical point-drying procedures permits us to observe cervical mucus by scanning electron microscopy (Chrétien 1975). The cervical secretion after de-hydratation appears to be composed of a framework of interlacing filaments forming a network whose meshes vary in size at different phases of the menstrual cycle (Chrétien et al. 1973, 1976).

Visible changes occurring in the rheological properties of the mucus throughout the menstrual cycle, appear to be closely related to variations in the length of the glycoproteic filaments which, in turn, cause the meshes to enlarge or to shrink (Chrétien and Psychoyos 1975, Chrétien et al. 1976). Fig.9.

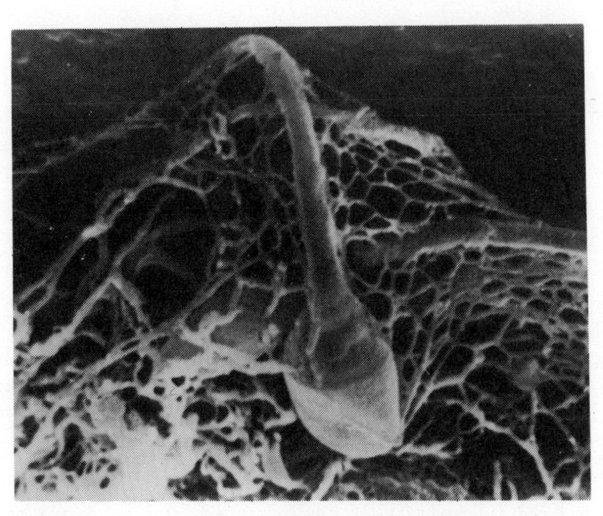

Fig. 9 - Example of human cervical mucus in non-ovulatory period. The magnification is equal to 11000.

These data were corroborated by other authors, using the same technique (Zaneveld et al. 1975, Jordan and Allen 1977, Gould et al. 1976). Pictures obtained by scanning electron microscopy correspond roughly to Odeblad's structural model which consequently appears extremely plausible.

However, a new model was recently proposed according to which cervical mucus is composed of "an ensemble of entangled random-coiled macromolecules" rather than a fibrillar macromolecular network. So, we decided to perform a systematic study of this medium by light scattering techniques. It has been shown (Yeh and Cummins 1964, Ford and Benedek 1965) that the relevant parameter in light scattering experiments is the scattering vector q :

$$q = \frac{4\pi}{\lambda_o} n \sin \frac{\theta}{2} \qquad (1)$$

where λ_o is the vacuum wavelength of the incident beam, n the refractive index of the medium and θ the scattering angle. From a physical point of view q^{-1} is the characteristic scale inside which the medium is "observed".

In our experiments, this length scale q^{-1} is to be compared with the mean size L of the meshes.

Equation (1) shows that by changing the scattering angle θ we can change the product qL - see table I

Table I

θ (degrees)	q cm^{-1}	q^{-1} cm	q^{-1} μm	Lμm	qL
5	1.42×10^4	7×10^{-5}	0.7	1	1.4
180	3.25×10^5	3×10^{-6}	0.03	1	33

Light scattering experiments allow study of the dynamical behaviour of this medium at very different length scales, from $qL \gg 1$ (local analysis) to $qL \sim 1$. It is important to know what kind of fluctuations one can study by light scattering techniques in this medium, at frequencies below 10 kHz.

1) From a physical point of view, we consider cervical mucus to be *a hydrogel* with non Newtonian mechanical properties (Odeblad 1968, Eliezer 1974) whose solid phase is composed mainly of weakly linked fibrillar glycoproteins. In fact, there are two types of fluctuation in such a medium.

a) Motions of the network itself ; according to the value of qL we can "see" local modes in scale with the size of the meshes, then more or less global modes and collective modes.

b) "Structural relaxation" of the medium : the bonds of weak energy existing in a hydrogel are not fixed, they can change with time. This evolution is defined by a distribution of characteristic life-times T_R of the local configuration. These times T_R are relaxation times.

2) In a medium composed of *entangled macromolecules*, which is a system very similar to hydrogel, there are also fluctuations of the two types previously described and a distribution of characteristic life-times T_R of the local configuration. But now T_R are regarded as times of "creep" of a macromolecule through the meshes. ("Reptation time". Adam et al. 1977 (1),(2), 1979. De Gennes 1976).

b - Theoretical analysis of the two models

1) Hydrogel with weak linkages.

. Each structural relaxation mode decays exponentially with time. The characteristic decay times τ_i do not depend on q (Ostrowsky 1973).

. The motions of the network give rise to overdamped modes due to the large friction forces between network and liquid phase. They also decay exponentially with time.

a) $qL \ll 1 (\theta \ll 5°)$. We will consider only the collective mode leading to a single exponential with decay rate $1/Dq^2$ where $D = G/f$ is the diffusion coefficient, G is the elastic modulus of the network, f the friction coefficient between the network and the liquid phase. (Tanaka et al. 1973, Geissler and Hecht 1976, Hecht and Geissler 1978, Candau et al. 1979).

b) $qL \gtrsim 1$ - This situation will give a superposition of modes with decay rates proportional to q^{-2}. The resulting temporal decay will be rather intricate (Geissler and Hecht 1976).

2) Entangled macromolecules

In this case, the changes in the local configuration arise from displacements of macromolecules themselves. Hence, evolution of configuration and movements of the network are not really distinct phenomena. Therefore, *all* decay times are *q dependent*.

a) $qL \ll 1$ "pseudo-gel domain" (Adam et al. 1977). As expected, one can observe a single exponential decay with rate $1/Dq^2$, as above.

b) $qL \gtrsim 1$ "internal motions domain" (Adam and Delsanti a, b 1977). Theory predicts non exponential decay with decay rate proportional to q^{-3} (Dubois-Violette and De Gennes 1967, De Gennes 1976) in good agreement with experimental results (Adam and Delsanti a, b 1977).

From a mechanical point of view, these two media have very similar behaviours, depending on t, the time of measurement (Ferry 1970).

- if $t \gg T_R$, a *viscous-like* response is obtained : structural rearrangements are dominant.

- if $t \ll T_R$, an *elastic* response is obtained. The order of magnitude of T_R is given (Geisser and Hecht 1977) by the relation $T_R \sim \frac{\eta}{G}$, where η is the macroscopic viscosity and G the elastic modulus of the medium. For cervical mucus, T_R is thus roughly equal to 0.1 s, according to data given by Meeyer et al. (1975).

By contrast, *light scattering* may allow one to decide which model must be chosen in order to describe cervical mucus taking into account the observed q-dependence of the experimental results. But, as in the mechanical studies, the response of the system will appear either viscous or elastic, according to the studied domain of times (or frequencies).

3) Materials and methods

a) Experimental device

The experimental set-up (fig.10) described in previous papers (Bergé and Dubois 1973, Dubois et al. 1975, Volochine and Bosq-Rolland 1979, Cazabat et al. 1979) is a classical two beam heterodyne device with variable scattering angle and adjustable heterodyne efficiency. The analysis of the scattered light is performed either by a real time wave analyzer (Spectral Dynamics) or by an analog PAR correlator. The cell containing the sample is placed in a thermostated holder, the temperature of which can be varied between ~ 20°C and 40°C.

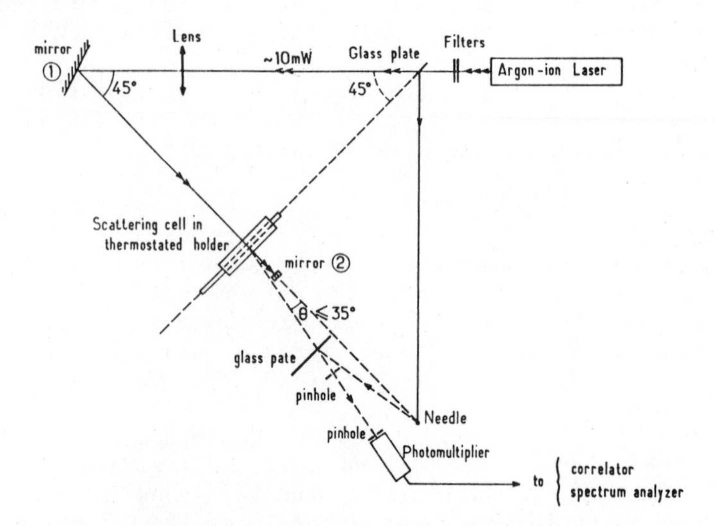

Fig.10 - Experimental device (heterodyne technique).

b) Biological material

The secretion of cervical mucus is known to be at its most intense throughout the ovulatory phase during which the mucus plays its main physiological roles. Moreover, at this time, the glycoprotein network is in a maximum state of looseness and the spatial structure thereof in an optimum state of organization. We have accordingly selected such ovulatory mucus for this physical study. The material was collected from 22 cyclic women, apparently healthy and gynaecologically symptom free. Their ages ranged from 24 to 35 years and 9 donors were undergoing ethinyl-oestradiol or Humegon therapy in order to improve their cervical factor. At least 2 samples per donor were taken. Cervical mucus was then drawn very carefully in a rectangular glass pipette (approximately 15 cm length) with inner square section (1 × 1 mm). A total of 47 samples were studied.

c - Results

All the experimental results presented here are obtained in *pure heterodyne detection ; in each case, samples obtained from the same donor give very similar results*. Table I shows $qL \gtrsim 1$ for all studied angles, thus the temporal decay $g_1(\tau)$ *will be non exponential* or, in other words, the frequency spectrum $S(\omega)$ *will not be lorentzian*.

Such features are quite obvious in figs. 11 and 12. Detailed analysis of these results leads to the following conclusions

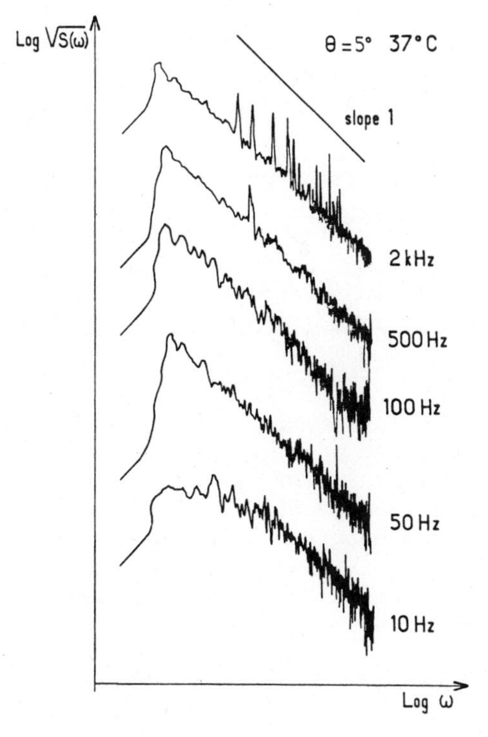

Fig.11 - Log-log plot of the amplitude spectrum at θ=5° for different frequency ranges of analysis.

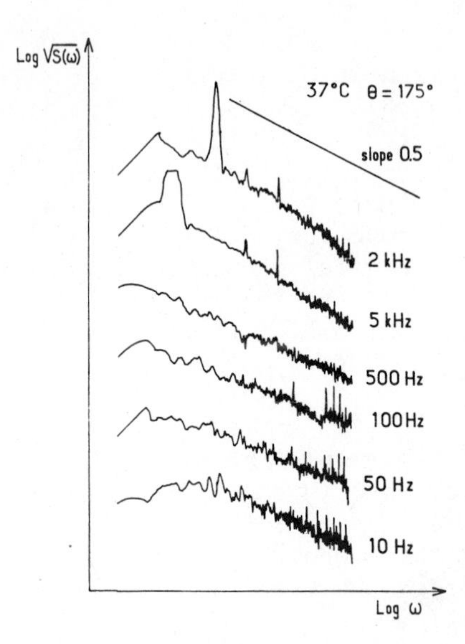

Fig.12 - Log-log plot of the amplitude spectrum at θ=175° for different frequency ranges of analysis.

- θ = 5°. The distribution of decay times includes a few short times (< 0.1 s). The major contribution appears to be around $\tau \sim 170$ ms (fig. 13).

The results at θ = 15° are identical.

- θ = 35°. The contribution of short decay times increases, an effect which is easily observed in temporal analysis. Frequency analysis indicates that the previous contribution remains unchanged.
- θ = 180°. The contribution of short decay times is now very important. The corresponding range (~ 4 ms) is very different from that of long decay times (170 ms), the contribution of which remains unchanged, (from frequency analysis).

All these results are summarized in figure 14, which suggests that two types of mode are present in this medium :

. relaxational structural modes with q-independent decay (long time contribution) and motions of the network with $1/q^2$ temporal decays, which are observed when they are shorter than the structural modes.
- Both types of contribution have a broad distribution of decay times (probably broader at large q because of the angular dependence of scattered light).
- The structural relaxation seems to be faster at higher temperatures (37°C) but further experiments are needed to measure the exact amount of this effect.
- From the measured time $\tau = 4$ ms at θ = 180°, we deduce

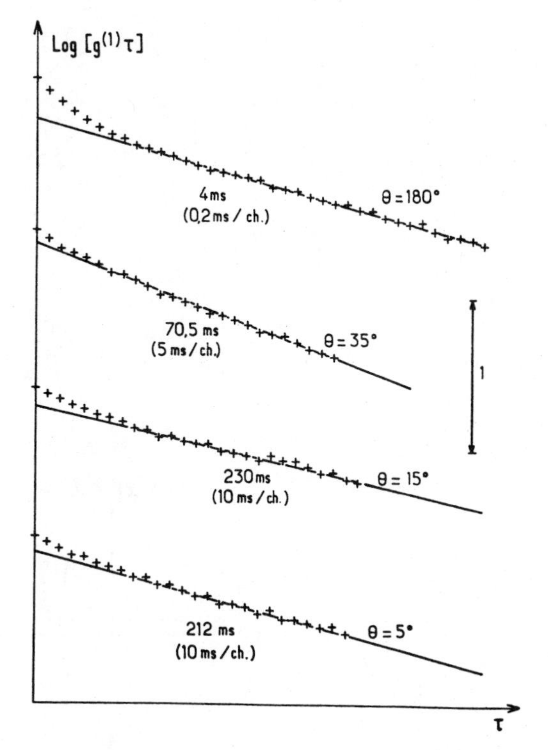

Fig.13 - Semi-log plot of the correlation function $g^{(1)}\tau$ versus τ for different scattering angles.

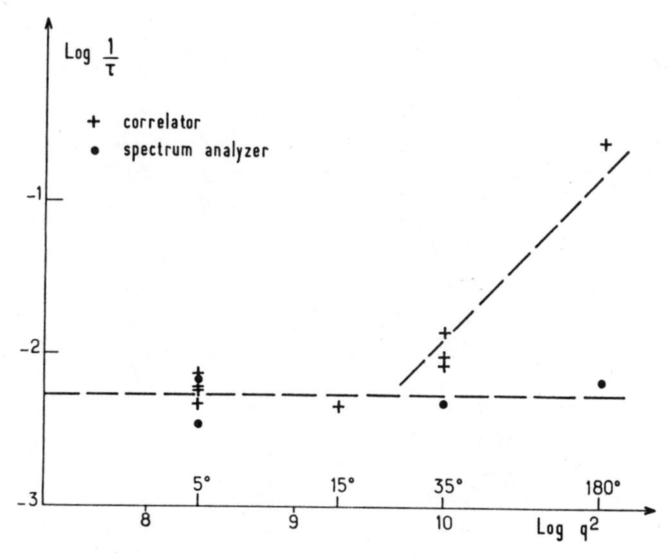

Fig.14 - Log-log plot of τ^{-1} versus q^2.

$1/\tau q^2 = 2.45 \times 10^{-9}$ cm^2/s, which would correspond to a diffusion coefficient D for particles having a mean size of 0.9 µm.
- With longer times, a very surprising behavior was observed : under very weak thermal or mechanical perturbations, strong oscillations appear in the temporal decay function $g_1(\tau)$. This corresponds to propagating modes and can only be explained by the non-Newtonian properties of this medium. (Such behavior is known to occur in non-Newtonian fluids. (Bordeaux 1979)). Usually, the perturbation (variation of laser intensity) induces a relatively rapid oscillation (period ~ 1s) which in turn excites a slower one, and so on (fig.15). This effect is very strong in the forward direction but its angular dependence is not clear (it is probably closely related to structural relaxation).

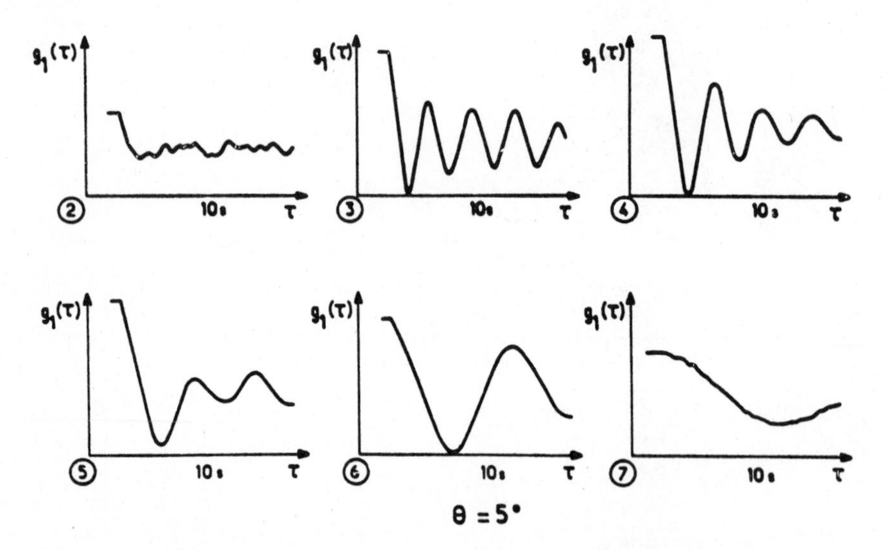

Fig.15 - Example of strong oscillations affecting the temporal decay function at θ=5° (the time interval between two consecutive plots is equal to ~ 30 seconds).

d - Conclusion

At the present time, our experimental results do not allow us to say whether these oscillations are those previously suggested by Odeblad (1968) and whether they really can assist the progress of spermatozoa. But it is interesting to note that they appear very readily during ethinyl-oestradiol or Humegon (mixture of pituitary hormones) therapy. These oscillations are not observed in old samples or under severely pathological conditions. Consequently, the way these large oscillations will appear in the cervical mucus and their correct frequencies could be parameters characterizing the healthy behavior of this important secretion, which plays an essential role in human reproduction. The results we obtained in cervical mucus by light-scattering techniques are in very good agreement with a hydrogel structure. That model, proposed by Odeblad (1968), was corroborated by electron scanning microscopy experiments whose conclusions remain highly plausible (Chrétien 1975). Our results also agree very well with extensive mechanical studies carried out in cervical or bronchial mucus by several authors (Eliezer 1974 Meeyer et al. 1975, Litt et al. 1976, Puchelle 1979).

By contrast, the angular dependence of the observed spectra is not compatible with the model composed of entangled macromolecules proposed by Lee et al. (1977).

A systematic study is now being undertaken in order to refine our experimental results and their theoretical interpretation. We consider that analysis of the non-Newtonian properties of this highly thixotropic medium is necessary in order to understand its physiological role.

REFERENCES

I°) Motility of human spermatozoa and applications

Bartak V., Sperm velocity test in clinical practice. Int.J.Fertil. 16, 107-112 (1971)

Bergé P., Volochine R. and Hamelin A., Mise en évidence du mouvement propre de microorganismes vivants grâce à l'étude inélastique de la lumière, C.R.Acad.Sci. 265, 889-892 (1967)

Bergé P., and Dubois M., Dispositif de mesures optiques adapté à l'étude du mouvement de microorganismes vivants, Rev.Phys.Appl. 8, 89-96 (1973)

Boon J.P., Nossal R. and Chen S.H., Light scattering spectrum due to wiggling motion of bacteria, Biophys.J. 14, 847-852 (1974)

Boselaar C.A. and Spronk N., A physical method for determination of the motility and concentration of spermatozoa, Nature 169, 19-19 (1953)

Castenholz A., Photokymographische Registriermethode zur Darstellung und Analyse der Spermatozoenbewegungen, Andrologia 6, 155-168 (1974)

Combescot R., J.Physique 31, 767 (1970)

Cummins H.Z., Knable N., Yeh Y., Phys.Mes.Lett. 12, 150 (1964)

D'Almeida, Belaisch M., Eyquem A., Soc.de Gynéco.C.Rendus 6, 2 (1969)

D'Almeida M., Palmer R., Eyquem A., Rev.Suisse Gynéco et Obst. 2, 315 (1971)

Dubin S.R., Lunacek J.H., Benedek G.B., Proc.Nat.Acad.Sci.USA, 57, 1164 (1967)

David G., Jouannet P., Bergé P., Dubois M., Volochine B., VIIIth World Congress Fert. Steril. (1974)

David G., Volochine B. and Bosq J., Cinétique de l'action immobilisante des anticorps antispermatozoïdes, 3rd Int.Symp.Immun. Reprod. Varna (1975)

Dubois M., Jouannet P., Bergé P. and David G., Spermatozoa motility in human cervical mucus, Nature 252, 711-713 (1974)
Dubois M., Jouannet P., Bergé P., Volochine B., Serres C. and David G., Méthode et appareillage de mesure objective de la mobilité des spermatozoïdes humains. Ann.Phys.Biol.et Méd. 9, 19-41 (1975)
Eyquem A., D'Almeida, 2nd Inst.Symp.on Immun. of Rep. Varna (1971) Acad.Sc.Sofia (1973)
Harvey C., The speed of human spermatozoa and the effect on it of various diluents, with some preliminary observations on clinical material. J.Reprod.Fertil. 1, 84-95 (1960)
Hynie J., A quick calculation of the velocity of spermatozoa. Int.J. Fertil. 7, 345-346 (1962)
Janick J., and Mac Leod J., The measurement of human spermatozoa motility. Fertil.Steril. 21, 140-146 (1970)
Jecht E.W., and Russo J.J., A system for the quantitative analysis of human sperm motility, Andrologie 5, 215-221 (1973)
Jouannet P., Volochine B., Deguent P., Serres C., David G., Andrologie 9, 36 (1977)
Mac Leod J., Gold R.Z., Fert.Steril. 2, 187 (1951)
Mac Leod J., Gold R.Z., Fert.Steril. 4, 10 (1953)
Munktell G., and Fjällbrant T., Electronic equipment for measuring the mobility of spermatozoa, 3rd Int.Conf.Med.Phys.Gothemburg (1968)
Nossal R., Chen S.H., J.Phys.Colloq. 33, cl.171 (1972)
Rickmenspoel R., Photoelectric and cinematographic measurements of the "motility" of bull sperm cells. Thesis Utecht 1957
Rickmenspoel R., Measurements of the "motility" of bull sperm cells under various conditions. J.Agric.Sci. 54, 399-409 (1960)
Rothschild, Lord, A new method of measuring activity of spermatozoa. J.Exp.Biol. 30, 178-199 (1953)
Timourian H., and Watchmaker G., Determination of spermatozoon motility. Develop.Biol. 21, 62-72 (1970)
Van Duijn jr. C., van Voorst C. and Freund M., Movement characteristics of human spermatozoa analysed from kenemicrograph. Europe J.Obst.Gynec. 4, 121-135 (1971)
Volochine B., Bosq-Rolland J., Revue de Phys.Appl. 14, 391 (1979)

II°) Cervical Mucus

Adam M., and Delsanti M., Macromolécules 10, 1229 (1977)
Adam M., Delsanti M., and Pouyet G., J.de Phys.Lettres 40, L435 (1979)
Adam M., and Delsanti M., J.de Phys.Lettres 38, L271 (1977)
Bergé P., and Dubois M., Revue de Phys.Appl. 8, 89 (1973), Bordeaux 1979, Colloque de la SFP, discussions
Candau S.J., Young C.Y., et Tanaka T., J.Chem.Phys. 70, 4694 (1979)
Cazabat A.M., Volochine B., Chrétien F.C. et Kunstmann J.M., CRAS B232 Paris (1979)
Chrétien F.C., J.Microsc.Biol.Cell., 24, 23 (1975)

Chrétien F.C., Gernigon G., David G., et Psychoyos A., Fertil. Steril. 24, 746 (1973)
Chrétien F.C., Cohen J., et Psychoyos A., J.Gynec.Obstet.Biol.Repr. 5, 313 (1976)
Chrétien F.C., and David G., Europ.J.Obstet.Gynec.Reprod.Biol. 816, 307 (1978)
Chrétien F.C., and Psychoyos A., Proc.Polish Gynec.Soc. Bialowieza, 30 june 1975
Cummins H.Z., in Photon correlation and light beating spectroscopy. H.Z.Cummins and Pike E.R., ed. Plenum Press New-York and London (1974)
De Gennes P.G., Macromolécules 9, 587 (1976)
Dubois M., Jouannet P., Bergé P., Volochine B., Serres C., et David G., Ann.Phys.Biol.et Méd. 9, 19 (1975)
Dubois-Violette E., and De Gennes P.G., Physics 3, 181 (1967)
Eliezer N., Biorheology 11, 61 (1974)
Ferry J.D., Viscoelastic properties of polymers,Wiley New-York (1970)
Ford N.C. Jr., and Benedek G.B., Phys.Rev.Lett. 15, 649 (1965)
Forrester T., J.Opt.Soc.Amer. 51, 253 (1961)
Geissler E., and Hecht A.M., J.Chem.Phys. 65, 103 (1976)
Gould K.G., Martin D.E., et Graham C.E., Proc.Workshop on S.E.M. in Reprod.Biol. ITT Research Institute, Chicago Illinois (1976)
Hecht A.M., and Geissler E., J.de Phys. 39, 631 (1978)
Jordan J.A., and Allen J.M., Workshop Conf.on the Uterine Cervix in Reproduction. Rottach-Egern, V.Insler and G.Bettendorf eds., Georg Thieme, Stuttgart, p.292 (1977)
Lee W., Blandau R.J., et Verdugo P., Workshop Conf. Rottach-Egern ed. by V.Insler, G.Bettendorf p.68 (1977)
Litt M., Khan M.A., et Wolf D.P., Biorheology 13, 37 (1976)
Meyer F.A., King M., et Gelman R.A., Biochemica et Biophysica Acta 392, 223 (1975)
Odeblad E., Acta Obstet.Gynec.Scand. 47, suppl.1, 59 (1968)
Odeblad E., Int.J.Fertil. 7, 313 (1962)
Ostrowsky N., in Photon correlation and light beating spectroscopy, ed. H.Z.Cummins and E.R. Pike Plenum Press New-York (1974)
Puchelle E., Thesis, Université de Nancy I., Faculté des Sciences Pharmaceutiques et Biologiques (1979)
Tanaka T., Hocker L.O., et Benedek G.B., J.Chem.Phys. 59, 5151 (1973)
Volochine B., and Cazabat A.M., Communication at the Bordeaux Conf. of the "Société Française de Physique" Octobre 1979
Volochine B., and Bosq-Rolland J.,Revue de Phys.Appl. 14, 391 (1979)
Yeh Y., and Cummins H.Z., Appl.Phys.Lett. 4, 176 (1964)
Zaneweld L.J.D., Tauber P.F., Port C., Propping D., et Schumacher G.F.B., Ann.J.Obstet.Gynec. 122, 650 (1975)

SYSTEMATIC ASSESSMENT OF SPERM MOTILITY

J.C.Earnshaw, G.Munroe, W.Thompson* and A.Traub*

Department of Pure and Applied Physics
The Queen's University of Belfast
Belfast BT7 1NN

*Department of Midwifery and Gynaecology

Light scattering is becoming increasingly popular as an objective method of evaluating sperm motility.[1-5] Studies on the effects of certain sperm stimulants necessitate continuous monitoring of control and treated samples for long periods of time.[6] Here we describe briefly how such an experiment has been automated using an Apple II microcomputer. The autocorrelation function of the scattered light is computer fitted by theoretically appropriate functional forms to extract various parameters related to the sample motility.

The sample (volume 0.4 ml) is contained in a rectangular glass cell (thickness 1 mm) within which a low power (1 mW) He-Ne laser is focussed (Fig 1). Light scattered at 15° is detected by a photomultiplier, the self-beat digital autocorrelation function being recorded with a 128-channel multi-bit correlator (Malvern). Up to 12 sample cells can be placed in a thermostatted sample block, which is moved across the laser path under computer control; a slotted opto-switch detects when a cell is in position. The computer also controls the correlator as required and after each correlator run the observed correlation function is transferred directly into the Apple memory. Here the data is reformatted to remove unnecessary characters (leading zeros etc) before transfer to floppy disc, where typically 120 correlation functions may be stored permanently on one side. Before commencement of an experiment suitable correlator settings are manually determined for each sample. These details are programmed into the computer together with the number of cells being used etc. From then on the experiment proceeds automatically to make the requested number of scans of the sample block (usually

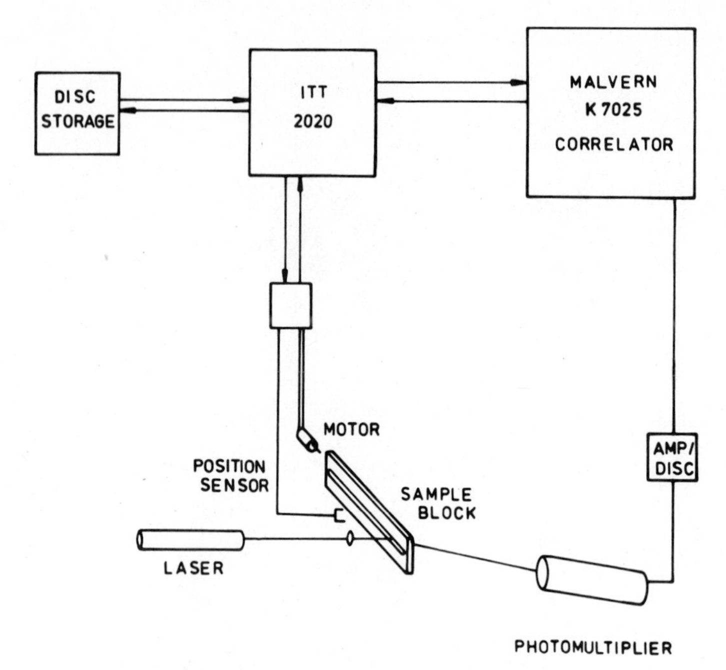

Figure 1 : Layout of apparatus

5-10). At least three correlation functions, using different sampling times, are recorded every time the motor stops and a cell is in position. With this variety in the recorded data it is possible to employ different analysis procedures.

Each recorded intensity correlation function is normalized (g^2) and the first order correlation function (g^1) derived using the Seigert relation. Now g^1 comprises contributions from both the motile and non-motile sperm:

$$g^1(\tau) = (1 - \alpha)\, g_{non-mot}(\tau) \quad + \quad \alpha\, g_{mot}(\tau)$$

The non-motile sperm are assumed to diffuse with a characteristic diffusion coefficient (D). The motile contribution is expressed as an integral over all possible swim speeds:

$$g^1{}_{mot}(\tau) = \int_0^\infty \frac{\sin K v \tau}{K v \tau}\, P(v)\, dv \qquad (1)$$

where K is the magnitude of scattering vector and P(v) the swim

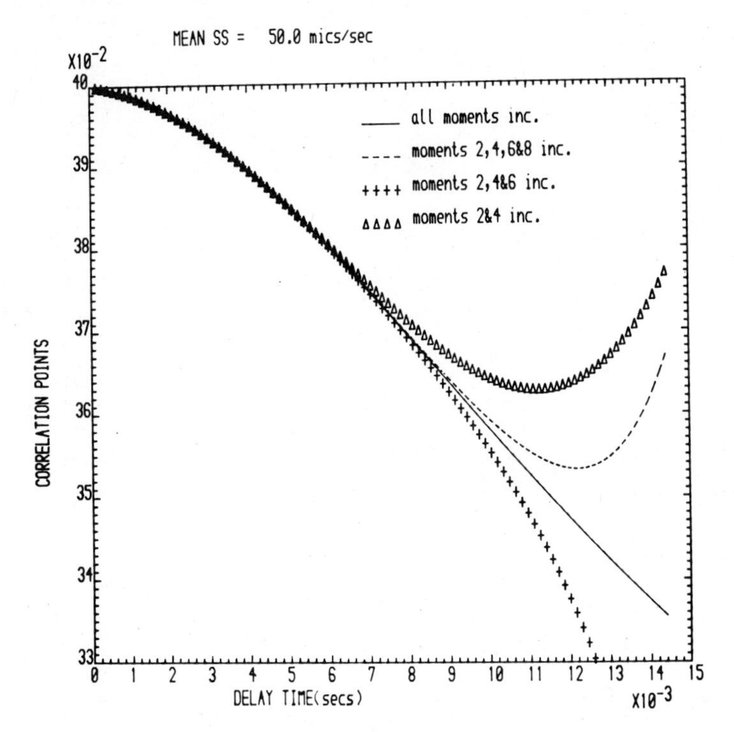

Figure 2 : The solid line shows the expected correlation function assuming a 'Saclay' distribution of $\bar{v}$ = 50 μm s^{-1} and the broken lines the approximations obtained using different numbers of terms in the moments expansion.

speed distribution. Two approaches to data analysis have been used. The first involves no assumptions about P(v) whereas the second considers a specific form for P(v).

The integral may be expanded as a power series in terms of the even moments of P(v):

$$g^1_{mot}(\tau) = 1 - \frac{(K\tau)^2}{3!} V_2 + \frac{(K\tau)^4}{5!} V_4 - \dots \qquad (2)$$

where

$$V_n = \int_0^\infty V^n P(v)\, dv,$$

so that

$$g^1_{mot}(\tau) \rightarrow e^{-DK^2\tau} (1 - \alpha + \alpha (1 - \frac{(K\tau)^2}{3!} V_2 + \dots)) \qquad (3)$$

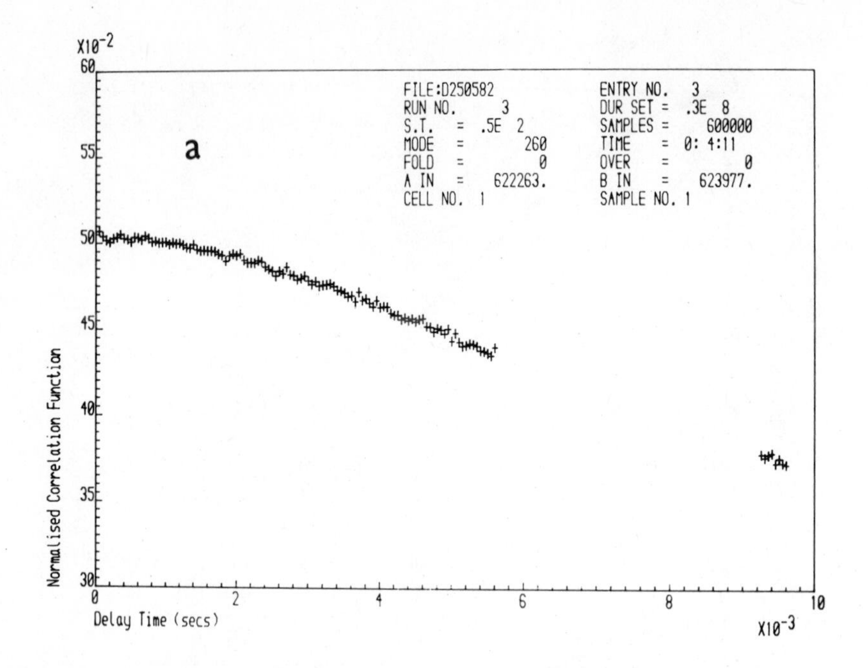

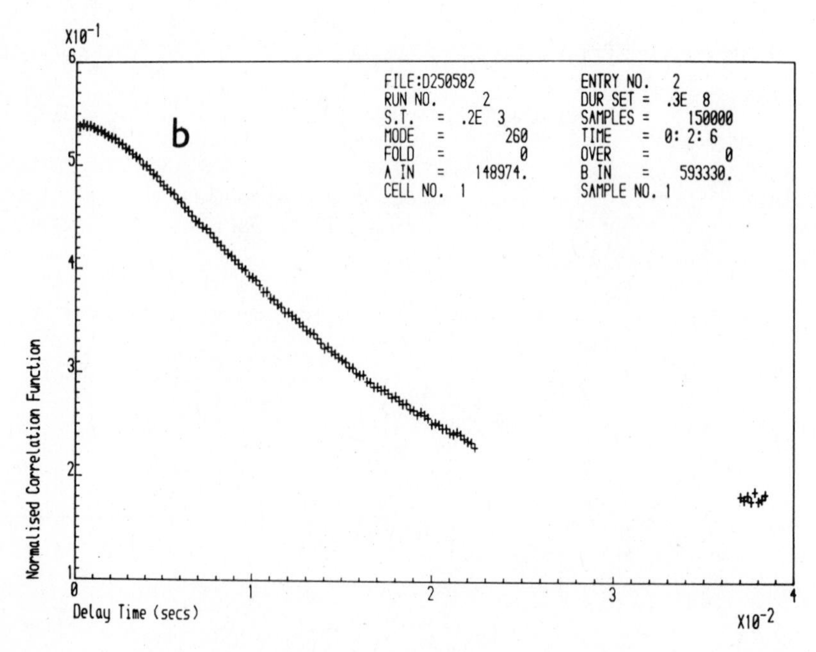

Figure 3 : Typical correlation functions for sample times of (a) 50 μs, (b) 200 μs, (c) 3 ms. In places the data of (c) fluctuate below zero; these are suppressed in the plot.

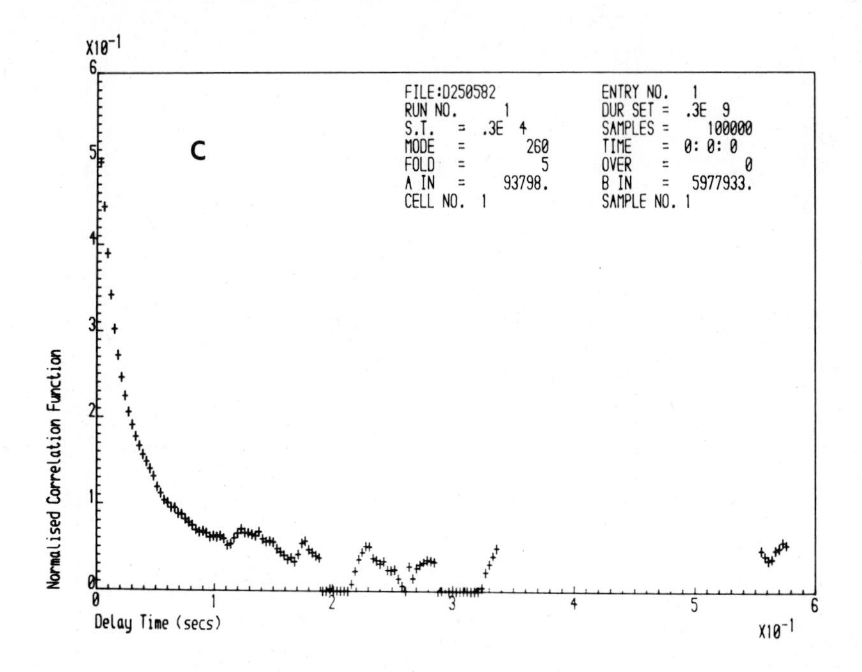

When fitting beyond the fourth moment problems arise due to noise; reduction of the number of terms in the expansion limits the range (of $K\tau$) of validity of eq 3 (cf method of cumulants[7]). Figure 2 suggests that eq 3 with two moments will adequately describe correlation data with sample times around 50 μs. An independant estimate of α is required to yield a unique solution. This can be obtained from longer sample time data (typically 1 ms) where it may be assumed that only the non-motile fraction is contributing to the latter part of the correlation function:

$$g^1_{tot}(\tau) = (1 - \alpha)\, e^{-DK^2\tau}$$

Thus two separate correlation functions provide estimates for α, D, V_2 and V_4 (where $\sqrt{V_2}$ = r.m.s. swim speed)

An example of the second type of analysis approach is to assume a 'Saclay' distribution of swim speeds[8]:

$$P(v) \;=\; \frac{4v}{\bar{v}^2} \exp\,(-2\, v/\bar{v}) \qquad (4)$$

in which case the integral reduces to a Lorentzian:

$$g^1_{mot}(\tau) = \frac{1}{1 + (K \bar{v} \tau/2)^2}$$

so that

$$g^1_{tot}(\tau) = e^{-DK^2\tau} \left(1 - \alpha + \frac{\alpha}{1 + (K \bar{v} \tau/2)^2}\right) \qquad (5)$$

Figure 2 shows that the full form of $g^1_{tot}(\tau)$ will require longer sample times than are appropriate to the moments analysis. Typically data for sample times of 200 μs permit estimation of α, D and $\bar{v}$ by fitting of eq 5 to the observed correlation functions.

Figure 3 shows a set of three correlation functions, recorded successively from the same sample, and Table 1 the results obtained from the analysis described. Accurate determination of the initial α from 3(c) is crucial to the reliability of the moments fit. However preliminary results would suggest that it is a feasible approach and has the advantage that no restrictions are forced on the swim speed distribution prior to computer fitting.

Table 1 : Fitted parameters for sample of figure 3

Deduced parameters	Saclay fit	Moments fit
α	0.71	0.64
D	$0.56\ 10^{-12}$	$0.99\ 10^{-12}$
$\bar{v}$	39.2 μm s^{-1}	-
V_2	$0.23\ 10^{-8}$*	$0.23\ 10^{-8}$
V_4	$0.18\ 10^{-17}$*	$0.62\ 10^{-17}$

* calculated from $\bar{v}$ and eq 4.

ACKNOWLDEGEMENTS

This work is supported by the M.R.C. GM wishes to thank the Company of Biologists, Cambridge and the Institute of Physics for financial aid in attending this meeting.

REFERENCES

1. Finsy, R., J Peetermans, H. Lekkerkerker; Optica Acta 27:25 (1980).
2. Shimizu, H., G. Matsumoto; IEEE Trans Biomed Eng BME-24:153 (1977).
3. Herpigny, B., J.P. Boon; J de Phys 40:1085 (1979).
4. Volochine, B., this volume.
5. Craig, T., F.R. Hallett, B. Nickel; Biophys J 28:457 (1979).
6. Traub, A.I., J.C. Earnshaw, P.D. Brannigan, W. Thompson; Fertil Steril 37:436 (1982).
7. Chu, B., this volume.
8. Jouannet, P., B. Volochine, P. Deguent, C. Serres, G. David; Andrologia 9:36 (1977).

THE APPLICATION OF LASER LIGHT SCATTERING TO THE STUDY OF PHOTO-RESPONSES OF UNICELLULAR MOTILE ALGAE

Cesare Ascoli, Carlo Frediani

Istituto di Biofisica del C.N.R.
56100 Pisa, Italy

INTRODUCTION

Quasi-elastic light scattering (QELS) has been used to investigate the mechanism of photosensory transduction in the unicellular motile algae *Euglena gracilis* and *Haematococcus pluvialis*.

Three kinds of photoresponses have been observed in unicellular motile algae:

i) photokinesis (dependence of swimming speed on light intensity);

ii) phototaxis (oriented motion towards or away from the light source);

iii) photophobic reactions (stop reactions followed by starts in random directions, induced by temporal changes in light intensity).

We have investigated photokinesis in *Euglena gracilis* and phototaxis in *Haematococcus pluvialis*.

The QELS measurements have been performed by using homodyne, or heterodyne detection, or heterodyne with frequency shift of the local oscillator, depending on the experiments and by analyzing the signal in the frequency domain.

QELS MEASUREMENTS OF *Euglena gracilis* MOTION PARAMETERS

Euglena gracilis has ellipsoidal shape and dimensions much greater than the laser wavelength (fig.1). The light scattering is highly anisotropic (fig. 2), so that cells moving in different directions scatter different intensities. Furthermore the motion is an helicoidal one (fig. 3), so the motion parameters to be taken into account are:

i) translational velocity,
ii) cell body rotation frequency,
iii) flagellar beating frequency.

Structure and motion of *Euglena* cause the frequency spectrum to be not easily interpretable. In fact cell body rotation and flagellar beating affect the spectrum due to translational motion.

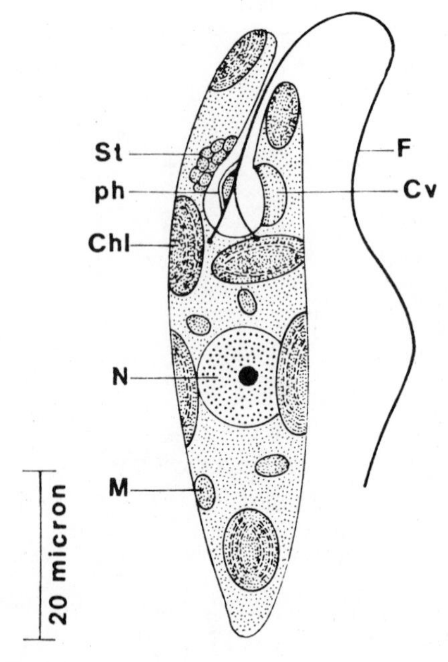

Fig. 1 - *Euglena gracilis* (schematic drawing). N, nucleus; St, stigma; F, flagellum; Chl, chloroplast; ph, photoreceptor; M, mitochondria; Cv, contractile vacuole.

Fig. 2 - Light scattering pattern from a single *Euglena*. The picture has been taken with the film placed perpendicularly to the laser beam, 20 cm away from the sample.

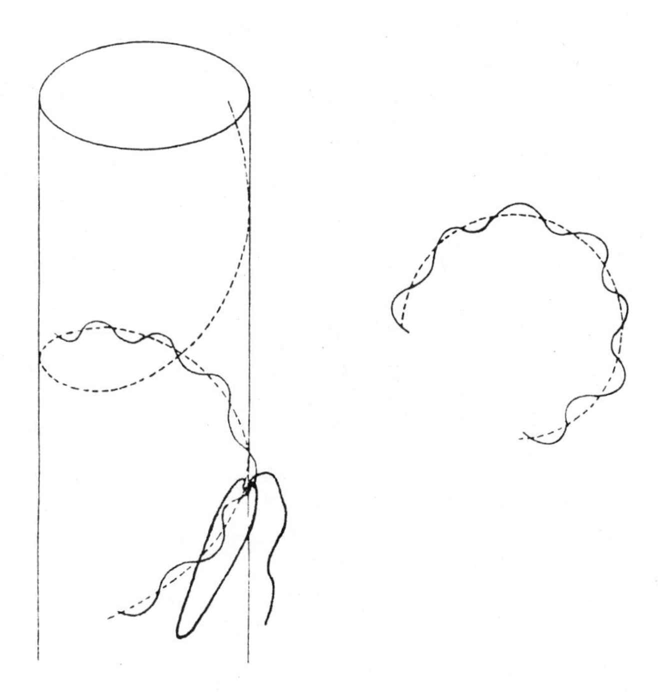

Fig. 3 - Schematic drawing of *Euglena* motion. After Ascoli et al.(1).

This influence occurs in two ways:

i) cell body rotation and flagellar beating frequencies appear in the spectrum as frequency modulation lines around the Doppler line. In fact it can easily be shown (1) that for a pure helicoidal motion the electric field scattered from a point scatterer is:

$$E_s \propto \exp \{ - j (\omega_0 t + q_z v t - q_x a \sin \omega_r t) \}$$

where ω_0 is the frequency of laser light, $\vec{q} = \vec{k}_s - \vec{k}_0$ is the difference between the wavevectors of the incident and the scattered beams, v is the swimming speed, ω_r is the body rotation frequency and a is the radius of the helicoidal pattern. The scattered field is a frequency modulated signal, where $\omega = \omega_0 + q_z v$ is the Doppler frequency modulated at the ω_r frequency. For a population, due to the heterogeneity of the population itself, that implies an enlargement of the Doppler spectrum (1).

ii) for the anisotropy of the light scattering pattern, the scattered light intensity is modulated at the frequencies of cell body rotation and of flagellar beating respectively. This effect is more evident for suitable positions of the photodetector.

In the general case these effects are mixed, but by using a suitable set up they can be separated. That tool is a radio-frequency field which orientates the *Euglena* cells (fig. 4), (2). In this case we can choose the photodetector position in such a way that the intensity of light scattered on it is constant and the second effect vanishes. So the heterodyne spectrum gives, by the Doppler relation, the swimming speed distribution (fig. 5). We have noted (1) that up to 20° scattering angle, the spectrum is not enlarged by frequency modulation lines. On the other hand, in ori- cells, we detect directly (homodyne detection) the intensity modulations at the rotation and flagellar beating frequencies by choosing a suitable photodetector position (1) that maximizes this effect (fig. 6). Homodyne detection is not really necessary; in fact we can see these intensity modulations in the spectrum of a single cell too.

Hence in *Euglena gracilis* populations we are able, by orienting the cells, to study effects which change scalar motion parameters, like swimming speed, cell rotation frequency and flagellar beating

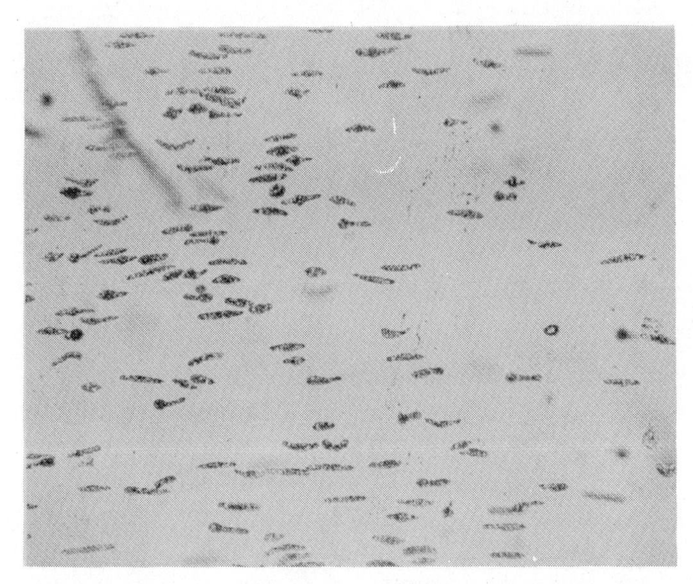

Fig. 4 - *Euglenae* moving parallel to the radio-frequency electric field.

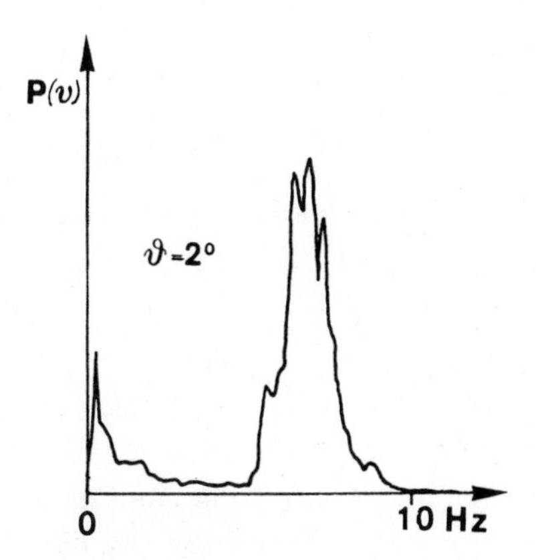

Fig. 5 - Doppler line obtained by heterodyne detection at small scattering angle; the Doppler relation gives a mean speed of 127 μ/s.

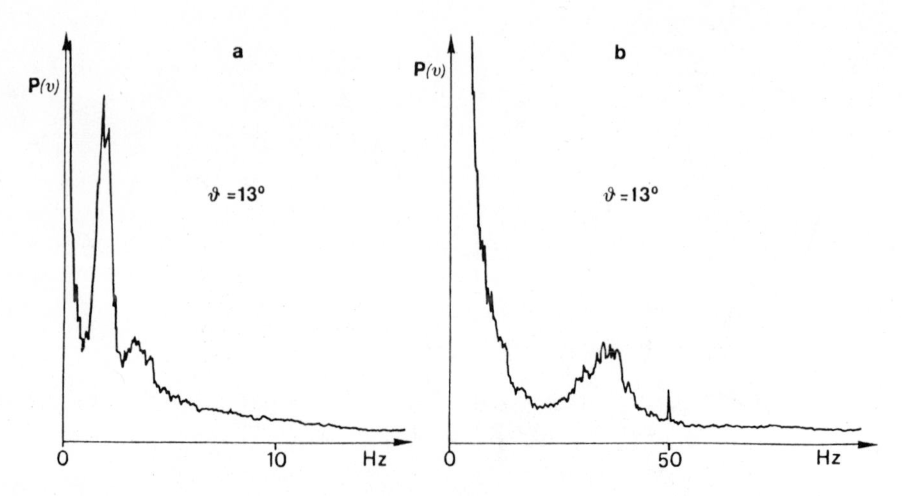

Fig. 6 - Cell body rotation (a) and flagellar beating (b) frequency distributions obtained by directly detecting the intensity modulations of the light scattered by oriented *Euglenae*. After Ascoli et al. (1).

frequency. In fact we have studied the effects on these parameters, of temperature and Ph (3), (4). We also investigated a photokinetic effect (fig. 7), (4); it is controlled by the photosynthetic apparatus and is present only in cells grown in an autotrophic medium.

APPLICATION TO PHOTOTAXIS OF *Haematococcus pluvialis*

Haematococcus pluvialis shows a quasi-spherical body of 20 μm diameter (fig. 8). It has been shown, by microscopical observations, to be phototactic (5). Phototaxis means a unidirectional motion, so it cannot be detected by an heterodyne detection in which the local oscillator has the same frequency as the incident beam. This is schematically illustrated in fig. 9. Therefore an heterodyne with frequency shifted local oscillator has been used (fig. 10). The phase modulator produces a phase difference $\Delta\phi$ between the two beams. This phase rises linearly in the time, so giving a frequency shift.

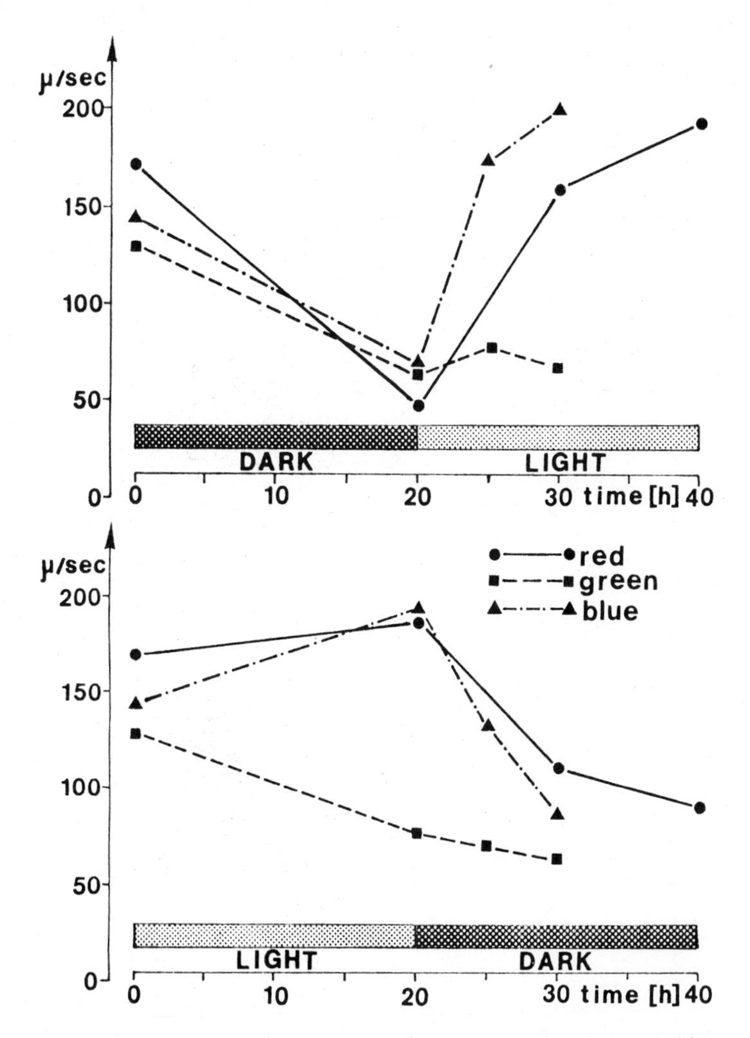

Fig. 7 - Mean speed of oriented *Euglenae* versus irradiation time. The wavelengths of the red (677 nm) and the blue (431) lights correspond to the chlorophyll absorption peaks, while the wavelength of the green light (512 nm) corresponds to a minimum of the chlorophyll absorption spectrum.

When $\Delta\phi = 2\pi$ the phase shift is suddenly reset to zero. By this detection technique an oriented motion can be revealed (fig. 11).

To see how the structure and the large dimensions of

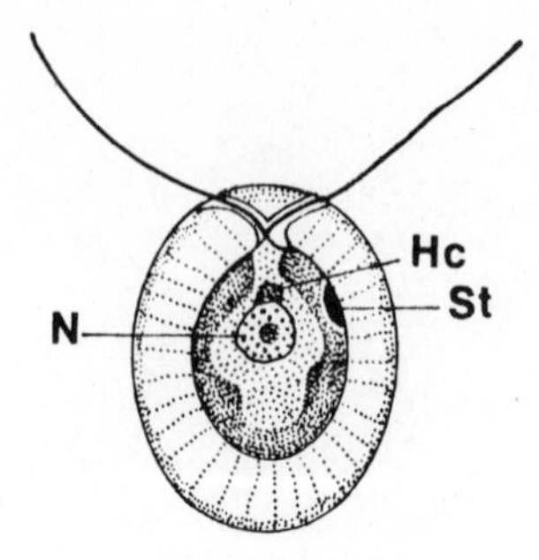

Fig. 8 - *Haematococcus pluvialis* (schematic drawing). N, nucleus; Hc, haematocrome; St, stigma.

Haematococcus pluvialis affect the Doppler spectrum, we have taken a spectrum of a single microorganism (fig. 12). This shows a very narrow Doppler line (without a meaningful structure) and two low symmetrical lines 30 Hz far away from the Doppler line (frequency modulation lines induced by the flagellar beating). So we can conclude that, at least in a first approximation, the Doppler spectrum of a population can be assumed to be homologous to the

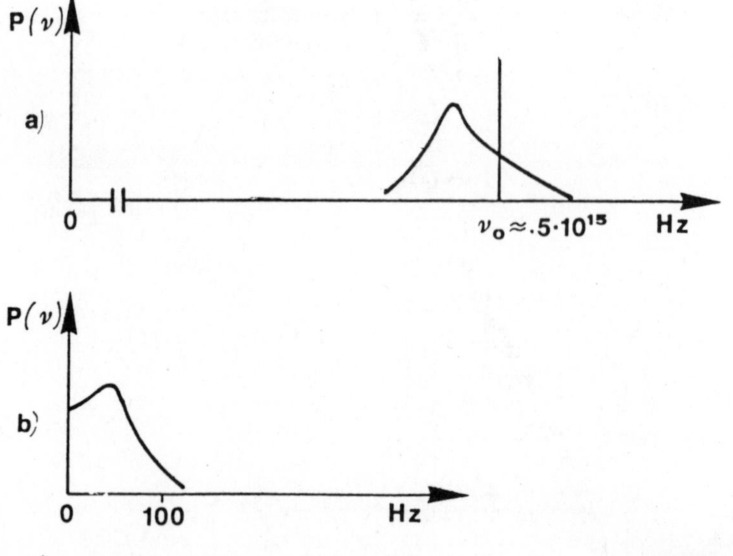

Fig. 9 - a) optical spectrum for anisotropically moving particles; b) heterodyne spectrum.

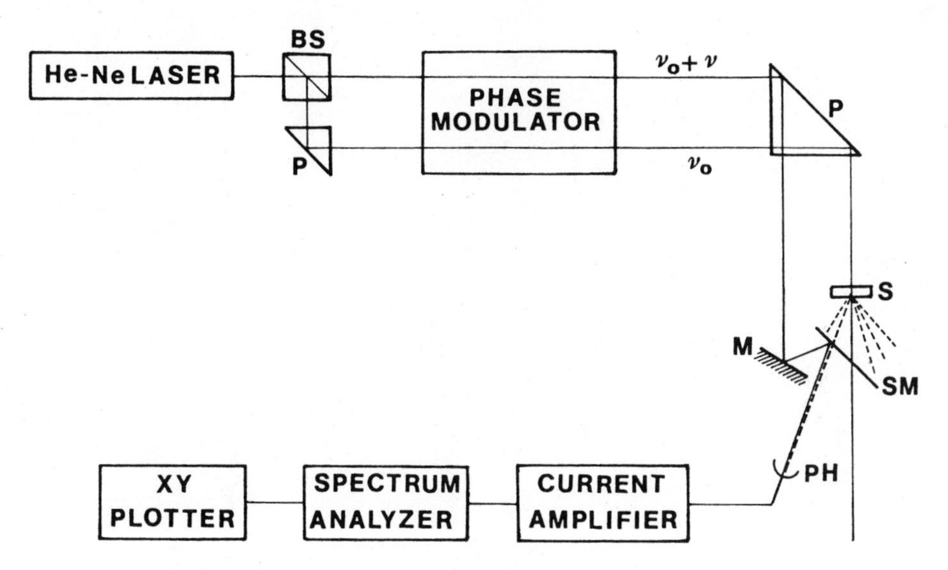

Fig. 10 - Block diagram of the frequency-shifted heterodyne spectrometer. Bs, beam splitter; P, prism; S, sample; SM, semireflecting mirror; M, mirror; PH, photodiode.

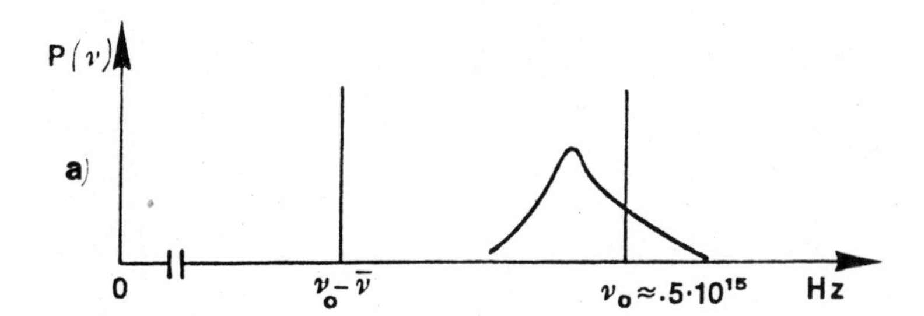

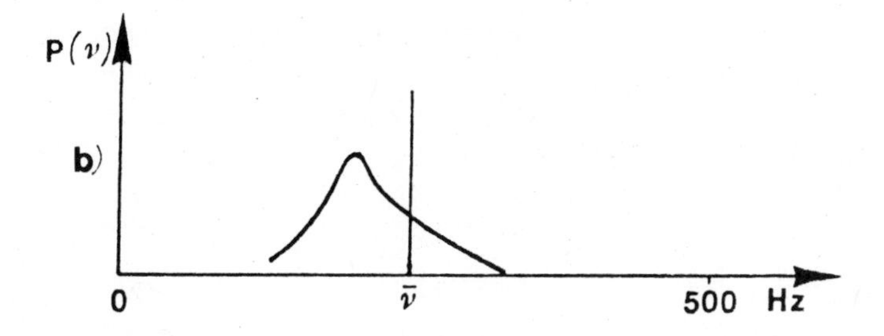

Fig. 11 - a) optical spectrum for anisotropically moving particles; b) frequency shifted heterodyne spectrum.

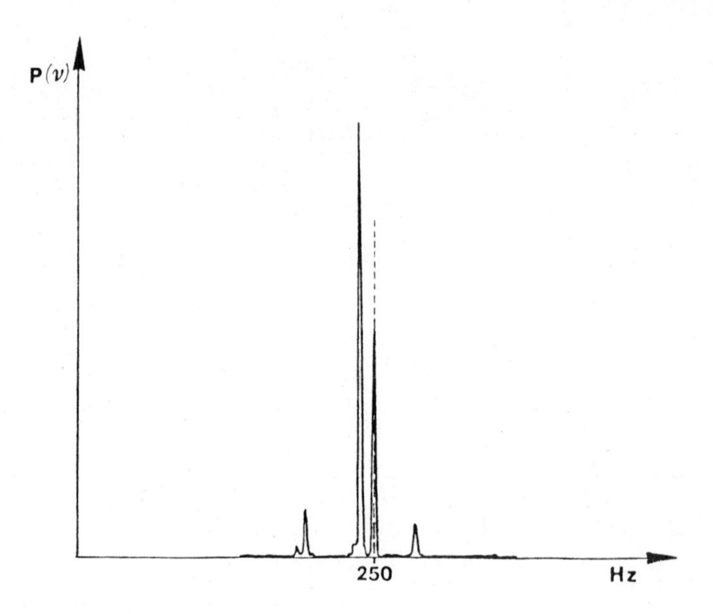

Fig. 12 - Frequency-shifted (250 Hz) heterodyne spectrum of a single *Haematococcus pluvialis*.

distribution of the $\vec{q}$ component of the translational velocity. The phototaxis measurements have been performed by laterally stimulating the cells with green light (480 nm). A large asymmetry clearly appears in the Doppler spectrum (fig. 13). The peak to the left of the frequency (250 Hz) of the local oscillator indicates that the mean component of the velocity is directed towards the light source.

CONCLUDING REMARKS

Quasi-elastic light scattering is a very fast method to obtain a quantitative description of the unicellular algae photobehaviour, but it must be adapted to the particular problem to be faced. However other complementary techniques must be used to clarify the mechanism of photobehaviour. To investigate the photoreceptor and the mechanism of phototaxis in *Haematococcus pluvialis* we recorded cell membrane potential. The preliminary results show that the photoreceptor is localized near the stigma. Our next step is to perform action spectra by both the above mentioned methods and compare the results.

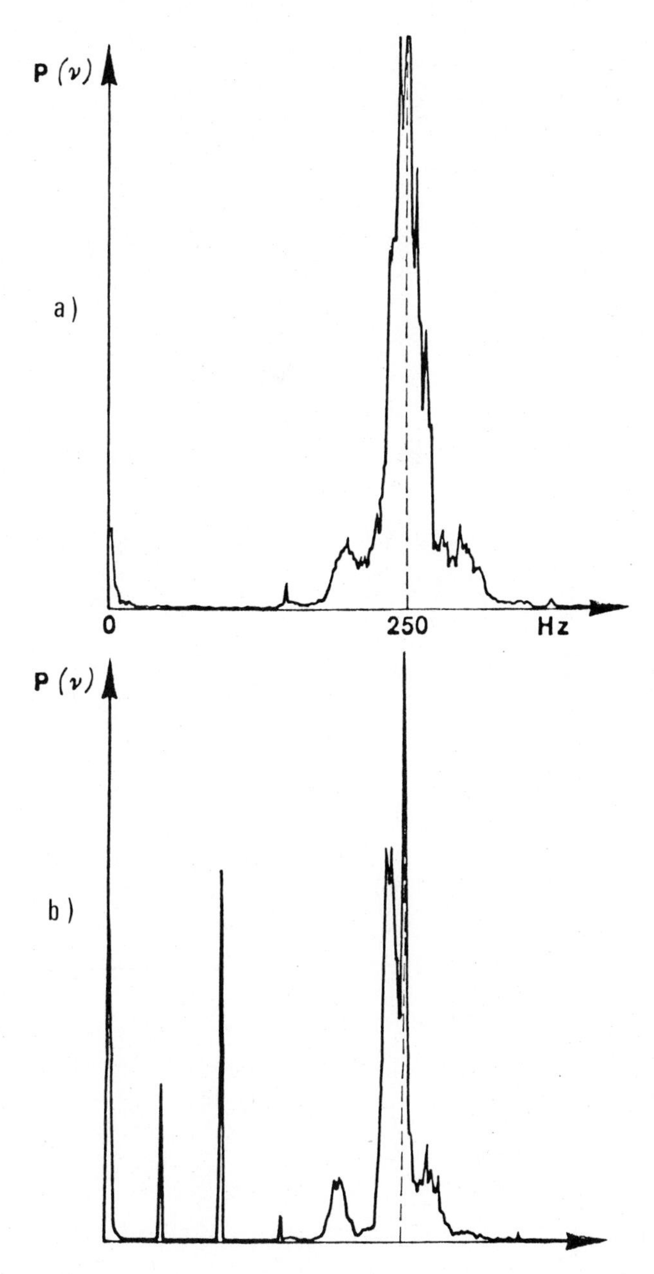

Fig. 13 - Frequency-shifted heterodyne spectra of a population of *Haematococcus pluvialis*: a) in dark; b) after 8 minutes uniform light.

REFERENCES

1. C. Ascoli, M. Barbi, C. Frediani, and A. Murè, Measurements of *Euglena* motion parameters by laser light scattering, Biophys. J. 24:585 (1978).
2. C. Ascoli, M. Barbi, C. Frediani, D. Petracchi, Effects of electromagnetic fields on the motion of *Euglena gracilis*, Biophys. J. 24:601 (1978).
3. C. Frediani, Misure della frequenza del battito flagellare con la tecnica del laser light scattering. Applicazioni allo studio della motilità di alghe flagellate, in:"Atti del 5° Congresso di Cibernetica e Biofisica",Pisa (1979).
4. C. Ascoli and C. Frediani, Quasi-elastic light scattering in the measurement of the motion of flagellated algae, in:"Light scattering in liquids and macromolecular solutions", V. Degiorgio et al., eds., Plenum, New York (1980).
5. F. F. Litvin, O. A. Sineshchekov, V. A. Sineshchekov, Photoreceptor electric potential in the phototaxis of the alga *Haematococcus pluvialis*, Nature, 271:476 (1978).

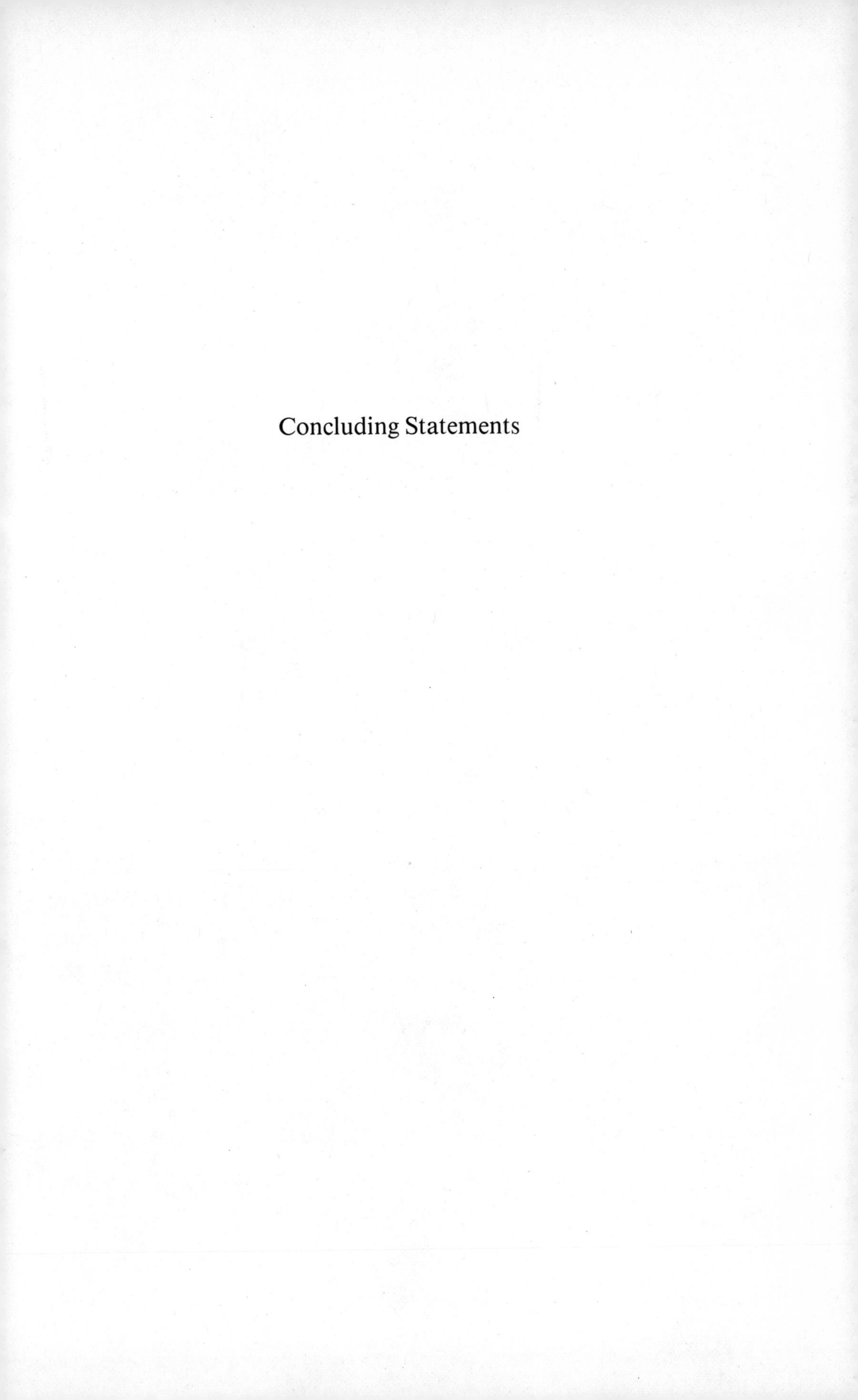

Concluding Statements

A BIOLOGIST SUMS UP

Richard P.C. Johnson

The Botany Department
University of Aberdeen
Old Aberdeen, Scotland, AB9 2UD

The aim of this most enjoyable Advanced Study Institute has been to bring biologists, chemists and physicists together to further the application of laser light scattering to the study of biological motion. We owe a very great debt indeed to John Earnshaw, Martin Steer, Ben Chu and their helpers for their outstanding organisation of it. The meeting has been great fun, but what has it achieved scientifically? Has there been any useful communication between us? Has there been any meeting of true minds? Or has physicist merely spoken to brother physicist, biologist to brother biologist and, of course,to sister? Have we done any more than have fun in the, remarkably, hot sun of this hospitable and scenic place? What fertilisation has occurred between disciplines?

Because I am a biologist interested in movement in cells I have, of course, learned much from the other biologists here and have understood most of what they have said. For me that alone would have made the meeting worth while. That has been an easy benefit because people in the same discipline share a common framework to hang new facts and ideas upon; they can relate terms and descriptions to similar practical experience. On the other hand, I estimate that I have understood, less easily, about 33.3% of what the physicists and chemists have had to say, which is rather more than I had expected. What we have all gained from the meeting is perspective and a better understanding of a different point of view; a better perception of another framework. This gain alone would also justify the meeting because the framework of a new field of study is the hardest part of it to acquire. As in the writing of an article, the grind lies in obtaining the outline, the overall shape and feeling. After that the arrangement of detail within it becomes a pleasure.

The framework for a subject consists not of facts joined together by logic, but of opinions and lines of thought put forward by people. This truth seems so often to be forgotten in the design of formal scientific courses, which is a pity because the true excitement of science lies in the personal aspects of it, in the opinions and argument. During the course of this meeting we have had enough time to meet people and appreciate their approach to their science, to hear their opinions and enjoy their arguments. Above all, we have met the people and put faces to their names.

What interactions have gone on here? I think that they may best be summarised with the aid of a pseudo-scientific diagram (Fig.1) suggested to me by Dennis Koppel during one of his lectures when he referred to a "two dimensional walk in the complex plane". The lines across the Figure represent partially

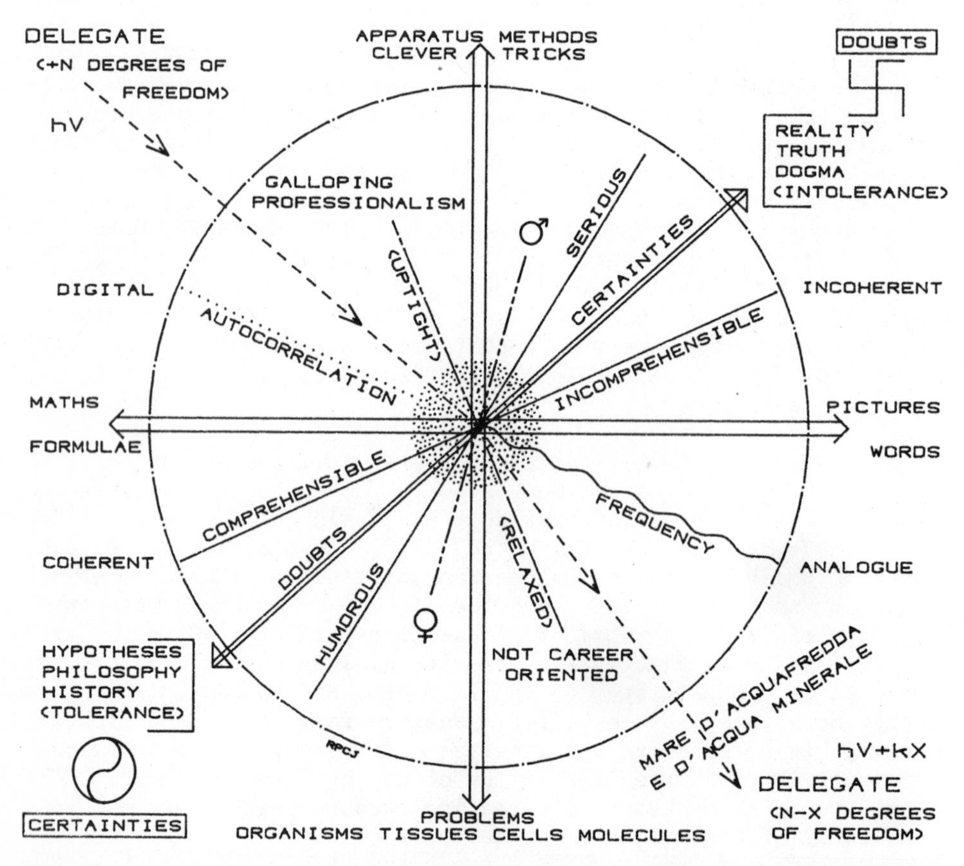

Fig. 1. A diagram to represent the deflection of a delegate by collisions in the multidimensional personal interaction space of this Advanced Study Institute.

orthogonal spectra with high and low at opposite ends. The cloud of dots round the point of interaction represents uncertainty of the second kind; the bonhomie due to good living. Also, it represents the affection which good lecturers seem to radiate towards their audiences.

I have spoken as though we have been sharply polarised into two camps, but of course, that is an over-simplification. A wide spectrum of points of view and interests has been evident and we have been fortunate to have amongst us some of those rare individuals who, like electrons in a covalent bond, seem entirely happy whichever sphere they are in orbit around. A large part of the charm of their attraction has been humour.They have contributed immeasurably by catalysing our meeting and holding it together. Nevertheless, there does seem to have been some polarisation here between those people who are happy with mathematics and those who deal more comfortably with pictures. Undoubtedly, this has hindered communication. The pictures, still or moving, presented by some biologists, are to them what symbols and formulae are to the mathematicians, a kind of shorthand. Both kinds of statement represent feelings and experience which may take many years to acquire. It is not enough to hear terms explained in other words. Experiments and practical experience are needed to grasp the meaning of terms and to place them in perspective. In my experience of learning and teaching biology it takes a new student at least a year to begin to be happy with the welter of new terms thrown at him. Therefore it is not to be expected that pictures or mathematical equations should be fully or even partly meaningful to anyone who has not soaked in them until their meaning is in his muscles.

At this meeting it has become clear that to some non-biologists the spectacular images which we have seen of movement in living cytoplasm present a horrid ill-defined mess, while to experts on cytoplasmic streaming Maxwell's elegant equations and even the practical problems of curve-fitting may have seemed so remote from their work as to be irrelevant. One major problem when one is trying to understand a lecture in an unfamiliar discipline is to realise the order or category of a concept. For example, what is the relation between first and second order electric field correlation functions; is the tonoplast a tissue or a kind of cell, or a part of a cell and when is it useful to think of it as being made of molecules? What levels of organisation or classification do these terms apply to? We need to grasp the relative importance, fluidity and novelty of terms. Are they essential or merely used as corroborative detail to add artistic verisimilitude? These barriers to understanding show up in the kind of atmosphere we have been able to generate here. They show up acutely when one finds ones self at lunch sitting opposite to the chap who was lecturing half an hour ago and one is still unable to formulate a question to ask him!

A special problem which deserves attention and which has become apparent during our discussions is the problem of scale and relative importance. We all need to construct mental Venn diagrams to see where our own area of study lies in the general scheme of progress. A major benefit of this meeting has been the opportunity to do this. However, it is too easy to conclude that one's own subject, or someone else's, is more important because it appears more "fundamental". The smaller a thing is the more scientific it may appear to be. For example,to the layman the study of the vibrations of DNA molecules may seem more scientific than the study of sap-flow in trees or the movement of particles in milk. Over dinner I have heard the argument that "biological observations should be explicable in terms of physics and chemistry because physics and chemistry are more fundamental". I suppose that one might try to explain the behaviour of Lady Macbeth in terms of physics and chemistry, but Shakespeare's level of discussion still has much to recommend it; a string of biochemical equations would not necessarily convey what he had to say. In other words, once we have chosen a level of organisation where we can do experiments we should not then seek to deny the validity of work at other levels. I have been much impressed in this matter by an argument put forward by P.B.Medawar[1] who points out that subjects can be arranged in a hierarchy as, for example:

1. Physics
2. Chemistry
3. Biology
4. Sociology

Here the subjects are arranged so that the higher up they are in the list the more "fundamental" they are in the accepted sense. Medawar points out that the lower a subject is in this hierarchy the richer it is in concepts. He points out that the terms used in physics and chemistry may be meaningful to the biologist, but not necessarily very important! Conversely, the terms used in sociology may have no meaning in biology, chemistry or physics. This explains the tendency for workers whose subjects are higher in the order of the list to dismiss those below as woolly and imprecise; biologists tend to look askance at sociology and theology is beyond the pale altogether. This kind of mistake (a version of the category mistake) leads to unhappiness. It is much exploited in the scramble for funds. Of course, in the end of the day, the terms used in any subject are imprecise; even photons are fuzzy. As Bronowski pointed out, awareness of the limits of resolution of our methods should lead us to realise the limits of error, of tolerance in the engineering sense, of our own knowledge and therefore to be more tolerant of other people's points of view.

It is evident that in order to study motion in biological systems we require approaches at all levels of biological and molecular organisation. We must beware of what Alexandrov calls "the tyranny of molecular biology". The question arises as to whether it is worthwhile, in order to be able to cope with biological mess while applying laser light scattering to the study of biological motion, for a physical scientist to learn much of what Professor Yu referred to as the wisdom of biology? Is it efficient for a biologist to learn much of the mathematics of light scattering? Do we have enough time to spend advancing ourselves sufficiently in another discipline to the point where we can have original thoughts in it and know when we have had them? Some rare spirits seem to have been able to do this, but life is short. Collaboration is better and the dialogue of collaboration may itself help to clarify problems. Collaboration can be fun, but it requires that collaborators understand each other. They must spend time cultivating their communal interests, even if these are not immediately to the point of their combined study. Their tolerances must overlap. They must work to develop a common feeling. This meeting has provided an excellent opportunity to start these processes. I believe that less than one and a half weeks would hardly have been long enough. Fortunately, we have had enough time to appreciate each others frameworks both in the lecture theatre and round the swimming pool; the good food and the sun have played their part and the ancient ruins have helped to keep things in proportion. The provision of these facilities reflects the wisdom of our patrons in NATO who have seen them to be necessary.

What other seeds may we have sown? I have spoken of some polarisation between biologists and physical scientists. It seems to me that mid way between their two poles of interest, between the study of molecules and photons on the one hand and of cytoplasm and cytoplasmic organelles on the other, there may lie a kind of no man's land, a level of organisation of structure and function which has not yet been explored. This is the region of scale where flow and diffusion are not clearly separate; where the concepts of temperature and of molecular motion overlap; where it is not clear whether molecules move or are moved; where ideas of "active" and "passive" lose their meaning. We may need to invent new terms to describe what happens in this ill defined region of study; lack of them may be fogging the raw reality on which we base models and explanations of how the cross bridges work in muscle and how the force is generated to drive cytoplasmic streaming. It may be that this region has not yet been recognised and defined because there have been no instruments acute enough or non-invasive enough to explore it. Perhaps the application of laser light scattering will open it up.

Now, sadly, it is time for us to go our separate ways. But

this too should be an essential part of the process of our Advanced Study Institute. Written over an archway in Marischal College at Aberdeen University and, no doubt, seen there by Robert Brown and James Clerk Maxwell in their time, are the words "They have said. What say they? Let them say". In other words, now that we have heard the professed experts we should all go home and cultivate our own points of view.

"When one is very young and knows a little, mountains are mountains, water is water, and trees are trees.
When one has studied and become sophisticated, mountains are no longer mountains, water is no longer water, and trees are no longer trees.
When one thoroughly understands, mountains are again mountains, water is again water, and trees are again trees."

An old Zen saying[3].

REFERENCES

1. P. B. Medawar, A geometric model of reduction and emergence, in: "Studies in the Philosophy of Biology",F. J. Ayala and T. Dobzhansky, eds., Macmillan, London (1974).

2. V. Ya. Alexandrov, The problem of behaviour at the cellular level (cytoethology), J.theor. Biol. 35:1-26 (1972).

3. D. A. Baker, "Transport Phenomena in Plants",p 72 Chapman and Hall, London, (1978).

EPILOGUE

Hyuk Yu

Department of Chemistry
University of Wisconsin
Madison, Wisconsin 53706, U.S.A.

Since the editors of this volume have requested that I provide the summarizing remarks from a "physicist's" perspective, I have taken on this assignment with express intent of sharpening the focus of the underlying feelings of the physicists during the conference with regard to how they should be involved in the study of the biological motions of relevance and significance. A corollary issue is how to interact with biologists in order to attain the most benefit to both.

Firstly, the preceding papers make it quite apparent that the laser light scattering technique has undergone substantial refinements since its theoretical inception by Pecora[1] and its experimental initiation by Cummins, Knable and Yeh[2] in 1964. If I were to make a conjecture, it may be said that the first decade of the technique was involved in exploring the range of application[3,4,5] together with the technical development in the time-domain detection method, i.e. photocurrent autocorrelation technique, by taking advantage of the digital nature of photons.[6] In the second decade up to this point, the applications have extended further into a wide variety of biological problems that is far beyond anyone's wildest expectations in the early seventies.[7,8] The range of applications extends from the diffusive motions of individual macromolecules to those of molecular and supramolecular aggregates such as viruses, gels, vesicles and membranes, to various subcellular and supracellular mechanisms such as cytoplasmic streaming, contractile mechanism and blood flow, to the intracellular motions of organelles and cellular movements, and finally to motility of bacteria and spermatozoa. Not only these applications are broadly scoped but also seemingly profound in their biological importance. At the same time, these applications entail a price of complexity. The practitioners of the technique no longer deal with simple hydrodynamic characterizations of a narrowly distributed collection of isolated particles or macromolecules. As the preceding papers by biologists attest, the problems being addressed are either very complex in biological functions or come with exceedingly heterogeneous compositions if their biological significance is straightforward; often the problems involve both biological complexity and compositional heterogeneity. In all cases, I have been struck by the common denominator that success of the application depends crucially on a thorough working knowledge of the systems being studied while the technical component of the method is becoming relatively routine; I do not mean by this provision that the laser light scattering has readied to the point of common accessibility of a method such as spectrophotometry that it can be provided in any chemical or biological laboratory. It is true that the technique still requires expertise in geometric and physical optics as well as in data analysis and interpretation even with a commercially available scattering goniometer and correlator. Nevertheless, research efforts without "good biology" mixed with a sensible application will likely end up in the deadly pile of inconsequential results. We all feel at times that each one of us is guilty of contributing to the pile as well as being victimized by it.

I should now like to discuss briefly the apparent disparity in scientific styles and approaches of the two groups. Instead of dividing them as "biologists" and "physicists" whose fruitful interactions and further colloboration are the objects of this institute, I would rather categorize the groups as "systems-intensive" and "technique-intensive"; the former may stereotypically be identified as biologists and the latter similarly as physicists, but with numerous exceptions. A systems-intensive group has its focus on a particular set of systems, be it physical, chemical or biological, and employs all possible probes at its disposal to examine the structure, properties, functions or any other aspects of the systems. On the other hand, a technique-intensive group has its emphasis on search for the broad range of application of a particular set of experimental methods while it may concentrate on a given system at any time. Ultimately, the goals of a technique-intensive group are to refine and perfect a set of techniques with use of a large number of systems. At this conference we have in effect a confluence of these two modus operandi of scientific endeavour; while we claimed that there existed a large difference in languages of the disciplines, I suspect that the tension we experienced arose more from the difference in these sorts of professional predilections. In any event, one rarely finds productive interactions of different disciplines without the attending tensions. Despite an ample middle ground between these two contrasting styles and rather forced division of the two groups by such a seemingly exaggerated difference in their modus operandi, I should now come to my point through this dichotomy. It is this: the biological systems are so intricate and complex that one cannot "learn" them without studying in detail by oneself. A technique-intensive group can rarely design an intelligent experiment without being converted in part into a systems-intensive group at the same time. Unlike in the last decade, it is no longer possible to study a significant biological problem by performing a scattering experiment "in a hurry of waste, and haste, and gloss, and glitter."[9] In more colorful language of American western movies, the attitude of "have gun, will travel" or "have machine, will scatter" can scarcely be appropriate or fruitful at this juncture. Biology, unlike mathematics or physics, seems to require wisdom in addition to logical rigor and precision, and this takes a time to acquire. Hence biologists are naturally systems-intensive because their discipline demands cognitive fidelity. I can cite several examples when certain physicists have succeeded in this mode by committing to and concentrating on a given biological problem for quite a number of years, though their introduction to the problem was through the technique-intensive route. Note the number of years that Carlson's group has been involved in the muscle contractile mechanism.[10,11] Volochine[12] was first engaged in the spermatozoan motility in the late sixties and note how much progress he has made in understanding some aspects of the human reproductive mechanism including the mechanical property of cervical mucus. I can cite other groups including that of Fujime[13] with the structure of F-actin & thin filaments of muscle, and those of Boon, Chen, Hallett and Nossal[14,15,16] with bacterial motility. Although these examples are hardly exhaustive and they reflect merely my ignorance of other works, the cited examples certainly serve to buttress my point. Without being converted in part into the systems-intensive mode, we, the physicists, may end up running the scattering experiments for others whose choice of the problem we must blindly trust. Colloborations with biologists are imperative in the conversion process, but a commitment of several years must be made by each one of us. Unless a physicist is willing to learn the necessary biology, he or she should avoid getting involved in biological applications of the laser light scattering.

REFERENCES

1. R. Pecora, J. Chem. Phys. 40: 1604 (1964).
2. H. Z. Cummins, N. Knable and Y. Yeh, Phys. Rev. Letts. 12: 1501 (1964).
3. G. B. Beneedek, in "Polarization: Matiere and Rayonnement", Les Presses Universitaires de France, Paris (1969).
4. H. Z. Cummins and H. L. Swinney, Prog. Opt. 8: 133 (1970).
5. B. J. Berne and R. Pecora, "Dyanamic Light Scattering", Wiley, New York (1976).
6. H. Z. Cummins and E. R. Pike, Eds., "Photon Correlation and Light Beating Spectroscopy", Plenum, New York (1974).
7. S.-H. Chen, B. Chu and R. Nossal, Eds., "Scattering Techniques Applied to Supramolecular and Non-Equilibrium Systems", Plenum, New York (1981).
8. D. B. Sattelle, W. I. Lee and B. R. Ware, Eds., "Biomedical Applications of Laser Light Scattering", Elsevier Biomedical,

Amsterdam (1982).

9. Lord Byron, in "Don Juan".

10. F. D. Carlson, B. Bonner and A. Fraser, Cold Spring Harbor Sym. Quant. Biol. 37: 389 (1972).

11. T. J. Herbert and F. D. Carlson, Biopolymers 10: 231 (1971).

12. P. Berge, B. Volochine, R. Billard and N. Hamelin, C. R. Acad. Sci. Ser D 265: 889 (1967).

13. S. Fujime, J. Phys. Soc. Jpn. 29: 75 (1970).

14. R. Nossal, S.-H. Chen and C. C. Lai, Opt. Commun. 4: 35 (1971).

15. J.-P. Boon, R. Nossal and S.-H. Chen, Biophys. J. 14: 847 (1974).

16. D. F. Cooke, F. R. Hallett and C. A. V. Barker J. Mechanochem. and Cell Motility 3: 219 (1976).

Participants

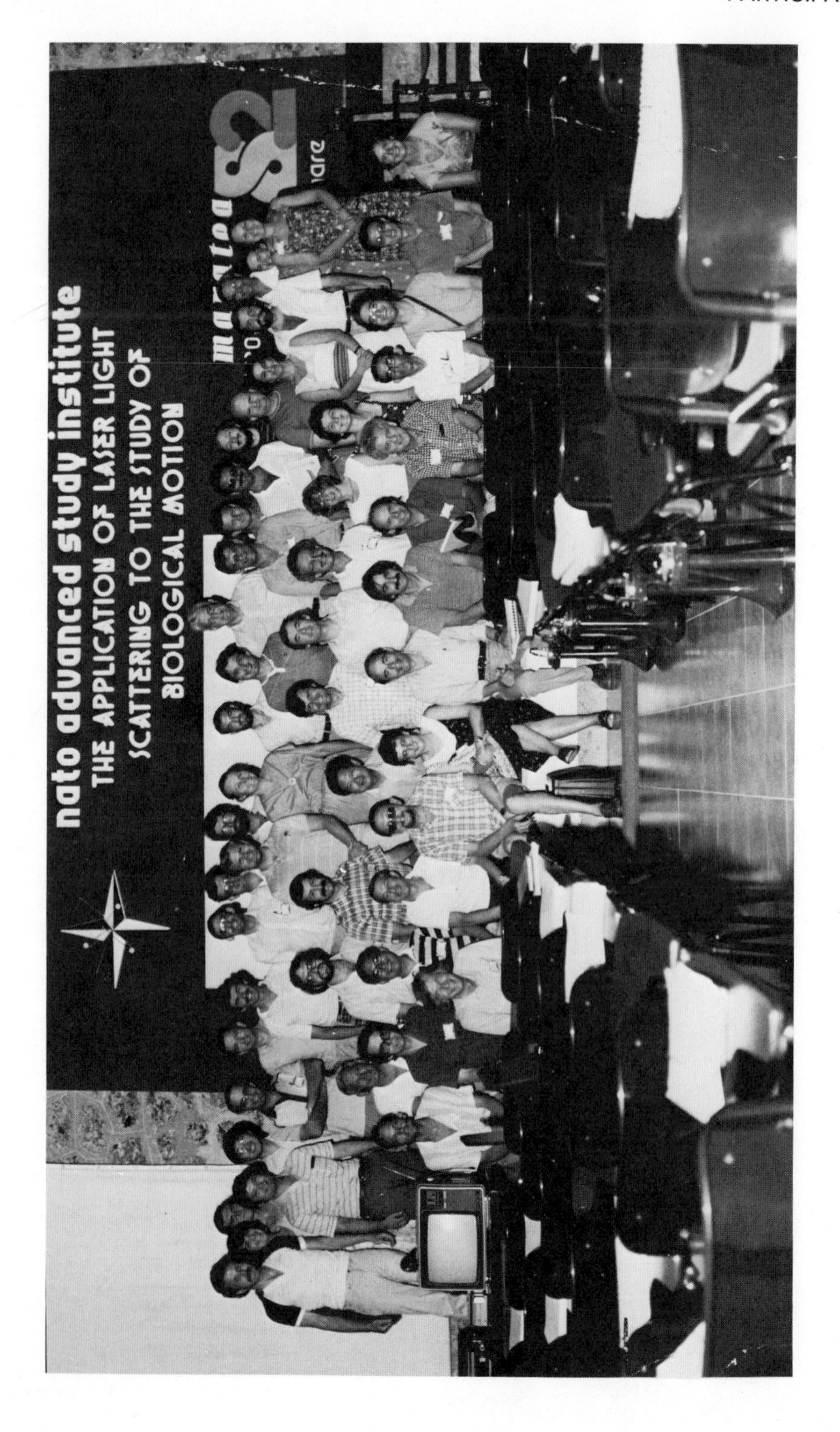
nato advanced study institute
THE APPLICATION OF LASER LIGHT
SCATTERING TO THE STUDY OF
BIOLOGICAL MOTION

1	Crilly	18	Woolford	35	Wohlfarth-Bottermann
2	Crawford	19	Milani	36	Chu
3	Munroe	20	Eastwood	37	Petracchi
4	Sato	21	Sellen	38	MacInnes
5	Wang	22	Giordano	39	Cummins
6	Fytas	23	Micali	40	Johnson
7	Ascoli	24	Jarosch	41	Duteil
8	Craig	25	Foissner	42	Eden
9	Geerts	26	Griffin	43	Bernengo
10	Klein	27	Steer	44	Maier
11	Ware	28	Volochine	45	Carlson
12	Norland	29	Earnshaw	46	Luzzati
13	Allen, N S	30	Taylor	47	Fujime
14	Holt	31	Yu	48	Blank
15	Minero	32	Boon	49	Wei
16	Allen, R D	33	Koppel	50	Picton
17	Frediani	34	Degiorgio		

PARTICIPANTS

Dr. N.S. Allen	Dept. of Biological Sciences, Dartmouth College, Hanover, NH 03755 USA
Prof. R.D. Allen	Dept. of Biological Sciences, Dartmouth College, Hanover, NH 03755 USA
Dr. C. Ascoli	CNR Istituto di Biofisica, Via S Lorenzo 26, 56100 Pisa, Italy
Dr. J.C. Bernengo	University of Nice, Laboratoire de Biochemie, Parc Valrose, 06034 Nice Cedex, France
Mr. P.S. Blank	The Johns Hopkins University, Jenkins Department of Biophysics, 3400 North Charles Street, Baltimore, Maryland 21218 USA
Dr. J.P. Boon	Service de Chimie Physique II, Code Postal No 231, Campus Plaine ULB, Boulevard du Triomphe, 1050 Bruxelles, Belgium
Prof. F.D. Carlson	The Thomas C Jenkins Department of Biophysics, The Johns Hopkins University, Baltimore, Maryland 21218 USA
Prof. B. Chu	Dept. of Chemistry, Suny at Stony Brook, Long Island, NY 11794 USA
Dr. T. Craig	Physics Dept., University of Guelph, Guelph, Ontario, Canada N1G 2W1
Miss. G. Crawford	Dept. of Pure and Applied Physics, Queen's University, Belfast BT7 1NN UK
Dr. J. F. Crilly	Dept. of Pure and Applied Physics, Queen's University, Belfast BT7 1NN UK

Prof. H.Z. Cummins	Dept. of Physics, The City College, City University of New York, NY 10031 USA
Prof. V. Degiorgio	Instituto di Fisica Applicata, Università di Pavia, Via Bassi 6, 27100 Pavia, Italy
Dr L. Duteil	Centre International de Recherches Dermatologiques, Sophia Antipolis 06565 Valbonne Cedex, France
Dr. J.C. Earnshaw	Dept. of Pure and Applied Physics, Queen's University, Belfast BT7 1NN, U.K.
Mr. J.C. Eastwood	Dept. of Physiology and Pharmacology Bute Medical Buildings, St Andrews, Fife, KY16 9TS Scotland
Prof. D. Eden	Dept. of Chem. San Francisco State Univ., San Francisco, CA 94132, USA
Dr. W.D. Eigner	Institut fur Physikalische-Chemie, University of Graz, Heinrichstrasse 28, A-8010 Graz, Austria
Dr. I. Foissner	Institut fur Botanik, Lasserstrasse 39, A-5020 Salzburg, Austria
Dr. C. Frediani	Istituto di Biofisica Del C.N.R., Via S. Lorenzo 24, 56100 Pisa, Italy
Dr. S. Fujime	Mitsubishi-Kasei Institut of Life Sciences, 11 Minamiooya, Machida, Tokyo 194, Japan
Dr. G. Fytas	University of Bielefeld, Faculty of Chemistry, P.O.B. 86 40, D-4800 Bielefeld 1, West Germany
Mr. H. Geerts	Dept. of Mathematics, Physics and Physiology, Limburgs Universitair Centrum, 3610 Diepenbeek, Belgium
Dr. R. Giordano	Istituto di Fisica, Universita Messina, Messina, Italy
Dr. M.C.A. Griffin	National Institute of Research in Dairying, Shinfield, Reading RG2 9AT, England
Dr. C. Holt	Hannah Research Institute, Ayr KA6 5HL Scotland

Prof. R. Jarosch	Institut fur Botanik, Universitat Salzburg, Lasserstrasse 39, A-5020 Salzburg, Austria
Dr. R.P.C. Johnson	Dept. of Botany, Aberdeen University, Aberdeen AB9 2UD, Scotland
Dr. R.A. Klein	Molteno Institute, Medical Research Council, University of Cambridge, Downing Street, Cambridge CB2 3EE England
Prof. D.E. Koppel	Dept. of Biochemistry, University of Connecticut Health Centre, Farmington, Connecticut 06032, USA
Miss. S. Luzzati	Polymer Research, Weizmann Institute of Science, Rehovot 76 100, Israel
Mr. D. MacInnes	MRC Brain Metabolism Unit, University Dept. of Pharmacology, 1 George Square, Edinburgh, Scotland
Mrs. K. Maier	Dept. of Chemistry, Stanford University, Stanford, Ca 94305, USA
Dr. N. Micali	Istituto di Fisica, Universita di Messina, Messina, Italy
Dr. M. Milani	Istituto di Fisica, Dell Universita, Via Celoria 16, 20133 Milano, Italy
Dr. C. Minero	Cise S.p.A. P.O.B. 12081, 20100 Milano, Italy
Mr. G. Munroe	Dept. of Pure and Applied Physics, Queen's University, Belfast BT7 1NN, N Ireland
Dr. S. Norland	Allegt 70, 5000 Bergen, Norway
Dr. D. Petracchi	Istituto di Biofisica, Via S. Lorenzo 26, 56100 Pisa, Italy
Dr. J.M. Picton	Dept. of Botany, Queen's University, Belfast BT7 1NN, N Ireland
Mr. M. Sato	Dept. of Biology, Gilman Hall, Dartmouth College, Hanover, NH 03755, USA
Dr. D.B. Sellen	Dept. of Biophysics, The University of Leeds, Leeds LS2 9JT, England

Dr. M.W. Steer	Dept. of Botany, Queen's University, Belfast BT7 1NN, N Ireland
Mr. T.W. Taylor	Dept. of Physics, Oklahoma State University, Stillwater, Oklahoma 74078, USA
Prof. B. Volochine	Dept. of Physics UER, EMB, 45, Rue Des Sts Peres, 75270 Paris Cedex 06, France
Mr. P.C. Wang	Rm 24-209 MIT, Cambridge, MA 02139, USA
Prof. B.R. Ware	Dept. of Chemistry, 108 Bowne Hall, Syracuse University, Syracuse, NY 13210 USA
Dr. J.G. Wei	Dept. of Biochemistry, University of Minnesota, St Paul, Mn 55108, USA
Prof. K.E. Wohlfarth-Bottermann	Institute for Cytology, Ulrich-Haberland Strasse 61a, D-5300 Bonn 1, Federal Republic of Germany
Dr. M.W. Woolford	Biophysics Group, Ruakura Agricultural Research Centre, Private Bag, Hamilton, New Zealand
Prof. H. Yu	Dept. of Chemistry, University of Wisconsin, 1101 University Avenue, Madison, Wisconsin 53706, USA

INDEX